Genetics and Advance Breeding Methods of Vegetable Crops

NIPA® GENX ELECTRONIC RESOURCES & SOLUTIONS P. LTD.
New Delhi-110 034

About the Authors

Prof. (Dr.) Pranab Hazra, M.Sc. (Ag), Ph.D., Former Professor. Department of Vegetable Science, Dean, Faculty of Horticulture and Dean, Post Graduate Studies, Bidhan Chandra Krishi Viswavidyalaya, Mohanpur-741252, West Bengal completed his illustrious professional career of 40 years in Agricultural University in 31 March, 2025. He has supervised doctoral level research works in Vegetable Science of 22 students and 15 Govt. sponsored research projects. He has so far written 14 books, 38 book chapters in 28 books and published 213 research papers with total citation of more than 3200. He visited Israel, Bulgaria and Greece in connection with International course and visit-oriented academic programme. He is Fellow, FAO, Rome, Indian Academy of Horticultural Science, Indian Society of Vegetable Science and International Society for Noni Science and Member of National Academy of Science, India. Recipient of Gold medal of Indian Academy of Horticultural Science, Dr. Kirti Singh Memorial Lifetime Achievement Award of Indian Society of Vegetable Science and Lifetime Achievement Award of the Academy for Advancement of Agricultural Science.

Prof. (Dr.) Arup Chattopadhyay, M.Sc. (Ag), Ph.D., Professor and Head, Department of Vegetable Science, Faculty of Horticulture, Bidhan Chandra Krishi Viswavidyalaya, Mohanpur-741252, West Bengal. He was Former Officer-in-Charge of All India Coordinated Research Project on Vegetable Crops, BCKV Centre. He has supervised 13 Ph.D students; written 6 books, 12 book chapters and published 240 research papers. Recipient of first Dwarika Nath Gold Medal Award for the best Ph.D. thesis in Vegetable Science conferred by Indian Society of Vegetable Science, Lt. Amit Singh Memorial Award for the Best Coordinating Centre of AICRP on Vegetable Crops in India conferred by ICAR, International Research Awards on Science, Health and Engineering SHEN. He is the Fellow, Indian Society of Vegetable Science and Crop and Weed Science Society, BCKV, Mohanpur, West Bengal.

Dr. Soham Hazra, M.Sc. (Ag), Ph.D., Assistant Professor in Genetics and Plant Breeding and Head, Department of Agriculture, Brainware University, Barasat, West Bengal. He pursued his doctoral degree programme as a Senior Research Fellow under the Department of Atomic Energy, Bhabha Atomic Research Centre- sponsored research project. His prominent research achievement includes isolation of YVMV resistant lines of okra through applied mutagenesis. He has published more than 20 research papers in SCI and SCOPUS indexed journal, co-authored 7 book chapters, 2 book and several popular articles. He has been awarded for 3 best paper presented in International and National Conferences. He has acted as peer reviewer to several scientific journals of repute. He is the recipient of 'Eminent Scientist' Award-2024 by the Society for Plant Research, Vegetos and "Young Scientist" award in Crop Science by the Academy for Advancement of Agricultural Science in 2025.

Genetics and Advance Breeding Methods of Vegetable Crops

Pranab Hazra
Former Professor, Department of Vegetable Science
Former Dean, Faculty of Horticulture and Dean, Post Graduate Studies
Bidhan Chandra Krishi Viswavidyalaya
Mohanpur – 741252, West Bengal

Arup Chattopadhyay
Professor and Head
Department of Vegetable Science
Former Officer-in Charge, AICRP on Vegetable crops
Bidhan Chandra Krishi Viswavidyalaya
Mohanpur – 741252, West Bengal

Soham Hazra
Head, Department of Agriculture
Brainware University
398, Ramkrishnapur Road, Near Jagadighata Market
Barasat, Kolkata- 700125, West Bengal

NIPA® GENX ELECTRONIC RESOURCES & SOLUTIONS P. LTD.
New Delhi-110 034

NIPA® GENX ELECTRONIC
RESOURCES & SOLUTIONS P. LTD.

101,103, Vikas Surya Plaza, CU Block
L.S.C.Market, Pitam Pura, New Delhi-110 034
Ph : +91 11 4386 0225, 9717133558, 9540816132
E-mail: newindiapublishingagency@gmail.com
Website: www.niparesources.com

Print ISBN: 978-93-5887-703-8

ebook ISBN: 978-93-5887-199-9

Composed and Designed by NIPA®.

Dedicated to:
Late Nabaghanashyam Hazra
Father of Pranab Hazra
and
Grand father of Soham Hazra

Preface

India is the second largest producer of vegetables after China. Spectacular increase in vegetable production was envisaged in the country from 15 million metric tonnes during 1950 to 204.83 million metric tonnes from an area of 11.35 million ha at present. India also ranks second after China in manually produced hybrid seeds of different vegetable crops. Productivity of different vegetable crops has been enhanced markedly during the last 60 years mostly due to genetic improvement and availability of improved varieties and hybrids of different vegetable crops. Advanced production and protection technologies also helped in enhancing the productivity.

The first part of the book "Genetics and Advance Breeding Methods of Vegetable Crops" has been structured with 17 chapters detailing history of vegetable breeding, origin and domestication, reproductive features and pollination control mechanisms, general breeding methods for sexually and asexually propagated crops, breeding for high temperature stress, disease resistance and product quality, heterosis breeding, mutation breeding, breeding through alteration of ploidy, ideotype breeding, procedure for release of a variety, intellectual property rights and biotechnological approaches for vegetable breeding. The second part has been framed on the details of 24 vegetable crops covering all the reproductive feature group pertaining to their origin, taxonomy, distribution, crop botany and floral biology, reproductive behaviour, genetics of important traits with published gene list, breeding objective, germplasm resource, breeding methods and breeding achievement highlighting the released varieties of the public sector. Important references have been provided at the end of each chapter which will be helpful for the post graduate and doctoral level students.

This book follows the syllabi prescribed by ICAR for B.Sc. Hons. (Agri.), B.Sc. Hons. (Hort.), M. Sc (Agri), M. Sc (Hort.) and Doctoral degree programmes in Vegetable Science particularly genetics and breeding of vegetable crops. Long experience of the authors in teaching and research on different aspects of vegetable breeding has been put together in laying out this book.

This book will be an excellent ready reference for the teachers and students pursuing different courses on crop breeding particularly, vegetable breeding. This book will also meet the requirement of vegetable breeders working in both public and private sector.

Pranab Hazra deeply feels from the core of his heart that this work would not have been possible without the blessings of his mother (Sandhya), teacher (M.G. Som) and sustained support of his wife (Antimoni), son (Soham) and daughter-in-law (Torsha).

The authors thank NIPA® Genx Electronic Resources & Solutions P. LTD. for bringing out this publication.

Kalyani, 2025

Pranab Hazra
Arup Chattopadhyay
Soham Hazra

Contents

1

Importance and History of Vegetable Breeding

History of plant breeding

The process of plant breeding takes care of changing or modifying the genetically controlled characters of a particular variety or genotype of a crop. The practice of age-old agriculture uses the process of selection from the existing variability to save a particular quality of seed to grow better crops in the next harvest.

Gregor Mendel who is known as the father of Genetics discovered the law of segregation in the middle of the 1800s. This theory forms the base for the process of plant breeding. The primitive ways used for plant breeding are very different than what modern technology has contributed to the process of plant breeding during the 20th century.

Plant breeding is the science and art of purposeful manipulation of characters of crop species to create new varieties with a set of desired characteristics. Significance of plant breeding has been highlighted especially after the "Green Revolution". Later, with the advancement in the process of plant breeding new technologies for developing varieties were discovered. Plant breeding is done not only to improve yield and quality of a particular variety but also to impart resistance in it for different biotic stresses inflicted by fungi, bacteria, virus, phytoplasma and nematodes and abiotic stresses like high temperature, soil salinity, flood or drought. However, to think globally, the main significance of plant breeding is to meet the increasing demands of the burgeoing population all over the world.

Basic objectives of vegetable breeding

- To develop high-yielding varieties being the main objective of plant breeding
- To develop high-yielding varieties with enhanced nutrient, antioxidants and important phytochemical contents and improved flavour
- To develop high-yielding varieties with enhanced shelf-life and processing qualities
- To develop varieties with resistance to different biotic stresses inflicted by fungi, bacteria, virus, phytoplasma and insect pests
- To develop varieties with resistance to different abiotic stresses like, high temperature, low temperature, drought, flood, soil salinity or soil alkalinity

Vegetables in global food economy

Vegetables play a major role in world agriculture by providing food and offering nutritional and economic security. The most dominant vegetables in the global food economy are potato, tomato, cucurbits (pumpkin, squash, cucumber and gherkin), allium (onion, shallot, garlic) and chilli. These vegetables are consumed in nearly all countries although there exists much variation in shape, size, colour and taste. However, traditional vegetables often have superior nutritional properties. For instance, 100 g leaves of vegetable amaranth or pods of vegetable cowpea can provide substantial quantity of vitamin A and protein in the diet. Globally, a one per cent increase in per capita income in developing countries is associated with a 0.5% increase in per capita availability of vegetables. Bulk of the global supply of fruits and vegetables (77% of total value) is produced in populous middle-income countries. China accounts for 45% of the global value of vegetable production. India comes second, accounting for 8.6% of global vegetable production.

Production and consumption of vegetables has expanded dramatically, with a global growth in the production of more than 50% in the last three decades, a rate of increase that is much higher than for other plant commodities. Vegetables constitute an important part of a varied and healthy diet and provide significant amounts of vitamins, antioxidants and other substances that prevent diseases and contribute to an improvement in the quality of life. In consequence, it is expected that in the coming years, production of vegetable crops will continue its expansion. Improved varieties have the main role not only to increase yield but also to enhance nutritional quality. In this respect, the vegetables seed market is very dynamic and competitive, and predominant varieties are quickly replaced by new varieties. Therefore, updated information on the science and art of the genetic improvement of specific crops is of interest to vegetable crops breeders, researchers and scholars. During the last three decades, immense knowledge on the genetic diversity of vegetable crops, utilization of genetic resources, breeding methods and techniques and utilization of modern biotechnologies in breeding of vegetables crops have been accumulated.

Diversity in vegetable species

It was recorded earlier that around 400 species contribute to global vegetable crop diversity (Arora, 1991). However, a much greater diversity of vegetables exists in traditional food systems, but many of these crops are poorly integrated in current markets and diets. A database of cultivated vegetable species by Bioversity International scientists in collaboration with the Food and Agriculture Organization of United Nations and compiled in the review of the Mansfeld Encyclopaedia of Agricultural and Horticultural Crops, 2001 (cited by Meldrum *et al.*, 2018) revealed that a total 1,097 vegetable species from 133 families, with a great variety of uses and growth forms, are cultivated worldwide. Of the total of 1,097 vegetable species cultivated worldwide, 495 species are used for leaves (leafy green and salad crop); 227 species are used for multiple vegetative parts (bulb, root, tuber, stem, leaf, etc.); 204 species are used for roots (root crops); 90 species are used for fruits or seeds and 80 species are used for other plant parts like flowers, inflorescences and stems. However, hardly, 80 species are known to be cultivated by the farmers (Kays, 2011). In India, range of 60 (Arora, 1991; Kochhar, 1998) to 90 (Nayar, 2003) species of higher plants are generally cultivated as vegetables. These reports amply suggest that hardly 7-8 % of the total plant species are at present used as vegetables around the world. Hence, ample opportunity is left to widen our vegetable basket through their integration in the diets and subsequent commercialization.

History and evolutionary aspects of vegetable breeding

Plant breeding / genetic improvement aims at altering plant genotypes in a way that they suit human needs for food, feed and industrial raw products and support sustainable plant production. In that sense, plant breeding may be regarded as the evolution of plant type by the will of mankind. Plant breeding / genetic improvement of crop is important because an improved cultivar is the most economic and efficient input for farmers.

Plant breeding is one of the early accomplishments of mankind. Many of the cultivated plants are the result of slow and involuntary improvement from wild progenitors by human being over centuries. In fact, the process of domestication of wild plants is one of the most fascinating chapters in the history of mankind. Concept of plant breeding came into existence through domestication as it involved selection of the most desirable types. Among the ancient domesticates, some of the vegetable crops are front-runners like, peas, radish, muskmelon, watermelon, etc. Other present day popular vegetable crops were domesticated later in the Neolithic times. Some species including certain vegetables are still undergoing the domestication process. However, for all practical purposes breeders have given up domestication of wild relatives of the cultivated species. Today, breeders often incorporate important characters like, resistance to biotic stresses, adaptability to environments etc. to the cultivated forms from their wild relatives.

Selection from the available variations is the oldest and the most effective breeding procedure. However, genetic recombination at present forms the basis of modern plant breeding approach. In this respect, Josef Kolreuter's extensive crossing experiments between 1755 and 1806 were essentially the first attempts to scientifically reconstruct naturally occurring species and types by combining their characters. Anyway, no matter what method is adopted in the breeding approach, it inevitably involves selection, which is the most difficult part of the overall breeding process.

The industrial revolution in Western Europe and concomitant flourish of capitalism and increasing commercial agricultural production stimulated breeding activities as a profitable venture of capitalistic economy. Throughout the 19th century thousands of large and small seed companies were developed in Germany, England and the United States out of this impetus.

Outstanding practical results in the breeding of horticultural crops (fruits, vegetables and ornamentals) were obtained through hybridization and multiple selections by American breeder Luther Burbank at the turn of last century. However, the first person who provided the theoretical basis of plant and animal breeding through major contribution to understanding the evolutionary process and progress in natural sciences was the prominent English scientist Charles Darwin. All these and many other research workers were careful experimenters and good biologists, but they could not explain their results because the basic principles of Genetics had not yet been established. The decisive factor of progress in scientific plant breeding was rediscovery of Mendel's work on law of heredity formulated way back in 1865 by three scientists independently, Hugo de Vries in Holland, Carl Correns in Germany and Erich Tschermak in Austria in 1900. This historical event established a new science "Genetics".

It was thereafter established that Genetics provided the theoretical basis for both plant breeding and seed production. Although both the disciplines of plant science are quite independent, they are inextricably intermingled and complementary to each other. Present day modern plant breeding is characterized by an integrated approach involving various disciplines *viz*., genetics, basic biology, physiology, biochemistry, phytopathology, entomology and biotechnology. The breeder performs the function of a chief designer working according to a carefully prepared long-term programme.

Conventional vegetable breeding

Vegetables not only have a pristine place in agricultural economy but also play an important role in ensuring food and nutritional security. Most of the vegetable crops are generally of short duration, high yielding, nutritionally and phyto-chemically rich and its successful farming generate substantial on-farm and off-farm employment. Most of the countries including India has witnessed a tremendous growth in vegetable production and productivity as a result of continual genetic improvement leading to the development of wide array of improved varieties and F_1 hybrids and framing of improved agro-techniques through systematic research coupled with their large scale adoption by the farmers and developmental policies of the respective Government.

Modern genetic improvement programmes focus on developing high yielding, biotic and abiotic stress resistant (such as insect pests, disease, heat tolerance, drought, flood and freeze damage, etc.) and high quality varieties. Maintaining a diversity of genes and germplasm pools is necessary for providing sources of useful genes. Aside from developing varieties with incorporated genetic resistance, breeding approaches also include improvement of different pigment, flavour, protein, starch, oil, phytochemicals and other feature characteristics of different vegetable crops.

Marker assisted breeding

Molecular marker or DNA markers are the short DNA sequences whose inheritance can easily be monitored and detected. Use of molecular markers in breeding has opened a new horizon called marker assisted breeding. Marker assisted selection is the selection for a trait based on genotype using associated markers rather than the phenotype of the trait. Marker assisted selection improve the efficiency and speed up the selection during breeding cycles. Gene pyramiding for the creation of multi-resistant varieties has been made possible with the development of molecular markers. Different potential molecular markers have been applied in breeding different vegetable crops.

Transgenic or Genetically modified crop

Genetic transformation technology is used for introgression of many important genes, such as drought and virus-resistant genes into elite germplasm and breeding populations. Pioneer public institutions which are working to strengthen the research under transgenic vegetable crops in India are National Institute for Plant Genome Research (NIPGR), ICAR- Indian Institute of Horticulture Research Hesaraghatta Bengaluru, Indian Institute of Vegetable Research, Varanasi and ICAR-Indian Institute of Agriculture Research. New Delhi.

Genome sequencing

Next generation genome sequencing provides opportunities of getting high density genetic maps which predict the linear arrangement of markers in a chromosome, identification of horticulturally important genes and development of millions of novel markers. Genes governing different traits can be identified through sequence data obtained from genome and transcriptomes along with their expression profile which are associated with different physiological conditions. Accumulation of data is also crucial for successful breeding programme by enabling the identification of allelic variation in candidate genes controlling important horticulturally important traits.

Cereal crops

Cereal crops belong to the same Monocotyledoneae family, Poaceae (grass family) cultivated for its edible grain. Cereals are the world's largest group of crops, and are therefore, staple foods. The eight principal cereal crops grown in the world are wheat (*Triticum* spp.), maize (*Zea mays*), rice (*Oryza sativa*), barley (*Hordeum vulgare*), oats (*Avena sativa*), rye (*Secale cereal*), pearl millet (*Pennisetum glaucum*) and sorghum (*Sorghum bicolor*). Edible grains from other plant families, such as buckwheat (*Fagopyrum esculentum* of the family Polygonaceae) and quinoa (*Chenopodium quinoa* of the family Amaranthaceae) are called pseudocereals.

Vegetable crops

Vegetable crops are defined as short-duration and widely diverse herbaceous plant species, grown in different season of the year in which the edible plant parts consist of leaves, roots, hypocotyls, stem, modified stem, root and inflorescence, petiole, flower bud, immature and ripened fruits. Vegetables being mostly low in fat and carbohydrates but high in vitamins, phytochemicals, minerals and dietary fibre play an important role in human nutrition.

Table 1: Major difference between Cereal and Vegetable breeding

Cereal breeding	**Vegetable breeding**
All the principal cereals grown in the world *viz.*, wheat, maize, rice, barley, oats, rye, pearl millet and sorghum belong to the same Monocotyledoneae family, Poaceae.	Wide array of vegetable crops grown in the world belong to large number of both Monocotyledoneae and Dicotyledoneae families
All these principal cereals crops are all sexually propagated and are both self- as well as cross-pollinated species.	Reproductive features and breeding system of the vegetable crops show wide variability with reference to floral biology and pollination control mechanisms. Vegetable crops are propagated sexually, asexually and through both the mode. They are self-pollinated, cross-pollinated and often cross-pollinated with varied floral and pollination control features (chasmogamy, dichogamy, unisexualism, self-incompatibility, male sterility, etc.)
The economic plant part of the principal cereals grown in the world is basically grain *i.e.*, dried seed or fruit. Hence, the breeding objectives for improvement of economic plant part are almost same.	The economic plant part of the wide array of vegetables vary widely *viz.*, fruit (brinjal, tomato, chilli, peas and beans, all cucurbits, okra, etc.), leaves (leaf lettuce, spinach, palak, amaranth, leaf coriander, fenugreek, Indian spinach, water spinach), growing point surrounded by overlapping leaves (cabbage, leaf lettuce), modified inflorescence (cauliflower), flower head (broccoli, globe artichoke), modified above ground stem (knolkhol, asparagus), modified underground stem (potato Jerusalem artichoke, yam, elephant foot yam, taro), bulb (onion, garlic), modified underground root (radish, carrot, beet, turnip, cassava, sweet potato, Chinese potato), etc. Hence, breeding objectives for improvement of economic plant part vary widely with the crop.
Basic quality traits of the cereals are starch, protein, micro-nutrient contents, milling quality, colour and flavour of flour.	The quality traits of different vegetable crops are mainly attributed to different carbohydrates, sugar, organic acid, vitamins, minerals, diverse pigments and antioxidant compounds such as carotenoids, anthocyanins, phenolics, alkaloids, nitrogen containing compounds, organo-sulphur compounds, etc.
All the cereals are utilized for almost the same basic purpose hence, the traits under consideration for breeding are almost same.	The vegetable crops are utilized in diversified culinary, raw, salad and processed purposes along with huge regional preference hence, the traits under consideration for breeding show wide variation.

References

Arora, R.K. 1991. Plant diversity in the Indian Gene centre. In: *Plant Genetic Resources-conservation and management* (Eds. Paroda, R.S. and Arora, R.K.), International Board for Plant Genetic Resources office for South Asia and South East Asia and National Bureau of Plant Genetic Resources Campus, New Delhi, pp.25-54.

Kays, S.J. 2011. *Cultivated vegetables of the world: a multilingual onomasticon.* Wageningen Academic Publishers, Wageningen. 828p. http: // dx.doi.org /10.3920 / 978-90-8686-720-2.

Kochhar, S.L. 1998. *Economic botany in the Tropics*, Second edition, Macmillan India Ltd., New Delhi.

Meldrum, G., Padulosi, S., Gaia Lochetti, G., Robitaille, R. and Diulgheroff, S. 2018. Issues and prospects for the sustainable use and conservation of cultivated vegetable diversity for more nutrition-sensitive Agriculture. *Agriculture*, 8: 112; doi:10.3390/agriculture8070112

Nayar, E.R., Pandey, A., Venkateswaran, K., Gupta, R., and Dhillon, B.S. 2003. *Crop Plants of India- A Checklist of Scientific Names*. National Bureau of Plant Genetic Resources, New Delhi. 48p.

2

Origin, Domestication, Evolution and Variability of Vegetable Crops

History of Agriculture

Cultivation of crops is one of the oldest occupations of mankind and probably began when it was discovered that certain seeds spilled on disturbed ground, grew in some mysterious way into new plants. The seeds were then collected and spread into different regions. This is how origin and introduction of crop plants started. Development of agriculture about 12,000 to13,000 years ago changed the way humans lived. They switched from nomadic hunter-gatherer lifestyles to permanent settlements and farming. However, scientists are unable to explain why man took so long to discover agriculture. Agriculture began independently in different parts of the globe and included a diverse range of taxa. At least eleven separate regions of the Old and New world were involved as independent centers of origin however, there was no single factor or combination of factors that led people to try farming in different parts of the world. Farming could have been brought on by climate changes at the end of the last ice age that favoured different annual plants like wild cereals to grow. But whatever the reasons for its independent origins, farming sowed the seeds for the modern age.

Archeological evidence for the origin of crops points to mountainous areas of both Old World (Asia, Europe and Africa) and New World (North and South America), which have pronounced wet and dry seasons. Under such climatic conditions seeds would need to germinate, grow quickly with the onset of rains, complete their reproductive cycles and go to seed formation before summer sets in such as we see in annual plants.

Prehistoric sites for ancient agricultural activity

All archeological and other evidences point to these prominent prehistoric sites for ancient agricultural activities

- **Indus valley civilization:** Well-watered hillocks and alluvial plain of the Indus river spanning an area from much of Pakistan (Harappa and Mohenjodaro of Punjab and Sindh provinces) to northeast Afghanistan, and northwestern India. Together with ancient Egypt and Mesopotamia, it was one of three early civilizations of the Near East and South Asia.Of the three, Indus valley civilization was the most widespread civilization.

- **Civilization in the Fertile crescent:** The fertile crescent is the transcontinental region near Euphrates, Tigris and Nile River having well-watered land stretches from the Nile river valley of Egypt northward along the coasts of the regions of Palestine, then eastward along the Tigris and Euphrates rivers to the Persian Gulf. This historical "Fertile Crescent" includes the areas of Mesopotamia (the historical region which included present day Iraq and parts of present- day Iran, Kuwait, Syria and Turkey), the Levant (present-day Israel, Jordan, Lebanon, Syria, the Palestinian territories and most of Turkey), Anatolia (the western-most extension of continental Asia in the region in Turkey), East Thrace (part of Turkey that is geographically a part of Southeast Europe), Cyprus and northern Egypt.
- **Mesoamerican civilization:** Mesoamerica was the region between central Mexico, Guatemala, El Salvador and a small section of Honduras. In Mesoamerica, several waves of civilizations developed, such as the Olmecs, Toltecs, Aztecs, Mayas, etc. . These civilizations were developed along the southwestern curve of the Gulf of Mexico, near the present city of Veracruz.
- **Andean civilization:** The Andean region includes what is now Venezuela, Colombia, Equador, Peru and Bolivia. In the Andes, populations in plains along the Pacific coast are thought to have grown into a large civilization dominated by the Inca. The Mesoamerican civilization, along with the comparable Andean civilization farther south, constitutes a New World counterpart to those of ancient Egypt, Mesopotamia and China.
- **Tehuacan valley:** Tehuacán Valley, part of the Mesoamerican region, is the arid or semi-arid zone with the richest biodiversity in all of North America. This valley is located in central-southern Mexico, at the southeast of the state of Puebla and north of the state of Oaxaca. It was the first place where maize was ever cultivated by humankind.
- **Chinese civilization:** This civilization, originated in Hwang Ho and Yangtze river basins, is one of the world's ancient civilizations, aged almost 5,000 years. The Hwang-Ho River is also known as yellow river because it brings the yellow alluvial soil and deposits it on its banks. This river bank is one of the prehistoric sites of ancient agricultural activity. Domestication of rice and millet took place in the Yangzi and Yellow river valleys.
- **Main land South East-Asia:** Many evidences suggest that the expansion of farming communities reached to mainland Southeast Asia (Myanmar, Thailand, Laos, Cambodia, and Vietnam) from the late third millennium BC. Its diverse vegetation to support a stable human population with fishing and hunting economy was ideal for agricultural beginnings. However, archaeological evidence for this view is meager.

Taking root around 12,000 to13,000 years ago, agriculture triggered such a change in society and the way in which people lived that its development has been dubbed the "Neolithic Revolution." Traditional hunter-gatherer lifestyles, followed by humans since their evolution, were swept aside in favour of permanent settlements and a reliable food supply. Out of agriculture, cities and civilizations grew, and because crops and animals could now be farmed to meet demand, the global population rocketed—from some 5.0 million people 10,000 years ago, to 8.1 billion at present.

Crop evolution

Biological evolution of crop is change in the properties of populations of organisms that transcend or go beyond the lifetime of a single individual. The ontogeny of an individual is not considered evolution because individual organisms do not evolve. The changes in populations that are considered evolutionary are those that are inheritable *via* genetic material from one generation to the next. Biological evolution may be slight or substantial embracing everything from slight changes in the proportion of different alleles within a population.

Advances in molecular genetics have ushered the exciting age for investigating where, when, and how crop plants arose. Progress in understanding crop evolution began with morphological studies and archaeological findings of early domesticates. Beginning in the mid-20th century, chromosome homology was used to investigate crop origins; later, allelic variants of enzymes (i.e., allozymes) were used to explore the origins and population genetics of crop species. The greatest technological leap for studying domestication, however, came with the development of DNA-based molecular markers beginning in the 1980s.

Domestication

Domestication is generally considered to be the end-point of a continuum that starts with the exploitation of wild plants, continues through cultivation of plants selected from the wild but not yet genetically different from wild plants.This process terminates in fixation, through human selection, of morphological characters which led to genetic differences distinguishing a domesticate from its wild progenitor. These differences constitute the domestication syndrome and generally render the domesticate less capable of survival in the wild, thus dependent on man for its growth and reproduction. Features of the domestication syndrome include loss of dispersal, increase in size (especially of the harvested part of the plant), loss of seed dormancy and loss of chemical or mechanical protection against herbivores.

Domestication of crops has been considered a key historical innovation that enabled the development of large, complex civilizations (Gepts, 2004). Crop plants have been deliberately selected by humans for desirable traits and to be adapted to the conditions of cultivation (Evans, 1993). There is accumulating evidence that crop domestication has profoundly altered tritrophic interactions between plants, insects, and their natural enemies. In general, domesticated crop plants tend to possess more exaggerated physical traits (Schwanitz, 1966; Evans, 1993), simpler morphologies (Rosenthal and Welter, 1995), altered nutritional content, and reduced plant defenses (Lindig-Cisneros *et al*, 1997) compared with their wild ancestors.

Crops vary within and between species in their degrees of domestication. As for example, all known accessions of *Capsicum pubescens* have large fruits that have lost their dispersal mechanism, and this species occurs only in cultivation. The four other domesticated species of chilli pepper includes a range of variation from wild peppers still capable of natural dispersal, to fully domesticated peppers with large fruits that remain firmly attached to the parent plant after maturity. 'Domesticated' may therefore describe an entire species, or just some of the variants within a species.

Clement (1999) proposed two intermediate categories, incipiently domesticated and semi-domesticated, to cover the spectrum of changes resulting from human interactions with the wild species. Cultivated populations are thus not genetically fixed for any characters distinguishing them from wild populations, but frequencies of alleles governing the character subject to human selection presumably differ.

Gepts (2004) considered cultivation a necessary but insufficient condition for domestication. However, domestication (or at least incipient or semi-domestication) without cultivation may occur by selective removal of undesirable phenotypes and/or enhancement of desirable phenotypes in wild populations, thus changing the proportion of phenotypes in the managed populations for various species (Casas *et al.*,1997 ; Anderson, 2004). Casas *et al.* (1997) noted that in open- or cross-pollinated species, selective retention of desirable phenotypes in managed populations promote mating between these plants. More progeny showed the desired characters, leading eventually to fixation of these characters. They termed this *in situ* domestication and suggested that, since *in situ* management of wild and weedy species is very common in Mesoamerica, *in situ* domestication may also have occurred frequently and may explain how various outbreeding species became some of the first domesticates in this region, in contrast to the Middle East, where most of the early domesticates are inbreeders (Zohary, 1984).

Despite difficulties in defining domestication, most workers agree that there were several independent regions of plant domestication. Quite frequently, different species of the same genus were domesticated independently, in different regions and by different peoples. More rarely, the same species was domesticated more than once, in different parts of the range of a widespread wild progenitor. These different, but congeneric (or conspecific) domesticate show similar changes resulting from human selection. This raises the question of whether their similar phenotypes are produced by similar genotypes or whether selection in different environments by humans of different cultural backgrounds, exerted on progenitor populations that differed genetically, resulted in different genetic pathways to the domesticated phenotype.

Domestication of crop is a good subject for the study of evolutionary process that happened about 12000 to 13000 years ago. The key role of selection in their origins and spread and diversification can also be studied from domestication. Recent studies such as, quantitative trait loci mapping, genome-wide association studies and whole genome resequencing studies have identified genes that are associated with initial domestication and subsequent diversification of crops. Together, these studies reveal the functions of genes that are involved in the evolution of crops that were under domestication, type of mutation that occurred during this process and parallelism of mutations that occurred in the same pathways and selective forces that were acting on these mutations.

Key concepts of domestication

- The most potent force influencing variation in cultivated populations was artificial selection, a specialized case of directional selection with humans acting as the agents of selection.
- The morphology and physiology of crop species were gradually changed through both conscious and unconscious selection by humans.
- Many traits associated with plant domestication are regulated by only a few genes that are close together on chromosomes.
- A high percentage of crop species are polyploid, with more than one set of chromosomes.
- Until recently, it was thought that once humans began cultivation, selection for the domestication genes would be rapid; however, new archaeological evidence indicates that pre-domestication cultivation may have actually taken 1000 years.
- During the process of crop domestication, genetic variability was greatly restricted by the selection of a limited number of desirable individuals.

Nature of domestication genes

Harlan *et al.* (1973) proposed that a distinct group of traits — later termed the "domestication syndrome" — would likely be selected during the initial stages of domestication. These traits are all associated with increasing ease of harvest and include reduced shattering, more determinate growth habit, greater inflorescence size, greater seed (or fruit) size, and loss of seed dormancy. In the time since Harlan first published this list in 1973, domestication genes have frequently been more loosely defined to include any gene selected on and consistently differing between a crop and its nearest wild relative (Weeden, 2007). Crop improvement after domestication can, however, also result in the fixation of traits that were likely not of interest to early farmers (e.g., dwarf growth habit of different crops).

In the past, domestication-related traits were widely believed to be conditioned by recessive, loss-of-function alleles (Lester, 1989); however, the results of QTL mapping studies as well as the lessons learned from the recent cloning of a handful of domestication genes show a less consistent pattern. While non-shattering appears to be a recessive trait in a number of cereals (Li *et al.*, 2006) and fruit weight is partially recessive in tomato and eggplant (Doganlar *et al.*, 2002), a number of QTL studies suggest that the genes underlying many other domestication-related traits act in a non-recessive manner (reviewed in Doebley *et al.*, 2006). Similarly, the view that domestication-related traits are typically conditioned by loss-of-function alleles has been challenged. In fact, only one of seven sequenced "domestication genes" reviewed by Doebley *et al.* (2006) is a loss-of-function allele (loss of red pericarp in rice according to Sweeney *et al.*, 2006). Other examples of loss-of-function alleles include the character affecting flowering time (Foucher *et al.*, 2003), dwarf habit in pea (Martin *et al.*, 1997), self-pruning habit in tomato (Pnueli *et al*, 1998) transition from two-rowed to six-rowed barley (Komatsuda *et al.*, 2007), etc.. Curiously, loss-of-function alleles appear to be relatively common for genes associated with crop improvement and/or varietal divergence. However, many "domestication genes" are amino acid substitutions and/or expression level differences caused by mutations in regulatory regions (Doebley *et al.*, 2006).

Impact of origin, domestication and cultivation of crop pants

- Survival of plants under natural conditions and alteration of crop plants to suit human needs (i.e. plant breeding) are contrary strategies.
- Domestication and cultivation of crop plants have been "unconscious plant breeding activities"
- Domestication and cultivation has created large crop diversity
- Most crop plants do not originate from the region where they are cultivated

Domestication and origin of crops

De Candolle (1886) was the first to develop the subject of origin of domesticated plant in his famous book, '*Origine des Plantes Cultivée'*. Nikolai I. Vavilov, a Russian scientist and pioneer of the plant genetic resource movement, showed that plant species are distributed unevenly on the earth. Some parts of the world are characterized by an astounding diversity of species. Northern countries and regions, such as Siberia, all of central and northern Europe and North America, on the other hand are poor in species composition. In certain areas of the world, like Costa Rica and El Salvador, the small Central American republics, tremendous concentrations of species are found.

Vavilov (1926) was the first to scientifically pursue the study of domestication and origins and proposed eight primary geographic "Centres of domestication" in the world, based on the regions having high crop varietal diversity.

All crop plants originated from wild plant populations. At the beginning of the domestication process, a number of changes began to appear in the genetic and physiological makeup of many crop species. Some of these changes were due to conscious selection such as increases in palatability and colour, but many of the others were the unconscious by product of the cultivation and harvesting. Changes in crop species during domestication were gradual, likely taking thousands of years. Many of the traits associated with plant domestication are regulated by only a few genes, and in many cases these genes have pleiotropic effects, where they affect a number of traits simultaneously. Many of the genes associated with domestication are also clustered close together on the same chromosome. The process of domestication severely limited the amount of genetic variation in crop species.

Wild grains were collected and eaten at least 105,000 years ago (Mercader, 2009). However, domestication did not occur until much later. The earliest evidence of small-scale cultivation of edible grasses is from around 21,000 BC with the Ohalo II people (an archaeological site in northern Israel) on the shores of the sea of Galilee (Snir, 2015). By around 9500 BC, the eight Neolithic founder crops – emmer wheat (*Triticum dicoccum*), einkorn wheat (*Triticum monococcum*), barley (*Hordeum vulgare*), peas (*Pisum sativum*), lentils (*Lens culinaris*), bitter vetch (*Vicia ervilia*), chickpea (*Cicer arietinum*), and flax (*Linum usitatissimum*) were cultivated in the Levant referring to a large area in the Eastern Mediterranean region of West Asia which include modern day Syria, Lebanon, Palestine, Israel, Jordan, Cyprus and most of Turkey (Melinda, 2011).Rice was domesticated in China by 6200 BCfollowed by soybean (*Glycine max*) and adzuki beans, *Vigna angularis* (Molina *et al*., 2011). Rice was cultivated in Indian sub-continent as early as 5000 BC. Rice was also independently domesticated in West Africa and cultivated by 1000 BC (Choi *et al*., 2019). Rye may have been cultivated earlier, but this claim remains controversial with the earliest known cultivation from 5700 BC, followed by mung bean (*Vigna radiata*), soybean and adzuki beans.

In sub-Saharan Africa, sorghum was domesticated in the Sahel region of Africa by 3000 BC, along with pearl millet by 2000 BC (Winchell *et al*., 2018). Yams (*Dioscorea* spp.) were domesticated in several distinct locations, including West Africa and cowpeas were domesticated by 2500 BC (Scarcelli *et al*., 2019). Other plant foods domesticated in Africa include watermelon, okra, tamarind and cowpeas, along with tree crops such as the kola nut and oil palm. Plantains were cultivated in Africa by 3000 BC and bananas by 1500 BC. In South America, agriculture began as early as 9000 BC, starting with the cultivation of several species of plants that later became only minor crops. In the Andes of south America, the potato was domesticated between 8000 BC and 5000 BC, along with beans, squash, tomatoes, peanuts and coca, Cassava was domesticated in the Amazon Basin no later than 7000 BC. Maize (*Zea mays*) found its way to South America from Mesoamerica, where wild teosinte was domesticated about 7000 BC and selectively bred to become domestic maize. Cotton (*Gossypium barbadense*)was domesticated in Peru by 4200 BC; another species of cotton (*Gossypium hirsutum*) was domesticated in Mesoamerica and became by far the most important species of cotton in the textile industry in modern times. Evidence of agriculture in the eastern United States dates to about 3000 BC. Several plants were cultivated, later to be replaced by the cultivation of "the three sisters" *viz*., maize, squash, and beans. These crops are called so because they nurture each other like family when planted together. The early inhabitants placed corn in small hills, planting beans around them and interspersing squash (*Cucurbita* spp.) throughout of the field. Sugarcane and some root vegetables were domesticated in New Guinea around 7000 BC. Bananas were cultivated and hybridized in the same period in Papua New Guinea.

Table 1: Eight independent Vavilovian centre of origin of crops

Centre of origin		Region and endemic species
1.	The Chinese Centre	This centre includes the mountainous regions of central and western China together with adjacent low lands. **Endemic species**: Soyabean, radish, turnip, pear, peach, plum, *Colacasia*, buckwheat, opium poppy, brinjal, apricots, oranges, tea, etc.
2.	The Indian (Hindustani) centre	This centre includes the Indian subcontinent (dominated by India, the area also includes present day Bangladesh, Nepal, Sri Lanka and Bhutan) including Burma (now Myanmar). **Endemic species**: Rice, red gram, chick pea, cowpea, mung bean, brinjal, cucumber, pointed gourd, sugar cane, black pepper, moth bean, rice bean, cotton, turmeric, indigo, millets, etc.
2a.	The Indo-Malay centre	This centre includes all of Malay Archipelago (part of Malaysia which hangs off the corner of mainland southeast Asia (stretching from the foot hills of the Himalayas in the north to the equator in the south), the Philippines and Indo-China. It is supplementary to Indian centre of origin.
3.	The Central Asia centre	This centre includes north-west India, Afghanistan, Uzbekistan and western China, western Punjab (Now in Pakistan), northern part of Pakistan, the Tadjik and Uzbek of former Soviet Republics (Now separate country Tajikistan, Uzbekistan) and western Tien shan (China). **Endemic species**: Bread wheat (*Triticum aestivum*), club wheat (*Triticum compactum*), sesame, linseed, muskmelon, carrot, onion, garlic, apricot, grape, hemp, cotton etc.
4.	The near east centre (Southwest Asia)	This centre includes transcontinental region around Iraq, Iran, Syria, Lebanon, Cyprus, Jordan, Israel, Palestine, Egypt, most of Turkey, Anatolia. **Endemic species**: Wheat, rye, pomegranate, almond, fig, cherry, walnut, alfalfa, Persian clover, etc.
5.	The Mediterranean centre	This centre includes Israel, Jordon, Syria, Cyprus, Turkey, Lybia, Lebanon. **Endemic species**: Durum wheat (*Triticum durum*), emmer wheat (*Triticum dicoccon*), oat, barley, lentil, pea, grass pea, broad bean, progenitor species of cole crops (broccoli, brussels sprouts, cabbage, cauliflower, collards, kale, and kohlrabi), asparagus, mint, etc.
6.	The Abyssinian centre	This region includes Ethiopia and parts of Somalia. **Endemic species**: Wheat, sorghum, bajra, safflower, castor, broad bean, okra, coffee, etc.
7.	South Mexican and central American centre	This centre includes southern parts of Mexico and small countries of central America which lie within a neck of land that joins north and south America (Guatemala, El Salvador, Costa Rica, Panama, Nicaragua, Hondurus, Haiti, Trinidad and Tobago, Cuba, etc.). **Endemic species**: Maize, French bean, lima bean, melon, pumpkin, sweet potato, arrow root, chilli, cotton, papaya, guava, avocado, etc.
8.	The South American centre	Mountainous areas and plateaus of Columbia, Ecuador, Peru, Paraguay, Bolivia and Chile **Endemic species**: Potato, sweet potato, lima bean, tomato, papaya, tobacco, quinine, cassava, rubber, ground nut, cocoa, pineapple etc.

The understanding of the number, size, and identity of the Vavilovian centres of domestication has changed substantially over the last 90 years of research. However, there is still no clear consensus as to the location and number of domestication sites (Gepts, 2008). Since Vavilov's time, eastern North America (Smith, 2006), New Guinea (Denham *et al.*, 2003), and the Amazon Basin (Olsen and Schaal, 2001) have been added as independent centres of domestication. China, the Andes and Ethiopia are later viewed as being less localized centres than was originally thought (Smith, 2006; Londo *et al.*, 2006). Fertile Crescent (Salamini *et al.*, 2002) and southern Mexico (Smith, 2006) remain as narrowly delimited centres of origin for multiple crops. The occurrence of several domestications in close proximity to one another suggests that agricultural technology arose with and followed the resident cultures that eventually migrated away from these regions (Willcox, 2005). It was also learned that high varietal diversity in a particular region is not a reliable indicator of a centre of domestication, as previously believed by Vavilov and his co-workers. Indeed, varietal diversity does not correlate well with archeological and molecular data regarding the origin of crops (Harlan, 1992), perhaps in part because elevated levels of diversity can result from secondary contact between crops and their wild relatives (Weissmann *et al.*, 2005). In fact, evolution of diverse landraces frequently involves introgression from wild relatives, as has been the case for cotton (Brubaker and Wendel, 1994), sorghum (Casa *et al.*, 2005), pea (Timmerman-Vaughan *et al.*, 2005), and sunflower (Burke *et al.*, 2005). Such evolution of landraces often occurs well outside the original site of domestication and such introgression can lead to false conclusions regarding the origin(s) of crops.

Depending on the domestication background, cultivated plants are usually divided into "Primary" and "Secondary" ones. The former include plants domesticated directly from wild state. Plants belonging to second category were initially weeds with respect to primary domesticates and started being cultivated later. According to Bioversity International and FAO, more than 1000 plant species constitute the global diversity in vegetable crops although 25-30 species are important commercially. The regions with rich diversity are tropical America (Central America, Mexico and South America), Asia and the Mediterranean. In the Asian region, India and China possess the maximum diversity in vegetable crops. Primary and secondary centers of diversity of major vegetable crops are presented in Table 2.

Table 2: Primary and Secondary centre of origin of important vegetable crops

Gene centre	Primary centre of origin and domestication	Secondary centre of origin
Chinese-Japanese	Eggplant, wax gourd, Chinese cabbage, welsh onion	Watermelon, amaranth
Indo-Chinese	Sponge gourd, ridge gourd, bottle gourd, sword bean, winged bean, taro, chayote, cucumber, yams, bitter gourd	Chinese cabbage, bottle gourd, yam bean, amaranth, cucumber, yard long bean, shallot
Hindustani Centre	Eggplant, wax gourd, cucumber, ridge gourd, sponge gourd, bitter gourd, hyacinth bean, pointed gourd, drumstick	Watermelon, melon, bottle gourd, amaranth, okra, vegetable cowpea
Central Asia	Onion, garlic, carrot, spinach, *Brassica* spp.	Eggplant, watermelon, melon, cauliflower
Near East	Onion, garlic, leek, *Brassica* spp.	Okra
Mediterranean	Cabbage, cauliflower, broccoli,kale, radish, beans	Sweet pepper, garlic, okra, onion, lima bean
African	Eggplant, watermelon, muskmelon, bottle gourd, cowpea, okra	Amaranth, shallot, etc.
European	Lettuce	Onion, cauliflower, cabbage, carrot, common bean, spinach
Central-American and Mexican region	Tomato, chilli, pumpkin, squashes (*Cucurbita pepo*), sweet potato, amaranth, etc.	–
South American	Potato, tomato, chilli, cassava, *Xanthosoma*, pumpkin and squashes, lima bean, chayote(*Sechium edule*)	Common bean, tomato, brinjal, muskmelon, watermelon, squashes, onion, lettuce, lima bean.

• Gupta *et al.* (1995)

Plant genetic resources in India

Plant genetic resources (PGR) are one of essential components of agro-biodiversity and defined as the genetic material of plants having value as a resource for present and future generations. PGR hold the key to the very foundation of agriculture as well as food and nutritional security for the world. Agricultural biodiversity, or agro-biodiversity (a subset of biodiversity) is defined as 'all crops and live stocks, their wild relatives, and all interacting species of pollinators, symbionts, pests, parasites, predators and competitors (Qualset *et al.*, 1995). Indian subcontinent has a rich and varied heritage of biodiversity in widely divergent habitats from tropical rainforests to alpine vegetation and from temperate forests to coastal wetlands. Indian subcontinent is one of the eight centres of origin (Vavilov, 1951) and is one of the 12 mega gene centres of the world. The 12 mega gene centres of the world are Brazil, China,

Columbia, Costa Rica, Ecuador, India, Indonesia, Kenya, Mexico, Peru, South Africa and Venezuela. India possesses 7-8% of the world's biodiversity and 11.9 % of world's flora and supporting 16% of the major forest types, varying from alpine pastures in the Himalayas to temperate, sub-tropical, tropical forests, and mangroves in the coastal areas. Agro-biodiversity of the Indian centre of origin (Hindustani centre of origin) are concentrated mainly in the North-East, Western Ghats, North West Himalayas and the Andaman and Nicobar Islands. Of the 49,219 higher plant species, 5,725 are endemic and belong to 141 genera under 47 families (Nayar, 1980). Of these recorded higher plant species in India, 3,500 are found in the Himalayas and adjoining regions and 1,600 in the Western Ghats alone (Arora, 1991). Earth's biologically richest and most endangered terrestrial ecoregions is termed as "Hotspots". Around the world, 34 biodiversity hotspots exist as of today and these sites support nearly 60 % of the world's plant, bird, mammal, reptile, and amphibian species, with a very high share of endemic species (Myers *et al.*, 2000).

Crop wild relatives

Crop wild relatives (CWR) are wild taxa closely related to crop plants, including wild progenitors and/or wild forms of crops. Maxted *et al.* (2006) defined a CWR as a wild plant taxon that has an indirect use derived from its close genetic relationship to a crop. The closer the species related, the more the possibility/ practicality to get their traits incorporated. They form an important source of useful traits such as agronomic/ horticultural quality, biotic and abiotic stresses, which forms the critical component for food security and environmental sustainability in the present era of climate change. CWRs are often associated with disturbed habitats and neither these habitats are offered adequate protection by ecosystem conservation agencies nor their diversity properly conserved *ex situ* (Maxted and Kell, 2009). Genetic diversity of CWR, like that for many species, is at a declining stage which necessitates the need to establish CWR inventories which is also an indispensable tool for exploration, surveys and collection of CWR (Maxted and Kell, 2009). Therefore, the need for novel genes for developing climate resilient varieties, increasing pressure on wild species populations and habitats and the present meagre *ex situ* collections, all accentuate the importance of locating and collecting germplasm of wild relatives. The Indian sub-continent, one of the Vavilovian centers of origin is endowed with rich diversity in several vegetable crops. Diversity of different vegetable crops in different phyto-geographical regions of India is presented in Table 3.

Table 3: Diversity of different vegetable crops in different phyto-geographical regions of India

	Phyto-geographical region	**Geographical ranges**	**Variability in vegetable crops and CWR**
1.	North-western Himalayan eegion	Jammu and Kashmir, Himachal Pradesh and Uttarakhand	*Allium* species (onion, garlic, leek, shallot), spinach and leafy Brassica, Chenopods, Amaranthus, *Beta vulgaris*; cucurbits (summer squash, pumpkin, cucumber (including hill type), chilli, sweet pepper, pea, faba bean, French bean, cowpea, drumstick, radish, artichoke, potato, colocasia, tomato and parsley; Underutilized vegetables like, cho-cho (*Sechium edule*), meetha karela (*Cyclanthera pedata*), tree tomato (*Solanum betaceum*), *Basella rubra, Flemingia vestita, Moghania vestita, Solanum khasianum,* etc.
2.	Eastern Himalayan region	Sikkim, Darjeeling district of West Bengal; Bhutan and the northeastern Indian states of Assam, Arunachal Pradesh, Manipur, Mizoram, Tripura, Meghalaya	*Allium* species (onion, garlic, leek, shallot), spinach and leafy *Brassica*, Chenopods, *Amaranthus, Beta vulgaris*; cucurbits (summer squash, pumpkins, cucumber (including hill type), brinjal, chilli, spinach beet, sweet pepper, pea, faba bean, French bean, cowpea, moringa, radish, artichoke, potato, colocasia, tomato, amaranth and parsley; Underutilized vegetables like, cho-cho, meetha karela, tree tomato, *Abelmoschus manihot* spp. *tetraphyllus, Solanum khasianum, Solanum torvum, Solanum sisymbriifolium, Solanum forex, Solanum verbascifolium, Cucumis histrix, Luffa echinata,* etc.
3.	The Northeastern region	Arunachal Pradesh, Assam, Manipur, Meghalaya, Mizoram, Nagaland, Sikkim and Tripura	*Cucumis* spp., cucumber, ridge gourd, pumpkin, summer squash, ash gourd, pointed gourd, leafy vegetables (amaranth, *Brassica* spp., *Malva verticillata* , etc.), French bean, chilli, tomato, brinjal, okra, taro, yams, bitter gourd, teasle gourd, etc Underutilized vegetables like, winged bean, sword bean, lima bean, meetha karela, tuberous vegetable, sohphlang (*Flemingia vestita*)
4.	Northern plains including Terai region	Uttar Pradesh, Bihar, West Bengal and the stretch of land from the Yamuna river in the west to the Brahmaputra river in the east	*Cucumis* spp., cucumber, snap melon (phoot), ridge gourd, smooth gourd, pumpkin, summer squash, ash gourd, Indian round gourd, pointed gourd, brinjal, tomato, chilli, okra, *Brassica* leafy vegetables, carrot, onion, garlic and taro/aroids, spinach beet, amaranth. Underutilized vegetables like, *Abelmoschus manihot, Solanum indicum, Solanum*

	Phyto-geographical region	Geographical ranges	Variability in vegetable crops and CWR
			khasianum, Solanum surattense, Cucumis sativus var. sikkimensis, Edgeria darjelingensis, Melothria assamica, Momordica cochinchinensis, Sechium edule, Tuladiantha cordifolia, Basella rubra.
5.	North western plain region	Punjab, Haryana, Delhi, Rajasthan (except Kota and Udaipur divisions) and Western UP (except Jhansi division), parts of J&K (Jammu and Kathua districts) and parts of Himachal Pradesh (Una district and Paonta valley) and Uttaranchal (Terai region)	Cucurbits (*Cucumis*, *Momordica* and *Citrullus* spp.), brinjal, amaranth, *Chenopodium*, okra, carrot, peas, chilli, beet, fenugreek, onion, garlic, *Vicia sativa*, *Vicia faba* Underutilized vegetables like, *Abelmoschus manihot* spp. *tetraphyllus* var. *pungens, Abelmoschus tuberculatus, Solanum indicum, Solanum torvum, Solanum khasianum, Solanum surrattense, Solanum hispidum, Cucumis hardwickii, Cucumis trigonus*
6.	Central region including plateau region	Central highlands includes the states of Maharashtra, Madhya Pradesh, Telangana, Andhra Pradesh, Karnataka, and Tamil Nadu	Pumpkins, summer squashes, cucumber, ash gourd, Indian round gourd, bitter gourd, ridged gourd, okra, brinjal, chilli, tomato; root, bulbous crops, leafy vegetables , fenugreek, amaranth, onion, Underutilized vegetables like, *Solanum surattense, Solanum torvum, Solanum nigrum, Solanum khasianum, Cucumis setosus, Luffa acutangula* var. *acutangula*
7.	Western and Eastern peninsular region	Rajasthan, Maharashtra, Gujarat, Telangana, Karnataka, Tamil Nadu, Kerala, and Odisha (Western peninsular region), Jharkhand, northern Chhattisgarh, and West Bengal's Purulia area (Eastern peninsular region)	Brinjal (eastern peninsular region), okra, chilli, cucurbits (pumpkin, cucumber, bitter gourd, bottle gourd, summer squashes, snake gourd, ridge gourd), amaranth, chenopods, spinach, *Beta vulgaris*, French bean, root and bulbous crops, Underutilized vegetables like, sword bean, mucuna bean, *Dolichos uniflorus, Dolichos purpureus, Basella rubra* (eastern peninsular region), *Abelmoschus crinitus, Abelmoschus angulosus, Abelmoschus ficulneus, Abelmoschus moschatus, Abelmoschus manihot* var. *tetraphyllus, Solanum trilobatum, Solanum indicum, Solanum insanum, Solanum pubescens, Solanum surattense, Solanum torvum, Luffa acutangula* var. *acutangula, Melothria angulate.*

Pandey *et al.* (2019)

Conservation of genetic resources

A species or a population sample of a particular part of its genetic variation can be maintained through *in situ* or *ex situ* conservation. *In situ* conservation is the preservation of species and populations of living organisms in a natural state in the habitat where they naturally occur. This method preserves both the population and the evolutionary processes that enable the population to adapt by managing organisms in their natural state or within their normal range. For example, large ecosystems may be left intact as protected reserve areas with minimal intrusion or alteration by humans. *Ex situ* conservation is the preservation and propagation of species and populations, their germ cell lines, or somatic cell lines outside the natural habitat where they occur. This method maintains the genetic diversity extant in the population in a manner that makes samples of the preserved material readily available. Different collections for *ex situ* conservation are mentioned below.

- **Base collection**: Collection of seed samples for long term conservation which are maintained at –18 to –20 °C temperature to ensure seed viability for maximum possible time period. The moisture content of seed to be stored as base collections should be between 3 and 7 % depending on the species.
- **Active collection**: Collection of seed samples for immediate use which are maintained in conditions that ensure at least 65 % viability for 10–20 years. The moisture content of seeds to be stored as active collections should be between 3 and 8 % for seeds having poor storability and between 7 and 11 % for seeds having good storability, depending on the temperature maintained for storage.

These collections are conserved in different types of storage facilities. Depending on the duration of storage, three basic types of storages are mentioned below.

- **Short term storage**: The period for short term storage of seeds is from one year up to 18 months. It requires a cool and dry atmosphere (20–22 °C and 45–50 % relative humidity) where the seeds can be conveniently stored for one to two years without much loss in their viability.
- **Medium term storage**: The active collection, which are generally larger than those meant for base collection are conserved in the medium term storage. The accessions are for regular distribution and therefore, the time period for storage is not more than 5 years. The active collections are stored at temperatures ranging from 0–10 °C coupled with of 20–30 % relative humidity.
- **Long term storage**: This storage facility is provided for base collection, where the seeds that meet all the above mentioned criteria for National Gene Bank (NGB) conservation are maintained at –18 to –20 °C with of 20–30 % relative humidity.

References

Anderson, M.K. 2004. Pre-agricultural plant gathering and management. In: *Encyclopaedia of plant and crop science* (Ed. Goodman R.) Marcel Dekker, New York, pp. 1055–1060.

Arora, R.K. 1991. Plant diversity in Indian gene centre. In: *Plant Genetic Resources – Conservation and Management* (Eds. Paroda, R.S., Arora, R.K.). New Delhi, India: IPGRI, Regional Office for South Asia, pp. 25-54.

Brubaker, C.L., and Wendel, J.F. 1994. Reevaluating the origin of domesticated cotton (Gossypium hirsutum; Malvaceae) using nuclear restriction fragment length polymorphisms (RFLPs). *Amer. J.Bot.,* **81**: 1309 – 1326 .

Burke, J. M., Knapp, S.J. and Rieseberg, L . H . 2005. Genetic consequences of selection during the evolution of cultivated sunflower. *Genetics*, **171** : 1933 – 1940 .

Casa, A.M. , Mitchell, S.E., Hamblin, M.T., Sun, H., Bowers, J.E., Paterson, A.H., Aquadro, C.F. and Kresovich, S. 2005 . Diversity and selection in sorghum: Simultaneous analyses using simple sequence repeats. *Theor. Appl. Genet.*, **111** : 23 – 30 .

Casas, A., Caballero, J., Mapes, C., and Zárate, S. 1997. Manejo de la vegetación, domesticación de plantas y origen de la agricultura en Mesoamérica. *Bot. Sci.* 61: 31–47. doi: 10.17129/botsci.1537

Choi, J. Y., Zaidem, M., Gutaker, R., Dorph, K., Singh, R.K., Purugganan, M.D. 2019. The complex geography of domestication of the African rice *Oryza glaberrima. PLOS Genetics*, **15** : 1007414. doi:10.1371 / journal.pgen.1007414.

Clement, C.R. 1999. 1492 and the Loss of Amazonian Crop Genetic Resources. I. The Relation between Domestication and Human Population Decline. *Econ. Bot.*, **53**: 188-202

De Candolle, A. 1886.Origin of Cultivated Plants ('Origine des Plantes Cultivée'), Cambridge University Press. https://doi.org/10.1017/CBO9781139107365

Denham, T. P. , Haberle, S. G. , Lentfer, C. , Fullagar, R. , Field, J. , Therin, M. , Porch, N. and Winsborough, B. 2003. Origins of Agriculture at Kuk Swamp in the Highlands of New Guinea. *Science*, 301: 189 (2003). DOI: 10.1126/science.1085255

Doebley, J., Gaut, B.S. and Smith. B.D. 2006. The molecular genetics of crop domestication. *Cell*, **127**: 1309 – 1321 .

Doganlar, S., Frary, A., Daunay, M.C., Lester, R.N. and Tanksley, S.D. 2002. A comparative genetic linkage map of eggplant (*Solanum melongena*) and its implications for genome evolution in the Solanaceae. *Genetics*, **161**: 1697 – 1711 .

Evans, L.T. 1993. *Crop Evolution, Adaptation, and Yield.* Cambridge Univ. Press, UK.

Foucher, F., Morin, J., Courtiade, J., Cadioux, S., Ellis, N., Banfield, M. and Rameau, C. 2003. DETERMINATE and LATE FLOWERING are two TERMINAL FLOWER1/ CENTRORADIALIS homologs that control distinct phases of flowering initiation and development in pea. *The Plant Cell*, **15** : 2742 – 2754

Gepts, P. 2004. Crop domestication as a long-term selection experiment. In: *Plant breeding reviews* (Ed. Janick, J.), Wiley, New York, New York, USA, pp. 1 – 44.

Gepts, P. 2008. Tropical environments, biodiversity and origin of crops. In: *Genomics of Tropical Crop Plants* (Eds. Moore, P.and Ming, R.), Springer, pp. 1-20.

Gupta, P.N., Rai, M. and Rana, R.S. 1995. Centres of origin and genetic variability of vegetable crops. In: *Genetic resources of vegetable crops; management, conservation and utilization* (Ed. Rana, R.S.). National Bureau of Plant Genetic Resources, New Delhi. pp. 52-62.

Harlan , J.R., DeWet, J. M. J. and Price, E. G. 1973 . Comparative evolution of cereals. *Evolution,* **27**: 311 – 325

Harlan, J.R. 1992 Agricultural Origins and crop domestication m the Mediterranean region. *Diversity*, 11(1/2): 14-16

Heiser, C. B. 1951. The sunflower among the North American Indians. *Proc. Amer. Philos. Soc.*, **95**: 432–448.

Komatsuda, T. , Pourkheirandish, M., He, C., Azhaguvel, P., Kanamori, H. , Perovic, D., Stein, N., Graner, A., Wicker, T., Tagiri, A., Lundqvist, U., Fujimura, T., Matsuoka, M., Matsumoto, T. and Yano, M. 2007. Six-rowed barley originated from a mutation in a homeodomain-leucine zipper I-class homeobox gene. *Proc.Nat. Acad.Sci.*, USA, **104** : 1424 – 1429

Lester , R . N . 1989 . Evolution under domestication involving disturbance of genic balance. *Euphytica*, 44 : 125 – 132 .

Li , C. , Zhou, A . and Sang, T. 2006. Rice domestication by reducing shattering. *Science,* **311**: 1936 – 193.

Lindig-Cisneros, R., Benrey, B. and Espinosa-García, F.J. 1997. Phytoalexins, resistance traits, and domestication status in *Phaseolus coccineus* and *Phaseolus lunatus*. *J. Chem. Ecol.* **23**:1997–2011.

Londo , J.P., Chiang, Y.C., Kuo, H. H., Tzen, Y. C. and Schaal, B.A. 2006. Phylogeography of Asian wild rice, *Oryza rufipogon*, reveals multiple independent domestications of cultivated rice, *Oryza sativa*. *Proc. Natl. Acad. Sci*., U S A, 103:9578-9583.

Martin , D.N., Proebsting , W. M. and Hedden, P. 1997. Mendel ' s dwarfing gene: cDNAs from the *Le* alleles and function of the expressed proteins. *Proc. Natl. Acad. Sci*., U S A, **94** : 8907 – 8911

Maxted, N. and Kell, S. 2009. Establishing conservation priorities for crop wild relatives. *In: Establishment of a Global Network for the in situ Conservation of crop wild relatives: Status and needs* (Eds. Maxted, N., Dulloo, E., Ford-Lloyd, B., *Kell*, S.P., Iriondo, J.M.), Back ground Study Paper No. 39, School of Biosciences, University of Birmingham, UK. http:// www.fao.org/docrep/013/i1500e/i1500e18a

Maxted, N., Ford-Lloyd, B.V., Jury, S., Kell, S. and Scholten, M. 2006. Towards a definition of a crop wild relative. *Biodivers. Conserv*., **15**: 2673-2685.

Melinda, Z. 2011. The Origins of Agriculture in the Near East. *Curr. Anthropol.*, **52** (S4): 221–235. doi:10.1086/659307.

Mercader, J. (2009). Mozambican grass seed consumption during the Middle Stone Age. *Science*. **326**: 1680–1683. doi: 10.1126/science.1173966.

Molina, J., Sikora, M. Garud, N. Flowers, J. M. Rubinstein, S. Reynolds, A. Huang, P. Jackson, S. Schaal, B.A. Bustamante, C.D. Boyko, A.R. and Purugganan, M.D. 2011. Molecular evidence for a single evolutionary origin of domesticated rice. *Proc. Nat. Acad. Sci*., USA, **108** : 8351–8356.

Myers, N., Mittermeier, R.A., Mittermeier, C.G., da Fonseca G.A.B. and Kent, J. 2000. Biodiversity hotspots for conservation priorities. *Nature*, **403**:853-858.

Nayar M.P. 1980. Endemism and pattern of distribution of endemic genera (angiosperm) in India. *J. Econ. Tax. Bot*., **1**: 99-110.

Olsen , K . and Schaal, B. A. 2001. Microsatellite variation in cassava (*Manihot esculenta*, Euphorbiaceae) and its wild relatives: further evidence for a southern Amazonian origin of domestication. *Amer. J. Bot*., **88** : 131-142.

Pandey, A., Panwar, N.S., Singh, R. and Ahlawat, S.P. 2019.*Vegetables: Status and Priorities for Exploration and Germplasm Collection in India*. ICAR-National Bureau of Plant Genetic Resources, New Delhi, India, 97p.

Pnueli , L., Carmel-Goren, L., Hareven, D., Gutfinger, T., Alvarez, J., Ganal, M., Zamir, D. and Lifschitz, E. 1998. The SELF-PRUNING gene of tomato regulates vegetative to reproductive switching of sympodial meristems and is the ortholog of CEN and TFL1. *Development*, **125**:1979-1989.

Qualset, C.O., McGuire, P.E. and Warbuton, M.L. 1995. Agrobiodiversity' key to agricultural productivity. *Calif. Agric.*, **49**: 45-49.

Rosenthal, J.P. and Welter, S.C. 1995. Tolerance to herbivory by a stemboring caterpillar in architecturally distinct maizes and wild relatives. *Oecologia,* **102**:146–55

Salamini, F., Ozkan, H., Brandolini, A., Schäfer Pregl, R. and Martin, W. 2002. Genetics and geography of wild cereal domestication in the near east. *Nat. Rev. Genet.*, **3**: 429-441.

Scarcelli, N., Cubry, P., Akakpo, R., Thuillet, A.C., Obidiegwu, J., Baco, M. N., Otoo, E., Sonké, B., Dansi, A., Djedatin, G., and Mariac, C. 2019. Yam genomics supports West Africa as a major cradle of crop domestication. *Science Advances,* 5 : eaaw1947. Bibcode:2019SciA....5.1947S.

Schwanitz, F. 1966. *The Origin of Cultivated Plants*. Harvard Univ. Press, Cambridge, MA, USA.

Smith, B.D. 2006. Eastern North America as an independent centre of plant domestication. *Proc. Nat. Acad. Sci.*, USA, **103**: 12223–12228.

Snir, A. 2015. The Origin of Cultivation and Proto-Weeds, Long before Neolithic Farming. *PLOS One*, **10**: e0131422.

Sweeney, M.T., Thomson, M.J., Pfeil, B.E. and McCouch, S. 2006. Caught red-handed: Rc encodes a basic helix-loop-helix protein conditioning red pericarp in rice. *The Plant Cell*, **18**: 283 – 294

Timmerman - Vaughan, G.M., Mills, A., Whitfield, C., Frew, T., Butler, R., Murray, S., Lakeman, M., McCallum, J., Russell, A. and Wilson, D. 2005 . Linkage mapping of QTL for seed yield, yield components and developmental traits in pea. *Crop Science,* **45** : 1336 – 1344.

Vavilov, N. I. 1926. *Studies on the Origin of Cultivated Plants*. Leningrad, 248p.

Vavilov, N.I. 1951. Phyto-geographical basis of plant breeding. In: *The Origin, Variation, Immunity and Breeding of Cultivated Plants* (Selected Writings of N.I. Vavilov and translated by Chester, K.S.), *Chronica Botanica*, **13**:1-364. Waltham Mass., USA.

Weeden, N . 2007. Genetic changes accompanying the domestication of *Pisum sativum* : Is there a common genetic basis to the ' domestication syndrome ' for legumes? *Ann.Bot.*, **100**: 1017 – 1025.

Weissmann, S., M. Feldman, and J . Gressel. 2005. Sequence evidence for sporadic intergeneric DNA introgression from wheat into a wild Aegilops species. *Mol. Biol. Evol.*, **22** : 2055 – 2062.

Willcox, G. 2005. The distribution, natural habitats and availability of wild cereals in relation to their domestication in the Near East: Multiple events, multiple centres. *Veg. Hist. Archaeobot.*, **14**:534-541

Winchell, F., Brass, M., Manzo, A., Beldados, A., Perna, V., Murphy, C., Stevens, C. and Fuller, D. Q. 2018. On the Origins and Dissemination of Domesticated Sorghum and Pearl Millet across Africa and into India: a View from the Butana Group of the Far Eastern Sahel". *The Afr. Archaeol. Rev.,* **35** :483–505.

Zohary, D. 1984. Modes of evolution in plants under domestication. In: *Plant biosystematics* (Ed. Grant, W.), Academic Press, Toronto, pp. 579–596.

3

Reproduction, Mode of Pollination, Inbreeding Depression and Deviation of Sexual Reproduction

Mode of reproduction

All the living organisms produce individuls of their own kind for the survival of their species which is called reproduction. Hence, reproduction is the biological process by which new individual organism (offspring) are produced from their "parents". In vegetable crops, modes of reproduction may be sexual, asexual and both. Most of the vegetable crops *viz*., tomato, brinjal, chilli, sweet pepper, peas, cowpea, hyacinth bean, French bean, cluster bean, runner bean, fenugreek, lettuce, cabbage, cauliflower, knolkhol, Brussels sprouts, sprouting broccoli, radish, beet, turnip, carrot, onion, leek, amaranth, palak, spinach, asparagus, pumpkin, bottle gourd, watermelon, muskmelon, ridge gourd, snake gourd, wax gourd, summer squash, bitter gourd, cucumber, sweet corn, coriander, parsnip, etc. are exclusively propagated sexually for both commercial cultivation and breeding purpose. Some vegetable crops like, pointed gourd, teasle gourd, yam, taro, Jerusalem artichoke, globe artichoke, etc. are exclusively propagated through asexual mode of reproduction. On the other hand, some vegetable crops like, potato, sweet potato, cassava, annual moringa, etc. can be propagated by both sexual and asexual modes of reproduction however, they are generally propagated asexually for commercial cultivation and their sexual reproduction is employed for the purpose of breeding.

Sexually propagated crops

Sexual reproduction involves two phases in the life cycle: diplophase (2n), associated with the sporophyte, *i.e.*, the plant itself and haplophase (n), associated with the gametophyte which produce gametes. Two types of spores emerge in flowers: microspore, forming in the anther which gives rise to male gametophyte (pollen) and megaspores, developed in the ovules which give rise to female gametophyte. Formation of haploid mega and microspores by meiosis marks the onset of haplophase in the plants. Having found its way to the stigma of the pistil, the pollen grains germinate to produce pollen tube, which grows down through the style towards the ovule all the way into the female gametophyte, *i.e.*, the embryo sac. Out of the two male gametes formed in the pollen tube, one fuse with the egg cell contained in the embryo sac, thus effects fertilization which result in a diploid (2n) zygote and marks the diplophase (2n) of the plant. The other male gamete fuses with two fused polar nuclei (2n) which result in the triploid (3n) primary endosperm nucleus. Sexually propagated crops can be categorized according to their mode of pollination.

The process by which pollen grains are transferred from anthers to stigma is referred as pollination. Pollination is of two types: *viz.* i) autogamy or self-pollination and ii) allogamy or cross-pollination. Transfer of pollen grains from the anther to the stigma of the same flower is known as autogamy or self-pollination while, transfer of pollen grains from the anther to the stigma of different flower is known as allogamy or cross-pollination.

Grouping of vegetable crops according to mode of pollination

- **Self-pollinated crops:** In these crops, pollination involves the stigma of the same flower that produces the pollens which results natural self-fertilization. In these crops, natural cross-pollination is normally less than 5.0 percent but sometimes may be slightly higher but do not exceed 10 per cent. Different self-pollinated vegetable crops are potato, tomato, brinjal, peas, cowpea, hyacinth bean, French bean, cluster bean, runner bean, fenugreek, lettuce, etc.
- **Cross-pollinated crops:** In these crops, pollens from the flowers of one plant are transferred by some agents (insects, wind, etc.) to the stigma of the flowers of another plant resulting cross-pollination and natural cross-fertilization. In cross-pollinated vegetable crops, natural cross-pollination generally exceeds 50 to 60 per cent which reach even to the extent of 100 per cent. Different cross-pollinated vegetable crops are cabbage, cauliflower, knolkhol, Brussels sprouts, sprouting broccoli, radish, beet, turnip, carrot, onion, leek, amaranth, palak, spinach, asparagus, pumpkin, bottle gourd, watermelon, muskmelon, ridge gourd, snake gourd, wax gourd, summer squash, bitter gourd, cucumber, sweet corn, pointed gourd, teasle gourd, cassava, sweet potato, coriander, parsnip, etc.
- **Often-cross-pollinated crops:** These crops are predominantly self-pollinated but natural cross-pollination sometimes exceeds 10 per cent and may even reach 30% depending on the variety, location, presence of insect pollinators and favourable environmental condition. Different often cross-pollinated vegetable crops are okra, lima bean, chilli and sweet pepper. Brinjal is usually a self-pollinated crop however, in favourable condition cross-pollination in the long-styled flowers may exceed 10 per cent.

Evolutionary advantages of cross and self-pollination

Cross-pollination is indicative of high efficiency of reproduction systems based on cross-fertilization. Cross-pollination clearly has certain evolutionary advantages. The seeds formed by outbreeding due to cross-pollination may combine the hereditary traits of both the parents, and the resulting offspring generally are more varied than would be the case after inbreeding through self-pollination. In a changing environment, the genetic variability within a cross-pollinated population may enable some individuals to be adapted to their new situation, ensuring survival of the species, whereas the individuals resulting from self-pollination might all be unable to adjust.

Vegetatively propagated crops

The process in which new plants are grown from the old parts of another plant like roots, shoots, leaves and other vegetative plant parts without involving sexual reproduction is termed as vegetative propagation. Plants produced in this way have characteristics identical to the parent plants (true to the type) which is the main and most important advantage of vegetative propagation. These vegetable crops generally reproduce asexually and plants regenerate from different vegetative parts of the mother plant.

Crop	**Plant parts for regeneration**
Potato (*Solanum tuberosum*)	Underground stem tuber
Sweet potato(*Ipomoea batatas*)	Underground root tuber and above ground vines
Elephant-foot yam (*Amorphophallus paeoniifolius*)	Corm, corm pieces and cormel
Taro (*Colocasia esculenta*)	Corm, cormel, head sets (top couple of centimetres of the corm)
Giant swamp taro (*Cyrtosperma chamissonis)* Swamp taro (*Colocasia esculenta* var. *stolonifera*)	Sucker, corm
Garlic (*Allium sativum*)	Segment of bulb (clove), areal bulbil
Pointed gourd (*Trichosanthes dioica*)	Fleshy root and vine cuttings
Teasle gourd (*Momordica subangulata* Blume subsp. *renigera*)	Fleshy root
Cassava (*Manihot esculenta*)	Stem cutting
Drumstick (*Moringa oleifera*)	Limb cutting
Jerusalem artichoke (*Helianthus tuberosus*)	Root tuber
Globe artichoke(*Cynara cardunculus* var. *scolymus*)	Root suckers or offshoots and division of old crown
Yam (*Dioscorea* spp.)	Pieces of stem tuber and vine cuttings
Rhubarb (*Rheum rhabarbarum*)	Division of crowns

All these vegetable crops are perennial in growth habit excepting potato, which is annual. Sweet potato, taro, pointed gourd and cassava, though perennial, yet are cultivated as annual crop. The annual drumstick (*Moringa pterygosperma*) is propagated readily through seed, however seed propagation in other vegetatively propagated vegetable crops is not generally followed unless required for the breeding approaches or special purposes (like, true potato seeds for cultivation) because of some inherent difficulties and/or special requirements for induction of flowering in those crops.

- **Potato:** Potato flowers and set fruits only in cool temperature and long day condition. Hence, it set fruits in high hill condition when grown in summer months under long-day condition. In the short-day and cool temperature condition in the plains, flowering and fruit set can be induced with the arrangement of artificial light to increase the photoperiod for at least 4-5 hours beyond the normal day length.

- **Cassava:** It is highly heterozygous and monoecious in sex form *i.e.*, production of both male and female flowers on the same inflorescence which favours cross-pollination. Male sterility, which is commonly found in cassava, also enhances cross-pollination. Sexually propagated seedlings are highly variable in character due to heterozygosity and at the same time it takes long time to give potential yield. However, hybridization following clonal selection is an important method to develop improved variety.
- **Sweet potato:** Flowering itself is a major constraint in this crop and a number of elite clones do not flower. Flowers are bisexual but sterility of different intensities is found because of hexaploidy and accumulation of recessive sub-lethal mutations through long clonal propagation. The clones are heterozygous due to cross-pollination hence, the sexual seedlings are highly variable in nature. However, sexual reproduction is conventionally employed for breeding purpose.
- **Yam:** All the members of yams are dioecious hence, the male and female flowers are borne on separate plant. Flowering and seed production is greatly limited due to shy flowering and poor seed set.
- **Taro:** Generally, flowering is rare in most the taro types. Inflorescence, the spadix, is cylindrical with female flowers are borne at the base and a longer protruding male section of staminate flowers with aborted flowers in-between.
- **Elephant-foot-yam:** The crop cycle of two to three years is required for the emergence of inflorescence, although very rarely. The spadix is extended bearing male flowers at the top and female flowers at the bottom and the neuters are absent. There is complete absence of seed set under natural condition due to extreme protogyny coupled with the delayed opening of the spadix.
- **Pointed gourd:** Seeds are recalcitrant in nature, which lose viability very soon. The seedlings normally do not flower in the first year. The seedlings from the open pollinated seeds are highly variable due to heterozygosity and generally, low yielder than the mother clone. The seedling population resulting from open pollination with the pollens of the male clone gives 50 per cent each of male (unproductive) and female (productive) plants. Pointed gourd clones have two slightly different chromosomes, may be designated as X and Y which in combination determine the sex of the clones. Female clones have two X chromosomes in their cells (XX), while the male clones have one X and one Y (XY). During open pollination, the male gamete of the male clone having haploid chromosome (X or Y) have the equal opportunity of fertilizing the ovules in the ovary of the female clone having haploid chromosome (X). Such open pollination results in the development of seeds with either female (XX) or male (XY) sex form.

- **Garlic:** Most of the garlic types are non-bolting where inflorescence does not emerge. In most of the cases, the inflorescence, if emerge, contain sterile flowers and bulbils (aerial bulbs). Very few cultivars bear fertile flowers and form viable seeds that germinate.

Floral organization influencing mode of pollination and reproduction

The structural features of flowers, which are stable traits, are of great importance in selecting controlled pollination procedures without which breeding of any crop is impossible. In some plant species, the flower structure favours self-pollination, whereas, in others, it favours cross-pollination. Hence, both the breeders and seed producers must be fully aware of the flower structure of the crop.

Floral organization favouring self-pollination

Bisexual flower structure and homogamy are the main floral features which favour self-pollination.

- **Bisexuality:** In this condition both male and female reproductive organs are present in the same flower which favours self-pollination. In fact, without this condition, natural self-pollination is never possible.
- **Homogamy:** This is the condition where the anthers and the stigmas of a bisexual flower mature at the same time which favours self-pollination. In almost all the vegetable crops, normally homogamy occurs after the flowers have opened (Chasmogamy). Homogamy occurs in tomato, brinjal, chilli, sweet pepper, okra, peas, French bean, hyacinth bean, cluster bean, runner bean, lettuce, etc.
- **Cleistogamy:** In this condition, the bisexual flowers never open and therefore, the self-pollination is only the way of pollination. However, true cleistogamy is not generally found in vegetable crops. In peas and beans, stamens and pistil in opened flower remain enveloped inside a tube-like structure of jointed petals called "keel" which ensures strict self-pollination however, it cannot be termed as cleistogamy.

Floral organization favouring cross-pollination

a) Decliny: Decliny or unisexuality is a condition in which either staminate (male) or pistillate (female) / bisexual flowers are borne separately on the same plant or on different plants.

- **Monoecious sex form:** In this condition, staminate and pistillate flowers are borne separately on the same plant, *e.g.*, pumpkin, bottle gourd, water melon, some varieties of musk melon, ridge gourd, snake gourd, wax gourd, summer squash, bitter gourd, cucumber, amaranth, sweet corn, cassava, etc.
- **Andro-monoecious sex form:** In this condition, staminate and hermaphrodite flowers are borne separately on the same plant, *e.g.*, most of the varieties of muskmelon grown in India.

- **Androecious sex form**: In this condition, only staminate flowers are produced turning the plant as super male. It is a rare phenotype in muskmelon (*Cucumis melo* L.)
- **Gynoecious sex form:** In this condition, only pistillate flowers are borne on the plant, turning the plant as super female, *e.g.*, some forms of cucumber, muskmelon and bitter gourd.
- **Dioecious sex form:** In this condition, staminate and pistillate flowers are borne on separate plants meaning that the male and female plants are separate, *e.g.*, pointed gourd, asparagus, spinach, teasle gourd, ivy gourd, yam, etc.

b) Genetically conditioned pollination control mechanisms favouring cross-pollination In some vegetable crops, despite having bisexual flower structure, self-fertilization is hindered and cross-pollination coupled with cross-fertilization is favoured due to some genetically conditioned pollination control mechanisms.

- **Self-incompatibility:** It is genetically controlled physiological hindrance to self-fertilization, which marks the inability of the functional pollens of the same flower to fertilize the ovule of the same flower after self-pollination. It is found in cabbage, cauliflower, kohlrabi, Brussels sprouts, broccoli, leaf mustard, radish and turnip.
- **Male sterility:** In this phenomenon, male parts of the bisexual flower are non-functional without affecting the female fertility. It is controlled by nuclear genes or sterile cytoplasm or by the interaction between nuclear gene and sterile cytoplasm. It is found in tomato, eggplant, hot pepper, musk melon, onion, carrot, sprouting broccoli, Brussels sprouts, cabbage, cauliflower, sweet pepper, etc.
- **Dichogamy:** In this genetically controlled mechanism where the stamens and pistils of hermaphrodite flowers mature at different time which facilitate cross-pollination.
  - **Protandry:** In protandrous flowers, pollens are shed before the stigma becomes receptive, *e.g.*, onion, carrot, beet, parsley, leek, etc.
  - **Protogyny:** In protogynous flowers, stigma becomes receptive before the pollens are shed from the anthers of the same flower, *e.g.*, elephant-foot yam.

c) Different floral adaptations favouring cross-pollination

Different adaptations in the structural features of the flowers for easy dispersal of pollens by agents (wind and insects) also favour cross-pollination.

- **Floral adaptations favouring wind pollination:** Certain floral adaptations, which favour wind pollination in the primarily wind-pollinated (anaemophilous) vegetable crops, like beet, amaranth, palak or beet leaf, sweet corn, celery and Swiss chard, are mentioned below.
  - Exposed inflorescence, *e.g.*, beet leaf or palak, amaranth, beetroot, etc., which sways in the blowing wind.

- Apical arrangement of flowers, *e.g.*, sweet corn, amaranth, beet leaf, beet root, etc.
- Petals are either inconspicuous or absent, *e.g.*, beet leaf, amaranth, beet root, etc.
- Profuse production of pollens, *e.g.*, sweet corn, amaranth, beet leaf, beet root, etc.
- Dry, fine and smooth surfaced pollens with air sacs for easy dispersal by winds, *e.g.*, sweet corn, amaranth, beet leaf, beet root, etc.
- Special structures to intercept the windblown pollens, *e.g.*, tassel in the pistillate flower of sweet corn.

Floral adaptations favouring insect pollination: In most of the cross-pollinated vegetable crops, the pollens are carried by insect pollinators (entomophilous), most commonly by honeybees, bumblebees and houseflies. There is no guarantee that an insect visitor will visit another plant of the same species to perform cross-pollination. However, many plants evolved flowers adapted to attract only one or few kinds of insects. In this way, the chances of that insect visiting another of the same kind are considerably increased. Different floral adaptations to attract the insect pollinators to perform successful cross-pollination are mentioned below.

- Bright yellow coloured petals to attract the insects, *e.g.*, flowers of cole crops, okra, pumpkin, cucumber, watermelon, bitter gourd, ridge gourd, etc.
- Nectar to attract the insects, *e.g.*, flowers of cole crops, radish, turnip, onion, carrot, etc.
- Pollens for food to attract the insects, *e.g.*, flowers of cucurbits, okra, etc.
- Strong smell of the flowers to attract the insects, *e.g.*, flowers of onion, carrot, coriander, etc.
- Clustering together of small flowers to form a conspicuous head, *e.g.*, flowers of onion, carrot, etc.
- Sufficient mechanical strength of the flowers to withstand damage and deformation by insect pollinators, *e.g.*, flowers of cole crops, radish, turnip, etc.
- Structural arrangements ensuring reliable contact for anthers and stigma with the insect's body, *e.g.*, flowers of cole crops, onion, radish, carrot, etc.
- Stickiness of the pollens, which adhere easily to the different body parts of pollinators particularly bees, *e.g.*, flowers of different cucurbits.

Inbreeding and Inbreeding depression

Inbreeding is the production of offspring from the mating of individuals related by ancestry which includes self-fertilization, sib-mating, parent–offspring mating, etc. Inbred offspring are more likely to inherit recent copies of the same allele from both the parents, that is, alleles that are identical by descent (derived from a common ancestor of both the parents). Two alleles that are identical by descent are homozygous, but not all homozygous alleles are identical by descent. Outbreeding, the reverse case of inbreeding, means production of offspring through mating between individuals unrelated or distantly related by ancestral descendent. The degree of outbreeding can be measured with coefficient of inbreeding, $F = 0$ or tending to 0, depending on the genetic distance between the parents. Outbreeding is different from random mating in that the mating is limited between some unrelated genotypes or individuals rather than random in the population. Organisms exhibit a wide array of mechanisms that promote outbreeding.

Inbreeding increases homozygosity in the offspring generation, leading to a reduction in the genetic variation in the population and an increase in the expression of recessive deleterious alleles. Extreme inbreeding occurs naturally in self-pollinated crops like, tomato, brinjal, chilli, okra, sweet pepper, potato, peas and beans, etc. In each generation of selfing or inbreeding, heterozygosity in the populations is reduced to one-half. In self-pollinated vegetable crops, homozygous and true breeding lines are available and can be developed by continuous inbreeding. On the other hand, different cross pollinated vegetable crops like, cole crops, bulb crops, root crops, sweet corn, etc. are basically out-breeders although some degrees of inbreeding may occur depending on the inherent pollination control mechanism like, degree of self-incompatibility, stability of male sterility, dichogamy, insect foraging, environmental condition, etc. In cross-pollinated vegetable crops, development of homozygous individuals depends on the extent of inbreeding depression.

Inbreeding depression is the reduced biological fitness that has the potential to result from inbreeding (the breeding of related individuals) where biological fitness refers to an organism's ability to survive and perpetuate its genetic material. Inbreeding depression generally reduce the vigour of the progeny which generally occurs due to concentration of harmful recessive or partially recessive genes in the population of the progeny. In the heterozygous condition, the harmful effects of those recessive genes are masked by their dominant counterpart hence, depressive effect of the recessive genes is not expressed in the population. Extent of inbreeding depression depends on the type of the crop, mode of reproduction and several other factors.

a) **Self-pollinated crops:** The self-pollinated crops like, tomato, brinjal, chilli, okra, capsicum, potato, peas and beans, lettuce, etc. are adapted to continuous inbreeding hence, do not generally show inbreeding depression.

b) **Cross-pollinated crops:** In cross-pollinated crops, inbreeding usually leads to reduce the vigour of the progeny. However, the extent of inbreeding depression varies with the crops.

- **Very high depression**: In this crop, inbreeding should not be practiced for more than two continuous generations. If depressions due to inbreeding become highly pronounced, occasional out-crossing can restore the vigour. *e.g.*, carrot.
- **High depression**: In this crop, inbreeding can be practiced for 2-3 generations to render some extent of homozygosity. In this crop occasional out-crossing can restore the vigour. *e.g.*, Onion.

- **Moderately high depression**: In these crops, inbreeding can be practiced for 3-4 generations hence, considerable homozygosity and uniformity can be attained in these crops. *e.g.* Cabbage, Group I and II cauliflower, broccoli, Brussels, sprouts, turnip, sweet corn.
- **Low depression**: In these crops, inbreeding can be practiced for 4-5 generations to develop highly homozygous and uniform lines for use in different breeding purposes. *e.g.* Snowball group of cauliflower (Group IV), radish.
- **Insignificant depression**: In these crops inbreeding can be practiced as effectively as done in self-pollinated crops. *e.g.* all cucurbits.

Concept of breeding based on inbreeding depression

- **Single plant concept**: In naturally self-pollinated crops which normally do not show any inbreeding depression *viz.*, tomato, brinjal, chilli, sweet pepper, okra, peas, French bean, hyacinth bean, cluster bean, runner bean, lettuce, etc., breeding concept of these crops is based on development of single superior and homozygous plant which is the base material for the line-bred variety or parental line of a hybrid (single plant concept). Cucurbits are naturally cross-pollinated crops but normally do not show inbreeding depression. It has been suggested that the lack of inbreeding depression in cucurbits is due to the small population size used by the farmers during the domestication of the species.In monoecious sex form of cucurbits, the male (staminate) and female (pistillate) flowers on a single plant don't mature at the same time and the female flower has a very limited fertility window. The insect pollinators which perform cross pollination in the cucurbits often collect the pollens from staminate flowers of the same vine due to long vine length. Hence, cross-pollination often does not result cross-fertilization in cucurbits. This condition has made the cucurbits adaptable to continuous inbreeding. For this reason, cucurbits can be bred following "Single plant concept" like that of the naturally self-pollinated crops.
- **Population concept**: In the naturally cross-pollinated crops which show variable degrees of inbreeding depression *viz.*, cabbage, cauliflower, knolkhol, broccoli, Brussels sprouts, leaf mustard, radish, turnip, carrot, onion, etc. breeding concept of these crops is based on the development of improved population from the initial random mating base material which forms the basis of developing an open-pollinated variety (population concept). Lines with appreciably high homozygosity can also be developed from these naturally cross-pollinated crops by controlling the pollination and subsequently fertilization process leading to the development of either a variety or parental lines of a hybrid.

Characteristics of a variety / hybrid /clone

- Line-bred varieties of naturally self-pollinated crops are both homogeneous and homozygous.
- Hybrids of both self and cross-pollinated crops are homogeneous but heterozygous.

- Clones of both self and cross-pollinated crops are homogeneous but either heterozygous or homozygous.
- Population or open pollinated varieties of naturally cross-pollinated crops are heterogeneous and heterozygous in nature.

Various deviations from normal sexual reproduction

Apomixis

Apomixis is asexual development of seed or embryo without fertilization. However, other definitions include replacement of the seed by a plantlet or replacement of the flower by bulbils. However, propagation of plant from cuttings or leaves has never been considered to be apomixes. Apomictically produced offspring are genetically identical to the parent plant, except those developed by non-recurrent apomixis. This asexual reproduction can be commercially used to produce true to the type seeds from the hybrids in cost-effective manner. However, this deviation of normal sexual reproduction eliminates the long-term evolutionary advantage of genetic contribution from two parents.

Types of apomixis in flowering plants

Maheshwari (1950) classified apomixis in flowering plants as mentioned below:

- **Non-recurrent apomixis:** In this type of apomixis the megaspore mother cell undergoes usual meiotic divisions and a haploid embryo sac is formed. The new embryo may then arise either from the egg (haploid parthenogenesis) or from some other cell of the gametophyte (haploid apogamy). The haploid plants have half the chromosomes number of the mother plant, and the process is not repeated from one generation to another (for this reason it is called non-recurrent apomixis).
- **Recurrent apomixis:** In this type of apomixis, the embryo sac develops from the diploid egg mother cell or from some other diploid cells of the embryo sac without fertilization. In this type, the embryo sac has the same number of chromosomes as the mother plant because meiosis was not completed. It generally arises either from an archesporial cell or from some other part of the nucellus.
- **Adventive embryony:** It is also called sporophytic apomixis or sporophytic budding, or nucellar embryony. In this type, there may be an embryo sac (gametophyte) in the ovule, but the embryos do not arise from the cells of the gametophyte rather, they arise from the cells of nucellus or the integument. Adventive embryony is important in several species of Citrus, Garcinia, Euphorbia dulcis, Mangifera indica, etc.
- **Vegetative apomixis:** In this type, the flowers are replaced by bulbils or other vegetative propagules which frequently germinate while still on the plant. Vegetative apomixis is important in *Allium* spp., Dioscorea, *Fragaria*, some grasses, etc.

Some terms related to apomixis

Parthenogenesis

It refers to the development of an embryo directly from an egg cell without fertilization which is of two types:

- **Haploid parthenogenesis:** Offspring is developed from normal haploid egg cell to produce haploid plant. This type of apomixis has been recorded in *Solanum nigram, Lilium* spp., *Orchis maculate, Nicotiana tabacum*, etc.
- **Diploid parthenogenesis:** In this type of parthenogenesis, the embryo sac is developed without completing meiosis and it develops into a diploid plant.

Androgenesis

Androgenesis is a form of quasi-sexual reproduction in which a male is the sole source of the nuclear genetic material in the embryo. Two types of androgenesis occur in nature. Under the first type, females produce eggs without a nucleus and the embryo develops from the male gamete following fertilization. Evolution of this type of androgenesis is poorly understood as the parent responsible for androgenesis (the mother) gains no benefit from it. Ultimate factors driving the evolution of the second type of androgenesis are better understood. In this case, a zygote is formed between a male and a female gamete, but the female genome is eliminated.

Apogamy and Apospory

Apogamy is an asexual method of reproduction in plants where the gametophyte develops from the sporophyte. Apospory is also an asexual method of reproduction where sporophyte develops from the gametophyte.

Table 1: Difference between apogamy and apospory

Apogamy	Apospory
Apogamy is an asexual method of reproduction in plants where the gametophyte develops from the sporophyte	Apospory is also an asexual method of reproduction where sporophyte develops from the gametophyte.
It produces a diploid gametophyte	It produces a haploid embryo
Gametes are formed without the formation of gametophytes	The embryo is formed without the process of fertilization

Pseudogamy

When the egg cell develops into an embryo, the process is termed as pseudogamy, and when any other haploid cell of the embryo sac develops into an embryo, the process is called apogamy.

Adventive embryony

Adventive embryony is an asexual reproduction that forms seeds without fertilization and produce a maternal clone in plants. The embryo does not develop from the reproductive cells or the gametes rather the embryo is produced from the diploid nucellus or integuments of the ovule. Adventive embryos may occur together with a normal zygotic embryo. Hence, one seed may contain several embryos, as is often the case in *Citrus* seeds, called polyembryony.

Polyembryony

Polyembryony is a condition where two or more embryos develop from a single embryo sac. In polyembryonic seed formation, many non-zygotic, nucellar embryos are initiated directly from the nucellar cells surrounding the embryo sac containing a developing zygotic embryo. Nucellar embryos give rise to seedlings that are of the same genotype as the female parent. Polyembryony occurs regularly in *Citrus*, *Mangifera*, etc.

Parthenocarpy

Parthenocarpy is the natural or artificially induced production of fruit without fertilization of ovules, which makes the fruit seedless. The phenomenon has been observed since ancient times but was first scientifically described by German botanist Fritz Noll in 1902. Natural parthenocarpy is governed by either an incompletely dominant gene or 2-3 independent, isomeric major genes with additive action. Banana, pineapple, fig are good examples of natural parthenocarpy. These plant species can no longer sexually reproduce but able to propagate by vegetative means.

Development of genetically controlled parthenocarpy in some vegetable crops bear substantial value. Almost all slicing and pickling cucumber varieties cultivated in the greenhouses/ polyhouses in Europe are both genetically parthenocarpic in nature and gynoecious in sex form (constant femaleness) which obviate pollination process to set fruits. These parthenocarpic varieties are exclusively utilized in the processing industries.

Types of Parthenocarpy

- **Stimulative parthenocarpy**: In stimulative parthenocarpy, pollination or other form of stimulation is required to produce parthenocarpic fruit. This condition occurs when the ovipositor of a wasp is inserted into the ovary of a flower and can also be achieved by flowing air or plant growth regulators into the unisexual flowers that are present inside the syconium (type of fruit borne by figs). e.g., fig.
- **Vegetative parthenocarpy**: In vegetative parthenocarpy, plants do not require pollination or other stimulation to produce parthenocarpic fruit. e.g., banana, pineapple, seedless cucumbers

Table 2: Difference between apomixes and parthenocarpy

Apomixis	Parthenocarpy
Asexual mode of reproduction	Asexual mode of reproduction
It leads to formation of seeds without fertilization	It leads to formation of fruits without fertilization

Stenospermocarpy

Stenospermocarpy refers to abortion of incompletely developed seeds with normal development of the berry as found in seedless grape. In stenospermocarpic fruits, normal pollination and fertilization are required to ensure that the fruits are set and continues to develop on the plant. However, subsequent abortion of the embryo following fertilization leads to a near seedless condition. The remains of the undeveloped seed are visible in the fruit. The fruit of seedless grapes is smaller than normal because seed development process induces synthesis as well as transport of the plant growth hormone, gibberellin to the developing fruits, which causes fruit enlargement. In stenospermocarpic varieties of grape (seedless varieties), such gibberellin-regulated development of fruit does not occur naturally. For this reason, most commercial seedless grape varieties are sprayed with gibberellin to increase the size of the fruit and also to make the fruit clusters less tightly packed.

Reference

Maheshwari, P. 1950. *An introduction to the Embryology of the Angiosperms*. McGraw-Hill, New York.

4

Genetically Governed Pollination Control Mechanisms

Male-sterility

Normal development of the male reproductive organ (stamen) and male gametophyte (pollen grain) is essential for successful completion of sexual reproduction in angiosperms. Abnormalities at any stage of stamen and pollen development can result in male sterility. Male sterility in plants implies the inability to produce or to release functional pollen. Male sterility is caused due to failure in the formation or development of functional stamens, microspores or gametes and /or release of pollens from the anther. Male-sterility can result from mutations in nuclear genes alone or mitochondrial genes or mitochondrial genes coupled with nuclear genes. The resulting conditions are known as genic male sterility (GMS), cytoplasmic male sterility (CMS) and cytoplasmic-genic male sterility (CGMS), respectively.

Classification of male-sterility

1. Genetic male-sterility

This male-sterility is caused due to mutations in nuclear genes alone or mitochondrial genes or mitochondrial genes coupled with nuclear genes which results abnormalities at any stage of stamen and pollen development. This genetically governed male-sterility may be found spontaneously in nature or can be induced artificially. Heritable male sterility can be induced through treatment with mutagenic agents, in general, and other chemicals that target cytoplasmic genome. However, temporary male sterility can be induced through spraying of gametocides. Male sterility is manifested phenotypically in three different manners –

(a) **Pollen sterility:** Male sterile individuals differ from normal plant only in the absence or extreme scarcity of functional pollen grains in the anthers.

(b) **Structural or staminal male sterility:** In the male sterile individuals, stamens are either absent or present in rudimentary form and non-functional.

(c) **Functional male sterility:** In the male sterile individuals, pollen grains are functional but are trapped in indehiscent anthers and thus prevented from functioning.

Long term interest of genetic male-sterility

- For the plant breeders it is an important pollination control mechanism for framing strategy of producing hybrid seeds in efficient and economic manner
- For the geneticists/molecular biologists it is an important system to study the differential/spatial expression of male sterility genes and interaction between the nuclear and cytoplasmic genome
- For the evolutionary biologists it is an important aspect for studying the endosymbiotic hypothesis (this theory states that some eukaryotic cell organelles, such as mitochondria and plastids, have evolved from free-living prokaryotes) which focuses on tracing the transmission mechanism of mitochondrial genes into the nuclear genome during evolution process.

2. Non-genetic male-sterility: This non-inherited type male-sterility may be induced by:

(a) Chemicals: Application of chemical hybridizing agent or gametocides may induce temporary male sterility.

(b) Exposing the plants in some ecological conditions: The major non-genetic factors which are known to cause temporary male sterility in plants are temperature, photo- period, duration and quality of light, and different soil-borne stresses.

Occurrence of genetic male-sterility

1. Male sterile genotypes arise as spontaneous mutants in most cases (high frequency mutation)
2. Male sterility can be artificially induced through
   (a) Applied mutagenesis
   (b) Distant hybridization
   (c) Protoplast fusion
   (d) Genetic engineering

Classification of genetic male-sterility

In plants, male sterility can be caused either by nuclear genes alone or mitochondrial genes or mitochondrial genes coupled with nuclear genes. The resulting conditions are known as genic male sterility (GMS), cytoplasmic male sterility (CMS) and cytoplasmic-genic male sterility (CGMS), respectively. Based on location of genes controlling male sterility, the genetic male sterility may be classified as:

1. Genic or nuclear male-sterility

This type of male-sterility (GMS) is controlled by the nuclear genes, and has been reported in many important vegetable crops. Most of the naturally occurring or induced male-sterile mutants are recessive in nature (*ms*) with few exceptions in some cole crops (cauliflower, cabbage and broccoli) and genetically transformed male sterile lines where male sterility is controlled by dominant (*Ms*) gene. Occurrence of GMS under predominantly recessive genetic control clearly indicates the result of mutation in any gene(s) controlling micro-sporogenesis (pollen development), stamen development or micro-gametogenesis (male gamete development) and accordingly, morphology of male-sterile mutants varies. It ranges from the complete absence of male reproductive organ to the formation of normal stamen with viable pollens that fail to dehisce. In general, male-sterile plants manifesting "Pollen sterility" are morphologically not distinguishable from the parent fertile plants with the exception of flower structure and sometimes size. In case of pollen less or pollen sterility condition, the male sterile flowers are generally smaller in size in comparison to the fertile flowers. In some cases, stamens may be modified as carpels, as in *stamenless-1* (*sl-1*) mutant of tomato. Functional male-sterile mutants produce viable pollens but are non-functional for mechanical reasons particularly due to indehiscent anther. Of the different types of functional sterility, (a) exerted stigma sterility (due to excessive length of the style that protrudes from the anther cone/anthers holding the stigma out and away from the concentration of pollens) and (b) positional sterility (presence of fertile pollens but indehiscent anthers) are important from their utilization point of view. Certain GMS lines are conditional mutants (mostly temperature sensitive) and male fertility is restored in them under certain environmental condition. Inheritance of genic male sterility follows Mendelian inheritance due to nuclear genes governing the sterility.

2. Cytoplasmic male-sterility

This type of male-sterility (CMS) is usually defined as "maternally inherited deficiency in producing viable pollen". Cytoplasmic male-sterility is caused due to any mitochondrial mutation which impairs the proper functioning of mitochondria. Hence, it is a maternally inherited trait with *S* (sterile) and *N* (normal) cytoplasm and reciprocal differences are observed. In most of the cases with few exceptions, mitochondria are not present in the pollen. The affected organs and tissues in CMS plants are also the stamens (anther and filament) and pollen grains (microspores). Pollen grains develop within the anther in a precisely timed, high-metabolic demand process, with the tapetal layer of the anther supplying enzymes, hormones, nutritive materials, nucleosides and nucleotides for micro-sporogenesis to proceed. It appears that CMS results because of a multitude of effects, all more or less related to insufficient or mistimed supply of necessary resources for developing microspores.

Cytoplasmic male-sterility besides being of spontaneous in origin may also result from distant hybridization because successive backcrossing between distantly related species or sub-species induces nuclear substitution (Hanson and Bentolila, 2004; Kubo *et al.*, 2011; Kaminski *et al.*, 2015), protoplast fusion, induced mutation or antibiotic effects on cytoplasmic genes. Protoplast fusion, where recombination between the two parental mitochondrial genomes occurs prior to cytoplasmic segregation in the somatic hybrids sometimes induce cytoplasmic male-sterility (Rambaud *et al.*, 1993; Carlsson *et al.*, 2007). Although CMS has been observed in more than 150 plant species, the mechanism involved in CMS has not been fully elucidated. Different evidence supports the roles for energy deficiency, programmed cell death (PCD) and reactive oxygen species (ROS) for the cause of male sterility (Hu *et al.*, 2014; Touzet and Meyer, 2014).

3. Cytoplasmic-genic male-sterility

In cytoplasmic- genic (CGMS) male-sterility both nuclear and cytoplasmic genes (*N* and *S* cytoplasm) are involved. This sterility is thought to have resulted as the product of a genetic conflict between two genomes that have different modes of inheritance. The expression of male sterility in such cases is the result of incompatibility between recessive nuclear allele (*r*) and cytoplasmic genome (*N* and *S* cytoplasm). Therefore, those CMS lines for which dominant nuclear gene (*Ms*) restorer male fertility have been identified are widely known as genic-cytoplasmic male-sterility (CGMS). The dominant nuclear allele which restores male fertility is called restorer gene (*Rf* gene). It represents an example of the crosstalk between mitochondrial and nuclear genomes. These fertility restorer genes (*Rf*) are distinct from genic male sterility genes. Male sterility factors, generally encoded by chimaeric mitochondrial genes, can be down-regulated by specific nuclear restorer genes. Hence, male sterility in CGMS is expressed under the presence of sterile *mt*-genome located in sterile cytoplasm (*S*-cytoplasm) and recessive allele of restorer (maintainer allele: *rfrf*) located in the nuclear genome. *Rf* requires to restore fertility in *S* cytoplasm where, *N rfrf* = male fertile genotype; *S Rf-* = male fertile genotype; *N Rf-* = male fertile genotype; *S rfrf* = male sterile genotype.

The Texas, or *T* cytoplasm (*cms-T*) of maize was first described in the Golden June line in Texas, USA. In *cms-T*; male sterility is characterized by the failure of anther exertion and pollen abortion. Two restorer nuclear genes, *Rf7* and *Rf2*, acting jointly, restore pollen fertility to *cms-T*. However, *Rf7* and *Rf2* do not restore pollen fertility to *cms-C* or *cms-S*; instead, different restorer genes are necessary to restore pollen fertility to these sterile cytoplasms. Cytoplasmic- genic (CGMS) male-sterility is most commonly found in maize, onion, carrot, beet, etc.

Spontaneous CGMS was first described in onion by Jones and Emsweller (1936), and its use for hybrid seed production was established by Jones and Clarke (1943) who reported that male sterility was originated by the interaction of the male sterile (*S*) cytoplasm with the homozygous recessive genotype (*msms*) at the male fertility restoration locus (*Ms*) in the nucleus. The dominant allele *Ms* restores male fertility in plants with *S* cytoplasm, while plants possessing the normal (*N*) cytoplasm are male fertile irrespective of the genotype of the restorer gene.

Limitations of CGMS system for utilization

- Adverse pleiotropic effects of the sterile cytoplasm on different horticultural traits (Kaminski *et al.* 2012)
- Poor cross-pollination of male sterile flowers by the insect pollinators due to changes in flower morphology and chemical composition of nectar (Soto *et al.*, 2013)

- Instability of male sterility in different environments, which causes contamination of the hybrid seeds with selfed seed (Weider *et al.*, 2009)

Maintenance of male sterility in different vegetable crops

• Genic male sterility

Genic male sterility causing pollen sterility is maintained by sib-crossing. Genetic constitution of the male sterile plant is '*msms*' and that of male fertile sister plants are *MsMs* (homozygous) and *Msms* (heterozygous). The male sterile plant can be maintained by sib-crossing it to the heterozygous male fertile plant because progeny of such crossing produce the population of male sterile and male fertile plant in 1:1 ratio. The male fertile plants are rouged out. At first, the male sterile (female line) is crossed with homozygous male fertile line (*MsMs*). The F_1 *i.e.*, heterozygous male fertile line (*Msms*) is then backcrossed with the male sterile line, and by repeated backcrossings male sterile line is maintained.

Male sterile (*msms*) × Male fertile (*MsMs*)

↓

msms x *Msms* (Male fertile)

↓

1 *Msms* : 1 *msms* x *Msms* (male fertilite plants are rogued out)

↓

Msms : *msms* (male sterile) (male fertilite plants are rogued out)

Functional male sterile plants with indehiscent anthers can easily be maintained by forced selfings. In tomato, the male sterile line can also be maintained vegetatively by cuttings.

• Cytoplasmic-genic male sterility

Male sterility due to sterile cytoplasm and single recessive gene (*Smsms*) is found in maize, onion, chilli and sweet pepper. The male sterile plants have the genetic constitution '*Smsms*'. The combination of sterile cytoplasm (*S*) and recessive nuclear genes in heterozygous condition (*SMsms*) produce male fertile plant. Normal cytoplasm (*N*) always produces male fertile plant. So, genetic constitutions of male fertile plants are *Nmsms*, *NMsms*, *NMsMs*, *SMsMs* and *SMsms*. The cross between male sterile (*Smsms*) and maintainer male fertile line (*Nmsms*) will produce male sterile plant and thereby restore male sterility because sterile cytoplasm is inherited from female (male sterile) parent to the progeny. The maintainer line (*Nmsms*) can be isolated following the breeding scheme.

The maintainer line (*Nmsms*) can be isolated from the population of all male fertile (*NMsMs*, *NMsms*, *Nmsms*) plants following progeny testing with the male sterile (*Smsms*) line.

Male sterile (*Smsms*) X Male fertile (*NMsMs*)

↓

F_1 *S Msms* (male fertile)

NMsMs x *SMsms* (Back cross)

(female) (male)

↓

NMsMs : *NMsms* (1:1 ratio, all male fertile)

↓ ↘

Selfing → Selfing

NMsMs: NMsMs: *NMsms*: ← *Nmsms* (1:2:1 ratio)

(breed true) (segregation) ↓

(Breed true)

The *Nmsms* plants (maintainer line) can be detected by examining the progeny. Only these plants will produce all male sterile progeny when crossed with the male sterile plants.

Male sterility by genetic engineering

Recombinant DNA techniques have made it possible to engineer entirely new systems of male sterility by disturbing any or a number of developmental steps specifically required for the production of functional pollens within the microspore or for the development of any somatic tissues supporting the microspores. Engineered male sterility can be achieved by using anther- or pollen-specific promoters fused to a ribonuclease gene to produce ablation of specific cell types that are essential for proper anther development.

- Barnase/barstar technique is a genetically engineered nuclear genetic male sterility. Barnase encodes a potent cytotoxic ribonuclease enzyme, which when expressed, kills a cell because this enzyme cleaves the RNA of RNA/DNA hybrids that form during replication and repair which could lead to DNA instability. The action of barnase is held in check by barstar, a strong repressor gene that prevents barnase action. Barnase and barstar are the extracellular ribonuclease and its intracellular inhibitor produced by *Bacillus amyloliquefaciens*. Both are small single-chain proteins and thus are suitable for application. It is only when barstar is not present that barnase can exert its lethal effect. By expressing the cytotoxic ribonuclease, barnase, under a tapetum-specific promoter in the anthers of targeted crop, it is possible to completely block pollen maturation, creating male sterile plants.
- *Pisum sativum ENDOTHECIUM 1* (*PsEND1*) is a pea anther-specific gene displaying very early expression in the anther primordium cells. This expression pattern has been assessed in both model plants and crops (tomato, tobacco, oilseed rape, rice, wheat) using genetic constructs carrying the *PsEND1* promoter fused to the *uidA* reporter gene. This promoter fused to the *barnase* gene produces full anther ablation at early developmental stages, preventing the production of mature pollen grains in all plant species tested. Additional effects produced by

the early anther ablation in the *PsEND1::barnase-barstar* plants are redirection of resources to increase vegetative growth, extension of flowering period, etc. Early anther ablation in transgenic *PsEND1::barnase-barstar* tomato plants promotes the developing of the ovaries into parthenocarpic fruits due to the absence of signals generated during the fertilization process and can be considered an efficient tool to promote fruit set and to produce seedless fruits. The *PsEND1::barnase-barstar* construct could also be useful to generate parental lines in hybrid breeding approaches to produce new cultivars in different legume species.

- Genetic engineering of the chloroplast genome using the β- *ketothiolase* gene, a component of the polyhydroxybutyrate pathway. Polyhydroxybutyrate (PHB) synthesis takes place by the consecutive metabolic action of β- ketothiolase (*phaA* gene), acetoacetyl-CoA reductase (*phaB*), and PHB synthase (*phaC*). The polymer, Polyhydroxybutyrate is primarily a product of carbon assimilation (from glucose or starch) and is employed as a form of energy storage molecule. The transgenic lines showed an accelerated pattern of anther development, affecting their maturation, and resulted in aberrant and collapsed tissue patterns of the pollens (Ruiz and Daniell, 2005).
- Accomplishment of mitochondrial rearrangements leading to CMS by disrupting the expression of *Mshl*, a nuclear gene involved in the suppression of mitochondrial DNA rearrangements in tomato and tobacco (Sandhu *et al.*, 2007). Disruption of the nuclear gene *MSH1*, which functions in maintaining mitochondrial and chloroplast genome stability, produces an array of unusual plant growth phenotypes including cytoplasmic male sterility

Chemical induced male sterility

Chemical induction of male sterility in plants has been of interest since 1950 when the potential for selective male sterility was first demonstrated. Since then, many attempts have been made to find substances effective to selectively disrupt pollen development without affecting the female functionality so that they can be used as the female parent in the production of hybrid seeds. These compounds, known as gametocides, pollen suppressors or chemical hybridizing agents (CHAs) belong to different chemical groups *viz.*, auxins and auxin inhibitors (NAA, IBA, 2,4-D, TIBA, MH), gibberellins, ethylene (ethephon–ethrel), halogenated aliphatic acids like, sodium 2, 3-dichlorobutyrate (FW450), Dalapon, arsenicals and brassinosteroids. Cytokinins, auxin, gibberellins, ethylene, jasmonic acid, Ethyl 4-fluorooxanilate (E4FO) and brassinosteroids play a role in anther development, and male sterility has been associated with changes in many plant growth regulators, suggesting that normal male development is controlled in concert by multiple hormones (Hirano *et al.*, 2008, Ye *et al.*, 2010). The sterility is non- genetic, non- heritable and induced with the help of chemicals that makes the male part abortive.

Characters of an ideal chemical hybridizing agent (CHA)

- Induction of male sterility but not female sterility
- Complete inhibition of pollen development
- Independent from environmental conditions
- Independent from genotypic differences
- Wide flexibility of rate and time of application
- Absence of phyto-toxicity or other adverse effects
- It should be non mutagenic in nature
- Environmentally safe and cause minimum side effects
- Cost effectiveness

Mode of action of chemical hybridizing agent (CHA)

- Disruption of meiosis leading to the arrest of anther development and degeneration of pollen mother cells (PMC) or microspores
- Disruption of exine formation resulting in thin walled, irregular and non-viable pollen
- Decrease in starch deposition and appearance of abnormal vacuoles in the microspores making them non-viable
- Persistence and abnormal growth of tapetal cell layers
- Indehiscence or delayed dehiscence of normally developed anthers with viable pollen

Stage of treatment

Selective disruption of pollen development by the chemical hybridizing agent (CHA) through different mode of action depends largely on the stage of treatments.

- Treatment at inflorescence initiation stage (pre-meiotic stage) affects normal development of anthers and pollen mother cells (PMCs).
- Treatment at flower initiation stage (meiotic phase) and flowering stage (post-meiotic) disturbs normal development and functioning the microspores.

Advantages of using chemical hybridizing agent (CHA)

- Ease of making and evaluating hybrid combinations
- It is labour efficient hybrid seed production method
- There is no need for developing male sterile and restorer lines

Disadvantages of using chemical hybridizing agent (CHA)

- Only certain specific developmental stages of plant are sensitive to gametocides
- It needs repeated application for getting male sterility which increase the cost of hybrid seed

- High doses of these chemicals have been found to be phytotoxic and incite harmful side effects like defoliation, stunting, wilting, etc.
- Gametocides are expensive and can potentially be a pollutant to the environment
- In many cases it has been found to cause female sterility to some extent
- The responses to the chemicals by the plants have been found to be genotype dependent.
- Responses to the chemicals by the plants is also dependent on environment as for example, heavy rainfall immediately after the spray nullify the gametocidal action
- Some chemicals show carryover effect
- Pollen abortion is incomplete and erratic and sometime cause incomplete sterility

Self-incompatibility

The self-incompatibility is defined as genetically controlled physiological hindrance to self-fertilization. It is a single multi-allelic locus, the *S-locus* that determines pre-zygotic mate-recognition system preventing self-pollination and is very common in angiosperms (Kao and McCubbin, 1996; Ferrer and Good, 2012). Self-incompatibility response is comprised of a self- and non-self-recognition process between pollen and pistil that is followed by selective inhibition of the pollen tubes of the self-pollens which is manifested in several forms: complete suppression of pollen germination, production of very short pollen tube which do not reach the ovule for fertilization, production of longer pollen tubes which coil over the surface of the stigmatic papillae but fail to penetrate or production of pollen tubes which penetrate the outer layer of the papillae wall but are inhibited within the layers of the papillae wall.

Mechanism of self-incompatibility

Many flowering plants have evolved self-incompatibility (SI) systems to prevent inbreeding. Self-incompatibility system is operative through self-recognition *via* signaling receptor complex and subsequent rejection of self-related pollens by the stigma during pollination, which is genetically controlled by the haplotypes of *S*-locus complex. Although, this recognition is not irreversible, it depends on the genotype and environment. Again genotypic variation exists within a crop for degree in self-incompatible reaction. Much of the research in the field of gametophytic self-incompatibility (GSI) has stemmed from the observation that within pistil extracts, certain proteins, called *S*-proteins with molecular weight ranging from 23 to 34 kD segregate with their respective *S*-alleles in genetic crosses. Temporal and spatial distribution of these proteins correlates with the onset of incompatibility in many species including *Solanum peruvianum, Solanum chacoense, Solanum tuberosum,* etc. The mechanism of self-incompability is quite complex and poorly understood. The various phenomena observed in self-incompability mating are grouped into three broad categories: (1) Pollen-stigma interaction, (2) Pollen tube-style interaction, and (3) pollen tube-ovule interaction.

Classification of self-incompatibility system

• Gametophytic self-incompatibility

It is the most widespread form of self-incompatibility, in which the incompatibility reaction of the pollen is controlled by its own haploid genotype (genetic architecture of the pollen itself) and is found in the members of the family Solanaceae, Rosaceae and Plantaginaceae. These *S* alleles do not show any dominance, so, the reaction is independent of other *S* alleles. In *S1S2* style, pollen tube growth of *S1* and *S2* pollen will be inhibited due to the presence of the same *S* allele in the pollen and stylar cells, but pollen tube growth of *S3* and *S4* pollen will not be inhibited due to the presence of dissimilar *S* alleles in the pollen and stylar cells. Therefore, there is possibility of 100% pollen non-functioning (pollens of *S1S2* plant), 100% pollen functioning (pollens of *S3S4* plant) or 50% pollen functioning (pollens of *S1S3*, *S1S4*, *S2S3* and *S2S4* plants).

• Sporophytic self-incompatibility

In this form of self-incompatibility, the incompatibility reaction of the pollen is controlled by the diploid (sporophyte) genotype of the parental anther on which it was produced (Hiscock and Tabah, 2003; Franklin-Tong and Franklin, 2003). This form of self-incompatibility is found in the members of the family Brassicaceae, Asteraceae, Convolvulaceae, Betulaceae, Caryophyllaceae, Sterculiaceae and Polemoniaceae. The phenomena associated with incompatible pollination vary according to the strength of the alleles involved (more than 50 alleles in *S*-locus have been identified in *Brassica*). The incompatible pollens even if germinate may succeed in penetrating the stigma cuticle but fail to invade stigma cell wall due to deposition of β-1, 3-glucan callose within the stigmatic papillae at the site of contact with incompatible pollen.

In the dominance relationship between *S*-alleles, one allele may be recessive in pollen only, in the stigma only or in both. If *S1* is dominant over *S2*, *S3*, *S4*, etc. incompatibility reaction of the diploid stigma (*S1S2*, *S1S3*, *S1S4*) and the diploid exine of the pollen (*S1S2*, *S1S3*, *S1S4*) will have the same incompatibility reaction like *S1*. Therefore, either pollens will not germinate or pollen tube growth in the style is inhibited if two parents have one or two *S* alleles are common. In dominance situation of this sporophytic incompatibility, there is chance of either 100% pollen functioning or non-functioning (*S1S1* x *S1S1*, *S2S2* x *S2S2*, *S1S2* x *S1S2*, *S1S2* x *S1S3*, *S1S2* x *S1S4*, *S1S3* x *S1S4*, etc. because *S1*>*S2*, *S3*, *S4*).

In co-dominance relationship between *S*-alleles, two alleles in an *S*-hetrozygote may be fully active in both pollen and stigma i.e., equal activity of both the alleles. In this *S* allele interaction (*S1* = *S2*, *S1* = *S3*, *S1* = *S4*, etc.), compatibility of pollens and pistil is manifested if the genotypes possess no corresponding *S* alleles. Hence, incompatibility reaction is manifested in the allelic situation: *S1S1* x *S1S1*, *S2S2* x *S2S2*, *S1S2* x *S2S3*, *S2S3* x *S3S4*, etc.

Sporophytic self-incompatibility system has been successfully utilized in the hybrid seed production in cabbage, cauliflower and broccoli.

General mechanisms of self-incompatibility

The best studied mechanisms of self-incompatibility act by inhibiting the germination of pollen on stigmas, or the elongation of the pollen tube in the styles. These mechanisms are based on protein-protein interactions, and the best understood mechanisms are controlled by a single multi-allelic locus, the *S*-locus, which has many different alleles in the species population. Despite their similar morphological and genetic manifestations, these mechanisms have evolved independently based on different cellular components. Therefore, each mechanism has its own, unique *S*-genes. The *S*-locus consists of at least two linked genes and having two basic protein coding regions - one expressed in the pistil, and the other in the anther and/ or pollen (referred to as the female and male determinants, respectively). Because of their physical proximity, these two genes are genetically linked, and are inherited as a unit. The units are called *S*-haplotypes. The translation products of the two regions of the *S*-locus are two proteins, which, by interacting with one another, lead to the arrest of pollen germination and/or pollen tube elongation, and thereby generate a self-incompatible response, preventing self-fertilization. The *SI* response occurs when both determinants are issued by the same *S*-haplotypes (Takayama and Isogai, 2005). However, when a female determinant interacts with a male determinant of a different haplotype, no self-incompatible reaction is created, and fertilization ensures. This is a simplistic description of the general mechanism of self-incompatibility, which is more complicated, and in some species the *S*-haplotype contains more than two protein coding regions. The identities of female and male determinants have been determined and molecular models for different types of *SI* have been developed (Kaothien-Nakayama *et al.* 2010, Serrano *et al.* 2015).

Mechanism of gametophytic self-incompatibility

• The RNase mechanism

In this mechanism, pollen tube elongation is halted when it has proceeded approximately one third of the way through the style. The female component ribonuclease, termed *S-RNase* probably causes degradation of the ribosomal RNA (rRNA) inside the pollen tube, in the case of identical male and female S alleles, and consequently pollen tube elongation is arrested, and the pollen grain dies (Franklin-Tong and Franklin, 2003).

• The S-glycoprotein mechanism

In this mechanism, the female determinant is a small, extracellular molecule, expressed in the stigma. Although, the identity of the male determinant remains elusive, which is probably some cell membrane receptor (Franklin-Tong and Franklin, 2003). The interaction between male and female determinants transmits a cellular signal into the pollen tube, resulting in strong influx of calcium cations which interferes with the intracellular concentration gradient of calcium ions which exists inside the pollen tube, essential for its elongation (Franklin-Tong *et al.*, 2002). The influx of calcium ions arrests tube elongation within 1–2 minutes. Subsequently, the cytosolic protein p26, a pyrophosphatase is inhibited by phosphorylation, possibly resulting in arrest of synthesis of molecular building blocks, required for tube elongation. Pyrophosphatase is an enzyme that catalyzes the conversion of one molecule of pyrophosphate to two phosphate ions. The functionality of this enzyme plays a critical role in lipid metabolism, calcium absorption and DNA synthesis, as well as other biochemical transformations. There happens de-polymerization and reorganization of actin filaments, within the pollen cytoskeleton (Geitmann *et al*., 2000; Snowman *et al*., 2002). Within 10 minutes from the placement on the stigma, the pollen is committed to a process, which ends in its death.

Mechanism of sporophytic self-incompatibility

In sporophytic self-incompatibility (SSI), the *SI* phenotype of the pollen is determined by the diploid genotype of the anther (the sporophyte) on which it was produced. Since SSI is determined by a diploid genotype, the pollen and pistil each express the translation products of two different alleles, *i.e.*, two male and two female determinants. In *Brassica* spp., the pollen coat, derived from the anther's tapetum tissue, carries the translation products of the two *S* alleles. These are small, cysteine-rich proteins. The male determinant is termed *SCR* or *SP11*, and is expressed in the anther tapetum as well as in the microspore and pollen (Takayama *et al*., 2000). The female determinant of the *SI* response in *Brassica* is a trans-membrane protein termed *SRK*, which has an intracellular kinase domain, and a variable extracellular domain. The trans-membrane protein SRK is expressed in the stigma, and probably functions as a receptor for the *SCR/SP11* protein in the pollen coat.

Another stigmatic protein, termed *SLG* is highly similar in sequence to the SRK protein, and seems to function as a co-receptor for the male determinant, amplifying the *SI* response. The interaction between the *SRK* and *SCR/SP11* proteins results in auto-phosphorylation of the intracellular kinase domain of *SRK*, and a signal is transmitted into the papilla cell of the stigma (Takayama *et al*, 2001). Another protein essential for the *SI* response is *MLPK*, a serine-threonine kinase, which is anchored to the plasma membrane from its intracellular side (Murase *et al*., 2004). The downstream cellular and molecular events lead eventually to pollen inhibition. There are possibly up to 100 polymorphs of the *S*-haplotype in *Brassica*, and within these, there is a dominance hierarchy. Dominance relationships results in complicated patterns of compatibility/self-incompatibility. These dominance relationships also allow the generation of individuals homozygous for a recessive *S* allele (Hiscock and Tabah, 2003).

Features of homomorphic system of self-incompatibility

- Flowers are morphologically identical
- Numerous mating types is controlled genetically by *S*-locus (*S*-haplophytes), containing two basic protein coding regions in close proximity- one expressed in the pistil and the other in the pollen
- The protein coded by the *S*-allele affects compatibility reaction
- Sporophytic self-incompatibility is controlled by the interaction between protein present within the sporophytic tissues i.e., pollen wall and stigma (diploid genotype)

Development of homozygous self- and sib-incompatible lines

In the cole crops and radish in which sporophytic self-incompatibility system is operative, all the plants are heterozygous for *S* alleles in the locus due to out breeding system. Homozygous self- and sib-incompatible but cross-compatible lines can be developed by selfing some selected superior plants of the elite variety following bud pollination.

S1S2 (genotype of the selected plants)

Selfing by bud pollination

↓

S1S1: *S1S2* : *S2S2* (segregation in 1:2:1 ratio)

In first selfing, 50% plants will be homozygous for *S* allele and rest 50% will be heterozygous. The *S1S1* and *S2S2* plants are self-incompatible but cross compatible. However, for producing hybrids utilizing self-incompatibility the degree of incompatibility alleles and their dominance is of prime importance. Furthermore, conditioning of incompatibility as influenced by environment must be properly considered before utilization.

Breakdown of self-incompatibility

The self-incompatibility system in *Brassica* does not fully functional until about two-day before anthesis. So bud pollination 1-2 days before maturity of stigma with the matured pollen of the same inflorescence (raceme) is very effective in overcoming self-incompatibility. Self-incompatibility can also effectively be broken down by various treatments –

(a) CO_2 gas treatment at 3-5% concentration for 8-24 hours at 100% relative humidity condition in a closed chamber like polyhouse,

(b) Spraying of 3% NaCl solution in the morning hours combined with bee pollination,

(c) Spraying of paraffin oil on the stigma just before self-pollination,

(d) Application of protein phosphatases inhibitor,

(e) Heat treatment at 70-80°C for 2 seconds before self-pollination

Assessment of self-incompatibility

Self-incompatibility can be assessed by determining the fertility index and through fluorescence microscopy.

$$\textbf{(i) Fertility index} = \frac{\text{Average number of seeds per siliqua from natural or compatible pollination}}{\text{Average number of seeds per siliqua from self-pollination in freshly opened flower}}$$

Note: Fertility index of 2.0 or more represent self-incompatibility.

(ii) Fluorescence microscopy: It is the technique of visualizing the pollen tube growth through stigma and style in UV light by fluorescence microscope. Penetration of 6-9 pollen tubes through the style is generally considered as operative incompatible reaction.

Advantages of self-incompatibility in hybrid seed production

- Only two self-incompatible lines carrying different *S* alleles are necessary and hybrid seed production requires maintenance of these two lines (Kucera *et al.*, 2006).
- Pollen and nectar production remain unaltered which ensure free open pollination between the plants of two self-incompatible but cross-compatible lines as recorded in cauliflower (Singh *et al.*, 2013)

Main disadvantages of utilizing self-incompatibility in hybrid seed production

- High cost of maintenance of self-incompatible (*SI*) lines
- Depression in self-incompatible (*SI*) lines due to continuous inbreeding
- Influence of environmental factors, particularly high temperature on the expression of self-incompatibility, pseudo compatibility, lack of synchronization of flowering, etc. poses hurdles for its utilization (Prakash *et al.*, 2015)

Gynoecious sex form in cucurbits

Monoecious is the predominant sex form in cucurbits. However, a wide range of variation in sex forms ranging from the primitive hermaphrodite to the advanced, gynoecious sex form occur in these vegetable crops. Inheritance of sex form plays an important role in breeding programme.

Cucumber (*Cucumis sativus* L.)

Sex determination in cucumber plants is genetically controlled by the *F* and *M* loci (Yamasaki *et al.*, 2002). The *F* and *M* loci interact to produce three different sex phenotypes *viz.*, gynoecious (*M-F-*), monoecious (*M-ff*) and andromonoecious (*mmff*). Both the loci play major roles in ethylene biosynthesis.

Musk melon (*Cucumis melo*)

Kubicki (1966) described a method for producing 100% genetically pure gynoecious lines. Such lines are derived from a cross of hermaphrodite (*aagg*) with a monoecious (*AAGG*) line and isolating gynoecious (*AAgg*) and hermaphrodite (*aagg*) lines. Crossing of gynoecious (*AAgg*) and hermaphrodite (*aagg*) lines produce all gynoecious (*Aagg*) plants. Gynoecious lines are generally maintained through induction of staminate or hermaphrodite flower in the gynoecious plant by spraying silver nitrate or silver thiosulphate @ 100-200 ppm at two to four leaf stage.

Bitter gourd (*Momordica charantia*)

Gynoecy in bitter gourd was reported to be governed by a single recessive gene, *gy-1* (Dey *et al.*, 2010) although, this gene might have partially dominant (Ram *et al.*, 2006) or semi-dominant (Iwamoto and Ishida, 2006) effect.

Labour intensive manual bagging and hand pollination commercially produce hybrid seeds of the cucurbits, which increase the cost of seed production. Use of gynoecious parental lines that produce only pistillate flowers reduce the labour requirement for either bagging or emasculation and at the same time, helps to maintain hybrid genetic purity.

References

Carlsson, J., Leino, M. and Glimelius, K. 2007. Mitochondrial genotypes with variable parts of Arabidopsis thaliana DNA affect development in *Brassica napus* lines. *Theor. Appl. Genet.*, **115**: 627- 641.

Dey, S., Behera, T., Munshi A.D. and Pal, A. 2010. Gynoecious inbred with better combining ability improves yield and earliness in bitter gourd (*Momordica charantia* L.). *Euphytica*, **173**: 37-47.

Ferrer, M. M., and Good, S. V. 2012. Selfsterility in flowering plants: preventing self-fertilization increases family diversification rates. *Ann. Bot.*, **110**: 535- 553.

Franklin-Tong, V. E. and Franklin, F. C. H. 2003. The different mechanisms of gametophytic self-incompatibility. *Philosophical Transactions of the Royal Society* B. **358** (1434): 1025-1032.

Franklin-Tong, V. E., Holdaway-Clarke, T. L., Straatman, K. R., Kunkel, J. G and Hepler, P. K. 2002. Involvement of extracellular calcium influx in the selfincompatibility response of Papaver rhoeas. *Plant J.*, **29** : 333–345.

Geitmann, A., Snowman, B. N., Emons, A. M. C. and Franklin-Tong, V. E. 2000. Alterations to the actin cytoskeleton of pollen tubes are induced by the selfincompatibility reaction in Papaver rhoeas. *Plant Cell.*, **12** : 1239-1252.

Hanson, M. R. and Bentolila, S. 2004. Interactions of mitochondrial and nuclear genes that affect male gametophyte development. *Plant Cell*, **16**: S154-S169.

Hirano, K., Aya, K., Hobo, T., Sakakibara, H., Kojima, M., Shim, R.A. , Hasegawa, Y., Tanaka, M. U. and Matsuoka, M. 2008. Comprehensive transcriptome analysis of phytohormone biosynthesis and signalling genes in microspore/ pollen and tapetum of rice. *Plant Cell Physiol.*, **49**: 1429-1450.

Hiscock, S. J., and Tabah, D. A. 2003. The different mechanisms of sporophytic self-incompatibility. Philos. *Trans. Roy. Soc. Lond.* B, 358: 1037-1045.

Hu, J., Huang, W., Huang, Q., Qin, X., Yu, C., Wang, L., Li, S., Zhu, R. and Zhu, Y. 2014. Mitochondria and cytoplasmic male sterility in plants. *Mitochondrion*, **19**: 282-288.

Iwamoto, E. and Ishida, T. 2006. Development of gynoecious inbred line in balsam pear (*Momordica charantia* L.). *Hort. Res.*, **5**: 101–104

Jones, H. A. and Emsweller, S. L. 1936. A male sterile onion. *Proc. Amer. Soc. Hort. Sci.*, **34**: 582-585.

Jones, H.E. and Clarke, A.E. 1943. Inheritance of male sterility in the onion and the production of hybrid seed. *Proc. Am. Soc. Hort. Sci.*, **43**: 189-194.

Kaminski, P., Dyki, B. and Stezpowska, A. A. 2012. Improvement of cauliflower male sterile lines with *Brassica nigra* cytoplasm, phenotypic expression and possibility of practical application. *J. Agric. Sci.*, **4**: 190-200.

Kaminski, P., Podwyszynska, M., Starzycki, M. and Starzycka-Korbas, E. 2015. Interspecific hybridization of cytoplasmic male-sterile rapeseed with Ogura cytoplasm and *Brassica rapa* var. *pekinensisas* a method to obtain male-sterile Chinese cabbage inbred lines. *Euphytica*, **208**: 519-534.

Kao, T. H., and McCubbin, A. G. 1996. How flowering plants discriminate between self and non-self pollen to prevent inbreeding. *Proc. Natl. Acad. Sci.* USA, **93**: 12059-12065.

Kaothien-Nakayama, P. K., Isogai, A. and Takayama, S. 2010. Self-incompatibility systems in flowering plants. In: *Plant Developmental Biology-Biotechnological Perspectives* (Eds. Pua, E. C. and Davey, M. R.), Springer Verlag, Berlin, Heidelberg, pp. 459-486.

Kaul, M.L.H. 1988. *Male Sterility in Higher Plants*. Monograph on Theoretical and applied Genetics, Volume 10, Springer-Verlag, Germany, 1005p.

Kubicki, B. 1966. Genetic basis for obtaining gynoecious muskmelon lines and the possibility of their use for hybrid seed production. *Genet. Polonica*, **7**: 27- 29.

Kubo, T., Kitazaki, K., Matsunaga, M. , Kagami, H. and Mikami, T. 2011. Male sterility-inducing mitochondrial genomes: how do they differ? *Crit. Rev. Plant Sci.*, **30**: 378-400.

Kucera, V., Chytilova, V., Vyvadilova, M. and Klima, M. 2006. Hybrid breeding of cauliflower using self-incompatibility and cytoplasmic male sterility. *Zahradnictvi Hort. Sci.*, **33**: 148-152.

Murase, K., Shiba, H., Iwano, M., Che, F. S., Watanabe, M., Isogai, A. and Takayama, S. 2004. A membrane-anchored protein kinase involved in Brassica self-incompatibility signaling. *Science*, **303**: 1516–1519.

Prakash, C., Dey, R. B., Dey, S. S. and Dhiman, M. 2015. Use of selfincompatibility and cytoplasmic male sterility in F_1 hybrid seed production of temperate vegetable crops. In: *Emerging Trends in Hybrid Vegetable Seed Production for Temperate Regions* (Eds. Kumar, R., Prakash, C., Kumar, V., Suman, R. S., Dhiman, M. R. , Dey, S. S., Dey, R. B. and Kumar, S. S.), ICAR, Katrain, pp. 21-30.

Ram, D., Kumar, S., Singh, M., Rai, M. and Kalloo, G. 2006. Inheritance of gynoecism in bitter gourd (Momordica charantia L.). *J. Heredity*, **97**: 294-295.

Rambaud, C., Dubois, J. and Vasseur, J. 1993. Male-sterile chicory cybrids obtained by intergeneric protoplast fusion. *Theor. Appl. Genet.*, **87**: 347-352.

Ruiz, O.N. and Dan V. and Mackenzie, S.A. 2007. Transgenic induction of mitochondrial rearrangements for cytoplasmic male sterility in crop plants. *Proc. Nat. Acad. Sci.* USA, **104**: 1766-1770.

Serrano, I., Romero-Puertas, M.C., Sandalio, L. M. 2015. The role of reactive oxygen species and nitric oxide in programmed cell death associated with self-incompatibility. *J. Exp. Bot.*, **66**: 2869-2876.

Singh, P.K., Pandey, V., Singh, M. and Sharma, S.R. 2013. Genetic improvement of cauliflower. *Veg. Sci.*, **40**: 121-136.

Snowman, B.N., Kovar, D.R., Shevchenko, G., Franklin-Tong, V.E. and Staiger, C.J. 2002. Signal-mediated depolymerization of actin in pollen during the selfIncompatibility response. *Plant Cell.*, **14**: 2613–2626.

Soto, V.C., Maldonado, I. B., Gil, R. A., Peralta, I. E., Silva, M.F. and Galmarini, C.R. 2013. Nectar and flower traits of different onion male sterile lines related to pollination efficiency and seed yield of F_1 hybrids. *J. Econ. Entomol.*, **106**: 1386-1394.

Takayama, S. and Isogai, A. 2005. Self-ncompatibility in plants. *Ann. Rev. Plant Biol.*, **56**: 467-489.

Takayama, S., Shiba, H., Iwano, M., Shimosato, H. , Che, F.S. , Kai, N., Watanabe, M., Suzuki, G., Hinata, K. and Isogai, A. 2000. The pollen determinant of self-incompatibility in *Brassica campestris*. *Proc. Natl. Acad. Sci.* U.S.A. **97**: 1920–1925.

Takayama, S., Shimosato, H., Shiba, H., Funato, M., Che, F.E., Watanabe, M., Iwano, M. and Isogai, A. 2001. Direct ligand–receptor complex interaction controls *Brassica* self-incompatibility. *Nature*, **413**: 534–538.

Touzet, P. and Meyer, E. H. 2014. Cytoplasmic male sterility and mitochondrial metabolism in plants. *Mitochondrion*, **19**: 166-171.

Weider, C., Stamp, P., Christov, N., H€usken, A., Foueillassar, X., Camp, K. H. and Munsch, M. 2009. Stability of cytoplasmic male sterility in maize under different environmental conditions. *Crop Sci.*, **49**: 77-84.

Yamasaki, S., Fujii, N., Takahashi, H., Mizusawa, H. and Matsuura, S. 2002. Molecular approach to the study of sex determination in cucumber plants. *Acta Hort.*, **588**: 309-312.

Ye, Q., Zhu, W., Li, L., Zhang, S., Yin, Y. , Ma, H. and Wang, H. 2010. Brassinosteroids control male fertility by regulating the expression of key genes involved in *Arabidopsis* anther and pollen development. *Proc. Nat. Acad. Sci.* USA, **107**: 6100-6105.

5

Genetic Basis of Plant Breeding

Plant breeding is the purposeful manipulation of characters in plants to create new genotype with a set of desired characters. Hence, it may basically be referred to as designed manipulation of the heritable variation for the character concerned. Living organisms possess the remarkable capacity to convert the non-living materials into living staff, which results in movement, growth, metabolism and reproduction, and thus continuity of living organisms from one generation to the next is accomplished. The continuity of life through transmission of characters is the subject of the study of heredity, while the resemblances and differences between individuals of the same or different generations is the concern of the study of variation. These two universal processes of continuity and of change, which are referred to as "Heredity" and "Variation" are primary consequences of all organisms. Organized study of these two biological processes is the science of Genetics.

Plant breeding, which may be termed as evolution of plant type directed by the will of mankind, is basically the process of selection of character(s) from heritable variation. Desirable heritable variations can also be induced through some genetic consequences like, hybridization, segregation, mutation, changing the number of chromosomes, etc. So frame of plant breeding lay on three basics:

- Methods of creating heritable variation
- Methods of predicting the outcomes
- Methods of selection from the variation

The methods of creation and subsequent prediction of heritable variation necessitates clear understanding of the basic biology and Genetics. It is clear that the hereditary contribution to the offspring or genetic materials contained in the single cell by which the new life is created. The same genetic information is then transmitted to billions of dividing cells for the development of the organism. Nuclei of each cell possess an organization in very condensed form, the chromosome and the genes in the chromosome depict the instructions transmitted from parents to offspring. These instructions *i.e.* gene expression through transcription and translation (the process through which information encoded in messenger RNA or mRNA directs the addition of amino acids during protein synthesis) guide the metabolism, which may also be environment dependent, so that the character is expressed. Success of plant breeding depends on the understanding of the magnitude of variation in the characters (particularly quantitative characters) occurred due to environmental influence because such variation is not heritable and therefore, not transmitted to the next generation. Heredity and environment do not operate together rather, a propensity is related to the environment while actual development requires genetic architecture or heredity. So, relative contribution of both heredity and environment for the manifestation of characters must be considered together.

Basic concepts of Genetics

- Heredity (architecture of the genotype) refers to the genetic information that is passed down from parents to offspring while different environmental factors that affect plant growth and development include light, temperature, water, humidity, nutrition, etc. Heredity and environment are both important factors that influence the manifestation of the phenotype.
- DNA is a long molecule (polymer) of two linked strands of paired chemical units (nucleotides) that wind around each other. The nitrogenous bases in the nucleotides come in four different types (the purine adenine (A) always pairs with the pyrimidine thymine (T); the pyrimidine cytosine (C) always pairs with the purine guanine (G). It carries information organized into units called genes. Genes typically provide instructions for synthesis of proteins, which give cells and organisms their functional characteristics.
- Ribonucleic acid (RNA) is a polymeric molecule (nucleotides) that is essential for most biological functions, either by performing the function itself (non-coding RNA) or by forming a template for the production of proteins (messenger RNA). Unlike double-stranded DNA, RNA is usually a single-stranded molecule in many of its biological roles and consists of much shorter chains of nucleotides. Unlike DNA, the nitrogenous bases in the nucleotide, purine adenine (A) always pairs with the pyrimidine Uracil (U) and the pyrimidine cytosine (C) always pairs with the purine guanine (G).
- A cell's set of DNA is called its genome.
- Some specific parts of the DNA molecule (coding region) in the chromosome and some other organelles (mitochondria, plastids, etc.) are the genetic material (gene) which provides instructions for the growth and functioning of living organisms and replication of DNA molecule marks the basis of reproduction.
- Each chromosome includes a single, linear molecule of DNA with its own set of genes, and genes are arranged along them in a specific order.
- Genes are the specific coding regions of the DNA molecule hence, are said to have been arranged in linear order on chromosomes.
- According to simple account, a gene is a sequence of nucleotides in DNA that is transcribed into a sequence of nucleotides making up a messenger RNA molecule that is in turn translated into sequence of amino acids that synthesis a protein.
- Gene expression consists of two major steps together: transcription and translation. During the process of transcription, the coded messages stored in

a gene's DNA is passed to RNA (ribonucleic acid) in the cell nucleus. The purpose of translation is to synthesize proteins by ribosomes from an mRNA template, which are the "workhorses" of the cell, carrying out all the functions necessary for life.

- Each species has its own characteristic number of chromosomes. In the diploid (**2*n***) species, the chromosomes remain in matched sets known as homologous pairs.
- The chromosomes are paired in fertilized egg and in somatic cells that are developed from fertilized eggs.
- Members of a pair of genes and chromosomes segregate to different reproductive cells and chromosomes are generally single units in gametes.
- Members of different gene pairs are assorted independently with respect to those of other gene pairs during the formation of gametes.
- Mutation is an alteration in the nucleic acid sequence of the genome of an organism which can result from errors in DNA replication during cell division, exposure to different mutagens, environmental factors such as UV light, etc.
- Multiple genes or polygenes control the inheritance of quantitative traits.
- Gene interaction (both allelic and non-allelic) is a common phenomenon for the expression of phenotype.
- Linkages are the close location of non-allelic genes or other DNA markers to each other on chromosomes which are inherited together from parents to offspring, while crossing over produces recombinants.
- Inheritance patterns of different characters are associated with different mating systems. The main mating systems in plants are outcrossing (cross-fertilization), autogamy (self-fertilization) and apomixis (asexual reproduction without fertilization). Mixed mating systems, where plants use two or three mating systems, are not uncommon.
- Genetic equilibrium is the condition of an allele or genotype in a gene pool (such as a population) where the frequency does not change from generation to generation. Genetic equilibrium describes a theoretical state that is the basis for determining whether and in what ways populations may deviate from it.
- Genes in population establish an equilibrium, the level of which can be changed by mutation, migration and selection, which provide the basis for race and species formation.

Mendelian Genetics

Various basic concepts have been established by observation and experimentation of Mendel as "Principles of Genetics". His success was the consequence of brilliant experimental design, main feature of which may be outlined as –

- Garden peas had been developed into pure lines through many generations of self-fertilization hence, single alteration in a trait was demonstrated by a visible difference between two genotypes.

- All the seven pairs of contrasting characters Mendel chose to study, one form was dominant over a well-defined, contrasting alternative *viz*., shape of pea (round *vs* wrinkled), colour of pea (green *vs* yellow), shape of pod (constricted *vs* inflated), colour of pod (green *vs* yellow), flower colour (purple *vs* white), plant size (tall *vs* dwarf) and position of flowers (axial *vs* terminal)
- All the seven major genes conditioning the characters segregated independently hence, perfect recombination occurred.
- All the crosses he made between the genotypes were fully fertile.

Mendel's hypothesis of the genes

- The unitary factors he referred (genes of the present day concept) were assumed to occur in alternative form (concept of allele) which were responsible for the expression of alternative characters in the genotypes.
- Each gamete possesses only one allele for a particular gene and are said to be haploid, so that the fertilized egg, happened to be diploid must carry the pair of this kind of gene.

However, this hypothesis was in fact, paralleled by the behaviour of chromosomes during meiosis and fertilization.

• The Principles of segregation

This principle depicts "The separation of allelic pairs of genes from one another and their distribution to different sex cells (gametes)"

- The diploid organisms possess paired genes and their gametes have these unpaired; members of gene pair must segregate and each are present 50% of the gametes.
- The gene present in a gamete is uninfluenced by the gene, whatever the allele, which was its partner in the adult.
- The hybrids must possess one allele of a gene contributed by each parent.
- Union of gametes during fertilization is random.

Table 1: Features of a monohybrid cross

Mating combinations	Gametes		Progeny		
	♂ parent	♀ parent	Genotypic frequency	Phenotypic ratio when dominance is	
				Complete	Intermediate
AA × aa	A	a	all Aa	all d	all intermediate
Aa × Aa (F_2)	A a	A a	¼ AA, ½ Aa, ¼ aa	3:1	1:2:1
Aa × AA (BC_1)	A a	A	½ AA ½ Aa	all d	1:1
Aa × aa (BC_2)	A a	a	½ Aa, ½ aa	1:1	1:1

Note: d = expression of dominant character; Intermediate = expression of character in intermediate magnitude

Allele: One of a pair, or series of alternative alleles, one contrasting form of a gene that occur at a given locus in a chromosome. The major forms of interaction between the alleles are:

(a) Dominance: The trait is expressed when the controlling gene is either homo or heterozygous meaning that AA and Aa individuals are alike phenotypically.

(b) Recessive: The trait is expressed only when the controlling gene is homozygous (aa) in the genotype.

(c) Co-dominance: Each allele produces independent effect in heterozygous condition.

• The principle of independent assortment or segregation

This principle depicts "Segregation of each pair of genes independently of other pairs" which does mean that "the chance of two or more independent events occurring together is the product of the chances of their separate occurrences"

The principle of independent assortment is the basic to the understanding of transmission genetics which results in genetic recombination. Assuming independent segregation of genes at gamete formation has occurred, the dihybrid may be derived by random fertilization of gametes. This law is in conformity to the law of probability (Probability is the chance of occurrence of an event). It may be stated that "Dihybrid ratio is the product of two-monohybrid ratio (3:1)".

Dihybrid ratio is the product of two monohybrid ratio (3:1)

P_1 (Round "RR" and Yellow "YY" seed) × P_2 (Wrinkled "rr" and green "yy" seeds)

Round > wrinkled; Yellow > green

F_1hybrid (Round and Yellow, RrYy)

F_2 segregation (RrYy × RrYy)

↓

Of the 4 F_2 segregates from monohybrid cross

¾ Round		¼ Wrinkled	
¾ yellow	¼ green	¾ yellow	¼ green
↓	↓	↓	↓
9/16 Ro Yel.	3/16 Ro gr.	3/16 wr. Yel.	1/16 wr. gr.

Dihybrid phenotypic ratio= $(3:1)^2$ or $(3+1)^2$ = 9:3:3:1

Table 2: Features of dihybrid F_2segregation

Phenotypes	Genotypes	Genotypic frequency	Phenotypic ratio
Round yellow	RRYY	1	9
	RRYy	2	
	RrYY	2	
	RrYy	4	
Round green	RRyy	1	3
	Rryy	2	
Wrinkled yellow	rrYY	1	3
	rrYy	2	
Wrinkled green	rryy	1	1

Trihybrid ratio is the product of three monohybrid ratio (3:1)

(P_1) DDGGWW × ddggww (P_2)

Tall vine, Yellow, Round seeds ↓ dwarf vine, green, wrinkled seeds

DdGgWw ⟶ ↓F_1 hybrid (Tall vine, yellow and round seeds)

F_2 segregation (DdGgWw × Dd Gg Ww)

Table 3: Features of trihybrid F_2 segregation

Characters			Phenotypic ratio
Height	**Seed colour**	**Seed surface**	
3 tall	3 yellow	3 round=	27 T, Y, R
		1 wrinkled=	9 T, Y, w
	1 green	3 round=	9 T, g, R
		1 wrinkled=	3 T, g, w
1 dwarf	3 yellow	3 round=	9 d, y, R
		1 wrinkled=	3 d, y, w
	1 green	3 round=	3 d, g, R
		1 wrinkled=	1 d, g, w

Notation of phenotype: T = Tall, Y = Yellow, R= Round, d=dwarf, g = green, w = wrinkled

Of the three gene pairs under study, one pair of genes has no influence on which way the other pair segregates.

Significance of Independent assortment

- It is the first proof that the germplasm is not composed of a single hereditary unit but consists of a number of units (genes) separable in heredity.
- It greatly contributes to our understanding of evolution of species reproducing sexually. If the new combination of genes is advantageous it might produce a line, which in time displaces the old line in competition with them. In this way, genetic recombination can speed up evolutionary processes.

- This principle has great practical significance in both plant and animal breeding. Desirable traits carried in different genotypes can be combined through hybridization and maintained in a single line.

Table 4: Relationship among pairs of independent alleles, gametes, F_2 genotype and phenotype

No. of heterozygous gene pairs	No. of kinds of gametes	No. of F_2 genotypes	% of homozygous genotypes in F_2	No. of F_2 phenotypes*
1	2	3	50	2
2	4	9	25	4
3	8	27	12.5	8
4	16	81	6.25	16
10	1024	59049	0.09	1024
n	2^n	3^n	$½^n \times 100$	2^n

*Considering dominance-recessive interaction of alleles

Table 5: Relationship of homo and heterozygous plants in segregating generations

If the character is controlled by					
One independently inherited gene				Two independently inherited genes	
	AA × aa	Homo to Hetero form ratio (%)		AABB × aabb	
	↓			↓	
F_1	Aa	0:100	0:100	AaBb	F_1
	↓			(AABB+AAbb+aaBB+aabb) :(AABb+AaBb+AaBB+ Aabb+aaBb)	F_2
F_2	(AA + aa) : Aa	50:50	25:75		
	↓				
				↓	
F_3	Same	75:25	56.2:43.8	Same	F_3
	↓			↓	
F_4	Same	87.5:12.5	76.5:23.5	Same	F_4

Table 6: Possibility of getting homozygous genotypes in F_2

No. of genes for control of character	Number of homozygous genotype
1	2/4
2	4/16
3	8/64
n	$2^n/(2^n)^2$

Probability of getting new genetic combinations in F_2 population

- The greater the number of loci heterozygous in F_1, the later the generations in which the plants in the population of advanced generations become homozygous. As for example, in a monohybrid, 75% plants in the F_3 generation are homozygous whereas, the F_3 generation of a F_1 heterozygous for 20 loci will have only 0.3% homozygous plants; its F_4 will have only 6.9% homozygous plants.

- Propensity of new genetic combinations in F_2 generation through recombination depends on the number of loci heterozygous in F_1. As for example, in a dihybrid 12.5% plants in F_2 are homozygous new combiners (RRyy, rrYY) and 25.0% plants are heterozygous new combiners (Rryy, rrYy), whereas in F_2 segregation of trihybrid 9.4% plants are homozygous new combiners (DDGGww, DDggWW, DDggww, ddGGWW, ddGGww, ddggWW) and 46.9% plants are heterozygous new combiners (DDGgww, DdGGww, DdGgww, DDggWw, DdggWW, DdggWw, Ddggww, ddGGWw, ddGgWW, ddGgWw, ddggWw).
- Genetic linkage changes the ratio between different homozygous genotypes, *i.e.*, their frequency varies with linkage.
- Percentage of homozygous plants in a population is decisive in the development of recombinant genotype. If a character concerned for breeding is governed by two independent genes, AABB/aabb, only 12.5% of homozygous recombinants (AAbb and aaBB) will be available in the F_2 population. On the other hand, if the character is governed by three independent genes, DDGGWW/ddggww, only 9.4% of homozygous recombinants (DDGGww, DDggWW, DDggww, ddGGWW, ddGGww and ddggWW) will be available in the F_2 population.
- Selection in the later generations is advantageous when the characters are under the control of more number of genes because percentage of heterozygous recombinants in F_2 generation increases with the increase in the number of genes controlling the character (25% and 46.9% in case the character is governed by 2 and 3 genes, respectively). With the increase in percentage homozygosity in a freely segregating population of a self-pollinated crop more number of homozygous recombinants will be available in later generations for selection of desirable type (s).
- Parents manifesting clear-cut characters must be selected for the breeding programme based on recombination of characters.
- It is better to produce as many F_2 seeds as possible to achieve favourable gene combinations particularly when many genes control the characters.
- Clear-cut selection criteria must be framed on the basis of expected recombination of characters.

Chromosomal Theory of Inheritance

The "Sutton-Boveri chromosome hypothesis" attempted to establish a parallelism between cytological chromosome behaviour and the principles followed by Mendelian factors. Today, this hypothesis is well accepted; however, it was not completely accepted at its proposal (1902-03). Several relevant ideas on chromosome hypothesis of Mendelian heredity was put forward by Sutton, Boveri, Montgomery, Cannon and Hugo de Vries, but Sutton alone should be credited with the first chromosome explanation of Mendel's laws.

The speculation that chromosomes might be the key to understanding heredity led several scientists to examine Mendel's publications and re-evaluate his model in terms of the behaviour of chromosomes during mitosis and meiosis. The Chromosomal Theory of Inheritance was consistent with Mendel's laws and was supported by the following observations:

- Bridge between one generation and the next is made through male and female gametes.
- The two gametes carry all the hereditary characters and both contribute equally in the heredity of the offspring.
- Homologous chromosome pairs are independent of other chromosome pairs.
- During meiosis, homologous chromosome pairs migrate as discrete structures that are independent of other chromosome pairs.
- The sorting of chromosomes from each homologous pair into pre-gametes appears to be random.
- Each parent synthesizes gametes that contain only half of their chromosomal complement.
- Even though male and female gametes differ in size and morphology, they have the same number of chromosomes, suggesting equal genetic contributions from each parent.
- The gametic chromosomes combine during fertilization to produce offspring with the same chromosome number as their parents.

Linkage and Recombination

All the seven pairs of genes which Mendel studied showed independent segregation. Although all the 7 characters governed by 7 genes were not present in different chromosomes, linkage was not encountered because genes present in the same chromosome were far apart ensuring free crossing-over. In fact, a chromosome contains hundreds or thousands of genes. Therefore, the genes being transmitted *en block* are linked corresponding to the chromosome they lay. So, linkage refers to the tendency for non-allelic genes in the same chromosome on 'linkage group' to enter the gametes together (parental combination) in higher proportion than expected from independent assortment. Mendel's law of independent assortment is applicable only when the genes are located either in different chromosomes or wide apart in the same chromosome so that genetic material (DNA) is exchanged between paired homologous chromosomes (one from each parent) that occurs during the development of male and female gametes during meiosis. This process results in new combinations of alleles in the gametes formed, which ensures genomic variation in any offspring produced.

A process of crossing over between homologous chromosomes explains the failure of linkage. Actually no more than 50% recombination is expected between any two loci because only two of the four chromatids in a meiotic tetrad are involved in any particular cross over event. The recombination value usually appears less than 50% because a) not every bivalent may possess a pair of crossover chromatids and b) double or multiple crossing over between gene pairs that are far apart along a chromosome reduce the apparent recombination products. Linkage group in a plant or animal species is equivalent to the haploid number of chromosomes.

The phenomenon of linkage was first reported by Bateson and Punnet in 1906. They studied flower colour and pollen shape in sweet pea involving two varieties. Plants with red flowers and long pollen grains (double dominant) were crossed to plants with white flowers and short pollen grains (double recessive). Linkage in garden pea was later studied by de Vilmorin and Bateson (1911). A pure double dominant (round seed, tendrils) plant was crossed with double recessive plant (wrinkled seed, no tendrils)

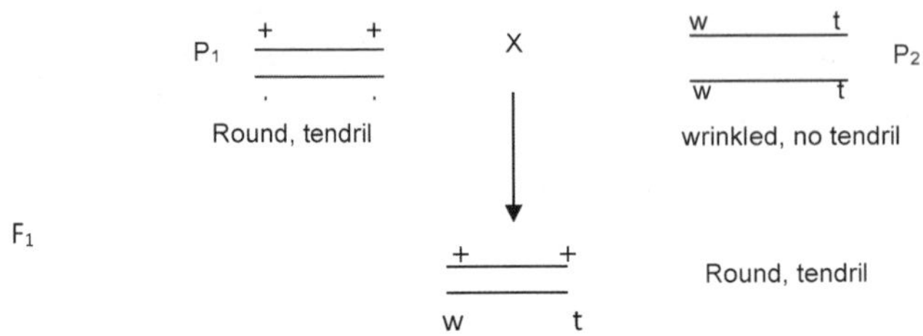

F_2 segregation

Pair character	Genotype	Phenotype		Expected ratio
		Observed	Expected	
Round, tendril	+ +	319	252.6	9
Round, no tendril	+ t	4	84.18	3
Wrinkled,tendril	w +	3	84.18	3
Wrinkled, no tendril	w t	123	28.06	1

- If the two pairs of genes inherited independently, there would have been in F_2 a 9:3:3:1 ratio. However, each gene pair showed monohybrid segregation in F_2-
- Round: Wrinkled gave 3:1 (323:126) ratio
- Tendril: no tendril gave 3:1 (322:127) ratio
- But, the two 3:1 ratios were not produced independently thereby did not confirm 4 phenotypic classes of 9:3:3:1 ratio in F_2 generation. Hence, there were too many F_2plants segregated phenotypically like parents (round, tendril/wrinkled, no tendril) indicating that the non-allelic genes did not inherit independently rather inherited together in a package. Consequently, there were too few new or recombination types (only seven exceptions to complete linkage).
- This deviation from the expected ratio is due to linkage of the character pairs.
  - round seed and tendrils are linked because both the genes are located on the same chromosome
  - wrinkled seed and no tendrils are linked together because the genes are located in the same chromosome (which is the homologue of the previous one).
- Few recombinations were obtained due to crossing over between the two concerned genes in some of the spore mother cells.

Characteristic features of Linkage

- Linkage involves two or more genes which are located in the same chromosome close to each other in a linear fashion
- Linkage may involve "Coupling" and "Repulsion" phase. Coupling refers to the condition where dominant alleles are on the same homologous chromosome and both recessive alleles are on the other homologuous chromosome(AABB × aabb). Repulsion refers to the condition where each homologous chromosome from the parents has one dominant and one recessive allele from the two genes (AAbb × aaBB).
- It may involve either all desirable traits or all undesirable traits or some desirable and some undesirable traits.
- It is observed for oligo-genic as well as polygenic traits.
- The strength of linkage depends on the distance between the linked genes; lesser the distance, higher the strength and *vice-versa.*
- Presence of linkage leads to higher frequency of parental types than recombinants in test cross.
- When two genes are linked the segregation ratio of dihybrid test cross progeny deviates significantly from 1:1:1:1 ratio.
- Linkage can be determined from test cross progeny data.
- If crossing over does not occur, all genes located on one chromosome are expected to be inherited together.
- Linkage reduces variability because linked genes contradict the independent assortment law thus reducing genetic variation.
- The maximum number of linkage groups possible in an organism is equal to the haploid chromosome number.
- Linkage can be broken by repeated inter-mating of randomly selected plants in segregating population for several generations or by mutagenic treatment.
- Besides pleiotropy, linkage is an important cause of genetic correlation between various plant characters.

Expression and interaction of genes

Mendel studied independent segregation of gene pairs affecting different characters. In his study, effect of one pair on the phenotype produced by another gene pair was absent which produced unambiguous results. Thus, the probability model between genotypes and phenotypes was perfect fit. But Bateson and his associates first showed that typical phenotypic results did not appear in conformity to the genotypic expectation. This was due to different gene pairs affecting the same character, and the phenotype is governed by an intricate interaction of a number of non-allelic genes. However, each allelic pair follows Mendelian rules during segregation.

Gene interaction is a common phenomenon for polygenes or multiple genes governing quantitative characters. It would be a big mistake to consider hybridization method of breeding as mere recombination of parental traits in the offspring. The interaction of genes in the hybrids sometimes gives rise to entirely new characters distinct from the parental forms. So, in addition to the parental characters, the hybrid genotype displays its own features resulting from a particular combination of genes. Occurrence of such interactions suggests that the phenotype we observe is not due to individual genes but arises from complex developmental processes although the genes themselves follow the regular laws of inheritance. Overall phenotype of an organism, therefore, is the result of epigenetic interactions of a network of genes controlled physio-biochemical reactions.

Principles of gene expression and interaction

(a) Many genes can affect the same trait.

(b) The expression of genes affecting a given character is diverse.

(c) Combinations of non-allelic genes affecting the same trait may produce phenotypic expression which

- superpose
- show epistasis
- are new to the series of crosses.

(d) It gives abnormal ratios and character expression, exceptions to Mendel's principles.

(e) Gene interaction frequently appears in polygenes.

Epistasis

Bateson (1990) used the word "epistasis," which translates as "standing upon," to define the masking action of one gene by another.Epistasis is a circumstance where the expression of one gene is modified (e.g., masked, inhibited or suppressed) by the expression of one or more other genes. Epistasis is an interaction at the phenotypic level of organization. When a dominant allele masks the expression of both dominant and recessive alleles at another locus, it is referred to as dominant epistasis or simple epistasis. When it is a recessive allele that masks the expression, it is called recessive epistasis. The non-allelic gene is epistatic to that gene and the gene suppressed is said to by hypostatic.The genes that are involved in a specific epistatic interaction may still show independent assortment at the genotypic level. In such cases, however, the phenotypic ratios may appear to deviate from those expected with independent assortment.

Explanation of epistasis at molecular level

The concept of epistasis originated in genetics but is now used in biochemistry, computational biology and evolutionary biology. Epistatic genes block particular reactions and mask the effects of normal or hypostatic genes. The phenomenon arises due to interactions, either between genes (such as mutations in regulators of gene expression) or within them (multiple mutations is needed before the gene loses function), leading to non-linear effects. In the depiction at molecular level, epistatic genes may be referred to the inactive genes that produce defective enzymes or no enzymes at all.

Examples of Epistasis

1. Gene for Rose comb (RRpp) × Gene for Pea comb (rrPP) in chicken ⟶ F_1 Walnut comb (RrPp); F_2 ⟶ 9 Walnut (R-P-): 3 Rose (R-pp): 3 pea (rrP-): 1 single (rrpp). The genes R and P were non-allelic but each was dominant over its allele (*i.e.* R over r and P over p). When R and P were together they produce different character (Walnut comb). So, the two non-allelic genes R and P acted independently in different ways, similar to the act of codominant allele. They are thus called co-epistatic genes.
2. White flower (CCpp) × purple flower (ccPP) in sweet pea ⟶ F_1 Purple (CcPp); F_2 ⟶ 9 Purple (CcPP, CcPp, CcPP, CcPp): 7 white (CCpp, Ccpp, ccPP, ccPp, ccpp). Two non-allelic gene pair C-P- are necessary to produce the purple colour. Thus flowers of the original parents were white in the absence of either one of two such complementary genes. The double recessive genotype (ccpp) also produced white flowers because of the absence of both the dominant genes.
3. White fruit (WWYY) × green fruit (wwyy) in summer squash (*Cucurbita pepo*) ⟶ F_1 white (WwYy); F_2 ⟶ 12 white (W- - -) : 3 yellow (- - Y -) : 1 green (wwyy). The dominant gene W conditions white colour and no other gene for fruit colour is expressed in its presence, thus, W is referred to dominant epistatic gene. Another colour gene Y is expressed in producing yellow fruit colour in presence of the other gene in homozygous recessive (ww) condition. Homozygous recessive plant for both the genes (wwyy) bears green fruits.
4. Yellow endosperm ($Y_1Y_1Y_2Y_2$) × white endosperm ($y_1y_1y_2y_2$) in maize ⟶ F_1 yellow ($Y_1y_1Y_2y_2$); F_2 ⟶ 15 yellow (Y_1 – Y_2 -, Y_1 - - -, - - Y_2 -) : 1 white ($y_1y_1y_2y_2$). Two different genes determine the same or nearly same phenotype, called duplicate genes such as Y_1 and Y_2 determine yellow endosperm colour in maize. When both or either of the genes is dominant, endosperm will be yellow. Presence of both the genes in recessive condition produce white endosperm colour.
5. Epistatic gene action is almost mandatory in the inheritance of quantitative characters governed by multiple genes or polygenes. This is a type of gene action in which the genotypic values associated with genetic phase at one locus are influenced by the genetic phase at the other locus. In polygenic inheritance, epistasis cannot occur of its own. It always occurs along with either additive or dominance gene action or both. This interaction could be additive × additive, additive × dominance or dominance × dominance type.

Multiple alleles

The two partners of a gene pair are called alleles occupying a specific spot called locus on a chromosome and they remain in corresponding positions on homologous chromosomes. Typically, there are only two alleles for a gene in a diploid organism. They may be alike, as in homozygote, or different, as in heterozygote. It is generally believed that dominant genes evolved first and the recessive counterpart originated as a result of changes in the dominant gene due to mutation or other factors. Different types or degrees of changes in the same gene might have resulted into a series of recessive genes. Degrees of recessiveness *i.e.*, intensity of gene expression in both homozygous and heterozygous conditions vary in these alleles. Thus a single gene may have more than two possible kinds of alleles.

Multiple alleles are the alternative forms of the same gene so they influence the same trait. The wild-type allele is mostly dominant over the mutant alleles. The wild type is considered the standard and all other alleles are considered variants.

Characteristics of multiple alleles

- Multiple alleles of a gene occupy the same locus on the chromosome.
- Multiple alleles are the alternative forms of the same gene so they influence the same trait.
- The wild-type allele is mostly dominant over the mutant alleles.
- The wild type is considered the standard and all other alleles are considered variants.
- The variant or mutant allele may be dominant, recessive or have an intermediate phenotypic effect.
- Multiple alleles exist at the population level and an individual can have only two alleles for the given gene.
- An individual can have only one type of allele (homozygous) or two different alleles (heterozygous).
- A series of mutant alleles can exist in a population. They are generated by a spontaneous mutation in the wild type and the mutant alleles.
- Multiple alleles lead to the formation of varied phenotypic traits for a character in the population.

Examples of multiple allele

1. **ABO blood group in human**: Blood type of human refers to certain proteins called antigens which are found on the red blood cells. The 4 main blood groups A, B, AB, O are due to 3 allelic genes A, A^B, a, although any one person may carry, of course, any two.

Phenotype	Genotype
A	AA, Aa
B	A^BA^B, A^Ba
AB	AA^B
O	aa

3. **The coat colour of rabbits**: Coat colour of rabbits is determined by the C gene. The C gene has multiple allelic forms that lead to varied phenotypes. The different phenotypes for coat colour in rabbits are agouti (full colour), chinchilla, Himalayan and albino.The four common allele of C gene are C, c^{ch}, c^{h} and c. A rabbit with genotype CC has black or brown fur, $c^{ch}c^{ch}$ have chinchilla or greyish coat colour, $c^{h}c^{h}$ have Himalayan pattern and cc have white coat colour, i.e. albino.

   Due to multiple alleles, many combinations of alleles are possible.

   - The black C allele is dominant over the other three alleles.
   - The c^{ch} (chinchilla) allele shows incomplete dominance over Himalayan and albino alleles.
   - The c^{h} (Himalayan) allele shows dominance over the albino allele (c).

4. **The eye colour in *Drosophila***: It is another trait that is determined by multiple alleles. There are a dozen different alleles that show phenotype between the wild-type red colour (w^{+}) and white colour (w). e.g., coral, blood, eosin, cherry, apricot, etc. The wild-type red colour is dominant over all other mutant alleles and the white colour is recessive to all the alleles.

5. **Sporophytic self-incompatibility in *Brassica* species**: Some *Brassica* species like cauliflower and cabbage have more than 50 different '*S*' alleles responsible for self-incompatibility in different intensities. The alleles produce their effects through production of enzymes, which is less abundant or efficient as one proceeds in the allelic series. In this way, self-incompatibility expression is manifested in different intensities. In fact, complexity and variability in self incompatibility response in different genotypes of *Brassica* species may be ascribed to large receptor kinase family and multiple *SLG* related sequences encoded by *SRK* and *SLG* gene, respectively located in *S* locus.

Pleiotropy

In genetics, pleiotropy is defined as the expression of multiple traits by a single gene and is often identified as a single mutation that affects two or more wild-type traits. Presence of a gene is inferred from alteration in the concerned character. It is generally assumed that one or two major genes control the qualitative characters. In the analyses of inheritance, emphasis is laid on the character whose change is most perceptible. However, close examination reveals that change in a single gene very often affect many characters. So pleiotropy refers to multiple effects on the phenotype, which a gene may have. This activity would be of a biochemical, perhaps enzymatic nature which would affect many varied chemical reactions involved in the production of different, at first appearance unrelated, phenotypic traits.

Example of pleiotropy

1. **Sickle cell anemia in human:**A classic example of pleiotropy is sickle cell anemia. The gene mutation that results in sickle-shaped red blood cells also leads to other affected traits like, anemia, spleen enlargement, heart failure, brain damage causing paralysis, kidney damage and skin lesions, either single or in combinations.
2. **Flower color, seed coat color, and axil color of sweet pea plants:** The gene for seed coat color is responsible for all three traits in sweet pea (*Lathyrus odoratus*). If the trait for colored seed coats is inherited, then the flowers and axils will also be colored.

Multiple gene inheritance or quantitative inheritance

Basic laws of genetics were postulated on the basis of inheritance patterns of characters, called qualitative characters which show discontinuous variation *e.g.*, flower or seed colour in peas, albinism *vs* pigmentation, blood group types in humans, etc. In each case, an individual belongs clearly to one class or another. The mutation theory also proposes that discontinuous variation occurs due to genetic change.

Controversy arose with the quantitative characters showing continuous variation, for which sharp classification of individuals as to type is not possible e.g., plant height, seed and fruit size, number, etc. The continuous range of phenotypes requires each individual be measured. Hence, these are often called quantitative traits. The controversy was resolved with the hypothesis that the quantitative traits which are controlled by multiple genes. Multiple gene inheritance occurs when more than one gene works together to control a trait. One prominent example of quantitative character is skin color. Skin color in humans is controlled by several genes in humans and the number of dominant alleles a person inherits is directly linked to how dark their skin is.

Many of the traits that plant breeders strive to improve are quantitatively inherited. Breeding efforts targeting quantitative traits have allowed major increases in crop yield during the past 80 years or so. A quantitatively inherited trait is controlled by many non-allelic and independent genes at different loci, with each gene — known as a polygene — contributing a small but additive effect to the expression of the single quantitative character. Polygenes are also known quantitative trait loci (QTL). QTLs involved in the expression of a quantitative character act cumulatively to determine the phenotype of the trait. Their mode of inheritance is called quantitative genetics. Quantitative genetics describes the connection between phenotype and genotype and provides tools to show how phenotypic selection of complex characters changes allele frequencies.

Each polygene involved in the inheritance of quantitative character segregate in Mendelian fashion however, in a Mendelian segregation variation is discontinuous but in polygenic inheritance, variation is continuous. In the inheritance of qualitative characters, each allele may show dominance/recessive or co-dominance. In contrast, each gene in polygenic inheritance has a quantitative effect and generally do not show dominance or co-dominance and such gene has equal and cumulative or additive effect on the phenotypic manifestation of a quantitative character (*i.e.*, phenotypically AAbb = AaBb = aaBB). Dominance is also manifested in polygenic inheritance where each gene shows dominance and having opposite effect on a trait. Multiple genes should not be confused with multiple alleles. Multiple alleles are many alternatives of the same gene and only two of these alleles exist at a time in a diploid organism. It is believed that mutations in different genes might have given rise to genes with similar effects. Tandem duplication of genes and chromosomes, which are found frequently in eukaryotes, may points towards the hypothesis of polygenes conditioning quantitative characters. Reasons of continuous variation in quantitative characters may be outlined in two distinct points.

1. Each gene involved in polygenic inheritance segregate in Mendelian fashion but because of many genes control the same characters, the familiar phenotypic ratio is modified and the classes of offspring increase. Usually so many gene pairs are involved in the inheritance of quantitative characters that the effect of any one is difficult to distinguish. Intra and inter allelic gene interaction coupled with environmental influence makes the phenotypic manifestation too complex to determine straight forward inheritance pattern.
2. The quantitative traits are rather strongly influenced by the environment, which has a large effect as compared to one of the polygenes governing the concerned trait. As for example, effect of fertilizer, light on plant height, fruit number, etc. As the number of gene pairs involved increases, the classes of offspring increase but if the number is very high as in the case of most of the quantitative characters, environmental influence causes loss of discrete classes.

Table 7: Main differences between the qualitative and quantitative traits

Qualitative or discontinuous trait	Quantitative or continuous trait
Inheritance depends on major gene(s) which is an all or none phenomenon depend on one gene or few interacting genes	Inheritance depends on polygenes and the pattern gives a graded series extending from one parental extreme to the other due to additive action producing a small proportion of the total effect.
The character differs in kind	The character differs in degree
Individual belongs to distinct phenotypic classes.	The characters do not permit sharp classification of individuals as to type.
No need to measure each individual as simple frequencies in each discrete class needs to be counted.	The continuous range of phenotypes requires each individual to be measured.
Effects of individual genes are identifiable and the number of genes controlling the characters can be assessed.	Individual effect of any gene is too small to be identified particularly when more number of genes are involved in the inheritance.
The characters are not or very little affected by environment. So environmental effect on the character expression is negligible.	The characters are rather strongly influenced by the environment, which has a large effect as compared with one of the polygene. Such environmental influences are the most confusing factors for analysis.

Qualitative or discontinuous trait	Quantitative or continuous trait
It is simple arithmetic concept, and based on fitting frequencies of individuals in each phenotypic class observed to the expected ones.	Polygenic inheritance is a statistical concept. Only mean and variances of the population are considered. Such factors as epistasis, cytoplasmic influences, interactions among genes and gene products and interactions with environment are reflected in the mean variance.
Parental characters are recovered in F_2 excepting in case of epistasis.	Parental extremes frequently do not occur in F_2 segregation because of large number of polygenes controlling the quantitative character

Example of polygenic inheritance

Simple cases of polygenic inheritance involving few pairs of genes are considered to understand polygenic inheritance and its properties. Two active gene pairs showing no dominance on each other control hereditary variation for size of ear in Maize.

AA BB × aabb

(16.8 cm) ↓ (6.6 cm)

F_1 AaBb (12.1 cm)

F_2 genotype	Frequency	Phenotype (cm)	Phenotype ratio
AABB	1	16.8	1
AaBB	2	14.2	4
AABb	2	14.2	
AaBb	4	11.7	6
aaBB	1	11.7	
AAbb	1	11.7	
aaBb	2	9.1	4
Aabb	2	9.1	
aabb	1	6.6	1

Genotypic frequency: 1: 2: 2: 4: 1: 1: 2: 2: 1

Phenotype ratio: 1 : 4 : 6: 4 : 1

Each gene, A- and B- produces an equal effect on the size of ear beyond the size that is dependent on residual genotype 6.6 cm. So individual contribution may be determined as —

$$\frac{\text{Maximum ear size - minimum ear size}}{\text{presence of active allele}} = \text{relative contribution of each allele}$$

$$\frac{16.8-6.6}{4} = 2.55\text{cm}$$

AABB = 6.6 + 4 x 2.55 = 16.8 cm

AaBB / AABb = 6.6+ 3 x 2.55 = 14.2 cm

AaBb / aaBB / AAbb = 6.6+2x 2.55 =11.7 cm

Aabb / aaBb = 6.6 + 2.55 = 9.1 cm

aabb = 6.6 + 0 = 6.6 cm

Thus, the 9: 3: 3: 1 phenotype ratio expected from a dihybrid cross involving qualitative character is modified to 1 : 4 : 6 : 4 : 1 ratio with continuous variation showing no dominance in expression and on each other.

Table 8: Probability of occurrence of F_2 individuals as extreme of either parent

Pairs of segregating alleles	Fraction of F_2 as extreme as either parent
1	¼
2	1/16
3	1/64
4	1/256
5	1/1024
n	$1/(2^n)^2$

Illustration of continuous variation in polygenic inheritance

In polygenic inheritance showing continuous variation, the genes are considered to segregate in the typical Mendelian fashion, however, their phenotypic results differ from the familiar ratios. Thus a 9:3:3:1 ratio expected from a dihybrid cross is modified to 1: 4: 6: 4: 1 ratio with continuous variation with no dominance on each other (*i.e.*, AAbb = AaBb = aaBB phenotypically). If three pairs of independent alleles are involved in a cross, the ratio of 1:6:15:20:15:6:1 will replace the familiar trihybrid ratio of 27:9:9:9:3:3:3:1 based on discontinuous variation and complete dominance. The situation of continuous phenotypic variation may be described through expansion of a binomial: $(p + q)^n$ where $p + q = 1$, p and q are alleles present in the population and their frequency of gamete combination is ½. It can be illustrated as:

$(p+q)^2 = p^2\,(AA) + 2pq\,(Aa) + q^2\,(aa)$

$$= \left(\frac{1}{2}\right)^2 + 2x\frac{1}{2}x\frac{1}{2} + \left(\frac{1}{2}\right)^2$$

$$= \frac{1}{4} + \frac{2}{4} + \frac{1}{4} \qquad \text{[A and a are alleles of a gene]}$$

$$= 1$$

Table 9: Expansion of binomials with respect to number of alleles

Case	Equation	Number of independent alleles with no dominance
1.	$(p+q)^2 = p^2+2pq + q^2$	2
2.	$(p+q)^3 = p^3 + 3p^2q + 3pq^2 + q^3$	3
3.	$(p+q)^4 = p^4 + 4p^3q + 6\,p^2q^2 + 4pq^3 + q^4$	4
4.	$(p+q)^5 = p^5 + 5p^4q + 10\,p^3q^3 + 10\,p^2q^3 + 5pq^4 + q^5$	5
5.	$(p+q)^6 = p^6 + 6p^5q + 15\,p^4q^2 + 20p^3q^3 + 15\,p^2q^4 + 6\,pq^5 + q^6$	6

Such binomial expansions are actually the product of the chances of their separate occurrences. Phenotypic frequency of 3 pair of genes (6 alleles) = Eq. 3 × Eq.1.

Practical application of binomial expansion in quantitative genetics

Example 1: Two active gene pairs (four alleles) controlling hereditary variation of a quantitative trait.

AABB ×aabb

↓

F_1 → AaBb

↓

F_2 phenotype frequency

$(p+q)^4 = p^4 + 4\,p^3q + 6p^2q^2 + 4pq^3 + q^4$ $\qquad (p+q) = 1$

$$= \left(\frac{1}{2}\right)^4 + 4\left(\frac{1}{2}\right)^3 \text{x} \frac{1}{2} + 6\left(\frac{1}{2}\right)^2\left(\frac{1}{2}\right)^2 + 4\,\text{x}\frac{1}{2}\text{x}\left(\frac{1}{2}\right)^3 + \left(\frac{1}{2}\right)^4$$

$$= \frac{1}{16} + \frac{4}{16} + \frac{6}{16} + \frac{4}{16} + \frac{1}{16}$$

So of the 16 possible segregating genotypes –

1 will be p^4 *i.e.* AABB

4 will be p^3q *i.e.* AaBB (2) + AABb (2)

6 will be p^2q^2 *i.e.* AaBb (4) + AAbb (1) + aaBB (1)

4 will be pq^3 *i.e.* aaBb (2) + Aabb (2)

1 will be q^4 *i.e.* aabb

The 'A' and 'B' being the active alleles of the two genes influence equally on the metric trait and designated as 'p' allele. Similarly, 'a' and 'b' alleles are designated as 'q' allele.

Example 2: Three active gene pairs (six alleles) are controlling hereditary variation of a metric trait.

AABBCC ×aabbcc

↓

F_1 → AaBbCc

↓

F_2 phenotype frequency

$(p+q)^6 = p^6+6p^5q+15p^4q^2+20p^3q^3+ 15p^2q^4 + 6\,pq^5 + q^6\,(p+q) = 1$

$$= \left(\frac{1}{2}\right)^6 + 6\left(\frac{1}{2}\right)^5 \times \frac{1}{2} + 15\left(\frac{1}{2}\right)^4 \times \left(\frac{1}{2}\right)^2 + 20\left(\frac{1}{2}\right)^3 \times \left(\frac{1}{2}\right)^3 + 15\left(\frac{1}{2}\right)^2 \times \left(\frac{1}{2}\right)^4 + 6 \times \frac{1}{2} \times \left(\frac{1}{2}\right)^5 + \left(\frac{1}{2}\right)^6$$

$$= \frac{1}{64} + \frac{6}{64} + \frac{15}{64} + \frac{20}{64} + \frac{15}{64} + \frac{6}{64} + \frac{1}{64}$$

So of the 64 possible segregating genotypes –

1 will be p^6 *i.e.* AABBCC

6 will be p^5q *i.e.* AABBCc(2) + AABbCC (2) + AaBBCC (2)

15 will be p^4q^2 *i.e.* AABbCc (4) + AaBBCc (4) + AaBbCC (4) + AABBcc(1) + AAbbCC (1) + aa BBCC(1)

20 will be p^3q^3 *i.e.* AaBbCc (8) + AABbcc(2) + AaBBcc (2) + aaBbCC(2) + aaBBCc (2) + AAbbCc (2) + AabbCC (2)

15 will be p^2q^4 *i.e.* AaBbcc (4) + AabbCc (4) + aaBbCc (4) + AAbbcc (1) + aaBBcc (1) + aabbCC (1)

6 will be pq^5 *i.e.* aabbCc(2) + aaBbcc (2) + Aabbcc (2)

1 will be q^6 *i.e.* aabbcc

The 'A', 'B', 'C' being the active alleles of the 3 genes or loci and having equal influence on the metric trait are designated as 'p' allele and 'a', 'b' and 'c' alleles of those loci are designated as 'q' allele.

Components of Genetic Variance

Polygenic inheritance causes measurable genetic variation. Different components of genetic variation are discussed below.

Additive genetic variance

It refers to that portion of genetic variance which is produced by the deviations due to average effects of the alleles of genes at all segregating loci. It is a mechanism of quantitative inheritance such that the combined effects of the genetic alleles at two or more gene loci are equal to the sum of their individual effects. Additive genes are gene which are neither dominant nor recessive alleles, although gradations may exist between extremes. The resulting phenotype does not follow a pattern of Mendelian inheritance but instead is determined by the combined effect of the alleles.

Thus, this component of variation arises from differences between two homozygotes of a gene *i.e.*, AA and aa. The genes act in such a way that substitution of one allele for another produces same plus or minus effect regardless of other gene(s) and the effect is the same whether replacement occurs in a homozygote or heterozygote, which is illustrated in Additive gene action model.

	AA	**Aa**	**aa**	**Mean**
BB	10	8	6	8
Bb	8	6	4	6
bb	6	4	2	4
Mean	8	6	4	

This illustration shows that Aa = $\frac{AA+aa}{2}$ = 8, implying absence of dominance. This is true at Bb and bb levels also, Replacement of 'a' by 'A' produces a plus effect of 2 units disregarding the presence of B gene and genetic phase at B locus showing no interaction between A and B. Further, totals of any two diagonal values are equal showing no interaction between two polygenes.

The additive genes show lack of dominance, *i.e.* intermediate expression. The additive genetic variance is associated with homozygosity and, therefore, it is expected to be the maximum in self-pollinating crops and minimum in cross-pollinating crops. Additive variance is fixable and therefore, selection for the traits governed by such variance is very effective. Additive genetic variance is important for following major reasons.

(a) It is required for estimation of heritability in narrow sense and response to selection is directly proportionate to narrow sense heritability.

(b) It is pre-requisite for selection because this is the only variance, which responds to selection.

(c) Breeding value of an individual is measured directly by the additive gene effects. The general combining ability (gca) effect of a parent is a measure of additive gene effects.

(d) In natural plant breeding populations, additive variance is the predominant one closely followed by dominance variance.

Dominance variance

In contrast, the second component of genetic variance i.e., dominance genetic variance involves deviation due to interactions between alternative alleles at a specific locus. It arises due to the deviation from the additive scheme of gene action and such deviation is related to intra-allelic interaction, *i.e.* in between alleles of same gene or same locus. This gene action is detected due to the deviation of heterozygote (Aa) from the average of two homozygotes (AA and aa).

The genes act in such a way that the substitution of 'A' for 'a' in the genotype 'aa' is not the same as in the genotype 'Aa', which is illustrated in the dominance gene action model assuming complete dominance at both A and B loci and without interaction.

	AA	**Aa**	**aa**	**Mean**
BB	9	9	7	8.3
Bb	9	9	7	8.3
bb	7	7	5	6.3
Mean	8.3	8.3	6.3	

This model shows that Aa = AA disregarding whether at other locus there is BB or Bb. Likewise, Bb = BB. Thus, there is complete dominance for both the genes and interaction is absent. No interaction between A and B genes is further reflected by the fact that total of any two diagonal values is the same.

Such gene actions are associated with heterozygosity. Therefore, this gene action is expected to be maximum in cross-pollinating crops and minimum in self-pollinating species. Dominance variance is not fixable and, therefore, selection for traits controlled by such variance is not effective. Heterosis breeding is the best option in such situation.

Table 10: Difference between additive and dominance variance

Additive variance		Dominance variance
1.	It refers to difference between homozygotes (AA/aa)	It refers to deviation of Aa from the mean of AA and aa
2.	Genes show lack of dominance	Genes show incomplete, complete or over dominance
3.	Associated with homozygosity and is more in self-pollinated crops	Associated with heterozygosity and is more in cross-pollinated crops
4.	It is fixable	It is non-fixable
5.	Selection is very effective as it is fixable	Selection is ineffective as it is non-fixable
6.	It is the chief cause of transgressive segregation	It is the chief cause of hybrid vigour

Epistatic variance

Epistasis is the deviation of two-locus genotypic values from the sum of the contributing single-locus genotypic values. Function of this parameter leads to statistical tests for epistasis. Epistatic variance arises due to the deviations as a consequence of their allelic interaction, *i.e.*, interaction between alleles of two or more different genes or loci. The epistatic variance is of three types, *viz.*, additive × additive, additive × dominance and dominance × dominance.

- **Additive × Additive (A × A):** In this case, both the interacting loci have homozygous condition. Thus this type of variance results due to interaction of alleles AA/aa with BB/bb, which can be illustrated in the following model

	AA	**Aa**	**aa**	**Mean**
BB	9	6	3	6
Bb	6	4	2	4
bb	3	2	1	2
Mean	6	4	2	

In this given model, Aa = mean of AA and aa and similarly Bb = (Bb + bb)/2 show no dominance. Further substituting 'a' by 'A' has equal effect from aa to Aa and from Aa to AA (3 units with AA, 2 units with Aa and 1 unit with aa). Same is true for B gene. This shows additive gene action. However, the amount of shift at 'A' locus is not independent of genetic phase at 'B' locus and *vice-versa* implying interaction which is also reflected by the fact that the diagonal totals are not matched.

- **Additive × Dominance (A × D):** In this type of epistasis, one locus is homozygous and the other is heterozygous for the interacting genes. Thus this type of variance results due to interaction of alleles AA/aa with Bb or BB/bb with Aa.

- **Dominance × Dominance (D × D):** In this type of epistasis, both the interacting loci have heterozygous condition. This type of variance results due to interaction between Aa and Bb.

Table 11: Difference between Dominance and Epistatic variance

Dominance variance		Epistatic variance
1.	It is due to interaction between genes of the same locus	It is due to interaction between genes of two or more different loci
2.	It is of three types *viz.* incomplete, complete and over-dominance	It is also of three types, *viz.* additive × additive, additive × dominance and dominance × dominance
3.	It is unfixable	Additive × Additive type of epistatic variance is fixable but other types are not fixble
4.	Magnitude is higher than epistatic variance	Magnitude is lower than dominance variance

The A × A type of epistasis is fixable and, therefore, selection is effective for traits governed by such variance. The A × D and D × D types of epistatic variances are unfixable and therefore, heterosis breeding may be rewarding for the traits under the control of such genetic variance. In the natural plant breeding populations, epistatic variance has the lowest magnitude. The overall epistatic interaction of polygenes may be categorized under two types.

i) **Complementary type:** This involves interaction between two non-alleles complementing each other to produce a new phenotype, which is not ascribable to them individually. This is created by the interaction of two homozygotes, each acting additively (additive × additive *i.e.* A × A). Epistasis related to expression of some qualitative characters revealing 9:7 (complementary), 9:3:4 (recessive epistasis) ratios fall in this category.

ii) **Duplicate type:** This involves two non-allelic genes, which tend to cancel or weaken the effect of each other in hybrid combination. Such a variation arises from homozygote-heterozygote (additive × dominance) or heterozygote-heterozygote (dominance × dominance) combination. Epistasis related to the expression of some qualitative characters revealing 13:3 (inhibitory gene expression), 13:3 (duplicate gene epistasis) and 12:3:1(dominant gene epistasis) ratios fall in this category.

In quantitative characters, when dominance (D) and dominance × dominance variance components have same sign (+ or -), it depicts complementary type of epistasis and otherwise it marks duplicate types epistasis.

Partitioning of phenotypic variance

The total phenotypic variance may be partitioned as Fixable (additive and additive × additive epistatic components) and Non-fixable (dominance, additive × dominance and dominance × dominance type of epistatic components) and environmental fraction, which may be summarized as follows:

VP = VG + VE; VG = VA + VD + VI; and VI = VAA + VAD + VDD

where, VP = Phenotypic variance, VG = Genotypic variance, VE = Environmental variance; VA = additive variance, VD = dominance variance, VI = epistatic variance, VAA = additive × additive variance, VAD = additive × dominance variances and VDD = dominance × dominance variance.

In homozygous genotypes, the genetic variance is of additive (A) and additive × additive epistasis (A x A) types while, all the three types of genetic variances, *viz.* additive, dominance and epistasis are present in the segregating populations.

Estimating the components of genetic variance may also assess the magnitude of variation. This involves crossing of a number of genotypes of strains in definite mating designs like, diallel, partial diallel, line × tester, generation mean analysis, etc. and subsequent evaluation of progenies thus obtained in replicated trials.

Concept of Heritability

The genes cannot cause a character to develop unless they have the proper environment, and conversely, no amount of manipulation of the environment will cause a character to manifest unless the necessary genes are present. So, in quantitative inheritance the phenomenon, genotype × environment interaction for the expression of a character is ubiquitous. Phenotypes that vary between the individuals in a population do so because of both environmental factors and the genes that influence traits, as well as various interactions between genes and environmental factors. Unless they are genetically identical (e.g., monozygotic twins in humans, dihaploids, inbred lines in experimental populations, or clones), the individuals in a population tend to vary in the genotypes they have at the loci affecting particular traits. The combined effect of all loci, including possible allelic interactions within loci (dominance) and between loci (epistasis), is the genotypic value. This value creates genetic variation in a population when it varies between individuals.

Heritability is a statistic used in the fields of breeding and genetics that estimates the degree of variation in a phenotypic trait in a population that is due to genetic variation between individuals in that population. In fact, heritability is formally defined as the proportion of phenotypic variation (V_P) that is due to variation in genetic values (V_G). Depending on the components of variance in the calculation, heritability is of two types, *viz.*, broad sense heritability and narrow sense heritability.

• Broad sense heritability

It captures the proportion of phenotypic variation due to genetic values that may include effects due to dominance and epistatic variance. It is calculated using the following formula suggested by Burton and Devane (1953).

Broad sense heritability (H) = Vg/VP = Vg/(Vg + Ve); where, Vg, VP and Ve are genotypic, phenotypic and environmental components of variances, respectively.

Broad sense heritability plays an important role in animal breeding than in plant breeding because in plants the environmental effects cannot be controlled. However, this heritability estimate can be used in selecting superior lines from the homozygous materials.

• Narrow sense heritability

Narrow-sense heritability is defined as the fraction of phenotypic variance that can be attributed to variation in the additive effects of genes. If the breeder is interested in predicting the phenotype of an offspring from a cross of two parents, the portion of additive variance is important which depicts relative contributions that the parents will make to the F_1 population. Narrow-sense heritability is always less than or equal to broad-sense heritability. Hence, it is the ratio of additive or fixable genetic variance to the total or phenotypic variance. It is calculated from the following formula.

Narrow sense heritability = VA/ VP; where, VA is the additive genetic variance and VP is the phenotypic variance. Since narrow sense heritability is estimated from additive (heritability fixable) genetic variance, it plays important role in the selection of elite genotypes from the segregating populations.

Table 12: Difference between Broad sense and Narrow sense heritability

Broad sense heritability		**Narrow sense heritability**
1.	It is estimated from total genetic variance	It is estimated from additive genetic variance
2.	It can be estimated from both parental and segregating material	It requires development of segregating population through crossing in definite mating design
3.	It is more useful in animal breeding	It is useful in both plant and animal breeding
4.	It is useful in selection of elite types from homozygous lines	It is useful in selection of elite types from segregating material

Important points about heritability

- The heritability estimate is specific to the population and environment under analysis.
- The estimate is a population, not an individual parameter.
- Heritability does not indicate the degree to which a trait is genetic, it measures the proportion of the phenotypic variance that is the result of genetic factors.

Factors affecting phenotypic expression

With the historical "pure line" experiments in genetics of Wilhelm Johannsen in 1903, the distinction between genotype and phenotype was demonstrated (Cited by Peirson, 2012). Experiment of Nilsson-Ehle and East demonstrated that quantitative characters were inherited according to the law of Mendel however, the variation arises from the joint action of the genotype and environment. In describing the phenotype, it is convenient to express this joint action in a linear fashion so that the phenotypic expression of any particular character denoted by A, can be represented as: $A= \mu + a + e + (ae)$

In this expression, the numerical value of the phenotype is visualized as the sum of a general population mean (μ), a genotypic effect (a), an environmental effect (e), and an interaction effect of genotype and environment (ae). If it was possible to catalogue all genotypes and all environments, the value obtained would represent the general population mean (μ). Any particular genotype (or particular environment) adds or subtracts from the mean depending on whether its effect on yield is above or below the general population mean. The interaction term (ae) will be zero only when all genotypes behave or perform consistently in all environments, that is, there are no genotypic x environmental interactions. Since all measurements are necessarily on phenotypes, any practical definition of μ, a, e, and ae must also be in phenotypic terms. In this system, the effect of a genotype, therefore, is defined only relative to other genotypes, and this relationship must be averaged over a number of environments. If the relationship is not constant from environment, this variability will be reflected by the interaction term.

From the point of view of the plant breeder, genotypic values must be measured with reference to some particular group of environments, usually the ones occurring over a period of years at a number of locations within some comparatively homogeneous geographical area. Experimental measurements of genotypic values, no matter how extensive (within practical limits), can only sample the population of environments occurring within this geographical area. Hence, the effects of the environmental components that are likely to be important in determining the agricultural value of genotypes, for example, soil fertility, rainfall, temperature, etc. are usually identified in plant breeding approach. In practice, these items are usually identified with years and locations. If the analysis is conducted in terms of different locations and different seasons, the first equation can be expanded to:

$A = \mu + a + r + l + y + (al) + (ay) + (lg) + (alg) + e.$

In this expression, μ and a are the population mean and the genotypic effect, respectively; r, l, and y are the direct effects of replicates, locations, and years, respectively, and e is a composite error term of the remaining effects (including the effect of the plot compared to others in the same replication, error due to sampling among the plants in the same family, and errors of measurement) and the interaction effects are al (genotype × location), ay (genotype × year), aly (genotype × location × year).

Genetic advance under selection

The essential aspects of most, if not all, breeding programmes are: (1) selection with a base population of genetically variable individuals or families, and (2) utilization of the selected material for the creation of new populations to be employed either as potential new commercial varieties or as the base for a new cycle of selection. It is apparent from the two equations on phenotypic expression that genetic advance under selection (as represented by improvement in ‘a’ the genotypic value) in the new population as compared to the base population will depend on a) the amount of genetic variability, that is, the magnitude of the differences in ‘a’ among different individuals (or families) in the base population, and b) the magnitude of the masking effect of the environmental and interaction components of variability on the genetic variability. Hence, “Genetic advance” explains the degree of gain obtained in a character under a particular selection pressure. High genetic advance coupled with high heritability estimates offers the most suitable condition for selection. Genetic advance can be calculated using the formula suggested by Johnson *et al.* (1955).

$G = (k)(\sigma_A)(H)$

where, G = Expected genetic advance under selection. It measures the difference between the mean genotypic values of the selected lines 'q' and the mean genotypic value of the lines or families (n) in the base population.

σA = Phenotypic standard deviation of the character mean of the lines or families in the base population.

H = Heritability in broad sense i.e., ratio formed by dividing genotypic by the phenotypic variance.

K = Selection differential which takes into account the mean phenotypic value of the 'q' selected lines, the mean phenotypic value for all the lines or families (n) tested, the phenotypic standard deviation (σA), and the stringency of selection q/n. Since K is expressed in standard deviation units i.e., in terms of a unit normal curve, k does not vary except as the selection intensity, q/n is varied. Thus, if 1% highest performing lines are selected, K can be shown to have the value of 2.64. In selection intensities of 2, 5, 10, 20 and 30 percent, K takes the values of 2.42, 2.06, 1.76 and 1.16, respectively.

Selection index

Careful analysis of the characters reveals a greater complexity of the processes than simple and free recombination of characters. The discovery of genetic linkages and better understanding of the close relationship between functions and organs in a living organism have established solid and scientific basis of correlation theory. In fact, genotypic correlations may derive from genetic linkage, pleiotropy or developmentally induced relationships between components that are only indirectly the consequence of gene action. Correlation study will help to know the suitability of various characters for indirect selection because selection for one or more traits results in correlated response in several other traits. In crop improvement programmes, several traits with varying degrees of agronomic and economic importance are usually considered at the same time. Because genetic correlations exist between traits, their selection cannot be considered separately. Improvement of one will provoke changes in others, usually in a sense contrary to that desired. In tomato, for example, if only number of fruits per plant is selected, the best families will surely have fruits of less weight.

The alternative methodology available to plant breeders to counteract this type of problem comprises selection indices (Hazel, 1943). The importance of these lies in evaluating the merits of individuals or families in terms of several traits. Selection based on indices permits maximizing the response to selection for one or a group of traits. In reality, selection based on indices reflects not only the response with direct selection, but also the correlated response as selection is practised for other traits.

To construct the selection indices, the magnitude of the genetic and phenotypic correlations between the traits need to be determined first. For the specific case of population improvement through recurrent selection, the additive components of genetic variance and covariance need to be estimated because these are directly related to the response to selection.

Population genetics

Genes exist in association with other genes in an individual and many individuals comprise a population. The sum total of all genes in a population is called the gene pool. So, a gene pool is the total genetic information possessed by reproductive members in a population of sexually reproducing organisms. Mendelian population is an interbreeding group of organism sharing a common gene pool. Understanding of population genetics lay the basis of breeding cross pollinated vegetable crops manifesting inbreeding depression like,cabbage, cauliflower, broccoli, Brussels sprout, radish, carrot, turnip, onion, etc. because breeding approaches for these crops is basically lay on the improvement of their existing population.

Principle of genetic equilibrium

Population genetics is a field of biology that studies the genetic composition of biological populations, and the changes in genetic composition that result from the operation of various factors, including natural selection. Population geneticists developed abstract mathematical models of gene frequency dynamics, extract predictions about the likely patterns of genetic variation in actual populations, and test the predictions against empirical data. The Hardy-Weinberg principle, discovered independently by them is one of the simplest and most important principles in population genetics (Hardy, 1908; Weinberg, 1908). According to this basic law of population genetics, equilibrium in the population is maintained as long as population size is large, mating is at random and mutation, selection and migration are negligible. Genes in the pool have dynamic relations with other alleles and with the environment in which the organisms are exposed. According to the principle, in a large interbreeding population in which mating is random and any factors for changing the gene frequency like selection, mutation and migration do not occur then –

1. Both gene frequencies and genotype frequencies remain constant from generation to generation.
2. Only one generation of random mating is required to reach genetic equilibrium which will be true regardless of what the initial frequencies of the two genes in the population may be.
3. This principle holds good for both multiple allele and multiple loci.
4. Neither dominance nor recessiveness will change the gene frequencies in subsequent generations.

Concept of genetic equilibrium principle

A single locus is assumed to have two alleles 'A' and 'a' with gene frequencies p and q, respectively, where $p + q = 1$ because the heterozygote Aa contribute either to 'A' or 'a' only half of their frequencies. The segregation of a single pair of alleles Aa, in a monohybrid cross may be represented by the simplest binomial expansion $(a+b)^n = (A+a)^2 = 1AA + 2Aa + 1aa$. At equilibrium the frequencies of genotypic classes are: p^2 (AA), $2pq$ (Aa) and q^2 (aa).

Table 13: Combination of gametes from heterozygous individual

Female gamete	Male gamete	
	A (p)	**a (q)**
A (p)	AA (p^2)	Aa (pq)
a (q)	Aa (pq)	aa (q^2)

Note: Allele frequency relation: p^2 (AA) + 2 pq (Aa) + q^2 (aa) = 1

In a large population including parental genotypes (AA and aa) as well as the heterozygotes (Aa), equilibrium is established for a single pair of alleles after one generation of random mating. The genetic proportion of an equilibrium population is entirely determined by its gene frequencies.

Table 14: Equilibrium population and gene frequencies

Genotypes in the random mating population	Random mating	Mating may be summarized as
AA	AA × AA	AA × AA
Aa	AA × Aa	2(AA × Aa)
	Aa × AA	
	AA × aa	2(AA × aa)
aa	aa × AA	
	Aa × Aa	Aa × Aa
	Aa × aa	2(Aa × aa)
	aa × Aa	
	aa × aa	aa × aa

Table 15: Maintenance of genetic equilibrium in a randomly mating population with similar frequency of two alleles 'A' and 'a' present in the population (p+q=1) and one generation random mating is required to reach the equilibrium

Random matings		**AA × AA**	**2 (AA × Aa)**	**2 (AA × aa)**	**Aa × Aa**	**2(Aa × aa)**	**aa × aa**
Mating frequencies		$p^2 \times p^2$	$2\ (p^2 \times 2\ pq)$	$2(p^2 \times q^2)$	$2pq \times 2\ pq$	$2(2pq \times q^2)$	$q^2 \times q^2$
Offspring frequencies	AA	p^4	$2p^3q$	-	p^2p^2	-	-
	Aa	-	$2p^3q$	$2p^2q^2$	$2p^2q^2$	$2pq^3$	-
	aa	-	-	-	p^2q^2	$2pq^3$	q^4
Total offspring frequencies		$p^4 +$	$4p^3q$	$+ 2p^2q^2$	$+ 4p^2q^2$	$+ 4pq^3$	$+q^4=\Sigma=1$
		It can be written as – $(p^4 + 2p^3q + p^2q^2) + (2p^3q + 4p^2q^2 + 2pq^3) + (p^2q^2 + 2pq^3 + q^4)$ or, $p^2 (p^2+2pq+q^2) + 2pq\ (p^2 + 2pq + q^2) + q^2 (p^2 + 2pq + q^2)$ or, $p^2 \times 1 + 2\ pq \times 1 + q^2 \times 1$ or, $p^2 + 2pq + q^2$ or, $(p + q)^2$ $= 1$					
		Deductions in the equation					
$2(p^2 \times 2\ pq)$		$2\ pq \times 2\ pq$			$2\ (2pq \times q^2)$		
$= 2\ (2p^3q)$		$= 4p^2q^2$			$= 2\ (2pq^3)$		
$= 2\ (p^3q + p^3q)$		$= p^2q^2 + 2p^2q^2 + p^2q^2$			$= 2\ (pq^3 + pq^3)$		
$= 2\ p^3q + 2\ p^3q$					$= 2pq^3 + 2\ pq^3$		

Note: Genotypes in random mating population: AA, Aa, aa

Existence of the same gene and genotype frequencies in a large free mating population (variety of a cross pollinated crop) is maintained generation after generation following this genetic equilibrium principle so long the population is not contaminated with other genes. For this reason, a particular open pollinated variety of the random mating cross pollinated vegetable crops like, cabbage, cauliflower, broccoli, Brussels sprouts, carrot, beet, radish, onion, etc. remain stable through maintenance of genetical purity generation after generation owing to this equilibrium if seed production is done in strict isolation.

References

Bateson, W. 1909. *Mendel's Principles of Heredity.* Cambridge University Press, UK.

Boveri, T.H. 1901. Das Problem der Befruchtung. Fischer, Jena. Partial translation In: *Classics of Biology* (Ed. Suñer, A.P.). Sir Isaac Pitman & Sons, London.

Burton, G.N. and Devane, E.M. 1953. Estimating heritability in tall Fescue (*Festuca arundiancea* L.) from replicated clonal materials. *Agronomy Journal*, **45**: 478-481.

De Vilmorin, P. and Bateson, W. 1911. A case of gametic coupling in *Pisum. Proc. Royal Soc.*, London, Series B, 84:9-11

Hardy, G.H., 1908. Mendelian Proportions in a Mixed Population, *Science*, **28**: 49–50.

Hazel, L.N. 1943. The genetic basis for constructing selection indexes. *Genetics*, **28**(6): 476–490.

Johnson, H.W., Robinson, H.F. and Comstock, R.E. 1955. Estimates of genetic and environmental variability in soybean. *Agron. J.*, **47**: 314-318.

Peirson, B.R.E. 2012. *Wilhelm Johannsen's genotype-Phenotype distinction.* Embryo Project Encyclopedia, Centre for Biology and Society, Arizona State University, USA.

Sutton, W.S. 1903. The chromosomes in heredity. *Biol. Bull.* **4**: 231-251.

Weinberg, W. 1908. Über den Nachweis der Vererbung beim Menschen. *Jahreshefte des Vereins für vaterländische Naturkunde in Württemberg*, **64**: 368-382.

6

Methods of Vegetable Breeding: Introduction, Selection, Hybridization and Selection

Vegetable breeding is the purposeful genetic manipulation of qualities in vegetable crops to create new varieties with a set of desired characteristics that are better suited for cultivation, give better yield, quality and are resistant to biotic and abiotic stresses. Vegetable breeding can be accomplished through many different techniques ranging from plant introduction, simply selecting plants with desirable characteristics, to more complex molecular techniques. Classical plant breeding includes hybridization followed by selection and mutation to develop the desirable plant types. With the advancements in genetics, molecular biology and *in vitro* culture, different molecular genetics tools are also being used in vegetable breeding.

Sexually propagated vegetable crops

1. Plant Introduction

The term "Plant introduction" describes the movement of crop plants, species, varieties, wild species, etc. from their original growing sites to other locations where it was not grown earlier. Plant introduction is essentially the method of cultivating crop plants in a new environments. The agro-climatic conditions can differ greatly depending on the origin of the plant species and the area where they are introduced. Plant introduction is a time-honored strategy for improving crop diversity.

Methods of plant introduction

- **Procurement of germplasm**: The new germplasm is procured through National Bureau of Plant Genetic Resources (NBPGR), New Delhi. After receiving the indent or request of the indenter (individual of Institution) through NBPGR, the sender (individual or Institution) of the foreign countries sends the required genetic material to the indenter scientist or organization through NBPGR. The NBPGR after checking the phyto-sanitary certificate attached with the consignment send the material to indenter after receiving the requisite fee for handling and processing the germplasm. The Bureau also keeps some seed sample of the germplasm for evaluation at their end. The Bureau also requests the indenter scientist or Institution to send evaluation report pertaining to the introduced material. Individual scientist or Institution can also submit their requirement to the Director, NBPGR, Pusa Complex, New Delhi. If the required germplasm is available in the stock of NBPGR, the Bureau sends the genetic material to the indenter. If the Bureau is unable to meet the request from its own stock or from known source, it attempts to procure them from the counterparts in other countries.

- **Quarantine:** All introduced material is thoroughly inspected for contamination with diseases, weeds and insects. The material is kept in isolation to prevent spreading of diseases, insect pests, etc. Plants that are suspected to be contaminated are grown in isolation for observation of insect pests and diseases. The entire process is known as quarantine and the rules which are used for this are known as "Quarantine rules". It is essential that all the material(s) being introduced must be accompanied by an authentic phytosanitary certificate. The phytosanitary certificate is thoroughly inspected and returned back to the sender or owner of the germplasm. The plant material being introduced or exported must confirm to certain quarantine regulations and quarantine control is exercised by NBPGR at different points of entry.
- **Cataloging**: All the plant material which is introduced is given an entry number and information regarding agency, place of origin, adaptation, etc is well documented. Plant material is classified in three categories
  - Exotic collection (EC)
  - Indigenous collection(IC)
  - Indigenous wild collection
- **Evaluation**: The plant material is sent to sub-stations of the Bureau and evaluated with respect to various characters to assess the potential of new introductions.
- **Acclimatization**: It is the process that leads to the adaptation of a new species or variety to the new environment with respect to altitude, temperature, humidity, light, soil salinity and reaction, presence of certain chemicals, etc.within the lifetime of an individual. This ability allows plants to cope with the continuous variation in ambient conditions to which they are exposed as sessile organisms. It is a kind of natural selection.
- **Multiplication and distribution**: Plant material which is introduced is to be multiplied and further tested at various locations. The suitability of cultivation in different regions of the country should be assessed before using it as a commercial variety.

Importance of Plant Introduction

Plant introduction is not only an important method of developing a variety but also set the initial stage of breeding programme for both sexually and asexually propagated vegetable crops through the source of new germplasm. Advances in the development of new varieties and/or hybrids are largely dependent on diversity of the source material at breeder's disposal. When a well-known commercial variety or a true breeding material is introduced from outside country, it can be utilized directly as a new variety after proper evaluation trials. Introduced materials can also be developed through acclimatization through a genetic shift based on natural and artificial selection. A good number of varieties of self- and cross-pollinated vegetable crops successfully cultivated in India are either directly introduced varieties or developed after introduction.

Introduced varieties released for cultivation

- **Tomato**: Sioux, Roma, Marglobe, etc.
- **Sweet pepper**: California Wonder ,Yolo Wonder, etc
- **Cabbage**: Golden Acre, September, etc
- **Knolkhol**: White Vienna , Purple Vienna, etc
- **Garden pea**: Arkel , Bonneville, etc.
- **French bean**: Contender, Giant Stringless, etc.
- **Carrot**: Nantes, Chanteny, etc.
- **Garden beet**: Detroit Dark Red, Crimson Globe, etc.
- **Turnip**: Purple Top White Globe, Early Milan Red Top, etc.
- **Radish**: White Icicle, Japanese White, etc.
- **Water melon**: Asahi Yamato, Sugar Baby, etc.
- **Summer squash**: Australian Green, Patty Pan, etc.
- **Cucumber**: Japanese Long Green, Poinsett, etc.

Varieties developed through selection from the introduced material

Introduction of heterozygous materials of both cross- and self-pollinated and non-uniform materials of cross-pollinated vegetable crops also form the base for selecting desirable types. Progenies of desirable types are increased, purified, evaluated and released as a new variety.

Cabbage: Pusa Drum Head (selection from a hybrid material introduced from Japan).

Tomato: Arka Alok (selection from an introduced line (CL-114-5-1-0) from AVRDC, Arka Ashis (selection from an introduced line (UC 83B) from California), CO-2 (selection from a Russian introduction)

Chilli: Punjab Guchhedar (selection from an introduction of Indonesia), Hot Portugal (selection from an introduction of Portugal)

Sweet pepper: Arka Baswant (selection from an introduced line Soroksari from Hungery); Arka Gaurav (selection from an introduced variety Golden Calwonder from USA); Arka Mohini (selection from an introduced variety Titan from USA)

Garden pea: Harbhajan (selection from an exotic stock)

Carrot: Solan Rachna (selection from an introduction of Netherlands)

Radish: Pusa Chetki (selection from an introduction of Denmark)

Varieties developed through hybridization with the introduced material

Introduced material can be utilized effectively in the hybridization programme to incorporate some desirable characters to the progenies. Some promising varieties have been developed in this manner where one or both the parents involved in the hybridization programme were exotic.

- **Tomato**: Pusa Ruby [Sioux (exotic) × Improved Meeruti] , Pusa Sheetal [Balkan (exotic) × Jemnorrosnej (exotic)], Pusa Early Dwarf [Improved Meeruti × Red Cloud (exotic)], Punjab Chhuhara [EC55055 (exotic) × Punjab Tropic (exotic)], Pusa Sheetal [Balka (exotic) × Jemnoroshnese (exotic)], Pusa Gaurav [Glamour (exotic) × Watch (exotic)]
- **Chilli**: Prakash [G-3 × Huntaka (exotic)]
- **Garden pea**: Jawahar Matar-3 [T-19 × Early Badger (exotic)], Hisar Harit [Bonneville (exoic) × P-23]
- **Cowpea**: Pusa Dofasli [Pusa Phalguni × Philippines Bush (exotic)]
- **Cauliflower**: Pusa Snowball-1 [EC-12012 (exotic) × EC-12013 (exotic)]
- **Radish**: Pusa Himani [Radish Black × Japanese White (exotic)]
- **Carrot**: PusaYamdagni [EC-9981 (exotic) × Nantes (exotic)], Pusa Kesar [Local Red × Nantes Half Long (exotic)]
- **Turnip**: Pusa Chandrima [Snow Ball (exotic) × Japanese White (exotic)], Pusa Kanchan [Local Red Round × Golden Ball (exotic)],
- **Muskmelon**: Punjab Sunheri [Hara Madhu × Edisto (exotic)]
- **Watermelon**: Arka Manik [HR 21 × Crimson Sweet (exotic)]

Varieties developed through hybridization with exotic wild relatives

Introduction of wild relatives and other cultivated species often serves as a gene complex reservoir for resistance to some insect-pests and diseases. There are some instances where introduced wild relatives have been utilized in the hybridization programme to develop resistant varieties.

- **Okra**: Yellow vein mosaic virus resistant varieties *viz.*, Arka Anamika and Arka Abhay (Okra variety × *Abelmoschus manihot* var. *tetraphyllus*), Parbhani Kranti and Punjab 7 (Pusa Sawani × *Abelmoschus manihot* var. *manihot*), Punjab Padmini (Reshmi × *Abelmoschus manihot* var. *manihot*), etc.
- **Tomato**: Leaf curl resistant variety Hisar Anmol (Hisar Arun × *Solanum habrochaites*); Leaf curl resistant variety Kashi Vishes (Tomato × *Solanum habrochaites*); High fruit quality variety Pusa Red Plum (Tomato × *Solanum pimpinellifolium*)
- **Chilli**: Leaf curl virus tolerant variety Pusa Sadabahar (Pusa Jwala × *Capsicum frutescens* (IC 81339)

2. Selection

Selection is the process of choosing plants with desirable characteristics. This is the oldest breeding method and is the basis of all crop improvement programmes. In this process of selection, plants with useful and desirable characteristics are chosen and maintained. In simple breeding method based on only selection, the source material for selection usually constitute breeding line, farmers' variety, land races, other collections including introduced materials. In this method, improved and more homogeneous varieties are developed within a rather short period. Selection is highly effective for improving simply inherited traits.

Determination of genetic variability for the concerned character

- **Data collection and record keeping:** In the breeding materials and trials comprising of different germplasm (varieties, cultivars, strains and segregating populations of various crosses), data are recorded on both qualitative and quantitative characters. The record is thus, maintained to keep a complete history or pedigree of various strains from the time these are introduced into the trials at an experimental station up to a particular stage in the breeding programme.
- **Analysis of variance for the character:** The statistical procedure which separates or splits the total variation into different components is known as an analysis of variance. It is useful in estimating the different components of variance. It also provides basis for the test of significance (F value).
- **Components of genetic variability parameters:** The germplasm assemblage which shows wide range of variability through determining the coefficient of variation, heritability and genetic advance for the concerned characters, selection would be rewarding for improvement of these characters by selection. However, high heritability coupled with high genetic advance is more effective and reliable in predicting the results and the effect of selection (Dudley and Moll, 1969).

Considering that all the line/varieties tested in randomized block design are genetically uniform, the expected error mean sum of squares or E (MSE) $\delta^2 e$ i.e. will be purely a random environmental variance. So, the mean sum of squares between varieties/lines will consist of variances.

(I) Attributable to varietal difference (i.e. genotypic differences) and

(II) Due to environmental variation among the individuals of each genotype. Thus the expected mean sum of square are as follows:

E (MST) - $\delta^2 e + r.\ \delta^2 g$ (r = no. of replication)

E (MSE) = $\delta^2 e$

Therefore, Genotypic variance, $\delta^2 g = \frac{MST - MSE}{r}$

Now, the genotypic variance being $\delta^2 g$ and environmental variance being $\delta^2 e$. The phenotypic variance, i.e. $\delta^2 p$ will be equal to $\delta^2 p = \delta^2 g + \delta^2 e$

The genotypic (GCV), phenotypic (PCV) and environmental (ECV) coefficients of variations are calculated using the following formulae (Burton, 1952; Burton and De Vane, 1953)

$$PCV = \sqrt{\frac{\delta^2 p}{\overline{X}}} \text{ x } 100 = \frac{\text{Phenotypic standard deviation}}{\text{Grand Mean}} \text{ x } 100$$

$$GCV = \sqrt{\frac{\delta^2 g}{\overline{X}}} \text{ x } 100 = \frac{\text{Genotypic standard deviation}}{\text{Grand Mean}} \text{ x } 100$$

$ECV = \sqrt{\delta^2 e/X} \text{ x } 100$ = Environmental standard deviation / Grand mean × 100

Heritability in broad sense (h^2) can be estimated by the following formula (Allard, 1960)

$$h^2 = \frac{\delta^2 g}{\delta^2 p} \text{ x } 100 = \frac{\text{Genotypic variance}}{\text{Phenotypic variance}} \text{ x } 100$$

The estimated genetic advance can be calculated by the following formula (Johanson *et al.*, 1955)

Genetic advance (G_A) = K x $h^2 \times \delta p$

Where, K = a constant (The value equal to 2.06 at 5% selection intensity)

h^2 = heritability in broad sense

δp = phenotypic standard deviation

Genetic advance as percentage of mean = GA/Grand mean × 100

Selection in self-pollinated Crop

The purpose of selection in self-pollinated crops is to isolate superior lines from the population of genotypically different materials. In India, most of the varieties of brinjal, chilli, cucurbits, hyacinth bean, cluster bean, etc. have been developed utilizing wide range of variation available in these crops.

Single plant selection

In this method a genetically heterogeneous population of plants is inspected, and "superior" individual plants with the most desired characters are selected for further multiplication. After necessary multilocation trial, homogenous progeny of this selected plant may be released as a new variety. The principal sources of this selection are the assemblage of material consisting of farmers' variety, improved or relatively unimproved lines, exotic materials, strains or varieties found in cultivation over a wide range of conditions. Well-adapted varieties that are found to be variable and containing a composite of a number of biotypes can also be used as source material. Typical self-pollinated crops are prone to some degree of natural cross-pollination and consequently, the recombination of genes in the population is possible. Spontaneous mutations also serve the sources of variations in the population. Thus, the natural out-crossing, spontaneous mutation and mechanical mixture supply the source material for single plant selection. Adaptability of a genotype is controlled by a combination of multiple genes with respect to which the genotype may not be homogeneous. In unusual growing condition, such inhomogeneity can be identified and used in selection. So, single plant selection is employed to select chance variants, off types or superior types from the source material. In India, several varieties of tomato, brinjal, chilli, cowpea, hyacinth bean, etc have been developed by single plant selection.

- **Tomato:** Sonali, CO-1, Pant Bahar, HS-110, Arka Saurav, etc.
- **Brinjal**: Pusa Purple Long, Azad B-1, Bidhan Supreme, etc.
- **Chilli**: Sabour Arun, Phule Jyoti, Bidhan Chilli-4, etc.
- **Cowpea**: Kashi Shyamal, Kashi Gouri, Swarna Suphala, etc.
- **Hyacinth bean**: Pusa Early Prolific, Phule Gouri, JDL-79, etc.

Pure line selection

A pure line is the progeny of a single self fertilized homozygous plant. The concept of pure line was proposed by Johannsen, the Danish botanist in 1903 (cited by Churchill in 1974) on the basis of his studies with the bean (*Phaseolus vulgaris*) variety called Princess. He obtained the seeds from the market and observed variation in seed size. The lot consisted of a mixture of larger as well as smaller size seeds. Johannsen selected seeds of different sizes and grew them individually. Progenies of larger seeds produced larger seeds and progenies from smaller seeds produced small seeds only which clearly showed that there was variation in seed size in the commercial lot and it has a genetic basis. He studied 19 lines altogether. He concluded that the market lot of the bean seeds is a mixture of pure lines. He also concluded whatever variation observed with in a line was due to environment only.

Confirmatory evidence was obtained in three ways. In line 13 which was having average 450 mg seed weight he divided the seeds on weight basis. He divided the line into seeds having 200, 300, 400 and 500 mg weights and studied the progenies. Ultimately he got lines having weight ranging from 458 to 475mg. Thus the variation observed was purely due to environment. The second evidence was that selection with in a pure line was ineffective.

After six generations of selection, the line for large seed as well as for small seed gave progenies having 680-690 mg weight which did not change further. Thus it was proved that selection within a pure line was ineffective. In third evidence when parent - offspring regression was worked in line 13 found be to zero indicating that variation observed is non-heritable and it was due to environment only.

According to the pure line theory, the population of self-fertilized species consists of several homozygous genotypes. Variation in such a population has genetic base and therefore, selection is effective. Therefore, pure line theory furnishes the basis for the isolation of types that differ appreciably in heritable characters. Natural crossing, mechanical mixture and spontaneous mutations may be the source of variation among the plants within a variety or line.

A pure line which is generated through repeated self fertilization has become homozygous for a particular trait. The trait is passed on to the progeny if bred with another true breeding line. Pure-line selection generally involves three more or less distinct steps:

- Several superior appearing plants are selected from a genetically variable population.
- Selfed progenies of the individual plant selections are grown and evaluated for important quantitative characters and again selfed which is repeated over a period of several years until homozygosity is attained. Detailed pedigree record of each generation can also be recorded to understand the genetic control for different characters.
- When selection can no longer be possible on the basis of observation alone, extensive trials are undertaken, involving careful recording of data on different characters to determine whether the remaining selections are superior in yielding ability and other aspects of performance. Any progeny superior to an existing variety is then released as a new "pure-line" variety.

In India, varieties developed following this method of selection are not many. However, few varieties of tomato, brinjal, chilli, French bean, cowpea, etc. have been developed by pure line selection.

- **Tomato**: Swarna Naveen, Swarna Lalima, Pant T-3, etc.
- **Brinjal**: Swarna Abhilamb
- **Chilli**: K-1, G-4, RHRC Pendant Cluster, etc.
- **French bean**: Arka Komal, Swarna Priya, etc.
- **Cowpea**: Swarna Harita

Selection in Cross-pollinated crop

Usually, the cross-pollinated crops are highly heterozygous to such a degree and so many new gene combinations emerge in the course of continuous cross-pollination that systematic selection combined with controlled pollination always makes it possible to develop a new variety with new properties. Selection in cross-pollinated crops may also be termed as "population improvement schemes." Different selection methods for cross-pollinated vegetable crops are discussed below.

Mass selection

In mass selection, the oldest breeding procedure, seeds are collected from desirable appearing individuals in a population (usually a few dozen to a few hundred), and the next generation is sown from the stock of mixed seed. This procedure, sometimes referred to as phenotypic selection, is based on how each individual looks. Mass selection is based on the concept that the observed character *i.e.*, phenotype of a plant may be interrelated with its genotype. In reality, however, such one to one interrelationship often does not exist. Mass selection has been widely used to improve old indigenous cultivars that have been passed down from one generation of farmers to the next over long periods and is common in vegetable crops.

Mass selection, with or without progeny test, is perhaps the simplest and least expensive of plant-breeding procedures. In this method, the best individual plants are selected from the base population and their seeds are mixed for raising the following generations. This selection is repeated in uniform growing condition before treating the population as a new variety. Size of the base population and selection intensity in each generation are dependent on the crop, character under selection, etc. In this selection method, there is no control over the pollination. Since mass selection is done exclusively on the phenotypic performance of the mother plant without any progeny test, the success of selection depends largely on the heritability of the character under selection. This is why mass selection, particularly when it is used to improve quantitative characters like, fruit number, fruit weight, yield, etc., success is meagre. On the other hand, mass selection is effective in improving simply inherited and highly heritable qualitative characters like shape of root in radish, carrot, beetroot and turnip, shape and colour of bulb in onion, shape of head in cabbage, sugar content in muskmelon and watermelon, disease resistance, etc.

Most of the varieties of onion, a good number of Asiatic radish and a few of cauliflower and carrot have been developed in India by mass selection.

- **Onion:** Pusa Red, Nasik Red, Arka Pragati, Agrifound Dark Red, etc.
- **Radish**: Pusa Desi, Pusa Mridula, Punjab Ageti, etc.
- **Carrot**: Shalimar Carrot-1, Hisar Gairik, Hisar Selection-1, etc.
- **Cauliflower**: Early Kanwari, Kashi Kanwari, Pusa Kartiki, etc.

Refinement of mass selection

In refinement of mass selection, best plants are harvested separately and grown in separate lines to compare their progenies. The poorer progenies are discarded and the seeds of the remainder are harvested. It should be noted that selection is now based not solely on the phenotype of the parent plants but also on both phenotype and performance of their progeny. Progeny selection is usually more effective than phenotypic selection when dealing with quantitative characters of low heritability. It should be noted, however, that progeny testing requires an extra generation; hence gain per cycle of selection must be double that of simple phenotypic selection to achieve the same rate of gain per unit time.

a) Line breeding

It is an important population improvement scheme for improvement of cauliflower, cabbage, Brussels sprouts, broccoli, kohlrabi, radish, carrot, turnip, onion, amaranth and sweet corn. In this method, the best plants are selected from the base population. Progeny of the selected plants, raised by controlled pollination, are grown separately in rows for evaluation. After several cycles of selection, the seeds of some superior and almost similar progeny lines based on progeny tests are composited and treated as a new variety. Several varieties of cauliflower, cabbage, radish, amaranth, etc., have been developed in India by this method.

- **Cauliflower**: Pusa Meghna, Pusa Aswini, Pant Shubhra, Pusa Sharad, Pusa Himjyoti, etc.
- **Cabbage**: Pride of India, Pusa Drumhead, etc.
- **Radish**: Pusa Mridula, Pusa Sweta, etc.
- **Amaranthus**: Arka Samraksha, Kashi Suhavaani, etc.
- **Palak**: Pant Composite, Pusa Bharati, etc.

b) Family breeding

This method is almost similar to line breeding except that in family breeding, the progenies are tested more elaborately in more number of generations. This method consists of i) individual plant selection from base population, ii) raising the progeny of selected plant by controlled pollination (open pollination, selfing, pair cross, diallel cross or poly cross), iii) progeny testing and selection of promising families, and iv) compositing 4-6 or more almost similar families. In this method, progenies are tested not only in first generation but also in later 2 or 3 generations. This population improvement scheme is very effective for the improvement of carrot, radish, beet and turnip. This method can also be adopted for improvement of cabbage, cauliflower, knolkhol, Brussels sprouts, kale and broccoli.

Recurrent selection

This population improvement scheme is designed to concentrate favourable genes scattered among the number of individuals in the base population. This method is based on selection of superior plants followed by controlled mating to produce the new base population. This method is effective in improving yield and other quantitative characters by accumulation of additive genetic variance in the population. The source population for recurrent selection may be an open pollinated variety, F_1 hybrid, synthetic variety, composite variety or inter-crossed progenies of selected inbred lines. Period for selection can be reduced by employing suitable molecular marker. The new and improved genetic base after recurrent selection can be used in developing hybrids, synthetics or composites. Three types of recurrent selection are generally practiced.

- **Simple recurrent selection:** In this method, number of desirable plants are selected and self pollinated. Separate progeny rows are grown from the selected plants in the next generation. The progenies are intercrossed in all possible combination by hand. Further selections are made in the intercross progenies. In India, good number of varieties of carrot and few of cauliflower, cabbage, radish, etc. has been developed by simple recurrent selection.
    - **Carrot**: Pusa Rudhira, Pusa Asita, Punjab Black Beauty, Kashi Krishna, Arka Suraj, etc.
    - **Cauliflower**: Pant Gobi-4
    - **Cabbage**: Pusa Ageti
    - **Radish**: Punjab Pasand

- **Recurrent selection for gca and sca:**In this population improvement scheme, two heterozygous populations are involved and each population serves as a tester for the other population. Through this scheme, both GCA and SCA variances are exploited. In recurrent selection for general combining ability (gca), the test cross is made with a heterozygous tester stock having a broad genetic base, while in the recurrent selection for specific combining ability (sca), a homozygous genetic base is used for test cross. The promising families are selected for inter-crossing of selfed progenies and further selections are made in the intercross composite population.
- **Reciprocal recurrent selection:**It is a cyclical population improvement scheme designed to improve the cross of two populations from different heterotic groups by using both general and specific combining ability. In this method, several plants with desirable characters are selected from two heterozygous and genetically diverse populations which are selfed and test crossed each other. Based on the test cross trial, the superior progenies are selfed and grown next year in the intercross block. The superior inbreds are intercrossed. The two resulting intercross composite populations 'A' and 'B' are used as source populations for next cycle of selection.

Individual plant selection and inbreeding

In this method, an individual plant from the source population consisting of heterozygous and non-uniform materials is selected only on individual basis and unlike mass selection, the seeds of the selected plants are not mixed even if they are equally good and look similar. This method of selection is based on the principle that the actual worth of the selected plant may be masked by the environmental variation and, therefore, the only way to recognize it is by growing the progeny of each selected individual plant for re-evaluation. In cucurbitaceous vegetable crops, *viz.* pumpkin, musk melon, water melon, bitter gourd, bottle gourd, etc., progenies of individual plant for evaluation can be developed by selfing as these vegetable crops do not show loss of vigour due to inbreeding. Homogzygosity for the concerned characters can also be attained in the individuals of the progeny by such repeated selfings. After necessary evaluation, the best selected individual may be treated as a new variety. Several varieties of cucumber, muskmelon, watermelon, pumpkin, bitter gourd, ridge gourd, bottle gourd, wax gourd, snake gourd, round gourd, sponge gourd and long melon have been developed in India by individual plant selection.

- **Muskmelon**: Lucknow Safeda, Durgapura Madhu, Pusa Madhuras, etc.
- **Watermelon**: Arka Muthu, Thar Manak, PKM-1, etc.
- **Pumpkin**: CO-1, Pusa Vikas, Narendra Amrit, etc.
- **Cucumber**: Sheetal, Phule Subhangi, Swarna Sheetal, etc.
- **Bitter gourd**: Coimbatore Long, Pusa Do Mousami, Arka Harit, etc.
- **Bottle gourd**: Pusa Summar Prolific Long, Arka Nutan, Pusa Samridhi, etc.
- **Ridge gourd**: Pusa Nasdar, Pusa Nutan, Punjab Sadabahar, etc.

- **Sponge gourd**: Pusa Sneha, Kashi Shreya, Kashi Divya, etc.
- **Long melon**: Pusa Utkarsh, Arka Sheetal, etc.
- **Snap melon**: Pusa Shandar
- **Leaf amaranthus**: Pusa Kirti, Pusa Lal Chauli, Arka Arunima, Arka Varna, etc

Inbred selection

Inbreeding is the mating of closely related individuals, *i.e.*, either selfing or sibmating. Inbreeding increases homozygosity and reduces the proportion of heterozygosity in the population. Recessive, lethal and deleterious genes usually remain concealed in the heterozygous system of cross-pollinated crops which are increased in proportion upon rendering to homozygosity by inbreeding. The consequential reduction of vigour is called inbreeding depression. Inbreeding is the way to get necessary alleles in homozygous condition. Hence, uniformity in the population is attained. Inbreeding depression varies with the crop, which is drastic in carrot, somewhat lesser in onion and still lesser in cabbage. Late cauliflower varieties grown in India also show less depression upon inbreeding. On the other hand, cucurbitaceous vegetable crops do not show loss of vigour due to inbreeding, therefore, in cucurbits, inbred selection can be practiced and maximum uniformity can be attained due to this advantage. On the contrary, inbreeding is not advocated for more than 2 generations in carrot, 2-3 generations in onion and broccoli and 3-4 generations in cabbage and cauliflower. If loss of vigour is pronounced in these crops, the vigour can be restored by occasional out-crossing and sibmating among the best inbred lines in the population. Important uses of inbreeding in a cross pollinated vegetable crops are i) to attain uniformity in plant characters, ii) to exercise plant selection in cucurbits which do not show inbreeding depression, iii) to select strong and stable self-incompatible lines in cole crops and iv) to develop inbred lines for the development of synthetics or hybrids. A few varieties of cauliflower, muskmelon, pumpkin, summer squash, ridge gourd, sponge gourd and palak have been developed by inbred selection method.

- **Cauliflower**: Pusa Deepali
- **Muskmelon**: Hara Madhu
- **Pumpkin**: Arka Chandan, Pusa Viswas, Ambili
- **Summer squash**: Punjab Chappan Kaddu, Pusa Pasand
- **Bitter gourd**: Pusa Vishes
- **Bottle gourd**: Pusa Naveen, Arka Bahar
- **Ridge gourd**: Arka Prasan, Kashi Shivaani
- **Sponge gourd**: Pusa Supriya
- **Palak**: All Green
- **Leaf amaranthus:** Arka Suguna, Arka Varna

Selfing and massing

This population improvement scheme was suggested by Jones and Mann (1963) for onion which shows considerable loss of vigour due to inbreeding. This method can also be employed in carrot, beetroot, radish and turnip, which are also vulnerable to inbreeding depression. In this method, improvement programme is carried out by selfing followed by massing according to the following procedure outlined for onion.

- First year (bulb crop): Selection of 100 best bulbs of desired type from open pollinated material.
- Second year (seed crop): Growing of selected bulbs, selfing one or more umbels per plant to initiate a separate line each.
- Third year (bulb crop): Growing the progenies of each inbred line separately. The poor performing lines are discarded during the growing season, at harvest or in storage. Selection of at least 25 best lines and keeping 15-20 bulbs of each for selfing and open pollination for next year.
- Fourth year (seed crop): Self pollination of 1-2 umbels in each plant and allow others to open pollinate. Large-scale production of open pollinated seeds for more than two generations to avoid much inbreeding depression.
- Fifth year (bulb crop): Growing of selfed progenies separately. Selection of the best 25 lines again and 15-20 bulbs of each lines as done in the third year.
- Sixth year (seed crop): Composite and planting of bulbs of all selections in a field or in a cage for free open pollination among the unrelated lines. The open pollinated seed can be massed and increased as foundation seed.

In India, one variety of onion has so far been developed by this population improvement scheme,

- **Onion:** Pusa Madhavi

Development of Synthetic variety

In cauliflower, cabbage, turnip, kohlrabi, radish and beetroot, where considerable heterosis is manifested but commercial production of F_1 hybrid seeds needs utilization of inherent pollination control mechanisms like self-incompatibility and male sterility. In these crops, synthetic variety can be developed with advantage by intercrossing a number of genotypes *viz.*, inbred lines, mass selected varieties and other lines, commercial variety of known superior combining ability i.e., genotypes that are known to give superior hybrid performance when crossed in all combinations. (By contrast, a variety developed by mass selection is made up of genotypes bulked together without having undergone preliminary testing to determine their performance in hybrid combination) Synthetic varieties are known for manifesting hybrid vigour and for their ability to produce usable seed for succeeding seasons. Because of these advantages, synthetic varieties have become increasingly favoured in the production of many vegetable crops in which expense prohibits the development or use of hybrids. In this population improvement scheme, additive genetic variance is basically exploited in developing a synthetic variety. Synthetic varieties of cauliflower and cabbage have been developed in India.

- **Cauliflower**: Pusa Early Synthetic, Pusa Early Synthetic 78-1
- **Cabbage**: Pusa Synthetic

3. Hybridization and selection

Hybridization results in a combination of valuable parental characters in the hybrid progeny. Among the characters whose recombination usually is the purpose is framed in the breeding objectives. The recombination of characters inherent in one variety with those of another during hybridization is a consequence of the discrete nature of inheritance. However, this process should not be oversimplified. If the concerned characters are determined by a small number of genes, their recombination is relatively easy. But when complex quantitative characters controlled by polygenes are dealt with, recombining them is extremely difficult. In such cases, breeders can achieve only very modest improvement.

Hybridization has much farther going consequences than simple recombination of the traits inherent in the parent being crossed. Along with the parental characters, the hybrid also exhibits its own properties as a result of a particular combination of genes as well as manifestation of the hereditary potential passed on through the parental gametes. Sometimes, union of the interacting genes during crossing gives rise to entirely new traits and characters making the hybrid radically distinct from the parental forms. Therefore, it would be a big mistake to consider hybridization to mere recombination of parental traits in the progeny. Due consideration should be paid in the fact that the linkage of genes and some other genetic phenomena may render the recombination of some traits most difficult or even impossible. Such situation is often encountered in most of the distant hybrids. Successful outcome of hybridization is to a great extent dependent on how well the breeder is familiar with the genetic processes occurring in the segregating progenies of hybrid population. Therefore, the genetic consequences of crossing must be considered.

With the accessibility of precise information on genetic conditioning of characters and advanced crop improvement programmes, the method of analytical breeding based on only selection has somewhat become obsolete as an independent practical procedure. It has been replaced by the method of synthetic breeding, or in other words, hybridization aimed at combining in the future variety the useful characters inherent in each parent separately.

During the 20th century planned hybridization between carefully selected parents has become dominant in the breeding of self-pollinated species. The objective of hybridization is to combine desirable genes found in two or more different varieties and to produce pure-breeding progeny superior in many respects to the parental types. Pedigree breeding starts with the crossing of two genotypes, each of which has one or more desirable characters lacked by the other. If the two original parents do not provide all of the desired characters, a third parent can be included by crossing it to one of the first generation hybrid progeny (F_1). Superior types are selected in successive generations following pedigree method, and a record is maintained of parent–progeny relationships.

The F_2 generation (selfed progeny of F_1 individuals) offers the first opportunity for selection in pedigree programmes. In this generation, emphasis is laid on the elimination of individuals carrying undesirable major genes. In the succeeding generations, heterozygosity is reduced and gives way to pure breeding as a result of natural self-pollination, and families derived from different F_2 plants begin to display their unique character. Usually one or two superior plants are selected within each superior family in these generations. By the F_5 generation the pure-breeding condition (homozygosity) is extensive, and emphasis shifts almost entirely to selection between families. The pedigree record is useful in making these eliminations. At this stage, each selected family is usually harvested in mass to obtain the larger quantity of seeds needed to evaluate the families for quantitative characters. This evaluation is usually carried out in plots grown under conditions that simulate commercial planting practice as closely as possible. When the number of families has been reduced to manageable proportions by visual selection, usually by the F_7 or F_8 generation, precise evaluation for performance and quality begins. The final evaluation of promising lines involves a) observation, usually in a number of years and locations, to detect weaknesses that may not have appeared previously, b) precise yield testing, and c) quality testing.

Types of crosses and selection of parents for hybridization

Different types of crosses are used in plant breeding. Their selection depends on the breeding objectives. Types of cross are divided into two groups, *i.e.*, simple and compound. However, selection of an appropriate crossing scheme depends on a number of factors, primarily the objective of the breeding programme, diversity of source material, inheritance pattern of character(s), etc. The greatest success may be achieved through combination of various crossing systems at different stages of the breeding process.

Simple	**Compound**
Single crosses	Three-way cross
Reciprocal crosses	Back crosses
Multiple crosses	Convergent crosses
Top crosses	Step-by-step crosses
Diallel crosses	Inter-hybrid crosses

Selection of parental material is the first, and at the same time, most critical step in any breeding programme. It is very important to understand the inheritance pattern of different characters in the crop under improvement. It is rather difficult task in many cases because one must take into consideration the entire complex of characters inherent in the parents being crossed, for which all possible kinds of interaction between genes are expected.

a) Simple crossing programme

In simple crossing programme, selection is carried out directly in the first advanced generations (F_2 generation) of the hybrid.

- **Single cross**

It is the first generation hybrid between two inbred lines. Breeding based on single crosses offers selection of elite plants through evaluation of their progenies in the segregating generations. However, in some breeding approaches particularly based on distant hybridization, single crossing may not be sufficient for obtaining the necessary combination of characters in the segregates.

- **Reciprocal crossing**

Reciprocal cross is a kind of crossing strategy which make crosses between a pair of parents (A and B) by using them in turn as female parent and male parent to obtain two reciprocal crosses of A×B and B×A. Reciprocal crossing is used to elucidate the interaction between the genetic materials present in the nucleus and cytoplasm of the parental forms. It is especially important in distant hybridization when direct and inverse combinations may give dissimilar results in terms of both setting of seeds and character manifestation in the hybrids, which depends on the interaction between the nucleus and cytoplasm.

- **Multiple crossing**
  - **Top crossing:** In top cross, crosses are made between the inbreds and either a tester or an open-pollinated variety. This is also known as an inbred–variety cross. Top crosses are most often used to determine the overall combining ability of lines and varieties and manifestation of heterosis.
  - **Diallel crossing:** Diallel crossing is intended to produce hybrids between all varieties or lines under investigation. The number of all possible combinations in crossing a definite number of lines or varieties can be calculated using the formula $n(n\text{-}1)$ for reciprocal diallel crosses and $n(n\text{-}1)/2$ for direct combinations where, n being the number of parental forms involved in hybridization. Diallel crossing scheme has been extensively applied to estimate the combining ability as well as the effect of reciprocal crosses in plants. In statistical terms, the general combining abilities are main effects and the specific combining ability is an interaction. Griffing (1956) defines diallel crosses in terms of genotypic values where the sum of general combining abilities for the two gametes is the breeding value of the cross (i, j). The analyses proposed by Jinks and Hayman (1953) and described in more detail by Hayman (1954a, 1954b) are robust in terms of genetic information generated and provide information on additive and dominance effects of genes, average degree of dominance, proportion of dominance, direction of dominance, distribution of genes, maternal and reciprocal effects, number of groups of genes that control a trait and exhibit dominance, ratio of dominant to recessive alleles in all the parents, and broad-sense (H^2) and narrow-sense (h^2) heritability.

b) Compound crossing programme

In compound crossing programme, the hybrid is crossed again with another inbred line or a third variety, one of the parents, or another hybrid. In compound cross, an elaborate system of crosses is involved followed by selection from the segregating generation.

• Three-way cross

It is the first generation hybrid between a single cross and an inbred line. Selection of the desirable segregates is exercised in the advanced generations.

• Backcross method of crossing

The back cross method of breeding proposed by Harlan and Pope (1922) is the second most important method of breeding a new variety or development of parental lines of a hybrid after pedigree method of handling the advanced generations. In this breeding method, the resulting hybrid is crossed again with one of the parental lines. Backcross breeding enables breeders to transfer a desired trait such as a transgene or disease resistant gene from one variety (donor parent) into the desirable genetic background of another (recurrent parent). Backcrosses are used when it is necessary to impart to a variety "A"one or several characters inherent in another variety "B" without detriment to the integrity of the first variety. In this case, the first variety "A" is referred to as "recurrent parent" and the variety "B" is referred to as "donor". The characters and genes of the variety "B", introduced into "A" by way of backcrossing, are known as characters and genes under transfer. Backcrosses may be denoted BC_1, BC_2, and so on. Backcrossing is most often used to eliminate a defect, such as high susceptibility to certain diseases, from a variety or line with a desirable set of characters. Since the donor parent, in most of cases, possessing limited other worthy characters, the hybrids have to be time and again crossed with the productive recurrent parent in order to get rid of the negative properties of the donor. In the course of this process, the recurrent parent's genotype is restored in the hybrid progenies and the undesirable genes of the donor are eliminated. Each generation of back crossing reduce the proportion of the donor parent present in the population by half. If, for example the recurrent parent was employed in the backcrossings for six times (A_6XB), the proportion of the germplasm obtained from the donor parent "B" will be $\left(\frac{1}{2}\right)^6 = \frac{1}{64}$. This cycle of back crossing progeny to the recurrent parent continues until a new line that is identical to the recurrent parent but with the designed character(s) from the donor parent is developed. However, requirement of number of back crossings and generations vary depending on the inheritance pattern of the trait under transfer, recovery of characters of the recurrent parent, selection during back crossing, genetic similarities between recurrent and non-recurrent parent and linkages.

- **If the trait of interest is governed by a dominant gene**: The process of developing a new line of the recurrent parent with the desirable character of the donor parent involves 4-5 rounds of backcrossing (BC_4 to BC_5)within 7 generations.

- **If the trait of interest is governed by a recessive gene:** Back-cross method for transfer of recessive gene is complex and need elaborate evaluation. The recurrent parent (desirable variety lacking in resistance) is crossed with the donor parent possessing the resistance governed by recessive gene. The F_1 plant which do not show any resistance are back crossed to the recurrent parent. In the BC_1 generation, all the plants are susceptible hence, there is no test for resistance. All the plants are self-pollinated. In the BC_1 F_2 generation, resistant plants which emerge in the segregating generation are selected and back-crossed with the recurrent parent. Selection is made for the plant type and other characteristics of the recurrent parent. In BC_2 generation no plant will show resistance and selection will be based on the characters of the recurrent parent. The selected plants are again back crossed with the recurrent parent. In this manner of back-crossing and selfing, final selection can be done in BC_5F_5 generation after completion of 11 generations. The new line of the recurrent parent with the resistance character of the donor parent can be tested in replicated trial after 12 generations.

  Marker assisted fore ground selection is particularly useful, if the target gene is recessive, or for combining redandantly acting target genes. Also foreground selection is useful for environmentally sensitive genes and in case of expensive phenotyping for example, some quality traits. Marken-based fore ground selection enables early selection and elimination of undesirable plants, thus reducing the cost related to growing and managing plants

• Convergent method of crossing

The principle of convergent improvement is necessarily concerned only with favourable genes which are completely or partly dominant, and can be recognizable in the heterozygous condition. Richey (1927) proposed convergent improvement that aims at parallel improvement through back crossings of two inbred lines by the reciprocal addition of dominant favourable genes present in one but lacking in the other. The method of convergence is based on double backcrossing of different donors with the same recurrent parent for transferring to the latter several valuable characters at a time. The desired characters become combined within a variety at final stage of the breeding process through crossing of the lines produced concurrently and recombination of the corresponding genes. Its importance in the development of varieties with resistance to several diseases is exceptionally great. Usually, two series of backcrosses are performed at a time to produce eventually two lines, which are virtually identical as far as the genes of the recurrent parent are concerned, the only difference being primarily in the genes introduced into the variety being improved from the donors. This method makes it possible to transfer not only two but also more genes and characters to the variety being improved.

• Step-by-Step method of crossing

During step-by-step hybridization, the initial hybrids are crossed again with a third variety, and if necessary, a fourth variety or species and so on. Thus, step-by-step crossing yields a hybrid material of several parental forms including the germplasm of several varieties or even species. As for example, by selecting varieties for successive step-by-step hybridization of which one is early ripening, another is high pigmented and the third is resistant to a disease, the breeder can produce a genotype combining all these three characters in advanced generations. Step-by-step crossing can also be used in distant hybridization. Step-by-step crossing is sometimes combined with backcrossing to enhance, in the progeny, the valuable characters of a particular variety. Sometimes, step-by-step crossing is used at final stage of breeding. All such modifications depend on the breeding objective and concerned characters under manipulation.

• Inter-hybrid crossing

In inter-hybrid crossing, the characters inherited from several parents are combined not consecutively, as in step-by-step hybridization, but in parallel, by producing simple hybrids at first and then crossing them. Double-cross hybrid is the first generation hybrid between two single-cross hybrids.

For instance, four lines may be combined in the hybrid progeny within two years using the following crossing scheme:

Parental lines	A	B	C	D
Single crosses in the first year	A × B		C × D	
Final crossing in the second year	(A × B)	×	(C × D)	
		↓		
Hybrid progeny		(A, B, C, D)		

In this manner, more number of genotypes can be combined in the hybrid progeny within shortest period. It takes three years to combine the genes of eight parental forms in a hybrid progeny. Step-by-step hybridization would take seven years to accomplish the same.

In such inter-hybrid crossing, segregation begins with the second hybrid generation, which calls for many crosses both in first and in second year to attain all possible gene combinations. This method can successfully be applied not only in producing heterotic hybrids but also developing improved varieties. In some modification, the best productive variety as for example, "A" may be used in all single crosses to reinforce its influence in the hybrid generations, *i.e.*, (A × B), (A × C), (A × D) and (A × E), which makes it possible to retain a large number of genes of the variety 'A' in the hybrid progeny in spite of the segregation that takes place. Inter-hybrid crossing can also be combined with backcrossing. This approach also permits combination, within a shortest period, of the valuable genes inherent in several lines resulting from convergent improvement.

Distant hybridization

Wild relatives of different vegetable crops are the great reservoir of genetic variation. Breeding programmes for the development of new varieties and parental lines of the hybrids in many vegetable crops resistant to different biotic and abiotic stresses have been increasingly dependent on distant hybridization covering wild and weedy species. Distant or wide hybridization refers to the crosses made between distantly related species or genera. Crossability barriers are very frequent in inter-specific and inter-generic hybridization programmes. These barriers along with linkage drag of undesirable characters limit effective utilization of crop wild relatives in breeding vegetable crops.

Post-fertilization barriers and methods of overcoming the barriers

Post-fertilization barriers result in the failure of the fertilized ovules to develop into mature seeds. Many causative factors have been identified for the hindrance of embryo development at different stages which include the presence of lethal genes, genic disharmony in the embryo and failure or early breakdown of endosperm. Arrest of normal development of embryo and endosperm may also result from unfavourable interactions between embryo sac and surrounding ovular tissues. Often, embryo sac is associated with the proliferation of the nucellar or integumentary cells. A range of *in vitro* methods has been developed to overcome post-fertilization barriers in a number of plant species.

• Ovary culture and ovary-slice culture

When embryo abortion occurs at a very early stage and maternal tissue has no negative influence on development of seeds, ovary culture can be applied. Young fruits can be grown *in vitro* to a stage at which dissection of embryos is possible. Ovary culture has been applied in many species, *viz. Brassica*, *Lilium*, *Nerine*, *Tulipa* and *Phaseolus*. Ovaries are harvested 7-10 days after intra-stylar pollination depending on the species and after surface sterilization, sliced into 2 mm thick disks. By these methods, plantlets are obtained from very small embryos. *In vitro* cut-style pollination and *in vitro* grafted style technique were developed and applied on various interspecific crosses of *Lilium longiflorum, L. dauricum, L. henryi.*

• Embryo rescue technique

Embryo rescue is an *in vitro* culture technique that is used to save weak, immature and hybrid or sometimes inviable embryos to prevent its degradation. The procedure involves excising weak, immature plant embryos while still inside their ovules and culture them on specially devised culture medium (Rogo *et al.*, 2023).Embryo culture after its rescue from the immature fruits can be applied successfully in crosses in which pollinated flowers stay on the plant for a considerable time before natural abscission occurs. This method has been applied in large number of crops, *viz. Allium*, *Alstroemeria*, *Howea*, *Lilium*, and *Solanum*. To overcome the problems involved in isolating the young embryos from ovules and providing suitable conditions for their growth, embryo culture has been modified in some systems. The ovule is cut in half and the cut halves or only the halves containing the embryo are cultured in a liquid medium. Out of these half ovules, the germinating embryos emerge, which could be raised to plantlets on a medium. In vegetable crops, embryo rescue technique has been employed to overcome inter-specific crossing barriers of the crosses *viz.*, (*Solanum melongena* × *Solanum sisymbrifolium*, *Solanum gilo* × *Solanum integrifolium*),

(*Solanum lycopersicum* × *Solanum habrochiates*), (*Solanum lycopersicum* × *Solanum peruvianum*), (*Solanum lycopersicum* × *Solanum chilense*), (*Capsicum. annuum* × *Capsicum baccatum* var. *pendulum*), etc.

Handling of the advanced generations

The hybrid developed by either simple or compound crossing schemes according to the breeding objectives is advanced to the subsequent generations by selfings or controlled random mating to select desirable segregates, which forms the base of a new variety. Such handling of advanced generations and subsequent selection varies with the reproduction system of the crop.

1. Self-Pollinated Crop

These crops are adapted to natural self-fertilization and normally do not show inbreeding-depression. Hence, advancement of the hybrid population is based on isolation of desirable homozygous segregates, which show improvement over the existing variety. In self-pollinated vegetable crops, the hybrid progeny derived from the planned hybridization is handled by pedigree or bulk method.

• Pedigree method

In the pedigree method, superior types are selected in successive generations, and a record is maintained regarding parent–progeny relationships. The F_2 generation offers the first opportunity for selection from the segregating population. In the F_2 generation, individual plants are selected for raising F_3 families of each selection and at the same time, undesirable segregates are discarded. Such selection and advancement of generations are generally continued until F_7 or F_8 generation when almost all families will have become uniform.

• Bulk or mass method

The bulk method of breeding differs from the pedigree method primarily in the handling of generations following hybridization. The F_2 generation is sown at normal commercial planting rates in a large plot. At maturity, the crop is harvested in mass and the seeds are used to establish the next generation in a similar plot. No record of ancestry is kept. During the period of bulk propagation, natural selection tends to eliminate plants having poor survival value. Two types of artificial selection also are often applied: a) destruction of plants that carry undesirable major genes and b) harvesting when only part of the seeds are mature to select for early maturing plants.

In this method, the hybrid populations are advanced up to F_5 generations in bulk when desired level of homozygosity has been attained. At this level, single plant selections are then exercised and evaluated in the same way as in the pedigree method of breeding. The chief advantage of the bulk method is that it allows the breeder to handle very large numbers of individuals inexpensively.

Various modifications of bulk methods are single seed descent method (hybrid populations are advanced by taking single seed from each plant), double seed descent method, etc. Pedigree selection is advantageous for characters having high heritability, while single seed descent (SSD) method is useful for improving low heritability traits since a broader genetic base is maintained to advance the segregating generations. Depending on the traits under improvement, pedigree with SSD, mass-pedigree, backcross-pedigree and mass-sib-crossing methods are generally adopted in the development of varieties of tomato, eggplant, chilli, okra, garden pea, vegetable cowpea, hyacinth bean, etc (Hazra *et al.*, 2022).

- **Tomato:** Hisar Arun (Pusa Early Dwarf × K1), Kashi Hemant (Sel-18 × Floradade), Hisar Lalit (HS 101 × R2), Arka Meghali (Arka Vikas × IIHR 554), Utkal Urbashi (Punjab Chhuhara × LE-79)
- **Brinjal**: Pusa Kranti [(Pusa Purple Long × Hydaerpur) × Wynad Giant], Pusa Anupam (Pusa Kranti × Pusa Pruple Cluster), Punjab Sadabahar (Japanese Long × R-34), Punjab Barsati (Pusa Purple Cluster × PH-4), Bidhan Suphala (Bidhan Supreme × Punjab Sadabahar), etc.
- **Chilli**: Kovilpatti-2 (K-1 × Sattur Samba), Pusa Jwala (NP46A × Puri Red), Arka Suphal (Pant C-1 × IIHR-517), Pant C-1 (NP46-A × Kandhari), Phule Sai (Pant C-1 × Kamandalow)
- **Garden pea**: Kashi Udai (Arkel × FC-1), Jawahar Matar-3 (T-19 × Early Badger), Azad P-2 (Bonneville × 6587), Punjab 88 (Pusa 2 × Morrasis-55), Arka Priya (Arka Ajit × IIHR 562)
- **Vegetable cowpea**: Pusa Komal [(Pusa Dofasli × EC 26410) × P-426], Arka Garima (TUV-762 × *Vigna unguiculata* subsp. *sesquipedalis*), Bidhan Barbati-1 (EC243954 (*Vigna unguiculata* subsp. *unguiculata*) × EC 305827 (*Vigna unguiculata* subsp. *sesquipedalis*), Bidhan Barbati-2 (V-70 (*Vigna unguiculata* subsp. *cylindrica*) × Sel TM-3 (*Vigna unguiculata* subsp. *sesquipedalis*), Bidhan Sadabahar (Bidhan Barbati-1 × Kashi Kanchan), etc.
- **Hyacinth bean**: Arka Krishna (IIHR 178 × Arka Swagath),Arka Amogh [(Arka Jay × Arka Vijay) × Konkan Bhusan], Arka Jay (Hebbel Avare 3 × IIHR 99), Arka Vijay (Hebbel Avare 3 × Pusa Early Prolofic), etc.
- **Okra**: Pusa Sawani (Pusa Makhmali × IC 1541), Utkal Gaurav (Pusa Sawani × Gouripada Local), Varsha Uphar (Lam Selection-1 × Parbhani Kranti), Hisar Unnat (Selection 2-2 × Parbhani Kranti), Kashi Satdhari [(Parbhani Kranti × IC 111542) × IIVR 20], Bidhan Sahebmukto (VRO-6 × Mukta), etc.

2. Cross-Pollinated Crop

In most of the cross-pollinated vegetable crops except cucurbits, induced selfing results inbreeding depression although the severity of depression depends on the crop. Hence, in cauliflower, cabbage, broccoli, radish, carrot, turnip, onion, etc., improvement is not based on isolation of homozygous genotypes as done in self-pollinated crops. As a consequence of out-breeding, any variety of these cross-pollinated crops has heterozygous genotypes, each being to some extent different from the other ones in a given population. Hence, a number of methods and huge variants in the population are available for handling the hybrid generations for developing a new variety.

On the other hand, in cucurbits, which do not show inbreeding depression, the generations can be handled based on continued individual selection through self-pollination as done in self-pollinated crops. So, the hybrid progenies of cross-pollinated crops are handled by two different approaches, *i.e.*, population improvement scheme and individual plant selection.

- **Population improvement approach:** In cauliflower, cabbage, broccoli, radish, carrot, turnip, turnip, palak, etc., the hybrid progenies are allowed to pollinate openly among themselves in isolation followed by selection of desirable plants to increase uniformity of the population and at the same time, to increase desirable gene frequency in the genotypes and eventually in population. In hybrid population, such a change in the frequency of genes must be achieved that would increase the probability of emergence of the best genotypes. Mass selection can be adopted following high selection intensity. However, handling of the progenies is generally based on family-group and individual family selection methods.
    - **Family-group method:** Seeds from the selected elite plants from the population are sown separately thus, making families. These families are grouped according to similar characters. This is followed by free open-pollination within each group.
    - **Individual-family method:** In this method, seeds from the selected elite plants from the population are sown separately by families in isolation and then, the generations are advanced by free open-pollination among the plants of each family.

Several varieties of cauliflower, radish, carrot, turnip, palak and amarnthus have been developed following these population improvement approaches.

- **Cauliflower**: Pusa Shubhra [(MG 2-3 × 15-1-1) × D-96], Pusa Snowball (EC 12013 × EC 12012), Pusa Snowball-1 (EC 12012 × EC 12013), Pusa Snowball K-53 (Lawyana Old × 13-85), etc.
- **Radish**: Pusa Reshmi (Green Top × Desi Type), Punjab Safaid (White 5 × Japanese White), etc.
- **Carrot**: Pusa Yamdagni (EC 9981 × Nantes Half Long), Pusa Kesar (Local Red × Nantes Half Long), Pusa Meghali (Pusa Kesar × Nantes),
- **Turnip**: Pusa Chandrima (Snow Ball × Japanese White), Pusa Swarnima (Japanese White × Snow Ball), Pusa Kanchan (Local red Round × Golden Ball), etc.
- **Palak**: Pusa Harit (Sugar beet × Local palak), Banerjee's Giant (Local palak × Garden beet), Arka Anupama (IIHR 10 × IIHR 8)
- **Amaranthus**: Pusa Kiran (*Amaranthus tricolor* × *Amaranthus tristis*)

• Individual plant selection approach

In cucurbitaceous vegetable crops, like watermelon, muskmelon, cucumber, pumpkin, bottle gourd, bitter gourd, ridge gourd, etc., pedigree method of handling the segregating generation is applicable because inbreeding depressing is not pronounced in these crops. Hence, the hybrid generations can be advanced by induced selfings as is done in the self-pollinated crops and superior homozygous plants are selected in the advanced generations. Several varieties of different cucurbits have been developed following this method.

- **Musk melon:** Punjab Sunheri (Hara Madhu × Edisto), Hisar Madhur (Pusa Sharbati × 75-34), Pusa Sharbati (Kutana × PMR 6), Punjab Rasila (WMR 29 × Hara Madhu), etc.
- **Watermelon:** Arka Manik (HR 21 × Crimson Sweet)
- **Pumpkin:** Thar Kavi (CM 16 × CM 19)
- **Cucumber:** Himangi (Poinsett × Kalyanpur Ageti)
- **Bitter gourd:** Kashi Urbasi (IC 85650B × IC 44435A), Kashi Ganga (IC 92465 × DVBG 151), Pusa Komal (LC 11 × LC 5), etc.
- **Ridge gourd:** Arka Sujat (IIHR 54 × IIHR 18), Arka Sumeet (IIHR 54 × IIHR 24), etc.

Asexually propagated vegetable crops

Crops which are propagated asexually or by vegetative means are known as asexually or vegetatively propagated or clonal crops. Many vegetable crops *viz*., potato, sweet potato, cassava, elephant foot yam, yam, taro, swamp taro, garlic, pointed gourd, teasle gourd, Jerusalem artichoke, globe artichoke, etc. are propagated by asexual means. General characteristic features of these crops are given below.

- Most of these crops are perennial but grown as annual crop. e.g., potato, sweet potato, yam, cassava, pointed gourd, etc.
- Many of them shows reduced flowering and seed set and many varieties do not flower at all e.g., sweet potato, yam, cassava, elephant foot yam, garlic, etc.
- Some crops require special climatic condition for induction of flowering and fruit set e.g., potato (long day and cool temperature condition as found in high hill situation)
- Some crops have recalcitrant seeds which are very short lived and 50% seedlings are non-productive due to dioecious sex form. e.g., pointed gourd, teasle gourd, etc.
- Almost all the crops are cross-pollinated and show high inbreeding depression.

Plant introduction

Plant introduction is not only an important method of developing a variety but also set the initial stage of breeding programme for many asexually propagated vegetable crops through the source of new germplasm. A good number of varieties of asexually propagated vegetable crops successfully cultivated in India have been developed through selection from exotic material. Introduced materials have also been utilized in planned hybridization programme and subsequent clonal selection for development of new variety.

Clonal selection

Different vegetative plant parts usually propagate these crops. Clone is referred to a group of plants produced from a single plant through asexual propagation. Clones within which individual plants are genetically identical and phenotypically uniform even if they are heterozygous represent the varieties of these crops. Therefore, the chances to develop a new variety from a single clone by selection are minimal because a clone may undergo changes only by somatic mutation.

The most important step in clonal selection is to assemble wide range of clones from different source, as chance of getting variability within a clone is remote. Superior clones are selected based on phenotype and released as a new variety. A number of varieties of sweet potato, cassava, yam, taro, garlic, pointed gourd and elephant-foot yam have been developed by clonal selection from both indigenous and exotic materials.

Selection from indigenous material

- **Sweet potato:** Samrat, Kalmegh, etc.
- **Cassava**: CO-1, Sree Swarna, Sree Pavitra , etc.
- **Greater yam**: Sree Keerthi, Sree Roopa, Sree Shilpa, Sree Karthika, etc.
- **Lesser yam:** Sree Latha,Konkan Kanchan
- **Taro:** Sree Rashmi, Sree Pallavi , Muktakeshi,Pani Saru-1, Bidhan Chaitanya, Bidhan Joydev
- **Elephant foot yam:** Sree Padma, Sree Athira, Gajendra, Bidhan Kusum
- **Yam bean:** Rajendra Mishrikand-1, Rajendra Mishrikand-2
- **Chinese potato:** Sree Dhara
- **Pointed gourd:** Rajendra Parwal-1,Swarna Alaukik,Kashi Suphal, etc.
- **Ivy gourd**: Indira Kundru-5, Indira Kundru-35

Selection from exotic material

- **Sweet potato:** Pusa Safed (exotic material, FA-17-White from China), Pusa Lal (exotic material, Norin from Japan), Sree Bhadra (exotic material from IITA, Nigeria)
- **Cassava**: M 4 (exotic material from Malaysia)
- **Lesser yam:** Sree Kala (exotic material from Puerto Rico)

Clonal breeding utilizing their sexual mode of reproduction

Several important asexually propagated vegetable crops *viz*., potato, sweet potato and cassava have been bred effectively utilizing their sexual mode of reproduction.

The possibility of combining sexual and asexual reproduction systems in almost all vegetatively propagated crops is the key strategy to develop new cultivars, because hybridization is the best approach to increase genetic diversity and combine traits in an individual (Hancock, 2004). This group of vegetable crops is quite diverse in nature due to perpetuation of clonal heterozygosity through vegetative mode of reproduction. Sexual mode of reproduction either in the form of open pollination or controlled hybridization invariably create variability in the progeny, may it be hybrid or seedlings from open-pollinated seeds. Autopolyploidy in some vegetable crops, like autotetraploidy in potato, autohexaploidty in sweet potato produce variability in gamete formation, and thereby, throw segregates upon both selfing and hybridization.

Cause of occurrence of variation upon both selfing and hybridization

In an autotetraploid crop like potato, four chromosomes are homologous to each other, thus, each gene has four copies. A *simplex* individual has one dominant and 3 recessive alleles (Aaaa), a *duplex* has 2 dominant and recessive alleles each (AAaa), a *triplex* has 3 dominant and 1 recessive alleles (AAAa), a *quadruplex* has all dominant alleles (AAAA), while a *nulliplex* has all recessive alleles (aaaa). Let the situation of gamete and genotype production be conceived in a simplex (Aaaa) individual.

- **Random chromosome segregation:** In the case where, the gene (A/a) is located closed to the centromere, which inhibit crossing over between the centromere and the gene. Hence, the sister chromatids carrying the dominant alleles would move to the opposite poles. Therefore, the two dominant alleles do not reach the gametes, *i.e.*, AA gamete is not produced. So two types of gametes, Aa and aa in 1:1 ratio is produced, and upon selfing, three genotypes AAaa, Aaaa and aaaa in the 1:2:1 ratio are produced, assuming formation of only quadrivalents and random separation of chromosomes.
- **Random chromatid segregation:** In the case where the gene (A/a) is located far apart from centromere ensuring crossing over between the centromere and the gene. Hence, two chromosomes involved in the crossing over each carrying a dominant allele after crossing over may move to the same pole, and thus, AA gamete is produced. Therefore, three types of gametes AA, Aa and aa are produced in the ratio of 1:12:15. Selfing in such simplex individual produces AAAA, AAAa, Aaaa, Aaaa and aaaa genotypes in the ratio of 1:24:174:360:225, respectively.

These two situations of gamete and genotype formation are related to the assumptions of a) regularity in quadrivalent formation and b) complete or no linkage between centromere and the gene in question. However, these two assumptions are rarely, if ever, fulfilled. Therefore, the actual gametic and zygotic frequencies are usually intermediate between the two extremes. This feature amply visualizes the complexity in gamete production and variability in genotype formation upon selfing and crossing. The advantage lay on the selection of desirable segregates at any stage of segregation and reinforcement of the characters through vegetative propagation.

Hybridization and selection from variable F_1 and segregating population

Breeding of clonal variety by this method is comparatively easy and fast. In this scheme, F_1 clones may be the potential varieties. Total amount of genetic variation can be used directly in F_1 generation. However, the main breeding task is the selection of appropriate parental lines and the identification of the best F_1 clone. This important breeding method is based on the advantage of highly heterozygous nature of the parental clone, which creates genetic variation in the F_1 seedlings. Ample scope is left for promising selection in F_1 progeny. However, the selected F_1 progeny may be advanced to later generations for selection. Success in hybridization and selection method can be achieved more easily and quickly in potato, cassava and sweet potato and other vegetatively propagated crops because of their vegetative propagation. This enables selection of valuable elite plants at any stage of segregation and reinforcement of their properties through vegetative propagation. Therefore, rendering to homozygosity in the desirable segregates is not required because vegetatively propagated plants, may it be homo- or heterozygous, always show uniformity in the expression of characters.

The main feature of this clonal breeding lay on the framing of specific objectives and selection of parents, development of widely divergent cross combinations, growing of as many F_1 seedlings as possible, selection of seedlings by visual observations, reselection of the seedlings for specific characters like disease resistance, final evaluation for different horticultural characters, multilocational trials and release of a variety.

Among the vegetatively propagated crops, potato, cassava and sweet potato has been improved extensively by hybridization and subsequent selection from the variable F_1 population. In potato, distant hybridization has successfully been practiced with the utilization of the wild relatives *viz.*, *Solanum tuberosum* ssp. *andigena*, *Solanum demissum*, *Solanum acule*, *Solanum curtilobum*, etc. to develop resistant varieties mainly against late blight and viruses. In potato, hybridization followed by clonal selection is mostly employed to improve quality and to incorporate resistance.

- **Potato**: Kufri Chamatkar [Phulwa × Ekishirazu (exotic)], Kufri Chipsona-4 [Atlantic (exotic) × MP/90-94], Kufri Khyati (MS/82-638 × Kufri Pukhraj), Kufri Jyoti [(3069d (4) × 2814d (1)], Kufri Lalima [Kufri Red × AG 14 (Wis × 37)], Kufri Pukhraj (Craigs Defiance × JEX/ B-687).
- **Sweet potato**: H-42 [Vella Damph × Triamph (exotic)], H-268 (H-74 × H-42), Rajendra Shakarkand-5 (Cross 4 × H 5), Pusa Sunheri [Puerto Blanco × Wannop) × Australian Canner], H 41 [Norin (exotic) × Local], Kishan (poly cross with female accession No.1016)
- **Cassava**: H 97 [(Manjavella × Acc. No. 300 (exotic)], H-165 (Chadayamangalam Vella × Kalikalan), H 226 [M 4 (exotic) × Ethakkakaruppam], Sree Visakham [(Acc. No. 1501 × S 2312 (exotic)], Sree Athulya [Op 4 (2x) × Sree Visakham (4x)], Sree Apoorva [Ambakkadam (2x) × Sree Sahya (4x)],etc.
- **Taro**: Sree Kiran (C 303 × C 383)

Selection of seedling segregates

Heterozygous nature of the asexually propagated crops and virtually obligatory out-crossing in sweet potato, pointed gourd, cassava, yam, etc., allow a wide range of genetic recombinants in natural seed selfing. Complexities in gamete formation in the polyploids (potato, sweet potato, yam, etc.) also serve a means of getting variable segregates upon setting. By selection from the seedling segregates, few varieties of sweet potato, cassava and white yam has been developed.

- **Sweet potato**: Sree Varun (seedling selection of the seeds from CIP, Peru), Sree Nandini (seedling selection of the seeds from Sudan),
- **Cassava**: CO-2 (seedlings from open pollinated seeds), CO3 (seedlings from open pollinated seeds from IITA, Nigeria),
- **White yam** : Sree Subhra, Sree Priya, Sree Dhanya

References

Allard, R.W. 1960. *Principles of Plant Breeding*, John Willey and Sons, Inc., New York, 485p.

Burton, G.W. 1952. Quantitative inheritance in grasses. *Proc. 6th Int. Grasslands Cong. J.*, **1**: 227-283.

Burton, G.W. and De Vane, E.H. 1953. Estimating heritability in tall fescue (*Fistvea arundiancea*) from replicated clonal material. *Agron. J.*, 45: 284-291

Churchill, F.B. 1974. William Johannsen and the genotype concept. *J. His. Biol.*, **7** : 5-30.

Dudley J.W. and Moll, R.H. 1969. Interpretation and use of estimates of heritability and genetic variances in plant breeding. *Crop Sci.*, **9**: 257-262.

Griffing B. 1956. Concept of general and specific combining ability in relation to diallel crossing systems. *Aust. J. Biol. Sci.*, **9**: 463–493.

Hancock, J.F. 2004. *Plant evolution and the origin of crop species.* CABI, Cambridge, 313p.

Harlan, H.V. and Pope, M.N. 1922. The use and value of back crossing in small grain breeding. *Journal of Heredity*, **13** : 319–322

Hayman B.I. 1954a. The analysis of variance of diallel tables. *Biometrics,* **10**: 235–244.

Hayman B.I. 1954b. The theory and analysis of diallel crosses. *Genetics,* **39**: 789–808.

Hazra, P., Banerjee, M.K. and Chattopadhyay, A. 2022. *Varieties of vegetable crops in India.* Kalyani Publishers, Ludhiana, 215p.

Jinks J.L., and Hayman B.I. 1953. The analysis of diallel crosses. *Maize Genet. Co-op. Newsl.* **27**: 48–54.

Johanson, N., Robinson, H. and Comstok, R. 1955. Estimates of genetic and environmental variability in soybean. *Agron. J.*, **47**: 314-318

Jones, H.A. and Mann, L.K. 1963. *Onion and their Allies*. Leonard Hill, (Books) Ltd., London, 169p.

Richey, F.D. 1927. The Convergent Improvement of selfed lines of Corn. *The Amer. Nat.*, **61**: 430-449.

Rogo, U., Fambrini, M. and Pugliesi , C. 2023. Embryo Rescue in Plant Breeding. *Plants*, **12**: 3106; https://doi.org/10.3390/plants12173106

Van Tuyl, J.M., Van Diën, M.P., Van Creij, M.G.M., Van Kleinwee, T.C.M. , Franken, J. and Bino, R.J. 1991. Application of *in vitro* pollination, ovary culture, ovule culture and embryo rescue for overcoming incongruity barriers in interspecific *Lilium* crosses. *Plant Sci.*, **74**: 115-126.

7

Mutation Breeding

Mutation breeding is a method that uses physical radiation or chemical means to induce genetic variation in plants to develop new crop varieties. Mutation can be defined as a sudden heritable change in the character of an organism which is not due to either segregation or recombination. Mutation is the source of most genetic variation and the motor of evolution. It is a natural process, which occurs spontaneously and slowly over generations in plants, animals and all living beings. It involves alteration of their DNA, leading to the development of changes within the organism. Mutation may be sped up through the treatment of physical or chemical mutagens to achieve alteration in characteristics that are useful for the development of new variety of a plant species.

Spontaneous mutations

Spontaneous mutations are the net result of all that can go wrong with DNA during the life cycle of an organism (Glickman *et al.*, 1986). Thus, the types and amounts of spontaneous mutations produced are the resultant of all the cellular processes that are mutagenic and those that are anti-mutagenic. Spontaneous mutations occur naturally without any apparent cause however, there are two possible sources of origin of these mutations.

- Due to error during DNA replication.
- Due to mutagenic effect of natural environment *e.g.*, UV rays from sunlight

The rate of spontaneous mutations is very low, 1 in 10 lakhs *i.e.* 10^{-6}. However, different genes may show considerably different mutation rates. Different spontaneous mutants have successfully been utilized in vegetable breeding programme.

Induced mutations

Exploitation of induced mutations for crop improvement is called mutation breeding. The induced mutation is a mutation that arises from exposure of an organism's DNA to the physical or chemical mutagens. Mutagens can insert, delete or substitute a nucleotide. For instance, point mutation occurs when one half of a base DNA pair is changed. The rate of induced mutations is comparatively higher than spontaneous mutation. However, the induced mutations did not differ from spontaneous mutations in expression.

The term mutation was first used by Hugo de Vries while working with *Oenothera lamarckiana* to describe the sudden phenotypic changes which were heritable. However, the systematic studies on mutations started by Morgan (1910) who used *Drosophila melanogaster* for his studies. He identified the role chromosomes in heredity. That year, Morgan was breeding *Drosophila*, or fruit flies. After observing thousands of fruit fly offspring with red eyes, he obtained one that had white eyes. Morgan began breeding the white-eyed mutant fly and found that in one generation of flies, the trait was only present in males. Muller (1927) demonstrated for the first time the artificial induction of mutations by using X-rays in *Drosophila.* Stadler (1928) demonstrated an increase in the rate of mutations due to X-rays in barley and maize. Steinberg (1939) demonstrated induction of mutations in *Aspergillus niger* by chemicals. Later, Auerbach and Robson (1946) induced mutations in *Drosophila* with the use of chemicals.

In 1928, the Swedish geneticists Hermann Nilsson-Ehle and Åke Gustafsson started on their suggestion experiments with induced mutations using the barley crop. Limited research was also conducted in Russia and Germany. It was only after World War II, with the advent of atomic reactors, the general expansion in radiobiology and the expanded genetics and breeding research, mutation breeding research developed in many countries. In 1953, at the instigation of the Swedish Government, the 'Group for Theoretical and Applied Mutation Research' was established. Mutation breeding has so far developed more than 4000 varieties of different crop species including many vegetable crops.

The physical basis of heredity and mechanisms of mutation

By far, the most outstanding development in molecular biology has been the discovery of the physical basis of heredity and the genetic code. The information encoded in the structural gene (cistron) in the DNA is transcribed from one of the two strands of the double helix to the messenger RNA (m-RNA) by complementation. The m-RNA then moves across the nuclear membrane into the cytoplasm and there with the ribosomes and the appropriate transfer-RNA (t-RNA) molecules, the right amino acids are inserted in the correct sequence to form the specified polypeptide *i.e.*, long chains of amino acids. A protein is composed of one or more long chains of amino acids, which are linked by covalent bonds called peptide bond. A peptide bond is basically an amide-type of the covalent chemical bond. This bond links two consecutive α-amino acids from C1 (carbon number one) and N2 (nitrogen number two) of another. This linkage is found along a peptide or protein chain (a peptide bond forms when the – NH_2 ends of amino acid joins to the – COOH end of another amino acid).

Thus, there can be several potential sites for mutation:

- A change in the code in the DNA molecule and this change, if not repaired, will be permanent.
- A change of one base in the codon triplet, particularly the first, is sufficient to affect a mutation.

- A change in the m-RNA molecule but such change will be transient and not significant.
- A change in the other components (in a given t-RNA, for example) due to a change in their code and this type of change will affect more than one trait and may act as a suppressor.
- The smallest unit of mutation, called the muton, is one nucleotide.

Table 1: The genetic code: m-RNA codons for amino acids

Second base

First base		U		C		A		G			Third base
	U	UUU	phe	UCU	ser	UAU	tyr	UGU	cys	U	
		UUC		UCC		UAC		UGC		C	
		UUA	leu	UCA		UAA	non	UGA	non	A	
		UUG		UCG		UAG	non	UGG	try	G	
	C	CUU	leu	CCU	pro	CAU	his	CGU	arg	U	
		CUC		CCC		CAC		CGC		C	
		CUA		CCA		CAA	gln	CGA		A	
		CUG		CCG		CAG		CGG		G	
	A	AUU	ileu	ACU	thr	AAU	asn	AGU	ser	U	
		AUC		ACC		AAC		AGC		C	
		AUA		ACA		AAA	lys	AGA	arg	A	
		AUG	met	ACG		AAG		AGG		G	
	G	GUU	val	GCU	ala	GAU	asp	GGU	gly	U	
		GUC		GCC		GAC		GGC		C	
		GUA		GCA		GAA	glu	GGA		A	
		GUG		GCG		GAG		GGG		G	

The m-RNA triplets for histidine are CAU and CAC, whereas those for tyrosine are UAU and UAC. Obviously, U replaced the first nucleotide C in the m-RNA. In terms of the DNA this meant a change from G, which would have given C in the m-RNA to A, *i.e.*, a *transition* as shown in Table 2. Similar analyses can be made where the exact nature of the mutation is known on the polypeptide level. Thus, much better understanding of the effect of chemical mutagens has been obtained in studies utilizing biochemical mutations. Such studies led to the identification of chemical mutagens with different modes of action, *e.g.*, base analogs, alkylating agents, acridine dyes, etc.

Table 2: Types of base substitution in the DNA by the effect of chemical mutagens

Initial base	**Final base**	**Type**	**Bases**
Purine	Other purine	Transition	Adenine → Guanine
Pyrimidine	Other pyrimidine	Transition	Cytosine → Thymine
Purine	Any pyrimidine	Transversion	A,G → C,T
Pyrimidine	Any purine	Transversion	C,T → A,G

Two-way transitions, A,T ← G-C can be induced by 5-bromouracil, 2-amimopurine and nitrous acid. One-way transitions, G-C → A-T can be induced by hydroxylamine. Ethyl ethane sulfonate can induce one-way transition as above and transversion G-C

→ T-A. and G-C → C-G. In addition, *frame shift* mutations can occur due to nucleotide deletions or insertions, as induced by acridine dyes.

Induction of mutation

Physical mutagens

Physical mutagens include various types of radiations, *viz.*, X-rays, γ rays, α rays, ß-rays, fast neutrons, thermal or slow neutrons, UV rays, etc. The physical mutagens are classified into two types: Ionizing radiations and Non-ionizing radiations.

• Ionizing radiations

Ionizing radiations include X-rays, γ rays, α rays, ß-rays, fast neutrons. X-rays, being of shorter wavelength (0.1 to 10Å). These radiations are very penetrating and produce ions by colliding with atoms. Ionizing radiations can also be produced by protons, neutrons as well as by α, β and γ rays from radioactive sources like cobalt-60, radium, etc. When a highly energetic wave is captured in the cell, enough energy is transferred to an electron to cause it to become a secondary β particle, leaving behind a positively charged ion. Such an electron is ejected from the atom at a high speed, and in turn, it can cause other atoms to be ionized. All atoms that capture free electrons become negatively charged ions in this process. Due to a change in the number of electrons, a stable atom or molecule changes to a highly reactive ionic state. X-ray produces a series of chemical reactions along its track. All ionizing radiations leave energy in cells in the form of electrically charged particles, besides leaving back some energy in the form of heat and excitation of molecules.

A gamma ray, also known as γ radiation, is a penetrating form of electromagnetic radiation arising from the radioactive decay of atomic nuclei. It consists of the shortest wave length electromagnetic waves, typically shorter than those of X-rays. Gamma radiations have been used very extensively for the induction of mutations in different vegetable crops, like tomato, pea, sweet pepper, French bean, broad bean, cucumber, okra, etc.

Frequency of mutation is generally proportional to the dose of radiation, greater the dose, higher the rate of mutation. The linear relationship between the frequency of mutation and the dose of irradiation had led to the "target theory of mutation" which means that ionizing particles produced by irradiation or the rays they impinge on the target to hit the genes, thereby causing mutation. However, at higher doses, lethality is increased and regeneration capacity of the organism is adversely affected.

Primary effect of irradiation is the breakage in DNA strand. Ion clusters can produce breaks by attacking the DNA molecule itself. Breaks may occur in single chromatids or in both the adjacent chromatids, depending on the degree of ionization. If the breaks are repaired, no damage is done. Unrepaired breaks appear as chromosomal aberrations. Sometimes, the repair is not precise and results into deletion, addition or substitution of one or a few bases, thereby, causing point mutations. Continuous dose of radiation is more effective than an intermittent one. Besides the dose of irradiation, the physiological state of cell too is important for effective mutagenesis. Irradiation is more effective as a mutagen at higher oxygen concentrations. It may act indirectly by attacking other molecules, thereby, producing ions that are free radicals. The latter may react directly with the DNA or may initiate chain reactions, the end product of which ultimately interacts with the DNA, *e.g.*, irradiation may ionize water molecules into hydroxyl ions:

$$H_2O + \text{Irradiation} \rightarrow H + OH$$

In the presence of oxygen, hydrogen peroxide is easily produced from hydrogen atoms:

$$H + O_2 \rightarrow H_2O \text{ (water)}$$

$$HO_2 + H \rightarrow H_2O_2 \text{ (Hydrogen peroxide)}$$

Hydrogen peroxide, being highly reactive, may be responsible for some of the mutations. Irradiation, therefore, may be highly mutagenic because of its direct as well as indirect effects. It can cause chromosome breakage, oxidation of deoxyribose, deamination and dehydroxylation of bases and formation of peroxides.

In seed propagated vegetable crops, generally, dry seeds are subjected to irradiation treatment. In vegetatively propagated crops, all kinds of plant parts like tubers, rhizomes, bulbs and other actively growing parts can be treated with mutagenic agents. However, mutation-breeding utilizing vegetative plant parts as the base material is of limited use because it is difficult to trace the hidden variation that happens in the mutated cells of the somatic tissue.

• Ultraviolet radiation

Seven years after the pioneering work of Muller (1927), Altenburg (1934) found that ultraviolet (UV) irradiation too is mutagenic. DNA and its bases, particularly cytosine and thymine, absorb ultraviolet light. Ultraviolet hydrates cytosine by inserting a water molecule into the C=C double bond and joins together two thymine bases after disrupting their double carbon bonds and this thymine dimerization (Cyclobuthane thymine dimer) is the main cause of ultraviolet mutagenesis. Cyclobuthane thymine dimer is a photo-lesion produced by UV radiation in sunlight and is considered as a potential factor causing skin cancer. It is formed as a covalently bonded complex of two adjacent thymines on a single strand of DNA. Pollen grains are generally treated with ultraviolet irradiation.

Chemical mutagens

Soon after the discovery of mutagenic effects of radiations, a vigorous search was made for chemicals that could induce mutations. Mutagenic effect was first identified in nitrous acid and subsequently in nitrogen and sulphur mustards and a variety of carbonyl compounds. Chemical mutagen can be divided into four groups.

a) **Alkylating agents:** This is the most powerful group of mutagens. Alkylating agents are highly reactive chemical that binds to certain chemical groups (phosphate, amino, sulfhydryl, hydroxyl, and imidazole groups) commonly found in nucleic acids and other macromolecules, bringing about changes in the DNA and RNA of cells causing induction of mutation in different crops. They induce mutations especially transitions and transversions by adding an alkyl group (either ethyl or methyl) at various positions in DNA. Alkylation produces mutation by changing hydrogen bonding in various ways. Different alkylating agents are Dimethyl sulphonate (DMS), Ethyl methane sulphonate (EMS), Nitrosomethyl Urea (NMU), Nitrosoethyl Urea (NEU), Methyl methane sulphonate (MMS), nitrosoguanidine (NTG), hydroxylamine, etc.

Ethyl methane sulfonate ($CH_3SO_3C_2H_5$) is widely used organic compound which produces random mutations in genetic material by nucleotide substitution; particularly through G:C to A:T transitions induced by guanine alkylation. EMS typically produces only point mutations.

b) **Base analogues:** These are chemicals which are very similar to DNA bases. Such chemicals are sometimes incorporated in DNA in place of normal bases during replication. Thus they can cause mutation by wrong base pairing. An incorrect base pairing results in transitions or transversions after DNA replication *e.g.,* 5-bromouracil, 3-bromodeoxy uridine, 2-amino purine, etc.

c) **Antibiotics:** A number of antibiotics like, mitomycin and streptomycin have been found to possess chromosome breaking properties. However, their usefulness for induced mutation purposes is very limited.

d) **Acridine dyes:** Acridine dyes *e.g.,* proflavin, acriflavin, acridine orange, etc. are very effective mutagens. These are positively charged and they insert themselves between two base pairs of DNA. This is known as intercalation. Replication of intercalated DNA molecules results in addition or deletion of one or few base pairs which produces frame shift mutations.

e) **Miscellaneous:** Hydroxyl amine produces chromosomal aberrations. Nitrous acid (deaminating agent) has strong mutagenic activity in a variety of viruses and micro-organisms. However, it is not useful in higher plants.

Chemical mutagens have been used extensively to induce mutations in potato, tomato, chilli, peas, broad bean, summer squash, French bean, carrot, etc. The effectiveness of chemical mutagens depends upon their solubility, toxicity and reactivity. The widely used chemical mutagen, EMS is generally used at concentrations ranging from 0.1 to 0.5 per cent. Dry seeds are generally treated with EMS solution. The application protocol is 8-12-hour pre-soaking of seeds to activate germination followed by treatment with 0.1 to 0.5% EMS solution for 3-6 hours depending on the crop. Chemical mutagens are less effective for treating the vegetative plant parts.

Table 4: Features of mechanisms of induced mutations

Mutagenic agent	Mutation mechanism
Physical mutagen	Mechanical breaks in DNA strands
	Breaks caused by ionizing radiations or P^{32} decay
	Chromosomal non-disjunction
	Elevated temperature
Chemical mutagen	Removal or alteration of DNA bases
	Incorporation of altered bases
	Intercalation of oligocyclic aromatic compounds
	Alteration of DNA backbone
Enzymatic agent	Mistakes in the DNA replicating system
	Alterations in the DNA replicating system
	Mistakes during recombination
	Mistakes during repair
	Production of chemicals that affect DNA

Classification of mutations

Mutations can be classified in several ways.

Based on direction of mutations

- **Forward mutation:** Forward mutations confer a gene sequence and phenotype different from that conferred by the wild-type gene. It means any change from wild type allele to mutant allele.
- **Backward mutation or reverse mutation:** Reverse mutation, also called reversion, denotes any mutational process or mutation that restores the wild-type phenotype to cells already carrying a phenotype-altering forward mutation. It means a change from mutant allele to wild type allele.

Based on source/cause of mutations

- **Spontaneous mutation:** Mutation that occur naturally. Thus, the types and amounts of spontaneous mutations produced are the resultant of all the cellular processes that are mutagenic and those that are antimutagenic.
- **Induced mutation:** The induced mutation is a mutation that arises from exposure of an organism's DNA to a mutagen. It means a mutation that originates in response to mutagenic treatment.

Based on tissue of origin

- **Somatic mutation:** A somatic mutation describes any alteration at the cellular level in somatic tissues occurring after fertilization. These mutations do not involve the germline and consequently do not pass on to offspring.
- **Germline mutation:** A mutation in germline cells or in reproductive tissues. Germline mutations are congenital genetic mutations in germ cells that originate from sperm or ovum and are generally incorporated into every cell of the offspring's body.

Based on effect on survival

- **Lethal mutation:** A gene mutation whose expression results in the premature death of the organism carrying it.
- **Sub-lethal mutation:** A gene mutation whose expression results in the mortality of more than 50% of individuals that carry mutation.
- **Sub-vital mutation:** A gene mutation whose expression results in the mortality of less than 50% of individuals that carry mutation.
- **Vital mutation:** When all the mutant individuals survive (survival 100%).

Based on trait or character effected

- **Morphological mutation:** A mutation that alters the morphological features of an individual. It affects the outward appearance of an individual. Plant height mutations could change a tall plant to a short one, or from having smooth to rough seeds.
- **Biochemical mutation:** A mutation that alters the biochemical functions of an individual. Biochemical mutations have a lesion in one specific step of an enzymatic pathway.

Based on visibility or quantum of morphological effect produced

- **Macro-mutations:** A mutation which produce a distinct morphological change in phenotype (which can be detected easily without any confusion due to environmental effects). This mutation has a profound effect on the resulting organism, as a change in a regulatory gene that controls the expression of many structural genes. Such mutation is generally found in qualitative characters, *e.g.* colour of flowers, colour of fruit, inflorescence type, growth habit, etc.
- **Micro-mutations:** A mutation with invisible phenotypic changes, (which can be easily confused with effects produced due to environment). Micro-mutations are small, affecting at the most a single gene and usually involving fewer than ~103 nucleotides. Such mutation is generally observed in quantitative characters.

Based on the site of mutation or on cytological basis

- **Chromosomal mutations:** Chromosomal mutations involve changes in the number or structure of chromosomes. These mutations can change the location of genes on chromosomes and can even change the number of copies of some genes. There are four types of chromosomal mutations: deletion, duplication, inversion, and translocation.
- **Gene or point mutations:** A point mutation occurs in a genome when a single base pair is added, deleted or changed. While most point mutations are benign, they can also have various functional consequences, including changes in gene expression or alterations in encoded proteins. Point mutations can be subdivided into the following three classes on the basis of molecular change associated with them *viz.*, (a) Base substitution, (b) Base deletion, (c) Base addition.

Base substitution: When a single base in a DNA molecule is replaced by another base it is known as base substitution which can be of two types.

a) **Transition:** Replacement of a purine base by another purine base or a pyrimidine base by another pyrimidine base in DNA or RNA is known as transition (A G or C T).

b) **Transversion:** Replacement of a purine base by a pyrimidine base and *vice versa* in DNA or RNA is known as transversion (A or G C or T or U).

Base deletion: In base deletion, one or more bases are altogether deleted.

Base addition: There is insertion of one or more bases. If the number of bases added or deleted is not a multiple of three, a frameshift mutation is obtained, as the reading frame in such case is shifted from the point of addition or deletion onwards. Hence, in a frameshift mutation, all the amino acids of a polypeptide chain located beyond the site of mutation are substituted / altered.

Cytoplasmic mutations

A mutation which is associated with the changes in chloroplast DNA (cpDNA) and mitochondrial DNA (mtDNA). Mitochondrial genomes are inherited *via* the maternal cytoplasm.

Characteristic features of mutations

1. Mutations are mostly recessive and very rarely dominant. This mutation generally does not lead to visible phenotypic effects when present in heterozygous combination with the wild-type allele.
2. The majority of mutations are neutral in their effects on the organisms in which they occur however, beneficial mutations may become more common through natural selection.
3. Mutation rates in humans have been estimated to be on the order of 10^{-4} to 10^{-6} per gene per generation. The rate of nucleotide substitutions is estimated to be 1 in 10^{8} per generation, implying that 30 nucleotide mutations would be expected in each human gamete, assuming no back mutation.
4. Mutations result either from errors in DNA replication or from the damaging effects of mutagens, such as chemicals and radiation, which reacts with DNA and changes the structures of individual nucleotides.
5. All cells possess DNA-repair enzymes that attempt to minimize the number of mutations that occur.
6. If mutation occurs at both the loci simultaneously it is dominant in nature and the mutants can be identified in M_1 generation. This consequence is really rare. However, if it is restricted to one locus only (recessive in nature), which mostly happens the effect can be seen only in M_2 generation when recessive mutant genes become homozygous in the locus.

7. Macro-mutations are visible and can be easily identified, while micro-mutations do not have discernable phenotypic expression and statistical analyses are required to establish the variation over the parental genotype.
8. Many mutants show male sterility which is classified into three groups: 1) genetic, 2) cytoplasmic, and 3) cytoplasmic - genetic.
9. Most mutants are of negative in selection value because more DNA changes are harmful than are beneficial. Hence, negative selection plays an important role in maintaining the long-term stability of the biological structures by removing deleterious mutations. Thus, negative selection is sometimes also called purifying selection or background selection.
10. Mutations are random *i.e.* they can occur in any tissue or cell of an organism however, some genes show higher mutation rate than others.
11. Mutations can be sectorial and the branches arising from mutated sector show mutant characters.
12. Mutations are recurrent *i.e.* the same mutation may occur again and again.
13. Induced mutations commonly show pleiotropy often due to mutation in closely linked genes.

Approaches in applied mutagenesis and isolation of mutants

Mutation is a method for the creation of genetic variability for further selection or for utilization in hybridization and selection programme of breeding. The nature of the problem, the mode of reproduction of the plant, the availability of efficient screening methods as well as other factors should lead to a decision as to which breeding method or combination of methods is employed. In asexually propagated crops, mutation breeding is an effective method for generating variability. In sexually propagated crops, utilization of mutation breeding in self-pollinated crops is fairly well known whereas in out crossing species its application is more difficult and identification of the origin of the desirable genotypes is doubtful.

The primary step in mutation breeding approach is to select the crop and the genotype. Generally, an otherwise good variety lacking in one or few characteristics like growth habit, resistance etc. are selected for mutagen treatment. Plant materials used for treating with mutagens are seeds, pollen, vegetative buds, whole plants, bulbils, tubers, suckers etc. However, mutagenic treatment of seeds is the most convenient and, therefore, the standard method in seed propagated crops.

Gamma radiation is the most effective electromagnetic radiation for crop improvement as it penetrates deeply into target tissue and is less destructive (Mba *et al.*, 2010).

Optimization of mutagen dose

An optimal dose is that which induces a higher rate of mutation with little biological damage. Optimization of mutagen dose requires a median lethal (LD50) dose, which kills 50% of the total population, and median growth reduction (GR50) dose, which reduces growth by 50% in M_1 generation (Songsri *et al.*, 2011; Angeles-Espino *et al.*,

2013). Both are based on the concept that a low dose fails to produce an impact on the genome resulting in no phenotypic change; a high dose increases mutation frequency but lowers the chance of obtaining desired and utilizable mutants.

Probit analysis according to Finney (1971) is employed to determine optimum gamma radiation or chemical mutagen dose through evaluation of dose-response to compare mutagen tolerance of the plant species. It helps to identify a target dose for large-scale testing. Generally, variation in germination percent after treating with the mutagen is statistically tested after angular transformation. The differences in fertility of M_1 plants are considered as the measure of sensitivity of a genotype to mutagen (Walter *et al.*, 1987).

Damage in M_1 generation

Plants grown from the seeds treated with mutagen are called M_1 generation. The impact and tolerance level of any genotype to the mutagen is manifested in M_1 generation itself in terms of lethality (reduction in seed germination), injury (reduction in seedling growth) and sterility i.e., reduction in pollen viability (Gaul, 1977). Inhibitory effect of mutagens is more pronounced in seed germination compared to seedling growth ((Sikder *et al.*, 2013; Hazra *et al.*, 2021). All the mutagenic treatments showed inhibitory effect on seed germination which happens due to a number of factors *viz.*, adverse effect on cytochrome oxidase content, thus reducing the respiration (Swaminathan, 1962), drastic distortion of the actively dividing phase (Singh and Singh, 1989), damage to cell constituents at the molecular level, altered enzyme activity and/or block in formation of cellular DNA causing the plant growth to slow or stop (Khan and Goyal, 2009; Roslim *et al.*, 2015), interference in the synthesis of enzyme and acceleration in the degradation of existing enzyme (Yusuf and Nair, 1974) , altered enzyme activity (Khan and Goyal, 2009), block in synthesis of growth hormone (Gunckel and Sparrow, 1961) and/or nature and extent of chromosomal damage (Monica and Seetharaman, 2016). These physiological consequences may explain inhibition of seed germination, seedling emergence and seedling growth parameters and pollen viability in different crops.

Selfed or open pollinated seeds are harvested from each plant of the M_1 generation to raise the next generation, M_2 following single seed descent or plant to row method. The recessive mutations, which are heterozygous in M_1, can be recovered in the M_2 generation. Thus, mutants can be identified and selected from single plant progenies showing segregation in M_2 generation.

Mutation frequency, efficiency and effectiveness

- **Chlorophyll mutation frequency**: It is determined as percentage of mutated M_2 progenies for chlorophyll deficiency (both viable and non-viable).
- **Total mutation frequency (Mf):** It is determined as percentage of mutated M_2 progenies for both chlorophyll deficient and other viable macro-mutants.
- **Biological damage, lethality**: Percent reduction in seed germination in M_1 generation

- **Biological damage, injury**: Percent reduction in seedling growth parameters in M_1 generation
- **Biological damage, sterility**: Percent reduction in pollen viability in M_1 generation

Data on biological damage (lethality, injury and sterility) in M_1 generation and total mutation frequencies are used to calculate the mutagenic efficiency and effectiveness according to standard formula given by Konzak *et al.* (1965).

- **Mutagenic efficiency**: It is determined as (Mf/L); (Mf/I) and (Mf/S) where Mf, L, I and S denote total mutation frequency in M_2 generation, lethality in seed germination, seedling injury and pollen sterility, respectively in M_1 generation.
- **Mutagenic effectiveness**: It is determined by using the formula Mf × 100 / KR or Mf × 100/C x T where, Mf, KR, T and C indicates total mutation frequency, dose of radiation in kilo rad, duration of treatment in hours and percent concentration of EMS solution, respectively.

Spectrum of mutation

The spectrum of mutation is essentially a parameter for the index of mutation frequency. Spectrum of mutation varied with the mutagen, its dose and genotypes indicating variation in allelic mutability of different genotypes. The spectrum of different chlorophyll deficient mutants include:

- **Albina**: Entirely white leaves
- **Chlorina**: Light green leaves
- **Maculata**: Yellowish or white dots on leaves
- **Tigrina:** Yellow and green patches on leaves
- **Viridis**: Viridine green colour of leaves
- **Xantha**: Bright yellow leaves

All these chlorophyll deficient mutants are lethal except maculata and viridis mutants. Different macro-mutants showing altering growth habit, branch number, inflorescence, leaf characters, male sterility, flower and fruit characters may emerge in the M_2 generation which can be isolated and later purified. These putative mutants may form the base for development of new variety. It was clearly evident from the studies on different crops that the physical and chemical mutagens induced different mutation spectrum and the type of mutant depends not only on the type of mutagen but also on the genotype employed in the applied mutagenesis programme (Konzak *et al.*, 1965; Gottschalk and Wolff, 1983; Walter *et al.*, 1987; Sakin and Sencar, 2002; Sikder *et al.*, 2012; Hazra *et al.*, 2022).

Total mutation frequency

Total mutation frequency does not necessarily reflect the spectrum of mutation hence, both mutation spectrum and frequency are important to ascertain the genetic variation that is available for selection in M_2 and onward generations.

Mutagenic efficiency

The efficient mutagenesis is the production of desirable changes free from association with undesirable ones (Konzak *et al.*, 1965). It has been generally observed that the mutagen dose that gives the higher mutation rates also induces lethality, sterility and other undesirable effects. In mutation breeding programme, such treatments are not desirable as they result in multiple mutations and other drastic effects reducing the yield potentials of the mutants. It is therefore, desirable to induce high mutation rates with least accompanying deleterious effects. Hence, mutagenic efficiency gives an idea of the proportion of mutation in relation to deleterious effects of the mutagen (Konzak *et al.*, 1965). The higher efficiency of a mutagen indicates relatively less biological damage in relation to mutations induced (Jain and Khandelwal, 2009). It seemed that strong mutagens reach their saturation point even at a lower dose in the genotypes having highly mutable allelic sites, and any further increase in the mutagen dose add to their mutation frequency. However, all treatments which yield the highest mutagenic frequency might not confer the highest mutagenic efficiency (Sakin and Sencar, 2002; Sikder *et al.*, 2013). It has been recorded in several experiments that the efficiency of lower doses/ concentrations is more compared to higher doses in producing desirable mutants.

Mutagenic effectiveness

Mutagenic effectiveness is defined as a measure of frequency of mutation induced by a unit of mutagen. The degree of effectiveness of mutagen and the response of the genotypes vary.

Table 3: Doses of γ radiation generally recommended for different vegetable crops

Crop	Dose (Gray, Gy)
Gamma radiation treatment in seeds	
Tomato	150-250
Brinjal	200-300
Chilli	100-200
Peas	150-250
French bean	100-250
Broad bean	100-250
Onion	100-150
Carrot	300-500
Brassica sp.	600-800
Radish	700-800
Okra	300-600
Gamma radiation treatment in vegetative plant parts	
Potato	50-60
Sweet potato	30-100

Application of mutagens (both physical and chemical) causes macro-mutations (major, mono or oligogenic qualitative alternations) and/or micro-mutations (minor, polygenic or quantitative changes) in the genotypes. It is quite easy to detect and select macro-mutations through apostatic selection (*i.e.*, selection of rare genotypes). Different macro-mutations recorded in the vegetable crops are dwarf mutants (French bean, pea, summer squash, muskmelon etc.), leaf mutants (pea), flower mutants (tomato, cabbage, cowpea), male sterile mutants (tomato, muskmelon, brinjal, chilli, sweet pepper, etc.), disease resistant mutants (tomato, brinjal, French bean, potato, okra, etc.), mutants with high nutritional qualities (potato, sweet potato, tomato, peas, chilli, etc.), etc.

It is not easy to detect minor polygenic changes for quantitative traits. Such changes are characterized by continuous variation, which can be measured on a suitable scale in terms of means, range, variance, coefficient of variation and heritability. It is often assumed that M_2 is analogous to F_2 generation so far as segregation events and relative degree of genetic variability are concerned, although M_1 cannot be compared with F_1 because heterozygosity is much more, even 100% in F_1 while mutant loci in M_1 are random and not so frequent. Handling of different generations after applied mutagenesis for the development of variety becomes, by and large, resembles pedigree method of breeding.

Few varieties of different vegetable crops have been developed by mutation breeding.

- **Tomato**: CO3 (Maruthan), EMS mutant of the variety CO1
- **Okra**: Punjab 8 resistant to yellow vein mosaic virus, EMS (1%) mutant of the variety Pusa Sawani
- **Bitter gourd**: MDU-1, Gamma radiation (100 Gy) mutant of the variety MC 103
- **French bean**: Pusa Parbati, X-ray mutant of the variety Wax Pod
- **Cucumber**: Swarna Ageti, Gamma radiation mutant

Some unique macro-mutants like, "Multiparus cyme mutant" manifesting branched cyme and bearing increased number of flowers and fruits isolated from the M_2 population of the exotic variety of tomato, Berika from Bulgaria irradiated by 50Gy gamma radiation and "Dark green fruit mutant"having high chlorophyll and lycopene contents in the fruit isolated from the M_2 population ofthe tomato cultivar, Patharkutchi irradiated by 150 Gy gamma radiation(Sikder et al, 2013; Das et al, 2019), a mutant of white-fruited bitter gourd having comparatively high charantin content isolated from the M_2population of the bitter gourd cultivar BG1346501 irradiated by 100 Gy gamma radiation(Hazra *et al.*, 2022) and a yellow vein mosaic virus resistant mutant isolated from the M_5 population of the okra cultivar Pusa Swani irradiated by 350 and 450 Gy gamma radiation (Hazra *et al.*, 2023) may form the basic material for development of new variety of tomato, okra and bitter gourd.

Mutation creates slightly different versions of the same genes, called alleles. These small differences in DNA sequence make every individual unique which are called mutants posessing such specific characters. Through induced mutagenesis, breeders are able to develop mutant varieties across the globe, which has helped in uplifting the agricultural economy of many developed and developing countries.

References

Altenburg, E. 1934.The Artificial Production of Mutations by Ultra-Violet Light. *American Naturalist*, **68**: 491-507.

Angeles-Espino, A., Valencia-Botin, A. J.,Virgen-Calleros,G. , Ramirez-Serrano, C.,Paredes Gutierrez, L. and Hurtado-de La Pena, S. 2013. Lethal doses (LD_{50}) determination using Co_{60} on *Agave tequilana* var. Azul *in vitro* plantlets. *Revista Fitotecnia Mexicana*, 36:381–386.

Auerbach, C. and Robson, J.M. 1946. Chemical production of mutations. *Nature*, **157**: 302.

Das, I., Hazra, P., Longjam, M., Bhattacharjee, T., Maurya, P. K., Banerjee, S. and Chattopadhyay, A. 2019. Characterization of induced mutants and their hybrids of tomato(*Solanum lycopersicum* L.) for growth, yield and fruit quality traits to explore the feasibilityinfuturebreeding.*Genetic Resources and CropEvolution*, **66**:1421-1441.

Finney D. J. 1971. *Probit Analysis* (3rd edition). Cambridge University Press, Cambridge, UK.

Gaul H. 1977. Mutagen effects observable in the first generation. I. Plant injury and lethality, II. Cytological effects, II sterility In: *Manual on Mutation Breeding* (second edition). IAEA technical report series No. 119, IAEA, Vienna, Austria, pp. 85-99.

Glickman, B.W., Burns, P.A. and Fix, D.F. 1986. Mechanisms of spontaneous mutagenesis: clues from altered mutational specificity in DNA repair-defective strains, In: *Antimutagenesis and Anticarcinogenesis Mechanisms* (Eds. Shankel, D.M., Hartman, P.E., Kada, T. and Hollaender, A.), Plenum Press, New York, pp. 259-281

Gottschalk W. and Wolff G. 1983. Induced mutations in plant breeding. *Monographs on Theoretical and applied Genetics*, Heidelberg, Berlin, 10-14.

Gunckel J. K. and Sparrow A. H. 1961. Ionizing radiations: biochemical, physiological and morphological aspects of their effects on plants. *Encycl. Plant Physiol.*, **16**: 555-611.

Hazra ,S., Gorai ,S., Umesh Kumara ,V., Bhattacharyaa , S., Majia ,A., Jambhulkarb , S., Ali ,N. and Chattopadhyay, A. 2021. Optimization of gamma radiation dose for induction of mutations in okra. *Intern. J. Veg. Sci.*, **27**: 574-584.

Hazra, P., Hazra, S., Acharya, B., Dutta, S., Saha, S., Mahapatra, P., Pradeepkumar, P., Pal, H., Chattopadhyay, A., Chakraborty, I., Jambhulkar, S., Chatterjee, S. and Ghosh, S.K. 2022. Diversity of nutrient and nutraceutical contents in the fruits and its relationship to morphological traits in bittergourd (*Momordica charantia* L.). *Sci. Hort.*, https://doi.org/10.1016/j.scienta.2022.111414).

Hazra, S., Gorai, S., Bhattacharya, S., Bose, S., Hazra, P., Chattopadhyay and Maji, A. 2022. Radio-sensitivity of diverse tomato genotypes with respect to optimization of gamma irradiation dose. *Brazilian Journal of Botany*, https://doi.org/10.1007/s40415-022-00823-2.

Hazra, S., Gorai, S., Roy, S., Bose, S., Hazra, P., Chattopadhyay, A., Nasim Ali, Md., Jambhulkar, S. and Maji, A. 2023. Isolation of YVMV resistant mutants of okra (*Abelmoschus esculentus* L.) through applied mutagenesis. *Plant Breed.*, 2023;1–14, DOI: 10.1111/pbr.13151.

Jain S. K. and Khandelwal R. 2009. Induced polygenic variability in blackgram [*Vigna mungo* (L.) Hepper]. *Indian J. Genet.*, **69**: 72-75.

Khan S. and Goyal S. 2009. Improvement of mungbean varieties through induced mutations. *African J. Plant Sci.*, **3**: 174-180.

Konzak C.F., Nilan R.A., Wagner J. and Foster R.J. 1965. Efficient chemical mutagenesis. *Radiation Bot.*, (Suppl.), **5**: 49-70.

Mba, C., Afza, R., Bado, S. and Jain. M. 2010. Induced mutagenesis in plants using physical and chemical agents, In: *Plant cell culture: Essential methods* (eds. Davey, M.R. and Anthony, P.). John Wiley and Sons, Ltd., Chichester, U.K. pp. 111–130.

Monica, S., and N. Seetharaman. 2016. Effect of gamma irradiation and ethyl methane sulphonate (EMS) mutagenesis in early generation of garden bean [*L. purpurens* (L.) Sweet var. *typicus*]. *Int. J. Adv. Sci. Tech. Res.*, **6**: 398–410.

Morgan, T. H. 1910. Chromosomes and heredity. *The Amer. Nat.,* **44** : 449–96.

Muller, H. J. 1927. Artificial Transmutation of the Gene. *Science,* **66**: 84–87

Roslim, D.I., A. Herman, and I. Fiatin. 2015. Lethal dose 50 (LD50) of mungbean (V. radiata L. Wilczek) cultivar kampar. *SABRAO J. Breed. Genet.,* **47**: 510–516

Sakin M.A. and Sencar O. 2002. The effects of different doses of gamma ray and EMS on formation of chlorophyll mutations in durum wheat (*Triticum durum* Desf.). *Tarim Bilimleri Dergisi*, **8**: 15-21.

Sikder, S., Biswas, P., Hazra, P., Chattopadhyay, A., Badigannavar, A.M. and D'Souza, S.F. 2013. Induction of mutation in tomato (*Solanum lycopersicum* L.) by gamma irradiation and EMS. *Indian J. Genet*., **73**: 392-399.

Singh, R. R. and Singh, J. 1989. Genotypic variability relationship with EMS in a *Trigonella* species complex. *Indian J. Hort*., **46**: 111-114.

Songsri, P., Suriharn, B., Sanitchon, J., Srisawangwong, S. and Kesmala, T. 2011. Effects of gamma radiation on germination and growth characteristics of physic nut (*Jatropha curcas* L.). *J. Biol. Sci*., **11**: 268–274.

Stadler, L.J. 1928. Genetic effects of X-rays in maize. *Proc. Natl. Acad. Sci.*, U S A. 14: 69–75.

Steinberg, R.A. 1939. Relation of carbon nutrition to trace element and accessory requirements of *A. niger*. *J. Agric. Res*., **59**: 749.

Swaminathan, M.S., Chopra, V.L. and Bhaskaran, S. 1962.Chromosome aberrations and the frequency and spectrum of mutations induced by ethyl methane sulphonate in barley and wheat. *Indian J. Genet*., **22**: 192-207.

Walter, R.F., Elinor, L.F. and Holly, J.J. 1987. *Mutation Breeding Principles of Cultivar Development, Theory and Technique*. Macmillan Publishing Co. New York, pp. 287-303.

Yusuf, K.K. and Nair, P.M. 1974. Effect of gamma irradiation on the indole acetic acidsynthesizing system and its significance in sprout inhibition of potatoes. *Radiation Bot*., **14**: 251-256.

8

Breeding Through Alteration of Ploidy Level

The term genome refers to a complete set of DNA, including all of its genes. Each genome contains all of the information needed to build and maintain that organism. In the nucleus of each cell, the DNA molecule is packaged into thread like structures called chromosomes and each species has specific chromosome number. Most of the plants are diploid and they show in each one of its body cell, twice the number of haploid chromosomes (n) found in the gametes.

There are two main types of chromosomal variation that involve changes in the number. These are euploidy (in Greek "eu"means true or even and "ploid"means unit), which involves the variation of whole sets of chromosomes and aneuploidy("aneu" means uneven), where a single chromosome within a set varies. This variation may be either increase or decrease in the chromosome number. The euploids include diploid (2n), triploid (3n), tetraploid (4n), hexaploid (6n) or even higher degree of polyploidy. Polyploidy, thus involves more than two genomes and is, therefore, a genometic mutation involving increase in the dose of whole genome. This condition of polyploidy is more common in angiosperms.

A number of polyploids have been evolved spontaneously in nature. Polyploids arise when a rare mitotic or meiotic catastrophe, such as non-disjunction, causes the formation of gametes that have a complete set of duplicate chromosomes. Diploid gametes are frequently formed in this way. When a diploid gamete fuses with a haploid gamete, a triploid zygote forms, although these triploids are generally unstable and can often be sterile. If a diploid gamete fuses with another diploid gamete, this gives rise to a tetraploid zygote, which is potentially stable as for example, potato. Many types of polyploids are found in nature, including tetraploids (four sets of chromosomes e.g., potato), hexaploids (six sets of chromosomes e.g., sweet potato), and other chromosome-pair multiples.

Induction of polyploidy

Polyploidy can also be induced artificially and such variation in the chromosome number is associated with the phenotypic changes in the organisms. Such phenotypic variation may result in the development of a new variety having unique character.

Polyploidy can be induced due to aberration in cell divisions. This may occur both in the mitosis as well as in meiosis. In mitosis, failure to form cell wall after the division of nucleus leads to doubling of chromosome number. If a shoot arises from such a cell, the plant will be tetraploid and if such failure occurs in germinal tissue shortly before reduction division, diploid gametes are formed from such cells. This type of irregularity may also occur at anaphase, when the spindle mechanism may fail to separate the sister chromatids to opposite poles. Consequently, the chromosome number is doubled.

Table 1: Types of variation in chromosome number

Types	Number	Description
Euploids		
Monoploids	n	Genetically hemizygous
Diploids	2n	Two homologous sets
Polyploids		
Triploids	3n	Three homologous sets
Tetraploids	4n	Four homologous sets
Hexaploid	6n	Six homologous sets
Aneuploids		
Hypoploids	2n-x	Loss of one or more chromosomes
Nullisomics	2n-2	Loss of homologous pair
Monosomic	2n-1	Loss of one member of a pair
Double monosomic	2n-1-1	Loss of two non-homologous chromosomes
Hyperploids	2n+x	Addition of one or more chromosomes
Trisomic	2n+1	Addition of one chromosome of a pair
Tetrasomic	2n+2	Addition of two chromosomes of a pair
Double tetrasomic	2n+2+2	Addition of two chromosomes of two separate pair

Failure in meiosis produces a diploid gamete, instead of a haploid and a chance fusion of such a gamete with other types of gametes produces different types of polyploidy. Such chance fusion of two such gametes produces a tetraploid (4n) plant. Blackslee and Avery (1937) and Nebel (1937) demonstrated the method of inducing polyploidy in plants (*Portulaca* and *Datura*) artificially by the application of the alkaloid colchicine, the extract from the autumn crocus plant *Colchicum autumnale* (Liliaceae family). Colchicine brings about the doubling of chromosome number by acting on the spindle mechanism and tetraploid plants are produced. Colchicine produces some characteristic change in the movement of the chromosome during cell division. It temporarily inhibits the formation of spindle, so there is a slight change of normal mitosis. Such changed mitosis is called "C-mitosis". The polyploidy formed through application of colchicine is called "colchiploidy".

Cotton swab method is very effective method of using colchicine for the induction of polyploidy due to low mortality rate. Sterilized cotton balls of suitable size are placed on the emerging apical tips between two cotyledonary leaves of 3–4-day-old seedlings. Then aqueous solution of colchicine is applied drop by drop on the cotton ball with the help of Pasteur pipette. Colchicine treatment is generally applied 4–8 hour continuously for 2–3 days (Dar *et al.*, 2017).

Apart from the widely used anti-mitotic and mutagenic agent colchicine, different other mutagenic agents such as, oryzalin, nitroxide, and epoxomicin are used for polyploidy induction in plants.

The triploids, containing three sets of chromosomes (3n) may arise due to fertilization between diploid gametes (2n) from the tetraploid plant and a haploid gamete (n) from the normal diploid plant. Though the triploid plants are viable, the gametes formed are of defective types and so are completely sterile. Normal synapsis between odd numbers of chromosomes at meiosis is not possible. Pairing will occur with only two homologous at a time thus, the third remains unpaired, thereby, forming a univalent. Even, if they pair to form a trivalent, it may be randomly distributed in the gametes. Thus, in either of the cases, abnormal gametes will be formed which will consequently fail to fertilize resulting sterility.

Methods of determining higher ploidy level

Conventionally ploidy levels are determined by counting the chromosomes of meristematic tissue (*i.e.* root tip cells) from individual plants. However, to screen large number of plants by this method requires trained skills and is labour intensive and time consuming as well. Several alternative and indirect methods are also used for determination of ploidy level.

- **Chloroplast number in the stomatal guard cell:** Positive correlation exists between the number of chloroplast in the guard cell and ploidy level in many plant species. This technique involves simply stripping a piece of epidermis from a leaf (preferably lower epidermis because of higher stomata concentration) and counting chloroplast number in a pair of guard cells surrounding a stomatae using a microscope. Chloroplasts can be viewed with or without staining using a bright field, phase contrast or fluorescence microscope.
- **Stomatal length and pollen grain diameter:** These two characters have also been commonly used as the alternative methods for determining the ploidy of plants. Differences in these parameters between the ploidy levels have been reported to be less than in the numbers of chloroplasts in stomatal guard cells. However, induced polyploids of pointed gourd (*Trichosanthes dioica*) could be identified by having larger but less viable pollen grains, reduced stomatal frequency and increased size of stomata (Hazra and Ghosh, 2001).
- **Flow cytometry:** It is a technique used to detect and measure physical and chemical characteristics of a population of cells . In this process, a sample containing cells is suspended in a fluid and injected into the flow cytometer instrument. The sample is focused to ideally flow one cell at a time through a laser beam, where the light scattered is characteristic to the cells and their components. Cells are often labeled with fluorescent markers so light is absorbed and then emitted in a band of wavelengths. Tens of thousands of cells can be quickly examined and the data gathered are processed by a computer.

Flow cytometry can quickly and accurately determine the nuclear DNA content of cells which is widely used to identify chromosome ploidy of wide range of plant species. For both the analysis of ploidy and genome size, external standard method is used wherein test samples and reference standards are analyzed separately on the flow cytometer with the same instrument gain settings. The samples are generally prepared as per the methodology given by Dolezel *et al.* (2007). The modified HPI buffer containing propidium iodide (PI) is used to isolate intact nuclei. The young leaf (approximately 20 mg) is taken in an ice-cold plastic petri plate and chopped in 1 ml of modified HPI buffer using surgical blades under dark condition. The 50 ml modified HPI buffer is prepared in distilled water by adding trisodium citrate dihydrate (0.05 gm), triton X-100 (150 μl), RNase (2.5 ml) and PVP (0.5 gm). The solution is sterilized using filter sterilization, and β-mercaptoethanol (15 μl) and PI (1.25 ml) are added to the filtrate. The samples are then filtered through the 40-μm filters. The homogenate is incubated for 15 minutes on ice under the dark condition with occasional shaking to allow the PI to bind to the DNA of the isolated nuclei. Then, the samples are analyzed for estimation of DNA ploidy and genome size using reference DNA standards. As for example, diploid potato species, *Solanum chacoense* has been used for ploidy estimation of potato (Das *et al.*, 2023). The stained nuclei of the test sample and reference standard are run through the flow cytometer with an excitation wavelength of 480-nm band pass filter. Ten thousand PI cell signals are detected per sample, and PI-fluorescence histograms are generated to estimate the ploidy and genome size.

Role of ploidy manipulation in vegetable breeding

Ploidy manipulation has been used successfully in vegetable breeding as a means of increasing yield, enhancing product quality and to impart seedlessness in the fruits of many vegetable crops.

Tetraploids

Induced tetraploids generally show gigantism and thereby show high yield compared to the normal diploids. After induction of tetraploidy through treating the active plant parts with colchicine, the manipulated plants need to be stabilized through growing and selection in different selfed generations so that the selected tetraploids produce normal diploid (2n) pollen and ovule because all the plants of the progeny do not always have the expected ploidy level.

Diploid plant

2n ← Colchicine treatment

↓

Tetraploid plant (4n)

↓ Selfings and selections

Stable tetraploid plant (4n)

(normal meiosis)

2n x 2n

↓

4n

↓

Perpetuation of tetraploidy through sexual cycle

Tetraploids of some vegetable crops like, radish, pumpkin, watermelon, muskmelon, palak, asparagus, etc. are highly productive and have improved quality.

- **Palak:** Pusa Jyoti (Induced tetraploid plant from the variety, All Green by application of 2% colchicine).

Tetraploid muskmelon plants could be induced successfully from diploid inbred M01-3 (2n = 24) by colchicine treatment which showed high fruit weight and better agronomical characters than the diploids and could be used as the improved variety (Zhang *et al.*, 2010).

Triploids

Triploids contain 3 complete sets of chromosome and can be developed by hybridizing normal diploid with induced tetraploid. Triploidy arise by the fusion of gametes containing haploid (n) and dihaploid (2n) chromosomes. In triploids, only trivalents are produced in meiosis instead of normal bivalents in the diploids. These trivalents do not disjoin in equal number of chromosomes to the two poles. Therefore, the number of chromosomes in the gametes would vary from n to 2n. Most of these gametes are unbalanced leading to high degree of sterility. In triploid watermelon, frequency of viable gametes is only 0.1 percent. So seeds are not produced in the fruits of the triploids. This phenomenon has been commercially utilized in developing seedless triploid watermelon.

Schematic diagram of producing and utilizing triploid watermelon

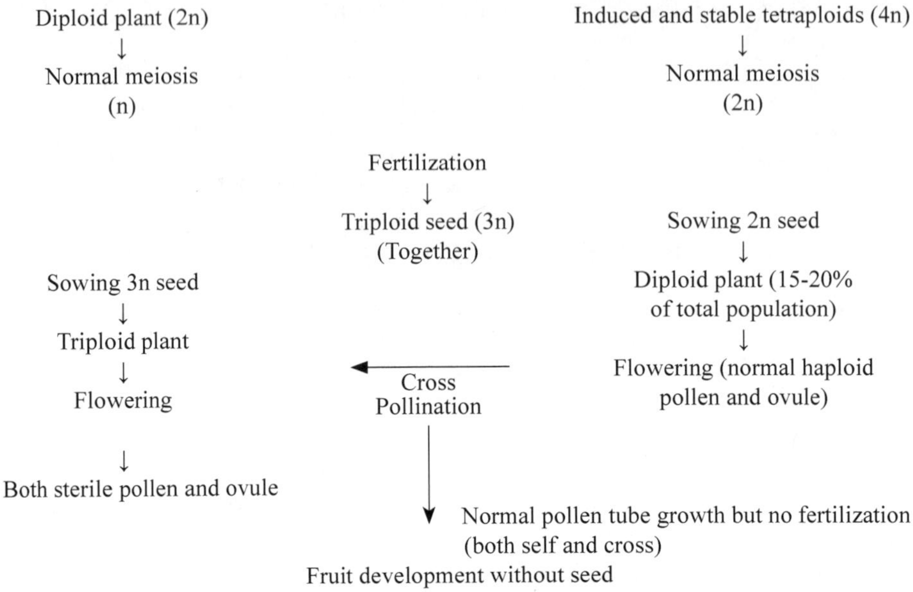

In watermelon, natural parthenocarpy does not occur. Therefore, it is essential to artificially pollinate the flower of 3x plant with diploid viable pollens to induce fruit set. Germination of pollens and development of pollen tubes on the stigma incite hormonal movement for the development of ovary into fruit. Since ovules are mostly sterile, generally no fertilization occurs. Generally triploids and diploids should be planted in 3-5 : 1 ratio. Triploid seeds are to be produced every year by hybridizing normal diploid × induced tetraploids. Triploid watermelon varieties have so far been developed in India at IARI, New Delhi, IIHR, Bengaluru and Kerala Agricultural University, Kerala. These triploid and seedless varieties have become popular due to high sugar content, good transportability and seedless character. First triploid hybrid of cassava, Sree Harsha was developed by crossing the diploid clone OP-4 x induced tetraploid of Sree Sahya. Triploid plants of cassava can be maintained vegetatively through clonal propagation or by tissue culture.

- **Watermelon:** Pusa Bedana, Arka Madhura, Swarna, Shonima
- **Cassava:** Sree Harsha

References

Blakeslee, A.F. and Avery, A.G. 1937. Method of inducing doubling of chromosomes in plants by treatment with colchicine. *J. Hered.*, **28**:393–411.

Dar, T.H., Raina, S.N. and Goel, S. 2017. Cytogenetic and molecular evidences revealing genomic changes after autopolyploidization: a case study of synthetic autotetraploid *Phlox drummondii* hook. *Physiol. Mol. Biol. Plants*, https://doi.org/10.1007/s12298-017-0445-8.

Das, B., Hazra, P., Chattopadhyay, A., Chakraborty, A.K., Hazra, S., Kardile, H.B., Maji, A. and Chakrabarti, S.K. 2023. Characterization of traditional potato cultivars of North-Eastern region of India for nutritional and quality traits and ploidy level. *Indian J.Genet.,* **83**: 407-413.

Hazra, P. and Ghosh, R. 2001. Induced polyploidy as a breeding approach in pointed gourd. SABRAO *J. Breed. Genet.*, **33**: 47-48.

Nebel, B.R. 1937. Mechanism of polyploidy through colchicine. *Nature*, **140**: 1101

Zhang, W., Hao, H., Ma, L. and Zhou, C. 2010. Tetraploid muskmelon alters morphological characteristics and improves fruit quality. *Sci. Hort.,* **125**:396-400

9

Heterosis and Development of Hybrids

The effect of inbreeding and outbreeding was found in ancient time in animals and human beings since civilization. Hence, the phenomenon "Heterosis" was conceived since pre-Mendelian era. But in plants this phenomenon was recognized only 200 years ago. Koelreuter (1766) first published his scientific investigation on hybridization. Knight (1799) first recorded increased vigour following crossing. Mendel (1865) also observed increased vigour in pea hybrids. Charles Darwin (1876) suggested that cross-fertilization was beneficial whereas self-fertilization was injurious. However, the scientific concept of heterosis phenomenon was established in post-Mendellian era by the monumental work by East (1908) and Shull (1908).

General concepts of Heterosis

- Heterosis is the widely occurring genetically governed biological phenomenon involving both plants and animals.
- It frequently occurs in plants and is found mostly in cross-pollinated plant species than in self-pollinated plants.
- It is built in evolutionary mechanism favouring better survival of heterozygotes in nature.
- Not all hybrids show heterosis (*i.e.* superiority over parents) rather it is exhibited by a few and specific ones.
- It shows increase in size, vigour, seed producing capacity, disease and pest resistance, metabolic activity, stability, and many other characters.
- Heterosis is expressed in F_1 only and reduced in advanced generation, *i.e.* from F_2 onwards due to breakdown of specific gene combination.
- Expression of heterosis generally becomes more in the hybrids those are developed involving genetically diverse parental lines than genetically related lines.
- Heterosis is associated with specific combining ability of the cross combination.
- Heterosis manifested in the hybrids is unfixable from generation to generation however, it can be fixed through vegetative propagation of the hybrids.

Definition of Heterosis

Shull (1914) proposed the term "Heterosis" to replace the phrase "the special stimulus of heterozygosis". Shull (1948) later defined heterosis as the interpretation of increased vigour, size, fruitfulness, speed of development, resistance to diseases and to insect pests or to climatic rigors of any kind manifested by crossbred organism as compared to corresponding inbreds as the specific result of unlikeliness in the constitution of the uniting parental gametes.

Classical genetic basis of heterosis

Shull coined the term heterosis with the intention to provide a denomination free from any underlying theoretical concept. By doing so, he accounted for the controversy on the genetic principle of hybrid vigour. At that time, the dominance and over-dominance models had been suggested as alternatives to explain the genetic basis of hybrid vigour. The debate on these models is still ongoing to this day.

Dominance model of heterosis

The dominance model of heterosis is based on the hypothesis that in many genes slightly deleterious recessive alleles are complemented by superior dominant alleles in hybrids. The basic assumptions involved in this hypothesis are:

1. Genes governing vigour are dominant, beneficial in nature and the recessives are usually deleterious.
2. No epistasis *i.e.* there is complete additivity of gene effects between loci.
3. No linkage *i.e.* there is complete recombination between alleles at various loci.

According to this model, heterozygosity *per se* is not a prerequisite for heterosis. Instead, hybrid vigor is based on the combination of superior alleles at as many loci as possible.

For example: In a cross of two inbreds A and B having the genetic constitution AAbbCCdd and aaBBccDD, respectively the cross would be:

AAbbCCdd × aaBBccDD

(Inbred A) ↓ (Inbred B)

F_1 ---- AaBbCcDd (manifestation of heterosis)

As is apparent, the hybrid containing more number of favourable dominant genes and would be according to dominant hypothesis, more vigorous than either of the parent, which has comparatively less number of dominant genes.

The dominance model explains inbreeding depression as the accumulation of homozygous slightly deleterious recessive alleles by subsequent cycles of self-pollination. In this scenario, if deleterious alleles are purged over time and replaced by favorable alleles, as recently demonstrated for modern inbred lines compared to traditional landraces, a slight decrease of relative heterosis would be expected. However, the degree of heterosis (in maize, for instance) remained constant over the last decades despite the significant improvement of the vigour of the inbred line. An explanation for this, one that is not necessarily in conflict with the dominance model, is that linkage of loci representing deleterious and beneficial alleles might substantially hamper the enrichment of favorable alleles. Moreover, in modern inbred lines of maize it has been demonstrated that deleterious alleles are fixed in regions with low recombination rates, such as the centromere. However, the situation is further complicated when the analysis of heterosis is extended to polyploids, which exhibit progressive heterosis that cannot be explained by the dominance model. Progressive heterosis describes the scenario in which hybrid vigor generally increases with the number of distinct genomes present in allopolyploids. A similar observation has been made in wheat, where tetraploids and hexaploids display progressive increases in heterosis, even though the ratio of positive and negative alleles has not been altered in polypoid *versus* diploid plants. Moreover, autotetraploids in maize and the corresponding diploids display a similar rate of inbreeding depression over several generations, although homozygosity is reached much faster in the diploids than in the tetraploids. It has been demonstrated in maize that most deleterious alleles display incomplete dominance. Although the effects observed in polyploids cannot be explained by complementation of purely recessive alleles, the action of incompletely dominant alleles and gene dosage effects can contribute to the understanding of the genetic patterns observed in polyploids.

Principal shortfall of the Dominance hypothesis to explain heterosis

- Failure of selection to identify inbred lines as good as hybrids. Heterosis due to dominance should, in principle, be fixable by inbreeding. However, in practice, heterosis due to dominance is fixable only if the number of dispersed complementary genes is small. If the number of favourable and unfavourable allele pairs differentiating the parents is large, the probability of recovering an F_2 segregating line with all the desired favourable dominant alleles is extremely small. It has been argued that if only the accumulation of dominant genes dispersed in parents is required for heterosis, plant breeders would have been successful in assembling such dominant blocks in a single variety long ago, which is actually the basic shortfall of this hypothesis.
- Absence of the skewed F_2 frequency distribution of the heterotic character because in such a trait, the dominant and recessive loci would be distributed according to the expression of binomial $(3/4 + 1/4)^n$ where n is the number of genes involved in the expression of the character.
- It has been found that dominant traits are not always linked to productivity. Wild species possessing more dominant traits than cultivated crop plants are generally low yielders, indicating that accumulation of dominant genes alone cannot explain the heterosis phenomenon.

Over-dominance model of heterosis

This hypothesis was proposed independently by Shull (1908) and East (1908) and later advocated by Gustafson (1938), Stadler (1939) and Hull (1945). The over-dominance model assumes that the heterozygous condition of the alleles at a single locus is superior to either of the homozygous condition of the alleles at that locus. Thus, heterozygosity is the cause of heterosis, while homozygosity, resulting due to inbreeding, causes inbreeding depression. The theory also envisages that the alleles 'A' and 'a' supposedly produce favourable but different effects. A combination of these diverse effects in the hybrid 'Aa' tends to be more favourable than the single effect produced by either of the homozygotes 'AA' and 'aa'. Several examples of over-dominance have been reported in *Arabidopsis* species (*erecta* and *augustifolia*), maize and tomato where a heterotic trait was conditioned by a single gene. In these examples, the trait was superior in its heterozygous over its homozygous state.

Jinks (1983) however, opined that apparent over-dominance due to non-allelic interaction and linkage diseqilibrium is a common contributor to heterosis. Milborrow (1998) suggested that the growth of a plant may be limited by the genes that regulate certain metabolic pathways to lower levels than the maximum possible. Heterozygotes may partially escape this regulation since they have two slightly different alleles for these genes, allowing greater flow on these pathways. This is not over-dominance but, like the over-dominance hypothesis it predicts that heterozygotes have an inherent advantage in vigour that cannot be duplicated by any amount of selection in open-pollinated homozygous lines.

Principal shortfall of the Over-dominance hypothesis to explain heterosis

- The major shortfall to over dominance theory stems from the fact that most of the support and evidences obtained from this hypothesis have come from the cases of single locus heterosis.
- Heterozygosity cannot solely explain heterosis because in some species, for instance, highly homozygous modern inbred lines perform better than highly heterozygous hybrids developed some decades ago.
- Evidence of genuine over-dominance for quantitative characters is much more difficult to establish.

Epistatic interaction model of heterosis

Neither the dominance nor the over-dominance model can exclusively explain heterosis. However, they also do not exclude each other. This epistatic interactions hypothesis suggests that interlocus genetic interactions between different genes or epistasis play an important role as the genetic basis of heterosis and inbreeding depression. Jinks and Jones (1958) observed a significant association between epistasis and heterosis in 80 per cent of the crosses in *Nocotiana rustica.* However, they pointed out that not all heterotic crosses showed non-allelic interactions, and conversely, not all crosses showing non-allelic interactions were heterotic. Wright (1968) visualized that each gene replacement may have effects on many characters and based on this concept, epistasis should be one of the most important genetic components in the inheritance of quantitative characters. According to the concepts of physiological epistasis and statistical epistasis, the former occurs when genotypic values at one locus vary depending on the alleles present at other loci and the latter occurs when there are quantitative genetic variance components that can be attributed to interactions among genes. In this context, "compensatory gene complex" (CGC) hypothesis put forward by Strunnikov (1983) may be referred where high level of heterosis was recorded involving a parent exhibited genetically controlled low vitality.

Molecular biology related to heterosis

It is important to keep in mind that the genetic hypotheses were suggested before the era of molecular biology and cannot be directly associated with molecular principles

• Complementation at genomic level / level of gene expression

Heterosis has been extensively exploited but, despite a century of investigations, its genetic basis is not completely understood, yet. The main strategy of exploiting heterosis in maize and other crops is based on the "genetic distance" model where heterosis defined and measured as the superiority of the hybrid over the mid-parent is related to the genetic divergence between its parental lines. A number of studies have shown that the genetic distance of the parental lines determined either through multivariate analysis or molecular markers was correlated with heterosis in yield but the relationship in most of the cases was not strong enough for getting regression of heterosis on genetic distance of the parents to confidently predict the level of heterosis based on a given value of genetic distance of the parents as recorded in cowpea (Hazra *et al.*, 1993), brinjal (Hazra *et al.*, 2010) and rice (Zhang *et al.*, 2010). These findings indicate that there may be optimum level of genetic divergence between parents to obtain heterosis in the F_1 generation (Hazra *et al.*, 2010).

However, for the maize breeders of particular interest is to identify genetic factors contributing to heterosis as well as a suitable method that could predict heterosis with some accuracy before field evaluation of test hybrids based on genetic divergence of the parents. It has been shown in maize that hybrid vigour correlates with the genetic distance between the parental inbred lines based on SSR markers (Drinić *et al.*, 2012). In several other studies, the genetic diversity of maize genotypes is based on intensive structural diversity between different genotypes including copy number variation and presence/absence of variation.

• Single parent expression complementation

It has been demonstrated that ~10% of genes present in either of the maize inbred lines B73 and Mo17 are absent in the other genotype. The activity of genes absent in one parental genotype can be complemented in the hybrid by the second genotype. It has been shown that complementation occurs not only on the genomic level but also on the level of gene expression. In a panel of distinct inbred lines and their hybrids, ~1,000 genes that were active in the hybrid were only expressed in one of the two parental inbred lines. This phenomenon was designated as 'single parent expression complementation' (Baldauf *et al.*, 2018). Remarkably, only about 10% of these genes are accounted for by genomic presence/ absence variations. Likely, other classes of genomic variations such as single-nucleotide polymorphisms and insertion/deletion polymorphisms in the regulatory sequences of these genes might account for this observation. As a consequence of single parent expression complementation, it was demonstrated that hybrids expressed, on average, ~600 additional genes compared to the average of their parental inbred lines in the surveyed tissue and the hybrids benefit from single parent expression (SPE) complementation (Baldauf *et al.*, 2022). Intriguingly, the class of evolutionarily younger non-syntenic genes (that is, genes that emerged after the evolutionary separation of the lineages that led to maize and sorghum) were significantly over-represented (~60–65%) among the genes displaying single parent expression patterns relative to their genomic prevalence (38%) and prevalence among all expressed genes (~18–20%). It was hypothesized that the activity of these additionally active genes could be beneficial for hybrids, for instance in the adaptation to changing environmental conditions, because it has been demonstrated before that non-syntenic genes are often involved in such processes.

- **Gene expression complementation**

These instances of gene expression complementation of hundreds of genes fit into the dominance model of heterosis. However, one needs to keep in mind that these global gene expression patterns have not yet been linked to the manifestation of phenotypic traits. There are several examples in which single-gene over-dominance has been observed. For instance, heterozygosity of loss-of-function alleles of the SINGLE FLOWER TRUSS (SFT) gene in tomato increases yield by up to 60%. The observed effects are a consequence of the suppression of growth termination mediated by the SFT antagonist SELF PRUNING (SP) gene. Moreover, the expression of the maize cell number regulator genes ZmCNR1 and ZmCNR2 is negatively correlated with plant vigour for many traits related to plant growth. Hence, these genes display significantly higher expression levels in homozygous inbred lines compared to their heterozygous parents. If it could be demonstrated that these dosage-dependent genes and the molecular networks in which they are interwoven play a role in heterosis, they could be considered as examples for negative over-dominance.

- **Dosage effects and gene balance hypothesis**

Several observations including the experiments with polyploids and the identification of dosage-dependent regulators of plant growth suggest that gene dosage likely plays a role in heterosis manifestation. The theoretical framework underlying these observations is known as the gene balance hypothesis which was developed by Birchler (2003) and Birchler and Veitia (2007). According to this molecular model of heterosis, combination of different alleles produces an interaction that causes gene expression in the hybrid to deviate relative to the midparent predictions (*e.g.*, by an upregulation of many housekeeping genes). Regulatory genes in multicellular organisms often function as part of complexes, and exhibit some measure of dosage dependence whereas housekeeping genes or those that encode metabolic functions are less likely to show dosage effect. Quantitative traits are expected to be controlled in large part by multiple dosage-dependent regulatory loci which has led to this proposal that heterosis is the result of different alleles being present at loci that contribute to the regulatory hierarchies that control quantitative traits.

This concept does not oppose the dominance and over-dominance hypotheses, but it does provide an additional molecular explanation for observations that are not covered by these classical genetic concepts. In brief, the gene-balance hypothesis suggests that regulatory genes, such as transcription factors, often function as part of complexes that are sensitive to changes in the abundance, and thus dosage, of their different subunits. Non-additive gene expression as well as protein and metabolite patterns related to regulatory genes are frequently observed in hybrids and might be associated with dosage-sensitive molecular complexes. Since these complexes are often involved in multi-level regulatory hierarchical cascades, alteration of expression of such regulatory genes can result in the altered expression of a large number of downstream genes. A similar concept proposed by Goff (2011) suggested that hybrids achieve greater cellular energy efficiency by selectively regulating transcription and translation of the more stable, energy-efficient alleles in hybrids. Despite the large number of heterosis-related gene expression studies that have revealed non-additive, differential and allele-specific gene activity patterns, no consensus set or enriched functional categories related to heterosis have been identified. This can be attributed in part to the different experimental platforms used, as well as to tissue- and genotype-specific gene expression differences. Nevertheless, even if there are a limited number of regulatory genes that control heterosis manifestation, it could be very difficult to identify them because their activities might be masked by the plethora of downstream genes affected by them. Moreover, the interpretation of heterosis-related transcriptome, proteome and metabolome data sets is hampered by the fact that hybrids and their parental inbred lines are often compared at developmental stages when the hybrids are substantially more vigorous than their parents. Hence, a substantial number of the expression differences observed might be the result of growth differences rather than directly related to processes involved in the manifestation of heterosis. In such analyses, it is therefore virtually impossible to separate heterosis-related and growth-difference conditioned effects.

• Epigenetic factors for heterosis

Although many hypotheses have been proposed to explain the mechanisms of heterosis in plants at the genetic level, studies have shown that the genetic mechanisms of heterosis cannot be fully explained by one or even several hypotheses at the genetic level.Genotype-specific epigenetic variation and unique epigenetic constitutions of hybrids have their possible role in heterosis manifestation. Through the intensive study of epigenetics, epigenetic factors such as DNA methylation, small RNAs, and histone modifications have been found to be involved in the development of heterosis in plants (Zhao *et al.*, 2007; Shivaprasad *et al.*, 2012). Epigenetic modifications play an important role in the formation of plant phenotypes by regulating gene transcription and gene expression (Kenchanmane and Niederhuth, 2018). The degree of DNA methylation as a possible molecular regulator of heterosis suggests a gradual accumulation of methylation during inbreeding or selfing, which is then released and/or re-patterned when these lines are crossed to generate hybrids (Stuber, 1992; Tsaftaris and Kafka, 2008). Such epigenetic modifications could be heritable and influence diverse developmental processes in plants, including flowering and seed development.

- **Enzyne activities and protein polymorphisms**

Stuber (1992) discussed several cases where heterosis was analysed with respect to gene and enzyme activities. Demonstrations of multimeric hybrid enzymes or isozyme spectrum due to non-allelic or intergenomic complementation provide evidence for a buffering mechanism at the subcellular level in the hybrids. Leonardi *et al.* (1991) suggested that genes controlling protein amounts, particularly those with multiple effects, directly influence the expression of hybrid vigour.

- **Inter-genomic interactions**

It has been demonstrated with corn, barley and wheat that in single-cross hybrids oxidative phosphorylation activity of mitochondria is similar to that of mixtures of parental mitochondria. In most examples studied where these hybrids exhibited heterosis for a productive trait there was also heterosis for oxidative phosphorylation of mitochondria. It has been shown that heterosis in a hybrid of two *Drosophila* lines is correlated with heterosis of mitochondria. Mechanisms for explaining mitochondrial heterosis include formation of "hybrid" mitochondria, production of heterotic mitochondrial enzymes and introduction of missing cytochromes. Oxidative phosphorylation information on mixtures of mitochondria may be useful in assessing lines for cross performance (McDaniel and Sarkissian, 1970). The phenomenon of chloroplast complementation was attributed to higher photosynthetic rates in isolated chloroplasts of hybrids of several crop species observed at the seedling stages.

Conclusion

It is clear from the foregoing discussion on the various aspects of heterosis that there is yet no simple or single explanation of heterosis. According to Schnell (1982), this phenomenon is widely used yet little understood phenomenon in plant breeding. However, the introduction of hybrids has been one of the most important innovations in modern agriculture. The exploitation of heterosis by constant improvement of hybrid yield and environmental stability is one pillar of global food and nutritional security. Despite the agronomic and thus economic success of hybrids, to date no unifying genetic or molecular theory can explain hybrid vigor completely. All modes of gene action—additivity, dominance, over-dominance, and epistasis could be perceived at the molecular level, but the interpretation of those molecular effects to final phenotype remains complex and largely undefined. One explanation for this is simply that there is no single principle underlying this phenomenon. Alternatively, this principle might be masked by the genetic diversity between and within plant species or by trait-specific networks of gene expression, proteome accumulation and metabolome activity. Today, heterosis is extensively exploited without detailed knowledge of the underlying molecular principles. The identification of the genetic and epigenetic framework underlying heterosis would be a landmark discovery revolutionizing modern plant breeding by allowing the optimal exploitation of crop productivity and stability.

Exploitation of heterosis in vegetable crops

Exploitation of heterosis in crop plants is considered to be one of the landmark innovations in modern agriculture and horticulture, and is of outstanding economic importance.

- It is the unique way of translating biological science to industry
- Incredible value of the hybrid in enhancing yield and adaptability
- Income and employment generation in the rural sector through manual production of hybrid seeds in many vegetable crops particularly, tomato, brinjal, sweet pepper, watermelon, musk melon, bottle gourd, ridge gourd, okra, etc.
- Earning of foreign exchange through export of hybrid seeds

Maize, which provides the highest annual global kernel yield of all crop species, is exclusively hybrid based in the major corn-producing countries in the world. In vegetable crops, exploitation of hybrid vigour was first recorded in cucumber (Hayes and Jones, 1916) but first commercial hybrid was developed in brinjal in Japan (Kakizaki, 1931). By 1930, breeders successfully introduced the first commercial hybrid of maize in the United States, and by the early 1940s, every commercial operator in the country was planting hybrid maize. Whereas maize yield had remained essentially unchanged for decades, a constant increase has been observed since the introduction and improvement of hybrids, accompanied by more efficient agricultural practices. From 1950 onwards, in addition to maize, hybrids of other crops including different vegetable crops have substantial market shares in different regions of the globe. Different major genetic and physiological consequences have led the way to establishing "Hybrid vegetable Industry" in different countries of the world.

- First record to exploit hybrid vigour in vegetable crop, Cucumber: Hayes and Jones (1916)
- First commercial hybrid in vegetable crops, Brinjal : Kakizaki (1931)
- First utilization of self-incompatibility system, Cabbage: Pearson (1931)
- First utilization of cytoplasmic - genic male sterility mechanism, Onion: Jones and Clarke (1943)

- First identification of gynoecious sex form, Cucumber: Peterson and Andher (1960)
- First identification of cytoplasmic male sterility, Radish: Ogura (1968)
- First ethylene-induced femaleness in cucurbits: Rudich *et al.* (1969)
- First identification of cytoplasmic - genic male sterility mechanism in carrot: Morelock *et al.* (1996)

Data Bridge Market Research analyses reported that the global hybrid fruits and vegetable seeds market which was USD 19.34 billion in 2022, is expected to reach USD 42.15 billion by 2030, growing at a CAGR of 10.2% during the forecast period of 2023 to 2030. The Indian vegetable seed market size is estimated at USD 700.99 million (Equivalent INR 6000 crore) in 2024, and is expected to reach USD 970 million (Equivalent INR 8000 crore) by 2030, growing at a CAGR of 5.56% during the forecast period (2024-2030).

Source:https://www.mordorintelligence.com/industry-reports/vegetable-seed-market-india

Main reasons for rapid growth of vegetable hybrid seed industry

- Continuous flow of new hybrids with high product quality, better adaptability and fortified with resistance to biotic and abiotic stresses.
- Continuous development of the parental lines with unique traits through marker assisted selection, utilization of crop wild relatives, mutants, etc.
- Development and utilization of different pollination control mechanisms for facilitating hybrid seed production.
- Usage of new molecular technologies in breeding like, gene insertion, gene editing, etc.
- Adoption of different seed technologies like, seed coating, priming, pelleting, etc.

Estimation of Heterosis

The phenomenon of heterosis is simply the reverse phenomenon of inbreeding depression. Inbreeding leads to a dispersal of gene frequencies among the lines. The lines becoming differentiated in gene frequencies as inbreeding proceeds and the coefficient of inbreeding is a means of expressing the degree of differentiation. Heterosis, just like inbreeding depression depends on degree of dominance. Loci without dominance cause neither inbreeding depression nor heterosis. The amount of heterosis following a cross between particular lines or populations depends upon the square of the difference of gene frequencies between two populations. If the populations crossed do not differ in gene frequency, there will be no heterosis and the heterosis will be greatest when one allele is fixed in one population and other allele in other populatiopn. If some loci are dominant in one direction and some in other direction, their effects will tend to cancel out and no heterosis may be observed in spite of the dominance at the individual loci. Thus, heterosis depends on directional dominance. In case of F_2 of a particular cross of a two parent populations, the genotypic frequency will follow the Hardy Weinberg principle since the F_2 is made by random mating among individuals of F_1. The gene frequency of F_1 being the mean of the gene frequency in two parent populations hence the amount of heterosis shown by F_2 is the difference between the F_2 and the mid parent values.

In fact, heterosis shown by F_2 is only half that is manifested by the F_1. The change of mean from F_1 to F_2 may, therefore, be regarded as inbreeding depression. In a particular cross, heterosis is usually measured in terms of two parameters namely:

(i) Heterosis over the mid parent (MP)

(ii) Heterosis over the better parent (BP)

In plant breeding programme it is also estimated in terms of heterosis over the check or standard variety/hybrid. This value is the most important from the point of view of commercial exploitation of heterosis for a particular crop.

1. Heterosis over the MP is commonly known as relative heterosis (RH). It can be calculated by $\frac{F_1 - MP}{MP} \times 100$ where, F_1 and MP are the average performances of F_1 and Mid parental values, respectively.
2. Heterosis over better parent is commonly known as Heterobeltiosis and can be calculated by $\frac{F_1 - BP}{BP} \times 100$ where, F_1 and BP are the average performances of F_1 and better parent, respectively.
3. Heterosis over the check or standard variety /hybrid is known as standard heterosis and can be calculated by $\frac{F_1 - CV}{CV} \times 100$ where, mean of F_1 and the standard variety /hybrid are used for calculation.

Relative heterosis, heterobeltiosis and standard heterosis are measured as proportions of deviation of F_1 values from mid parent, better parent and check variety/hybrid, respectively. After measuring by any of these methods, the variation needs to be established statistically.

1. $SE(MP) = \sqrt{\frac{3\,Me}{2r}}$
2. $SE(BP) = \sqrt{\frac{2\,Me}{r}}$
3. $SE(CV) = \sqrt{\frac{2\,Me}{r}}$

Where, Me = error mean squares emanated from the trial using Randomised Block Design and r = number of replications.

Manifestations of hybrid vigour

Hybrid vigour, *i.e.*, superiority of the hybrids over the parents is manifested in various aspects, however, some of the important characters with respect to acceptability of the hybrids are mentioned below.

- **High productivity:** The major attribute of the F_1 hybrid is their increased yield as compared to the best homozygous/open-pollinated varieties. In vegetable crops, like tomato, brinjal, sweet pepper, cucumber, muskmelon, watermelon, cabbage, cauliflower, summer squash, etc. heterosis in yield generally ranges between 40-50 per cent and sometimes it is even more than 60 per cent. Yield of a crop is the cumulative expression of different yield components hence, heterosis in yield is manifested through heterosis in different yield components. As for example, high yield in the hybrids of tomato, brinjal and chilli is realized mainly through heterosis in plant vigour, fruit set percentage, fruit number per plant, fruit weight and fruiting duration. Similarly, high yield in the hybrids of carrot, radish, onion, cabbage, cauliflower, cucumber, muskmelon, watermelon, etc. is manifested through heterosis in the yield components, like root length and girth in carrot, uniform bulb size and bulb weight in onion, head/curd size and weight in cabbage and cauliflower, sex ratio and number of fruits per plant in cucumber, muskmelon, watermelon, etc.
- **Uniformity:** Uniform produce of the hybrids is highly valued, especially in cross-pollinated crops like onion, cabbage, cauliflower, carrot, etc. Large variations in shape, size and colour of bulb/root and shape, size and compactness of curd/head are common features of most of the improved open-pollinated varieties of onion, cabbage, cauliflower, carrot, etc. Total yield as well as marketability of the produce is reduced due to these variations. Uniformity is attained because the population of a hybrid is heterozygous in the similar loci and for this reason, is homogeneous. Uniform products of the hybrids are marketed with greater advantage.
- **Improved quality:** Hybrid vigour in most of the vegetable crops is exploited for higher early and total yield and uniformity of the produce. Some hybrids also show worth mentioning nutritional and processing qualities. Sometimes, better quality may not be accompanied by increased yields. Some other quality criteria like fruit with thick pericarp, high lycopene content and high total soluble solids in tomato, medium-long, cylindrical fruit in bottle gourd, high lycopene and sugar contents in watermelon, high β carotene content in carrot and high total soluble solids contents in onion are of much value as far as processing and marketing of the produce are concerned. Enhanced shelf-life and high transportability are also considered as important quality characters of these vegetable crops
- **Built-in resistance:** Some hybrids exhibit greater resistance to insect pests and diseases. Resistance to several diseases in different vegetable crops is conditioned by single dominant gene, *e.g.* resistance to *Fusarium* wilt, *Verticillium* wilt, early blight, late blight, tobacco mosaic virus and nematode in tomato. Disease resistance, even multiple disease resistance can be realized in the hybrids due to supplementary allelic action in them. Some of the genes are even expressed in better way under heterozygous than homozygous condition at high temperature condition (*e.g.*, *Mi* gene for resistance to nematode and *Tm* gene for tobacco

mosaic virus in tomato). Resistant lines derived by introgression from wild relatives are also used as parents to develop resistant hybrids.

- **Environmental adaptation:** Hybrids of some vegetable crops are found more adaptive to different environmental conditions than the improved open-pollinated varieties. This manifestation is relevant to the hypothesis "Interaction of gene products" in the hybrids. Tropical cabbage and cauliflower hybrids have special qualities of high temperature tolerance and long stay in the fields, and such hybrids are very much suitable for milder climatic condition of eastern India. High and cold temperature tolerant tomato hybrids have also been developed.
- **Earliness:** Earliness is a critical character of both open-pollinated varieties and hybrids. Early flowering or early maturity judges this highly desirable character. Some hybrids of tomato, brinjal, muskmelon, watermelon, onion, and cabbage show conspicuous earliness.

Important steps to be taken to popularize vegetable hybrids in India

At present, the hybrid seed business is totally controlled by different multinational corporate sectors and few big National companies. They have been marketing the hybrids either directly imported and relabeled or produced indigenously. Capitalizing the advantages of "open general license" for the import of both hybrid seeds and seeds of the parental lines, hybrid seeds are also produced indigenously involving exotic parents. For this reason, these indigenously produced hybrids often show lack of adaptation and at the same time, susceptibility to some diseases. Many Corporate and National companies have developed strong research and development sector in India to develop well-adapted hybrids resistant to biotic and abiotic stresses. However, infrastructural facilities, allocation of technical manpower and development of hybrid seed production technology should be promoted in the public sector agencies. A long-term arrangement may also be made between the research institutes (both ICAR Institutes and Agricultural Universities) and private seed companies for collaborative research and development work. In other words, coordinated efforts with well-planned strategies by the breeders, vegetable scientists, private seed companies and extension service agencies are essentially needed to develop indigenous hybrid seed industry in vegetable crop.

- **Cost-effective technology:** Hybrid seed production technology should be cost-effective, so that the cost of the hybrid seed remains within the reach of the average farmers of our country. In the present situation, where hybrid seed industry is in the grip of a few corporate and private organizations, Government should take necessary steps, so that economic methods of hybrid seed production with the involvement of indigenous parental lines is promoted in the public sector seed organizations.
- **Training programmes:** Hybrid seeds of most of the vegetable crops are produced manually by the farmers with the use of specific parental lines under the control of different seed companies. Hence, training programmes for

the production and management of hybrid seeds for the farmers need to be strengthened to bring more area under hybrid seed production.

- **Government check on the import of hybrid seeds:** Under the new seed import policy, Union Government is allowing free import of hybrid seeds and exotic parents for the production of hybrid seeds. Total dependence on such imports may strangle indigenous hybrid breeding research. The new seed policy can serve only as a catalyst. Indiscriminate import may not be worthy in the long run as far as adaptability and resistance attributes point of view. Import of the hybrids resistant to the diseases and pests not important in India may be discouraged. Seed companies should be asked to test the new hybrids under Indian condition before marketing. There must be a close surveillance regarding performance of imported hybrids. It is to be mentioned that black rot disease (caused by the bacteria *Xanthamonas campestris* pv *campestris*) and sclerotinia rot disease (caused by the fungus *Sclerotinia sclerotiorum*) of cabbage and cauliflower once came through the seeds is now assuming serious proportions.
- **Intensification of research:** Research works on heterosis breeding of vegetable crops should be intensified and at the same time, well-coordinated in the different research institutes and Agricultural Universities. Some of the focused research areas are:
    - Identification of superior hybrid combinations involving indigenous and adaptable parental lines.
    - Integration of heterosis and resistance breeding for not only resistance to diseases and pests but also resistance to environmental stresses.
    - Identification of stable self-incompatible and cytoplasmic male sterile lines in cabbage and cauliflower.
    - Identification of suitable genic-cytoplasmic male sterile lines of chilli to develop suitable hybrids.
    - Identification of seedling markers to maintain sporogenous genic male sterility in different vegetable crops.
    - Development of cytoplasmic-genic male sterile lines in pink bulb cultivars of onion to develop onion hybrids with pink bulb colour, pungency, uniformity in bulb size and good storage capacity.
    - Utilization of functional male sterile lines to economize hybrid seed production of tomato and brinjal.
    - Development of suitable parental lines with monoecious sex form in place of usual andromonoecious sex form in muskmelon to reduce the cost of hybrid seed production.
    - Commercialization of the use of growth regulators for the suppression of staminate flowers for considerable period of time to reduce the cost of hybrid seed production in cucurbitaceous vegetable crops.

- Development of tropical gynoecious sex form for the production of promising hybrids of cucumber.
- Identification of suitable locations where fertility could be restored in an otherwise sterile line (such as in carrot).

Table 1: Targeted characters in the hybrids of different vegetable crops

Hybrid	Targeted characters
Tomato	Determinate growth habit, dual purpose and processing quality, good transportability, high yield, resistant to high temperature + TLCV+ early blight + bacterial wilt Indeterminate growth habit with extended harvest span Determinate growth habit , flattish-round, high acid fruits
Okra	Attractive light green to green fruit, tender, thin, medium long, smooth, 4-5 ridged, free from conspicuous hairs, resistant to BYVMV and OELCuV disease
Cabbage	High temperature tolerant, firm, semi-flat heads, weighing within two kg and with good field holding capacity, resistant to black rot + diamond back moth
Chilli	Very pungent (>70,000 Schoville Heat Units) for fresh market and as dry powder, medium pungent (30,000 SHU) for pickle making, resistant to leaf curl + thrips + mites + anthracnose Low pungency coupled with attractive deep red colour
Cauliflower	High temperature tolerance, non-riceyness, compact, bract free and self-blanching curd with retentive cream/white colour, resistant to black rot
Watermelon	Jubilee pattern type, large, elongated melons with finely-textured, bright pink-red flesh and sweet, resistant to WBN + Fusarium wilt Sugar Baby and Icebox type with 3 to 3.5 Kg fruit weight, blackish green colour skin, bright pink-red flesh very good transportability, TSS 13°Brix and above, fine texture
Cucumber	Early maturing, nice green or dark green poinsett type fruit, uniform long cylindrical shape without crook-neck , prolific yielder, resistant to CMV + PM + DM
Bottle gourd	Attractive green fruit of round/cylindrical shaped with sparse hairs persisting on skin and non-fibrous flesh at edible stage, resistant to DM + PM
Bitter gourd	Spindle shaped green fruit with medium prickle on fruit surface with extended harvest span, resistant to Mosaic + DM
Muskmelon	Consistent yield, netted cantaloupe, honeydew, sutured melon and Galia types, better adaptability and shipping qualities, resistant to *Fusarium* wilt + CMV + PM + DM
Brinjal	Round with purple and white stripes, green long, purple long, green round and large purple oval fruits, green calyx, resistant to Phomopsis blight + Bacterial wilt
Pumpkin	Fruits with high flesh recovery and β carotene content

TLCV: Tomato leaf curl virus; RKN: Rot knot nematode, WBN: Watermelon bud necrosis, PM: Powdery mildew, DM: Downey mildew, CMV: Cucumber mosaic virus, BYVMV: Bhindi yellow vein mosaic virus; OELCuV = Okra enation leaf curl virus.

- **Development of infrastructure for vegetable production:** Hybrids of several vegetable crops, like cucurbits, tomato, sweet pepper, cabbage, etc. can be grown successfully in off season. Development of infrastructures like, polyhouse, poly tunnel, net house, etc. with drip irrigation system is required to capitalize the advantages of high production of the hybrids.
- **Improvement in marketing of vegetables:** The farmers should be given the opportunity of capitalizing the advantage of enhanced production by improving marketing, storage and processing facilities of vegetables.

Methodology in hybrid breeding

Hybrid is the first generation (F_1) progeny of a cross between two selected and genetically diverse parents and the superiority of the performance of a hybrid over its parents (s) is termed as heterosis.

- **Combining ability of the parental lines:** Choice of the parental lines is the most critical component in any plant breeding programme including hybrid breeding. Value of an inbred line in hybrid breeding programme ultimately depends on its ability to combine well with other lines to produce superior hybrids.
- **Development of heterotic pools:** Heterotic pools can be identified by evaluating testcross performance among available germplasm in the field trials. However, the large number of parental lines available in any breeding programme makes the evaluation of all the possible crosses impractical. Hence, to overcome this problem, Melchinger (1999) suggested the following steps

  (i) Clustering germplasm based on genetic similarities using molecular markers

  (ii) Selecting representative genotypes from each subgroup

  (iii) Evaluating crosses among the representative genotypes in field trials

  (iv) Identification of heterotic pools based on the *per se* performance, combining ability and heterosis.

Heterotic pool need to be made as divergent as possible, because heterosis depends on the differences in allele frequency between two populations. Genes should be concentrated for contrasting traits in two heterotic pools.

Hybrid seed production technology

- **Hand emasculation and hand pollination:** In crops having bisexual flowers like tomato, brinjal, chilli, sweet pepper and okra and in andromonocious varieties of muskmelon male part in bisexual flower is emasculated and pollinated manually by hand. It involves different steps like emasculation, bagging or cotton wading, removal of bags/cotton wads, hand pollinations and subsequent bagging.
- **Hand pollination without emasculation:** This method is followed in the crops where staminate and pistillate flowers are borne separately in the same plant as in monoecious cucurbits or in self-pollinated crops where male part of bisexual flowers is non-functional due to different forms of male-sterility. This method

of hybrid seed production has been exploited commercially in watermelon, summer squash, bottle gourd, bitter gourd, ridge gourd, pumpkin, cucumber, etc. For production of hybrid seed in monoecious cucurbits, pistillate flowers of the intended female parent and staminate flowers of the intended male parents are bagged/enveloped separately before opening and hand pollination is done subsequently after opening of the pistillate and staminate flowers inside the respective bag/envelope.

- **Emasculation and wind pollination:** This method is generally employed in monoecious crop like sweet corn. Hybrid sweet corn seed production involves deliberately crossing a female parent population with a male parent in isolated fields. Thus, from the very start of hybrid seed production, the identity and arrangement of the two parent populations determine the outcome. The field management of the two parents is also important and requires attention to timing of planting, elimination of off-types, removal of tassels (male flower) near the apex of the plant from the female line before pollen shedding, separate harvesting of the seeds from the female line and careful extraction and processing of the seeds to maintain seed quality
- **Removal of male plants and wind pollination:** This method is generally employed in dioecious crops like spinach and asparagus. In this method, male plants from the intended female population are uprooted before opening of flowers and free wind pollination in open is allowed to produce the hybrid seeds. This hybrid seed production method is easy and economical as no labour is required for pollination.
- **Hand emasculation and free insect pollination:** This method is used mostly in monoecious cucurbits like cucumber, watermelon, summer squash, pumpkin, bitter gourd, ridge gourd, etc. where intended male and female parents are planted alternatively in commercial hybrid seed production plots in a 1: 2 ratio and staminate flowers of the intended female parents are pinched off before opening regularly and free insect pollination in isolation is allowed.
- **Use of chemicals and free insect pollination:** Hybrid seeds can also be produced in different cucurbits by the application of chemicals for suppression of staminate flowers in the vine for at least a month. Generally, spraying of ethrel (2-choloro-ethylphosphonic acid) @ 200-300 ppm at two and four true leaf stage and another at flowering is useful for suppression of staminate flowers and induction of pistilate flower successively in first few nodes on the intended female parent in bottle gourd, pumpkin and summer squash. The row of the untreated vine which is regarded as intended male parent is grown side by the side of the intended female in isolation and natural cross pollination by insect pollinators is allowed. Hand pollination can also be done to produce hybrid seed.

Free insect pollination without emasculation

It is one of the most important methods particularly in the crops where emasculation is tedious as in onion and carrot due to umbellate inflorescence and in crops where each pollination results in only a small number of seeds as in cabbage, cauliflower, Brussels sprouts, broccoli, kale, radish, beet and chilli. Hybrid seed production in these crops is accomplished through utilization of inherent pollination control mechanisms like, self-incompatibility, male-sterility and gynoecious sex form. Self-incompatibility and cytoplasmic male sterility in cabbage, cauliflower, Brussels sprouts, broccoli, kale and radish; cytoplasmic-genic male sterility in onion, carrot, beet and chilli and gynoecious sex form in cucumber have been exploited commercially to overcome laborious manual emasculation process and cross-pollination is accomplished mainly by insect pollinators. In chilli, if considerable natural cross-pollination does not take place on the male sterile plants supplementary hand pollination is often needed to produce sufficient quantity of hybrid seeds.

Table 2: Commonly utilized mechanisms/methods for developing commercial hybrids

Mechanism	Commercially exploited crops
Hand emasculation + MP	Tomato, eggplant, okra, sweet pepper, muskmelon
Bagging/protection of staminate and pistillate flowers + MP	Bottle gourd, pumpkin, watermelon, bitter gourd, cucumber, ridge gourd
Pinching of staminate flowers + MP/NP	Bottle gourd, pumpkin, watermelon, bitter gourd, cucumber, ridge gourd
Male sterility + MP	Tomato, sweet pepper, chilli
Male sterility + NP	Onion, cabbage, cauliflower, carrot, radish, muskmelon, chilli
Self-incompatibility +NP	Cabbage, cauliflower, broccoli
Gynoecism + NP	Cucumber
PGR and pinching of staminate flowers + NP	Summer squash, cucumber, bottle gourd, etc.

MP: Manual pollination, NP: Natural pollination, PGR: Plant growth regulator application (generally ethephon)

Utilization of male sterility in hybrid seed production

Male sterility, the otherwise undesirable phenomenon is considered to be most important for commercial hybrid seed production of many cross-pollinated vegetable crops like onion, carrot, radish, cabbage, cauliflower, etc. which produce numerous but very small flowers and each pollination results few hybrid seeds. The male sterile lines eliminate the tedious emasculation procedure and make it possible to produce hybrid seeds of the cross-pollinated crops in open pollinated condition. Male sterile lines with marker characters further facilitate identification of the undesirable male fertile plants in the population even at seedling stage.

The review of Sanjeet Kumar *et al* (2000) describes histological, biochemical and molecular mechanisms of nuclear (GMS) and nuclear-cytoplasmic (CGMS) male sterility in tomato, chilli, sweet pepper, cole crops, radish, carrot, onion, muskmelon, etc. The roles of tapetum, callase, esterase and plant growth substances (PGSs) have also been discussed in the light of male sterility expression. Similarly, the role of sterile cytoplasm specific ORFs, namely T-cytoplasm of maize (orf-13) and Ogura cytoplasm of radish (orf-138) has also been described. Nuclear male sterility in tomato, brinjal, chilli, muskmelon and sweet pepper and nuclear-cytoplasmic male sterile lines in chilli, onion and carrot are available, which are being utilized either to produce seeds of commercial hybrids or develop cost effective experimental crosses.

Table 3: Nuclear male sterility (GMS) in vegetable crops

Crop	Salient features of male sterility
Tomato	More than 55 recessive genes have been reported; *sl-2, ms-13 and ms-15* are temperature sensitive, *ps-2* gene for functional sterility has been exploited at commercial scale
Brinjal	Monogenic recessive gene has been reported; monogenic recessive functional sterility is available.
Chilli	More than 12 recessive genes have been reported, *MS-12* (*ms-509/ms-10*) and *ms-3* genes are commercially utilized in India and Hungary, respectively. The *ms-10* gene is linked with taller plant height, erect growth and dark purple anther.
Cauliflower	Both recessive and dominant genes have been reported.
Cabbage	Both recessive and dominant genes have been reported
Brussels sprout	Recessive male sterile mutant has been reported.
Broccoli	Six recessive non-allelic genes have been reported; linkage of *ms* gene with bright green hypocotyls.
Watermelon	Recessive mutants have been reported; linkage of *ms* gene with delayed-green (*dg*) seedling marker gene.
Muskmelon	Five recessive non-allelic genes have been reported; *ms-1* is commercially utilized in India.
Cucumber	Monogenic recessive gene has been reported; limited scope of utilization because of the availability of gynoceous lines.
Summer squash	Monogenic recessive gene has been reported; very limited scope of utilization because of the availability of sex regulating mechanism using plant growth regulator, particularly ethephon
Onion	Monogenic recessive gene has been reported, but not utilized commercially because of availability of CGMS lines
Carrot	Recessive male sterile genes have been reported but not utilized at commercial scale because of the availability of CGMS lines.
Radish	Three recessive mutants have been reported but commercially not utilized.

In tomato, approximately 19 genic male sterile based hybrids have been released in Europe and 17 of them are based on functional sterility controlled by positional sterility (*ps*-2) gene. The *ps*-2 mutant is being utilized at commercial scale in few European countries like Czech Republic, Moldova and Bulgaria. However, under tropical climatic condition of West Bengal, India both *ps*-2 and *ps* functional male sterile lines showed inconsistent functional male sterility character under open condition (Hazra *et al.*, 2009). The *ms-10*^{35}gene is linked with a recessive marker gene (*a*) responsible for absence of anthocyanin. Hence, *ms-10*[35] sterile plant can be identified at seedling stage and fertile plant can be rouged out in the nursery itself. The functional male sterility in brinjal line, UGA-1MS is manifested through anther non-dehiscence particularly under poly house condition but in open condition, only 63.5% plants on an average exhibited anther non-dehiscence character which restricts its use in hybrid seed production following no emasculation and hand pollination method (Hazra *et al*., 2008).

Table 4: Cytoplasmic-genic male sterility (CGMS) in vegetable crops

Crop	**Salient features of male sterility**
Tomato	Sterile cytoplasm has been derived from the distant wild species through protoplasm fusion; restorer gene (*Rf*) is not available.
Chilli and sweet pepper	First reported in an Indian accession; most of the lines are temperature sensitive; occurrence of *Rf* allele is common in small fruited (chilli) and *rf* in large fruited (usually sweet pepper) lines.
Cauliflower, Cabbage, Brussels sprout, Broccoli	In cole crops, sterile cytoplasm has been derived from *B. nigra* and *Raphanus sativus* (Ogura type); problem of seedling yellowing at low temperature is associated with Ogura based CGMS lines of broccoli, cauliflower, cabbage, Brussels sprout has been solved using protoplast fusion. Hybrid lines of cabbage and cauliflower are being utilized by different seed companies; protoplast fusion has been utilized to transfer Ogura cytoplasm from broccoli into cabbage.
Onion	Two types of sterile cytoplasms, *viz*., *S* and *T* are available; *S*-cytoplasm is most widely exploited commercially
Carrot	Two types (petaloid and brown anther) of male sterile lines are available; genetics of fertility restoration is complex because structural variants of mitochondrial (mt) DNA are numerous.
Radish	Sterile cytoplasm is widely distributed in wild radish; occurrence of *Rf* allele is frequent in European and Chinese cultivars and *rf* in Japanese cultivars.

Utilization of male sterile lines in the hybrid seed production involves the following steps

- Identification or development of male sterile line
- Transfer of male sterility character into the otherwise desirable variety by repeated back-crossings which will be used as female parent in hybrid seed production
- Propagation of male sterile lines by developing maintainer line
- Production of hybrid seeds in open or by hand pollination

Problems in the utilization of male sterility in vegetable crops

- Cytoplasmic-genic male sterile lines are often sensitive to environment. Thus change of environment may lead to restoration of fertility in the male sterile line. So, hybrid seed production needs to be carried out in suitable set of environments.

- Maintenance of male sterile stock, particularly genic male sterile lines is cumbersome because of the rouging of male fertile plants from the population following progeny test.
- Cytoplasmic-genic male sterility is maintained by crossing male sterile line with the male fertile maintainer line but, with the continuation of this process, male sterility may break partially or completely because normal cytoplasm from male fertile plant is incorporated into the male sterile line, although in very small amount, in each crossing.
- In self-pollinated crops, like tomato, brinjal, sweet pepper, etc., very poor fruit set is resulted in the male sterile line (female line) by natural outcrossing which necessitates hand pollination. Even hand pollination in the male sterile line of self-pollinated crops sometimes do not result satisfactory fruit set.

Utilization of self-incompatibility in hybrid seed production

Self-incompatibility provides a natural mechanism against self- or sib-fertilization. It allows natural hybridization between two dissimilar lines without either emasculation or employing male sterility mechanism. Self- incompatibility is sometimes more advantageous than male sterility because visits of the pollinators to the male sterile plants are limited because of less attraction due to non-functional pollens in the male sterile plants resulting less hybrid seed set. On the other hand, normal and functional pollen grains are produced in the self-incompatible plants, which amply attract the pollinators, so that natural hybridization is accomplished successfully. Self-incompatibility system is generally utilized in hybrid seed production of cole crops and radish and it involves the following steps:

• Development of homozygous self- and sib-incompatible lines

In cole crops and radish, in which this self-incompatibility system is operative, all the plants are heterozygous for *S* alleles in the locus. Homozygous self- and sib-incompatible but cross-compatible lines are developed by selfing some selected superior plants of the elite variety following bud pollination.

S_1S_2 (genotype of the selected plants)
selfing by bud pollination
↓
$S_1S_1 : S_1S_2 : S_2S_2$ (segregation in 1:2:1 ratio)

Hence, in first selfing, 50% plants will be homozygous for *S* allele and rest 50% are heterozygous. The S_1S_1 and S_2S_2 plants are self-incompatible but cross compatible. However, for producing hybrids utilizing self-incompatibility the degree of incompatibility alleles and their dominance is of prime importance. Furthermore, conditioning of incompatibility by environment must be considered properly before utilization.

• Testing the combining ability

On the basis of combining ability, *i.e.*, relative ability of the parents to transmit desirable characters to its cross, the inbred lines are selected as parents for the hybrid seed production. Specific combining ability is more reliable for selecting parental inbred lines for hybrid seed production. Generally single, double, three way crosses, top cross, diallel cross and polycross mating designs are adopted for testing the combining ability.

• Production of hybrid seeds

The homozygous and strong self-incompatible but cross-compatible lines are grown in isolation for natural crossing by insect pollinators. The hybrid seeds may be produced by:

Single cross: Two self-incompatible lines are grown in alternate rows and hybrid seeds are harvested from both the parents. The hybrid seeds produced in both the lines are identical genetically excepting the maternally inherited variations.

$$S_1S_1 x S_2S_2$$
$$\downarrow$$
$$S_1S_2 \text{ (hybrid)}$$

Double cross: Four homozygous self-incompatible lines are required for double cross or four-way cross.

$$S_1S_1 \text{ x } S_2S_2 \qquad\qquad S_3S_3 \text{ x } S_4S_4$$
$$\downarrow \qquad\qquad\qquad \downarrow$$
$$S_1S_2 \text{ x } S_3S_4$$
$$\downarrow$$

Double cross hybrid seed

Triple cross: Here six self-incompatible inbred lines are required for the hybrid seed production.

$$S_1S_1 \text{ x } S_2S_2 \qquad\qquad S_4S_4 \text{ x } S_5S_5$$
$$\downarrow \qquad\qquad\qquad \downarrow$$
$$S_1S_2 \text{ x } S_3S_3 \qquad\qquad S_4S_5 \text{ x } S_6S_6$$
$$\downarrow$$
$$S_1S_3, S_2S_3 \text{ x } S_4S_6, S_5S_6$$

Triple cross hybrid seed

• Maintenance of self-incompatibility

i) **Bud pollination:** Young flower buds are initially self-compatible and become self-incompatible only 1-2 days before maturity of the anthers. So, pollination on the stigma of the immature buds with the mature pollens does self-fertilization effectively and thereby self-incompatible lines are maintained.

ii) **CO_2 gas treatment:** Treatment of 3-4 % CO_2 gas at 100% relative humidity condition under closed poly house for 8-24 hours is widely used to produce selfed seeds of the self-incompatible lines for their maintenance in pure form.

It has been found that the failure in self-incompatible reaction due to increased concentration of CO_2 gas causes a structural alteration of the papillae cell and thereby the self-incompatibility is removed.

Problems in the utilization of self-incompatibility in vegetable crops

- In some cases, particularly in Indian cauliflower and Brussels sprouts, sib-incompatibility is weak in certain inbred lines, especially at high temperature condition. In this situation, sib seeds are produced due to fertilization among the plants within each inbred (sister × brother crossing). The sibs are weak and very poor yielder. This disadvantage may be overcome by producing hybrid seeds under favourable environment, particularly at moderate temperature and humidity condition, so that the self-incompatibility reaction is not adversely affected.
- Continuous inbreeding may lead to severe loss of vigour of the inbred lines. So, it is desirable to propagate the parental lines by tissue culture techniques.
- Restriction of pollination within parental lines by bees also hampers cross-pollination because bees do not always visit the plants at random. This problem may be overcome by increasing morphological similarity of the parental inbred lines and by using other pollinators, such as, flies.
- Pseudo-incompatibility may lead to the production of selfed as well as sib seeds in place of hybrid seeds. This problem may be overcome by using strong self-incompatible lines.

Unisexualism and its utilization in hybrid seed production

- **Monoecious sex form:** In this sex form, staminate and pistillate flowers are produced separately in the same plant. Different cucurbitaceous vegetable crops *viz.*, pumpkin, summer squash, winter squash, bitter gourd, sponge gourd, ridge gourd, bottle gourd, watermelon, muskmelon, cucumber, etc and other crops like, sweet corn, amaranthus, etc., manifest this sex form. Sex expression *i.e.*, production of staminate and pistillate flowers in the plants though is genetically controlled yet amply influenced by environmental conditions.
- **Gynoecious sex form:** In this sex form, only pistillate flowers are produced in the plants of some cucurbits so that the plants can be treated as super female. Gynoecious lines have been identified in cucumber, muskmelon and bitter gourd. Gynoecy is conditioned genetically (*e.g.*, single dominant gene in cucumber; single recessive gene in bitter gourd), but subject to considerable influence by modifying genes and environmental factors.

Maintenance of Gynoecious line

- The gynoecious lines of cucumber can be maintained in homozygous condition by inducing staminate or hermaphrodite flowers in them. Staminate flowers can be induced by spraying the plants with 1500 ppm GA or 250-500 ppm silver thiosulphate ($Ag_2S_2O_3$) or 200-250 ppm silver nitrate ($AgNO_3$) at 2-3 true leaf stage and before opening of first flower in muskmelon. Application of silver thiosulphate is most effective in inducing staminate flowers.

- Solution of silver thiosulphate is prepared as per the method given by Reid *et al.* (1980). The stock solution of each of 4mM silver nitrate and sodium thiosulphate ($Na_2S_2O_3.5H_2O$) is prepared by dissolving 680 mg silver nitrate and 1000 mg of sodium thiosulphate separately in one litre of distilled water. For safer use, each bottle is kept in dark and stored in refrigerator till final use. The solution of silver thiosulphate is prepared by slowly pouring the calculated volume of silver nitrate into sodium thiosulphate, stirring rapidly as the solutions are mixed as needed on the day of experiment. To prepare 0.5mM (500 ppm) solution of silver thiosulphate from 4mM stock solution of silver nitrate and sodium thiosulphate, 125ml of each stock solution is mixed and the final volume is brought to one litre. Similarly, using the normality equation, silver thiosulphate in concentration of 250 ppm and 1000 ppm are prepared accordingly.
- In cucumber, the gynoecious lines are maintained by selfing the pistillate flower with the pollens from the staminate or hermaphrodite flower of the gynoecious line.

Hybrid seed production utilizing unisexualism

- **Bagging and hand pollination:** This method is applicable to monoecious cucurbits where both staminate and pistillate flowers are covered prior to opening and hand pollination is carried out for hybridization.
- **Emasculation and open-pollination:** In monoecious cucurbits, staminate flowers on the intended female line are removed before opening and open-pollination in isolation is allowed. Seeds set in the fruits of the intended female line are the hybrid seeds.
- **No emasculation and open pollination:** This method of hybrid seed production is applied with the utilization of:
  - **Gynoecious line:** Planting gynoecious and monoecous or andromonoecious lines in the ratio of 4:1 produces hybrid seeds. Seeds set in the fruits of the gynoecious line (female line) are the hybrid seeds. It is a commercial hybrid seed production method of cucumber.
  - **Chemical suppression of staminate flowers:** Suppression of staminate flowers on the intended female parent of monoecious cucurbits, like cucumber, pumpkin, summer squash, watermelon, etc., can be achieved for considerable period of time (about one month) by spraying Ethrel or Ethephon at 2- and 4-true leaf stage. Effective concentration is 250 ppm for cucumber, 600 ppm for summer squash and 250-300 ppm for pumpkin, bottle gourd and ridge gourd. Another growth substance, 2-CPA (2-chlorophenoxy acetic acid) at 100-200 ppm is effective for bottle gourd and bitter gourd. However, the concentration will have to be adjusted according to the climatic condition. Three to four fruit set at initial nodes are sufficient to produce enough quantity of hybrid seed.

Problems in the utilization of unisexualism in hybrid seed production

- Bagging and hand pollination is the commercial hybrid seed production method of cucurbits using monoecious sex form. Hybrid seed production in open condition is handicapped because there is always a large mixture of sib mated seeds in hybrid seeds collected from seed parent because considerable extent of self-pollination occurs in cucurbits in nature, which sometimes go up to 40 per cent.
- The gynoecious lines, which are the only effective solution for successful hybrid seed production in open condition in cucumber are stable only under moderate temperature and photoperiodic conditions. Stability of the gynoecious lines is affected when temperature goes beyond 30°C. Developing tropical gynoecious lines in cucumber can solve this problem.
- Exogenous application of growth regulators, particularly Ethrel, distinctly suppresses the staminate flowers for considerable period of time, although the efficacy depends on environments. So, this technique needs to be adopted under definite set of environments.

References

Baldauf, J.A., Liu, M., Vedder, L., Yu, P., Piepho, H.P., Schoof, H., Nettleton, D. and Hochholdinger, F.2022. Single-parent expression complementation contributes to phenotypic heterosis in maize hybrids. *Plant Physiol.*, **189**:1625-1638.

Baldauf, J.A., Marcon, C., Lithio, A., Vedder, L., Altrogge, L., Piepho, H.P., Schoof, H., Nettleton, D., and Hochholdinger, F. 2018. Single-parent expression is a general mechanism driving extensive complementation of non-syntenic genes in maize hybrids. *Curr. Biol.*, **28**: 431–437.

Birchler J.A. and Veitia R.A. 2007. The gene balance hypothesis: From classical genetics to modern genomics. *Plant Cell*, **19**: 395–402

Birchler, J. A., Donald, L. A. and Riddle, N. 2003. In search of molecular basis of heterosis. *Plant Cell*, **15**: 2236-2239

Brewbaker, J. L. 1964. *Agricultural Genetics*. Prentice Hall Inc. New Jersey.

Darwin CR. 1876. *The effects of cross and self-fertilisation in the vegetable kingdom*. John Murray, London, UK.

Drinić, S.M., Kostadinović, M., Ristić, D. and Stevanovic, M.2012. Correlation of yield and heterosis of maize hybrids and their parental lines with genetic distance based on SSR. *Genetika,* **44**:399-408

East, E.M. 1908. Inbreeding in corn. In: *Report of the Connecticut Agricultural Experiment Station for the Year 1907*. Conn. Agric. Exp. Stn., New Haven, pp. 419–428

Goff, S.A. 2011. A unifying theory for general multigenic heterosis: energy efficiency, protein metabolism, and implications for molecular breeding. *N. Phytol.,* 189:923–937.

Gustafsson, A. 1938. Studies on genetic basis of chlorophyll formation and mechanism of induced mutations. *Hereditas*, **24**: 33-93.

Hayes, H.K. and Jones, D.F. 1916. First generation crosses in cucumber. *Cann. Agric. Stu. Newsl.*, **3**: 34-35

Hazra P., Roy, M., Akhtar, S., Balacheva, E. and Atanassova, B. 2009. Expressivity oftwo genes controlling functional male sterility (*positional sterile* (*ps*) and *positional sterile-2*(*ps-2*) in tomato under the climatic conditions of Eastern India. *Genetics and Breeding*, Bulgaria **38**: 10-16.

Hazra, P., Roy, T. and Choudhury, J. 2008. Characterization of genic functional malesterility in brinjal and its expression in different genetic background. *Indian Journal of Genetics and Plant Breeding*, **68** : 47-51.

Hazra, P., Sahu, P. K., Roy, U., Dutta, R., Roy, T. and Chattopadhyay, A. 2010. Heterosis in relation to multivariate genetic divergence in brinjal (*Solanum melongena*). *Indian J. Agric. Sci.*, **80** : 119 – 124.

Hazra, P., Som, M. G.and Das, P. K.1993. Analysis of heterosis for pod yield and itscomponents in relation to genetic divergence of the parents and specific combining ability of the crosses incowpea. *Indian J. Genet.*, **53**: 418-423

Hull, F. H. 1945. Recurrent selection for specific combining ability in corn. *J. Amer. Soc. Agron.*, **37**: 134-145.

Jinks, J. L. 1983. Biometrical genetics of heterosis. *Theor.Appl. Genet.*, **6**: 1-46.

Jinks, J.L. and Jones, R.M. 1958. Estimation of the Components of Heterosis. *Genetics*, **43**: 223-234.

Jones, H. and Clarke, A. 1943. Inheritance of male sterility in the onion and the production of hybrid seed. *Proc. Amer. Soc. Hort. Sci.*, **43**: 189-194.

Kakizaki, Y. 1924. Flowering habit and natural crossing in the eggplant. *Jap. J. Genet.*, **3**: 29-36.

Kenchanmane, R. S. K. and Niederhuth, C. E. 2018. Epigenetic diversity and application to breeding. *Adv. Bot. Res.*, **88**: 49–86.

Knight, T.A. 1799. An Account of some Experiments on the Fecundation of Vegetables. In a Letter from Thomas Andrew Knight, Esq. to the Right Hon. Sir Joseph Banks, K. B. P. R. S. Royal Society, London, https: // royalsocietypublishing.org › pdf › rstl.1799.0013.

Koelreuter, J.G. 1766. *Vorlaufigennachircht von einigen das Geschlecht der Pflanzen betreffenden.* Versuchen und beobachtungen, Leipzig, 266p.

Leonardi, A., Damerval, C., Herbert, Y., Gallais, A., and de Vienne, D. 1991. Association of protein amount polymorphisms (PAP) among maize lines with performances of their hybrids. *Theor. Appl. Genet.*, **82:** 552–560.

McDaniel, R. G., and Sarkissian, I. V. 1970. Kinetics of mitochondrial complementation. *Physiol. Plantarum*, **23**: 335-342.

Melchinger, A.E. 1999. Genetic diversity and heterosis In: *Genetic and exploitation of heterosis in crops* (Eds. Coors, C.G. and Pandey, S.). *American Society of Agronomy*, Madison, USA, pp 99–118.

Mendel, Gregor. 1865. Experiments in Plant hybridization. The paper read at the February 8th, and March 8th, 1865, meetings of the Brünn Natural History Society

Mendel, Gregor. 1866. Versuche über Plflanzenhybriden. Verhandlungen des naturforschenden Vereines in Brünn, Bd. IV für das Jahr 1865, Abhandlungen, pp. 3–47.

Milborrow, B.V. 1998. A biochemical mechanism for hybrid vigour. *J. Exp. Bot.*, **49:** 1063–1071

Morelock, T.E., Simon, P.W. and Peterson, C.E. 1996. Wisconsin Wild: another petaloid male-sterile cytoplasm for carrot. *HortScience*, **31**: 887-888.

Ogura, H. 1968. Studies on the new male-sterility in Japanese radish, with special reference to the utilization of this sterility towards the practical raising of hybrid seeds. *Mem. Fac. Agric. Kagoshima Univ.*, **6**: 39–78

Pearson, O. H.1931. Methods for determining the solidity of cabbage heads. *Hilgardia*, **5**: 383-393.

Peterson C. E., Andher E. D. 1960. Induction of staminate flowers on gynoecious cucumbers with gibberellin A3. *Science,* **131**: 1673–1676.

Reid, M.S., Paul, J.L., Farhoomand, M.B., Kofranek, A.M. and Staby, G.L. 1980. Pulse treatment with the silver thiosulphate complex to extend the vase life of cut carnations. *Journal of American Society of Horticultural Science*, **105**: 25-27

Rudich J., Halevy A. H., Kedar N. 1969. Increase in femaleness of three cucurbits by treatment with Ethrel, an ethylene releasing compound. *Planta,* **86**: 69–76.

Sanjeet Kumar, Banerjee, M.K. and Kalloo, G. 2000. Male sterility: Mechanisms and current sttus on identification, characterization and utilization in vegetables. *Veg. Sci.*, 27: 1-23.

Schnell, F.W 1982. A synoptic study of the methods and categories of plant breeding. *Z. Pflanzenziichtg*, **89**: 1-18.

Shivaprasad, P. V., Dunn, R. M., Santos, B. A., Bassett, A. and Baulcombe, D. C. 2012. Extraordinary transgressive phenotypes of hybrid tomato are influenced by epigenetics and small silencing RNAs. *EMBO J.*, **31**: 257–266.

Shull, G. H. 1914. Duplicate genes for capsule form in Bursa bursa pastoris. *Ztschr. Ind. Abst. Ver.* **12**: 97-149.

Shull, G. H. 1948. What is Heterosis? *Genetics*, **33**: 439-446.

Shull, G.H. 1908. The composition of a field of maize. *Amer. Breed. Assoc. Rep*., 4: 296–301.

Stadler, L. J. 1939. Some observations on gene variability and spontaneous mutation. - Spragg Memorial Lectures, Michigan State College, 15 p.

Strunnikov, V.A. 1983. New hypothesis of Heterosis: its scientific and practical significance (in Russian). *Vestn Skh Nauki* (Moscow), **1**: 34-40.

Stuber CW, Lincoln SE, Wolff D, Helentjaris T, Lander E. 1992. Identification of genetic factors contributing to heterosis in a hybrid from two elite maize inbred lines using molecular markers. *Genetics*, **132**: 823-239.

Tsaftairs, A.S. and Kafka, M. 2008. Mechanisms of Heterosis in crop plants. *J. Crop Prod.*, DOI: 10.1300/J144voln01_05.

Wright, S. 1968. *Evolution and the Genetics of Populations. Vol. 1: Genetic and Biometric Foundations,* University of Chicago Press, Chicago, USA.

Zhang, T., Lin, X., Jiang, K.F., Deng, H.F., He, Q., Yang, Q.H., Yang, L., Wan, X.Q., Cao, Y.J. and Zheng, J.K. 2019. Relationship between heterosis and parental genetic distance based on molecular markers for functional Genes related to yield traits in rice. *Rice Science*, **17**: 288-295.

Zhao, X., Chai, Y. & Liu, B. 2007. Epigenetic inheritance and variation of DNA methylation level and pattern in maize intra-specific hybrids. *Plant Sci.,* (Amst., Neth.) **172**: 930–938.

10

Breeding Vegetable Crops for Quality Traits

In the changing Global scenario, food, nutrition, healthcare and livelihood security are the matter of great concern to human, social and economic development. The global community is increasingly looking towards finding means of sustainable nutrition for the growing population across all countries. The emphasis is on identifying low-resource strategies acceptable to the communities that do not put an unnecessary burden on the environment (FAO, 1993).

Nutritional security denotes the consumption and equitable access to adequate quantities of healthy, safe, affordable and nutritious food by every member of the family essential for optimal health and well-being. Concern for improvement of nutritional quality of vegetables has arisen in the context of providing nutritional security to our ever-increasing population because poor nutrition is linked with increased risk of obesity, diabetes, and heart disease as well as broader impacts including higher health care costs and decreased productivity.

More than 1.0 billion people globally are hungry due to food insecurity problem and about 3.0 billion people are malnourished due to poor nutrition causing mainly micronutrient deficiency. Of the total 3.0 billion malnourished people in the world, 70% live in South and Southeast Asia, 53% of children of the world are malnourished and underweight, and 40% of them live in India. COVID-19 pandemic pushed over 130 million more people into chronic hunger across the globe because of increase in food price and widespread food insecurity. People suffering from hidden hunger due to lack of vitamins and minerals in the food may appear healthy due to the absence of the classic symptoms of hunger (Uchendu and Atinmo, 2010; Burchi *et al.*, 2011). Micronutrients refer to vitamins, minerals and different phyto-chemicals essential for the body's physical and mental development, immune system functioning and various metabolic processes (Burchi *et al.*, 2011). This would entail the need to ensure a varied food intake, comprising all the essential macro- and micronutrients, plant pigments and other functional phyto-chemicals through a diversified and balanced diet.

Vegetables are considered essential for well-balanced diets since they supply natural macro-nutrients (protein, carbohydrates, fat), vitamins (vitamin A, vitamin B, vitamin C, folate, etc.), plant pigments (lycopene, β-carotene, anthocianin, lutein, capsanthin, zeaxanthin, etc.), minerals (potassium, calcium, iron, zinc, iodine, selenium, etc.), antioxidants (β carotene, lycopene, lutein, anthocyanin, ascorbic acid, polyphenol, flavanoids, etc.) and different functional phyto-chemicals (isothiocyanates, sulpharophane, thiosulfides, polyphenols, flavonols, flavones, triterpenoids) and dietary fibre (Dias and Ryder, 2011). Phyto-chemicals are secondary metabolites, which often play a crucial role against different stresses but are not important for basic metabolic processes of the plant (Sevindik *et al.*, 2017). Moreover, these phyto-chemicals are vital basis for various medicines and pharmaceutical industry (Sevindik, 2019). In the daily diet, these nutrients, minerals and phyto-chemicals have been strongly associated with improvement of gastrointestinal health, good vision, and reduced risk of coronary heart disease, stroke, diabetes and some form of cancer (Rimm, 1996; Prior and Cao, 2000; Mullie and Clarys, 2011). Vegetables are the most affordable source of vitamins and minerals needed for good health and can be a powerful agent in malnutrition management through their dietary diversity and association in balanced and healthy diet (Singh *et al.*, 2020).

Increasing evidence suggests that a healthy eating strategy with increased consumption of plant-based foods plays important roles in the prevention of chronic diseases, such as heart disease, cancer, stroke, diabetes, Alzheimer's disease, cataracts, and age-related function decline (Liu, 2004). This suggests that changes in dietary patterns and lifestyle, such as increasing the consumption of fruits and vegetables and more balanced intakes of meat and plant foods is a practical and effective strategy for reducing the incidence of chronic diseases. It is recommended that consumers should eat a wide variety of fruits and vegetables from different sources and including all forms, fresh, frozen, canned, dried, and 100% juices. One of the hypotheses about the health benefits of fruits and vegetables is attributed to the synergy or interactions of bioactive compounds and other nutrients in whole foods (Liu, 2004).

High-income group in the world conceived the implication of vegetables and its diversity in the diet very well since human health in the recent years has assumed an unprecedently important status. Foods have now assumed the status of "functional foods", which should be capable of providing additional physiological benefits, such as preventing or delaying the onset of chronic diseases as well as meeting basic nutritional requirements. Hence, with high-income elasticity, the demand for vegetables and perception of diversified diet have been growing steadily, hence, both access and consumption of vegetables have been increasing in the urban and peri-urban population groups. However, meeting the requirements for macro- and micronutrients for most of the population groups in households and communities of South Asia, Latin America and African countries seems far off because of meager presence of vegetables in the diet.

Nutrient and Antioxidant potential of vegetables

Since last three-decades nutritional studies are concentrating on examining food for their protective and disease preventing potential instead of negative attributes such as microorganism count, fatty acids, inorganic pollutant concentration, etc. (Nicoli *et al.*,1999). Vegetables contain fair amount of different primary nutrients *viz.*, carbohydrate, protein and fat and supply 10% calories, 7.0% protein and adequate dietary fibre to the daily diet (Table 1). However, vegetables are particularly important source of micronutrients (calcium, phosphorus, iron, magnesium, potassium, etc.), vitamins (provitamin A, vitamin C, vitamin B_6, thiamine and folic acid, etc.), different bioactive phyto-chemicals (lycopene, ß-carotene, anthoyanin, lutein, capsanthin, zeaxanthin *etc.*), different secondary metabolites (flavonoids, isothiocyanates, glucosinolates etc.) which provide colour, flavour and texture of different vegetables. About 90% of the human dietary vitamin C and 50% of vitamin A requirement is fulfilled from fruits and vegetables (Table 1). Though vegetables do not contain active vitamin A, they are rich source of provitamin A (β carotene), which is readily converted into active retinol (vitamin A) in the liver and intestine.

Table 1: Nutritional contribution of fruits and vegetables

Major Nutrients	Percent of total food supply	Major Nutrients	Percent of total food supply
Ascorbic acid	90	Thiamine	20
Vitamin A	50	Niacin	20
Vitamin B_6	35	Calories	10
Magnesium	25	Protein	7
Iron	20	Fat	1

Grouping of vegetables according to the types of major nutrients and bioactive phyto-chemicals they provide (Table 2).

Table 2: Vegetable groups according to the primary type of nutrients and bioactive phyto-chemical they provide

Nutrients	Vegetables
Carbohydrate	Potato, sweet potato, dry beans, elephant foot yam, yam, taro, plantain, cassava, tapioca and sweet corn
Protein	Beans, peas, leafy crucifers and sweet corn
Fat	Mature seeds of some legumes and cucurbits
Pro-vitamin A (β-carotene)	Carrot, pumpkin, muskmelon, sweet pepper, chilli, green peas, green beans, drumstick and green leafy vegetables
Lutein and zeaxanthin	Dark green leafy vegetables, including spinach, kale, turnip greens, broccoli, Brussels sprouts, collards, etc
Lycopene	Tomato, watermelon, muskmelon, etc.
Betalains	Garden beet
Anthocyanin	Red cabbage, red onions, red radishes, purple cauliflower, purple corn and the skin of purple brinjal
Vitamin C	Bitter gourd, sweet pepper, chilli, cabbage, broccoli and other crucifers, leafy vegetables, melons, tomato, immature bean seed, bean sprouts, etc.
Minerals	All leafy vegetables, cole crops and other crucifers, salad crops
Glucosinolates (isothiocyanates, sulpharophane)	Cabbage, cauliflower, broccoli, Brussels sprout, kale, radish, knolkhol, etc

Nutrients	Vegetables
Organo-sulfuric compounds (allicin, diallyl sulfide, diallyl disulfide)	Onion, garlic, shallot, chive, welsh onion, etc.
Dietary phenolics	Spinach , red sweet pepper, beets, broccoli, Brussels sprouts, eggplant, asparagus, green chilli, yellow onion, cauliflower, cabbage, radish, mushroom, sweet potato, carrot, sweet corn, potato, squash, white onion, green pea, tomato, green bean, celery, lettuce, romaine lettuce, cucumber, etc.
Vitamin E	Green leafy vegetables
Dietary fibre	Green leafy vegetables, root vegetables, beans, brinjal, cucurbits particularly, pointed gourd and ridge gourd, etc

Antioxidants

Free radical is a molecule or atom with an unpaired electron. Hence, they are unstable and quite reactive. The free radicals like super oxide (O_2^-), hydroxyl (OH^-), hydroperoxyl (HOO), alkyl peroxyl (ROO), etc. and reactive oxygen species (ROS) like hydrogen peroxide (H_2O_2), nitric oxide (NO), etc. are produced during cellular metabolism by oxidation. These free radicals or reactive oxygen species abstract an electron from the stable compounds that, in turn, are transformed into a new free radical. Therefore, one free radical can initiate a self-propagating chain reaction of destructive process by removing electrons from the stable compounds and forming many free radicals or reactive oxygen species.

Antioxidants are naturally occurring phytonutrients and different secondary metabolites, *viz.* carotenoids (β-carotene, lycopene, zeaxanthin, cryptoxanthin, lutein), polyphenols (phenolic acid, flavonoids,lignans, etc), flavonoids (quercetin, rutin, macluraxanthone, genistein, kempferol, myricetin, luteolin, anthocyanin, etc.) ascorbic acid, vitamin K, etc. The antioxidants neutralize the free radicals by donating one of their own electrons, ending the electron stealing reaction. However, they do not themselves become free radicals by donating electrons since they are stable in either form. Epidemiological studies have consistently shown that there is a clear significant positive association between consumption of fruits and vegetables and reduced rate of heart diseases mortality, common cancers and other degenerative diseases as well as ageing (Prior and Cao, 2000; Wargovich, 2000, Pandey and Rizvi, 2009). Vitamin C is a strong enhancer of nonheme iron absorption, as is an unidentified factor in meat, commonly known as the meat–fish–poultry factor. Other organic acids (e.g., citric acid), alcohol, and fermented foods also enhance nonheme iron absorption (Allen and Ahluwalia, 1997; Beck, 2016).

Phytochemicals

Phytochemicals are bioactive non-nutrient plant compounds in fruits, vegetables, whole grains, and other plant foods that have been hypothesized to reduce the risk of major chronic diseases (Liu, 2004). More than 5000 individual dietary phytochemicals have been identified in fruits, vegetables, whole grains, legumes, and nuts, but a large percentage of them still remain unknown. These phytochemicals need to be isolated and identified before we can fully understand the health benefits of bioactive compounds in whole foods (Liu, 2003). The most important groups of dietary phytochemicals can be divided into general categories as phenolics, alkaloids, nitrogen-containing compounds, organosulfur compounds, phytosterols, and carotenoids (Liu, 2004). The most studied groups of dietary phytochemicals related to human health and well-being are phenolics and carotenoids.

1. Phenolics

Phenolics are a group of naturally occurring phenols which are products of secondary metabolism in plants with ≥ 1 aromatic rings possessing ≥ 1 hydroxyl groups. Phenolics are generally classified as subgroups of phenolic acids, flavonoids, stilbenes, coumarins and tannins. High total phenolics containing fruits are wild blueberry, blackberry, pomegranate, cranberry, blueberry, plum, raspberry, strawberry, red/black grape, apple, pear, pineapple, peach, grapefruit, nectarine, mango, kiwifruit, orange, banana, lemon, avocado, cantaloupe, honeydew melon, and watermelon. Among the common vegetables, spinach has the highest phenolic content, followed by red sweet pepper, beets, broccoli, Brussels sprouts, eggplant, asparagus, green chilli, yellow onion, cauliflower, cabbage, radish, mushroom, sweet potato, carrot, sweet corn, potato, squash, white onion, green pea, tomato, green bean, celery, lettuce, romaine lettuce and cucumber (Wolfe *et al.*, 2008; Song *et al.*, 2010). It clearly showed that phenolic compounds are not found only in fruits and vegetables with bright colours. For example, potatoes are also good sources of dietary phenolics and contain 36 mg of gallic acid equivalents per 150 g of fresh potatoes (Song *et al.*, 2010).

Phenolics are generally considered as anti-nutrients, but recently, the recognition of antioxidative properties of these phenolic compounds has evoked a rethinking towards the health benefits of these secondary metabolites (Bravo, 1998; Dai and Mumper, 2010). Antioxidant properties of phenolics are mainly because of their redox properties, which allow them to act as reducing agents, hydrogen donators and singlet oxygen quenchers (Rice-Evans *et al.*, 1997).

• Flavanoids

Flavonoids are a major group of phenolic compounds that commonly have a generic structure consisting of 2 aromatic rings (A and B rings) connected by 3 carbons that are usually in an oxygenated heterocycle ring or C ring. This class of compounds ubiquitous in plants usually occurs as glycosides (a large important class of sugar derivatives in which the sugar is combined with a non-sugar; when glucose is the sugar, they are called glucosides). Fruits, vegetables, and other plant foods are rich sources of flavonoids. More than 5000 individual flavonoids have been isolated and identified. Structural differences in the heterocycle C ring categorize them as flavonols (quercetin, kaempferol, and myricetin), flavones (luteolin and apigenin), flavonols (catechin, epicatechin, epigallocatechin, epicatechin gallate, and epigallocatechin gallate), flavonones (naringenin), anthocyanidins (cyanidin and malvidin), and isoflavonoids (genistein and daidzein). These are common flavonoids in the diet. The estimate of human intake of all flavonoids is about few hundred milligrams to 650 mg/day. Flavonols (quercetin, myricetin, and kaempferol) and flavones (luteolin and apigenin) are major flavonoids in human diet, which is estimated as 23 mg/day, of which quercetin contributed about 70%, kaempferol 17%, myricetin 6%, luteolin 4%, and apigenin 3% (Hertog *et al.*, 1993).

Vegetables contain a variety of flavonoids, such as flavonols, flavones and anthocyanidins, which are associated with anticarcinogenic activity in animal cell systems (Papas, 1999; Panche *et al.*, 2016). Flavonoids are now considered as an indispensable component in a variety of nutraceutical, pharmaceutical, medicinal and cosmetic applications. Numerous *in vitro* studies have revealed diverse biochemical actions of flavonoids and their antioxidant activity. Quercetin (3,3',4',5,7-pentahydroxyflavone), a flavonol-type flavonoid, is an anti-atherosclerotic food factor. Quercetin and other flavonoids frequently exist as their glycoside forms in vegetables such as in onion. Glucoside-bound quercetin is absorbed and hydrolyzed in small intestine through glucose-transport system or passive transport after lactase phlorizin hydrolase (LPH)-dependent deglucosidation. Quercetin glycosides other than glucoside forms are likely to be absorbed in large intestine after hydrolysis with enterobacteria. Efficiency of the absorption is considerably affected by the coexisting food components. Lipids and emulsifiers are found to enhance the absorption. In any cases, resulting quercetin aglycone is converted to conjugated glucuronides and/or sulfates during the process of their absorption. Some metabolites circulating in the blood stream can act as free radical scavengers, as well as the inhibitors of lipoxygenase and xanthine oxidase. It is, therefore, likely that biological effect of quercetin is originated from conjugated metabolites. Metabolic conversion of flavonoids to their conjugates may be a process for regulating their action on cellular oxidative stress (Saito *et al.*, 2013). Vegetables such as kale, spinach, Brussel's sprouts, sprouting broccoli, beets, red bell-pepper, onion, corn, eggplant, cauliflower, tomato, garden beet, cucumber, lettuce, etc. are rich source of flavonoids hence, have potent antioxidant activity (Prior and Cao, 2000). Garlic extracts are also used as potential cardiovascular and anticancer agents. Allicin, produced when onion bulb or garlic cloves are crushed, spontaneously decomposes to form sulphur-containing compounds having chemo-preventive activity. *Allium* vegetables contain high levels of flavonoids and flavanols, which scavenge free radicals (Kothari *et al.*, 2020).

• Phenolic acids

Phenolic or phenolcarboxylic acids are the most widely distributed plant non-flavonoid phenolic compounds present in the free, conjugated-soluble and insoluble-bound forms. Phenolic acids can be divided into two major subgroups: hydroxybenzoic acid and hydroxycinnamic acid derivatives. Benzoic acid derivatives include p-hydroxybenzoic acid, salicylic acid, gallic acid, and ellagic acid. The common cinnamic acid derivatives include p-coumaric acid, caffeic acid, and ferulic acid (Liu, 2004). They are usually present in the bound form in foods as components of complex structures such as lignins and hydrolyzable tannins or attached to cell walls and proteins. They can also be found as derivatives of sugar and organic acids in fruits, vegetables, and whole grains. Caffeic, p-coumaric, ferulic, protocatechuic, and vanillic acids are found in virtually all plant-based foods.

Phenolic acids are found in the seeds, skins of fruits and leaves of the variety of vegetable and fruit crops. Phenolic acids possess phenol moiety and resonance stabilized structure, which causes the H-atom donation results in antioxidant property through radical scavenging mechanism (Naresh Kumar and Goel, 2019).

Table 3: Quantitative distribution (mg/100 g fresh weight) of different flavonoids in some vegetables

Vegetable	Quercetin	Kaempferol	Myricetin	Luteolin
Lettuce	0.7-3.0	0.2	0.1	0.1
Onion	28.4-48.6	0.2	0.1	0.1
Red Pepper	0.1	0.2	0.05	0.7-1.4
Broad bean	2.0	0.2	2.6	0.1
French bean	1.9-18.35	0.56-1.48	-	-

Plant pigments

Plant pigments are edible colours found in tissue of plants which includes anthocyanins, betalains, carotenoids and chlorophylls. These colours are produced by several biochemical pathway. These colours are changed according the growth stage of the plant parts or vegetable product. These pigments play important ecological and metabolic functions in the plants. These pigments are more frequently exploited as the source of nutraceuticals to address a number of human ailments (Grotewold, 2006).

• Carotenoids

Carotenoids are the second most abundant pigments in nature, and consist of more than 700 members (Britton, 1998; Takaichi, 2011). Carotenoids are divided into two groups: carotenes and xanthophylls. Carotenes, such as α-carotene, β-carotene, γ-carotene and lycopene are hydrocarbons and about 50 kinds of carotenes are present in nature (Britton *et al.*, 2004). On the other hand, xanthophylls, such as β-cryptoxanthin, lutein, zeaxanthin, astaxanthin, fucoxanthin, and peridinin are carotenoids containing oxygen atoms as hydroxy, carbonyl, aldehyde, carboxylic, epoxide, and furanoxide groups in these molecules. Some xanthophylls are present as fatty acid esters, glycosides, sulfates and protein complexes. Structures of xanthophylls show marked diversity and about 800 kinds of xanthophylls have been reported (Britton *et al.*, 2004; Maoka, 2009). For photosynthesis, phototrophic organisms necessarily synthesize not only chlorophylls but also carotenoids. Carotenoids play an important role in plant reproduction, through their role in attracting pollinators and in seed dispersal, and are essential components of human's diets. Carotenoids provide protection to vision and eye function, and against macular degeneration and cataracts. Carotenoids are credited with biological promotion of immune system response. Carotenoids are associated with inhibition of several types of cancers including cervical, esophageal, pancreatic, lung, prostate, colorectal and stomach.

Carotenoids are especially powerful in scavenging singlet oxygen generated from light induced lipid oxidation or radiation. Astaxanthin, zeaxanthin, and lutein are excellent in scavenging free radicals because of the unique end functional groups. β-carotene, α-carotene, and β-cryptoxanthin have pro-vitamin A activity and can be converted to retinol (vitamin A) in the intestine in humans. Zeaxanthin and lutein are the essential carotenoids in the macular region (yellow spot) of the retina of eyes in humans. A diet rich in zeaxanthin and lutein has been associated with a reduced risk of the development of cataract and macular degeneration. Orange and yellow vegetables and fruits, including carrots, spinach, pumpkins, papaya, sweet potatoes, winter squash, mangoes, cantaloupes and red peppers, are rich sources of β-carotene. Dark green leafy vegetables, including spinach, kale, turnip greens, broccoli, Brussels sprouts, and collards, are rich sources of lutein and zeaxanthin. Tomatoes, watermelons, pink grapefruits, apricots, and pink guavas are the most common sources of lycopene. It has been estimated that 85% of lycopene intake come from fresh and processed products of tomato such as ketchup, tomato paste, and tomato soup. The most abundant carotenoids in potatoes are lutein and zeaxanthin, followed by β-carotene and β cryptoxanthin (Andre *et al.*, 2007).

- **Anthocyanins**

Anthocyanins are natural pigments belonging to the flavonoid family. They are responsible for the blue, purple, red and orange colour of many fruits and vegetables. Anthocyanins are capable of acting on different cells involved in the development of atherosclerosis, one of the leading causes to cardiovascular dysfunction. Anthocyanins and the aglycone cyanidin are found to inhibit cycloxygenase enzymes, which can initiate the stage of carcinogenesis. On one hand, they can interfere with glucose absorption and on the other hand they may have a protective effect on pancreatic cells. The most extensively documented phytomedicinal role of anthocyanin pigments is in improving eyesight including night vision.

- **Chlorophylls**

Chlorophyll is the most important plant pigment, which play a crucial role in photosynthesis and are abundantly found in green fruits and vegetables that form an integral part of our diet. Although limited, existing studies suggest that these photosynthetic pigments and their derivatives possess therapeutic properties. These bioactive molecules exhibit a wide range of beneficial effects, including antioxidant, antimutagenic, antigenotoxic, anti-cancer, and anti-obesogenic activities (Martins *et al.*, 2023).

- **Betalains**

Betalains are unique nitrogen-containing pigments found exclusively in families of the Caryophyllales order (Chenopodiaceae, Amaranthaceae, Polygonaceae, Caryophyllaceae, etc.) and some higher order fungi, where they replace anthocyanin pigments. Betalains, consisting of betacyanins (reddish to violet) and betaxanthins (yellow) are generally used as colour additives in food. Its edible sources consist primarily of garden beet (*Beta vulgaris*) and prickly pears (fruit of the *Opuntia* genus of cacti). Betalains have been widely used as natural colorants for many centuries but their attractiveness for use as colorants of foods (or drugs and cosmetics) has increased recently due to their reportedly high anti-oxidative free radical scavenging activities, and concerns about the use of various synthetic alternatives (Rahimi *et al*., 2019).

Other functional phyto-chemicals

Since last three decades, special attention has been given to vegetables that are rich in plant secondary metabolites responsible for the induction of detoxifying enzymes, e.g., glutathione-S-transferase, quinone reductase, and epoxide hydrolase, which inactivate reactive carcinogens by destroying their reactive centers or by conjugating them with endogenous ligands, thereby triggering their elimination from the body and this inducer activity of detoxifying enzymes is principally because of highly reactive thiocyanates e.g., broccoli, cabbage, cauliflower, etc. (Zhang *et al*., 1992). The anticarcinogenic activity of isothiocyanate, a sulphoraphane present in cole crops was demonstrated in a rat mammary tumour model (Zhang *et al*., 1992). Glucosinolates, which are thought to be anticarcinogenic are very stable precursors of isothiocyanates, typically present in cole crops and other vegetable crucifers at very high levels and their hydrolysis by myrosinase is the prerequisite for biological activity (Fahey *et al*., 2002; Maina *et al*., 2020). Several studies confirmed that many leafy spices, especially those belonging to the Labiatae family such as sage (*Salvia officinalis* L.), rosemary (*Rosmarinus officinalis* L.), oregano (*Origanum vulgare* L.) and thyme (*Thymus vulgaris*) show strong antioxidant activity (Nakatani, 1997).

Dietary phytochemicals in the prevention of cardiovascular disease

Consumption of flavonoids has been significantly and inversely correlated with mortality from coronary heart disease (CHD) with the incidence of myocardial infarction (Hertog *et al*., 1995). Total intake of flavonoids (quercetin, myricetin, kaempferol, luteolin, and ficetin) was inversely associated with the LDL cholesterol and plasma total cholesterol concentrations (Arai, 2000).

In a randomized and controlled trial conducted by Watzl *et al*. (2005), a group of nonsmoking men consumed a diet of <2 servings per day of fruits and vegetables for 4 week. After this time period, the men were then randomly assigned to consume either 2 servings per day, 5 servings per day, or 8 servings per day of carotenoid-rich fruits and vegetables for an additional 4-week period. Compared with the low intake group (2 servings per day), the high-intake group was found to have significantly increased total carotenoid concentrations in plasma as well as significantly reduced C-reactive protein levels at week 8 (Watzl *et al*., 2005).

Consumption of approximately one serving per day of green leafy vegetables and foods rich in phylloquinone (vitamin K), lutein, folate, α-tocopherol, and kaempferol may help to slow cognitive decline with aging (Morris *et al*., 2018).

Dietary phytochemicals in the prevention of cancer

It has been estimated that one third of all cancer deaths could be prevented through the dietary modification (Anand *et al.*, 2008). Increasing bioactive compounds and antioxidant defenses through dietary phytochemicals, present in fruits, vegetables, whole grains, and other plant foods, may prevent, reduce, or delay the oxidation of DNA and affect cellular signal transduction pathways controlling cell proliferation and apoptosis (Liu, 2004). Consumption of 28 servings of vegetables per week showed a 35% reduction in the risk of prostate cancer compared with an intake of <14 servings per week (Cohen *et al.*, 2000). In another study, fruit and vegetable consumption was correlated with the reduced risk of pancreatic cancer (Chan et al, 2005). β-Carotene, an excellent dietary antioxidant in fruits and vegetables is thought to prevent lung cancer based on several epidemiological studies.

Dietary fibre

Dietary fibre is that part of plant material in the diet which is resistant to enzymatic digestion. The plant cell walls comprising of cellulose, pectin, xylolucean, xylan, mannan, lignin, free arabinan, etc. are the main source of dietary fibre, which ranges from 0.2 to 3 g/100 g edible portion of vegetables. Leafy vegetables, root vegetables and cucurbits particularly, pointed gourd and ridge gourd are high in dietary fibre content. High dietary fibre content in the food probably aid in digestion and utilization of high energy foods like meat, cheese, etc. and also neutralize the acid substances produced in the course of their digestion. Dietary fibre has attracted global attention due to their role in protection against certain types of cancer, regulation of transit, lowering of blood cholesterol, etc. (Dhingra *et al.*, 2012).

Antinutrient factors

Plants produce many defense strategies to protect themselves from predators and many of these, such as resveratrol and glucosinolate, which are primarily pathogen-protective chemicals also have demonstrated beneficial effects for human health. Many phyto-chemicals cause hindrance to nutrition and uptake of nutrients. For example, phytate, a plant phosphate storage compound is an antinutrient as it strongly chelates iron, calcium, zinc and other divalent mineral ions making them unavailable for uptake. Different antinutrient compounds (phytates, oxalates, trypisn inhibitors, lectins etc.), food allergens (albumins, globluins etc.) and toxins (glycoalkaloids, cynogenic glucosides, phyto-hemaglutinins) in different vegetables need to be reduced to enhance nutrient potential of the vegetables.

Role of whole food in the prevention of chronic diseases

Dietary antioxidants or phytochemicals lower the risk of the development of chronic diseases. However, these isolated compounds as dietary supplements in pure form may not work in the same way as the compounds in whole foods. Increasing evidence suggests that the health benefits of fruits, vegetables, whole grains, and other plant foods are attributed to the synergy or interactions of bioactive compounds and other nutrients in whole foods. Therefore, consumers should obtain their nutrients, antioxidants, bioactive compounds, or phytochemicals from their balanced diet with a wide variety of fruits, vegetables, whole grains, and other plant foods for optimal nutrition, health and well-being, not from dietary supplements.

Breeding vegetables for quality improvement

Vegetables are valued for their extrinsic and intrinsic quality traits. Diet rich in vegetables provide micronutrients and health promoting phytochemicals that alleviate malnutrition. The beneficial health effects are mainly attributed to diverse antioxidant compounds such as vitamins, carotenoids, phenolics. alkaloids, nitrogen containing compounds, organouslphur compounds, etc. Chief breeding objective will continue to be increasing yield to meet the food requirement of ever-increasing population. However, in order to ensure their health security and multipurpose utility of the varieties for fresh market and industry suitability, it is imperative that nutraceutical, edible colour and bioactive compound rich vegetable varieties should be bred which also ensure high remuneration to the farmers. Quality in vegetables is a complex character influenced by both genetical and environmental factors. Breeding for quality has been unsystematic and often empirical but significant progress has been made in several vegetable crops. Conventional breeding in conjunction with molecular biology has bright prospects of developing vegetable varieties high in phyto-chemicals, edible colours and bioacticve compounds suitable for fresh market as well as developing functional food industry. Breeding for quality attribute is a tricky job because not all compositions of vegetable crops are considered nutritious for the human being (Table 4). Consequently, quality improvement must be performed concurrently with improvement of productivity.

Table 4: Quality attributes of different vegetables along with their nutritional and anti-nutritional factors

Crop	Cosmetic and taste qualities	Nutritional and antioxidant factors	Anti-nutritional factors
Tomato	Firmness, number of locules, pericarp thickness, alcohol insoluble solids (AIS), dry matter, total soluble solids (TSS), titratable acidity, TSS: acid ratio, pH, colour, shape, etc.	Carotenoids particularly lycopene, β-carotene, ascorbic acid, vitamin K, potassium, phenolic compounds, such as flavonoids, hydroxycinnamic acid, chlorogenic, homovanillic acid, ferulic acid,etc.	Oxalates, caffeic acid derivatives, etc.
Brinjal	Texture, apparence, colour, flavour, etc.	Dietary fibre, carbohydrates, potassium, manganese, calcium, ascorbic acid, phenolic acids (chlorogenic acid in particular) and anthocyanins	Steroidal alkaloids (solanine, solasonine, solamargine and solasodine), trypsin inhibitors (in the fruit peel), phenols, amide proteins, etc.
Chilli	Colour, fruit smoothness, capsaicin content, oleoresin content, etc.	Carotenoids particularly capsanthin and β-carotene, ascorbic acid, vitamin B_6, dietary fibre contents, etc.	No major anti-nutritional factors however, phytates and oxalic acid may present during ripening stages
Sweet pepper	Fruit shape and firmness, sugar, organic acids, etc.	Carotenoids particularly capsanthin and β-carotene contents, ascorbic acid, vitamin B_6, fibre contents	Phenolic compounds such as chlorogenic acid content
Potato	Flavour and texture of tuber; commercial and industrial traits like tuber shape, colour, cold sweetening, dry matter, reducing sugar etc.	Proteins, carbohydrates, minerals and ascorbic acid contents	Alkaloids (alpha-solanine and alpha-chaconine), lectins and trypsin inhibitors
Cauliflower	Colour, flavour, texture and shape of curd	Vitamin C, β-carotene, anthocyanin, vitamin B_6, calcium, magnesium, vitamin K, folate, and fibre	Glucosinolates, the group of sulphur-containing glucosides contents
Cabbage	Compactness, crispness, fruity flavour, shape, appearance of the head	Vitamin C, β-carotene, anthocyanin, vitamin B_6, calcium, manganese protein, vitamin K, folate, and fiber	Phytic acid, tannic acid and trace amount of oxalic acid

Crop	Cosmetic and taste qualities	Nutritional and antioxidant factors	Anti-nutritional factors
Chinese cabbage	Crispness, flavour, shape and appearance of the head	Protein, β-carotene, vitamin C, vitamin B_6 folate and fibre contents	Phytate and oxalate in a trace amount
Onion	Flavour, colour, TSS, pungency, and shape of the bulb	Allicin, sulfur compounds, flavonoids, and fibre contents	Tannins, phytate, saponins, oxalates and cynogenic glycosides
Garlic	Flavour, colour, pungency and shape of the cloves and bulb	Allicin, sulfur compounds, flavonoids, and fibre contents	Tannins, Phytate, saponins, oxalates and cynogenic glycosides
Okra	Less fibre, high chlorophyll, less mucilaginous substances, high dry matter for dehydration purpose; low dry matter for canning and freezing purposes	High protein, minerals, carotenoids, dietary fibre, vitamin C, vitamin K, thiamine and iodine contents	Phytate and saponin contents
Pumpkin	Appearance, thick rind, thick flesh, deep yellow colour and high total soluble solids content	Carotenoids particularly β-carotene, sugar, vitamin C and vitamin K contents	Stachyose, raffinose, verbascose, trypsin inhibitor, phytic acid and tannin contents
Melons	Appearance, flesh colour, flavour, odour, sweetness, texture, soluble solids, TSS: titratable acids ratio and flesh firmness	β-carotene, vitamin C, niacin, folic acid, sodium, magnesium, protein, carbohydrate, dietary fibre and potassium contents	Tanins and saponins
Bitter gourd	Unique flavour, taste, shape, presence of tubercles on fruit surface, etc.	Vitamin C, iron, protein, phosphorus contents, hypoglycemic chemical like charantin	Phytic acid, polyphenol and saponin
Bottle gourd	Appearance, firmness, total soluble solids, shape, colour, etc.	Carbohydrates, dietary fibre, proteins, vitamin C and minerals like sodium, iron, etc.	Phytic acid, phytin, traces of lead in the seeds
Cucumber	Flesh colour, flesh firmness and crispy, fresh smell of the fruit, flesh juiciness, sweet taste	Carbohydrates, sugars, dietary fibre, thiamin (vitamin B_1), riboflavin (vitamin B_2), niacin (vitamin B_3), pantothenic acid (vitamin B_5), vitamin B_6, folate (vitamin B_9) vitamin C, calcium, iron, magnesium, phosphorus, potassium and zinc contents	Trypsin inhibitor (low level in seed meal), phytate, lectin and tannin

Crop	Cosmetic and taste qualities	Nutritional and antioxidant factors	Anti-nutritional factors
Carrot	Flavor attributes of raw carrot which include sweetness, harshness, and bitterness	Carotenoids particularly β-carotene, vitamin C, vitamin K, sugar, fibre and potassium contents	Traces of oxalate, saponin, alkaloids tannins, cyanide and phytate
Cowpea	Green colour, tenderness and glossiness of pods	High protein and methionine content and minerals contents	Trypsin inhibitors, tannins, phytic acid and haemagglutinin
Hyacinth bean	High sugar to polysaccharide ratio, less fibre, stringlessness and fleshy pod pericarp	Protein, flavonoids, sugar, vitamin C, vitamin K, manganese, fibre, potassium and folate contents	Proteolytic inhibitors, phytohemagglutinins, lathyrogens, cyanogenetic compounds and saponins
Garden pea	Sensory quality, texture, color and sweetness	Protein, vitamin C, manganese, fibre, vitamin B_1, folate contents.	Trypsin inhibitors and chymotrypsin
Lettuce	Greenish intensity of leaf, turgidity, crispyness, etc.	Carbohydrates, dietary fibre, protein, β-carotene, folate (vitamin B_9), vitamin C, vitamin K, vitamin E and iron contents	Nitrate, oxalate and saponin
Sweet potato	Flavour, starch content, cooking qualities and colour	Carbohydrates, carotenoids particularly β-carotene, vitamin C and manganese contents	Saponin, oxalate, tanin, phytate, trypsin inhibitor
Colocasia	Starch content, corm and cormel shape, cooking qualities, etc.	Ascorbic acid, reducing sugar, non-reducing sugar, starch, vitamin B_6, niacin, potassium, copper and manganese contents	Oxalic acid, trypsin inhibitors.
Leaf amaranth	Flavour, colour, leaf tenderness and texture of stem tissues.	β-carotene , vitamin C, fibre and minerals like iron, calcium, phosphorus, sodium, potassium, etc.	Oxalates and nitrates

Conventional breeding strategy

Conventional breeding uses inherent properties of the crop, having far-reaching impact on communities and has no regulatory constraints compared to genetically modified varieties. Breeding efforts targeting improved micronutrient content and composition began in the 1940s and 1950's, with research describing the inheritance and development of tomato breeding stocks high in β- carotene and vitamin C contents. Similar research leading to the development of darker orange and consequently high β carotene content in carrots began in the 1960s. Since then genetic improvement to increase levels of specific micronutrient has been pursued in several vegetable crops.

Several investigations on the genetic control for iron and zinc content of edible plant parts has been recorded, but such report for other minerals like, calcium and magnesium are meagre. As progress happened in improvement for crop yield, mineral content usually is reduced. Furthermore, breeding for improved mineral use efficiency usually does not alter mineral content of edible plant parts. Breeding for higher mineral and vitamin contents and their bioavailability along with reduction of anti-nutrient contents must ne considered.

Quality of vegetables is defined as a "dynamic composite of the physicochemical properties pertaining to horticultural commodities and consumer perception". Therefore, the concept of quality is strictly related to the specific vegetable and its consumption, and it depends on genetic, environmental conditions (open field, greenhouse, polyhouse conditions), and crop management (low input system production, pesticide residue free, etc.). Now-a-days, consumers are becoming extremely health-conscious regarding the system adopted to produce vegetables and the quality of the edible part, focusing on phytochemicals and bioactive compounds. Basic quality parameters of different vegetable crops are discussed below.

Tomato

On a worldwide scale, tomato continues to increase in importance for consumption as fresh vegetable as well as a major constituent in many prepared food products. Quality components of tomato can broadly be classified into two groups, *viz.,* i) flavour, taste and textural components and ii) nutritional components. Total soluble solids, sugar, acidity, viscosity, flavour and texture are the main processing quality parameters. Total soluble solids, alcohol insoluble solids, acidity, sweetness (sugar), pericarp thickness and number of locules per fruit basically constitute flavour and textural components. Different carotenoids particularly, lycopene and β-carotene, ascorbic acid, hydroxycinnamic acid, chlorogenic, homovanillic acid, ferulic acid are nutritional components.

Total soluble solid (TSS) content in tomato or its different products is expressed as °Brix which is a measure of the dissolved solids in a liquid. The TSS content in tomatoes is mainly sugars (fructose). Tomatoes for processing require a minimum °Brix of 4.5. High total soluble solids are the main quality components for the manufacture of different processed products like paste, puree, sauce, ketchup, etc. Alcohol insoluble solids constitute 20% of the total soluble solids and consist of pectin, polysaccharides including cellulose and structural proteins and determine the viscosity of fruit juice and thick consistency of the processed products. Titratable acidity (expressed as percent anhydrous citric acid) provides a measure of organic acids in the fruits (mainly citric and malic acid). In tomato fruits, acidity ranges widely from 0.32 to 0.84%. Sweetness is largely determined by sugar content (mainly glucose and fructose content). Interaction between sugar and acidity is important with regard to taste characteristics. Flavour is generally related to the relative concentrations of sugars and acids in the fruit, mainly fructose and citric acid. Combination of high sugar and high acid content give the most flavorsome product.

The benefits of tomato and tomato products have been contributed mostly by their carotenoid content. Different carotenoids present in tomato fruits are lycopene, β-carotene, phytoene, phytofluene, zeta-carotene, etc. Lycopene content in ripe fruits exceeds 80% of total carotenoids and next important carotenoid is colourless phytoene (Raffo *et al.*, 2002). β-carotene content in tomato at full ripe stage contain 10-12% of total carotenoids. Different tomato genotypes varied not only in the total carotenoid content but also in the qualitative distribution, which provides ample scope for improvement of lycopene and β-carotene in the fruits. Tomato is a good source of vitamin C and in different wild relatives of tomato, ascorbic acid content per 100 g fresh fruit ranges widely from 10 to 120 mg (Hobson and Davis, 1971). Small and soft-fruited types have more vitamin C content than large and firm fruited cultivars. It has also been found that fast ripening varieties have higher ascorbic content than slow ripening varieties. High locule number in the fruits is also associated with high ascorbic acid content. Different research reports indicated that most of the important fruit quality characters were governed overwhelmingly by non-additive gene action suggesting heterosis breeding as the best possible conventional breeding option for improving the quality traits of tomato.

Number of locules in the fruit is of paramount importance for the selection of varieties for processing purpose. Round and flattish-round fruited cultivars having higher locule number are suitable for fresh table purpose. Fruit firmness and the ratio of fruit wall to locular contents refer to fruit texture and increase in fruit pericarp thickness adds to increased firmness. Firmness of fruits is a desirable character for obtaining better keeping quality of fruits during storage and transportation leading to better transportation ability. Consumers of the western countries prefer firm fruits because they do not like too much juice when sliced. Pericarp thickness is correlated with fruit weight, compression tolerance, soluble solids content, fruit shape index and cracking resistance.

Brinjal

Size, shape, external colour, smoothness and softness of the fruit, green calyx, non-spinyness, etc. are the major cosmetic quality characters of brinjal. The biochemical constituents such as anthocyanins, total phenols, polyphenol oxidase activity and glycoalkaloid contents are important for taste. Anthocyanin is the main pigment giving mainly purple and violet colour and chlorophyll content has darkening effect (Nothman *et al.*, 1976), and colour development of the growing fruits has no relation to initial ovary colour (Nothman *et al.*, 1976; Hazra *et al.*, 2003). Nine different fruit colours are generally found, namely deep green, light green, purple, dark purple, light purple, pink, purple with green stripes or streaks, green with violet streaks or blush and white (Hazra *et al.*, 2003), although purple and dark purple fruit colour is mostly preferred by the consumers.

Different proximate compositions of the fruits, like sugar, protein and phenol contents widely vary with the genotypes which give ample scope for selection (Hazra *et al.*, 2005). However, these are also greatly influenced by stage of fruit development and environmental condition. Generally, the elongated-type varieties are rich in dry matter, crude protein, anthocyanin, phenolics, and glyco alkaloid contents, while the oblong-type varieties are rich in total sugars (Hazra *et al.*, 2004).

Brinjal has a very low caloric value and is considered among the healthiest vegetables for its high content of protein, various vitamins, minerals like, iron, calcium, potassium, magnesium, dietary fibre and different bio-active phyto-chemicals which are desirable for body growth (Docimo *et al.*, 2016). Brinjal contains different bioactive substances including secondary metabolites such as vitamins, minerals, carbohydrates, phenolics, glycoalkaloids and other antioxidants which appear to be the primary source of the plant's health benefits (Sharma and Kaushik, 2021) and for this reason, brinjal is regarded as one of the most important vegetable crops renowned for its nutritional benefits (Samtiya *et al.*, 2021). It contains huge number of polyphenols, such as anthocyanins (nasunin and delphinidin- 3-(coumaroyl-rutinoside)-5-glucoside) and phenolic acids, which are natural antioxidants (Li *et al.*, 2017; Sinanoglou *et al.*, 2019). Phenolic acids which are mostly present in fruit flesh (Nino-Medina *et al.*, 2017) show antioxidant, anti-carcinogenic, anti-inflammatory, anti-obesity, and anti-diabetic (type 2) properties (Samtiya *et al.*, 2021). Wide genetic variation was recorded in total phenol content from 3.26 – 8.77 mg GE equivalent /100 g fresh in different brinjal genotypes (Hazra et al., 2003) and from 16.95 – 19.46 mg GE equivalent /100 g fresh in other *Solanum* species *viz.*, *Solanum aethiopicum* cv.gr. *gilo*, *Solanum anguivi* and *Solanum macrocarpon* (Chattopadhyay *et al.*, 2009). Variations for total anthocyanin content in fruit ranging from 0.010 to 120.23 mg /100 g fresh weight (Bhanushree *et al.*, 2019) and from 0.00 to 945.34 mg /100 g fresh weight (Nandi *et al.*, 2021) was very wide. Brinjal is regarded as one of the top ten vegetables in terms of oxygen radical absorption capability due to its high anthocyanin and chlorogenic acid content (Todaro *et al.*, 2009; Braga *et al.*, 2016). However, there should be a threshold level for phenolic acids content to maintain palatability of the fruits.

Chilli and sweet pepper

Both chilli and sweet pepper belong to *Capsicum annuum*. The fruits provide a particular taste, colour and pungency as a spice, while as a vegetable they can be eaten in several ways, e.g., raw, pickled, fried, roasted and cooked vegetable. One of the most important quality attributes of chilli is pungency. The pungent principle is a mixture of closely related nitrogenous metabolites called capsinoids and chief among these is the potent, fat-soluble white crystalline alkaloid named capsaicin, which is chemically (E)-N- [(4-hydroxy-3-methoxyphenyl) methyl]- 8-methyl-6-nonenamide (Wood, 1987).

Capsaicinoids synthesis initiates from the amino acids phenylalanine and valine, from which the phenolic and acyl moieties are formed, respectively (Baas-Espinola *et al.*, 2016). The ability to synthesize capsaicinoids is related to the presence of at least one functional acyltransferase allele (*AT3*) at the *Pun1* locus (Stewart *et al.*, 2005). Recently, capsaicinoid synthase (CS) activity has been ascribed to *AT3* (Ogawa *et al.*, 2015). Non-pungent peppers carry a non-functional allele, which lacks a 2.5 Kb region at the 5' end, bearing part of the promotor and the first 735 bp of the open reading frame (Stewart *et al.*, 2007). In addition to pungency, aroma is one of the most important and typical flavour traits of chilli and sweet pepper. According to the sensory perception, 2-methoxy-3-isobutypyrazine is considered one of the most important volatiles for the final flavour "Bell pepper smell" or "pepper pyrazine" in *C. annuum* (Bosland and Votava, 2000).

Cauliflower

Cauliflower produces hypertrophic structure of the immature inflorescence termed as curd (Zhao *et al.*, 2016). Quality in cauliflower is related to the characters related to the external inflorescence (colour, firmness, tissue turgidity, absence of undesirable disorders) and chemical attributes such as glucosinolates, vitamin C, polyphenols and carotenoid contents (Raja *et al.*, 2011). The main pigments are made up of several phytochemicals such as β-carotene (orange), lutein (yellowish-green), chlorophyll (green), and anthocyanin (purple) (Singh *et al.*, 2020). In the curd of the purple cauliflower, it has been reported that the biosynthesis of anthocyanin is regulated by a single semi-dominant gene *Pr* (Chiu *et al.*, 2020). Carotenoids are secondary plant compounds, and they constitute lipid soluble yellow, orange, and red pigments, with the evidence on their positive action to prevent cancer and other chronic diseases (Johnson, 2000).

Cabbage

Cabbage is one of the most important vegetable crops grown in many countries of the world. White cabbage belongs to the variety of *capitata*, whose name comes from the Latin word 'capita' which means 'head'. Cabbage heads vary in shape, colour and texture of the leaves, resulting in a wide number of cabbage cultivars with cultivation potential under different climatic conditions (Björkman *et al.*, 2011). From ancient times, it is known to have many medicinal properties. Its juice reduces constipation, act as an antidote to mushroom poisoning, laxative, antidote to drunkenness, used to treat hangovers, headaches, sunstroke, treatment of swelling, protection of breast inflammation, etc. (Alexandra and Andreea-Daniela, 2020). Cabbage is rich in phytochemicals such as polyphenolics, glucosinolates, enzyme myrosinase, carotenoids and vitamins that have shown antioxidant, anticancer, and anti-obesity properties (Oloyede *et al.*, 2021).

Main objectives of cabbage breeding are crop uniformity, high yield, field appearance and adequate size and shape of head, non-splitting and good colour. It is difficult to achieve a high level of uniformity in case of open-pollinated varieties and in F_1 hybrids using conventionally developed inbred lines due to the cross-pollination nature of the crop. Production of F_1 hybrids with a great extent of uniformity and in a reduced time span could be possible by use of doubled haploid (DH) lines. At present, most of the cabbage varieties grown in the world are F_1 hybrids (Tanaka and Niikura, 2006). Breeding for quality traits are gaining importance in cabbage breeding. Red-colored cabbage is gaining attraction due to its high anthocyanin content. A selective approach for glucosinolates is mainly required because some have harmful health effects. Hanschen and Schreiner (2017) reported that broccoli and red cabbage were mainly rich in 4- methylsulfonyl butyl glucosinolate (glucoraphanin), whereas cauliflower, savoy cabbage and white cabbage contained mainly allyl glucosinolate (sinigrin). The sources of high levels of nutritive and biologically active substances were found, mostly within central Russian and Dutch white cabbage accessions (Solovyova *et al.*, 2014). Development of tropical hybrids has gained immense popularity in tropical and sub-tropical parts of the world.

Muskmelon and watermelon

Melons refer to both dessert melons of the genera *Cucumis* (musk melon) and *Citrullus* (water melon). The *reticulatus* group of musk melon includes the netted or rough skinned melons and the *inodorous* group includes the smooth skinned types. Fruit quality of the melons can be divided broadly into three categories: appearance (shape, size and skin colour), flesh colour and storability.

Muskmelons show enormous diversity of fruit types varying in flesh colour from green, pink to orange, skin colour from green, white, yellow, orange to grey, and in skin texture from smooth to net. White, green, and orange colour of fruit flesh is due to a combination of chlorophyll and carotenoid pigments (Burger *et al.*, 2009). β-carotene is the major carotenoid accumulated in orange-flesh varieties of musk melon (Nuñez-Palenius *et al.*, 2008).

Watermelon shows less diversity than muskmelons. The main pigment components in different flesh colour in watermelon genotypes are lycopene in red-flesh, β-carotene in orange flesh and xanthophylls and β carotene in yellow flesh (Tadmor *et al.*, 2005).

Fruit quality in both muskmelon and watermelon is generally related to high internal sugar levels and good flavour which is determined by total soluble solids. Total soluble solids should be at least 9 or 10% for good dessert quality however; high quality melons contain TSS in the range of 12-15% or even more. In watermelon, the principal requirements for market are flesh crispness, good flesh colour and sweetness. The principal sugar accumulated within muskmelons and in most watermelons is sucrose (Yativ *et al.*, 2010).

Muskmelon fruits have little or no starch so they require a concurrent supply of photoassimilates from leaves for sugar accumulation during development and ripening (Hubbard *et al.*, 1990). Also high total soluble solids, and concomitantly, the sugars are associated with slow ripening and high leaf area hence, sugar content of ripe muskmelons is very much dependent on allowing sufficient time for sugar accumulation in fruits prior to harvest.

Sucrose accumulation in melons occurs mainly from carbohydrates translocated into the fruits during ripening. Accumulation of sucrose is regulated mainly by the activities of two enzymes, i.e., acid invertase and sucrose phosphate synthase (SPS). In the phase of rapid sucrose accumulation in muskmelon, the acid invertase activity declines with a concomitant increase in sucrose-phosphate synthase activity thus, higher activity of sucrose phosphate synthase is an indicator of high sucrose accumulation and it may be utilized in specific breeding programme for enhancing sweetness.

Good aroma is one of the important quality attributes in melons particularly muskmelon. The volatile compounds particularly various ethyl and acetate esters are responsible for aroma development. Thioesters probably derived from methionine contribute to the unique aroma of Golden Crispy cultivars of muskmelon (Wyllie and Leach, 1990).

Muskmelon contains enormous genetic diversity for size, shape, skin and flesh colour, skin texture, low and high sugar levels and aroma, as well as climacteric and non-climacteric patterns of ethylene production, hence, considerable potential exists for improving the different quality and post-harvest attributes of musk melons by utilizing this natural variation in conventional breeding programmes. However, improved analytical methods and identification of volatile aromatic compounds enhance breeding programmes concerned with aroma.

Cucumber

Cucumbers, either fresh or pickled, are highly regarded as a crunchy and low-calorie food with little nutritional value. Cucumber germplasm has relatively limited genetic variation but the abundance of monogenic traits, the ease of inbreeding and rapid life cycle has yielded alteration of broad range of often-useful traits, like plant architecture, sex expression, disease resistance, etc. by intensive breeding (Fanourakis and Simon, 1987). Linkage maps in cucumber (*Cucumis sativus*) have been constructed using morphological traits, isozymes, restriction fragment length polymorphisms (RFLPs), and random amplified polymorphic DNAs (RAPDs). However, lack of polymorphism in cucumber has led to the construction of relatively unsaturated maps (13 to 80 points) (Stub *et al.*, 2000). Many individual factors, such as skin colour, spine colour, presence or absence of warts and spines, fruit shape and fruit firmness together constitute fruit quality and fortunately, most of these factors are under simple genetic control.

Fruit texture traits, such as firmness and crispness, are one of the most important characteristics in cucumber because these are directly related to consumer preference and demand. Wide range of cucumber cultivars grown and bred around the world differ greatly in firmness and crispness. Quantitative differentiation has occurred in the fruit texture and shape traits during domestication and subsequent breeding which has resulted extensive diversity in these characters (Shimomura *et al.*, 2016). Different breeding programmes in conjunction with new processing methods had yielded improved crunchy texture in fresh and pickled cucumbers.

Different nutrition elements, such as soluble solids, vitamins, and minerals constitute the nutrient quality in cucumber. Flavour quality contains both volatile and non-volatile compounds (Zhang *et al.*, 2021). Cucumbers are a fair source of vitamin C. Furthermore, fruit peel (exocarp) does contain carotenes and different oriental types have orange flesh. β carotene is the major carotene present in cucumber peel and flesh (Simon and Navazio, 1997). Common cucumber is low in carotenoid content with greenish white flesh colour while orange-fleshed cucumber germplasm from northeast region of India has high carotenoid content and thus, appears to be promising for utilization in breeding programme (Ranjan *et al.*, 2019). The orange flesh characteristic has been transferred to United States pickling and slicing cucumber types to yield immature fruit with 25 ppm carotenoids. Significant genetic variation exists for carotene content in cucumber peel. Dark green peel contains 25 to 45 ppm carotene, whereas, pale green peel contains 8 to 15 ppm carotene. This translates to 3 to 5 ppm carotene content on a whole fruit basis for dark green cucumbers. Increased consumer acceptability of dark green cucumber provides a means for plant breeding efforts to improve nutrient availability. Bitterness in fruit and foliage is due to the presence of cucurbitacins (terpenoids) which is under monogenic dominant control. Several genes have been described that control the trait, with *bi (bi-1)* making fruit and foliage bitter free and *Bt (Bt-1)* making the fruit highly bitter (Zhang *et al.*, 2013). For breeding purpose, non-bitterness of the whole plant is advantageous because genotypes that have non-bitter fruits but leaves containing bitter principles may produce bitter fruits under certain conditions.

Pumpkin

Pumpkin is well-known for multifunctional vegetable in the diet, full of nutrients. The fruits including the flesh, seed, and peel are a rich source of primary and secondary metabolites, including proteins, stored carbohydrates (mainly glucose), mono-unsaturated fatty acids, polyunsaturated fatty acids, carotenoids (74% β-carotene), tocopherols, tryptophan, delta-7-sterols, and many other phytochemicals with moderate quantity of ascorbic, nicotinic, pantothenic and folic acids and different minerals (Batool et al, 2022). These multifunctional ingredients have resulted the average nutritive value of pumpkin (2.68) higher than brinjal (2.41), tomato (2.09) and cucumber (1.69) (Gupta and Rai, 1990). In general, the soluble solids, soluble protein and soluble sugar contents were higher in ripe fruits than in green fruits.

A wide range of total sugar content from 1.18 to 6.0%, of β carotene content from 1.6 to 45.6 mg/100 g fresh pulp and of ascorbic acid content from 0.83 to 5.49 mg/100 g fresh weight in different genotypes sufficiently justifies the possibility of improving these quality characters of pumpkin through appropriate breeding approaches (Hazra *et al.*, 2007). Ascorbic acid and β- carotene content registered positive but non-significant association with fruit yield (Pandey *et al.*, 2003). Both β-carotene and total sugar content of pulp exerted no negative direct effect on fruit yield (Hazra *et al.*, 2007). These associations amply justified the possibility of simultaneous selection for sugar, ascorbic acid and βcarotene content of pulp without a concomitant depression in fruit yield. There were high correlations between sensory sweetness and total ethanol-soluble sugars, and between sensory texture and solids content of fruit pulp (Daniel *et al.*, 1996).

Pumpkin being a cross-pollinated in nature due to monoecious sex form generally shows no loss in vigour due to inbreeding. Hence, inbreeding and individual plant selection from the open pollinated and heterozygous base population could be an effective breeding approach for pumpkin. Marked improvement was recorded for fruit yield per plant, marketable fruits per plant and β-carotene content of the pulp showing 89.4, 85.9 and 54.2% increase, respectively over the base population (Hazra *et al.*, 2007). Heterosis breeding emerged as an important breeding method for increasing yield, sweetness, carotenoid and other antioxidant contents (Hazra *et al.*, 2007; Hosen *et al.*, 2022).

Carrot

Extensive genetic variation is available in carrot germplasm for many production and quality attributes useful for improvement. Furthermore, carrot is easily manipulated in tissue culture with well-developed genetic transformation systems and extensive diversity of molecular markers, e.g., RFLPs, RAPDs and isozymes so that biotechnological approaches can readily be applied.

Carotenoids, including β-carotene are abundant in carrot and they account for the familiar orange colour. Significant genetic variation occurs for carotenoid levels in adapted carrot germplasm. Phenotypic recurrent selection has yielded breeding stocks averaging 500 ppm β-carotene content in contrast to 100-150 ppm in widely grown cultivars (Simon *et al.*, 1993). Visual selection is moderately successful for improving β-carotene content up to 200 ppm but laboratory analysis is necessary for accurate selection at higher levels. Genetic selection for improved β-carotene content, sweetness and harshness levels has no deleterious effect on productivity of carrots (Simon *et al.*,1993).

In addition to the nutrients provided by carrots, flavour is also an important component of overall quality. The major flavour attributes of raw carrot include sweetness, harshness and bitterness. Consumers generally prefer sweet carrots without harsh, turpentine aftertaste or bitterness (Simon *et al.*, 1993).

The compounds described as most important for carrot flavour are sugars, bitter substances and volatile terpenoids. The main sugars associated with the sweetness perception are glucose, fructose and sucrose, which account for more than 95% of the free sugars and 40–60% of the stored carbohydrates in the carrot root (Cavagnaro, 2019). Analysis of volatile terpenoids requires gas chromatographic evaluation (Gaggiotti *et al.*, 2019). Sugars and volatile terpenoids account for sweetness and harshness, respectively. Volatile terpenoid content varies from 5 to 330 ppm among the inbreds and is quantitatively inherited with incomplete dominance for low terpenoid levels. Carrot roots containing 30 to 50 ppm volatile terpenoids are mild-flavored, unlike those with a higher content although this range of volatile terpenoids gives adequate flavour. In contrast to sweet and harsh flavour, bitterness is only found in stored carrots exposed to ethylene and is thought to be caused by isocoumarin, although this is not well established. Another important component of raw carrot quality is texture. Consumers prefer succulent juicy carrots. Texture is related to fibre and pectate levels but the relationships are not clear. Selection for succulent texture, however, can yield roots which are prone to cracking.

Onion and garlic

In onion, important bulb quality traits include size, shape, colour, pungency, firmness, dormancy and amount of soluble solids (carbohydrates concentration). Most of the characters have a genetic base and can be manipulated by breeding. Bulbs of various shapes and colours are preferred in different parts of the world. Onion bulbs may be white, red (deep or light), brown or yellow, round, pyriform or flat and soft or firm. Higher soluble solids are important for dehydration products of onion (flakes or powder). Onion and garlic flavour is an important component of quality.

S-Alk(en)yl cysteine sulfoxides are odourless, non-protein sulfur amino acids typically found in the members of the family Alliaceae and are the precursors to the lachrymatory and flavour compounds found in the genus *Allium*. The pungent flavour (volatile sulfur compounds) typical of these crops is the result of an interaction between alkyl cysteine sulfoxides in the cytoplasm with the vacuolar enzyme alliinase when cells are ruptured. Open-pollinated populations of onion show marked differences in pungency. Consumers of Europe and United States of America often prefer less pungent onion, whereas, many tropical onions are characterized by strong pungency. The sulfur containing cysteine derivatives (ajoene, allicin, aliin, and diallyl sulfide) are the breakdown products of more complex molecules found in intact plant tissue and they have been implicated in the reduction of blood cholesterol levels, hypertension, cancer, atherosclerosis and a wide range of other health problems (Sobolewska *et al.*, 2013). As the healthful derivatives of flavour may be expected to be correlated with intensity of pungency, development of pungent onions may yield onions with greater health benefits to consumers than mild onions. Improved and rapid methods for pungency evaluation and analytical methods to evaluate compounds beneficial to human health will improve the knowledge of onion quality genetics. Between genetic and environmental factors, genetic factors are the primary determinant of onion pungency (81.3% of total variation), with environmental factors controlling the remainder (Yoo *et al.*, 2006). Preliminary evaluation of onion pungency genetics indicates incomplete dominance for low levels and selection for low pungency has been exercised in the varietal development in Europe (Kopsell and Randle, 1997). A large, less pungent, soft onion may retain more water at maturity, diluting the alkyl sulfides responsible for pungency. Therefore, this onion would be softer and more likely to sprout during storage. The quality of the dehydrated onion product is mainly dependent on uniform white colour, globe shape for convenient handling, high solids (soluble and insoluble), more non-reducing sugars and high degree of pungency (pyruvic acid 0.50-0.70%) and its better retention after dehydration.

Sweetness, determined by free fructose and glucose, is another important component of onion and garlic flavour. The fructose polysaccharide fructans are the major storage carbohydrates of *Alliums*. Fructans are fructose polymers of four to several hundred residues that possess a sucrose core. Fructans are found widely in plants and appear to act as reserve carbohydrates, and improve cold and drought tolerance. The dynamic equilibrium between the nearly flavourless fructans and their very sweet component fructose monomers is not well understood during bulb development and storage. For this reason, the sugar or soluble solids levels are not always well correlated with sweetness. Furthermore, high pungency levels can mask sweetness. Potential for improvement of garlic quality is less understood than that for onion. However, analytical methods for determination of different quality parameters are more or less same for both onion and garlic. Opportunities for garlic quality improvement will greatly be expanded as true seed production and its concomitant sexual recombination becomes routine.

Garden peas and beans

Garden peas are known for its pivotal nutritional value with high-quality proteins, dietary fibres, starch, carbohydrates and micronutrients including vitamins and minerals. It is also enriched with phytochemicals, antioxidants, flavonoids, tannins and other phenolic compounds under non-nutritional category. Dietary fibre extracted from its cotyledon of the cell wall, seed coats and pods proved to accelerate the gastrointestinal activity by developing intestinal microflora. Intermediate amylose maintains blood glucose level by lowering the glycaemic index and reducing starch digestibility of peas. Pea proteins derived peptides have opportunities in nutraceuticals. Furthermore, pea-derived non-α-galactooligosaccharides could mitigate the gas production by colonic bacteria during the fermentation. Anti-acetylcholinesterase (anti-AChE) was also traced in the pea pod with some antioxidant properties (Kumari and Deka, 2021).

In garden pea, the seed quality is highly dependent on dark green colour of the skin of pods and higher soluble solids content. Protein content is an important nutritional quality in peas and beans. Wide range of variation among the genotypes for protein content has been reported in French bean, pea, faba bean, cowpea and other beans, which favours selection of genotypes with enhanced protein content. In peas, beans and soybean, seed protein content is negatively correlated with yield (Evans and Gridley, 1979) which indicated less likelihood of developing high yielding variety with high seed protein content. However, in vegetable cowpea genetic studies of pod yield and seed protein content suggested the encouraging possibility of enhancing pod yield without concomitant decrease in protein content (Hazra *et al.,* 1996). In broad bean (*Viciafaba* L.), non-additive gene action for ascorbic acid and protein content (Kalia and Pathania, 2007) indicated less likely hood of developing these quality traits by selection.

Leaf amaranth

Amaranth leaves do not differ significantly with other green vegetables in terms of appearance, texture and flavour except for nutritive value. Additionally, amaranth leaves have been found to have low levels of tannin and phytates which lower the bio-availability of minerals unlike in other vegetables (Mwangi, 2003). Dry matter content of amaranth plants showed positive correlation with crude fibre and phosphorous content at all stages of harvest, whereas, calcium, magnesium and iron content had negative correlation with dry matter content in most of the stages of harvest (Saha *et al.*, 2003). Rangarajan and Kelly (1999) reported that *Amaranthus tricolor* and *A. lividus* had the highest total and bioavailable iron but *A. hypochondriacus* had the lowest levels of total and bioavailable iron. Though field-grown amaranth accumulated higher levels of total and bioavailable iron, a greater proportion of the total iron was sequestered in insoluble and unavailable forms. These analyses indicated potential for genetic improvement of iron nutritional quality from *Amaranthus*, especially within the species *Amaranthus tricolor*.

Chinese cabbage

Chinese cabbage (*Brassica rapa* L. ssp. *pekinensis*) is an economically important leaf vegetable of the *Brassica* genus, growing worldwide, especially in Asian countries. The quality of Chinese cabbage includes flavour and nutritional quality (Gong *et al.*, 2016). Commodity quality is external quality which includes freshness, leaf ball tightness, etc. and flavour quality refers to the unique smell and taste, which need to be identified by tasting and analyzing the flavour-related components. Nutritional quality is determined by the content and distribution ratio of nutrients, such as crude protein, soluble sugar, fiber, organic acid, amino acid and mineral elements, and is related to human health (Wu *et al.*, 2012). Sugar and fibre contents had much greater influence on flavour than protein content. At least three pairs of genes controlled inheritance of fibre content but one pair of genes controlled sugar and protein contents (Zhao *et al.*, 2005).

Lettuce

Quality of lettuce is determined by several external and internal quality parameters *viz.*, appearance, colour, texture and flavour. Different quality parameters are dry weight, percentage of cellulose, vitamin C, soluble sugars and nitrates contents (Dimitrov *et al.*, 2006). Wild accessions (*Lactuca serriola* L., *Lactuca saligna* L., *Lactuca virosa* L. etc.) had higher β-carotene and lutein concentrations than cultivated lettuces, mainly due to the lower moisture content in the wild lettuces. Among the major types of cultivated lettuce, carotenoid concentration followed the order of green leaf or romaine > red leaf > butter head > crisp head. There is significant genetic variation in carotenoid concentration within each of these lettuce types. Crisp head lettuce generally accumulates more lutein than β-carotene, while other lettuce types had more β-carotene than lutein. Carotenoid concentration is higher in summer than in the fall but was not affected by the position of plant. β-carotene and lutein concentrations were highly correlated, suggesting that their levels could be enhanced simultaneously. β-carotene and lutein concentrations are both highly correlated with chlorophyll '*a*', chlorophyll '*b*' and total chlorophyll contents, suggesting that carotenoid content could be selected indirectly through chlorophyll or color measurement. Different research findings amply suggest that genetic improvement of carotenoid levels in lettuce is feasible (Mou, 2005).

Potato

Potato is a world food crop having a significant contribution towards food and nutritional security, especially in the developing world. It is considered as a balanced food due to the presence of high-quality proteins, vitamins and minerals, trace elements with lesser energy (Yadav *et al.*, 1994). It contains some important essential amino acids like leucine, tryptophan and isoleucine. Potatoes are also good sources of dietary phenolics and contain 36 mg of gallic acid equivalents per 150 g of fresh potatoes (Song *et al.*, 2010).

Demand for processed potato products like chips, French fries, lachcha, flakes etc. is ever increasing due to rapid urbanization, inclination of the young generation towards fast foods, higher income, increasing number of working women with their preference towards ready-cooked food and tourist demand. Even though India is the third-largest producer in the world, mostly table purpose potato is grown in the country with a little share of processing type potatoes. The table purpose cultivars had low dry matter (17–19%) and high reducing sugar (>250 mg 100 g^{-1} fresh tuber weight) content which in turn reflect poor chips colour score and crispiness. Processing quality potatoes should possess more than 20% dry matter content and reducing sugar level of less than 150 mg per 100g of tuber fresh weight. Higher dry matter content of potatoes results in higher yield of chips and lower consumption of the frying medium. However, the processing quality cultivars should be low in glyco-alkaloids and phenols, lesser undesirable colour (<5%) and total chips defects (<15%) (Kumar *et al.*, 2014).

Sweet potato

Sweet potato roots have high nutritional value and sensory versatility in terms of taste, texture, and flesh colour (white, cream, yellow, orange, purple). The varieties with high dry matter (>25%), white or cream flesh colour, and mealy firm texture after cooking are preferred by the consumers in the tropics.

Sweet potato is recognized for its nutritional value, presenting complex carbohydrates and a moderate glycemic index and is also a good source of vitamins, minerals and antioxidants (Severo *et al.*, 2021). In addition, sweet potato may have significant levels of bioactive compounds, like anthocyanins and carotenoids, depending on the cultivar (Truong *et al.*, 2018). Furthermore, recent work showed that flours from different sweet potatoes cultivated in Northeastern Brazil are potential prebiotic ingredients, suggesting their use in the formulation of functional foods or dietary supplements (De Albuquerque *et al.*, 2020).

Conventional breeding for vitamins, minerals and bioactive compounds

Vitamins, minerals, different bioactive compounds and edible colours are natural compounds which are regulated by several biochemical pathways and controlled by genetic and environmental factors. From early times people knowingly or unknowingly selected several vegetable crops for their food purpose. There are many vegetables which are rich in these beneficial compounds. The biochemical pathway and synthesis of these compounds are controlled by one or many genes which are scattered in the available or unknown germplasm of the particular vegetable crop. There is plenty of diversity in different vegetable crops which can be exploited for development of special varieties. Through conventional breeding approaches, it is possible to develop new vegetable varieties or integrate the favorable genes for vitamins, minerals, nutraceuticals, bioactive compounds and edible colour into the cultivated varieties. Advances in molecular biology and recombinant technology have paved the way for enhancing the pace of special trait variety development using marker assisted breeding and designing new vegetable crop plants following transgenic approach.

- **Utilization of genetic resources**

Diversity in genetic resources is considered essential for food security and nutrition. It can contribute to the achievement through improved dietary choices and positive health impacts. Dietary diversity depends not only on a diversity of crops but also on diversity within crops. Several research reports documented wide variation in nutrient contents within species, but data are lacking on nutrient composition and dietary intake for many underutilized species. Such information is needed both to enhance the use of more nutritious cultivars in diets and to make them available for use in breeding programme aimed at increasing the nutrient content of more commonly used varieties, eliminating the need for transgenic modifications.

Genetic resources are the foundation block that are essentially required for evolving improved crop varieties when the breeder aim at adding more desirable traits to an otherwise acceptable variety. This necessitates availability of the desired variability to the breeder within the land races, putative ancestral form, primitive cultivars and obsolete cultivars, heirloom cultivars of these crops or its wild forms and other related species constituting primary, secondary and tertiary gene pools.

Substantial genetic variation frequently exists for many quality traits in different vegetable crops although non-genetic influences also play role in the determination of vegetable quality. Plant genetic resources with wide and utilizable genetic variability for plant pigments are available. As for example, enhanced carotenoid profiles in sweet potatoes (Wu *et al.,* 2008; Afuape *et al.,* 2019), potato (Lachman *et al.,* 2016; Das *et al.,* 2023), carrot (Baranski *et al.,* 2012; Bhandari *et al.,* 2022), etc. and enhanced anthocyanin profile in radish (Singh *et al.,* 2017). Pavithra *et al.* (2014) found tomato lines rich in Zn content. Studies on Andean potato cultivars have exhibited a broad range of calcium, iron, and zinc contents in them (Andre *et al.,* 2007). Total anthocyanin concentrations in Andean potatoes ranged very widely from 14 to 16, 330 µg/g dry weight (Andre *et al.,* 2007). Usually, cultivars high in anthocyanins are low in carotenoids and *vice versa.* These widely divergent lines of different vegetable crops with high pigment and mineral contents form the base for development of nutritionally enriched variety. These elite germplasms can also be used to identify the prospective candidate gene through biotechnological tools for improving the nutritional value.

Breeding techniques

Breeding method in any crops depends upon the breeding system and genetic architecture resulting from natural selection as well as human selection during the course of cultivation. The genetic architecture or the pattern of inheritance of characters is another important consideration while determining the most appropriate breeding procedure applicable to any particular crops. The choice of breeding method would largely be guided by nature of gene action and relative magnitude of additive genetic variance, dominance variance and epistasis in a breeding population for quantitative characers. Development of F_1 hybrid is highly suitable for enhancing nutraceutical contents and edible colours in several vegetable crops. β-carotene content increased manifold in the F_1 hybrids of muskmelon (Moon *et al.,* 2002) and pumpkin (Pandit *et al.,* 2006). The best performing cytoplasmic male sterile (CMS) lines, *viz*. RRMA, RJA, PMA, RCA, 208A and 5A with higher concentrations of different nutritional compounds can be utilized in future breeding programs for quality F_1 hybrid development in cabbage.

Advanced breeding techniques

• Utilization of wild relatives to enhance quality traits

The wild relatives include all other species, which are related to crop species by descent during their evolution. Renowned Russian plant geneticist Nicolai Vavilov first recognized their potential as gene donors for crop improvement during the 1930s. Crop wild relatives are gaining increasing attention to increase the genetic base of the elite varieties because they maintain much higher level of genetic diversity compared with domesticated cultivars. High nutrient and bioactive content of many wild relatives has made them valued contributors for balanced diet and play significant role in human nutrition. Their use in plant breeding is essential to broaden the genetic basis of crops as well as to meet industrial demands, for global food security, sustainable production and enhanced quality.

Breeding tomato sets a unique example where different wild relatives (*Solanum pimpinellifolium*, *Solanum chmielewskii*, *Solanum habrochaites, Solanum pennelli*, *Solanum cheesmanii*, and *Solanum peruvianum*) being rich source of useful genes have been used successfully in developing several quality traits in cultivated tomato (Stommel, 2007). Segregation patterns for fruit pigmentation and percentage β carotene in F_1, F_2 and backcross populations emanated from *Solanum esculentum* × *Solanum cheesmanii* cross fit the expected models for a single dominant gene conditioning a high percentage of β-carotene and resultant orange fruit pigmentation. The inheritance of β carotene in *Solanum cheesmanii* paralleled that of the *B* gene described in *Solanum habrochaites*. Variation in colored carotenoid content indicated a parental influence on total lycopene and β-carotene content and suggested independent genetic control of these carotenoid levels (Stommel and Haynes, 1994).

An accession LA722 was identified within *Solanum pimpinellifolium*, which exhibited a number of desirable desirable characteristics including salt tolerance, disease resistance and high fruit quality (Chen and Foolad, 1999). Bernacchi *et al.* (1998) produced improved processing quality tomato lines by the molecular breeding strategy of advanced backcross quantitative trait loci (AB-QTL) analysis. These near-isogenic lines (NILs) contained unique introgressions of wild alleles originating from two donor wild species, i.e., *Solanum habrochaites* (LA1777) and *Solanum pimpinellifolium* (LA1589). Inter-specific hybridization with *S. pimpinellifolium* has been found encouraging in increasing the total soluble solids content of the fruit (Hazra *et al.*, 2001) although the alleles of *S. pimpinellifolium* reduced the fruit size considerably, suggesting remote likelihood in recovering big-fruited segregates from such cross (Grandillo *et al.*, 1996; Hazra *et al.*, 2001). Segregation pattern of the inter-specific hybrids of cultivated tomato x *Solanum pimpinellifolium* and cultivated tomato × *Solanum lycopersicum* var. *cerasiformae* suggested that there was more likelihood of getting the recombinant lines towards the wild relatives for fruit size traits. However, there is possibility of recovering promising segregates having medium fruit size with tolerance to leaf curl virus and enhanced fruit quality through advancement of these inter-specific hybrids. (Acharya *et al.*, 2024). A gene controlling sucrose accumulation in fruit, *sucr*, was introgressed from the wild species *Solanum chmielewskii* into the genetic background of a hexose accumulating cultivated tomato (Chetelat and Ji, 2006).

Total soluble solids content is a key determinant of tomato fruit quality for processing. Using the ILs developed with the wild species *Solanum pennellii*, 23 OTLs connected with increased total soluble solids content could be revealed and mapped (Eshed and Zamir, 1995). Charles *et al.* (2005) also developed several tomato lines carrying defined introgressions from *S. pennellii* in a *S. lycopersicum* background, which produced fruit with elevated Brix value, a refractive index measure of soluble solids. Based on interspecific hybridization between cultivated tomato and *S. peruvianum*, tomato cultivar "Double Reach" possessing high vitamin C content was developed (Stevens and Rick, 1986).

To evaluate components of fruit metabolic composition, Schauer *et al.* (2008) used previously metabolically phenotyped tomato introgression lines containing segmental substitutions of wild species chromosome in the genetic background of a cultivated variety. They studied the hereditability of fruit metabolic products by analyzing an additional year's harvest and evaluating the metabolite profiles of lines heterozygous for the introgression (ILHs), allowing the evaluation of putative quantitative trait locus (QTL) mode of inheritance. These studies revealed that most of the metabolic QTLs (174 of 332) were dominantly inherited, with relatively high proportions of additively (61 of 332) or recessively (80 of 332) inherited QTL and a negligible number displaying the characteristics of over-dominant inheritance.Comparison of the mode of inheritance of QTL revealed that several metabolite pairs displayed a similar mode of inheritance of QTL at the same chromosomal loci. Evaluation of the association between morphological and metabolic traits in the ILHs revealed that this correlation was far less prominent due to a reduced variance in the harvest index within this population.

Different wild relatives of potato (*Solanum chacoense* Bitter, *S. pinnatisectum* Dunal, *S. tuberosum* subsp. *andigena* Hawkes, *S. commersonii* Dunal, *S. hondelmannii* Hawkes and Hjerting, *S. jamesii* Torrey, *S. microdontum* Bitter, *S. sparsipilum* (Bitt.) Juz. and Bukasov, *S. stenotomum* Juz. and Bukasov, and *S. tarijense*) appeared superior to cultivars in terms of tuber starch contents and nitrogen utilization efficiency which can be utilized in breeding programme of potato for quality (Bachmann-Pfabe and Dehmer, 2020).

Traditional small potato cultivars of North Eastern region of India were tetraploid with an average genome size of 837.38 Mb based on ploidy level study by flow cytometry method although they were suspected to be diploid based on the tuber characters and overall plant morphology. These cultivars were characterized by having anthocyanin pigmentation of newly emerged sprouts and eye on tuber, small and narrow leaf, long stolon length (average 7.38 cm) and high number (average 33.90/plant) of very small (average 7.41 g) tubers compared to the commercial potato varieties. However, these cultivars surpassed India's commercial tetraploid potato varieties for dry matter, starch, ascorbic acid, β-carotene and total phenol contents. Appreciably rich nutritional profile coupled with matching levels of ploidy not only enhanced the biodiversity of cultivated potato but also suggested the possibility of utilizing these cultivars in the breeding programme to enhance nutrient and antioxidant contents (Das *et al.*, 2023).

• Utilization of mutants for quality traits

In a simple way, mutation is a random or directed change in the structure of DNA or the chromosome, which often result in a visible or detectable change in specific character or trait. In self-pollinated crops, it is well known whereas in cross-pollinated crops, its application is more difficult and identification of the origin of the desirable genotypes is difficult.

Spontaneous mutants

Tomato

Enhanced pigmentation is of a major economic importance in tomato contributing to their visual and functional properties. At the best of present knowledge, there are 64 tomato mutants with altered pigmentation of flowers and fruits (Hirschberg *et al.*, 2008).Few widely utilized genes, which result in enhanced carotenoid accumulation have been identified and the prominent mutants are *high pigment* (*hp-1*, *hp-1*^{w}, *hp-2*, *hp-2*^{j}, *hp-2*^{dg} and *hp-3*), *old gold crimson* (*og*^{c}) and *dark green* (*dg*) which have been extensively summarized (Lieberman *et al.*, 2004; Levin *et al.*, 2006). The high pigment mutants have higher pigmentation compared to normal wild tomato at all developmental stages.Tomato high pigment (*hp*) mutants are characterized by their exaggerated photo-responsiveness. Light-grown *hp* mutants display elevated levels of pigmentation, are shorter and darker than wild-type plants and have dark green immature fruits due to the overproduction of chlorophyll pigments (Mustilli *et al.*, 1999). The *old gold crimson* (*og*^{c}) gene increases lycopene content in tomato fruit by about 50%, while reducing β carotene by the same amount (O'Hare *et al.*, 2005).

The third mutant *dark green* (*dg*), characterized by exaggerated photoresponsiveness, as well, was found to be allelic to *hp 2* (Levin *et al*., 2003). The different genotypic combinations at the loci *alc*, og^c and *hp* had no effect on fruit yield. The alc^+/alc hybrid genotype significantly increased fruit firmness and significantly delayed the development of red color in maturing fruit. Simultaneous usage of og^{c+}/og^c and hp^+/hp promoted an increase in the red colour and lycopene content of alc^+/alc hybrids but did not have any additional effect on fruit firmness. Increasing both contents of β-carotene and intensity of red colour in tomato fruits may be achieved by coupling "crimson" og^c and *hp* pigment characters in one genotype (Faria *et al.*, 2003). A number of mutants, which controls delayed fruit ripening *viz*., *Never ripe* (*Nr*), *ripening inhibitor* (*rin*), *alcobaca* (*alc*) and *non-ripening* (*nor*) have been characterized. Fruits of the mutant fruits were typically non-climacteric. They lacked an obvious respiratory peak and concomitant rise in ethylene at the time when normal fruit undergo ripening and colour development occurred slowly during this period. Incorporation of particularly *rin* and *nor* alleles into standard cultivars has been extensively explored as a way of reducing the loss of fruit firmness during and after ripening, therefore, extending the shelf life of fruits (Tucker, 1990). According to Mutschler *et al.* (1992), the *alc* mutation affects the ripening and storability of tomato fruit due to a reduction in total pigment and a reduction in lycopene relative to total carotinoids and polygalacturonase (PG) activity was reduced to less than 5% of normal. However, the *nor* F_1 hybrids had the best fruit colour, firmness and shelf life (Seroczynska *et al.*, 1998) although it was observed that environmental factors exert influence on redness and lycopene level in tomato fruits (Potaczek and Michalik, 1998).

In tomato, different lycopene enhancing mutants *viz*., *dg*, *hp-1* and og^C ; anthocyanin expressing mutant *viz*., *Aft* in Alisa Craig background and different fruit ripening manipulator mutants *viz*., *rin*, *nor* and *Nr* of tomato have been fully characterized at both homozygous and heterozygous level for different quality characters (Biswas *et al*., 2016). These mutants have also been characterized at the level of bioactive compound (Siddiqui *et al.,* 2014), metabolic behaviour (Pal *et al.,* 2019) and postharvest physico-chemical changes in the fruits (Siddiqui *et al*., 2014; Siddiqui *et al.,* 2015). Immense possibility exists to utilize these mutants in the breeding programmes of tomato for enhanced fruit quality and shelf life (Biswas *et al*., 2016).

Breeding strategy has been framed to combine the lycopene enhancing gene *dg* or $hp\text{-}2^{dg}$ present in chromosome 1 and anthocyanin-expressing gene, *Aft* present in chromosome 10 in one genotype taking advantage of independent assortment in the ratio of 1:15 in the F_2 population based on morphological markers to develop "Purple tomato" line through conventional method. This breeding scheme took only one crossing Alisa Craig *Aft/Aft* × BCT-115 *dg/dg* and two selfings to select the desirable double homozygous genotype *AftAft dg/dg*. Ripe fruits of this novel breeding line contain both high lycopene (6.13 mg/100 g fresh) and anthocyanin (20.73 mg/100 g fresh) pigments (Hazra *et al*., 2018).

Applied mutagenesis

Tomato

In tomato, macro-mutants could be isolated from gamma-radiation induced M_2 generation based on wide array of distinct, unique and stable characters *viz.*, plant stature, leaf morphology, inflorescence type, floral morphology, pedicel character, fruit morphology and chlorophyll content of leaf and fruit. Two mutants *viz.*, "Multiparus cyme mutant" from Berika, the line bred variety of Bulgaria and "Dark green fruit mutant" from the cultivar, Patharkutchi emerged as most promising for further use in breeding programme (Sikder *et al.*, 2013; Das *et al.*, 2019; Das *et al.*, 2020). Multiparus cyme mutant induced by 50 Gy gamma radiation manifests a unique character where the primary axis of the inflorescence "cyme" produced lateral branches with increased flower as well as fruit number per plant. Fruits are pyriform, three-loculed with joint-less pedicel, dented blossom end, and are suitable for processing purpose. High chlorophyll containing "Dark green fruit mutant" induced by 150Gy gamma radiation produces dark green leaf (total chlorophyll content 318.52 mg/100 g fresh) as well as dark green fruit at unripe stage (total chlorophyll content 21.93 mg/100 g fresh. This induced "Dark green fruit mutant" resembling the spontaneous mutant gene *dg* or $hp\text{-}2^{dg}$ present in chromosome 1 bear dark red ripe fruits with high lycopene content (7.49 mg/100 g fresh).

Bitter gourd

In bitter gourd, exposure of dry seeds to 200-250 Gy gamma radiation was the most efficient in inducing desirable macro-mutants based on variation in sex form, fruit colour, fruit shape, prickliness, fruit length, girth, weight and charantin content (Dutta *et al.*, 2021). Radicle scavenging activity in the fruits of two macro-mutants of the white fruited cultivar, BG-1346501 (BG-1346501- 100 Gy) and (BG-1,346,501-200 Gy) improved significantly by 9.12 and 16.12 percent, respectively and total charantin content was maximum (1.26 mg /100 g dry weight) in the mutant line BG-1346501-100 Gy (Hazra *et al.*, 2022).

Linking of morphological traits with quality of bitter gourd

Predominant bitter gourd types of Indian origin (dark green, spindle shaped fruits with sharp tubercles) were associated with high nutrient and most of the nutraceutical contents (β carotene, ascorbic acid, total phenol and flavonoid contents) but comparatively low in charantin content. The predominant types in the region of China, Japan, Indonesia, Korea (whitish yellow, yellowish white, obtuse or roundish-oblong fruit with light ridges or roundish blunt tubercles) were associated with comparatively low nutrient and most of the nutraceutical contents but high in charantin content. This study confirmed the diversity among the genotypes, effectiveness of combining morphological characters with nutritional, and nutraceutical contents of fruit in future breeding programmes (Hazra *et al.*, 2022).

Alteration of ploidy for enhancing quality traits

Polyploid breeding

Polyploidy refers to the presence of more than two complete sets of chromosomes per cell nucleus. It is responsible for increasing genetic diversity and producing species with increased size, vigour, quality and resistance to diseases. Thus, production and application of polyploidy breeding have brought remarkable economic and social benefits.

Kihara (1951) first developed the tetraploidization process for developing seedless watermelons. By interbreeding a tetraploid (female parent) and a diploid (male parent) inbred line, seedless varieties ($2n = 3x = 33$) were created. The triploid hybrid fruit is seedless since the triploid hybrid female is sterile. Since triploids lack viable pollen, at least 20% of the diploid cultivar must be grown in the production area in order to provide pollen that promotes fruit set (Miri, 2020). Due to its special qualities, including its small size, vigorous growth, high fruit number per plant, high sugar content, flesh firmness, thin rind, and potential extended shelf life, triploid seedless watermelon is well-liked commercially and commands a premium price in the global market (Noh *et al.,* 2012).

Tetraploids in radish, pumpkin, and watermelon are highly productive and have improved quality. Zhang *et al.* (2010) developed tetraploid muskmelon, which is rich in vitamin C than those in the diploid fruit.

Polyploid *Capsicum frutescens* fruits were larger and contained more capsaicin than diploid fruit (Pliankong *et al.,* 2017). Vitamin C concentration has also been reported to be high in polyploids caused by the application of colchicine (Miri, 2020). Tetraploid parsley (*Petroselinum crispum*) plants treated with colchicine had larger leaves and stems than diploid plants (Nasirvand *et al.,* 2018). Induced tetraploid radish plants have huge leaves and a taproot in addition to having high levels of soluble sugar and protein, vitamin C, and antioxidant enzyme activity (peroxidase). Additionally, qRT-PCR results revealed that FLC1.1, a gene known to suppress flowering, was highly expressed in tetraploid plants (Pei *et al.,* 2019).

Utilization of haploids and double haploids

Haploid plants are the sporophytes with a gametophytic chromosome number and doubled haploids (DH) are haploids that have undergone chromosome duplication. Haploids and double haploids occur spontaneously in nature. Haploids can also be induced artificially by various techniques such as wide hybridization, pollination with irradiated pollens, androgenesis, gynogenesis and finally further chromosome doubling by the application of colchicine results in double haploids. Production of double haploids in vegetables shortens the breeding cycle. The DH technology is also useful in reverse breeding, CMS line production, gene stacking and various other genetic studies. Doubled haploidy continue to be a very efficient tool for the production of completely homozygous lines from heterozygous donor plants in a single step. Furthermore, the haploid induction technique can now-a-days be efficiently combined with several other plant biotechnological techniques, enabling several novel breeding achievements, such as improved mutation breeding, backcrossing, hybrid breeding and genetic transformation.

Biotechnological approaches for improvement of quality

In past five decades, conventional breeding has contributed significantly for the improvement of vegetable yield, quality, post-harvest life and resistance to biotic and abiotic stresses. However, there are many constraints in conventional breeding, like pre/post fertilization barriers in distant hybridization programme, linkage drag, difficulty in identification of target characters in large population, etc. Different biotechnological tools can successfully be employed to overcome the impediments of breeding. With the development of molecular markers, genetic maps have been constructed in some vegetable crops including potato, tomato, pea, cucumber, etc. which allow for the localization of major loci and QTLs controlling different traits. Molecular markers have been widely used for the introgression of major loci. However, marker-assisted selection for quantitative trait breeding is used less. In recent years, significant progress has been made to manipulate vegetable crops for abiotic stress tolerance, quality improvement and pharmaceutical and industrial applications. Though the progress in commercialization of transgenic vegetable crops has been very slow, transgenic vegetables engineered for nutraceutical and pharmaceutical use will contribute significantly to the value added agriculture in near future. Some examples of the utilization of different biotechnological tools in breeding for quality traits are cited, although such studies are many.

Molecular markers and marker assisted selection (MAS)

Molecular markers such as RAPD, ISSR, SSR, SCAR, CAPS are used to study linkage with gene(s) responsible for high neutraceuticals, bioactive compounds and edible colors using mapping population. Use of Single nucleotide polymorphic (SNP) marker is becoming more common. In marker assisted selection, a marker (morphological, biochemical or DNA) is used for indirect selection of a trait of interest. The mapping populations such as Near Isogenic Lines (NILs), Recombinant Inbred Lines (RILs) are employed to identify the molecular markers linked to genes of interest. Ripley and Roslinsky (2005) identified an ISSR Marker for 2 propenyl glucosinolate content in *Brassica.*

Investigation on the metabolic profile of *Or* gene mutants reported that β-carotene is the principal carotenoid compound in edible portion. AFLP markers, tightly linked with *Or* locus and developed functional SCAR markers to simplify positional cloning of *Or* gene which is very benificial for improving the quality. In broad bean, *zt 2* gene specific SCAR marker identified increase in protein content and reduced fiber content in the seeds (Gutierrez *et al.,* 2008). Song *et al.* (2010) reported that the common cucumbers have low level of carotenoid i.e., only 22 to 48 μg/100 g, but Xishuangbanna gourd having high level of carotenoid quantified as 700 μg/100 g on flesh weight basis has unique advantage as a germplasm to be used for enhancing nutritive value of the cucumber. The gene regulating β-carotene content was incorporated into cultivated background of cucumber from *Cucumis sativus* var. *xishuangbannesis*. Inheritance pattern of β-carotene in endocarp showed that the trait was controlled by a single recessive gene. Seven SSR makers specific to endocarp carotene content were identified on linkage group three and mapped in cucumber chromosome along with the putative candidate gene. QTL associated with orange color fruit flesh showed two genetic linkage maps with the markers of RAPD, SCAR, SSR, EST, SNP, AFLP and SSAP, which defined a common collinear region containing four molecular markers on linkage group LG6 inMAP1 and LG 3 in Map 2. Sinclair *et al.* (2004) identified two QTL affecting Vitamin C in melon and they were responsible for 14 and 12 % phenotypic variation, respectively. Four out of the nine molecular markers steadily linked with sweetness of fruits have been reported. Seven SSR makers were identified which can be utilized in marker assisted breeding for developing cucumber germplasm with high β-carotene content in the fruits.

In Chinese cabbage, SCAR markers linked to the *Or* gene were identified based on random amplified polymorphic DNA (RAPD) and amplified fragment length polymorphism (AFLP) through bulk segregant analysis (BSA) using a double haploid (DH) population derived from the F_1 cross between 91 and 112 (white head leaves) and T12–19 (orange head leaves) *via* microspore culture. On the basis of linkage analysis, the *Or* gene was mapped in a region conversing a total interval of 4.6 cM between two SSR markers derived from BAC clones AC172873 and AC189246 at the end of linkage group 9, which matches with chromosome I of A genome. A genetic map of the '*Or*' locus was constructed by using five SSR markers and two morphological markers. Three SSR markers were tightly linked to *Or* and two of them, *sau* (C) 586 and *syau* 19, were located on the same side at distances of 1,6 and 1,3 cM, respectively. The other marker, *syau* 15, was located on the other side at a distance of 3.3 cM. Ripley and Roslinsky (2005) identified ISSR marker for 2- propenyl glucosinolate content in *Brassica*. Efforts are also being made to use the genetic and molecular approaches for increasing tocopherol contents in potato tubers through metabolic engineering tools and techniques.

Quantitative trait loci (QTL) analysis

QTL analysis is the study of the alleles that occur in a locus and the phenotypes. Most traits of interest are governed by more than one gene, defining and studying the entire locus of genes related to a trait gives hope of understanding the effect genotype of an individual. The advent of molecular maps and the derived quantitative trait locus (QTL) mapping technology has provided strong evidence that despite the inferior phenotype, exotic germplasm is likely to contain QTLs that can increase the quality of elite breeding lines. Bin 3-C has previously been described as harboring a single gene mutation *r* yellow flesh in tomato (Fray and Grierson, 1993). QTL analysis of fruit antioxidants in tomato using *S. pennellii* introgression lines detected a total of 15 QTL including, six for ascorbic acid, and nine for total phenolics (Rousseaux *et al.*, 2005). Cervantes-Flores *et al.* (2010) have reported QTL for dry matter, starch content and ß-carotene content, opening up the possibility of genetic manipulation and further nutritional enhancement of sweet potato.

Advanced backcross QTL analysis

The AB-QTL strategy has so for been tested in tomato, chilli and sweet pepper. The most extensive experiments have been conducted in tomato, where populations involving crosses with five wild *Solanum* species have been genotyped and field tested in a number of locations around the world for numerous traits important for the tomato processing industry. Through the application of marker and phenotypic analysis of segregating generations of the cultivated tomato and wild relatives of tomato, QTLs for improved fruit colour could be identified. QTLs that improved fruit colour originating from red-fruited *S. pimpinellifolium* and green fruited S. *habrochaites* and *S. parviflorum* wild relatives could be detected in segregating populations of crosses of these species and the cultivated tomato. Quantitative trait loci associated with carotenoids and tomato fruit colour using introgression populations of *S. pennellii, S peruvianum* and *S. habrochaites* have been described in detail (Berancchi *et al.,* 1998).

Introgression line (IL) libraries

Introgression line (IL) libraries contain homogenous genetic backgrounds, only differing from one another by the introgressed donor segment. A tomato introgression line population that combines single chromosomal segments introgrossed from the wild, green fruited species *Solanum pennellii* in the background of the domesticated tomato was used to identify QTL for nutritional and antioxidant contents. Liu *et al.*(2003) applied the candidate gene approach to link sequences that have known functional roles in carotenoid biosynthesis to QTLs that are responsible for the variation of the tomato red fruit color.

Marker assisted backcross breeding has been used successfully to incorporate genes or QTL for both qualitative and quantitative traits in a number of crop species especially tomato, cucumber, potato leading to the development of improved cultivars. Indian cauliflowers have been introgressed with semi-dominant mutant *Or* gene to enhance their β-carotene content in an attempt to tackle malnutrition problem by making diverse β-carotene rich food available to consumers.

The dominant gene *Anthocyanin fruit* (*Aft*), which induces limited anthocyanin pigmentation upon stimulation by high light intensity was introgressed into domesticated tomato plants by an interspecific cross with *S. chilense* and the gene *Aubergine* (*Abg*) from *Solanum lycopersicoides*. Furthermore, the recessive gene *atroviolacea* (*atv*), derived from the interspecific cross with *Solanum cheesmaniae* stimulate strong anthocyanin pigmentation in the entire plant, particularly in vegetative tissues. Fruits with either *Aft* and *atv* alleles or *abg* and *atv* alleles have been obtained with higher production of anthocyanins in the peel, ranging in total amount from 1 to 4 mg/g fresh weight of peel. Anthocyanins were found in the skin and flesh of certain cultivars of potato. The fruit colour of red chili is genetically determined by three loci *y*, *cl*, and *c2*. Recently the gene for capsanthin-capsorubin synthase (CCs) has been considered as candidate gene for the *y* locus. The relationship between the phytoene synthase and carotenoid content in chilli was tested with interval mapping using QTL analysis revealed that they were detected only at the *PSY* locus.

Singh *et al.* (2002) observed enormous diversity in pigmentation of European and Asiatic carrot. The Asiatic types are mostly yellow and purple. The Asiatic type collection Local Rewari Black and Local Jaipur Black have higher anthocyanin content.

Few molecular markers linked to major genes or QTL have been developed for carotene content in carrot (Santos and Simon, 2002). To date, seven monogenic traits have been mapped for carrot viz., *yel, cola, Rs, Mj-1, Y,* Y_2 and P_1. QTL have been mapped for total carotenoids and five component carotenoids: phytoene, x-carotene, ß-carotene, zeta-carotene, and lycopene in carrot (Santos and Simon, 2002) and the majority of the structural genes of the carotenoid pathway are placed into this map. Anthocyanin accumulation in the carrot phloem is conditioned by the P_1 locus, with purple (P_1) dominant to non-purple (p_1). From the inheritance studies it was concluded that the P_1 and Y_2 loci are unlinked.

Transgenic approach for nutritional enrichment of vegetable crops

Genetic engineering facilitates vegetable breeders to integrate preferred transgene(s) into desirable back ground which may be well established varieties for enhancing their nutritive value significantly. Nevertheless, it offers distinctive prospects to improve nutritive quality and bringing various health benefits. Different vegetables were genetically transformed to enhance their nutritional quality by increasing nutritive composition or improve flavour, and by reducing anitinutritional factor including bitterness.

Enhancing β-carotene content: One of the most obvious benefits of enhancing carotenoid levels is the increase in pigmentation, which can lead to more deeply colored vegetables that are often preferred by the consumers. Thus, increasing levels of carotenoid content is doubly beneficial, both in terms of nutrition and esthetics. Three genes, encoding *phytoene synthase* (*CrtB*), *phytonene desaturase* (*Crt1*) and lycopene β-cylase (*CrtY*) from *Erwinia* have been introduced in potato to produce β-carotene. Romer *et al.* (2000) developed transgenic tomato to enhance the carotenoid content with the bacterial carotenoid gene (*crtl*) encoding the enzyme phytoene desaturase, which converts phytoene into lycopene. Lu *et al.* (2006) suggested that transgenic cauliflower with *Or* transgene is associated with a cellular process that triggers the differentiation of proplastids into chromoplasts for carotenoid accumulation and *Or* can be used as a novel genetic tool to induce carotenoid accumulation in a major staple food crops. Transformation of carrot with bacterial *crtB* gene reported 93 time rise in carotenoid content (Maass *et al.,* 2009).

There are a range of other approaches to enhance the carotenoid levels in potatoes and other root vegetables. Diretto *et al.* (2006) have silenced the first step in the ß-epsilon branch of carotenoid biosynthesis, *lycopene epsilon cyclase* (LCY-e) in potato which is low in carotenoids. This antisense tuber-specific silencing of the gene results in significant increases in carotenoid levels, with up to 14-fold more ß-carotene.

Enhancing anthocyanin and flavonoid contents: Potato does not normally produce anthocyanin, but germplasm expressing anthocyanin pigment has been developed which has attracted interest from the consumers. One of these genes which encode a novel single-domain MYB transcription factor has the potential to influence anthocyanin pigment production in potato. The resulting purple potato might offer both novelty and health functionality to consumers, who can also benefit from native Andean potatoes that do not always show desired tuber shapes for both table and processing industry. Substantial increase in the flavonoid content was reported in tomato genetically transformed with petunia *CHI-A* gene that encrypts *chalcone isomerase* (Muir *et al.,* 2001). In tomato 10-time enhanced flavonoid content was reported through ectopic expression of the transcription factors *LC* and *C1* from corn (Le Gall *et al.,* 2003)

Enhancing folate content: Diaz de la Garza *et al.* (2004) developed transgenic tomatoes by engineering fruit specific over expression of *GTP cyclohydrolase I* that catalyzes the first step of pteridine synthesis, and *aminodeoxychorismate synthase* that catalyzes the first step of PABA synthesis. Vine ripened fruits contained on average 25-fold more folate than controls by combining PABA and pteridine overproduction traits through crossbreeding of transgenic tomato plants.

Enhancing glucosinolate content in *Brassica* vegetable: The extent to which vegetable brassicas show cancer protection properties probably depend on the genotype possessing the allele present at the *GSTM 1* locus. This gene codes for the enzyme *glutathione transferase*, which catalyzes the conjugation of glutathione with isothiocyanates. Approximately, 50% of humans carry a deletion on the *GSTM1* gene which reduces their ability to conjugate process and excrete isothiocyanates. Individuals with two null alleles for *GSTM1* might gain less protection from these cultivars. *Brassica. rapa* contains different isothiocyanates compared to *B. oleracea* and evidence suggested that individuals who are null for *GSTM1* can gain a protective benefit from *B. rapa* (Gasper, 2005).

A 10-fold increase in the level of 4-methylsulphinylbutyl glucosinolate was obtained by crossing broccoli cultivars with selected wild taxa of the *Brassica oleracea* (chromosome number, n = 9) complex. Tissue from these hybrids exhibited 100-fold increase in the ability to induce quinone reductase in Hepa1clc7 cells over broccoli cutlivars, due to an increase in 4-methyl sulphinyl butyl glucosinolate content.

Reducing lachrymatory factor in *Allium*: Vegetables of the *Allium* genus such as onion, garlic, leek and chive are among the oldest crops associated with health-related properties. Three sets of transgenic onion plants containing antisense alliinase gene constructs (a CaMV 35S-driven antisense root alliinase gene, a CaMV 35S-driven antisense bulb alliinase, and a bulb alliinase promoter-driven antisense bulb alliinase) have been produced (Eady *et al.,* 2008). Transgenic hybrid onion seed from these transgenic lines has been developed by crossing a non-transgeneic open-pollinated parental line with a transgenic parental plant carrying a single transgene in the hemizygous state.

Miraculin rich vegetables: For reduction of bitterness in lettuce, the gene for sweetness and taste modifying protein miraculin, from the pulp of berries of West African shrub *Richadella dulcifica* was cloned (Sun *et al.,* 2006). This gene, with the CaMV 35S promoter, was introduced into the lettuce cultivar "Kaiser" using *A. tumefaciens* GV2260. Expression of this gene in transgenic plants led to the accumulation of significant concentrations of the sweet enhancing protein.

Fruit quality improvement through induction of parthenocarpy: In brinjal fruit quality could be improved through genetic manipulation for parthenocarpy (seedless fruits) by altering auxin levels in the developing ovules of the transgenic brinjal using *Agrobacterium tumefasciens* transformation of DefH9-iaaM gene (Acciarri *et al.*, 2002).

Protein rich potato: The genetically modified potato developed at CPRI in collaboration with NIPGAR "Protato" contains 60% enhanced protein content. This has been achieved by introducing *AmA1* gene (*Amaranth Albumin 1*) from edible amaranth (*Amaranthus hychondriacus*) into seven commercial varieties of potatoes. The *AmA1* gene has been reported to have potential for the nutritional improvement of other food crops as well (Delgado-Vargas *et al.,* 2003). The GM potato plants were tested in India and the results demonstrated greater harvest and moderate increase in tuber yield. Safety evaluation indicated that the transgenic potatoes are suitable for commercial cultivation and have no negative effects on animal health. In addition, the concentration of several essential amino acids increased significantly in transgenic tubers which are otherwise limited in potato (Chakraborty *et al.,* 2010).

References

Acciarri, N., Restaino, F., Vitelli, G., Perrone, D., Zottini, M., Pandolfini, T., Spena, A. and Rotino, G. 2002. Genetically modified parthenocarpic eggplants: improved fruit productivity under both greenhouse and open field cultivation. *BMC Biotechnol*, doi: 10.1186/1472-6750-2-4.

Acharya, B., Hazra, S., Gorai, S., Bose, S., Hazra, P., Chattopadhyay, A., Pal, H. and Maji, A. 2024. Genetic control of fruit quality characters in the segregating generations of interspecific crosses and metabolic profiling of the fruits of promising segregates of tomato. *South Afr. J.Bot.*, (in press).

Afuape, S.O., Tongoona, P.B., Asante, I.K., Egesi , C.N., Nwaigwe, G. and Samuel K. Offei, S.K. 2019. Breeding new sweetpotato genotypes combining elevated content of two antagonistic traits of high dry matter and beta-carotene content in a high root yield background, *Euphytica*, 215: DOI:10.1007/s10681-019-2532-0

Alexandra, S. I. M., and Andreea D, O. 2020. Cabbage (*Brassica oleracea* l.). Overview of the health benefits and therapeutical uses. Hop and Medicinal Plants, Year, XXVIII(1–2): 150- 169.

Allen, L.H. and Ahluwalia, N. 1997. Improving Iron Status through Diet. In: *Relative Impact of Potential Interventions to Improve Dietary Iron Availability, Opportunities for Micronutrients Intervention* (ed. Wagner, L.V.), Washington DC, USA, pp.1-86.

Anand, P., Kunnumakkara, A.B., Sundaram, C., Harikumar, K.B., Tharakan, S.T., Lai, O.S., Sung, B.and Aggarwal, B.B. 2008. Cancer is a preventable disease that requires major lifestyle changes. *Pharm Res.*, **25**:2097–116.

Andre, C.M., Ghislain, M., Bertin, P., Qufir, M., Del Rosario Herrera, M., Hcffmann, L., Hausman, J.F., Larondelle, Y. and Evers, D. 2007. Andean potato cultivars (*Solanum tuberosum* L.) as a source of antioxidant and mineral micronutrients. *J.Agril. Food Chem.*, **55**:366-378

Andre, C.M., Oufir, M., Guignard, C., Hoffmann, L., Hausman, J.F., Evers, D. and Larondelle Y. 2007. Antioxidant profiling of native Andean potato tubers (*Solanum tuberosum* L.) reveals cultivars with high levels of beta-carotene, alpha-tocopherol, chlorogenic acid, and petanin. *J. Agric. Food Chem.*, **55**:10839–49.

Arai, Y., Watanabe, S., Kimira, M., Shimoi, K., Mochizuki, R., Kinae, N.. 2000. Dietary intakes of flavonols, flavones and isoflavones by Japanese women and the inverse correlation between quercetin intake and plasma LDL cholesterol concentration. *J. Nutr.*, **130**:2243–2250.

Baas-Espinola , F.M., Castro-Concha , L.A., Vázquez-Flota , F.A. and Miranda-Ham , M.L. 2016. Capsaicin Synthesis Requires *in situ* Phenylalanine and Valine Formation in *in Vitro* Maintained Placentas from *Capsicum chinense. Molecules*, **21**: 799doi: 10.3390 / molecules 21060799

Bachmann-Pfabe, S. and Dehmer, K.J. 2020. Evaluation of Wild Potato Germplasm for Tuber Starch Content and Nitrogen Utilization Efficiency. *Plants, (Basel)* **9**: 833; https://doi.org/10.3390/plants9070833

Baranski , R., Allender, C. and Klimek-Chodack, M. 2012. Towards better tasting and more nutritious carrots: Carotenoid and sugar content variation in carrot genetic resources. *Food Res. Intern.,* **47**: 182-187

Batool, M., Roobab,U., Manzoor, M.F., Farooq, U., Nadeem, H.R., Nadeem, M., Kanwal, R., Abd Elgawad, H., Al Jaouni, S.K., Selim, S., and Ibrahim, S.A. 2022. Nutritional Value, Phytochemical Potential, and Therapeutic Benefits of Pumpkin (*Cucurbita* sp.). *Plants* (Basel), **11**: 1394.doi: 10.3390/plants11111394

Beck, K.L. 2016. Anemia: Prevention and Dietary Strategies. In: *Encyclopedia of Food and Health* (Editor-in-Chief. Caballero, B., Finglas, P.M. and Toldra, F.), Academic Press, USA.

Bernacchi, D., Beck-Bunn, T. , Eshed, Y. , Lopez, J. , Petiard, V. , Uhlig, J. , Zamir, D. and Tanksley, S. 1998. Advanced backcross QTL analysis of tomato. I. Identification of QTL for traits of agronomic importance from *Lycopersicon hirsutum. Theor. Appl. Genet.*, **97**:381–397.

Bhandari, S.R., Rhee, J., Choi, C.S., Jo, J.S., Shin, Y.K., Song, J.W., Kim, S.H. and Lee, J.G. 2022. Morphological and Biochemical Variation in Carrot Genetic Resources Grown under Open Field Conditions: The Selection of Functional Genotypes for a Breeding Program. *Agronomy, 12*, 553; https://doi.org/10.3390/agronomy12030553

Bhanushree, N., Saha, P., Tomar, B. S., Lyngdoh, Y. A., Gopala, K. S. and Gurung, B. 2019. Genetic variability study for yield components and anthocyanin in eggplant (*Solanum melongena* L.). *Indian J. Agric. Sci.*, **89**:1082-1085.

Biswas, P., Chattopadhyay, A.and Hazra, P. 2016. Genetic control of fruit yield and quality characters in tomato genotypes possessing *hp-1*, *og*^{c}, *dg*, *Aft* and *rin*genes. *Indian J. Genet. Plant Breed.*, **76**: 326-332

Björkman, M., Klingen, I., Birch, A. N., Bones, A. M., Bruce, T. J., Johansen, T. J., Meadow, R., Mølmann, J., Seljåsen, R., Smart, L. E., and Stewart, D. 2011. Phytochemicals of Brassicaceae in plant protection and human health–Influences of climate, environment and agronomic practice. *Phytochemistry*, **72**:538–556. https://doi.org/10.1016/j.phytochem.2011.01.014

Bosland, P. and Votava, E. 2000. *Peppers*: *Vegetable and Spice Capsicums*. CABI Publishing, Cambridge, UK.

Braga, P. C., Lo Scalzo, R., dal Sasso, M., Lattuada, N., Greco, V. and Fibiani, M. 2016. Characterization and antioxidant activity of semi-purified extracts and pure delphinineglycosides from eggplant peel (*Solanum melongena* L.) and allied species. *J. Funct. Foods*, **20**: 411–421.

Bravo, L. 1998. Polyphenols Chemistry, Dietary Sources, Metabolism, and Nutritional Significance. *Nutr. Rev.*, **56**: 317-333.

Britton G. 1998. Overview of carotenoid biosynthesis. In: *Carotenoids: Biosynthesis and Metabolism*. Vol. 3. (Eds. Britton, G., Liaaen-Jensen, S. and Pfander, H.), Birkhäuser; Basel, Switzerland, pp. 13–147.

Britton, G., Liaaen-Jensen, S. and Pfander, H. 2004. *Carotenoids Hand Book*. Birkhäuser, Basel, Switzerland.

Burchi, F., Fanzo, J. and Frison, E. 2011. The role of food and nutrition system approaches in tackling hidden hunger: Review. *Int. J. Environ. Res. Public Health,* **8**: 358-373.

Burger, Y., H. Paris, R. Cohen, N. Katzir, Y. Tadmor, E. Lewinsohn and A.A. Schaffer, 2009. Genetic diversity of *Cucumis melo*. *Hort. Rev*., (*Amer. Soc. Hort. Sci.*) **36**: 165-198.

Cavagnaro P.F. 2019. Genetics and genomics of sugars and polacetylenes. In: *The Carrot Genome*. Springer; Berlin/Heidelberg, Germany, pp. 295–315.

Cervantes-Flores, J.C., Sosinski, B., Pecota, K.V., Mwanga, R.O.M., Catignani, G.L., Truong, V.D., Watkins, R.H., Ulmer, M.R., Yencho, G.C. 2010. Identification of quantitative trait loci for dry matter, starch, and beta carotene content in sweet potato. *Mol. Breed.*, doi:10.1007/s11032-010-9474-5

Chakraborty, S., Chakraborty, N., Agrawala, L., Ghosh, S. and Narula, K. 2010. Next generation protein-rich potato expressing the seed protein gene *AmA1* is a result of proteome rebalancing in transgenic tuber. *Proc.Nat. Acad. Sci.*, USA, **107**:17533-17538

Chan, J.M., Wang, F. and Holly, E.A. 2005. Vegetable and fruit intake and pancreatic cancer in a population-based case-control study in the San Francisco bay area. *Cancer Epidemiol. Biomarkers Prev.*, **14**:2093–7.

Charles , M.T., Kalantari , S., Corcuff , R. and Arul , J. 2005. Postharvest Quality and Sensory Attributes of UV-Treated Tomato Fruit. *Acta Hort.*, **682**:537-542

Chattopadhyay, S.B., Sarkar, A. and Hazra, P. 2009. Study on the crossability of three species of *Solanum* in Gangetic plains of West Bengal. *J. Crop Weed,* **5**: 53-56

Chen, F Q and Foolad, M R. 1999. A molecular linkage map of tomato based on a cross between *Lycopersicon esculentum* and *L. pimpinellifolium* and its comparison with other molecular maps of tomato. *Genome*, 42, https://doi.org/10.1139/g98-103

Chetelat, R.T. and Ji, Y. 2006. Cytogenetics and evolution, In: *Genetic Improvement of Solanaceous Crops*. Vol 2: Tomato (Eds. Razdan, M. and Matoo, A.K.), Science Publishers, New Dehli, India.

Chiu, L.W., Zhou, X., Burke, S., Wu, X., Prior, R.L., Li, L. 2010. The Purple Cauliflower Arises from Activation of a MYB Transcription Factor. *Plant Physiol.,* **154**: 1470–1480.

Cohen, J.H., Kristal, A.R., Stanford, J.L. 2000. Fruit and vegetable intakes and prostate cancer risk. *J. Natl Cancer Inst.*, **92**:61–8.

Dai, J. and Mumper, R.J. 2010. Plant Phenolics: Extraction, Analysis and Their Antioxidant and Anticancer Properties. *Molecules,* **15**: 7313–7352.

Daniel, A.L., Brecht, J.K., Sims, C.A., Maynard, D.N. 1995. Sensory analysisof bush and vining types of tropical pumpkin. *Proc. Fla. State Hort. Soc.*, **108**:312–31

Das, B., Hazra, P., Chattopadhyay, A., Chakraborty, A.K., Hazra, S., Kardile, H.B., Maji, A. and Chakrabarti, S.K. 2023. Characterization of traditional potato cultivars of North-Eastern region of India for nutritional and quality traits and ploidy level. *Indian J. Genet. Plant Breed.*, **83**: 407-413.

Das, I., Hazra, P., Longjam, M., Bhattacharjee,T., Maurya, P.K., Banerjee, S. and Chattopadhyay, A. 2020. Genetic control of reproductive and fruit quality traits in crosses involving cultivars and induced mutants of tomato (*Solanum lycopersicum* L.). *J. Genet.*, 99: 56 https://doi.org/10.1007/s12041-020-01209-7

Das,I., Hazra,P., Longjam, M., Bhattacharjee,T., Maurya, P.K., Banerjee, S. and Chattopadhyay, A. 2019. Characterization of induced mutants and their hybrids of tomato (*Solanum lycopersicum* L.) for growth, yield and fruit quality traits to explore the feasibility in future breeding. *Genet. Resour.Crop Evol.*, **66**:1421-1441.

De Albuquerque, T.M.R., Borges, C.W.P., Cavalcanti, M.T. , Lima, M.S., Magnani, M. and de Souza, E.L. 2020. Potential prebiotic properties of flours from different varieties of sweet potato (*Ipomoea batatas* L.) roots cultivated in Northeastern Brazil. *Food Biosci.*, 36, Article 100614

Delgado-Vargas, F and Paredes-Lopez, O. 2003. *Natural colorants for Food and Nutraceutical Uses*. CRC Press, Boca Raton, Florida, USA

Dhingra, D., Michael, M., Rajput, H. and Patil, R.T. 2012. Dietary fibre in foods: a review. *J. Food Sci. Technol.*, **49**: 255–266.

Dias, J.S. and Ryder, E. 2011. World vegetable industry: production, breeding, trends. *Hort. Dev.*, **38**: 299-356.

Diaz de la Garza, R., Quinlivan, E.P., Klaus, S.M., Basset, G.J. and Gregory, F. F, 2004. Folate biofortification in tomatoes by engineering the pteridine branch of folate synthesis. *Proc. Nat. Acad. Sci.*, USA, **101**: 13720-13725.

Dimitrov, I., Stancheva, I., Mitove, I., and Atanasova, E. 2006. Comparative Study of Some Quality Parameters of Lettuce in Dependence on Way of Cultivation. *Bul. J. Agril. Sci.*, **12** : 421-427 .

Diretto, G., Tavazza, R, and Giuliano, G 2006. Metabolic engineering of potato tuber carotenoids through tuber-specific silencing of lycopene epsilon cyclase. *BMC Plant Biol.*, Dol: 10.1186/1471-2229:6-13

Docimo, T., Francese, G., Ruggiero, A., Batelli, G., De Palma, M. and Bassolino L. 2016. Phenylpropanoids accumulation in eggplant fruit: characterization of biosynthetic genes and regulation by a MYB transcription factor. *Front. Plant Sci.*, 6: 1233.doi: 10.3389/fpls.2015.01233.

Dutta, S., Hazra, P., Saha, S., Acharya, B., Bhattacharjee, T., Maurya, P.K., Banerjee, S., Chakraborty,I. and Chattopadhyay, A. 2021. Applied mutagenesis could improve economically important traits in bitter gourd (*Momordica charantia* L.). *J. Genet.*, 100:43 https://doi.org/10.1007/s12041-021-01283-5

Eady, C.C., Kamoi, T and Kato, M. 2008. Silencing onion lachrymatory factor synthase causes a significant change in the sulfur secondary metabolitie profile. *Plant Physiol.*, **147**:296-2106

Eshed, Y. and Zamir, D. 1995. An introgression line population of *Lycopersicon pennellii* in the cultivated tomato enables the identification and fine mapping of yield associated QTL. *Genetics*, **141**:1147–1162.

Evans A. M. and Gridley H. E. 1979. Prospects for the improvement of protein and yield in food legumes. *Curr. Adv. PI. Sci.*, **3**: 1-17.

Fahey , J.W., Haristoy , X., Dolan , P.M., Kensler , T.W., Scholtus , I., Stephenson , K.K., Talalay , P. and Lozniewski , A. 2002. Sulforaphane inhibits extracellular, intracellular, and antibiotic-resistant strains of Helicobacter pylori and prevents benzo[a]pyrene-induced stomach tumors. *Proc. Natl. Acad. Sci.*, US A, 28;99(11):7610-5 doi: 10.1073/pnas.112203099.

Fanourakis, N.E. and Simon, P.W. 1987. Analysis of genetic linkage in cucumber. *J. Hered.*, **78**: 238-242.

Faria, M.V., Maluf, W. R., de Resende, J.T.V., Andrade-Junior, V.C., do Nascimento, I.R., Benites, F.R.G., de Menezes, C.B. and Azevedo, S.M. 2006. *rin*, *nor*, *ogc* and *hp* mutants in tomatoes with different genetic backgrounds. *Pesqui. Agropecu Bras*., **41**: 793-800.

Food and Agriculture Organization of the United Nations. 1993. Indigenous knowledge and biodiversity. http://www.fao.org/docrep/004/v1430e/V1430E03.htm.

Fray, R. G, and Grierson, D. 1993. Molecular genetics of tomato fruit ripening. *Trends in Genet.*, **9**: 438-443 doi: 10.1016/0168-9525(93)90108-t.

Gaggiotti , S., Mascini , M., Pittia , P., Della Pelle , F and Compagnone , D. 2019. Headspace Volatile Evaluation of Carrot Samples—Comparison of GC/MS and AuNPs-hpDNA-Based E-Nose. *Foods,* **8**: 293, doi: 10.3390/foods8080293

Gasper, A.V. 2005. Glutathione-S transferase M1polymorphism and metabolism of sulforaphane from standard and high –glucosinolate broccoli. *Amer. J. Clinical Nutr.*, 82:1283-1291

Gong, Z.P., Yu, S.C., Zhang, F.L., Yu, Y.J., Zhao, X.Y., Zhang, D.S., Wang, W.H., Su, T.B. 2016. Evaluation of Chinese Cabbage Sensory Quality and Its Relationship with Contents of Main Nutrient Components. *Agric. Sci. Technol.,* **17**: 1592–1596.

Grandillo, S., Ku, H. M. and Tanksley, S.D. 1999. Identifying the loci responsible for natural variation in fruit size and shape in tomato. *Theor.Appl. Genet.*, **99**: 978-987

Grotewold, W .2006. The genetics and biochemistry of floral pigments. *Annu. Rev. Plant Biol.,* **57**: 761-780

Gupta, P. N. and M. Rai. 1990. Genetic divergence in cucurbit-pumpkin. *Indian Hort*., 36:1.

Gutierrez, N., Avila, C.M., Moreno, M.T., Torres, A.M. 2008. Development of Scar markers linked to zt 2, one of the genes controlling absence of tannins in faba bean. *Australian J. Agril. Res.*, **59**: 62- 68

Hanschen, F. S., and Schreiner, M. 2017. Isothiocyanates, nitriles, and epithionitriles from glucosinolates are affected by genotype and developmental stage in Brassica oleracea varieties. *Front.Plant Sci.*, **8**: 1095. https://doi. org/10.3389/fpls.2017.01095

Hazra, P., Das, R. K., Pandit, M. K.and Sahu, P. K.2001. Genetical studies on flower andfruit characters in an interspecific cross of tomato. *Indian J. Genet. Plant breed.,* **61**: 143-146.

Hazra, P., Dutta, R., Roy, T., Mandal, A. K. and Roy, U. 2005. Biochemical basis of eggplant susceptibility to shoot and fruit borer and their genetic control. *Egyptian J. Plant Breed.*, **9**: 43-51.

Hazra, P., Dutta, R., and Maity, T.K 2004. Morphological and biochemical characters associated with field tolerance of brinjal to shoot and fruit borer (*Leucinodes orbonalis*) and their implication in breeding for tolerance. *Indian J. Genet. Plant breed.,* **64**: 255-256.

Hazra, P., Hazra, S., Acharya, B., Dutta, S., Saha, S., Mahapatra, P., Pradeepkumar, P., Pal, H., Chattopadhyay, A., Chakraborty, I., Jambhulkar, S., Chatterjee, S. and Ghosh, S.K. 2022. Diversity of nutrient and nutraceutical contents in the fruits and its relationship to morphological traits in bittergourd (*Momordica charantia* L.). *Sci. Hort.*, https://doi. org/10.1016/j.scienta.2022.111414.

Hazra, P., Longjam, M. and Chattopadhyay, A. 2018. Stacking of mutant genes in the development of "purple tomato" rich in both lycopene and anthocyanin contents. *Sci. Hort.*, **239**: 253-258.

Hazra, P., Mandal, A. K., Dutta, A. K., Sikder, D. and Pandit, M. K. 2007. Breeding pumpkin (*Cucurbita moschata* Duch ex Poir.) for high yield and carotene content. *Acta Hort.*,**752**: 431-435.

Hazra, P., Rout, A., Roy, U., Nath, S., Roy, T., Dutta, R., Acharya, S. and Mandal, A. K. 2003. Characterization of brinjal (*Solanum melongena* L) germplasm. *Veg. Sci.*, **30**:145-149.

Hazra, P., Som, M. G. and Das, P. K.1996. Combining ability for pod yield and seed protein in cowpea. *Indian J. Genet. Plant breed.*, **56**:553-556.

Hertog, M.G.L., Feskens, E.J.M., Hollman, P.C.H., Katan, M.B. and Kromhout, D. 1993. Dietary antioxidant flavonoids and risk of coronary heart disease: The Zutphen Elderly Study. *Lancet,* **342**:1007–11.

Hertog, M.G.L., Kromhout, D., Aravanis, C., Blackburn, H., Buzina, R., Fidanza, F., Giampaoli, S., Jansen, A., Menotti, A. and Nedeljkovic, S. 1995. Flavonoid intake and long-term risk of coronary heart disease and cancer in the Seven Countries Study. *Arch. Intern. Med.*, **155**: 381–386.

Hirschberg, J., Galpaz, N., Kachanovsky, D., Ronn, I., Pankratov, I., Asaacson, T. and Zamir, D. 2008. Regulation of carotenoid biosynthesis: lessons from mutational analysis in tomato. Paper presented at the 15th International Symposium on Carotenoids, 22-27 June, 2008, Okinawa, Japan, *Interdisciplinary J. Res. Carotenoids*, **12**: 63 (Abst)

Hobson, G.E. and Davies, J.N. 1971. *The biochemistry of fruits and their products*. Academic Press, New York.

Hosen, M., Rafii, M.Y., Mazlan, N., Jusoh, M., Chowdhury, M.F. N., Yusuff, O., Ridzuan, R., Karim, K.M.R., Halidu, J. and Ikbal, M.F. 2022. Estimation of heterosis and combining ability for improving yield, sweetness, carotenoid and antioxidant qualities in pumpkin hybrids (*Cucurbita moschata* Duch. Ex Poir.). *Horticulturae*, *8*10, 863; https://doi. org/10.3390/ horticulturae 8100863

Hubbard, N. L., Huber, S. C. and Pharr, D. M., 1989. Sucrose phosphate synthase and acid invertase as determinants of sucrose concentration in developing muskmelon (*Cucumis melo* L.) fruits. *Plant Physiol.*, **91**: 1527-1534.

Johnson, E.J., Hammond, B., Yeum, K.J., Qin, J., Wang, X.D., Castaneda, C., Snodderly, D., Russell, R.M. 2000. Relation among serum and tissue concentrations of lutein and zeaxanthin and macular pigment density. *Am. J. Clin. Nutr.*, **71**: 1555–1562.

Kalia, P. and Pathania, N.K. 2007. Genetic variability and relationships for quantitative and quality characters in winter bean (*Vicia faba*). *Acta Horticulturae*, DOI:10.17660/ ActaHortic. 2007.752.72

Kihara, H. 1951. Triploid Watermelons. *Proc.Amer. Soc.Hot. Sci.*, **58**: 217–230.

Kopsell, D.E. and Randle, W. 1997. Onion cultivars differ in pungency and bulb quality changes during storage. *Hort Sci.*, **32**: 1260-1263.

Kothari, D., Lee, W.D. and Kim, S.K. 2020. *Allium* Flavonols: Health Benefits, *Molecular Targets and Bioavailability*. **9**: 888, doi: 10.3390/antiox9090888

Kumar V., Luthra S.K., Bhardwaj V., Singh B.P. 2014. *Indian Potato Varieties and Their Salient Features. Technical Bulletin No. 78* (Revised), ICAR-Central Potato Research Institute; Shimla, India.

Kumari, T. and Deka, S.C. 2021. Potential health benefits of garden pea seeds and pods: A review. *Legume Science*, DOI: 10.1002/leg3.82

Lachman, J., Hamouz, K., Orsák, M. and Z. Kotíková, Z. 2016. Carotenoids in potatoes - a short overview. *Plant Soil Environ.*, **62**:474-481

Le-Gall, G., DuPont, M.S., Mellon, F.A., Davis, A.L., Collins, G.J., Verhoeyen, M.E., Colquhoun IJ. 2003. Characterization and content of flavonoids glycosides in genetically modified tomato (*Lycopersicon esculentum*) fruits. *J. Agril.Food Chem.*, **51**:2438-2446

Levin, I., Frankel, P., Gilboa, N., Tanny, S. and Lalazar, A. 2003. The tomato dark green mutation is a novel allele of the tomato homolog of the DEETIOLATED1 gene. *Theor. Appl. Genet.*, **106**: 454-460.

Li, D., Wang, P., Luo, Y., Zhao, M. and Chen, F. 2017. Health benefits of anthocyanins and molecular mechanisms: Update from recent decade. *Crit. Rev Food Sci.Nutri.,* **57**: 1729–1741.

Liu, R.H. 2003. Health benefits of fruits and vegetables are from additive and synergistic combination of phytochemicals. *Am. J. Clin. Nutr*., **78**: 517S–20S.

Liu, R.H. 2004. Potential synergy of phytochemicals in cancer prevention: mechanism of action. *J Nutr*., **134**: 3479S–85S.

Liu, Y.S., Gur, A., Ribeb, A., Causes, M., Damidaux, R., Buret, M., Hirschberg, J. and Zamir, D. 2003. There is more tomato fruit colour than candidate carotenoid genes. *Plant Biotechnol. J.*, **1**: 131-240

Lu, S, Eck, J.V., Zhou, X., Lopez, A.B, Halloran, D.M., Cosman, K.M., Conlin, B.J., Dominick, J.P., Garvin, D.F., Vrebalov , J., Kochian , L.V., Küpper , H., Earle , E.D., Cao, J. and Li, L. 2006. The cauliflower *Or* gene encodes a DnaJ cysteine-rich domain-containing protein that mediates high levels of beta-carotene accumulation. *Plant Cell*, **18**: 3594-3605

Maass D, Arango J, Wust F, Beyer P, Welsch R. 2009. Carotenoid crystal formation in Arabidopsis and carrot roots caused by increased phytoene synthase protein Levels. *PLoS One*, **4**: e6373.

Maina, S., Misinzo, G., Bakari, G. and Kim , H.Y. 2020. Human, Animal and Plant Health Benefits of Glucosinolates and Strategies for Enhanced Bioactivity: A Systematic Review. *Molecules ,* 2020 Aug; 25: 3682.doi: 10.3390/molecules25163682

Maoka T. 2009. Recent progress in structural studies of carotenoids in animals and plants. *Arch. Biochem. Biophys*., **483**:191–195. doi: 10.1016/j.abb.2008.10.019.

Martins , T., Barros , A.N., Rosa,E. and Antunes , L. 2023. Enhancing Health Benefits through Chlorophylls and Chlorophyll-Rich Agro-Food: A Comprehensive Review. *Molecules*, 28: 5344.doi: 10.3390/molecules28145344

Miri, S.M. 2020. Artificial polyploidy in the improvement of horticultural crops. *J. Plant Physiol*., Breed.,**10** : 1-28.

Moon, S.S., Verma, V.K. and Munshi, A.D. (2002). Gene action of quality traits in muskmelon (*Cucumis melo* L.). *Vegetable Science*, **29** (2) 134-136.

Morris , M.C., Wang , Y., Barnes , L.L., Bennett , D.A., Dawson-Hughes , B. and Booth , S.L. 2018. Nutrients and bioactives in green leafy vegetables and cognitive decline. *Neurology*, **90**(3): e214–e222doi: 10.1212/WNL.0000000000004815

Mou, B. 2005. Genetic variation of beta-carotene and lutein contents in lettuce. *J. Am. Soc. Hortic*., Sci., **130**: 870-876

Muir, S.R., Collins, G.J., Robinson, S., Hughes, S., Bovy, A., Ric, D. 2001. Overexpression of petunia chalcone isomerase in tomato results in fruit containing increased levels of flavonols. *Nature Biotechnol.*, **19**: 470- 474.

Mullie, P. and Clarys, P. 2011. Association between cardiovascular disease risk factor knowledge and lifestyle. *Food Nutr. Sci.,* **2**:1048-1053.

Mustilli, A.C., Fenzi, F., Ciliento, R., Alfano,F. and Bowler, C.(1999).Phenotype of the tomato highpigment-2 mutant is caused by amutation in the tomato homolog of DEETIOLATED1. *The Plant Cell*, **11**: 145-158.

Mutschler , M., Wolfe , D.W., Cobb , E.D. and Yourstone , K.S. 1992. Tomato Fruit Quality and Shelf Life in Hybrids Heterozygous for the *alc* Ripening Mutant. *Hort Sci.*, **27** DOI:10.21273/HORTSCI.27.4.352

Mwangi, D. 2003. *Introduction to Amaranth.* 2nd Edition, Joypet Services and Printers Nairobi - Kenya. pp. 51-56.

Nakatani, N. 1997. Antioxidants from Spices and Herbs. In: *Natural Antioxidants: Chemistry, Health Effects, and Applications* (Ed. Shahidi, F.), AOCS Press, Champaign, 64-75.

Nandi, L.L., Saha, P., Jaiswal, S., Lyngdoh, Y.A., Behera, T.K., Pan, R.S., Munshi, A.D., Saha, N.D., Hossain, F., Srivastava, M. and Tomar, B.S. 2021. Bioactive compounds, antioxidant activity and elements content variation in indigenous and exotic *Solanum sp.* and their suitability in recommended daily diet. *Sci. Hort.*, **287**: 110232.

Naresh Kumar and Goel, N. 2019. Phenolic acids: Natural versatile molecules with promising therapeutic applications. *Biotechnol. Rep.* (Amst), **24**: e00370doi: 10.1016/j.btre.2019.e00370

Nasirvand, S, Asghari ,Z.R., Zare, N. and Esmaeilpoor, B. 2018. Poly ploidy induction in parsley (*Petroselinum crispum* L.) by colchicine treatment. *Cytologia*, **83** : 393—396

Nicoli , M.C.,Anese , M. and Parpinel, M. 1999. Influence of processing on the antioxidant properties of fruit and vegetables. *Trends Food Sci. Technol.*, **10**: 94-100.

Nino-Medina, G., Urías-Orona, V., Muy-Rangel, M. D. and Heredia, J. B. 2017.Structure and content of phenolics in eggplant (*Solanum melongena*)-a review. *South Afr. J. Bot.*, **111**: 161-169.

Noh, J., Sheikh, S., Chon, H.G., Seong, M.H., Lim, J.H., Lee, S.G., Jung, G.T., Kim, G.M., Ju, H.J. and Huh, Y.C. 2012. Screening different methods of tetraploid induction in watermelon (*Citrullus lanatus* (Thunb.) Manst. and Nakai). *Hortic. Environ. Biotech.*, **53** : 521—529.

Nothmann, J., Rylski, I. and Spigelman, M. 1976. Color and variations in color intensity of fruit of eggplant cultivars. *Sci. Hort.*, **4**: 191–197.

Nuñez-Palenius, H.G., M. Gomez-Lim, N. Ochoa-Alejo, R. Grumet, G. Lester and D.J. Cantliffe, 2008. Melon fruits: genetic diversity, physiology, and biotechnology features. *Crit. Rev. Biotechnol.*, **28**: 13–55.

O'Hare, T.J., Wong, L.S. and McGrath, D.J. 2005. Increasing visual appeal and health benefit of fresh-cut tomato slices using high lycopene varieties. *Acta Hort.*, **687**: 395-397.

Ogawa, K., Murota, K., Shimura, H., Furuya, M., Togawa, Y., Matsumura, T. and Masuta, C. 2015. Evidence of capsaicin synthase activity of the *Pun1*-encoded protein and its role as a determinant of capsaicinoid accumulation in pepper. *BMC Plant Biol.*, **15**: 93. doi: 10.1186/s12870-015-0476-7.

Oloyede, O. O., Wagstaff, C., and Methven, L. 2021. Influence of cabbage (*Brassica oleracea*) accession and growing conditions on myrosinase activity, glucosinolates and their hydrolysis products. *Foods*, 10, 2903. https://doi. org/10.3390/foods10122903

Pal, H., Kundu, A., Sahu, R., Sethi, A., Hazra, P. and Chatterjee, S. 2019. Unravelling the metabolic behavior in Tomato high pigment mutants (*hp-1, hp-2dg,* og^c) and non ripening mutant (*rin*) during fruit ripening. *Sci. Hort.*, 246: 652-663.

Panche, A.N., Diwan, A.D. and Chandra, S.R. 2016. Flavonoids: an overview. *J. Nutr Sci.*, 5: e47, doi: 10.1017/jns.2016.41

Pandey, K.B. and Rizvi, S.I. 2009. Plant polyphenols as dietary antioxidants in human health and disease. *Oxid. Med. Cell Longev.*, **2**: 270–278

Pandey, S., Singh, J., Upashyay, A.K., Ram, D. and Rai, M. 2003. Ascorbate and carotenoid content in an Indian collection of pumpkin (*Cucurbita moschata* Duch Ex. Poir). *Cucurbit Genet. Coop. Rep.*, **26**: 51-53.

Pandit, M. K., Hazra, P.and Dutta, A. K. 2006. Study on heterosis for yield and quality characters of pumpkin (*Cucurbita moschata* Duch. Ex Poir). In: *Cucurbits Breeding andProduction Technology* (Eds. Gautam, P.L., Ram, H. and Singh, H.P.) *Proc. Natl. Seminar on Cucurbits*. G.B. Pant Univ. Agril. and Tech., Pantnagar, 22-23 September, 2005, pp.319-327.

Papas, A.M. 1999. Diet and Antioxidant Status. *Food Chem. Toxicol.*, **37**: 999-1007.

Pavithra, G.J., Mahesh, S. and Shankar, A.G 2014. Assessment of zinc variability in tomato germplasm lines: An option for biofortification by introgression breeding. *Agrotechnol.*, 2:229

Pei, Y., Yao, N., He, L., Deng, D., Li, W. and Zhang, W. 2019. Comparative study of the morphological, physiological and molecular characteristics between diploid and tetraploid radish (*Raphanus sativus* L.). *Scientia Hort.*, **257** : 1—10.

Pliankong, P., Suksa-Ard, P. and Wannakrairoj, S. 2017. Effects of colchicine and oryzalin on polyploidy induction and production of capsaicin in *Capsicum frutescens* L. *Thai J. Agric. Sci.*, **50** : 108—120.

Potaczek, H. and Michalik, H. 1998. Improvement of the nutritional value of tomato (*Lycopersicon esculentum* Mill.) by means of breeding for quality.*Veg. Crops Res. Bull.*, **49***: 13-20.*

Prior, R.L. and Cao, G. 2000. Antioxidant phytochemicals in fruits and vegetables: Diet and health implications. *Hort Sci.,* **35**: 588-592.

Raffo, A., Leonardi, C., Fogliano, V., Ambrosino, P., Salucci, M., Gennaro, L., Bugianesi, R., Giuffrida, F. and Quaglia, G. 2002. Nutritional value of cherry tomatoes (*Lycopersicon esculentum* Cv. Naomi F1) harvested at different ripening stages. *J.Agril. Food Chem.*, **50**: 6550-6556.

Rahimi , P., Abedimanesh , S., Mesbah-Namin , S.A. and Ostadrahimi , A. 2019. Betalains, the nature-inspired pigments, in health and diseases. *Crit. Rev. Food Sci. Nutr.*, **59**: 2949-2978

Raja, M.M., Raja, M., Imran, M., Rahmsn, H.A. 2011. Quality aspects of cauliflower during storage. *Inernt. Food Res. J.,* **18**: 427–431.

Rangarajan, A. and Kelly, J. F, 1999. Iron bioavailability from *Amaranthus* species: 1 *In vitro* dialysable iron for estimation of genetic variation. *J. Sci. Food Agril.*, https://doi.org/10.1002/(SICI)1097-0010(199810)

Ranjan,, P., Pandey, A., Munshi, A.D., , Bhardwaj, R., Gangopadhyay, K.K., Malav, P., Pandey, C.D., Pradheep, K., Tomar, B.S. and Kumar, A. 2019. Orange-fleshed cucumber (Cucumissativus var. sativus L.) germplasm from North-East India: agro-morphological, biochemical and evolutionary studies. *Genet. Resour. Crop Evol.*, DOI:10.1007/s10722-019-00778-6

Rimm, E.B., Ascherio, A., Giovannucci, E., Spiegelman, D., Stampfer, M.J. and Willett, W.C. 1996. Vegetable, Fruit and cereal fibre intake and risk of coronary heart disease among men. *JAMA J. Amer. Med. Assoc.*, **275**: 447-451.

Ripley, V.L. and Roslinsky, V. 2005. Identification of an ISSR marker for 2 propenyl glucosinolate content in *Brassica juncea* L. and conversion to a SCAR marker. *Mol. Breed.*, **161**: 57-66

Romer, S., Fraser, P.D., Kiano, J.W., Shiption, C. A., Misawa, N., Schuch, W., Bramley, P.M. 2000. Elevation of the pro-vitamin a content of transgenic tomato plants. *Nature Biotech.*, **18**: 666-669.

Rousseaux, M.C., Jones, C.M., Chetelat, D.A.R., Bennett, A. and Powell, A. 2005. QTL analysis of fruit antioxidants in tomato using *Lycopersicon pennellii* introgression lines. *Theor. Appl. Genet.*, **111**:1396-1408.

Saha, S.R., Nazim Uddin, M., Rahman, M.A., Sharifuzzaman, S.M. and Anwar Habib, A.K.M. 2003. Growth and Harvestable Maturity of Red Amaranth at Different Sowing Dates. Asian J.Plant Sci., **2**: 431-433.

Saito , K., Yonekura-Sakakibara, K., Nakabayashi , R., Higashi , Y., Yamazaki , M., Tohge, T. and Fernie , A.R. 2013.The flavonoid biosynthetic pathway in Arabidopsis: structural and genetic diversity. *Plant Physiol. Biochem.*, doi: 10.1016/j.plaphy.2013.02.001.

Samtiya, M., Aluko, R. E., Dhewa, T. and Moreno-Rojas, J. M. 2021. Potential health benefits of plant food-derived bioactive components: An overview. *Foods*, **10**: 839.doi: 10.3390/ foods10040839.

Santos, C.A.F. and Simon, P.W 2002. QTL analyses reveal clustered loci for accumulation of major provitamin a carotenes and lycopene in carrot roots. *Mol. Genet. Genomics*, **268**:122-129

Sapir , M., Oren-Shamir, M., Ovadia, R., Reuveni, M., Evenor, D., Tadmor, Y., Nahon, S., Shlomo, H., Chen, L., Meir, A. and Levin, I. 2008. Molecular aspects of Anthocyanin fruit tomato in relation to high pigment-1. *J. Hered.*, **99**: 292-303.

Schauer, N., Semel, Y., Balbo, I., Steinfath, M., Repsilber, D., Selbig, J., Pleban, T., Zamir, D. and Fernie, A.R. 2008. Mode of inheritance of primary metabolic traits in tomato . *The Plant Cell*, **20**: 509–523.

Seroczynska, A., Niemirowicz, S.K. and Korrenicwsk, I.A. 1998. Utilization of three non-ripening mutants in tomato breeding for prolonged shelf life. *Folia Hort.*, **1**:3-14.

Severo, J., Santos, F.N. , Samborski, T. , Rodrigues, R. and Mello, A.F.S. 2021. Biofortified sweet potatoes as a tool to combat vitamin A deficiency: effect of food processing in carotenoid content. *Rev. Chil. Nutr.*, **48**: 414-424

Sevindik, M. 2019. The novel biological tests on various extracts of *Cerioporus varius*. *Fresen. Environ. Bull.*, **28**: 3713-3717.

Sevindik, M., Akgul, H., Pehlivan, M., Selamoglu, Z. 2017. Determination of therapeutic potential of *Mentha longifolia* ssp. *longifolia*. *Fresen Environ. Bull.*, **26**: 4757-4763.

Sharma, M. and Kaushik, P. 2021. Biochemical composition of eggplant fruits: A review. *Appl. Sci.*, 11: 7078. https://doi.org/10.3390/app11157078

Shimomura , K., Horie , H., Sugiyama , M., Kawazu , Y. and Yoshioka, Y. 2016.Quantitative evaluation of cucumber fruit texture and shape traits reveals extensive diversity and differentiation. *Sci. Hort.*, **199**: 133-141.

Siddiqui, Wasim, Md., Chakraborty,I., Mishra, P. and Hazra, P. 2014. Bioactive attributes of tomatoes involving *dg*, *og*^{c}, and *rin* genes. *Food &Function,* **5**: 936-943.

Siddiqui, Wasim, Md., Chakraborty,I., Mishra,P., Hazra, P. and Ayala-Zavalad, J.F. 2015. Postharvest physico-chemical changes in mutant (*dg, og*^{c}, and *rin*) and non-mutant tomatoes. *Acta Physiol. Plant.*, **37**:1-11.

Siddiqui, Wasim, Md., Chakraborty,I., Hazra,P. and Ayala-Zavalad, J.F. 2014.Characterization of quality indices on storage of puree of mutant (*dg* and *og*^{c}) and normal tomatoes. *Acta Alim.*, **43**: 426-436.

Sikder, S., Biswas, P., Hazra, P., Chattopadhyay, A., Badigannavar, A.M. and D'Souza, S.F.2013. Induction of mutation in tomato (*Solanum lycopersicum* L.) by gamma irradiation, EMS and their combinations. *Indian J.Genet. Plant Breed.*, **73**: 392-399.

Simon, P.W. 1993. Breeding carrot, cucumber, onion and garlic for improved quality and nutritional value. *Hort. Bras.*, **11**: 171-173.

Simon, P.W. and Navazio, J.P. 1997. Early Orange Mass 400, Early Orange Mass 402, and Late Orange Mass 404: High-Carotene Cucumber Germplasm. *Hort Sci.,* **32**:144-145.

Sinanoglou, V. J., Kavga, A., Strati, I. F., Sotiroudis, G., Lantzouraki, D. and Zoumpoulakis, P. 2019. Effects of infrared radiation on eggplant (*Solanum melongena* L.) greenhouse cultivation and fruits' phenolic profile. *Foods*, **8**: 630.doi: 10.3390/foods8120630.

Sinclair, J.W., Park, O.S. and Crosby, K.M. 2004. Identification of QTL affecting Vitamin C in Melon. *Subtropical Plant Sci.*, **56**:10-15.

Singh, B.K., Koley, T.K., Karmakar, P., Tripathi, A., Singh, B. and Singh, M. 2017. Pigmented radish (*Raphanus sativus*): Genetic variability, heritability and interrelationships of total phenolics, anthocyanins and antioxidant activity. *Indian J. Agril. Sci.*, **87**: 1600–1606

Singh, J., Singh, B, and Kalloo, G. 2002. Root morphology and carotene content in Asiatic and European carrots (*Daucus carota* var. *sativa*). *Indian J. Agril. Sci.*, **72**:225-227

Singh, S., Selvakumar, R., Mangal, M. and Kalia, P. 2020. Breeding and genomic investigations for quality and nutraceutical traits in vegetable crops- A Review. *Indian J. Hort.*, **77**: 1-40.

Singh, S., Kalia, P., Meena, R.K., Mangal, M., Islam, S., Saha, S. and Tomar, B.S. 2020. Genetics and Expression Analysis of Anthocyanin Accumulation in Curd Portion of Sicilian Purple to Facilitate Biofortification of Indian Cauliflower. *Front. Plant Sci.,* **10**: 1766-1772.

Sobolewska, D., Podolak, I. and Makowska-Wąs , J. 2013. *Allium ursinum*: botanical, phytochemical and pharmacological overview. *Phytochem Rev*, 2015; **14**: 81–97

Solovyova, A. E., Sokolova, D. V., Piskunova, T. M., and Artemyeva, A. M. 2014. Nutrients and biologically active substances in vegetable crops and their role in improving nutrition. *Proc. Appl. Bot., Genet. Breed.*, **175**: 5–19

Song, H., Chen, J., Staub, J.E. and Simon, P.W. 2010. QTL analysis of orange color and carotenoid content and mapping of carotenoid biosynthesis gene in cucumber. *Acta Hort.*, 871:607-614.

Song, W., Derito, C.M., Liu, K.M., Liu, H. X., Dong, M. and Liu, R.H. 2010. Cellular antioxidant activity of common vegetables. *J. Agric. Food Chem.*, **58**: 6621–9.

Staub, Jack E. and Serquen, Felix C. 2000. Towards an integrated linkage map of cucumber: Map merging. *Acta Hort.*, 510: 357-366. DOI: 10.17660/ActaHortic.2000.510.57

Stevens, M. and Rick, C.M. 1986. Genetics and breeding, In: *The Tomato Crop. A Scientific Basis for Improvement* (Eds. Atherton, J.G. and Rudich, J.), Chapman & Hall, New York, pp. 35-109.

Stewart, C., Kang, B.C., Liu, K., Mazourek, M., Moore, S.L., Yoo, E.Y., Kim, B.D., Paran, I. and Jahn, M.M. 2005. The *Pun1* gene for pungency in pepper encodes a putative acyltransferase. *Plant J.* ,42:675–688. doi: 10.1111/j.1365-313X.2005.02410.x.

Stewart, C., Mazourek, M., Stellari, G.M., O'Connell, M., Jahn, M. 2007. Genetic control of pungency in *C. chinense via* the *Pun1* locus. *J. Exp. Bot.*, **58**: 979–991. doi: 10.1093/jxb/erl243

Stommel J.R. 2007. Genetic enhancement of tomato fruit nutritive value. In: *Genetic Improvement of Solanaceous Crops* Vol 2: Tomato (Eds. Razdan, M.K. and Mattoo, A.K.) Science Publishers, Enfield NH, USA. pp. 193-238.

Stommel, J. and Haynes, K. 1994. Inheritance of Beta Carotene Content in the Wild Tomato Species *Lycopersicon cheesmanii*. *J. Hered.*, DOI:10.1093/

Sun, H.J., Cui, M.L., Ma, B. and Ezura, H. 2006. Functional expression of the taste modifying protein, Miraculin, in transgenic Lettuce. *FEBS Letters*, 580:620-626

Tadmor, Y., King, S., Levi, A., Meir, A. Wasserman, B., Hirschberg, J and Lewinsohn, K. 2005. Comparative fruit colouration in watermelon and tomato. *Food Res. Intern.l,* **38**: 837-841

Takaichi , S. 2011. Carotenoids in Algae: Distributions, Biosyntheses and Functions. *Mar Drugs*, **9**: 1101–1118.

Tanaka, N., and Niikura, S. 2006. Genetic analysis of the developmental characteristics related to the earliness of head formation in cabbage (*Brassica oleracea* L.). *Breed. Sci.*, **56**: 147–153. https://doi.org/10.1270/jsbbs.56.147

Todaro, A., Cimino, F., Rapisarda, P., Catalano, A. E., Barbagallo, R. N. and Spagna, G. 2009. Recovery of anthocyanins from eggplant peel. *Food Chem.*, **114**: 434-439.

Truong, V.D. , Avula, R.Y. , Pecota, K.V. and Yencho, G.C. 2018. Sweetpotato production, processing and nutritional quality. In: *Handbook of Vegetables and Vegetable Processing* (Eds. Siddiq, M. and Uebersax, M.A.), John Wiley & Sons Ltd., USA.

Tucker, G.A. 1990. Genetic Manipulation of Fruit Ripening, *Biotech. Genet. Engr. Rev.*, 8:1, 133-160, DOI: 10.1080/02648725.1990.10647867

Uchendu, F.N. and Atinmo, T. 2010. The silent and neglected crisis of malnutrition: scientific evidence for taking decisive action. *Global J. Health Sci.,* **3**: 193-202.

Wargovich, M.J. 2000. Anticancer properties of fruits and vegetables. *Hort Sci.*, **35**: 573-575.

Watzl, B., Kulling, S.E., Möseneder, J., Barth, S.W. and Bub, A. 2005. A 4-wk intervention with high intake of carotenoid-rich vegetables and fruit reduces plasma C-reactive protein in healthy, nonsmoking men. *Am. J. Clin. Nutr.*, **82**: 1052–1058.

Wolfe, K.L., Kang, X.M., He, X.J., Dong, M., Zhang, Q.Y. and Liu, R.H. 2008. Cellular antioxidant activity of common fruits. *J. Agric. Food. Chem.*, **56**: 8418–26.

Wood, A.B. 1987. Determination of the pungent principles of chillies and ginger by reversed-phase high-performance liquid chromatography with use of a single standard substance. *Flavour and Fragrance J.1*, https://doi.org/10.1002/ffj.2730020102

Wu, C.Y., He, Q.W., Song, T.W., Deng, Y.L., Wang, C.H. and Xu, W.L. 2012. Selection of evaluation indexes in flavor quality of Chinese cabbage. *J. Northwest Sci. Tech. Univ. Agric. For. (Nat. Sci. Ed.),* **40**: 161–168.

Wu, X., Sun,C., Yang, L., Zeng, G., Liu, Z. and Zuyang Li, Y. 2008. β-carotene content in sweet potato varieties from China and the effect of preparation on β-carotene retention in the Yanshu No. 5. *Innov. Food Sci. Emerging Technol.*, **9**: 581-586

Wyllie, S.G and Leach, D.N. 1990. Aroma volatiles of *Cucumis melo* cv. Golden Crispy. *J. Agric Food Chem.*, **38**: 2042-2044.

Yadav S.K., Lal S.S., Srivastava A.K., Bag T.K., Singh B.P. 1994. Efficacy of chemical and non-chemical methods of weed management in rainfed potato (*Solanum tuberosum* L.). *Ind. J. Agric. Sci.*, **85**: 382–386.

Yativ , M., Harary , I. and Wolf , S. 2010. Sucrose accumulation in watermelon fruits: Genetic variation and biochemical analysis. *J. Plant Physiol.*, **167**: 589-596

Yoo , K.S., Pike , L., Crosby , K. and Jones , R. 2006. Differences in onion pungency due to cultivars, growth environment, and bulb sizes. *Sci. Hort.*, **110**:144-149

Zhang , W., Hoa, H., Ma, L. and Zhao, C. 2010. Tetraploid muskmelon alters morphological characteristics and improves fruit quality. *Sci. Hort.*, **125**: 396-400

Zhang , Y., Talalay , P., Cho, C.G. and Posner , G.H. 1992. A major inducer of anticarcinogenic protective enzymes from broccoli: isolation and elucidation of structure. *Proc. Natl. Acad. Sci.* U S A, 89 :2399-403. doi: 10.1073/pnas.89.6.2399.

Zhang, S., Miao, H., Sun, R., Wang, X., Huang, S., Wehner, T.C. and XingfangGu, X. 2013. Localization of a New Gene for Bitterness in Cucumber. *J. Hered.*, **104**: 134–139

Zhang, W, Hao, H, Ma, L, Zhoa, C and Yu, X 2010. Tetraploid muskmelon improves fruit quality. *Sci. Hort.*, **125** : 396-400

Zhang, J., Feng, S., Yuan, J., Wang, C., Lu, T., Wang, H. and Yu, C. 2021. The Formation of Fruit Quality in *Cucumis sativus* L. *Front Plant Sci.*, 2021; 12: 729448.doi: 10.3389/fpls.2021.729448.

Zhao, J., Wang, X., Deng, B., Lou, P., Wu, J., Sun, R., Xu, Z., Vromans, J., Koornneef, M. and Bonnema, A.B. 2005. Genetic relationships within *Brassica rapa* as inferred from AFLP fingerprints. *Theor. Appl. Genet.*, 110:1301-1314

Zhao, Z., Gu, H., Sheng, X., Yu, H., Wang, J., Huang, L. and Wang, D. 2016. Genome-Wide Single-Nucleotide Polymorphisms Discovery and High-Density Genetic Map Construction in Cauliflower Using Specific-Locus Amplified Fragment Sequencing. *Front. Plant Sci., 7*: 334-338.

Zuo, X., Zhang, Y., Wu, B., Chang, X. and Ru, B. 2002. Expression of the mouse metallothionein mutant ââ-cDNA in the lettuces (*Lactuca sativa* L.). *Chinese Sci. Bull.*, 47:558-562.

11

Breeding for Disease Resistance

Vegetable crops are highly susceptible to biotic stresses incited by different disease-causing micro-organisms (fungi, bacteria, phytoplasma, virus), insect pests, nematodes, arachnids and weeds. In contrast to abiotic stresses caused by the environmental factors such as drought, heat, soil salinity, alkalinity, etc., biotic stress-causing agents directly deprive their host of its nutrients leading to reduced plant vigour and, in extreme cases, death of the host plant. In vegetable crops, different biotic stresses are the major cause of pre- and post-harvest losses.

Indiscriminate use of chemicals for the control of diseases, insect pests and nematodes has several disadvantages particularly, increasing cost involved in chemical control, residual toxicity of the chemicals (fungicides, antibiotics, insecticides) in food chain and development of resistance against pathogens and insect pests to different chemicals. Management of diseases and insect pests involve the use of resistant varieties and integration of cultural and chemical control measures.

Of the biotic stresses, much of the breeding efforts have been directed towards developing resistance against diseases. In fact, breeding for resistance has been one of the most important objectives of vegetable breeding programme in the last half of the century. In the biotechnological approaches for vegetable breeding also, most of the achievements have been to generate transgenic plants resistant to diseases. However, development of transgenic plant in different vegetable crops through genetic engineering of *Bt* gene has also become very important achievement. Often breeding for disease-resistance has assumed greater importance in a number of vegetable crops. Consequently, there have been remarkable achievements in developing disease-resistant varieties and use of such resistant cultivars impedes epidemics of pathogen and maintains biological balance in the environment.

Mechanism of the attack of pathogens

Plant cells consist of cell wall, cell membranes and cytoplasm, which contains the nucleus and various organelles and all the substances for which the pathogens attack them. The cytoplasm and the organelles are separated from each other by membranes that carry various types of proteins embedded in them. The plant surfaces that come in contact with the environment either consist of cellulose, as in the epidermal cells of roots and in the intercellular spaces of leaf parenchyma cells, or consist of a cuticle that covers the epidermal cell walls, as is the case in the aerial plant parts. Often an additional layer, consisting of waxes, is deposited outside the cuticle, especially on younger plant parts. Pathogens attack plants because during their evolutionary development they have acquired the ability to live off the substances manufactured by the host plants, and some of the pathogens depend on these substances for survival. Many substances are contained in the protoplast of the plant cells. If the pathogens are to gain access to them they must first penetrate the outer barriers formed by the cuticle and/or cell walls. Even after the outer cell wall has been penetrated, further invasion of the plant by the pathogen necessitates the penetration of more cell walls. Furthermore, the plant cell contents are not always found in forms immediately utilizable by the pathogen and must be broken down to smaller units that the pathogen can absorb and assimilate. Moreover, the plant reacting to the presence and activities of the pathogen, produces structures and chemical substances that interfere with the advance or the existence of the pathogen. If the pathogen is to survive and to continue living off the plant, it must be able to overcome such obstacles.

Mechanical forces exerted by the pathogens on the tissues of the host plant

Plant pathogens are generally tiny microorganisms that cannot apply a voluntary force to the plant surface. Only some fungi which are parasitic to the higher plants, and nematodes appear to apply mechanical pressure to the plant surface they are about to penetrate. The amount of pressure, however, may vary greatly with the degree of pre-softening of a plant surface by enzymatic secretions of the pathogen. For fungi and parasitic higher plants to penetrate a plant surface, they must, generally, first adhere to it. Hyphae and radicles are usually surrounded by mucilaginous substances, and their adhesion to the plant seems to be brought about primarily by the intermolecular forces developing between the surfaces of the plant and pathogen on close contact with the adhesive substances and with one another. In some cases, an adhesion pad forms from the spore when it comes in contact with a moist surface, and cutinase and cellulase enzymes released from the spore surface help the spore to adhere to the plant surface. Spores of some fungi carry adhesive substances at their tips that, on hydration, allow spores to become attached to various surfaces. After contact is established, the diameter of the tip of the hypha or radicle in contact with the host increases and forms the flattened, bulb-like structure called the appressorium. This increases the area of adherence between the two organisms and securely fastens the pathogen to the plant. From the appressorium, a fine growing point, called the penetration peg, arises and advances into and through the cuticle and cell wall. In some fungi, such as *Alternaria, Cochliobolus*, *Colletotrichum*, *Gaeumannomyces*, *Magnaporthe*, and *Verticillium*, penetration of the plant takes place only if melanin (dark pigment) accumulates in the appressorial cell wall. It appears that melanin produces a rigid structural layer which causes water to be absorbed by trapping solutes inside the appressorium.

Chemical weapons of the pathogens

Although some pathogens may use mechanical force to penetrate plant tissues, the activities of pathogens in plants are largely chemical in nature. Therefore, the effects caused by the pathogens on plants are almost entirely the result of biochemical reactions taking place between substances secreted by the pathogen and those present in, or produced by, the host plant. The main groups of substances secreted by the pathogens in plants that seem to be involved in production of disease, either directly or indirectly, are enzymes, toxins, growth regulators and polysaccharides (plugging substances). These substances vary greatly as to their importance in pathogenicity, and their relative importance may be different from one disease to another. Thus, in some diseases, such as soft rots, enzymes seem to be by far the most important, whereas in diseases such as crown gall, growth regulators are apparently the main substances involved. Among the plant pathogens, all except viruses and viroids can probably produce enzymes, growth regulators and polysaccharides. Plant viruses and viroids are not known to produce any substances themselves, but they induce the host cell to produce either excessive amounts of certain substances already found in healthy host cells or substances completely new to the host. Some of these substances are enzymes and others may belong to one of the other groups mentioned earlier.

• Enzymes

Cell wall degrading enzymes (pectinase, ligninase, hemicellulase and cellulase) secreted by the pathogens break down the structural elements in primary and secondary plant cell walls and the medulla which leads to maceration of tissues and cause watery lesions typical of soft rot fungi. It also results in loss of cell turgor, disruption of the plasma membrane, leakage of water and electrolytes.

Different released enzymes *viz*., amylases, lipase and proteases further degrade the cytoplasm and break down inert food substances in the cell or affect components of its membranes and the protoplast directly, thereby interfering with its functioning systems. The enzymatic breakdown of cellulose results in the final production of glucose molecules. Cellulolytic enzymes secreted by the pathogens play a role in the softening and disintegration of cell wall material. They facilitate the penetration and spread of the pathogen in the host and cause the collapse and disintegration of the cellular structure, thereby aiding the pathogen in the production of disease. Some of the nutrient compounds, *e.g*., sugars and amino acids are molecules sufficiently small to be absorbed by the pathogen directly. However, some of the other plant cell constituents such as, starch, proteins and fats can be utilized only after degradation by enzymes secreted by the pathogen.

• Toxins

Toxins seem to act directly on the protoplast components and interfere with the permeability of its membranes and with its function. Several of these toxins, e.g., tabtoxin, phaseolotoxin, cercosporin, etc. inhibit normal host enzymes, thereby leading to increases in toxic substrates or to depletion of needed compounds. Host-specific toxins have been shown to be produced only by certain fungi *viz*., *Cochliobolus*, *Alternaria*, *Periconia*, *Phyllosticta*, *Corynespora*, *Hypoxylon*, etc. Certain bacterial polysaccharides from *Pseudomonas* and *Xanthomonas* have been reported to be host specific.

Certain phytotoxins which are generally non-specific secondary metabolites such as oxalic acid work synergistically with pectinases to degrade plant tissues as found in *Sclerotinia sclerotiorum* by-

a) binding to Ca ions that normally would be bound to pectins and hemicellulose which are the constituents of plant cell walls,
b) lowering the pH of tissues to enhance pectinase activity,
c) inhibiting the synthesis of orthodiphenol oxidases which are responsible for browning due to the formation of quinones from the phenols, hence the rotted tissue does not turn brown.

• Growth regulators

Plant growth is regulated by a small groups of naturally occurring compounds that act as plant hormones and are generally called growth regulators. The most important growth regulators are auxins, gibberellins and cytokinins, but other compounds such as, ethylene and growth inhibitors play important regulatory roles in the growth and development of the plant. Growth regulators act in very small concentrations and even slight deviations from the normal concentration may bring about strikingly different plant growth patterns. Concentration of a specific growth regulator in the plant is not constant, but it usually rises quickly to a peak and then declines quickly as a result of the action of hormone-inhibitory systems present in the plant. Growth regulators appear to act, at least in some cases, by promoting the synthesis of messenger RNA molecules. This leads to the formation of specific enzymes, which in turn control the biochemistry and the physiology of the plant. Plant pathogens may produce more of the same growth regulators as those produced by the plant or more of the same inhibitors of the growth regulators as those produced by the plant. They may produce new and different growth regulators or inhibitors of growth regulators. Alternatively, they may produce substances that stimulate or retard the production of growth regulators by the plant. Whatever the mechanism of action involved, pathogens often cause an imbalance in the hormonal system of the plant and bring about abnormal growth responses such as stunting, overgrowth, rosetting, excessive root branching, stem malformation, leaf epinasty, defoliation and suppression of bud growth. Some of the diseases caused by the imbalance in plant hormones are club root of cabbage caused by the fungus, *Plasmodiophora brassicae*, crown gall caused by the bacteria, *Agrobacterium tumefaciens*, corn smut caused by the fungus, *Ustilago maydis*, root knot caused by the nematode, *Meloidogyne* sp., foolish seedling disease of rice caused by the fungus, *Gibberella fujikuroi*, etc.

• Polysaccharides

Different fungi, bacteria and nematodes constantly release varying amounts of mucilaginous substances that coat their bodies and provide the interface between the outer surface of the microorganism and its environment. Slimy polysaccharides seem to play a role in the vascular disease like wilt, in which they cause a mechanical blockage of the vascular bundles and interfere passively with the translocation of water in the plants.

Development of disease symptom

Development of disease symptoms refers to Pathogenesis. Disease is a condition in which the physiological and biochemical functions of the host plant are not properly discharged and the malfunctioning process that is caused by continuous irritation of the pathogen is manifested through development of disease symptom. Disease symptoms can be divided into four broad categories such as coloration/texture changes, growth changes, wilting and necrosis.

1. Changes in coloration or texture of plant tissue

- **Yellowing or chlorosis**: Chlorosis is a visible disease symptom shown by the green plants where yellow patches or spots are found on the foliage. It happens due to the degradation of chlorophyll pigment or interruption of pigment formation in the leaves mainly caused by the phytotoxins produced by the fungal pathogen. Virus infections also cause interveinal yellowing and thickening of leaves.
- **Reddening and decline**: This symptom is often associated with loss of chlorophyll.
- **Brown/black pigmentation**: This symptom is often associated with death of plants cells and oxidation of phenolics.
- **Tar spot**: This symptom is associated with the formation of melanized fungal sclerotia, stroma and pycnidia or perithecia which are asexual/sexual reproductive bodies (Example: potato scab, phomopsis blight of brinjal, etc).
- **Smut**: These symptoms are seen only at the time of flowering and characterized by the presence of black and sooty or powdery teliospores of the fungus on the inflorescence/seeds of the plants. They are classified into loose smut and covered smut. In loose smut, the fungus infects and develops the spores on the grains while in case of covered smut, the spores remain inside the grains. Example: Smut of wheat caused by the fungus *Ustilago tritici.*
- **Rust**: Appearance of rusty coloured (brown, red and yellow) pin-head size pustules on the leaves and stem of the affected plants is known as rust symptom. A pimple like eruptive fruiting mass is called pustule which is the uredinia of the fungus. The pustules may be erupted, minute or spreading irregularly on the plant parts. These pustules are surrounded by chlorotic halos at the initial stages and later the whole leaf becomes yellow. Example: pea rust caused by the fungus *Uromyses pisi* and French bean rust caused by the fungus *Uromyses fabae*.
- **Mildew**: In case of mildew, the symptoms appear on the ventral surface of leaves as a mass of fungal bodies, downy and powdery sporangiophores and conidiophores of the fungus, respectively in the form of powder. The powdery mildew symptoms are expressed by the presence of white powdery mass on the ventral surface of leaves and are mostly caused by ascomycetous fungi. The downy mildew is characterized by the appearance of white to blackish-brown

powdery patches on ventral surface of leaves whereas the dorsal side of leaves exhibits chlorotic patches. Examples: Powdery mildew of pea caused by the fungus *Erysiphe polygoni*, downy mildew of cucumber caused by the fungus *Pseudoperonospora cubensis*, downy mildew of onion caused by the fungus *Peronospora destructor*, etc.

2. Growth aberrations

- Stunting or dwarfing is a symptom of sub-normal development of various plant parts, causing reduction in plant height, stem length, leaf size and other plant organs. The symptoms start in the form of curling of top leaves. Later, leaves are reduced in size and markedly wrinkled. Branches of plant show weak growth, shortening in length and irregular phyllotaxy. As disease progresses, axillary and accessory buds sprout early on the top producing many groups of small branches. Stunting or dwarfing is caused through production of abscissic acid, ethylene or disrupted nutrient uptake and translocation.
- Overgrowth, galls, warts or tumours (Example: potato wart caused by the fungus *Synchytrium endobyoticum*, club root of crucifers caused by the fungus *Plasmodiophora brassicae*), leaf curl (Example: tomato and chilli leaf curl due to infection of virus causing imbalance in auxins synthesis and breakdown), etiolation (Example: foolish seedling disease of rice incited by the fungus *Gibberella fujikuroi* causing imbalance in gibberellins bio-synthesis), excess branching (Example: little leaf of brinjal caused by the phytoplasma causing imbalance in cytokinin synthesis)

3. Wilting

Vascular wilt diseases caused by the fungi and bacteria are broadly defined as those that result from the blockage of the vascular tissues causing reduction in water absorption through the xylem vessel. Xylem vessels are blocked by tyloses, mycelium and spore masses as found in *Verticillium dahliae* in potato and *Fusarium oxysporum* in tomato. Tyloses are balloon-like swellings or projections that fill the vessels and subsequently plug it and as a result, the vessel can no longer conduct water. Vessels can also be blocked by polymers derived from the fragments of cell walls (polysaccharides, glycoproteins, pectins, phenolic material) which cause brown colouration as found in bacterial wilt of tomato, brinjal and potato caused by *Ralstonia solanacearum*.

4. Necrosis and decay (rot)

Necrosis refers to the death of plant cells due to fungal and viral infections hence, necrosis is also known as cellular death. It is characterized by localized death in living tissues. Apart from fungal and viral infections, nutrient deficiency can also lead to necrosis. It causes the leaves to dry out and the colour of the leaf changes from brown to black. This results in brown or black dots on the leaves. The leaf edges are the first to be affected by necrosis and later, it grows to other areas as well. It is not feasible to reverse necrosis, while chlorosis due to plant nutrient deficiency can be reversed in situations of early detection. When the tissues disintegrate, decay and peel-off easily from any part of the plants due to the action of pathogens, such symptoms are called rot. On the basis of symptoms expressed over the infected tissue, the rots are distinguished in to two namely, soft rot and dry rot. Soft rot leads to oozing of liquid on the infected tissues/ plant parts while in case of dry rots, there is no oozing of fluid. Further, depending upon the plant organ affected, different rots are called as root rot, collar rot, cutting rot, trunk rot, etc. The symptoms of collar rot include the appearance of brown to black discoloration on stem near the soil line (collar region) in the form of ring, which affects the transportation of nutrients.

The symptoms of blight and blast are very characteristic. Generally, they develop in the form of browning/burning either at the tips or at the edges of leaf lamina. The preliminary symptoms appear in the form of isolated irregular brown colour patches which spread gradually on entire leaf surface, resulting in the death of leaves. The blight also affects other aerial parts of the plants. On the basis of plant parts affected by the blight, it is called as leaf blight, twig blight, stem blight, bud blight, sheath blight, etc.

- Various forms of necrois depending upon the extent of cell and tissue death and where on the plant that it occurs. Example: leaf spot disease of amaranthus caused by the fungus *Cercospora canescens*; anthracnose disease caused by the fungus *Colletotrichum* sp.; damping-off caused by the fungi, *Fusarium* sp, *Rhizoctonia* sp.; rot disease caused by the fungi *Pythium* sp, *Phytophthora* sp.; blight disease caused by the fungi *Phytophthora infestans*, *Alternaria solani*, *Rhizoctonia solani,* etc.
- Browning of the dead and rotted tissues is caused by non-enzymatic oxidation of phenol or by the action of phenol oxidases that occurs in the late or degenerative phases of infection.

5. Mosaic

Mosaic symptom refers to alternating light and dark green areas on some leaves, especially the younger ones. Leaves may also be curled. Mosaic symptom generally appears due to infection of the virus. Plant viruses, as obligate intracellular parasites, induce significant changes in the cellular physiology of host cells to facilitate their multiplication. Viral virulence influence different cellular processes such as dysfunction of chloroplast proteins, hormone manipulation, accumulation of reactive oxidative species and control of cell cycle, which are critical for symptom development. On the basis of colour pattern, the mosaic can be recognized and given name as yellow and green mosaic. Examples: Bean common mosaic virus, Cucumber mosaic virus, Bottle gourd mosaic virus, Tomato mosaic virus, Yellow vein mosaic virus of okra, etc.

Physiological responses to biotic stress

Biotic stress is caused by infectious diseases that is developed in different plant parts and is usually caused by the bacteria, fungi, virus or yeasts. Plants respond to biotic stresses through a defence system. The defence mechanism is classified as an innate and systemic response. The defences to biotic stress include morphological and structural barriers, chemical compounds, proteins and enzymes. These confer tolerance or resistance to biotic stresses by protecting products and by giving them strength and rigidity. After infection, reactive oxygen species (ROS) are generated and oxidative bursts limit pathogen spread (Atkinson and Urwin, 2012). Resistance to biotic stress can be induced through specific chemical compounds such as β-aminobutyric acid (BABA) or benzothiadiazole (BTH).

Plant hormones, salicylic acid (SA), jasmonic acid (JA), and ethylene play central roles in biotic stress signalling. Several transcription factors (TFs) are mediators in multiple hormone signalling. Plant defences against biotic stresses involve numerous signal transduction pathways. Abscisic acid (ABA) is reflected as the main hormone involved in the perception of many abiotic stresses (Cramer *et al.*, 2011). However, ABA also has a positive effect on biotic stress resistance (Rejeb *et al.*, 2014). Under abiotic and biotic stress, ABA acts antagonistically with ethylene, which induces liability of the plant against disease attack. However, under abiotic stress ABA increases and induces stomatal closure. As a result, the entry of biotic attackers through stomata is prevented. Therefore, under such situations, the plant is protected from both abiotic and biotic stress (Rejeb *et al.*, 2014). Kinase protein signals also interact with ROS and ABA leads to plant defence enhancement (Rejeb *et al.*, 2014).

Pathogenesis-related (PR) proteins are critical for plant resistance against pathogens. When plants are attacked, their expression is strongly upregulated. In molecular biology and genetics, transcription factors (TFs) are proteins that bind to DNA sequences specifically, thereby regulating the transcription of genetic information from DNA to messenger RNA. It is suggested that with an increase in ABA expression, specific TFs like C-repeat binding factors (CBFs), and cup-shaped cotyledon mediated by ABA could be enhanced, which induces upregulation of Pathogenesis-related (PR) genes (Rejeb *et al.*, 2014). There is a need to understand how different plant tissues and their metabolic pathways respond to different stresses, applied alone or in combination with others to frame the strategy to combat the pathogens (Madani *et al.*, 2019).

The first layer of the plant immune system comprises plasma membrane-localized receptor proteins and intracellular receptors of the nucleotide-binding leucine-rich repeat protein superfamily. Together, these immune receptors act as a network of surveillance machines in recognizing extracellular and intracellular pathogen invasion-derived molecules, ranging from conserved structural epitopes to virulence-promoting effectors. Successful pathogen recognition leads to physiological and molecular changes in the host plants, which are critical for counteracting and defending against biotic attack.

Induced structural and biochemical defenses

Early recognition of the pathogen by the plant is very important if the plant is to mobilize the available biochemical and structural defenses to protect itself from the pathogen. The plant apparently begins to receive signal molecules, i.e., molecules that indicate the presence of a pathogen, as soon as the pathogen establishes physical contact with the plant.

Pathogen elicitors

Various pathogens, especially fungi and bacteria, release a variety of substances in their immediate environment that act as non-specific elicitors of pathogen recognition by the host. Such non-specific elicitors include toxins, glycoproteins, carbohydrates, fatty acids, peptides, and extracellular microbial enzymes such as proteases and pectic enzymes. In various host–pathogen combinations, certain substances secreted by the pathogen, such as *avr* gene products, *hrp* gene products and suppressor molecules act as specific pathogen elicitors of recognition by the specific host plant. In many cases, in which host enzymes break down a portion of the plant surface polysaccharides, the released oligomers or monomers of the polysaccharides act as recognition.

Mobilization of defenses

Once a particular plant molecule recognizes and reacts with a molecule (elicitor) derived from a pathogen, it is assumed that the plant "recognizes" the pathogen. Following such recognition, a series of biochemical reactions and structural changes are set in motion in the plant cell(s) in an effort to fend off the pathogen and its enzymes, toxins, etc. How quickly the plant recognizes the (presence of a) pathogen and how quickly it can send out its alarm message(s) and mobilize its defenses determine whether hardly any infection will take place at all (as in the hypersensitive response) or how much the pathogen will develop, i.e., how severe the symptoms (leaf spots, stem, fruit, or root lesions, etc.) will be, before the host defenses finally stop further development of the pathogen.

Host defense providers: Signal transduction

Several types of molecules have been implicated in intracellular signal transduction. The most common such signal transducers appear to be various protein kinases, calcium ions, phosphorylases and phospholipases, ATPases, hydrogen peroxide (H_2O_2), oligogalacturonides released from plant cell walls, salicylic acid, jasmonic acid, ethylene, etc. Once the pathogen-derived elicitors are recognized by the host, a series of alarm signals are sent out to the host cell proteins and to nuclear genes, causing them to become activated to produce substances inhibitory to the pathogen, and to mobilize themselves or their products toward the point of cell attack by the pathogen. Some of the alarm substances and signal transductions are only intracellular, but in many cases the signal is also transmitted to several adjacent cells and, apparently, the alarm signal is often transmitted systemically to most or all of the plant. These signaling pathway leads to systemic acquired resistance against several diverse types of plant pathogenic viruses, bacteria, and fungi.

Defense mechanisms of the plants

A breadth of significant insights and conceptual advances have been derived from decades of research in various model plant species regarding the structural complexity, functional diversity and regulatory mechanisms of plant immune receptors (Kim *et al.*, 2020). Defense mechanisms of the plants can be categorized as structural /physical or biochemical.

Structural defenses or barriers

- Pre-existing structural characteristic include wax and cuticle on plant surfaces, shape and size of epidermal cells, location and size of natural openings such as stomata and lenticels, presence of tissues containing thick-walled cells (sclerenchyma)
- Different structures formed in response to pathogen attack include
  a) Formation of cork layers and scabs by wound periderm which contain suberin. Suberin, cutin and lignins are complex, higher plant epidermis and periderm cell-wall macromolecules which form a protective barrier. Suberin, a complex polyester biopolymer is lipophilic and composed of long chain fatty acids called suberin acids and glycerol. It seals off the pathogen from obtaining water and nutrients and separates the host from any toxic excretions from the fungus.
  b) Formation of abscission layers and as a result of which infected tissues die and fall off (Kim and Kim, 2002)
  c) Formation of tyloses which are membranous growth into xylem vessels from adjoining parenchyma cells through the pits and depending on their size and number they may clog xylem vessels preventing pathogen spread.
  d) Production of intercellular and intracellular gums and resins. Various types of gums are produced by many plants around lesions after infection by pathogens or injury. Gum secretion is most common in stone fruit trees but occurs in most plants. The defensive role of gums stems from the fact that they are deposited quickly in the intercellular spaces and within the cells surrounding the locus of infection, thus forming an impenetrable barrier that completely encloses the pathogen.
  e) In response to pathogen attack, plants increase cell lignification and this mechanism blocks the invasion of parasites and reduces host susceptibility (Atkinson and Urwin, 2012).
  f) Deposition of amorphous callose which is a polysaccharide in the form of β-1,3-glucan with some β-1,6-branches, which helps to sheath off the pathogen.

Biochemical defenses or barriers to invasion

Various natural compounds, ranging from cell wall components to metabolic enzymes have been reported to protect plants from infection by pathogens and hence, provide specific resistance to hosts against pathogens. It is termed as induced resistance which involves various biochemical components, that play an important role in molecular and cellular signaling events occurring either before (elicitation) or after pathogen infection. Induction of reactive oxygen species, activation of defensive machinery of plants comprising of enzymatic and non-enzymatic antioxidative components, secondary metabolites, pathogenesis-related protein expression (e.g. chitinases and glucanases), phytoalexin production, modification in cell wall composition, melatonin production, carotenoids accumulation, and altered activity of polyamines are major induced changes in the host plants during pathogen infection. Hence, the altered concentration of biochemical components in host plants restricts disease development (Kaur *et al.*, 2022).

- Pre-existing antifungal chemicals exuded into the environment or present inside cells and tissues. As for example, toxic phenolic compounds (protocatechuic acid) from red onions prevent spore germination of the onion smudge disease causing fungus, *Colletotrichum* sp.; phenolic compounds and tannins in young fruits often protect them from fungal attack by degrading enzymes of the grey mold fungus, *Botrytis*; saponins (tomatine of tomato and avenacin of oats) are fungitoxic; different enzymes *viz*., glucanases and chitinases which break down large polysaccharides and chitin, respectively in the fungal cell walls.
- Elicited defense reactions include the hypersensitive reaction that is very common for incompatible host-fungus combinations and very effective against obligate pathogens which involves rapid death and collapse of infected cells and a few cells around them and different chemical events like, accumulation of phenols and production of phytoalexins (small, antifungal phenolics such as phaseollin in beans, pisatin in peas, rishtin in tobacco, capsidiol in chilli, etc.). Phytoalexin production is elicited by pathogen substances (glucans, chitosans, glycoproteins, polysaccharides as fragments from fungal cells walls) or fragments from host cell walls that have been attacked. Some virulent fungi have evolved ways of suppressing the phytoalexins. Some pathogens such as *Fusarium solani* that attacks peas have evolved phytoalexin detoxification enzymes such as pisatin demethylase in order to restore virulence. Although it is not quite clear whether the hypersensitive reaction (HR) is the cause or the consequence of resistance, this type of necrotic defense is quite common, particularly in diseases caused by obligate fungal parasites and by viruses, bacteria and nematodes.
- Resistance mechanisms through stimulation of phenol oxiding enzymes (polyphenol oxidases), which form fungitoxic quinones and protective melanin pigments and peroxidases leading to lignin formation.
- Increased activity of phenylalanine ammonia lyase (PAL), a key enzyme in the production of phenolic compounds, including phytoalexins and lignin.
- Defense through the deactivation of hydrolytic enzymes and phytotoxins.
- Systemic induced resistance (resistance at some distance from the infection site) by salicylic acid, fungal proteins, polysaccharides for the production of new soluble host proteins which is mediated by all processes of the hypersensitive reaction.
- Cross-protection which refers to the phenomenon in which plants infected with the mild strain of a virus are protected from subsequent infection by a virulent strain of the same virus. Cross-protection with a mild strain of a disease-causing virus is considered an effective control measure and this mechanism is comparable to immunization in animal systems.

Different levels of various plant defense phenomena

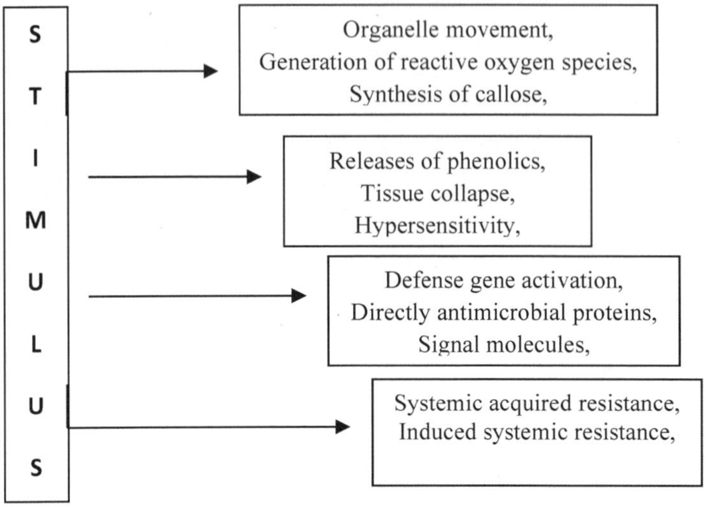

One concept that must be made clear at the outset is that whatever the kind of defense or resistance a host plant employs against a pathogen or against an abiotic agent, it is ultimately controlled, directly or indirectly, by the genetic material (genes) of the host plant and of the pathogen.

Nature of disease resistance and the plant genes involved

Plants lack an adaptive immune system, or the ability to adapt to new diseases and memorize past infections. Though lacking an adaptive immune system, plants have evolved a plethora of sophisticated strategies to counteract biotic stresses. The genetic basis of these defense mechanisms is stored in the plant's genetic code. Plant genomes encode hundreds of biotic stress resistance genes. With the completion of several plant genome sequences during the past decades a glimpse of the wealth of biotic stress resistance genes encoded within plant genomes are available in many crops including several important vegetable crops.

Resistance and susceptibility are the two ends of the disease spectrum. The disease reaction is complex since its intensity varies significantly depending on many factors. Three major factors namely pathogen, host plant and environment interact with each other in different ways, the outcome of which is disease resistance or susceptibility. Disease resistance is a cumulative response to a complex process in plants triggered by the interaction of the pathogen with the various plant components. The major mechanisms are either passive or active. The passive ones are the pre-existing resistance in the plant such as cuticle thickness, phenolics and other barriers to pathogen penetration. Active or induced resistance is a new development in response to the pathogen which includes the production of toxic substances such as phytoalexins, fungal wall degrading enzymes including chitinases, and 1,3 b-glucanases, etc.

Race-Specific, Monogenic, R Gene or Vertical Resistance

Vertical resistance refers to exhibition of high degree of resistance by a plant variety to a single race or strain of a pathogen and this ability usually is controlled by one or a few plant genes. In many plant–pathogen combinations, especially those involving biotrophic oomycetes (*Plasmopara viticola, Plasmopara halstedii, Pseudoperonospora cubensis, Peronospora tabacina, Bremia lactucae,* and *Albugo candida*) causing downy mildews, powdery mildews, rusts and many other fungi, e.g., *Cochliobolus, Magnaporthe, Cladosporium,* many bacteria, nematodes and viruses, defense (resistance) of a host plant against many of its pathogens is through the presence of matching pairs of juxtaposed genes (genes with contrasting effect) for disease in the host plant and the pathogen. The host plant carries one or few resistance genes (R) per pathogen capable of attacking it, while each pathogen carries matching genes for avirulence (Avr) for each of the R genes of the host plant. Avirulence (Avr) genes exist in many fungi that share a gene-for-gene relationship with their host plant. They represent unique genetic determinants that prevent fungi from causing disease on plants that possess matching resistance (R) genes. The avirulence gene of the pathogen serves to trigger the host R gene into action. This then sets in motion a series of defense reactions that neutralize and eliminate the specific pathogen that carries the corresponding (matching) gene for avirulence (Avr), while the attacked and a few surrounding cells die. Vertical resistance (VR) provides a population genetic framework that inevitably leads to the continuous appearance of new pathogen virulences and pathogenicity. This type of defense or resistance is known as race-specific, hypersensitive response (HR), major gene, R gene, or vertical resistance. However, some R genes as for example, the rice disease resistance gene, Xa21, do not induce a visible hypersensitive response (HR).

Horizontal or Quantitative resistance

Genetic resistance that is effective at preventing successful attack only by certain races of a pathogen is called vertical or specific resistance, whereas resistance that is effective at preventing successful attack by all races of a pathogen is called horizontal or general resistance. Van der Plank (1963) first advanced the current framework under which vertical and horizontal resistance were defined and contrasted. He opined that horizontal resistance (HR) and vertical resistance (VR) do not represent different kinds of resistance; they represent, rather, polygenic and oligogenic resistance that is based on the same underlying physiological mechanisms. Horizontal resistance can be explained by a polygenic system where the individual genes are vertical and operating on a gene-for-gene basis with virulence genes in the pathogen (Parlevliet and Zadoks, 1977). The frequencies of the resistance and virulence genes are such that the effective frequencies of resistance genes tend to be negatively related to the magnitude of the gene effect, explaining why major genes often occur at low frequencies, while minor genes appear to be frequent. It is in this way that the host and the pathogen, both as extremely variable and vigorous populations, can co-exist.

The concept of horizontal resistance invariably raises the consciousness of the distinction between "resistance" and "tolerance" (Parlevliet and Zadoks, 1977).

Another factor at play is the durability of HR relates to population dynamics and genetics. The absolute resistance conditioned by single-gene vertical systems presents a large population of pathogens with enormous selection pressure. Any mutation from avirulence to virulence within the population will be subject to tremendous positive selective forces as the newly virulent pathogen successfully enters the host and begins to consume energy resources and reproduce. In contrast, the selective pressures exerted by HR are much less severe. Experimental evidence has borne out the prediction that HR may last indefinitely, while VR breaks down relatively more frequently. Vertical resistance (VR) may be more absolute than HR but the host possessing VR genes is only one pathogen gene mutation away from a compatible disease response. In contrast, the breakdown of HR would require the loss of many independent genes simultaneously, or multiple mutations of the pathogen to overcome host barriers.

However, a clear definition of what vertical resistance is and basic ways it differs from horizontal resistance (Michelmore and Meyers, 1998; St Clair, 2010) are documented below.

- HR is controlled by several to many genes while VR is usually controlled by one or a few plant gene
- HR is more or less equally effective against all "races" of a pathogen while VR is specific to single race or strain of a pathogen.
- HR is more "durable" than vertical resistance while VR breaks down relatively more frequently.
- HR is also referred to as non-specific, polygenic, quantitative, durable or partial resistance.
- It is much easier to develop vertical resistance than horizontal resistance by breeding due to involvement of the number of genes and differences in their heritability.

Mode of inheritance of resistance

- **Vertical or Oligogenic inheritance**: Single or few major genes govern this disease resistance of the host plant, and both dominant and recessive genes may condition the resistance reaction. Such oligogenic resistance is specific to some pathotype of the pathogen and it breaks down relatively more frequently. However, it is much easier to develop vertical resistance following appropriate breeding methods.
- **Horizontal or Polygenic inheritance**: This disease resistance is governed by many genes (polygenes) with small effect expressed in cumulative manner thus showing continuous variation for disease reaction. In this case, several genes contribute to a quantitative trait loci (QTL) that show additive effect against different races of the pathogen. Horizontal resistance is more "durable" than vertical resistance. However, it is complicated to develop horizontal resistance due to the involvement of the number of genes and differences in their heritability.

- **Cytoplasmic inheritance**: Occasionally, a plant trait conferring disease resistance or susceptibility is associated with a cytoplasmic characteristic. In this case, the resistance trait will be maternally transmitted. Cytoplasmic inheritance is also known as non-Mendelian inheritance because the pattern of inheritance does not segregate in accordance with Mendel's laws. The cytoplasm consists of autonomous organelles like mitochondria and plastids that have their own genomes, which are circular like bacterial genomes. These genomes encode their own proteins that can have functions within and outside the organelle. Example: T-cytoplasm (cms-T) of maize is solely responsible for the susceptibility of the plants to Southern corn leaf blight caused by the fungus, *Bipolaris maydis* race T.

Table 1: Oligogenic and Polygenic resistance to different diseases

Crop	**Genetic control for oligogenic resistance to disease**
Tomato	Single dominant gene for Fusarium wilt (*I*), Verticillium wilt (*Ve*), early blight (*Ad*), late blight (*Ph-1*), leaf mould (*Cf*), leaf spot (*Sm*), mosaic virus (*Tm*), leaf curl virus (Ty_1, Ty_2, Ty_3, Ty_4, ty_5, Ty_6) Two independent genes for *Fusarium* wilt.
Potato	Single dominant gene for late blight (*R*), potato virus X, potato virus Y and few dominant major genes for bacterial wilt.
Brinjal	Single dominant gene for bacterial wilt (*Rps*); A pair of dominant genes for phomopsis blight.
Chilli	Single dominant gene for Phytophthora fruit rot, bacterial leaf spot, tobacco mosaic, single recessive gene for leaf curl virus (*lc*), two independent dominant genes for phytophthora root rot.
Okra	Two complementary dominant genes for yellow vein mosaic virus in *A. manihot* ssp.*manihot.*
Pea	Single dominant gene for Fusarium wilt (*Fnw*) and ascohyta blight and single recessive gene for powdery mildew (*er-1*), pea-seed-borne mosaic virus (*sbm*).
Cowpea	Single dominant gene for cercospora leaf spot (*Cls-1*) and bean yellow mosaic, single recessive gene for powdery mildew, bacterial blight and cercospora leaf spot.
Cucumber	Single dominant (*P*) for downey mildew; two pairs of duplicate genes and three complementary recessive genes for powdery mildew.
Muskmelon	Two incompletely dominant genes for downey mildew, single dominant gene for powdery mildew and watermelon mosaic, three recessive genes for cucumber mosaic.
Watermelon	Single dominant gene for fusarium wilt and anthracnose.
Crop	**Genetic control for polygenic resistance to disease**
Tomato	Bacterial wilt, bacterial spot, bacterial canker, leaf curl in *Solanum pimpinellifolium, Solanum habrochaites* (3-5 recessive gene); Early blight (two or more recessive genes).
Potato	Potato leaf roll virus.
Brinjal	Phomopsis blight.
Cauliflower	Black rot (polygenes with dominant effect).

Sources of disease resistance

The first requirement of resistant breeding programme is to find usable source(s) of resistance. Such sources may be present in the indigenous cultivars, land races, folk cultivars, semi-wild relatives and allied species of the crop. Source of resistance to diseases may be obtained from the following genetic resources

- Cultivated variety (improved variety, indigenous cultivars)
- Germplasm collection (Farmers' variety, heirloom variety, land race, folk cultivars, etc.)
- Wild and semi-wild relatives
- Mutants (spontaneous, induced)
- Somaclonal variants
- Allied species of the crop

Several adapted varieties/lines of several vegetable crops in India are found carrying resistance to indigenous races of pathogens (For example, resistance to black rot (*Xanthomonas campestris*) in acclimatized Indian cauliflower, powdery mildew (*Erysiphe polygoni*) in field pea, downey mildew (*Pseudoperonospora cubensis*) in snap melon, *Cucumis melo* var. *momordica,* chilli leaf curl virus in perennial chilli, *Capsicum frutescence*, etc. Different crop wild relatives and allied species of different vegetable crops are the prominent and reliable source of resistance to several diseases.

Table 2: Wild and allied species of the vegetable crops carrying resistance attribute

Crop	Disease	Pathogen	Species
Potato	Late blight	*Phytophthora infestans*	*Solanum tuberosum* ssp. *andigena, S. demissum, S. curtilobum, S. stoloniferum*
Tomato	Fusarium wilt	*Fusarium oxysporum* f.sp. *lycopersici*	*S. pimpinellifolium, S. habrochaites*
	Early blight Bacterial wilt	*Alternatia solani* *Pseudomonas solanacearum*	*S. pimpinellifolium, S. habrochaites* *S. pimpinellifolium*
	Leaf curl	Tomato leaf curl virus	*S. chilense, S. peruvianum, S. pimpinellifolium, S. habrochaites,*
Brinjal	Bacterial wilt	*Pseudomonas solanacearum*	*S. sisymbrifolium, S. integrifolium*
	Phomopsis blight	*Phomopsis vexans*	*S. xanthocarpum, S. indicum, S. gilo, S. khasianum, S. nigrum, S. sisymbrifolium*
	Little leaf	Phytoplasma	*S. viarum, S. incanum*
Chilli	Tobacco mosaic	Tobacco mosaic virus	*C. chacoense*
	Phytophthora rot	*Phytophthora capsici*	*C. frutescens*
	Leaf curl Anthracnose	Leaf curl virus *Colletotrichum capsici*	*C. frutescens* *C. baccatum, C. chinense*

Okra	Yellow vein mosaic	Yellow vein mosaic virus	*A.manihot* ssp. *manihot* *A. manihot* ssp. *tetraphylus*
Onion	Purple blotch	*Alternaria porri*	*Allium fistulosum*
Cabbage and cauliflower	Club root White rust	*Plasmodiophora brassicae* *Albugo candida*	*Brassica napus, B. campestris* *Brassica carinata*

Breeding techniques for disease resistance

Breeding techniques for disease resistance are categorized into conventional and molecular methods. Conventional methods include introduction, selection, hybridization and selection (Pedigree and backcross) and mutation (utilization of spontaneous and induced mutants), whereas different biotechnological interventions include *in vitro* culture (embryo rescue, protoplast fusion, somaclonal variation), marker assisted selections and genetic engineering.

The principles underlying the breeding for resistance to diseases are much the same as for other characters. However, in breeding for disease resistance, the breeder is to deal with two series of heritable factors: the heritable characters of the host plant and the heritable differences in the pathogen. Hence, proper screening method is of utmost importance. Breeding methods depend on the source of resistance and its inheritance pattern of host resistance.

Specific programmes need to be taken to integrate resistance breeding with heterosis breeding to develop promising disease-resistant hybrids. In this breeding programme, parental lines resistant to indigenous pathogens or races of pathogens need to be developed for their subsequent utilization as parents in breeding resistant hybrids.

Conventional breeding techniques

1. Plant Introduction

Introduction of resistant varieties for cultivation in new regions presents a relatively straight forward and swift approach to obtaining resistance. However, this method has certain limitations. Firstly, the introduced varieties may not perform optimally in the new environment. Secondly, they may become susceptible to the specific disease in the new region, and thirdly, they might prove vulnerable to other diseases prevalent in that area. However, introduction of wild relatives and other cultivated species often serves as a gene complex reservoir for resistance to some diseases. In India, some resistant varieties of okra, tomato, chilli, etc. have been developed through utilization of wild relatives in the hybridization programme followed by selection.

- **Okra**: Yellow vein mosaic virus resistant varieties *viz*., Arka Anamika and Arka Abhay (Okra variety × *Abelmoschus manihot* var. *tetraphyllus*), Parbhani Kranti and Punjab 7 (Okra variety Pusa Sawani × *Abelmoschus manihot* var. *manihot*), Punjab Padmini (Okra variety Reshmi × *Abelmoschus manihot* var. *manihot)*, etc.

- **Tomato**: Leaf curl resistant variety Hisar Anmol, Kashi Amrit, (Hisar Arun × *Solanum habrochaites*); Leaf curl resistant variety Kashi Vishes, Kashi Anupam (Tomato variety × *Solanum habrochaites*); Leaf curl resistant variety, Kashi Aman, Kashi Adarsh, Kashi Amul (introgression of *Ty-3* gene from introduced genotype from AVRDC, Taiwan)
- **Chilli**: Leaf curl virus tolerant variety Pusa Sadabahar (Pusa Jwala × *Capsicum frutescens* (IC 81339)

2. Selection

Selection is the process of choosing plants with desirable characteristics. This is the oldest breeding method and is the basis of all crop improvement programmes. The most cost-effective and expedient method for developing a resistant variety involves selecting resistant plants from a commercial variety. While this approach has proven valuable in many earlier cases, its utility is currently somewhat limited given the present state of crop improvement.

It is all the time efficient to exercise selection process in the hotspot area for a particular disease. Hotspot area is described as the areas of elevated incidence for prevalence of the particular disease or higher probability of disease emergence. Hotspots have been shown to have a higher potential for transmission risk and may be the source of infections, making them a priority for controlling epidemics. However, the role of edge areas of hotspots in disease transmission remains unclear.

Disease resistant varieties of some vegetable crops in India have been developed by straight selection from the germplasm assemblage.

- **Tomato**: Bacterial wilt resistant variety Arka Alok (from introduced line of AVRDC, CL-114-5-1-0)
- **Brinjal**: Bacterial wilt resistant variety, Pusa Purple Cluster (from indigenous genotype)
- **Chilli**: Chilli leaf curl virus resistant variety, Bidhan Chilli-4 (from indigenous genotype)
- **Pea**: Powdery mildew resistant variety, Phule Priya (from indigenous genotype)

3. Hybridization and selection

Hybridization results in a combination of valuable parental characters in the hybrid progeny. Among the characters whose recombination usually is the purpose is framed in the breeding objectives. The recombination of characters inherent in one variety with those of another during hybridization is a consequence of the discrete nature of inheritance.

Hybridization and selection is the most commonly employed approach in breeding for disease resistance which serve two primary purposes: (1) transferring disease resistance from an agronomically / horticulturally undesirable variety to a susceptible yet otherwise desirable variety through the backcross breeding method, and (2) combining disease resistance and other favourable traits from one variety with the superior characteristics of another variety *via* pedigree method of breeding.

For simply inherited resistance trait, generally backcross method of breeding is commonly employed to transfer the resistance from the donor parent to the commercial variety. However, in certain cases simple selection, pedigree method and combination of backcross and pedigree method are employed in breeding. In backcross method of breeding, the resulting new variety is agronomically / horticulturally similar to the susceptible variety but possesses disease resistance. In the pedigree method of breeding, the new variety is expected to excel in agronomic traits while retaining disease resistance. In both the cases, one of the parents is selected for its disease resistance, ideally exhibiting a high level of resistance against multiple pathogen races, governed by a few oligo genes. When the resistant variety is a non-adapted genotype and lacking in desirable agronomic traits, the backcross method becomes the obvious choice. However, when the resistant variety is well-adapted and possesses additional desirable characteristics, the pedigree method of breeding is preferred.

In polygenic control of resistance, mass selection, recurrent selection, controlled matings (among the resistant progeny) in the F_2 and succeeding generations and other breeding methods involving gene pyramiding are employed.

Depending on the genetic control of resistance of the host plant under improvement and reproductive system of the crop, back cross, pedigree, pedigree with SSD, mass-pedigree, backcross-pedigree, mass-sib-crossing methods for predominantly self-pollinated crops (tomato, eggplant, chilli, okra, garden pea, vegetable cowpea, hyacinth bean, etc.), different population improvement methods for cross pollinated crops (cabbage, cauliflower, carrot, radish, etc.) and clonal breeding method utilizing sexual mode of reproduction for asexually propagated crops (potato, sweet potato, cassava, etc.) are generally adopted for the development of disease resistant varieties (Hazra *et al.*, 2022).

- **Tomato**: *Fusarium* wilt and nematode resistant variety, Punjab NR 7 (S-12 × NMR-1)
- **Brinjal**: Bacterial wilt resistant variety, Utkal Tarini (Pusa Kranti × Gopa Local), Bidhan Suphala (Bidhan Suphala × Punjab Sadabahar)
- **Okra**: Yellow vein mosaic resistant variety, Varsha Uphar (Lam Selection × Parbhani Kranti), Hisar Unnat (Selection 2-2 × Parbhani Kranti), Kashi Pragati (VRO-6) (NIC 9303 × PK 20), Bidhan Sahebmukto (VRO 6 × Mukta)
- **Garden pea**: Powdery mildew resistant variety, Jawahar Pea 83 [(JM 1× JP 829) x (46c × JP 501)], Jawahar Peas 15 (JM 1 × R 98 B)
- **Vegetable cowpea**: Bacterial blight resistant variety, Pusa Komal [(Pusa Dofasli × EC 26410) × P 426)]
- **Cauliflower**: Black rot and curd blight resistant variety, Pusa Shubhra [(MG 2-3 × 15-1-1) × D-96)]
- **Potato**: Late blight resistant variety, Kufri Jyoti (3069 d (4) × 2814 a (1), Kufri Muthu (3046 (1) × M 109-3), Kufri Himalini (1-1062 × Tollacan), Kufri Sadabahar (MS/81-145 × PH/F-1545)

Mutation breeding

Spontaneous mutations are the ultimate source of genetic variation present in the biological world. Mutations either spontaneous or induced may help in developing resistant against pathogen. Very few disease resistant varieties have been developed in India through applied mutagenesis.

Yellow vein mosaic virus resistant mutant variety Punjab 8 was isolated from the okra cultivar Pusa Swani treated with 1% EMS solution. Similarly, yellow vein mosaic virus resistant mutant line 350//10///3-9////28 and 450//66///2-4////39 could be isolated from the M_5 generation of the okra cultivar Pusa Swani irradiated by 350 and 450 Gy gamma radiation (Hazra *et al.*, 2023).

Biotechnological approaches

In vitro culture

Different *in vitro* culture techniques like embryo rescue, protoplast fusion techniques are integrated with the conventional breeding methods. Crossing barriers are major hindrance to utilize resistant genes from the distantly related parental lines. These *in vitro* culture techniques are employed to overcome interspecific and even intergeneric crossing barriers as shown by the crosses: (*S. melongena* × *S. sisymbrifolium, S. gilo* × *S. integrifolium*), (*S. lycopersicum* × *S. peruvianum, S. lycopersicum* × *S. habrochaites*), (*C. annum* × *C. baccatum* var. *pendulum*), (*Sinapis alba* × *Brassica oleracea* var. *botrytis*) etc.

Isolation of disease-resistant somaclonal variants

Disease-resistant somaclonal variants can be generated through two distinct approaches. Firstly, plants that are regenerated from the cultured cells or their offspring are subjected to a disease test, and resistant plants are identified through screening against the particular pathogen. Secondly, cultured cells are chosen for their resistance to the toxin or culture filtrate produced by the pathogen, and plants are subsequently regenerated from these selected cells using cell selection. In the majority of instances, these plants also demonstrate resistance to the specific disease under consideration. Cell selection strategy is more successful in cases where toxin produced by the pathogen plays a significant role in disease development.

Marker assisted selection and Gene pyramiding

Marker assisted selection (MAS) increases genetic response by affecting effiency and accuracy of selection with the help of molecular markers component in backcross or pedigree breeding programmes for tracking the resistance genes in gene pyramiding programme.

Gene pyramiding, which aims to assemble multiple desirable genes into a single genotype, is a commonly used breeding method for self-pollinated crops. Gene pyramiding is gaining considerable importance as it would improve the efficiency of plant breeding leading to the development of genetic stocks and precise development of broad spectrum resistance capabilities. Traditionally, the main use of gene pyramiding is to improve an existing elite cultivar through introgression of one or two genes of large effects from other sources, since the presence of the target genes has to be monitored by phenotyping, which is only effective for major genes. Depending on the trait and inheritance of the targeted genes, gene pyramiding may require large labour, time and material resources. Development of modern plant molecular techniques and quantitative genetics in the last four decades has dramatically widened the applicability of gene pyramiding by providing enhanced knowledge of the genetics of the breeding traits and of the relative genomic location of functionally related as well as neutral markers associated with the genes responsible for the traits. Identification of genes with large effect for traits which are traditionally regarded as quantitative are not targeted by gene pyramiding programme. Marker-assisted selection reduces/eliminates extensive phenotyping, provides more effective options to control linkage drag, makes the pyramiding of genes with very similar phenotypic effects possible and reduces the breeding duration. Marker-assisted gene pyramiding now is the method of choice for inbred line development targeted at improving traits controlled by major genes. However, the success of gene pyramiding through MAS depends upon several critical factors, including the number of genes to be transferred, the distance between the target genes and flanking markers, the number of genotype selected in each breeding generation, the nature of germplasm, etc.

Genetic Engineering

Many crops including vegetable crops have been genetically modified to induce resistance to insects, plant pathogens (including viruses), and herbicides and for improved features, such as slow ripening, higher nutritional status, seedless fruit, and increased sweetness. Transgenic crops, commonly referred to as genetically modified (GM) crops enable breeders to bring favorable genes, often previously inaccessible, into already elite cultivars, improving their value considerably and offer unique opportunities for controlling insects and pathogens.

Conventional plant breeding that utilizes non-transgenic approaches remain the backbone of vegetable genetic improvement strategies. However, transgenic crop cultivars should not be excluded as products are capable of contributing to resistance and more nutritious and healthy food. Although development of transgenic was originally used for field crops, genetic engineering has the potential to address some of the most challenging biotic and abiotic constraints faced by vegetable farmers worldwide. These challenges are not easily addressed through conventional vegetable breeding alone. A successful application of biotechnological approaches is needed for the development of vegetable varieties possessing resistance to disease caused particularly by the viruses transmitted by the vectors.

Some important genetically modified vegetable crops include GM squash (*Cucubita pepo*) cultivars resistant to Zucchini yellow mosaic virus (ZYMV), Watermelon mosaic virus (WMV) and Cucumber mosaic virus (CMV).

Small RNA – regulator of biotic stress and pathogenesis in crops

Plants have myriad of defense mechanisms against biotic stress and small RNA (sRNA) constitute an important defense system against biotic stress. Small RNAs like siRNA, miRNA and circRNA play a main role in regulating the expression of plant genes at post transcriptional level, collectively termed as RNA interference (RNAi). The biogenesis and regulation of sRNAs are dependent on key proteins such as Dicer-like (DCL), Argonaute (AGO) and RNA dependent RNA polymerase (RDR). sRNAs responsible for the disease can originate either from the pathogen causing infection or from the infected plant to overcome the infection. Novel tRFs and rRFs from bacteria mimic miRNAs of plants. Extensive study on mechanisms and regulations of sRNAs in both pathogens and plants can provide a deep insight in to exploring an efficient way of disease management against biotic stress and pathogenesis in different food crops (Raja and Tennyson, 2023).

Benefits of breeding for disease resistance

Diseases inflict substantial economic losses and can often result in human suffering, including crop failures that may lead to famines. Resistant varieties play a critical role in averting such crises. Resistant varieties offer the most cost-effective means of disease control as they do not entail recurring expenses for farmers. Hence, they represent the most economically viable option for disease management, especially for small-scale farmers in developing countries. In certain cases, such as powdery mildew, downy mildew, rusts and smuts and virus diseases, resistance breeding is the sole practical method of disease control because the pathogens are obligate parasites.

Disease-resistant varieties can be effectively integrated with other disease control measures. Breeding for disease resistance focuses on maintaining productivity (productivity maintenance research) by reducing losses caused by diseases. In contrast, breeding for higher yields aims at achieving positive yield gains under disease-free conditions (productivity enhancement research). Disease-resistant varieties eliminate the need for pesticides, contributing to reduced environmental pollution and making them highly desirable. Effectiveness of disease-resistant varieties remains consistent regardless of environmental conditions, in contrast to chemical control methods, which can be heavily influenced by weather conditions.

Challenges in breeding for disease resistance

Major gene resistance is relatively straightforward to incorporate into a breeding programme. Unfortunately, in some cases, new resistance genes can become vulnerable to rapidly emerging virulent races, often within 2-5 years of the release of a new variety. This phenomenon leads to a cycle of success and decline for disease-resistant varieties, necessitating the continuous search for new resistance genes to base future varieties on. In particularly challenging situations, breeding programmes originally designed for major gene resistance can also incorporate polygenic resistance. Consequently, varieties carrying major R genes often exhibit limited polygenic resistance. This can lead to the well-known "Vertifolia effect" when the predominant virulent strains emerge. The "Vertifolia effect" is a well-documented phenomenon in the fields of plant breeding and plant pathology. It is characterized by the erosion of a crop's horizontal resistance to a disease during a breeding cycle due to the presence of strong vertical resistance, characterized by the presence of R genes. This effect was observed in late blight resistance of potato. Polygenic or partial resistance tends to be durable, but breeding for this type of resistance presents challenges in most cases. The primary difficulty lies in accurately and reliably assessing the level of resistance. In almost all crops, more than one disease and/or insect pest holds economic importance. Therefore, a successful variety should possess multiple resistances, i.e., resistance to several diseases and insect pests. Combining genes conferring these resistances into a single variety demands meticulous planning, elaborate execution, and significantly greater effort compared to breeding for single resistance. Marker-assisted gene pyramiding now is the method of choice for inbred line development targeted at improving traits controlled by major genes.

References

Atkinson, N.J. and Urwin, P.E. 2012. The Interaction of Plant Biotic and Abiotic Stresses: From Genes to the Field. *J. Expt. Bot.*, **63**: 3523-3543.

Cramer, G., Urano, K., Delrot, S and Pezzotti, M. 2011. Effects of abiotic stress on plants: a systems biology perspective. *BMC Plant Biol.*, 11:163DOI:10.1186/1471-2229-11-163

Hazra, P., Banerjee, M.K. and Chattopadhyay, A. 2022. *Varieties of Vegetable Crops in India.* 3rd Edition, Kalyani Publishers, Ludhiana, 215p.

Kaur, S., Samota, M.K., Choudhary, M., Choudhary, M., Pandey , A.K., Sharma , A. Thakur , J. 2022. How do plants defend themselves against pathogens-biochemical mechanisms and genetic interventions.*Physiol. Mol. Biol. Plants*, **28**: 485–504.

Kim, J.H., Danve , C. and Castroverde , C.D. 2020. Diversity, Function and Regulation of cell surface and intracellular immune receptors in *Solanaceae. Plants* (Basel), **9**: 434.. doi: 10.3390/plants9040434

Kim , Y.H. and Kim, K.H. 2002. Abscission layer formation as a resistance response of peruvian apple cactus against *Glomerella cingulate. Phytopath.*, **92**: 964- 969.

Madani, B., Mirshekari, A., and Imahori, Y. 2019. Physiological responses to stress, In: *Postharvest Physiology and Biochemistry of Fruits and Vegetables* (Ed. Yahia, E.M.), Elsevier, pp. 405-423.

Michelmore, R.W. and Meyers, B.C. 1998. Clusters of resistance genes in plants evolve by divergent selection and a birth-and-death process. *Genome Res.*, **8**: 1113-1130.

Parlevliet, J.E. and Zadoks, J.C. 1977. The integrated concept of disease resistance; a new view including horizontal and vertical resistance in plants. *Euphytica*, **26**: 5-21.

Raja, P. and Tennyson, J. 2023. Small RNA – regulator of biotic stress and pathogenesis in food crops. In: *Plant Small RNA in Food Crops* (Eds. Guleria, P., Kumar, V. and Mo, B.), Academic Press, Cambridge, USA.

Rejeb, I.B., Pastor, V. and Mauch-Mani, B. 2014. Plant responses to simultaneous biotic and abiotic stress: Molecular mechanisms. *Plants* (Basel), 15:458-475.

St.Clair, D.A. 2020. Quantitative disease resistance and quantitative resistance loci in breeding. *Ann. Rev. Phytopath.*, **48**:247-268

Vanderplank, J.E. 1963. *Plant Disease: Epidemics and Control.* Academic Press, New York, 349p.

12

Abiotic Stress Physiology and Management

Basic concepts of plant stress, acclimation and adaptation

Energy is the absolute requirement for the maintenance of structural organization over the lifetime of the organism. Maintenance of such complex structural organization over time requires a constant flow of energy through all biological organisms, which provides the dynamic driving force for the performance of important maintenance processes (the processes which are essential for an organism to remain alive and continue as the species) such as respiration, assimilation, nutrition, cellular biosyntheses and transport to maintain its characteristic structure and organization as well as the capacity to replicate and grow. Maintenance of such steady-state results in a meta-stable condition called homeostasis. Thus, "Homeostasis" is defined as a self-regulating process by which a living organism can maintain internal stability while adjusting to changing external conditions. It is the state of balance among all the living systems needed for the organism to survive and function properly.

Plant stress

Any change in the surrounding environment may disrupt homeostasis. Environmental modulation of homeostasis may be defined as biological stress. Thus, plant stress implies some adverse effect on the physiology of a plant induced upon a sudden transition from some optimal environmental condition where homeostasis is maintained to some suboptimal condition which disrupts this initial homeostatic state. Thus, plant stress is a relative term measured in comparison to survival, biomass accumulation or yield of a species. The experimental design to assess the impact of a stress always involves the measurement of a physiological phenomenon in a plant species under a suboptimal, stress condition compared to the measurement of the same physiological phenomenon in the same plant species under optimal conditions.

Different categories of plant stress

Plants respond to stress in several different ways. However, plant stress can be divided into two primary categories.

• Abiotic stress

Abiotic stress is imposed on plants by physical (e.g., light, temperature) or chemical insult that the environment may impose on a plant. The most commonly encountered abiotic stress factors are temperature extremes (such as heat stress, cold spells and freezing), drought (water stress), excessive watering (water logging, flooding or submergence), and soil ion content (salinity i.e., high sodium content and mineral toxicity) which negatively impact growth, development yield and seed quality of crop and other plants.

Climate change refers to a change in the state of the climate that can be identified (e.g. using statistical tests) by changes in the mean and/or the variability of its properties and that persists for an extended period, typically decades or longer. It has been estimated that CO_2 concentration in the atmosphere has increased drastically from 280 ppm to 370 ppm and is likely to be doubled in 21stcentury (IPCC, 2007). With the in°C higher than the current level by 2100 is predicted (IPCC, 2007) which is the threat to the biosphere. The Indian climate has undergone significant changes showing increasing trends in annual temperature with an average of 0.56°C rise over last 100 years (Rao *et al.,* 2009; IMD, 2010).

Heat stress has known effects on the life processes of organisms, acting directly or through the modification of surrounding environmental components. Plants are sessile organisms because they spend their lives anchored to the substrate and cannot move to more favourable environments as a result of which plant growth and developmental processes are substantially affected, often lethally, by high temperature stress (Lobell and Field, 2007). Rate of plant growth and development is dependent upon the temperature surrounding the plant and each species has a specific temperature range represented by a minimum, maximum, and optimum. The evidence of global warming has increased the interest in the cause of yield declines at temperatures only slightly above optimal in different crops such as rice (Baker and Allen, 1993; Wheeler *et al.*, 2000), groundnut (Vara Prasad *et al.*, 1999), sorghum (Craufurd *et al.*, 1998), cowpea (Craufurd *et al.*, 1998), tomato (Peet *et al.*, 2002; Hazra and Ansary, 2008), wheat (Barlow *et al.*, 2015), etc. Analysis by Meehl *et al.* (2007) revealed that daily minimum temperatures will increase more rapidly than daily maximum temperatures leading to the increase in the daily mean temperatures and a greater likelihood of extreme events and these changes could have detrimental effects on yield.

It has been evaluated that without the use of CO_2 fertilization for increased rate of photosynthesis and genetic transformation, each 1°C rise in the global mean temperature reduces global maize yield by 7.4%, wheat yield by 6.0%, rice yield by 6.2% , soybean yield by 3.1% (Zhao *et al.*, 2017; Xu *et al.*, 2021)

• Biotic stress

Biotic stress is imposed by biological agents (e.g., insects-pests and nematode, disease-causing organisms like, fungi, bacteria, phytoplasma and virus,) to which a plant may be exposed during its lifetime. Many biotic stresses affect photosynthesis as in case of the larvae of the lepidopteran insects having chewing type mandible reduce leaf area, bore inside the flower and fruit. Virus infections also reduce the rate of photosynthesis per leaf area. Despite lacking the adaptive immune system, plants can counteract biotic stresses by evolving themselves to certain physio-biochemical strategies. The defence mechanisms conditioned by resistant genes which act against these stresses are controlled genetically.

Acclimatization to the stress

Some sudden change in environmental condition generally imposes stress on plant species. However, in stress-tolerant plant species, exposure to a particular stress leads to acclimatization to that specific stress in a time-dependent manner. Thus, plant stress and plant acclimatization are intimately linked with each other. The stress-induced modulation of homeostasis can be considered as the signal for the plant to initiate processes required for the establishment of a new homeostasis associated with the acclimatized state. Plants exhibit stress resistance or stress tolerance because of their genetic capacity to adjust or to acclimatize to the stress and establish a new homeostatic state over time. Establishment of homeostasis associated with the new acclimatized state is not the result of a single physiological process but rather the result of many physiological processes that the plant integrates over the acclimatization period. Plants usually integrate these physiological processes over a short-term as well as a long-term basis. Acclimatization usually involves the differential expression of specific sets of genes associated with the exposure to a particular stress. The remarkable capacity to regulate gene expression in response to environmental change in a time-nested manner is the basis of plant plasticity.

Adaptation and phenotypic plasticity

Plants have various mechanisms that allow them to survive and often prosper in the complex environments in which they live. Adaptation to the environment is characterized by genetic changes in the entire population that have been fixed by natural selection over many generations. Development of high temperature tolerant cauliflower group "Indian cauliflower" (Swarup and Chatterjee, 1972) and short day onion group "Kharif onion" in India (Mahajan *et al.*, 2018) can be cited as important examples for the change in population through adaptation to the new environment.

Effect of imbalances of abiotic factors on plants

Plants may experience physiological stress when an abiotic factor is deficient or in excess (referred to as an imbalance). The deficiency or excess may be chronic or intermittent. Abiotic conditions to which native plants are adapted may cause physiological stress to non-native plants. For this reason, the new plant species or varieties introduced from their adapted environmental condition should be released for commercial cultivation in new environmental condition after necessary acclimatization. Imbalances of abiotic factors in the environment cause primary and secondary effects in plants. Primary effects such as reduced water potential and cellular dehydration directly alter the physical and biochemical properties of the cells, which then lead to secondary effects. These secondary effects, such as reduced metabolic activity, ion cytotoxicity and production of reactive oxygen species initiate and accelerate the disruption of cellular integrity and may lead ultimately to cell death. Different abiotic factors may cause similar primary physiological effects because they affect the same cellular processes. This is the case for water deficit, salinity and freezing, all of which cause reduction in hydrostatic pressure or turgor pressure that builds up against the cell wall of a plant cell due to the movement of water, stored inside. Turgor pressure which is generally caused by the osmotic flow of water is responsible for the enlargement of cells. Reduction in turgor pressure results in cellular dehydration and increases the cellular susceptibility to stress-induced damage. Secondary physiological effects caused by different abiotic imbalances may overlap substantially. It is evident that imbalances in many abiotic factors reduce cell proliferation, photosynthesis, membrane integrity and protein stability, and induce production of reactive oxygen species (ROS), oxidative damage and cell death.

Different imbalances of abiotic factors causing stress on plants

Abiotic stresses either individually or in combination cause morphological, physiological, biochemical and molecular changes that adversely affect plant growth and ultimately yield.

1. Excess light excitation stress

As photo-autotrophs, plants are dependent upon – and exquisitely adapted to – visible light for the maintenance of a positive carbon balance through photosynthesis. Higher energy wavelengths of electromagnetic radiation, especially in the ultraviolet range, can inhibit cellular processes by damaging membranes, proteins and nucleic acids. However, even in the visible range, irradiances far above light saturation point of photosynthesis (the point where the light intensity does not increase the photosynthesis rate) cause high light stress, which can disrupt chloroplast structure and reduce photosynthetic rates. This excess light excitation stress is known as photo-inhibition.

Photosystem II (PSII) has a P680 reaction centre containing chlorophyll 'a' that uses light energy to carry out the oxidation (splitting) of water molecules, and to produce ATP *via* a proton pump. Excess light excitation arriving at the PSII reaction centre can lead to its inactivation by the direct damage of the D1 protein. The D1 protein is encoded by the psbA gene residing in the chloroplast genome which represents a key subunit of photosystem II (PSII), a multicomponent pigment–protein complex of oxygenic photosynthetic organisms. Nicotinamide adenine dinucleotide phosphate ($NADP^+$) is a coenzyme that functions as a universal electron carrier, accepting electrons and hydrogen atoms to form NADPH, an essential electron donor in all organisms. Excess absorption of light energy by photosynthetic pigments also produces excess electrons outpacing the availability of NADP+ to act as an electron sink at PSI. The excess electrons produced by PSI lead to the production of reactive oxygen species (ROS), notably superoxide ($O_2^{\bullet -}$). In superoxide, O_2 is reduced to $O_2^{\bullet -}$ by NADPH.

Superoxide and other ROS are low-molecular-weight molecules that function in signalling and, in excess, cause oxidative damage to proteins, lipids, RNA, and DNA.

The oxidative stress generated by excessive ROS destroys cellular and metabolic functions and leads to cell death.

2. High temperature stress

Mesophytic plants (terrestrial plants which are neither adapted to particularly dry nor particularly wet environments) which include the majority of plants, trees, shrubs, and most crops including vegetable crops have a relatively narrow temperature range of about 10°C for optimal growth and development, the optimum range for those plants being 20°C and 30°C. Outside of this range, varying degrees of damage occur, depending on the magnitude and duration of the temperature fluctuation causing different types of stress: high temperatures, low temperatures above freezing and temperatures below freezing. Most plants with access to abundant water are able to maintain leaf temperatures below 45°C by evaporative cooling, even at elevated ambient temperatures. However, high leaf temperatures combined with minimal evaporative cooling causes heat stress. Leaf temperatures can rise to 4 to 5°C above ambient air temperature in bright sunlight near midday, when soil water deficit causes partial stomatal closure or when high relative humidity reduces the gradient driving evaporative cooling. Increases in leaf temperature during the day can be more pronounced in plants experiencing drought and high irradiance from direct sunlight.

Of the major forms of abiotic stress plants are exposed to in nature *viz*., heat, drought, cold and salinity, heat stress has an independent mode of action on the physiology and metabolism of plant cells. Susceptibility of the plants to high temperature stress varies with the stage of plant development however, it depends on the species and genotype, with abundant inter- and intra-specific variations (Barnabás *et al*., 2008; Sakata and Higashitani, 2008).

Different manifestations of high temperature stress

a) Physiological, cellular and molecular impact of high temperature stress

High temperatures negatively affect various physiological processes including photosynthesis, primary and secondary metabolism and hormonal signalling. Heat stress has negative effects on plant growth and development by disrupting the stability of various proteins, membranes and cytoskeleton structures.

Damaged membranes and enzymes

Plant membranes consist of a lipid bilayer interspersed with proteins and sterols, and any abiotic factor that alters membrane properties can disrupt cellular processes. The physical properties of the lipids greatly influence the activities of the integral membrane proteins, including H+ pumping ATPases, carriers and channel-forming proteins that regulate the transport of ions and other solutes. High temperatures cause an increase in the fluidity of membrane lipids and a decrease in the strength of hydrogen bonds and electrostatic interactions between polar groups of proteins within the aqueous phase of the membrane. High temperatures thus modify membrane composition and structure, and can cause leakage of ions. High temperatures can also lead to a loss of the three-dimensional structure required for correct function of enzymes or structural cellular components, thereby leading to loss of proper enzyme structure and activity. Misfolded proteins often aggregate and precipitate, creating serious problems within the cell.

Inhibition of photosynthesis

Photosynthesis and respiration are both inhibited by temperature stress. Typically, photosynthetic rates are inhibited by high temperatures to a greater extent than respiratory rates. Although chloroplast enzymes such as rubisco, rubisco activase, NADP-G3P dehydrogenase, and PEP carboxylase become unstable at high temperatures, the temperatures at which these enzymes began to denature and lose activity are distinctly higher than the temperatures at which photosynthetic rates begin to decline. This would indicate that the early stages of heat injury to photosynthesis are more directly related to changes in membrane properties and to uncoupling of the energy transfer mechanisms in chloroplasts.

Heat stress induces changes in respiration and photosynthesis and thus leads to a shortened life cycle and diminished plant productivity (Barnabás *et al.*, 2008). Heat stress induces major modifications at the sub-cellular level in chloroplasts, leading to significant reduction in photosynthesis by changing the structural organization of thylakoids (Karim *et al.*, 1997). Loss of grana stacking or its swelling, disruption of vacuoles and mitochondria has also been reported as the specific effects of high temperatures resulting in reduced photosynthetic and respiratory activities (Zhang *et al.*, 2005).

This imbalance between photosynthesis and respiration is one of the main reasons for the deleterious effects of high temperatures. On an individual plant, leaves growing in the shade have a lower temperature compensation point than leaves that are exposed to the sun (and heat). Reduced photosynthate production may also result from stress-induced stomatal closure, reduction in leaf canopy area, and regulation of assimilate partitioning.

Disturbed compartmentalization of metabolites

Homeostasis in general, including biosynthesis and compartmentalization of metabolites is disturbed in plant tissues due to high temperature stress (Maestri *et al.*, 2002). High temperature modifies the activities of carbon metabolism enzymes, starch accumulation and sucrose synthesis by down-regulating specific genes in carbohydrate metabolism (Ruan *et al.*, 2010). Among the primary metabolites accumulating in response to heat stress are proline, glycine betaine or soluble sugars (Wahid, 2007).

Plant hormone response

Under different abiotic stresses, hormones are very crucial for regulating many signaling pathways and responses such as salicylic acid (SA), abscisic acid (ABA), and ethylene (Kurepin *et al.*, 2017). Several key phyto-hormones including abscisic acid, ethylene and salicylic acid (a type of phenolic acid which functions as a plant hormone) increase their levels under heat stress, while others decrease, such as cytokinin, auxin, and gibberellic acids that ultimately cause premature plant senescence (Talanova *et al.*, 2003; Larkindale *et al.*, 2005). However, the most important and vital hormone for regulating the climatic stresses in the plant is ABA. ABA plays a major role in different stages of plant development particularly in stomata opening and closing, drought stress, seed germination, and dormancy. PYR/PYL/RCAR-PP2C-SnRK2 is recognized as a signaling cascade generated by ABA and controls the seed dormancy efficiently. Under drought conditions, the plant growth is severely retarded and it increases the ABA concentration in cells. ABA accumulation during drought stress controls transpiration and inhibit stomatal closure (Dong *et al.*, 2018).

Activation of silenced gene clusters

Heat stress also leads to the transient activation of repetitive elements or silenced gene clusters close to the centromeric regions as well as the transient loss of epigenetic gene silencing (Lang-Mladek *et al.*, 2010; Pecinka *et al.*, 2010). Such gene silencing mechanisms are thought to be involved in transcriptional repression by hetero-chromatinization of repetitive DNA regions in plants (Khraiwesh *et al.*, 2012). Heterochromatin is a tightly packed form of DNA which mainly consists of genetically inactive satellite sequences and both centromeres and telomeres are heterochromatic.

Oxidative damage

Heat induces accumulation of "Heat shock proteins" (HSPs) which prevent protein degradation and it also causes a state of metabolic imbalance and the build-up of toxic by-products, such as "Reactive oxygen species" (ROS) which are chemically reactive chemical species containing oxygen *viz.*, peroxides, superoxide, hydroxyl radical and singlet oxygen. Oxidative damage to proteins, lipids, RNA, and DNA ultimately affect plant vegetative and reproductive development, with negative consequences on fruit set, yield and product quality. The oxidative stress generated by excessive ROS destroys cellular and metabolic functions and leads to cell death.

b) Anatomical and morphological changes

Temperatures higher than the optimal can cause burning of twigs and leaves, senility of leaf and abscission, prohibition in the development of shoot and root, discoloration of fruit which lead to diminished production (Ismail and Hall, 1999; Vollenweider and Gunthardt-Goerg, 2005). High temperature considerably affects the anatomical structures not only at the tissue and cellular levels but also at the sub-cellular level. The cumulative effects of all these changes under high temperature stress may result in poor plant growth and productivity. A general tendency of reduced cell size, closure of stomata and curtailed water loss, increased stomatal and trichomatous densities and greater xylem vessels of both root and shoot has been observed at the whole plant level (A˜non *et al.*, 2004).

c) Negative impact on reproductive manifestation

Sexual reproduction and flowering in particular have been long recognized as extremely sensitive to heat stress, which often results in reduced crop plant productivity (Hedhly *et al.*, 2009; Thakur *et al.*, 2010). The most affected stage is the reproductive phase and the affected process is the physiology related to pollen grain development. However, several studies on different vegetable and other crops have documented that bud drop before pollination, undeveloped flowers, persistence of flower and calyx for a long time without fruit set, splitting of antheridial cone, lack of anther dehiscence, poor pollen production, pollen sterility, degeneration of embryo sac, browning and drying of stigma, reduction in stigma receptivity, style elongation, under-developed ovary, poor fertilization, slow pollen tube growth, poor pollen viability and germinability, poor ovule viability, ovule abortion and embryo degeneration, disruption in meiosis leading to prevention of pollen formation, hindered sugar metabolism and failure of viable pollen production, reduced carbohydrate availability for the fruits, reduced total soluble protein content, developmental abnormalities and poor fruit set, reduction in fruit size and seeds/fruit and inhibition of pathogen induced resistance mechanisms are some of the negative impact of high temperature stress on reproductive manifestations in the crops (Sato *et al.,* 2001; Hazra and Ansary, 2008; Hatfield *et al.*, 2011). All these changes lead to a single result: non setting of fruits/pods which causes significant reduction in yield.

Heat stress often accelerates rather than delays the onset of anthesis, which means that the reproductive phase of development will be initiated prior to the accumulation of sufficient resources (Zinn *et al.*, 2010). High temperature is most deleterious at the stage of flower bud initiation, and that this sensitivity is maintained for 10–15 days (Hedhly *et al.*, 2009; Nava *et al.*, 2009) most probably as result of reduced water and nutrient transport during reproductive development (Young *et al.*, 2004). The developmental pathway for the male gametophyte (pollen grains) starts with the separation of the reproductive tissues of the anther, continues with meiosis of the pollen mother cell, followed by mitosis and microspore maturation that results in the mature pollen grain. After initiation, the highly specialized anther tissues will acquire non-reproductive (e.g., the tapetum for support, the stomium for dehiscence) or reproductive functions (the pollen mother cell for pollen formation). Both tapetum and microspore development are essential for male fertility, as documented by numerous studies on male sterile mutants (Jaggard *et al.*, 2010; Zinn *et al.*, 2010). The male gametophyte is particularly sensitive to high temperatures at all stages of development, while the pistil and the female gametophyte are considered to be more tolerant (Hedhly, 2011). High temperature stress is possibly linked to tapetum degeneration and pollen sterility in several plant species (Oshino *et al.*, 2007; Endo *et al.*, 2009). Heat stress also induces early abortion of tapetal cells which cause the pollen mother cells to rapidly progress toward meiotic prophase and finally undergo programmed cell death (PCD), thus leading to pollen sterility (Sakata and Higashitani, 2008; Parish *et al.*, 2012).

Heat stress also reduces carbohydrate accumulation in pollen grains and in the stigmatic tissue by altering assimilates partitioning and changing the balance between symplastic and apoplastic loading of the phloem (Taiz and Zeiger, 2006). Heat stress down-regulates sucrose synthase and several cell wall and vacuolar invertases in the developing pollen grains and as consequence, sucrose and starch turnover are disrupted and thus soluble carbohydrates accumulate in reduced levels (Sato *et al.*,

2006). Continuing heat stress beyond a successful fertilization can also halt further development of the embryo (Barnabás *et al.*, 2008).

3. Stress due to drought (water shortage)

Nowadays climate has changed all around the globe by continuously increase in temperature and atmospheric CO_2 levels. The distribution of rainfall is uneven due to the change in climate which acts as an important stress as drought. The soil water available to plants is steadily decreased due to severe drought conditions and cause death of plants prematurely. Among the abiotic stresses, drought is a major environmental constraint as it limits plant growth and causes severe yield reduction (Morgan, 1984; Subbarao *et al.*, 1995) accounting for around $30.0 billion losses every year.

Different manifestations of drought stress

Drought stress alters the plant growth through changes in various morphological, physiological, and metabolic processes (Diaz *et al.*, 2010). Water shortage decreases both photosynthesis and the consumption of assimilates in the expanding leaves. As a consequence, water shortage indirectly decreases the amount of photosynthate exported from leaves. Because phloem transport depends on pressure gradients, decreased water potential in the phloem during water deficit may inhibit the movement of assimilates. As a results of drought stress, reactive oxygen species (ROS) is formed in plant tissues that induce damages in cellular membranes (Sairam and Saxena, 2000).

Avoidance of drought stress

The ability to continue translocating assimilates is a key factor in almost all aspects of plant resistance to drought. Negative impact of drought stress is mitigated to some extent through application of micro-irrigation, nano-technology and use of seawater after desalination. However, it requires high initial investment and is not feasible for most of the developing and low-income countries.

Physio-chemical basis of drought stress tolerance

Several physio-biochemical indicators *viz.* crop yield (Mwadzingeni *et al.*, 2016), relative water content (Szira *et al.*, 2008; Yadav *et al.*, 2019), root-shoot length, stomatal conductance (Munns *et al.*, 2010), seedling survivability percentage (Singh *et al.*, 1999), photosynthetic parameters (Lawlor and Cornic, 2002), accumulation of amino acids (Mwadzingeni *et al.*, 2016; Yadav *et al.*, 2019), etc. are important. However, determination of key associated traits should be carefully considered for screening of the genotypes for drought response (Sallam *et al.*, 2019).

Development of drought resistant genotypes

Identification of drought-tolerant genotypes suitable for water-limited environments as well as low-rainfall areas without having proper irrigation facilities is one of the ideal strategies to mitigate drought stress. The most critical factors for screening the drought tolerant lines are diverse genetic resources, selection of traits, precise phenotyping and determination of proper growth stage at which drought stress is imposed (Cattivelli *et al.*, 2008). Besides, understanding the physio-chemical basis of drought stress tolerance is equally important for achieving genetic gain for this extremely important trait.

Plants reduce their growth of shoots under drought conditions and reduce their metabolic demands. After that, protective compounds are synthesized by plants under drought by mobilizing metabolites required for their osmotic adjustment. Screening using small set of genotypes sometimes causes a misinterpretation of the results. However, screening of large set of genotypes under field conditions which is uncontrolled hampers the selection process as it is difficult to maintain uniform soil water potential throughout the field. At the same time, the genotypes may be prone to various biotic (e.g. soil-borne diseases) and abiotic factors (e.g. heat stress) (Munns *et al.*, 2010; Sallam *et al.*, 2018). Therefore, large-scale screening at seedling stage under controlled environment followed by validation under field condition increase the accuracy and efficiency of the selection process.

Many researchers tried with polyethylene glycol (PEG)-stimulated osmotic stress under controlled conditions to screen out drought tolerant genotypes but its practical uses is questioned as it causes hypoxia stress due to insufficient movement of O_2 around root surface area (Munns *et al.*, 2010; Dubois and Inzé, 2020). Moreover, it is very necessary to evaluate the same genotypes under multiple methods to increase the accuracy of screening process (Singh *et al.*, 2017; Cai *et al.*, 2020).

4. Chilling and freezing temperature stress

In temperate conditions plants are encountered by chilling and freezing conditions that are very harmful stress to the plants. Many important crops are incompetent to the process of cold acclimation. The abiotic stress caused by cold stress affect the cellular functions of plants in every aspect. Several signal transduction pathways are there by which these cold stresses are transduced like components of ROS, protein kinase, protein phosphate, ABA and Ca^{2+}, etc. and among these ABA proves to be the most important. Freezing temperatures result in intra- and extracellular ice crystal formation. Intracellular ice formation physically shears membranes and organelles. Extracellular ice crystals, which usually form before the cell contents freeze, may not cause immediate physical damage to cells, but they do cause cellular dehydration. This is because ice formation substantially lowers the water potential (Ψw) in the apoplast, resulting in a gradient from high Ψw in the symplast (continuous arrangement of protoplasts of a plant, which are interconnected by plasmodesmata) to low Ψw in the apoplast (non-protoplasmic components of a plant, including the cell wall and the intracellular spaces). Consequently, water moves from the symplast to the apoplast, resulting in cellular dehydration. Cells that are already dehydrated, such as those in seeds and pollen, are relatively less affected by ice crystal formation. Ice usually forms first within the intercellular spaces and in the xylem vessels, along which the ice can quickly propagate. This ice formation is not lethal to hardy plants, and the tissue recovers fully if warmed. However, when plants are exposed to freezing temperatures for an extended period, the growth of extracellular ice crystals leads to physical destruction of membranes and excessive dehydration.

5. Stress due to imbalance in soil mineral content

Several anomalies associated with the elemental composition of soils can result in plant stress, including high concentrations of salts (e.g., Na^+ and Cl^-) and toxic ions (e.g., As and Cd), and low concentrations of essential mineral nutrients, such as $Ca2^+$, $Mg2^+$, N, and P.

Salinity stress

The term salinity is used to describe excessive accumulation of salt in the soil solution. Genetically adapted salt-tolerant plants are termed "halophytes", while the plants that are not adapted to salinity are termed "glycophytes". In natural environments, there are many causes of salinity. Terrestrial plants encounter high salinity close to the seashore and in estuaries where seawater and freshwater mix or replace each other with the tides. The movement of seawater upstream into rivers can be substantial, depending on the strength of the tidal surge. Far inland, natural seepage from geologic marine deposits can wash salt into adjoining areas. Evaporation and transpiration remove pure water (as vapor) from the soil, concentrating the salts in the soil solution. Soil salinity is also increased when water droplets from the ocean disperse over land and evaporate.

Human activities also contribute to soil salinization. Improper water management practices associated with intensive agriculture can cause substantial salinization of crop lands. In many areas of the world, salinity threatens the production of staple foods. Irrigation water in semiarid and arid regions is often saline. Only halophytes, the most salt-tolerant plants, can tolerate high levels of salts.

Manifestation of salinity stress

Water deficit condition

Salinity stress has two components: non-specific osmotic stress that causes water deficits, and specific ion effects resulting from the accumulation of toxic ions, which disturb nutrient acquisition and result in cytotoxicity. Saline soils are often associated with high concentrations of NaCl, but in some areas Ca^{++}, Mg^{++} and $SO4^-$ ions are also present in high concentrations in saline soils. High Na^+ concentrations that occur in sodic soils (soils in which Na^+ occupies >10% of the cation exchange capacity) not only injure plants but also degrade the soil structure, decreasing porosity and water permeability. Salt incursion into the soil solution causes water deficits in leaves and inhibits plant growth and metabolism.

Cytotoxicity

High cytosolic Na^+ and Cl^- denature proteins and destabilize membranes. Cytosol is known as the matrix of the cytoplasm. It surrounds the cell organelles in eukaryotes. In prokaryotes, all the metabolic reactions occur here. Thus, cytosol is the fluid contained in the cell cytoplasm while cytoplasm is the entire content within the cell membrane.

The most widespread example of a specific ion effect is the cytotoxic accumulation of Na^+ and Cl^- ions under saline conditions. In contrast to extracellular fluid, cytosol has a high concentration of potassium ions and a low concentration of sodium ions. Under non-saline conditions, the cytosol of higher plant cells contains about 100 mM K^+ and less than 10 mM Na^+, an ionic environment in which enzymes are optimally functional. In saline environments, cytosolic Na^+ and Cl^- increase to more than 100 mM, and these ions become cytotoxic. High concentrations of salt cause protein denaturation and membrane destabilization by reducing the hydration of these macromolecules. However, Na^+ is a more potent denaturant than K^+.

At high concentrations, apoplastic Na^+ also competes for sites on transport proteins that are necessary for high-affinity uptake of K^+, an essential macronutrient. Further, Na^+ displaces Ca^{++} from sites on the cell wall, reducing Ca^{++} activity in the apoplast and resulting in greater Na^+ influx, presumably through non-selective cation channels. Reduced apoplastic Ca^{++} concentrations caused by excess Na+ may also restrict the availability of Ca^{++}in the cytosol. Since cytosolic $Ca^{++}$is necessary to activate Na^+ detoxification *via* efflux across the plasma membrane, elevated external Na^+ has the ability to block its own detoxification.

References

A˜non, S., Fernandez, J.A., Franco, J.A., Torrecillas, A., Alarc´on, J.J. and S´anchez-Blanco, M.J. 2004. Effects of water stress and night temperature preconditioning on water relations and morphological and anatomical changes of *Lotus creticus* plants. *Sci Hort.*,**101**: 333–342.

Ainsworth, E. A. and Ort, D.R. 2010. How do we improve crop production in a warming world? *Plant Physiol.*, **154**: 526–530.

Allakhverdiev, S. I., Kreslavski, V.D., Klimov, V.V., Los, D. A., Carpentier, R. and Mohanty, P. 2008. . Heat stress: an overview of molecular responses in photosynthesis. *Photosynthesis Res.*, **98**: 541–550.

Argyris, J., Truco, M. J., Ochoa, O., McHale, L., Dahal, P. and Van Deynze, A. 2011. A gene encoding an abscisic acid biosynthetic enzyme (LsNCED4) collocates with the high temperature germination locus Htg6. 1 in lettuce (*Lactuca* sp.). *Theor. Appl. Genet.,* **122**: 95–108.

Baker, J.T., and L.H. Allen, Jr. 1993. Contrasting crop species responses to CO_2 and temperature: rice, soybean and citrus. *Vegetatio*, **104**: 239–260

Barlow, K.M., Christy, B.P., O'Leary, G.J., Riffkin, P.A. and Nuttall, J.G. 2015. Simulating the impact of extreme heat and frost events on wheat crop production : a review. *Field Crops Res.*, **171**: 109-119.

Barnabas, B., Jager, K. and Fehér A. 2008. The effect of drought and heat stress on reproductive processes in cereals. *Plant Cell Environ.*, **31**: 11–38.

Cai, K., Chen, X., Han, Z., Wu, X., Zhang, S., Li, Q., Nazir, M.M., Zhang, G. and Zeng, F. 2020. Screening of worldwide barley collection for drought tolerance: The assessment of various physiological measures as the selection criteria. *Front. Plant Sci.*, **7**(11):1159.

Cattivelli, L., Rizza, F., Badeck, F.W., Mazzucotelli, E., Mastrangelo, A.M., Francia, E., Marè, C., Tondelli, A. and Stanca, A.M. 2008. Drought tolerance improvement in crop plants: an integrated view from breeding to genomics. *Field Crops Res.*, **105**: 1-14.

Challinor, A., Wheeler, T., Craufurd, P., Ferro, C. and Stephenson, D. 2007 . Adaptation of crops to climate change through genotypic responses to mean and extreme temperatures. *Agric. Ecosyst. Environ.*, **119**: 190–204.

Craufurd, P.Q., Bojang, M., Wheeler, T.R. and Summerfield, R.J. 1998. Heat tolerance in cowpea: Effect of timing and duration of heat stress. *Ann. Appl. Biol.*, **133**: 257–267

Craufurd, P.Q., Qi, A., Ellis, R.H., Summerfield, R.J., Roberts, E.H. and Mahalakshmi, V. 1998. Effect of temperature on time to panicle initiation and leaf appearance in sorghum. *Crop Sci.*, **38**: 942–947.

Diaz, P., Betti, M., Sanchez, D.H., Udvardi, M.K., Monza, J. and Marquez, A.J. 2010. Deficiency in plastidic glutamine synthetase alters proline metabolism and transcriptomic response in Lotus japonicus under drought stress. *New Phytol.*, **188**:1001–1013.

Dobra, J., Motyka, V., Dobrev, P., Malbeck, J , Prasil, I. T. and Haisel, D. 2010 .Comparison of hormonal responses to heat, drought and combined stress in tobacco plants with elevated proline content. *J. Plant Physiol.*, **167**: 1360–1370.

Dong, H., Bai, L., Chang J., and Song. S.P. 2018. Chloroplast protein PLGGI is involved in abscisic acid-regulates lateral root development and stomatal movement in *Arabidopsis, Biochem. Biophys. Res. Commun.,* **495**: 280-285.

Dubois, M. and Inzé, D. 2020. Plant growth under suboptimal water conditions: early responses and methods to study them. *J. Exp. Bot.*, **71**:1706–1722.

Ehlers J.D. and Hall, A.E. 1998. Heat tolerance of contrasting cowpea lines in short and long days. *Field Crops Res.*, **55**: 11–21.

Endo, M., Tsuchiya, T., Hamada, K., Kawamura, S., Yano, K., Ohshima, M. 2009. High temperatures cause male sterility in rice plants with transcriptional alterations during pollen development. *Plant Cell Physiol.*, **50**: 1911–1922.

Farooq, M., Basra, S., Wahid, A., Cheema, Z., Cheema, M. and Khaliq, A. 2008. Physiological role of exogenously applied glycine betaine to improve drought tolerance in fine grain aromatic rice(*Oryza sativa* L.). *J. Agron.Crop Sci.*, **194**: 325– 333.

Firon, N., Shaked, R. ,Peet, M., Pharr, D., Zamski, E. and Rosenfeld, K. 2006.Pollen grains of heat tolerant tomato cultivars retain higher carbohydrate concentration under heat stress conditions. *Sci. Hort.*, **109**: 212–217.

Foolad, M.R. 2005. Breeding for abiotic stress tolerances in tomato. In: *Abiotic Stresses: Plant Resistance Through Breeding and Molecular Approaches* (Eds. Ashraf, M. and Harris, P.J.C.), The Haworth Press Inc., New York, USA, pp. 613–684.

Gorantla, M., Babu, P., Lachagari, V.B.R., Reddy, A., Wusirika, R. and Bennetzen, J. L. 2007. Identification of stress-responsive genes in an indica rice (*Oryza sativa* L.) using ESTs generated from drought- stressed seedlings. *J. Expt. Bot.,* **58**: 253–265.

Gurley, W.B. 2000. HSP101: a key component for the acquisition of thermo-tolerance in plants. *Sci. Signal.,* **12**: 457

Hall, A. E. 2011. Breeding Cowpea for Future Climates, In: *Crop Adaptation to Climate Change* (eds. Yadav, S. S., Redden, R. , Hatfield, J. L., Lotze Campen, H. and Hall, A.J.W.), JohnWiley & Sons.USA.

Hatfield, J.L., Boote, K.J., Kimball, B.A., Ziska, L.H., Izaurralde, R.C., Ort, D., Thomson, A.M. and Wolfe, D.W. 2011. Climate impacts on agriculture: implications for crop production. *Agron. J.*, **103**: 351–370.

Hazra, P. and Ansary, S.H. 2008. Genetics of heat tolerance for floral and fruit set to high temperature stress in tomato (*Lycopersicon esculentum* Mill.). *SABRAO J. Breed. Genet.*, **40**: 117- 125.

Hedhly, A. 2011. Sensitivity of flowering plant gametophytes to temperature fluctuations. *Environ.Expt. Bot.*, **74**: 9–16.

Hedhly, A., Hormaza, J. I. and Herrero, M. 2009. Global warming and sexual plant reproduction. *Trends Plant Sci.*,**14**: 30–36.

IMD 2010. Annual Climate Summary, 2010. India Meteorological Department, Pune. Government of India, Ministry of Earth Sciences, 27p.

Ismail, A.M. and Hall, A.E. 1999. Reproductive stage heat tolerance, leaf membrane thermo-stability and plant morphology in cowpea. *Crop Sci.*, **39**: 1762-1768.

Jaggard, K.W., Qi, A. and Ober, E.S. 2010.Possible changes to arable crop yields by 2050. *Philosophical transactions of the Royal Society* of *London* Series B, *Biological Sciences,* **365**: 2835–2851.

Karim, M.A., Fracheboud, Y. and Stamp, P. 1997. Heat tolerance of maize with reference of some physiological characteristics. *Ann. Bangladesh Agril.*, **7**: 27–33.

Kaya, H., Shibahara, K.,Taoka, K., Iwabuchi, M., Stillman, B. and Araki, T. 2001. FASCIATA genes for chromatin assembly factor-1in *Arabidopsis* maintain the cellular organization of apical meristems. *Cell*, **104**: 131–142.

Khraiwesh, B., Zhu, J.K. and Zhu, J. 2012 . Role of miRNAs and siRNAs in biotic and abiotic stress responses of plants. *Biochimica et Biophysica Acta,* **1819**: 137–148.

Kotak, S., Larkindale, J., Lee, U., von KoskullDöring, P., Vierling, E. and Scharf, K.D. 2007. Complexity of the heat stress response in plants. *Curr. Opin.Plant Biol.*, **10**: 310–316.

Kuripin, L.V. Ivanov, A.G., Zaman, M., Pharis, R.P., Hurry, V. and Huner, N.P. 2017. Internation of glycine betaine and plant hormones: Protection of the photosynthetic appratus during abiotic stress. In: *Photosynthesis: Structures Mechanisms and Applications* (Eds. Hou, H.J.M. Najafour, M.M. Moore, G.F. and Allakhverdiev, S.I.), Springer, Germany. pp. 185-202.

Lang Mladek, C., Popova, O. , Kiok, K., Berlinger, M., Rakic, B. and Aufsatz, W. 2010. Transgenerational inheritance and resetting of stress-induced loss of epigenetic gene silencing in *Arabidopsis. Molecular Plant,* **3**: 594–602.

Larkindale, J., Hall, J.D., Knight, M.R. and Vierling, E. 2005.Heat stress phenotypes of *Arabidopsis* mutants implicate multiple signalling pathways in the acquisition of thermo-tolerance. *Plant Physiol.*,**138**: 882–897.

Larkindale, J. and Huang, B. 2004. Thermo-tolerance and antioxidant systems in *Agrostis stolonifera* : involvement of salicylic acid, abscisic acid, calcium, hydrogen peroxide, and ethylene. *J. Plant Physiol.*, **161**: 405–413.

Lawlor, D.W. and Cornic, G. 2002. Photosynthetic carbon assimilation and associated metabolism in relation to water deficits in higher plants. *Plant Cell Environ.*, **44**: 275–294.

Li, M., Ji, L., Yang, X., Meng, Q. and Guo, S. 2012.The protective mechanisms of CaHSP26 in transgenic tobacco to alleviate photo-inhibition of PSII during chilling stress. *Plant Cell Rep.*, **31**: 1969–1979.

Liu, X. and Huang, B. 2000. Carbohydrate accumulation in relation to heat stress tolerance in two creeping bent grass cultivars. *Amer. Soc. Hort. Sci.,* **125**: 442–447.

Lobell, D.B. and Field, C.B. 2007.Global scale climate–Crop yield relationships and the impacts of recent warming. *Environ. Res. Let.,* 2, doi:10.1088/1748-9326/2/1/014002.

Maestri, E. , Klueva, N. , Perrotta, C., Gulli, M. , Nguyen, H.T. and Marmiroli, N. 2002. Molecular genetics of heat tolerance and heat shock proteins in cereals. *Plant Mol. Biol.,* **48**: 667–681.

Mahajan, V., Gupta, A.J., Lawande, K.E. and Singh, M. 2018. Onion Improvement in India. *J. Allium Res.*, **1**: 1-15.

Meehl, G.A., Stocker, T.F., Collins, W.D., Gaye, A.J., Gregory, J., Kitoh. A., Knutti, R., Murphy, J.M., Noda, A., Raper, S.C.B., Watterson, J.G., Weaver, A.J. and Zhao, Z. 2007. Global Climate Projections, In: *The Physical Science Basis. Contribution of Working Group I to the Fourth Assessment Report of the Intergovernmental Panel on Climate Change* (eds. Solomon, S., Qin, D., Manning, M., Chen, Z., Marquis, M., Averyt, K.B., Tignor, M. and Miller, H.L.), Cambridge University Press, Cambridge, U.K. and New York, USA.

Morgan, J.M. 1984. Osmoregulation and water stress in higher plants. *Ann. Rev. Plant. Biol.*, **35**:299-319.

Morgante, M. and Salamini, F. 2003. From plant genomics to breeding practice. *Curr. Opin. Biotechnol.,* **14**: 214–219.

Munns, R., James, R.A., Sirault, X.R.R., Furbank, R.T. and Jones, H.G. 2010. New phenotyping methods for screening wheat and barley for beneficial responses to water deficit. *J. Expt. Bot.*, **61**: 3499–3507.

Mwadzingeni, L., Shimelis, H., Tesfay, S., Tsilo, T.J. 2016. Screening of bread wheat genotypes for drought tolerance using phenotypic and proline analyses. *Front. Plant Sci.*, **7**:1276. doi: 10.3389/fpls.2016.01276

Nagarajan, S. ,Jagadish, K.S. V., Prasad, A.S.H. , Thomarc, A. , Anand, A. and Pal, M. 2010. Local climate affects growth, yield and grain quality of aromatic and non-aromatic rice in north-western India. *Agric. Ecosyst. Environ.,* **138**: 274–281.

Nava, G.A., Dalmago, G.A., Bergamaschi, H., Paniz, R. , dosSantos, R.P. and Marodin, G.A.B. 2009. Effect of high temperatures in the pre-blooming and blooming periods on ovule formation, pollen grains and yield of 'Granada' peach. *Sci. Hort.*, **122**: 37–44.

Oshino, T., Abiko, M., Saito, R., Ichiishi, E., Endo, M. and Kawagishi-Kobayashi, M. 2007. Premature progression of anther early developmental programs accompanied by comprehensive alterations in transcription during high-temperature injury in barley plants. *Mol. Genet. Genom.*, **278**: 31–42.

Parish, R.W., Phan, H.A., Iacuone, S. and Li, S.F. 2012.Tapetal development and abiotic stress: a centre of vulnerability. *Func. Plant Biol.*, **39**: 553–559.

Pecinka, A., Dinh, H.Q., Baubec, T., Rosa, M., Lettner, N. and Scheid, O.M. 2010. Epigenetic regulation of repetitive elements is attenuated by prolonged heat stress in *Arabidopsis. Plant Cell*, **22**: 3118–3129.

Peet, M., Sato, S. and Gardner, R. 2002. Comparing heat stress effects on male-fertile and male-sterile tomatoes. *Plant Cell Environ.*, **21**: 225–231.

Pinto, R.S., Reynolds, M.P., Mathews, K.L. , McIntyre, C. L., Olivares-Villegas, J.J. and Chapman, S.C .2010. Heat and drought adaptive QTL in a wheat population designed to minimize confounding agronomic effects. *Theor. Appl. Genet.,* **121**: 1001–1021.

Prasad, P., Boote, K., Allen, L., Sheehy, J. and Thomas, J. 2006. Species, ecotype and cultivar differences in spikelet fertility and harvest index of rice in response to high temperature stress. *Field Crops Res.*, **95**: 398–411.

Rao, G.G.S.N., Rao, A.V.M.S. and Rao, V.U.M. 2009. Trends in rainfall and temperature in rainfed India in previous century. In: *Global climate change and Indian Agriculture case studies from ICAR network project* (Ed. Aggarwal, P.K.), ICAR Publication, New Delhi, pp.71 73.

Rivero, R.M. , Ruiz, J.M., Garcia, P.C., López-Lefebre, L.R., Sánchez, E. and Romero, L. 2001 .Resistance to cold and heat stress : accumulation of phenolic compounds in tomato and watermelon plants. *Plant Sci.*, **160**: 315–321.

Rodríguez, M., Canales, E. and Borrás-Hidalgo, O. 2005. Molecular aspects of abiotic stress in plants. *Biotechnol. Appl. Biochem.*, **22**: 1–10.

Roitsch, T. and González, M.C. 2004. Function and regulation of plant invertases: sweet sensations. *Trends Plant Sci.*, **9**: 606–613.

Ruan, Y.L., Jin, Y., Yang, Y.J., Li, G.J. and Boyer, J.S. 2010. Sugar input, metabolism, and signaling mediated by invertase: roles in development, yield potential, and response to drought and heat. *Mol. Plant.,* **3**: 942–955.

Sairam, R.K. and Saxena, D.C. 2000. Oxidative stress and antioxidants in wheat genotypes: possible mechanism of water stress tolerance. *J. Agron. Crop Sci.*, **184**:55–61.

Sakamoto, A. and Murata, N. 2002 . The role of glycine betaine in the protection of plants from stress: clues from transgenic plants. *Plant Cell Environ.*, **25**: 163–171.

Sakata, T. and Higashitani, A. 2008. Male sterility accompanied with a normal anther development in plants–genes and environmental stresses with special reference to high temperature injury. *Int. J. Plant Dev. Biol.*, **2**: 42–51.

Sallam, A., Alqudah, A.M., Dawood, M.F.A., Baenziger, P.S. and Börner, A. 2019. Drought stress tolerance in wheat and barley: Advances in physiology, breeding and genetics research. *Int. J. Mol. Sci.*, **20**:3137. doi: 10.3390/ ijms20133137

Sato, S., Kamiyama, M., Iwata, T., Makita, N., Furukawa, H. and Ikeda, H. 2006. Moderate increase of mean daily temperature adversely affects fruit set of *Lycopersicon esculentum* by disrupting specific physiological processes in male reproductive development. *Ann. Bot.*, **97**: 731–738.

Sato, S., Peet, M.M. and Gardner, R.G. 2001. Formation of parthenocarpic fruit, undeveloped flowers and aborted flowers in tomato under moderately elevated temperatures. *Sci. Hort.,* **90**: 243–254.

Savchenko, G. , Klyuchareva, E. , Abramchik, L. and Serdyuchenko, E. 2002. Effect of periodic heat shock on the inner membrane system of etioplasts. *Russian J. Plant Physiol.*, **49**: 349–359.

Scafaro, A.P., Haynes, P.A. and Atwell, B.J. 2010. Physiological and molecular changes in *Oryza meridionalis* Ng., a heat-tolerant species of wild rice. *J. Expt. Bot.,* **61**: 191–202.

Singh, B.B., Mai-Kodomi, Y. and Terao, T. 1999. A simple screening method for drought tolerance in cowpea. *Indian J. Genet. Plant Breed.*, **59**: 211- 222.

Singh, B., Reddy, K.R., Redona, E.D. and Walker, T. 2017. Screening of rice cultivars for morpho-physiological responses to early-season soil moisture stress. *Rice Sci.*, **24**: 322-335.

Subbarao, G.V., Johansen, C., Slinkard, A.E., Nageswara, R.R.C., Saxena, N.P., Chauhan, Y.S. and Lawn, R.J. 1995. Strategies for improving drought resistance in grain legumes. *Crit. Rev. Plant Sci.*, **14**: 469-523.

Swarup, V. and Chatterjee, S.S. 1972. Origin and Genetic Improvement of Indian Cauliflower. *Econ. Bot.*, **26**: 381-393.

Szira, F., Bá lint, A.F., Börner, A. and Galiba, G. 2008. Evaluation of drought related traits and screening methods at different developmental stages in spring barley. *J. Agron. Crop Sci.*, **194**: 334–342.

Taiz, L. and Zeiger, E. 2006. Stress physiology, In: *Plant Physiology* (eds. Taiz, L. and Zeiger, E.), Sinauer Associates, Sunderland, USA. pp. 671–681.

Talanova, V., Akimova, T. and Titov, A. 2003. Effect of whole plant and local heating on the ABA content in cucumber seedling leaves and roots and on their heat tolerance. *Russian J. Plant Physiol.*, **50**: 90–94.

Thakur, P., Kumar, S., Malik, J., Berger, J.D. and Nayyar, H. 2010. Cold stress effects on reproductive development in grain crops, an overview. *Environ. Expt. Bot.*, **67**: 429–443.

The Intergovernmental Panel on Climate Change (IPCC). 2007. *Climate Change 2007.* World Meteorological Organization and the United Nations Environment Programme.

Vara Prasad, P.V., Craufurd, P.Q. and Summerfield, R.J. 1999. Fruit number in relation to pollen production and viability in groundnut exposed to short episodes of heat stress. *Ann.Bot.*, **84**: 381-386.

Vierling, E. 1991.The roles of heat shock proteins in plants. *Ann. Rev.Plant Biol.*,42: 579–620.

Vollenweider, P. and Gunthardt-Goerg, M.S. 2005.Diagnosis of abiotic and biotic stress factors using the visible symptoms in foliage. *Environ. Pollut.*, **137**: 455–465.

Wahid, A., Gelani, S., Ashraf, M. and Foolad, M. 2007. Heat tolerance in plants: an overview. *Environ.Expt. Bot.,* **61**: 199–223.

Wahid, A. 2007. Physiological implications of metabolite biosynthesis for net assimilation and heat-stress tolerance of sugarcane (*Saccharum officinarum*) sprouts. *J. Plant Res.*,**120**: 219–228.

Wang, W., Vinocur, B., Shoseyov, O. and Altman, A. 2004. Role of plant heat-shock proteins and molecular chaperones in the abiotic stress response. *Trends Plant Sci.*, **9**: 244–252.

Wheeler, T.R., Craufurd, P.Q., Ellis, R.H., Porter, J.R. and Vara Prasad, V. 2000. Temperature variability and the yield of annual crops. *Agril. Ecosyst. Environ.*, **82**: 159-167.

Xu, Y., Chu, C. and Yao, S. 2021. The impact of high-temperature stress on rice: Challenges and solutions. *The Crop J.*, **9**: 963-976.

Yadav, A.K., Carroll, A.J., Estavillo, G.M., Rebetzke, G.J. and Pogson, B.J. 2019. Wheat drought tolerance in the field is predicted by amino acid responses to glasshouse-imposed drought. *J. Expt. Bot.*, **70**: 4931–4947.

Young, L.W., Wilen, R.W. and Bonham-Smith, P.C. 2004. High temperature stress of *Brassica napus* during flowering reduces micro-and mega-gametophyte fertility, induces fruit abortion, and disrupts seed production. *J Expt. Bot.*, **55**: 485–495.

Zhang, J.H., Huang, W.D., Liu, Y.P. and Pan, Q.H. 2005. Effects of temperature acclimation pre-treatment on the ultrastructure of mesophyll cells in young grape plants (*Vitis vinifera* L. cv. Jingxiu) under cross-temperature stresses. *J. Integrative Plant Biol.*, **47**: 959–970.

Zhang, J.X., Wang, C., Yang, C.Y., Wang, J.Y., Chen, L.and Bao, X.M. 2010.The role of *Arabidopsis* AtFes1 A in cytosolic Hsp70 stability and abiotic stress tolerance. *Plant J.*, **62**: 539–548.

Zhao, C., Liu, B., Piao, S., Wang, X., Lobell, D.B. and Huang, Y. 2017. Temperature increase reduces global yields of major crops in four independent estimates. *Proc. Nat. Acad. Sci.*, USA, **114**: 9326-9331.

Zinn, K.E.,Tunc-Ozdemir, M. and Harper, J.F. 2010. Temerature stress and plant sexual reproduction: uncovering the weakest links. *J. Expt. Bot.*, **6**:1959-1968.

13

Breeding for High Temperature Stress

High temperature stress

Abiotic stresses either individually or in combination cause morphological, physiological, biochemical and molecular changes that adversely affect plant growth and ultimately yield. The major forms of abiotic stress plants are exposed to in nature are heat, drought, cold and salinity. Of these abiotic stresses, heat stress has an independent mode of action on the physiology and metabolism of plant cells. High temperatures above optimal have generally damaging effects on plant development which has led to the prediction that Global warming have a negative effect on plant growth. High temperature regimes may cause severe yield reduction leading to even famine. High temperature stress has wide range of effects on plant in terms of physiology, biochemistry and gene regulation pathways. The susceptibility to high temperatures in plants varies with the stage of plant development and genotype.

Mechanisms of heat tolerance

Heat tolerance is generally defined as the ability of the plant to grow and produce economic yield under high temperature conditions. Several plant physiological parameters and mechanisms that contribute to heat tolerance in the field include:

- Amendments to essential processes such as photosynthesis, concomitant increases of transcripts coding for proteins involved in protection, protein stability, antioxidant activity, membrane lipid unsaturation, accumulation of compatible solutes, etc.,
- Increased affinity of Rubisco for CO_2 resulting improvement in photosynthetic capacity and efficiency. (Parry *et al.*, 2011; Cossani and Reynolds, 2012).
- Higher photosynthetic rates, increased membrane thermo-stability and heat avoidance.
- Changes in the organization of organelles, cytoskeleton, and membrane functions (Weis and Berry,1988) hence, increasing the saturation level of fatty acids appears to be critical for maintaining membrane stability and enhancing heat tolerance.
- Decrease in the synthesis of normal proteins and the accelerated transcription and translation of heat-shock proteins, HSPs (Bray *et al.*, 2000),

- Production of phyto-hormones such as abscisic acid (ABA), salicylic acid, ethylene, cytokinin and auxin and anti-oxidants and other protective molecules like, late embryogenesis abundant (LEA) proteins, osmo-protectants, antioxidant defense, etc.
- Accumulation of osmo-protectants like, proline, glycine betaine, and soluble sugars (e.g., glucose and sucrose) which is an important adaptive mechanism in plants to tolerate high temperature stress because accumulation of these osmo-protectants is necessary to regulate osmotic activities and protect cellular structures from increased temperatures by maintaining the cell water balance, membrane stability and by buffering the cellular redox potential.

In all plant species, the ability to sustain leaf gas exchange under heat stress is directly correlated with heat tolerance. However, plant responses to high temperature stress clearly depend on the genotypic parameters, as certain genotypes are more tolerant.

Sucrose is the principal end product of photosynthesis, which translocate from source leaves to sink organs through the phloem. Different studies suggest that high carbohydrate availability (e.g., glucose and sucrose) during heat stress represents an important physiological trait associated with heat stress tolerance (Liu and Huang, 2000). Sucrose not only plays the role of a carbon carrier between sources and sink tissues, but also acts as a long-distance signal for the control of many processes involved in plant development such as apical dominance and root growth. It also intervenes in the production/consumption balance of plant carbohydrates. For this reason, sucrose and its cleavage products regulate plant development and response to stresses through carbon allocation and sugar signaling (Roitsch and González, 2004). Studies on a heat-tolerant tomato genotype demonstrated that it is the high cell wall and vacuolar invertases activities and increased sucrose import into young tomato fruit that contribute to heat tolerance through increasing sink strength and sugar signalling activities. Sucrose phosphate synthase (SPS) is a key rate-limiting enzyme (*SlSPS*gene) mediates plant growth and thermos-tolerance in tomato (Zang *et al.*, 2022). Similarly, the carbohydrate content of developing and mature pollen grains may be an important factor in determining pollen quality, as heat-tolerant tomato genotypes appear to have a mechanism for maintaining the appropriate carbohydrate content under heat stress condition.

Enhanced synthesis of secondary metabolites under heat stress conditions also protects against oxidative damage. Several studies in tomato and watermelon indicated that thermal stress induces the accumulation of phenolics in the plant by activating their biosynthesis as well as inhibiting their oxidation, which could be an acclimation mechanism of the plant against thermal stress. Several plant growth regulators, such as abscisic acid, salicylic acid, ethylene, cytokinin and auxin may play important role in plant thermo-tolerance. It has been found that the accumulation of proline, anthocyanins, and furanocoumarins (secondary metabolites), but not of phenolic acids or flavonols, contributes to protection of *Heracleum sosnowskyi* plants against heat stress (Rysiak *et al.*, 2021).

The most important characteristic of thermo-tolerance is the massive production of heat-shock proteins, HSPs. Heat shock proteins (HSPs) are the molecular chaperones (proteins that guide proteins along the proper pathways for folding) that are not only expressed during the normal growth process of cell cycle consecutively, but also get induced in cells during various stressful conditions produced by cellular insult, environmental changes, higher temperatures, pH variation, hypoxic conditions, infections, tumors etc. In many plant species, thermo-tolerance of cells and tissues after a heat stress is pretty much dependent upon induction of HSP70, though HSP101 has also been shown to be essential. The enhanced expression of HSP70 was reported to assist in translocation, proteolysis, translation, folding, aggregation, and refolding of denatured proteins, while a methionine-rich chloroplast HSP has been shown to protect the thermolabile photosystem II and consequently the whole-chain electron transport during heat stress.

Breeding strategies for high temperature stress

With the rapid increase in the world's population, there is a corresponding increase in food demand owing to concerns about the stability of the global environment. To tackle these limitations and guaranteed food security there is a need for development of new climate-smart crop cultivars (Wheeler and Von Braun, 2013).

Conventional breeding strategy

In general, the negative impacts of abiotic stresses on crop productivity can be reduced by a combination of genetic improvement and cultural practices. Genetic improvement entails development of varieties which can produce economic yield under high temperature stress condition. Under various environmental stresses, plant breeding shows dynamic techniques in crop improvement. It gives a way to potentially guarantee food security and safety under harsh weather variations and help plants escape from various stresses through the crucial phase of plant growth by developing stress resistant cultivars (Blum, 2018).

Development of new vegetable cultivars tolerant to heat stress is a major challenge for vegetable breeders. The responses of crops to high temperature stress are dynamic depending upon the extremity and duration of the high temperature condition as well as the plant type and other environmental factors in the surroundings. However, the identification and confirmation of the traits that confer tolerance to heat stress still remain elusive in many vegetable crops (Wahid *et al.*, 2007).

Screening of genotypes for tolerance to high temperature stress

Tolerance to high temperatures is a multigenic character hence screening for heat tolerance in the field presents a challenge due to interactions with other environmental factors. Traditional breeding of heat resistant crops basically based on screening and selection. However, successful selection in the field condition for high temperature tolerance depends on the identification of a variety of screenable traits. Breeding for stress tolerance requires efficient screening procedures, identification of key traits in diverse donor or tolerant lines and understanding their inheritance and molecular genetics. The common technique of selecting crops for heat stress resistance has been to grow breeding materials in high temperature condition and detect individuals/lines with higher yield.

Several types of morphological and physiological charactersthat help in screening for heat tolerance in the conventional breeding approaches are-

- Long root length which has a good ability to uptake water and nutrients from the soil
- Short life-span which helps to minimize the temperature effect on plant
- Hairiness which provides partial shade to cell wall and cell membrane and repels sunrays
- Small size of leaf which resists evaporation due to reduction of stomata
- Leaf orientation which enhances the photosynthetic activity and produces tolerance against heat stress
- Leaf glossiness and waxiness which repel sunlight
- Different physiological characteristics that are influenced by high temperature efficiency of photosynthesis, assimilate translocation, mesophyll tolerance, and cellular membranes disorganization (Chen *et al.*, 1982)
- Increased membrane-thermostability and successful fruit set under high temperature conditions.

Wide array of germplasm should regularly be evaluated to identify sources of heat resistance by screening for fruit set under high temperature condition (Berry and Rafique-Uddin, 1988). In tomato, a potent positive correlation has been found between yield and fruit set under high temperature. Breeding to develop such characteristics under high temperatures can lead to improvement of varieties with heat resistance approaches.

However, detection of appropriate genetic resources is the major concern. Genetic divergence analysis based on genetic distance and similarities is considered a very important method for initiation of the development of new cultivars, particularly resistant to environmental stresses (Raza *et al.*, 2018). Landraces are valuable resource to initiate breeding for stress resistance as it contains cultivars adjustable to diverse environmental stress (Lopes *et al.*, 2015).

Various other concerns when applying conventional breeding protocols to promote heat resistant crops may arise. Detection of genetic resources with heat resistance approaches is one such concern. In most of the vegetables crops, for instance tomatoes and legumes, the genetic differentiations within the cultivated species are restricted which necessitates the detection and use of wild accessions (Foolad, 2005). However, photosynthesis and reproductive development are the most sensitive physiological processes to stress (Prasad *et al.*, 2008). Hence, a heat-tolerant variety is usually characterized by higher photosynthetic rates reflected in stay-green leaves, increased membrane-thermostability and successful fruit set under high temperature conditions.

Sometimes heat resistance is linked to various unfavorable agronomical or horticultural traits. In tomato, for instance, two unfavorable traits generally observed in heat resistant lines are small fruit size and limited foliar canopy (Scott *et al.*, 1997). The small fruit production is mostly because of improper impacts of high temperature on the creation of auxins in the fruit and the poor canopy is for the sake of the highly reproductive nature of the heat resistant varieties (Scott *et al.*, 1997).

AVRDC, Taiwan, has made significant contributions to the development of heat-tolerant tomato and Chinese cabbage lines (*Brassica rapa* subsp. *pekinensis* and *chinenesis*) adapted to hot and humid climate. The key to achieving high yields with heat-tolerant cultivars is the broadening of their genetic base through crosses between heat-tolerant tropical lines and disease-resistant temperate or winter varieties (Opena *et al.*, 1992). The heat-tolerant tomato lines were developed using heat-tolerant breeding lines and landraces from the Philippines (*viz.*, VC11-3-1-8, VC 11- 2-5, Divisoria-2) and the United States (*viz.*, Tamu Chico III, PI289309) (Opena *et al.*, 1992). However, lower yields in the heat-tolerant lines are still a concern.

More heat-tolerant varieties are required to meet the needs of a changing climate, and these must be able to match the yields of conventional, non-heat tolerant varieties under non-stress conditions. A wider range of genotypic variation must be explored to identify the additional sources of heat tolerance. For example, AVRDC's tomato breeding line, CL5915, has demonstrated high levels of heat ranges from 15% to 30%, while there is complete absence of fruit set in heat-sensitive lines in mean field temperatures of 35 °C. New breeding lines have been developed from CL5915 and other sources that exhibit increased heat tolerance. A CL5915 line is considered best combiners for percentage fruit set and total yield in hybrids developed for heat-tolerance (Metwally *et al.*, 1996). Similarly, EC-520061 (*Solanum habrochaites*) can set fruits under both high (40±2°C) and low (10±2°C) temperatures. Some heat tolerant breeding lines of tomato likeCLN-2413R, CLN 2116B and COML CR-7 have performed well under high temperature condition of West Bengal (Akhtar *et al.*, 2012). The Division of Vegetable Science, Indian Agricultural Research Institute, New Delhi has developed some varieties of vegetables which performed appreciably well under high temperature condition. Tomato variety, Pusa Sadabahar and the hybrid Pusa Hybrid-1 are tolerant to high temperatures. Radish variety, Pusa Chetki produce root of appreciable length under high temperature regime. Similarly, carrot variety, Pusa Vrishti can form root at high temperature and high humidity condition. Early cauliflower variety, Pusa Meghna has been developed which can form curd at high temperature condition.

Framing of conventional breeding strategy

In different fruit vegetable crops, physiological causes of fruit set reduction in heat tolerant lines must be well-understood to improve heat tolerance and address these constraints through breeding. The upper limit of heat tolerance in heat tolerant lines should be fully characterized before using them in combination breeding programme. Different studies based on fruit vegetable crops indicated that single plant selection in the F_2 for heat tolerance will not be effective. Hence, advancement of generation of the cross involving heat tolerant genotype and selection for some important characters *viz.*, flower/ plant, flower drop %, pollen germination, fruits / plant, fruit weight and fruit yield/ plant based on replicated family testing in F_3 and later generations may be effective for improving heat tolerance in different fruit vegetable crops, particularly tomato (Hazra and Ansary, 2008).

Integrated breeding strategy

Various ingredients of resistance, handled by various sets of genes are vital for heat resistance at various steps of crop growth or in diverse tissues. Therefore, use of genetic stocks with diverse levels of heat resistance, co-segregation and correlation analyses, molecular biology methods, molecular markers and quantitative trait loci (QTLs) are promising tools to detect the genetic source of thermo-resistance. However, regardless the screening method, the key objective for plant breeders is to develop an effective set of thermo- tolerance markers which can be used to further implement heat tolerance into various crop species. Since plants adapt to temperature stress by developing more appropriate morphological, physiological and biochemical characteristics, analyzing plant phenology in response to heat stress often gives a better understanding of the plant response and facilitates further molecular characterization of the tolerance traits. Molecular and integrated plant breeding are useful to develop the biotic and abiotic stress tolerance cultivars using genomics approaches like marker-assisted selection (MAS) and genome wide associated studies (GWAS).

Physiological approach

Photosynthesis and photosynthate partitioning

Reduction in carbon fixation and the consequent oxygen evolution result in generation of harmful reactive oxygen species (ROS). As a consequence, the repair mechanism of the damaged photosystem is inhibited. Hence, approaches to develop varieties with improved productivity in a high temperature environment could include manipulating leaf photosynthesis and even photosynthate partitioning (Ainsworth and Ort, 2010).

Table 1: Effect of temperature stress on physiological characters of different vegetable crops

Vegetable crop	Temperature	Effects	Reference
Pea	38°C	Reduced chlorophyll and carotenoid contents	Georgieva and Lichtenthaler (2006)
Potato	25°C	Reduced rate of photosynthesis and carotenoid content	Aien *et al.* (2011)
Radish	40°C	Changes in stomatal characteristics, and decreased photosynthetic rate	Chen *et al.* (2014)
Mungbean	40°C	Reduced relative water contents and damaged chlorophyll contents	Nahar *et al.* (2015)
Cauliflower	40 °C	Decrease in chlorophyll fluorescence	Lin *et al.* (2015)
Cabbage	40 °C	Reduced in photosynthesis activity	Chang *et al.* (2016)
Mungbean	40 °C	Decrease in the viability of pollen	Sharma *et al.* (2016)

Vegetable crop	Temperature	Effects	Reference
Okra	45 °C	Smaller sacs of pollen which affect germination of pollen	Hayamanesh (2018)
Tomato	36 °C	Decreased chlorophyll contents and CO_2 assimilation rate	Zhou *et al.* (2017)
Hot and sweet pepper	40 °C	Decreased chlorophyll contents	Haq *et al.* (2019)
Leafy radish	40 °C	Reduced chlorophyll contents	Yang *et al.* (2019)
Chickpea	40 °C	Decrease in the content of chlorophyll	Kaloki *et al.* (2019)
Carrot	38 °C	Decreased cell membrane stability, increased relative cell injury	Nijabat *et al.* (2020)
Water spinach	42 °C	Decreased chlorophyll content, photosynthetic rate, carbon fixation, and increased respiration rate	Guo *et al.* (2020)
Sweet potato	37 °C	Reduction in chlorophyll contents	Heider *et al.* (2021)
Cabbage	42 °C	Decreased stomatal conductance rate, and chlorophyll contents	Moradpour *et al.* (2021)
Potato	39 °C	Decreased gaseous exchange and damaged photosynthetic pigments	Şanlı and Öztürk Gökçe (2021)

Membrane lipid saturation

One of the typical heat stress symptom is tissue senescence, characterized by membrane damage associated with increased fluidity of membrane lipids, lipid peroxidation, and protein degradation in various metabolic processes (Savchenko *et al.*, 2002). It has been found that compared to tissues grown at 22°C, tissues grown at 32°C had significantly higher overall membrane lipid saturation level and when submitted to a short heat shock at 35°C, displayed a noticeably reduced calcium influx and a consequent reduced heat shock gene expression. Hence, membrane lipid saturation may, therefore, be considered as an important physiological indicator in high temperature tolerance (Balogh *et al.*, 2013).

Osmo-protectants

Osmo-protectants or compatible solutes are small organic molecules with neutral charge and low toxicity at high concentrations that act as osmolytes and help the organisms to survive in extreme osmotic stress condition. Osmo-protectants may include amino acids, polyamines, quaternary ammonium compounds and sugars. These non-toxic compounds stabilize cellular structures and enzymes, act as metabolic signals, and scavenge reactive oxygen species produced under stressful conditions (Singh, 2015). Glycine betaine (N,N,N - trimethylglycine betaine) and proline are two major osmo-protectant osmolytes, which are synthesized by many plants (but not all) in response to stress, including salinity stress, and thereby help in maintaining the osmotic status of the cell to ameliorate the abiotic stress effect (Ashraf and Foolad, 2007).Transgenic approaches have confirmed the beneficial effect of proline over-production during high temperature stress as enhanced proline production in transformed plants correlates well with a more negative leaf osmotic potential and higher production of protective xanthophyllic pigments under heat stress. Glycine betaine plays an important role as compatible solute in plants experiencing high temperature conditions. Glycine betaine production in chloroplasts maintains the activation of Rubisco by sequestering Rubisco activase enzyme near thylakoids and preventing its thermal inactivation (Allakhverdiev *et al.*, 2008). These osmo-protectants may, therefore, be considered as important physiological indicator in high temperature tolerance.

Plant phenomics

The difficulty of expanding agricultural land, along with the effect of climate change and the increase in world population are the current societal changes that make necessary to accelerate research to improve yield-potential and adaptation to stressful environments (Lobos *et al.*, 2014). Increasing yields will require implementing novel approaches in gene discovery and plant breeding that will significantly increase both production per unit of land area and resource use efficiency (Parry and Hawkesford, 2010; Tanger *et al.*, 2017). Hence, "Plant phenomics" is an approach to envisage complex traits (a phenome is the set of physical and biochemical traits belonging to a given organism) that are appropriate for selection (forward phenomics), and also provide explanations as to why the given genotypes stands out in a specific environment (reverse phenomics) (Lobos *et al.,* 2017).

Generally, tolerance to heat is characterized by consistent increases of transcripts involved in the biosynthesis of protective components. Emerging phenomics methodologies can be used to identify genes associated with the traits of interest by establishing functional relationships between genetics and the associated phenotype.

Antioxidant mechanisms to alleviate oxidative stress

The harsh environment, abiotic stress, and sessile nature of plants (permanently restricted to their site of germination) have given plants the mechanisms to fight against oxidative damage caused by ROS (Hasanuzzaman *et al.,* 2020; Mittler *et al.*, 2022).

Table 2: Production of Reactive oxygen species (ROS) in vegetable crops under heat stress

Vegetable crop	Temperature	ROS production	References
Radish	40 °C	H_2O_2	Chen *et al.* (2014)
Cucumber	40 °C	O_2•− and H_2O_2	Li *et al.* (2016)
Tomato	39 °C	H_2O_2	Sakhonwasee and Phingkasan (2017)
Chilli	45 °C	O_2•− and H_2O_2	Feng *et al.* (2019)
Potato	37 °C	O_2•− and H_2O_2	Xi *et al.* (2020)
Water Spinach	42 °C	O_2•− and H_2O_2	Guo *et al.* (2020)
Garlic	45 °C	H_2O_2	Ji *et al.* (2021)
Eggplant	45 °C	H_2O_2	Hannachi *et al.* (2022)
Mungbean	40 °C	H_2O_2	Kareem *et al.* (2022)

Note: Reactive oxygen species (ROS) is an umbrella term for an array of derivatives of molecular oxygen that occur as a normal attribute of aerobic life. Elevated formation of the different ROS leads to molecular damage, denoted as 'oxidative distress'.

The anti-oxidant mechanism of plants is generally divided into two categories, i.e., enzymatic and non-enzymatic mechanisms. The non-enzymatic mechanism consists of ascorbic acid, Glutathione (GSH), α-tocopherol, carotenoids, phenolics, favonoids, and osmolyte proline. The enzymatic system consists of Dehydroascorbate reductase (DHAR), Superoxide dismutase (SOD), Catalase (CAT), Ascorbate peroxidase (APX), Monodehydroascorbate reductase (MDHAR), Glutathione Peroxidase (GPX), and Glutathione reductase (GR) (Hasanuzzaman *et al.*, 2020).

Anti-oxidant mechanism of the enzymes

The enzyme Dehydroascorbate reductase, located in the chloroplast, plays a critical role in the ascorbate–glutathione recycling reaction in higher plants. Superoxide dismutase alternately catalyzes the dismutation (a reaction between two identical molecules in which one is reduced and the other oxidized) of the superoxide radical into ordinary molecular oxygen and hydrogen peroxide. Catalase catalyzes the decomposition of hydrogen peroxide to water and oxygen. Ascorbate peroxidase is also a $H_2O_2^-$ scavenging enzyme and is indispensable for the protection of chloroplasts and other cell constituents from damage by H_2O_2 and hydroxyl radicals (•OH). Ascorbic acid maintains redox homeostasis by scavenging reactive oxygen species from prokaryotes to eukaryotes, especially plants. The enzyme monodehydroascorbate reductase (MDHAR) regenerates ascorbic acid by catalyzing the reduction of monodehydroascorbate, using NADH or NADPH as an electron donor. Peroxidase catalyzes the reduction of hydrogen peroxide to water and oxygen. Glutathione reductases are NADPH dependent enzymes that catalyse the conversion of oxidised glutathione (GSSG) to reduced glutathione (GSH) thus efficiently maintaining reducing environment within the cells under extreme environments.

Influence of polyamines (PAs) to high temperature stress

Polyamines (an organic compound having more than two amino groups) are known as simple molecules which are present abundantly in different vegetable crops. They are simple in structure. There are two types of polyamines *viz.*, simple PAs like diamine putrescine (Put), cadaverine (Cad), spermidine (Spd), and higher PAs like triamine spermidine (Spr). These PAs are present constantly in all vegetable crops and play their part in different physiological and developmental processes (Bano *et al.*, 2020). Whenever plants face temperature stress, PAs help in increasing photosynthesis, antioxidant capacity and osmotic adjustment, which induces tolerance to heat stress (Shao *et al.*, 2015).

Marker assisted breeding

The search for molecular markers associated to phenotypic traits is one aspect of molecular genetics usually carried out with methods based on segregation mapping, genomic introgression, and association mapping (Morgante and Salamini, 2003). QTL mapping has become the method of choice to identify specific chromosome segments that contain candidate genes for heat tolerance. Plant responses under heat stress can be easily detected by using different approaches such as Genome-wide association studies (GWAS) and quantitative trait locus (QTL). QTL is a region of DNA associated with a specific phenotype or trait that varies within a population. GWAS help to identify genes associated with a particular trait. This method studies the entire set of DNA (the genome) of a large group of population, searching for small variations, called single nucleotide polymorphisms or SNPs. Typically, QTLs are associated with traits with continuous variance, rather than traits with discrete variance. These approaches help to understand phenotypic traits linked with different genetic variations under different climate conditions which ultimately help to develop resilient crops against extreme temperatures (Varshney *et al.*, 2019).

High temperature tolerance QTL identification is performed using different yield related traits, such as the thousand-grain weight, yield, etc or senescence related traits and oxidative damage (Vijayalakshmi *et al.*, 2010; Zhou *et al.*, 2021). In tomato, QTL analysis helped to identify heat stress related genes (Wen *et al.*, 2019). In bottle gourd, analysis of QTL-seq method identified seven heat-tolerant QTLs (qHT1.1, qHT2.1, qHT2.2, qHT5.1, qHT6.1, qHT7.1, and qHT8.1). In the qHT2.1 region three non-synonymous SNPs were potentially linked with heat stress tolerance. These SNPs were positioned in the genes that may have roles in signal recognition, intracellular transport, and pollen sterility (Song *et al.*, 2020). In cucumber seedlings, Dong *et al* (2020) detected two loci, qHT3.1 and qHT3.2, and five loci, namely qHT3.2, qHT3.3, qHT4.1, qHT4.2, and qHT6.1, that regulate short-term extreme or long-term mild thermos-tolerance.

In order to transfer these traits, classical breeding requires the establishment of rapid and cost effective screening procedures and implementing these using conventional non-transgenic breeding approaches such as "marker-assisted screening," "association mapping" or "genomic selection" procedures.

Heat-shock transcriptional factors (HSFs)

HSFs are the complex network of a transcriptional regulatory system that controls the downstream responses in plants during heat shock and other abiotic stress responses (Ohama *et al.*, 2017; Haider *et al.*, 2022). HSFA1 subfamily acts as a master regulator of thermo-tolerance in tomato during heat stress (Andrási *et al.*, 2021). HSFs in eggplant comprise 20 members and are subcategorized into 14 groups. These factors not only respond to heat shock, but qRT-PCR has revealed that these HSFs responded to cold, salinity, and drought stress (Wang *et al.*, 2020). In carrot, 35 HSFs were identified, and these factors responded to heat shock and other abiotic stresses (Huang *et al.*, 2015).

Conclusion

High temperature is a devastating environmental factor that influences crop growth and yield by affecting numerous physiological mechanisms. High temperature is an alarming concern that needs to be considered since it directly affects yield in a situation where realization of high yield need to be ensured to ensure global food security and eradicate hunger. High-temperature tolerance is not only important for the current situation, but also for the future since the simulated predictions show the elevating high temperature that enormously affects yield. New possibilities to resist or enhance high-temperature tolerance need to be investigated. Even though there are several studies related to high temperature on crops that reveal various factors that are up-regulated and down-regulated, there are abundant wild types that have not yet been explored. Touching the untouched wild types can reveal amazing outcomes on how high temperatures can be managed with novel tolerant genes. Advancements in molecular techniques provide rapid detection of traits in wild types. It is also important to implement or follow advanced agricultural practices across the globe towards raising the global temperature, by implementing climate-smart practices, high-throughput phenotyping methods, and revealing the traits that are tolerant can protect, safeguard and defend tomorrow.

References

Adli, M. 2018. The CRISPR tool kit for genome editing and beyond. *Nat. Commun.*, **9**:1911. https://doi.org/10.1038/ S41467-018-04252-2

Aien, A., Khetarpal, S.a and Pal, M. 2011. Photosynthetic characteristics of potato cultivars grown under high temperature. *American- Eurasian J. Agric. Environ. Sci.*, **11**:633–639.

Akhtar, S., Ansary, S. H., Dutta, A. K., Karak, C. and Hazra, P. 2012. Crucial reproductive char Crucial reproductive characters as screening indices for tomato (*Lycopersicon esculentum* Mill.) under high temperature stress. *J. Crop Weed*, **8**: 114–117.

Andrási, N., Pettkó-Szandtner, A. and Szabados, L. 2021. Diversity of plant heat shock factors: Regulation, interactions, and functions. *J. Expt. Bot.*, **72**:1558–1575

Ashraf, M. and Foolad, M.R. 2007. Roles of Glycine Betaine and Proline in Improving Plant Abiotic Stress Resistance. *Environ. Expt. Bot.*, **59**: 206-216.

Balogh, G., Peter, M., Gombos, I., Torok, Z., Horvath, I., Harwood, J. L. and Vigh, L. 2013. Key role of lipids in heat stress management. *FEBS Letters*, **587**: 1970-1980.

Bano, C., Amist, N. and Singh, N.B. 2020. Role of polyamines in plants abiotic stress tolerance: advances and future prospects. In: *Plant Life under Changing Environment* (Eds. Tripathi, D.K., Chauhan, D.K., Prasad, S.M., Ramawat, N., Singh, V.P., Sharma, S. and Dubey, N.K.). Elsevier, Academic Press, USA, Pp. 481–496.

Berry, S. Z. and Rafique-Uddin, M. 1988. Effect of high temperature on fruit set in tomato cultivars and selected germ plasm *Hort Sci.*, **23**: 606–608

Blum, A. 2018. *Plant Breeding for Stress Environments*, CRC Press, Boca Raton, FL, USA.

Bray, E.A., Bailey-Serres, J. and Weretilnyk, E. 2000. Responses to abiotic stress. In: *Biochemistry and Molecular Biology of Plants* (Eds. Gruissem, W. and Jones, R.), American Society of Plant Physiologists, Rockville, USA, pp. 1158-1203.

Cardi, T., Batelli, G. and Nicolia, A. 2017. Opportunities for genome editing in vegetable crops. *Emerg. Top. Life Sci.*, **1**:193–207

Chang, K.Y., Lin, K.H. and Lo, H.F. 2016. Physiology and proteomics of cabbage under heat and flooding stress. *Res. Rev. J. Bot. Sci.*, **5**: 44–53

Chen, T. H. H., Shen, Z. Y. and Lee, P. H. 1982. Adaptability of crop plants to high temperature stress. *Crop Sci.*, **22**: 719–725

Chen, W.L., Yang, W.J., Lo, H.F. 2014. Physiology, anatomy, and cell membrane thermostability selection of leafy radish (*Raphanus sativus* var. *oleiformis* Pers.) with different tolerance under heat stress. *Sci. Hort.*, **179**:367–375

Cossani, C.M. and Reynolds, M.P. 2012. Physiological traits for improving heat tolerance in wheat. *Plant Physiol.*, **160**: 1710–1718.

Craufurd, P.Q., Qi, A., Ellis, R.H., Summerfield, R.J., Roberts, E.H. and Mahalakshmi, V. 1998. Effect of temperature on time to panicle initiation and leaf appearance in sorghum. *Crop Sci.*, **38**: 942–947.

Dangol, S.D., Hashmi, M.H. and Saeed, F. 2021. Utilizing RNA-based approaches to understand plant-insect interactions. In: *RNA Based Technologies for Functional Genomics in Plants* (Eds. Tang,G.,Teotia, S., Tang, X. and Singh, D.), Springer, Cham. pp. 393–428.

Dong, S., Zhang, S. and Wei, S. 2020. Identification of quantitative trait loci controlling high-temperature tolerance in cucumber (*Cucumis sativus* L.) seedlings. *Plants*, **9**:1–11.

Feng, X.H., Zhang, H.X. and Ali, M. 2019. A small heat shock protein CaHsp25. 9 positively regulates heat, salt, and drought stress tolerance in pepper (*Capsicum annuum* L.). *Plant Physiol Biochem*., **142**:151–162

Foolad, M.R. 2005. Breeding for abiotic stress tolerances in tomato. In: *Abiotic Stresses: Plant Resistance Through Breeding and Molecular Approaches* (Eds. Ashraf, M. and Harris, P.J.C.), The Haworth Press Inc., New York, USA, pp. 613–684.

Georgieva, K. and Lichtenthaler, H.K. 2006. Photosynthetic response of different pea cultivars to low and high temperature treatments. *Photosynthetica*, **44**:569–578.

Gocal, G.F.W., Schöpke, C. and Beetham, P.R. 2015. Oligo□mediated targeted gene editing In: *Advances in New Technology for Targeted Modification of Plant Genomes* Chapter 5 (Eds. Zhang F., Puchta H. and Thomson J.G.), Springer Verlag, Heidelberg, Berlin, pp. 73–90.

Guo, R., Wang, X. and Han, X. 2020. Physiological and transcriptomic responses of water spinach (*Ipomoea aquatica*) to prolonged heat stress. *BMC Genomics*, 21. https:// doi. org/ 10. 1186/ s12864- 020- 06953-9

Haider, M.S., Jaskani, M.J. and Fang, J. 2021. Overproduction of ROS: underlying molecular mechanism of scavenging and redox signaling. In: *Biocontrol Agents second metabolites* (Ed. Jogaiah, S.), Elsevier, UK. pp. 347–382.

Hannachi, S., Signore, A., Adnan, M. and Mechi, L. 2022. Single and associated effects of drought and heat stresses on physiological, biochemical and antioxidant machinery of four eggplant cultivars. *Plants*, **11**: 2404

Haq, S.U., Khan, A. and Ali, M. 2019. Knockdown of CaHSP60-6 confers enhanced sensitivity to heat stress in pepper (*Capsicum annuum* L.). *Planta*, **250**: 2127–2145.

Hasanuzzaman, M., Bhuyan, M.H.M.B., Zulfiqar, F., Raza , A., Mohammad Mohsin , S., Al Mahmud, J. Fujita, M. and Fotopoulos, V. 2020. Reactive oxygen species and antioxidant defense in plants under abiotic stress: revisiting the crucial role of a universal defense regulator. *Antioxidants*, **9**:1–52

Hayamanesh, S. 2018. The effect of high temperature on physiological and metabolic parameters and reproductive tissues of okra (*Abelmoschus esculentus* (L.) Moench), Doctoral dissertation. The University of Sydney, Sydney, NSW, Australia. 181p.

Hazra, P. and Ansary, S.H. 2008. Genetics of heat tolerance for floral and fruit set to high temperature stress in tomato (*Lycopersicon esculentum* Mill.). *SABRAO J. Breeding Genet.,* **40**: 117- 125.

Heider, B., Struelens, Q., Faye, E. and Flores, C. 2021. Intraspecific diversity as a reservoir for heat-stress tolerance in sweet potato. *Nat. Clim. Chang*., **11**:64–69. https:// doi. org/ 10. 1038/ s41558- 020- 00924-4.

Huang, Y., Li, M.Y. and Wang, F.2015. Heat shock factors in carrot: genome-wide identification, classifcation, and expression profiles response to abiotic stress. *Mol. Biol. Rep*., **42**:893–905.

IMD, Annual Climate Summary, 2010. India Meteorological Department, Pune. Government of India, Ministry of Earth Sciences, 27p.

Ji, H.S., Bang, S.G. and Ahn, M.A. 2021. Molecular cloning and functional characterization of heat stress-responsive superoxide dismutases in garlic (*Allium sativum* L.). *Antioxidants*, **10**:815

Kaloki, P., Devasirvatham, V. and Tan, D.K. 2019. Chickpea abiotic stresses: combating drought, heat and cold. In: *Abiotic and Biotic Stress in Plants* (Eds. de Oliveira, A.B.), Intech Open, London. Pp.139–162.

Kareem, H.A., Saleem, M.F. and Saleem, S. 2022. Zinc oxide nano particles interplay with physiological and biochemical attributes in terminal heat stress alleviation in mungbean (*Vigna radiata* L.). *Front. Plant Sci.*, **13**: 842349–842349

Li, H., Ahammed, G.J. and Zhou, G. 2016. Unraveling main limiting sites of photosynthesis under below-and above-ground heat stress in cucumber and the alleviatory role of luffa rootstock. *Front. Plant Sci.*, **7**:746

Lin, K.H., Chen, L.F.O., Li, S.D., Lo, H.F. 2015. Comparative proteomic analysis of cauliflower under high temperature and flooding stresses. *Sci. Hort.*, **183**:118–129.

Lobos, G.A., Matus, I., Rodriguez, A., Romero-Bravo, S., Araus, J.L., and del Pozo, A. 2014. Wheat genotypic variability in grain yield and carbon isotope discrimination under Mediterranean conditions assessed by spectral reflectance. *J. Integr. Plant Biol.*, **56**: 470–479.

Lobos, G.A., Camargo, A.V., del Pozo, A., Araus, J.L., Ortiz, R. and Doonan, J.H. 2017. Plant phenotyping and phenomics for plant breeding. *Front. Plant Sci.*, 8, https://doi.org/10.3389/fpls.2017.02181.

Lopes, M.S., El-Basyoni, I., Baenziger, P.S., Singh, S., Royo, C., Ozbek, K., Aktas, H., Ozer, E., Ozdemir, F. and Manickavelu, A. 2015. Exploiting genetic diversity from landraces in wheat breeding for adaptation to climate change. *J. Exp. Bot.*, **66**: 3477–3486.

Metwally, E., El Zawily, A., Hassan, N. and Zanata, O. 1996. Inheritance of fruit set and yields of tomato under high temperature conditions in Egypt. *First Egypt-Hung. Hort Conf.*, **1**: 112– 122

Mittler, R., Zandalinas, S.I., Fichman, Y., Breusegem, F.V. 2022. Reactive oxygen species signalling in plant stress responses. *Rev. Mol. Cell Biol.*, **23**: 663–679.

Moradpour, M., Abdullah, S.N.A. and Namasivayam, P. 2021. The impact of heat stress on morpho-physiological response and expression of specific genes in the heat stress-responsive transcriptional regulatory network in *Brassica oleracea*. *Plants*, 10. https:// doi. org/ 10. 3390/ plant s1006 1064

Morgante, M. and Salamini, F. 2003. From plant genomics to breeding practice. *Curr. Opin. Biotechnol.,* **14**: 214–219.

Nahar, K., Hasanuzzaman, M., Alam, M.M. and Fujita, M. 2015. Exogenous glutathione confers high temperature stress tolerance in mung bean (*Vigna radiata* L.) by modulating antioxidant defense and methylglyoxal detoxification system. *Environ. Exp. Bot.*, **112**:44–54.

Nijabat, A., Bolton, A., Rehman M.U. 2020. Cell membrane stability and relative cell injury in response to heat stress during early and late seedling stages of diverse carrot (*Daucus carota* L) germplasm. *Hort Sci.*, **55**:1446–1452.

Ohama, N., Sato, H., Shinozaki, K. and Yamaguchi-Shinozaki, K. 2017. Transcriptional regulatory network of plant heat stress response. *Trends Plant Sci.*, **22**:53–65

Opena, R. T., Chen, J. T., Kuo, C. G. and Chen, H. M. 1992. Genetic and physiological aspects of tropical adaptation in tomato. In: *Adaptation of Food Crops to Temperature and Water Stress* (Ed. Kuo, C. G.), AVRDC, Shanhua, Taiwan, pp. 321–334

Parry, M.A.J., and Hawkesford, M.J. 2010. Food security: increasing yield and improving resource use efficiency. *Proc. Nutr. Soc.*, **69**: 592–600.

Parry, M.A.J., Reynolds, M., Salvucci, M.E., Raines, C., Andralojc, P.J., Zhu, X.G., Price, G.D., Condon, AG. and Furbank, R.T. 2011. Raising yield potential of wheat. II. Increasing photosynthetic capacity and efficiency. *J. Expt. Bot.*, **62**: 463-467.

Prasad, P. V. V., Staggenborg, S. A., and Ristic, Z. 2008. Impacts of drought and/or heat stress on physiological, developmental, growth, and yield processes of crop plants, In: *Response of Crops to Limited Water: Understanding and Modeling Water Stress Effects on Plant Growth Processes, Advances in Agricultural Systems Modeling* (Eds Ahuja, L. R., Reddy, V. R. , Saseendran, S. A. and Qiang, Y. U.), American Society of Agronomy, Madison, WI, USA. pp. 301–355.

Raza, A., Mehmood, S.S., Ashraf, F. and Khan, R.S.A. 2018. Genetic diversity analysis of Brassica species using PCR-based SSR markers. *Gesunde Pflanzen*,**71**:1–7

Ricroch, A. 2019. Global developments of genome editing in agriculture. *Transgenic Res.*, 282:45–52. https://doi.org/10.1007/ S11248-019-00133-6

Roitsch, T. and González, M.C. 2004. Function and regulation of plant invertases : sweet sensations. *Trends Plant Sci.*, **9**: 606–613.

Rysiak, A., Dresler, S., Hanaka, A., Hawrylak-Nowak, B., MaciejStrzemski, M., Kováčik, J., Sowa,I., Latalski, M., and Wójciak, M. 2021. High temperature alters Secondary metabolites and photosynthetic efficiency in *Heracleum sosnowskyi*. *Int. J. Mol. Sci.*, **22**: 4756.doi: 10.3390/ijms22094756

Sakhonwasee, S. and Phingkasan, W. 2017. Effects of the foliar application of calcium on photosynthesis, reactive oxygen species production, and changes in water relations in tomato seedlings under heat stress. *Hort. Environ. Biotech.*, **58**:119–126

Şanlı, B.A. and Ozturk Gokce, Z.N. 2021. Investigating effect of miR160 through overexpression in potato cultivars under single or combination of heat and drought stresses. *Plant Biotechnol. Rep.*, **15**:335–348.

Scott, J. W., Bryan, H. H. and Ramos, L. J. 1997. High temperature fruit setting ability of large-fruited, joint less pedicel tomato hybrids with various combinations of heat-tolerance. *Proc. Fla. State Hortic. Soc.*, **110**: 281–284

Shao, C.G., Wang, H., Yu-Fen, B.I. 2015. Relationship between endogenous polyamines and tolerance in *Medicago sativa* L. under heat stress. *Acta Agrestia Sin.*, **23**:1214–1219

Sharma, L., Priya, M., Bindumadhava, H. 2016. Influence of high temperature stress on growth, phenology and yield performance of mungbean [*Vigna radiata* (L.) Wilczek] under managed growth conditions. *Sci. Hort.*, **213**:379–391.

Singh, M., Kumar, J., Singh, S., Singh, V.P. and Prasad, S.M. 2015. Roles of osmoprotectants in improving salinity and drought tolerance in plants: a review. *Rev. Environ. Sci. Bio Technol.,* **14**: 407–426.

Song, H., Huang, Y., Gu, B. 2020. QTL-Seq identifies quantitative trait loci of relative electrical conductivity associated with heat tolerance in bottle gourd (*Lagenaria siceraria*). *PLoS One*, 15. https:// doi.org/10.1371/journal.pone.0227663

Tanger, P., Klassen, S., Mojica, J. P., Lovell, J. T., Moyers, B. T. and Baraoidan, M. 2017. Field-based high throughput phenotyping rapidly identifies genomic regions controlling yield components in rice. *Sci. Rep.,* **7**: 42839. doi: 10.1038/srep42839

Varshney, R.K., Pandey, M.K., Bohra, A., Singh, V.K., Thudi, M. and Saxena, R.K. 2019. Toward the sequence-based breeding in legumes in the post-genome sequencing era. *Theor. Appl. Genet.*, **132**: 797–816

Vijayalakshmi, K., Fritz, A., Paulsen, G., Bai, G., Pandravada, S. and Gill, B. 2010. Modeling and mapping QTL for senescence-related traits in winter wheat under high temperature. *Mol. Breed.,* **26**: 163–175.

Wang, J., Hu, H., Wang, W., Wei, Q., Hu, T. and Bao, C. 2020. Genome-wide identification and functional characterization of the heat shock factor family in eggplant (*Solanum melongena* L.) under abiotic stress conditions. *Plants*, 9. https://doi.org/10.3390/ plants9070915

Weis, E. and Berry, J.A. 1988. Plants and high Temperature Stress. In: *Plants and Temperature* (Eds. Long, S.P. and Woodward, F.I), The Company of Biologists Ltd., Cambridge, pp. 329-346.

Wen, J., Jiang, F. and Weng, Y. 2019. Identification of heat-tolerance QTLs and high-temperature stress-responsive genes through conventional QTL mapping, QTL-seq and RNA-seq in tomato. *BMC Plant Biol.*, 19. https://doi.org/10.1186/s12870-019-2008-3

Wheeler, T. and Von Braun, J. 2013. Climate change impacts on global food security. *Science*, **341**: 508–513.

Xi, Y., Han, X. and Zhang, Z. 2020. Exogenous phosphite application alleviates the adverse effects of heat stress and improves thermotolerance of potato (*Solanum tuberosum* L.) seedlings. *Ecotoxicol. Environ. Safety*, **190**:110048

Yang, Z., Li, W. and Su, X. 2019. Early response of radish to heat stress by strand-specific transcriptome and miRNA analysis. *Int. J. Mol. Sci.*, 20. https:// doi. org/ 10. 3390/ ijms2 01333 21.

Zang, Y., Zeng, D., Liu, Y. and Zhy, W. 2022. *SlSPS*, a Sucrose Phosphate Synthase Gene, mediates plant growth and thermotolerance in Tomato. *Horticulturae,* **8**: 491; https://doi.org/10.3390/horticulturae8060491

Zhang, J., Huang, W., Pan, Q.H. and Liu, Y. 2005. Improvement of chilling tolerance and accumulation of heat shock proteins in grape berries (*Vitis vinifera* cv. Jingxiu) by heat pre-treatment. *Postharvest Biol. Technol.,* **38**: 80–90.

Zhang, J., Huang, W., Liu, Y. and Pan, Q.H. 2005. Effects of temperature acclimation pretreatment on the ultrastructure of mesophyll cells in young grape plants (*Vitis vinifera* L. cv. Jingxiu) under cross-temperature stresses. *J. Integr. Plant Biol.*, **47**: 959–970.

Zhou, R., Kjar, K.H., Rosenqvist, E., Yu, X., Wu, Z., and Ottosen, C.O. 2017. Physiological response to heat stress during seedling and anthesis stage in tomato genotypes differing in heat tolerance. *J. Agron. Crop Sci.*, **203**:68–80.

Zhou, L., Tang, M., Hassan, M. J., Zhang, Y., Han, L. and Peng, Y. 2021. Adaptability to high temperature and stay-green genotypes associated with variations in antioxidant, chlorophyll metabolism, and γ-aminobutyric acid accumulation in creeping bent grass species. *Front. Plant Sci.*, **12**: 750728. doi: 10.3389/fpls.2021.750728

14

Ideotype Breeding

Concept of Ideotype breeding

In the breeding programme, elite crop varieties are developed through combination of relatively low-frequency alleles which are of known agronomic / horticultural importance (Lyu *et al.*, 2013). During breeding programme, these elite varieties are developed targeting different main objectives, such as yield, product quality, resistance to abiotic and biotic stresses and agronomic suitability.

A crop ideotype is defined as an idealised structural model, combining both morphological and physiological traits, with the potential to maximise yield and quality within a defined environment (Donald, 1968). Ideotype generally include factors that increase the survivability of the plant such as drought resistance, resource use efficiency, root architecture and disease resistance (Arraiano *et al.,* 2006; Rezzouk *et al.,* 2022). Definition of "Ideotype" may be broadened through combination of morphological and physiological traits (or their genetic bases) conferring to satisfying adaptation of a crop to a particular biophysical environment, crop management and end use.

The ideotype approach was designed to overcome limitations traditionally experienced by the breeders during selection. Primarily, the ideotype breeding approach relies on defining a theoretically efficient or ideal plant type, and then it is bred by targeting all the component traits that define it however, it can complement traditional breeding approaches (Reynolds *et al.*, 2009). As the end goal is defined for different locations and situations, characteristics of the ideotype are expected to vary across different environments.

According to this approach, breeders should select directly for the plant ideotype, rather than empirically for yield. Concrete guidelines have been framed to conduct the ideotype breeding process in view of several well-defined target traits besides final yield (Reynolds *et al.*, 2009). For instance, guided by the idea of improving light capture and assimilate partitioning in cereals, Donald (1968) identified short stature, strong stem, few small erect leaves, large erect awned ears, low tillering capacity, disease resistance, local adaptation and low plant competitive ability as important target traits for wheat crops sown at high density and under non-limiting conditions. Breeding for such an ideotype has resulted in improved lodging resistance and higher harvest index in wheat (Hamblin, 1993).

Ideotype breeding is a continuous process because ideotypes are developed to meet the changing needs. Plant breeders can complement traditional breeding methods with ideotype breeding to generate a better understanding on individual traits and to diversify the genetic variation for traits. Ideotype breeding also provides hypotheses on the paths to increase yield or other important end traits, and this approach is also expected to give greater quantity of useful end product in a specific environment.

Ideotype breeding was initially developed for mainly cereals and was later applied to oilseed crop, forest and fruit tree species and other crops. Since then, the ideotype approach was expanded and refined to include other concerns such as market, new outlets, climate change adaptation (Semenov and Stratonovitch, 2013; Semenov *et al.*, 2014), emerging pests and diseases, and changes in farming system (Jeuffroy *et al.*, 2013). Attempts have also been made to define traits at the biochemical and molecular levels. Overall, the level of knowledge to support the ideotype approach depends on the consideration of important characters as well as final trait (e.g. disease resistance, yield, product quality, etc.)

Ideotype for targeted environment

The initial ideotype of Donald (1968) was developed for low- or non-stress environments where light capture was the major limitation to grain yield. In water-limited environments, it is difficult to design an ideotype because water deficit affects crop growth and development to a different extent depending on the timing, severity and duration of stress and interactions between water deficit and other factors such as temperature and nutrient availability. Therefore, it is unlikely that a single trait will improve plant performance in all the scenarios of water deficit (Tardieu, 2012). Therefore, specific ideotypes should be conceived for targeted environments. For environments with large inter-annual weather variability, stability of performance is often considered an important varietal characteristic but selecting for more specific adaptation could be an alternative.

Eco-physiological models to develop ideotype

The targeted traits should reliably be estimated based on genetic information under eco-physiological models to develop the ideotype. Different levels of complexity found in the models concerning the integration of genetic details are:

- Complexity in the models dealing with quantitative trait loci (QTL) /genes effects
- Applications of modeling to support trait assessment in multi-environments through genotype × environment interactions
- Assessment of the potential particularly to decipher QTL × environment interactions
- Use of integrated models in breeding approaches

White and Hoogenboom (2003) identified six classes of models in relation to genetic detail:

1. Generic model with no reference to species
2. Species-specific model with no reference to cultivars
3. Genetic differences represented by cultivar specific parameters
4. Genetic differences represented by gene actions modeled through their effects on model parameters
5. Genetic differences represented by genotypes, with gene action explicitly simulated based on the knowledge of regulation of gene expression and effects of gene products
6. Genetic differences represented by genotypes, with the gene action simulated at the level of interactions of regulators, gene products, and other metabolites.

The 'generic' (class 1) and 'species specific' (class 2) models were developed first, in the 1970s. Progressively, basic genotypic information has been included, so that most eco-physiological models are now 'cultivar-specific' (class 3). Currently 'gene-specific' models (class 4) have been developed which include information from the major genes associated to 'simple' traits or from the quantitative trait loci associated to more complex traits. These models have been proposed for hypothetical genes when key genetic controls were unknown (Chapman *et al.*, 2003; Hammer *et al.*, 2005) and for genetic controls identified experimentally (Chenu *et al.*, 2009). Models based on the 'regulation of gene expression' (class 5) are rare as the understanding of gene action is restricted to particular physiological processes mostly in model species (Welch *et al.*, 2005; Chew *et al.*, 2012). Despite incomplete information, models of "class 5" can be proposed for crop species. Lastly, models that simulate gene action based on interactions of regulators, gene products, and other metabolites (class 6) have only been developed for unicellular organisms (Karr *et al.*, 2012; Sanghvi *et al.*, 2013).

In overall scenario, while models of classes 5 and 6 can be relevant for 'simple' traits associated to a restricted number of genes (e.g. flowering time, resistance to a pathogen, enhanced carotenoid content, etc.), more complex traits are usually modeled *via* QTL analysis. However, increase in the complexity of a model does not necessary lead to better predictions (Challinor et al., 2014). Typically, a trade-off between complexity and accuracy in prediction has to be addressed to avoid over-parameterization and to limit uncertainty (Reynolds and Acock, 1985). For this reason, well-known processes that affect traits of interest or key variables of the system might be worth including with further detail in a model.

Before addressing the question of how to translate the information of model simulations to knowledge that can be used by physiologists or geneticists, better understanding of the model properties and behaviour is needed. One of the best ways to do that is to conduct a global uncertainty and sensitivity analysis of the model to investigate its behavior in response to variations in inputs (Cariboni *et al.*, 2007). By perturbing model parameters associated with simple physiological traits, uncertainty and sensitivity analyses allow investigation of crop responses and can help identify those traits that lead to high and stable yield, particularly grain yield in the target environments.

Features of Ideotype breeding

- **Emphasis on individual trait :** In ideotype breeding, each character is specified before initiation of the work to achieve the goal. Emphasis is given on individual morphological and physiological trait which enhances the yield and other targeted characters.
- **Inclusion of yield components:** Yield is quantitative trait which is the cumulative expression of different yield components. Characters that enhance yield are identified through character association analysis. Only those traits which exhibit positive association with yield or the targeted character are included in this model.
- **Exploitation of physiological variation:** Genetic difference exists for various physiological characters such as photosynthetic efficiency, photo-respiration, nutrient uptake, etc. Ideotype breeding makes use of genetically controlled physiological variation in enhancing various characters of agronomic interest and yields.
- **Slow progress:** Ideotype breeding is very slow method to achieve the goals because incorporation of various desirable characters from different sources into a single genotype takes long time. Moreover, sometimes undesirable linkage affects the progress adversely.
- **Selection of genotype:** Selection of genotype is core component of ideotype breeding. Selection is focused on individual plant character which enhances the yield and other targeted character and efficient selection save a lot of time.
- **Designing of model:** In ideotype breeding, the phenotype of a new variety to be developed is specified as a "model" in terms of physiological and morphological traits in advance.
- **Interdisciplinary approach:** In true sense, it is an interdisciplinary breeding approach. It does not depend upon only single approach but also includes disciplines of genetics, breeding, physiology, soil science, pathology, entomology etc. to deal with different morphological, physiological and biochemical characters.
- **Continuous process:** It is long and continuous process of developing new ideotype to meet changing and increasing demands. Hence, development of ideotype is a moving target.
- **Genetic background:** Effectiveness of breeding depends on the genetic background that controls the phenotypes. Undesirable effect of the genes and genetic linkage should be taken into consideration before initiation of the breeding process.

Difference between Traditional and Ideotype breeding

In case of ideotype breeding, values for individual traits are specified whereas such values are not fixed in case of traditional breeding. In other words, first a conceptual plant model is fixed and then efforts are made to achieve such model. In traditional breeding such models are not developed before initiation of breeding programme.

Traditional breeding	**Ideotype breeding**
The main objective is defined before initiating the breeding work	The conceptual theoretical model is prepared before initiation of breeding work
Selection is mainly focused on yield and some other characters.	Selection is focused on different individual plant characters through which yield on other target character is enhanced.
It usually includes various morphological and economic characters of the crop	It usually includes various morphological, physiological and biochemical characters of the crop
Value of each character is not fixed in advance	Value of each trait is defined in advance
This is simple and rapid method of variety development	This is difficult and slow method of variety development
The phenotype of a new variety is not specified in advance	The phenotype of a new variety to be developed is specified in advance

Methods of developing an Ideotype

The breeder should be knowledgeable with several components or features related to yield and other important characters, such as inheritance pattern of important characters, correlations between yield and its related traits and interactions among different character for their expression. In order to develop an ideotype, association studies are crucial (Sharma *et al.,* 2012). If a trait and yield is positively correlated both these characters can be improved concomitantly but in case of negative correlation, both the characters cannot be developed at the same time.

While productivity improvement has slowed down in crops like wheat in recent years, the major benefits of ideotype design are seen as being conceptual and analytical rather than in direct yield improvements (Richards *et al*., 2019). Progress from using ideotype breeding has often not been as fast as hoped. However, it stimulated interactions among the breeders, physiologists, agronomists and more recently, crop modelers about what traits might be important and in which conditions. Andrivon *et al.* (2013) considered three views for the ideotypes based on formal or simulation experiments.

- Historical or 'genetic' view i.e., genetic architecture of the ideotype
- Agronomic view, where new genotypes are designed for specific cropping systems
- Modelling view i.e., the best combinations of traits (usually represented by model parameters)

Steps in Ideotype Breeding

- **Target definition:** Definition of the main goal (target) for the breeding process (e.g. breeding for resistance to particular disease, improved water-deficit tolerance, improved vitamin and mineral content, etc.)
- **Development of conceptual theoretical model (Ideotype design):** Ideotype consists of various morphological and physiological traits and is specified to develop a conceptual theoretical model with a combination of desired traits based on the objectives of ideotype development. Ideotype design is basically identification of morpho-physiological traits to reach the defined goal and the way to assemble them within an ideotype. As for example, specification of maturity duration, maintenance of photosynthesis under water stress condition, leaf size, leaf number, plant height, angle of leaf etc. and subsequently efforts are made to attain this design.
- **Selection of base material:** It is crucial to choose the right base material. In order to develop a climate resilient variety capable of giving consistent yield under a variety of climatic conditions, the genotype to be used in generating a model plant should have wider adaptability and broad genetic base. Active cooperation of the physiologists, pathologists, entomologists and soil scientists is needed for the selection of genotypes for plant stature, maturity duration, leaf size, and angle from the global gene pool of the concerned crop species. These genotypes would be employed as basis or source material for the desired characters.
- **Incorporation of desirable traits:** Finding the appropriate means to incorporate the required qualities from several sources into one type is the crucial next step after choosing the basic material. This can be accomplished through hybridization, mutation breeding, backcrossing, etc. Understanding the relationships between various characters is crucial before executing hybridization programme so that the efficient and potential trait combination can be created in the segregating population. The backcross approach is frequently used to introduce oligogenic characteristics from the chosen germplasm into the background of a genotype that has been modified. If there are reproductive barriers, different biotechnological interventions *viz.*, embryo rescue, protoplast fusion, etc. are sought to transfer the targeted character to new genetic background.
- **Selection of ideal plant type:** In ideotype breeding, selection of plant type is a crucial technique. The genotypes combining the features indicated in the conceptual model are chosen, replicated and evaluated across different locations before released for commercial cultivation.
- **Assessment of the Ideotype:** Multi-criteria assessment of the suggested ideotypes is done to prove the agronomic relevance of the trait integration under different environments through simulations or field experiments. Under controlled environmental settings, it is possible to check for resistance to soil salinity, alkalinity, drought, diseases and insects.

Generally, the ideotype is thought of in terms of crop improvement *via* breeding, but crop management (e.g. sowing density, row width, nitrogen fertilization, irrigation, etc.), diversified use of the product (e.g., use of tomato as fresh and processing purpose) may also produce the desired ideotypes by exploiting phenotypic plasticity. Different studies recorded that plant architecture traits and crop canopies may limit the epidemic development of pests and disease (Andrivon *et al.*, 2013; Desanlis *et al.*, 2013). So ideotyping not only result from breeding and varietal choice but also from crop management and cropping system strategies. While crop management is often added in a second step as an effective driver to complement genetic gains and greatest productivity improvement may arise when combining together both breeding and agronomic practices (Duvick *et al.*, 2004).

Andrivon *et al.* (2013) concluded that different breeding strategies should lead to development of ideotype. Emergence of new objectives (e.g. low input systems, dual purpose crops) and new constraints (e.g. increasing risk of extreme weather events, price volatility) is now arguing for both new breeding objectives and new design methods. Designing crop ideotypes for these new targets is an important consideration in the present scenario.

Tools to aid ideotype development

- ***Improvements to phenotyping approaches:*** Improvements to plant phenotyping approaches have led to methods to rapidly capture detailed descriptions for a variety of plant traits. These often rely on the use of non-invasive and/or digital technologies. Incorporating multiple sensors within the same system can permit high spatial and temporal data collection for a wide range of plant parameters simultaneously (Rao Mogili and Deepak, 2018; Barbedo, 2019).
- ***Computer modelling:*** Generating new ideotypes is confounded by the high level of uncertainty associated with rapidly changing environmental conditions. Improvements in phenotyping approaches provide one approach to capturing plant structural traits. Limitations arise for access to equipment, knowledge, sufficient variation and time. To overcome limitations associated with the time and resource intensive breeding of ideotypes, tools are being continually developed to narrow down target traits. For example, designing and testing potential ideotypes *in silico* (experimentation performed by computer) using simulation models enables a wider range of traits and environmental conditions to be tested.
- ***Genetic selection:*** Many desirable plant traits are quantitative and governed by multiple genes. Therefore, quantitative genetics provides a powerful framework for crop improvement. For ideotype design, appropriate parental lines should be selected for breeding programmes. Following trait recombination, genomic selection could be used as a tool to help avoid lengthy phenotypic evaluation of the progeny and to achieve greater genetic gains.

  When defining a genetic ideotype, the first stage is to accurately define the target environment and the corresponding impact on the phenotype (Hodson and White, 2007; Trethowan, 2014; Gerard *et al.*, 2020). Following this, known genetic information, gene x environment interactions and physiological response to stress can be integrated to predict the optimal plant type. For the areas of the genome where unknown genes may be located, association genetic analysis can be performed *via* random genotyping (Crossa *et al.*, 2007). Once the genetic ideotype has been defined, it can then be bred for. This can be aided by methods such as marker assisted recombinant selection (Ribaut and Ragot, 2007) and through software to predict the probability of obtaining the desired genotype.

- ***Combined approaches:*** Computer modeling, genetic selection and new breeding approaches have all resulted in improvements of crop traits however greater improvements could be achieved by combination of approaches. For example, combining simulation studies with new breeding approaches provides the platform for targeted ideotype development which significantly accelerates the generation of new varieties. Potential drawbacks of each individual approach may be overcome. For example, computer simulation approaches often do not consider the genetic basis of a trait of interest, not do they account for the available genetic variation in traits (Gu, 2022). Accurate quantitative genetic analyses require screening of large populations to obtain sufficient genetic resolution. However, accuracy of predicting genetic-based models may be reduced due to the complex gene x gene and gene x environment interactions. For this reason, quantitative trait loci (QTL) are often only detected under a limited set of environmental conditions (Zhou *et al.*, 2007). Combining both the approaches has therefore been proposed to overcome these issues (Letort *et al.*, 2008; Gu, 2022).

Ideotype features of important vegetable crops

Ideotyping was focused primarily on cereals and oilseed crops. Later, forest and fruit crops have come under the purview of ideotyping. Wheat was the first crop in which ideotype was first proposed by Donald (1968) which consisted of seven main features:

- Short and strong culm, called tillers
- Erect leaves
- Few and small leaves
- Large ears
- Erect ears
- Presence of awns
- Single culm

Later, New Plant Type (NPT) proposed for rice shared some characteristics in common with the wheat ideotype including, upright leaves, low tiller number and a strong culm, large panicles that sit below the top three leaves and a high harvest index (Peng *et al.*, 2008).

Designing plant architectures of different vegetable crops suitable for different environments, cropping systems, resistance to biotic and abiotic stress etc. has been considered for the development varieties to harness their maximum potential for different purposes. Different characteristic features of important vegetable crops, which the breeders should aim at for developing ideal line-bred variety or hybrid are mentioned below.

Tomato for fresh purpose

- Determinate growth habit, high yield
- Indeterminate hybrids with extended harvest span
- Dual purpose, good transportability, resistant to high temperature + tomato leaf curl virus + early blight + bacterial wilt
- Determinate growth habit, flattish-round, high acid fruits

Tomato for processing purpose

- Compact and determinate growth habit, high yield
- Indeterminate growth habit with extended harvest span
- Concentrated flowering and fruiting habit
- Jointless pedicel of fruit suitable for mechanical harvest
- High processing quality fruits with high total soluble solid content and high viscosity of fruit pulp
- High fruit pH and acidity
- Resistant to tomato leaf curl virus + early blight + bacterial wilt

Brinjal

- High yield and good fruit quality
- Round fruit with purple and white stripes, green long, purple long, green round and large purple oval fruits
- Shiny blackish purple round or oblong with dark green non-spiny calyx for export purpose
- Tolerance to abiotic stresses (drought, high or low temperature, salinity, etc.)
- Resistant to Phomopsis blight + Bacterial wilt

Chilli

- Early flowering and fruiting habit
- High yield, very pungent fruit (>70,000 Schoville Heat Units), deep red colour for fresh market and as dry powder
- High yield, medium pungent fruit (30,000 SHU), deep red colour for preparation of dry powder
- High yield, low pungent coupled with attractive deep red colour for pickle making

- High fruit weight and large fruit size for pickle purpose
- Resistant to Leaf curl + thrips + mites + Anthracnose

Sweet pepper

- Early flowering and fruiting habit
- High yield, high fruit weight and larger fruit size
- Blocky fruits of dark green, red, yellow, orange, violet or white colour
- Resistant to Leaf curl virus + thrips + mites + Anthracnose
- Wide adaptation particularly in high temperature condition

Cabbage

- Good heading under high temperature condition
- Early head formation / early maturity
- Firm, semi-flat heads, weighing within two kg
- Good field holding capacity
- Resistant to Black rot caused by *Xanthomonas campestris* + Diamond back moth

Cauliflower

- High temperature tolerance, non-riceyness, compact, bract free and self-blanching curd
- Early curd formation / early maturity
- Retentive cream/white colour of hemispherical shape
- Resistant to Black rot caused by the bacteria *Xanthomonas campestris*

Watermelon

- Vine with low ratio of staminate: pistillate flower (sex ratio)
- Jubilee pattern type, large, elongated melons with finely-textured, bright pink-red flesh and sweet
- Sugar Baby and icebox type with 3 to 3.5 Kg fruit weight, blackish green skin colour, bright pink-red flesh
- Very good transportability, TSS 13°Brix and above, fine texture
- Resistant to watermelon bud necrosis virus + Fusarium wilt

Muskmelon

- Vine of monoecious sex form with low ratio of staminate : pistillate flower (sex ratio)
- High number of hermaphrodite flower in andro-monoecious sex form
- Consistent early and total yield

- Netted cantaloupe, honeydew, sutured melon and Galia types
- Attractive round fruit shape, thick flesh with attractive orange or green colour
- Better adaptability and shipping qualities
- Resistant to *Fusarium* wilt + cucumber mosaic virus + powdery mildew + downy mildew

Pumpkin

- Plant with reduced vine length
- Earliness in fruiting and maturity
- Vine with low ratio of staminate: pistillate flower (sex ratio)
- Globular or flattened fruit shape with thick pericarp
- Fruits with high flesh recovery
- High sugar and β-carotene content

Cucumber

- Vine with low ratio of staminate: pistillate flower (sex ratio)
- First pistillate flower to appear at lower node
- Early maturing and prolific yielder
- Nice green or dark green Poinsett type fruit
- Uniform, long cylindrical fruit shape without crook-neck
- Resistant to cucumber mosaic virus + powdery mildew + downy mildew

Bottle gourd

- Vine with low ratio of staminate: pistillate flower (sex ratio)
- Earliness: first pistillate flower to appear at lower node
- High early and total yield
- Attractive green fruit of round/cylindrical shape with sparse hairs persisting on skin
- Non-fibrous flesh at edible stage
- Resistant to powdery mildew + downy mildew

Bitter gourd

- Vine with low ratio of staminate: pistillate flower (sex ratio)
- Earliness: first pistillate flower to appear at lower node
- High total and early yield
- Spindle shaped, green fruit with medium prickle on fruit surface
- Extended harvest span
- Resistant to Mosaic + downy mildew

Ridge gourd

- Vine with low ratio of staminate: pistillate flower (sex ratio)
- Earliness: first pistillate flower to appear at lower node
- High total and early yield
- Club shaped dark green fruit with extended harvest span
- Immature seeds for longer period during edible green stage of the fruit
- Resistant to mosaic + powdery mildew + downy mildew

Ash gourd

- Vine with low ratio of staminate : pistillate flower (sex ratio)
- Earliness: first pistillate flower to appear at lower node
- High total and early yield
- Uniform, cylindrical, light green fruit with soft bristles
- Resistant to mosaic + powdery mildew + downy mildew

Okra

- Medium tall and non-branching growth habit
- Short-statured and highly branching growth habit
- High early and total fruit yield
- Attractive light green to green fruit, tender, thin, medium long, smooth, 4-5 ridged, free from conspicuous hairs,
- Resistant to bhendi yellow vein mosaic and okra enation leaf curl virus disease

Garden pea

- Early or medium maturity
- High pod yield
- Long and attractive green pods with shelling percentage of around 45.
- Sweet seeds containing around 5% total sugar
- Resistant to powdery mildew disease

French bean

- High pod yield
- Bush/pole plant type
- High number of primary branches/plant
- Early flowering and harvesting of pod
- High number of green pods/plant
- Non-stringy, long pods, flat or round in shape

- Wide adaptability in tropical and sub-tropical condition
- Resistant to bean common mosaic and bacterial leaf spot disease

Radish

- Early rooting habit
- Higher and early root yield
- Uniformity in shape/size/color/maturity
- White or pink long / stump and non-pithy roots with non-branching habit
- Suitability to high temperature and rainfall conditions
- Pungency of roots as per consumers' preference
- Longer field stay and delayed bolting
- Resistant to Alternaria blight, Fusarium wilt, aphids and beetles

Carrot

- Early rooting habit
- Higher and early root yield
- Uniformity in shape/size/color/maturity
- Scarlet / orange coloured roots with uniformity in shape and size
- Smooth root surface and self-coloured core
- Non-branching habit and free from cracking
- High sugar, β-carotene, anthocyanin and dry matter content in the roots and free from harsh flavour
- Wide adaptation particularly to temperature condition

Potato (for cooking purpose)

- High yield
- Long-oval or round tuber shape
- Skin colour white, yellow or red
- Eye depth shallow or medium
- Flesh colour white or yellow
- Flesh texture waxy or mealy
- Dry matter content 18-20%
- Glycoalkaloid content less than 15 mg/ 100g fresh
- Good keeping quality
- Resistant to virus and late blight caused by *Phytophthora infestans*

Potato for processing purpose (French fries or Chips)

- High yield
- Long oval (> 7.5 cm) or round tuber shape (6.25 – 8.25 cm)
- Skin colour white or yellow
- Eye depth shallow
- Flesh colour white or yellow
- Flesh texture mealy
- Dry matter content > 20%
- Reducing sugar content 100-200 mg/100 g fresh
- Glycoalkaloid content less than 15 mg/ 100g fresh
- Good keeping quality
- Resistant to virus and late blight caused by *Phytophthora infestans*

Sweet potato

- High yield
- Early bulking character
- Good appearance and eating quality (sweet and fine texture)
- High β-carotene content in orange-fleshed variety
- Resistance to diseases and insect pest, particularly sweet potato weevil (*Cylas formicarius)*

Conclusion

Ideotype breeding is an inter-disciplinary breeding approach for increasing yield and other targeted characters by modifying a variety of physiological and morphological characteristics taking the advantage of morphological and physiological diversity. Utilization of physiological variation that is genetically regulated is an effective way to get through yield constraints. Ideotype breeding offers a time-sensitive solution for a number of issues, including disease, pest, maturity duration, lodging resistance, yield and quality by creating a single genotype from the combination of advantageous genes for certain qualities from several sources. It is a successful strategy for creating cultivars for specific situation or environment. However, there are number of difficulties, as mentioned below, that have hampered the greater acceptance and use of the concept of breeding model plants or ideotypes.

1. It is challenging to combine multiple desired morphological, physiological and resistance features from various sources into a single genotype. Due to the close relationship between desirable and undesirable characteristics, it is very difficult to combine different characters. Development of ideotype breeding is hindered by the presence of undesirable linkages.

2. Strong relationships among individual traits of importance may hinder progress in cultivar development. There are four different factors or types of trait inter-relationships which could slow breeding process:
   - Requirement of size symmetry among plant parts (fruit size is inversely correlated with fruit number)
   - Compensation among plant parts (increased fruit size leads to decreased fruit number)
   - Pleiotropy, the phenomenon where one gene affects more than one character. For example, seed weight affects many root characters of French bean (Singh *et al.*, 2019)
   - Genetic background effects (for example, linkage drag where desirable genes occur only in inferior plants).
3. Ideotype breeding is a slow method of cultivar development because it requires more time than traditional breeding, which only improves yield and one or two other characters.
4. Ideotype is a dynamic model that alters in response to new information, demands, National policies, etc. In order to satisfy the change and growing demands of commercial items, new ideotype must be developed.

Future prospects

The concept of ideotype proposed by Donald (1968) suggests that there is an ideal phenotype of a crop that can produce higher yield and better quality within a given environment. This concept revolutionized the breeding process by providing targeted traits that are known to underlie key physiological processes. Since its proposal, the ideotype concept has been applied to a variety of different crop species, expanded to specific environments and alternative cropping systems and applied to varying end uses. Ideotype should not be considered to be static, and will depend on a combination of biotic, abiotic and cultural variables. However, limitations in these approaches exist, restricting their predictive capacity. Regardless of potential limitations, there is still a role for the ideotype in crop breeding, but that this concept needs to be expanded to combine knowledge of the genetic and environmental interactions governing plant physiology.

Ideotype breeding is not a substitute for traditional or conventional breeding rather it is supplementary to the former. In addition to conventional breeding methods, different biotechnological interventions, particularly *in vitro* culture, marker assisted selection and transgenic technology must be used to create new plant ideotype. Crop ideotype as the frame of breeding objectives have been developed for many vegetable crops. However, the ideotype should now be developed for fresh market, product quality and processing purpose as well. There is ample scope for developing ideal plants or model plants for important vegetable crops *viz*., tomato, cherry tomato, brinjal, sweet pepper, chilli, cabbage, cauliflower, broccoli, watermelon, muskmelon, cucumber, bottle gourd, bitter gourd, ridge gourd, garden pea, French bean, radish, carrot, okra, potato, sweet potato, etc

References

Andrivon, D., Giorgetti, C. and Baranger, A. 2013. Defining and designing plant architectural ideotypes to controlepidemics, *Eur. J. Plant Pathol.*, **135**: 611–617.

Arraiano, L., Brading, P., Dedryver, F. and Brown J. 2006. Resistance of wheat to septoria tritici blotch (*Mycosphaerella graminicola*) and associations with plant ideotype and the 1BL–1RS translocation. *Plant Pathol.*, **55**: 54–61.

Barbedo, J.G.A. 2019. Plant disease identification from individual lesions and spots using deep learning. *Biosyst. Engin.*, **180**: 96-107

Cariboni, J., Gatelli, D., Liska, R. and Saltelli, A. 2007. The role of sensitivity analysis in ecological modelling. *Ecol. Model.*, **203**: 167–182.

Challinor, A., Martre, P., Asseng, A., Ewert, F., Thornton, P.K. 2014. Making the most of climate impacts ensembles. *Nat. Clim. Change*, **4**: 77–80.

Chapman, S., Cooper, M., Podlich, D. and Hammer, G. 2003. Evaluating plant breeding strategies by simulating gene action and dryland environment effects. *Agron. J.*, **95**: 99–113.

Chenu, K., Chapman, S., Tardieu, F., McLean, G., Welcker, C., Hammer, G. 2009. Simulating the yield impacts of organ-level quantitative trait loci associated with drought response in maize: a "gene-to-phenotype" modeling approach. *Genetics*, **183**: 1507–1523.

Chew, Y.H., Wilczek, A.M., Williams, M., Welch, S.M., Schmitt, J., Halliday, K.J. 2012. An augmented *Arabidopsis* phenology model reveals seasonal temperature control of flowering time. *New Phytol.*,**194**: 654–665.

Crossa, J., Pérez-Rodríguez, P. Cuevas, J., Montesinos-López, O., Jarquín, D., de los Campos ,G., Burgueño, J., González-Camacho, J.M., Pérez-Elizalde, S., Beyene, Y., Dreisigacker, S., Singh, R., Zhang, X., Gowda, M., Roorkiwal, M., Rutkoski, J. and Varshney, R.K. 2007. Genomic Selection in Plant Breeding: Methods, Models, and Perspectives. *Trends in Plant Sci.*, **22**: 961-975

Desanlis, M., Aubertot, J.N., Mestries, E. and Debaeke, P. 2013. Analysis of the influence of a sunflower canopy on *Phomopsis helianthi* epidemics as a function of cropping practices. *Field Crops Res.*, **149**: 63–75.

Donald, C. M. 1968. The breeding of crop ideotypes. *Euphytica*, 17: 385–403.

Duvick, D.N., Smith, J.S.C. and Cooper, M. 2004. Long-term selection in a commercial hybrid maize breeding program. *Plant Breed. Rev.*, 24: 109–151.

Gerard, G.S., Crespo-Herrera, L.A., Crossa,, J., Mondal, S., Velu, G., Juliana, P., Huerta-Espino,J., Vargas, M., Rhandawa, M.S., Bhavani, S., Braun, H. and Singh,, R.P. 2020. Grain yield genetic gains and changes in physiological related traits for CIMMYT's High Rainfall Wheat Screening Nursery tested across international environments. *Field Crops Res.*, 249: 107742.doi: 10.1016/j.fcr.2020.107742.

Gu, J. 2022. Ecophysiological crop modelling combined with genetic analysis is a powerful tool for Ideotype design. *Agronomy*, *12*, 215; https://doi.org/10.3390/agronomy12010215.

Hamblin, J. 1993. The ideotype concept: useful or outdated? In: *International crop science* I. (Eds. Buxton, D.R., Shibles, R. and Forsberg, A.), CSSA, Madison, pp. 589-597.

Hammer, G.L., Chapman, S., van Oosterom, E., Podlich, D.W. 2005. Trait physiology and crop modelling as a framework to link phenotypic complexity to underlying genetic systems. *Aust. J. Agric. Res.*, **56**: 947–960.

Hodson D.P. and White J.W. 2007. Use of spatial analyses for global characterization of wheat-based production systems. *J. Agric. Sci.*, **145**:115–125.

Jeuffroy, M.H., Casadebaig, P., Debaeke, P., Loyce, C., Meynard, J.M., 2013. Agronomic model uses to predict cultivar performance in various environments and cropping systems. A review. *Agron. Sustain. Dev.*, doi: 10.1007/s13593-013-0170-9.

Karr, J.R., Sanghvi, J.C. and Macklin, D.N. 2012. A whole-cell computational model predicts phenotype from genotype. *Cell*, **150**: 389–401.

Letort, V., Mahe, P., Cournède, P.H., de Reffye, P. and Brigitte Courtois, B. 2008. Quantitative Genetics and Functional–Structural Plant Growth Models: Simulation of Quantitative Trait Loci Detection for Model Parameters and Application to Potential Yield Optimization. *Ann Bot.*, **101**: 1243–1254.

Lyu, J., Zhang, S., Dong, Y., He, W., Zhang, J. and Deng, X. 2013. Analysis of elite variety tag SNPs reveals an important allele in upland rice. *Nat. Commun.*, **4**: 2138. doi: 10.1038/ncomms3138

Peng, S., Khush, G.S., Virk, P., Tang, Q. and Zou, Y. 2008. Progress in ideotype breeding to increase rice yield potential. *Field Crop Res.*, **108**: 32-38.

Rao Mogili, U.M. and Deepak, B.B.V.L. 2018. Review on Application of Drone Systems in Precision Agriculture. *Procedia Comp. Sci.*, **133**: 502-509

Reynolds, J.F. and Acock, B. 1985. Predicting the response of plants to increasing carbon dioxide: A critique of plant growth models. *Ecol. Model.*, **29**: 107–129.

Reynolds, M., Foulkes, M., Slafer, G., Berry, P., Parry, M. and Snape, J. 2009. Raising yield potential in wheat. *J. Exp. Bot.*, **60**: 1899–1918.

Rezzouk, F., Gracia-Romero, A., Kefauver, S., Nieto-Taladriz, M., Serret, M. and Araus, J. 2022. Durum wheat ideotypes in Mediterranean environments differing in water and temperature conditions. *Agric. Water Manag.*, **259**: 107257. doi: 10.1016 / j.agwat.2021.107257

Ribaut, J.M. and Ragot, M. 2007.Marker-assisted selection to improve drought adaptation in maize: the backcross approach, perspectives, limitations, and alternatives. *J.Expt. Bot.*, **58**: 351–360.

Richards, R., Cavanagh, C. and Riffkin, P. 2019. Selection for erect canopy architecture can increase yield and biomass of spring wheat. *Field Crops Res.*, **244**: 107649. doi: 10.1016/j.fcr.2019.107649

Sanghvi, J.C., Regot, S. and Carrasco, S. 2013. Accelerated discovery *via* a whole-cell model. *Nat. Meth.*, **10**: 1192–1195.

Semenov, M. and Stratonovitch, P. 2013. Designing high-yielding wheat ideotypes for a changing climate. *Food Energy Secur.*, **2**: 185–196.

Semenov, M., Stratonovitch, P., Alghabari, F. and Gooding, M. 2014. Adapting wheat in Europe for climate change. *J. Cereal Sci.*, **59**: 245–256.

Sharma, D., Arvind, S., Motiramani, N. K., Verulkar, S. B., Tuteja, S. S., Tiwari, P. K. and Verma, R. 2012. Breeding super hybrid rice for Chhattisgarh State in India, In: *Proc. 6th Intern. Hybrid Rice Symp.*, pp. 73-74.

Singh, J., Gezan, S.A. and and Vallejos, C.E. 2019. Developmental Pleiotropy Shaped the Roots of the Domesticated Common Bean (*Phaseolus vulgaris*). *Plant Physiol.*, **180**: 1467–1479.

Tardieu, F. 2012. Any trait or trait-related allele can confer drought tolerance: just design the right drought scenario. *J. Exp. Bot.*, **63**: 25–31.

Trethowan, R.M. 2014. Defining a genetic ideotype for crop improvement. *Methods Mol. Biol.*, **1145**:1-20. doi: 10.1007/978-1-4939-0446-4_1.

Welch, S.M., Dong, Z.S., Roe, J.L., Das, S. 2005. Flowering time control: gene network modelling and the link to quantitative genetics. *Aust. J. Agric. Res.*, **56**: 919–936.

White, J.W. and Hoogenboom, G. 1996. Simulating effects of genes for physiological traits in a process-oriented crop model. *Agron. J.*, **88**: 416–442.

Zhou L, Wang J, Yi Q, Wang Y, Zhu Y, Zhang Z. 2007. Quantitative trait loci for seedling vigor in rice under field conditions. *Field Crops Res.,* **100**: 294–301.

15

Procedures for Release and Notification of Varieties

Introduction

Improved varieties, hybrids and quality seeds are considered as the basic and key inputs for the development of agriculture and horticulture. High quality seed production has been the major concern in the Indian sub-continent till 1960s. Before that, India was mostly dependent on USA for food grain. After 1960, Government of India established State Agricultural Universities in different states following "Land grant model" of USA, different ICAR Research institutes and All India Coordinated Crop Research Projects in a systematic manner to give the boost to develop large number of improved varieties with assured seed quality in all major agricultural and horticultural crops. The ultimate intention was to introduce the newly developed high yielding varieties for broad spectrum cultivation in the area of their adoption.

Importance of variety and system of its release

There are several ways and means to increase the crop production and productivity, however using genetically pure and high-quality seed is first and prime objective in agriculture/ horticulture. Govt. of India acknowledged seed as an essential commodity under essential commodities Act, 1955. On October 1964, Variety release system (VRS) came into existence with the formation of Central Variety Release Committee (CVRC) at National level and State Variety Release Committee (SVRC) at each state level. A Central seed committee (CSC) was established under the Ministry of Agriculture, Cooperation and Farmers Welfare under the provision of the Seed Act, 1966.

Therefore, the variety which will be used by the farmers must have undergone several evaluations in order to ensure its stable yield potential, tolerance to biotic and abiotic stresses and these criteria are being fulfilled by a legal varietal release system. The main objective of the varietal release system in India is to introduce newly developed, high yielding varieties to the farmers for broad-spectrum cultivation in the area of their adaptation and only those varieties will be notified which are superior to existing one. It provides choice to the farmers to cultivate a specific variety, based on their need for crop diversification. In India, the systematic framework has helped farmers to get high quality seeds from market and production has increased many folds since the inception of AICRPs. Notification is mandatory to release a variety, though the release process itself does not have legal cover.

Functions of CVRC to ensure the quality of seeds on sale and notification of the kinds / varieties were taken over by the CSC in 1969. To perform the function at Central level to release / notification, provisional notification / de-notification of varieties, CSC constituted a Central sub-committee on crop standards, notification and release of varieties for agricultural and horticultural crops, while to perform similar functions at state level, State Seed Sub-Committee (SSSC) was constituted.

Development of plant genetic material

Entries (pure lines/open pollinated varieties/composites/synthetics/hybrids etc.) are developed by the concerned plant breeders/agencies through plant breeding which is the technique to improve the genetic material of plants for the benefit (food and nutritional security, industrial use, etc.) of human kind.Different conventional (Introduction, selection, hybridization followed by selection, mutation breeding and polyploid breeding) and advanced breeding (tissue culture-based techniques like, somaclonal variation, anther and pollen culture, marker assisted breeding, transgenic or genome editing techniques) methods are being used by the different agencies (ICAR or non-ICAR National research institutes, SAUs, private/ corporate or National or multinational companies, etc.) to generate elite material for high yield potential, nutritional quality, resistance to biotic and abiotic stresses and other associated traits. The elite genetic materials are tested by the concerned plant breeders at their research institutes / SAUs/ research station for three to four years in replicated manner for stability. The selected superior variety/varieties or hybrids are given entry in the All India Coordinated Research projects (AICRPs) trials for further testing in multi-environments across the country.

Variety

- Botanically a variety is a sub group of a species. According to seed act (1966, sub section 16 of section 2) it is a sub division of a kind identified by its growth, yield, plant, fruit, seed or other characters.
- According to seed technology, variety after its release and notification is considered as a group of plants uniform in their morphological, physiological, biochemical and other characters without any variation from generation to generation and can be differentiated from other groups of plants of the same species by some distinguishing characters.
- In India, new and improved varieties of different crops are developed by crop research institutes of Indian Council of Agricultural Research, State Agricultural Universities and few Corporate / Private seed companies.

All India Coordinated Research Project (AICRP) system of varietal testing

The AICRPs have been developed for all the major crops including vegetable crops and is implemented in both SAUs (through memorandum of understanding with respective state Government) and ICAR institutes. The AICRPs under the umbrella of ICAR has great role in the development of improved crop varieties as well as generation of production and protection technologies that directly benefit farmers for their economic upliftment. First All India Coordinated Research Project (AICRP) was started way back in 1957 by ICAR on maize for systematic testing of entries followed by onward release of high yielding new maize varieties. In general, the three-tier system of multi-location evaluation is used for three years except perennial fodder crops (which requires four years testing: one year for crop establishment and three years for evaluation) in India. Multilocation trials are conducted by the Project Coordinator (PC)/Project Director (PD) of AICRPs with the help of concerned principle investigators (Tandon *et al.*, 2015). All AICRP trials are well organized, systematic and conducted through a uniform testing procedure across the centres as per crop standard. It is a powerful system to screen large number of entries and recommend well-tested, superior, and adapted new varietie to the end users. The new entries must have the following requirements.

- The proposed entry must have undergone screening against major insect pests and diseases, crop based quality parameters and tolerance to key abiotic stresses. Before the coordinated trial, data on yield, trait stability and other related agronomic/ horticultural characters of the entry must be available to the PC/PD in support of the relevance of the entry (Singh, 1999).
- The entry must have a high degree of genotypic stability, phenotypic uniformity, germination percentage and the seed lot must have physical purity as per the minimum seed certification standards.
- The entry must have few distinct uniform and stable (DUS) characters which help to identification of the variety during legal infringement (Chopra, 1989).
- All the information related to the development of entry *i.e.* parentage or pedigree and method of breeding should be available to the PC/PD by the concerned plant breeder/agency. If the performance of the entries is same in the coordinated trials, then preference will be given to the variety which has been developed by using of diverse parents in breeding programme.
- Corporate /Private Companies can give entry of their variety/ hybrid into the coordinated trial system similar to public agencies however, prescribed fees for their entries as per guidelines of the Government of India need to be paid.
- All the material (product of selection, hybridization followed by selection, synthetics, composites and hybrids, etc.) shall be subjected to the same system.

Three tier system of All India Coordinated Research Project (AICRP)

The AICRP centers for various crops are located at ICAR institutes or State Agricultural Universities (SAUs) or other volunteer centres recommended in the AICRP workshop based on the covered crop area, adaptability, agro-climatic condition, etc. The entries are tested under AICRP through the following steps (Tandon *et al.*, 2015).

- **Initial Evaluation Trial (IET):** All the superior entries based on respective station trials would be introduced into the IET for testing in one year. These entries would be used for multi-location trials along with the checks. In general, three checks (National, zonal and local checks) are used for efficient evaluation of entries across the centres. The National check (a crop variety which was earlier released nationally) is generally used for a long period but zonal (a crop variety which was earlier released for a specific zone) and local checks (high yielding local variety) can be replaced based on the requirement. The IET trials are conducted in such a manner that minimum difference of yield (5–10%) and other ancillary characters can be recorded. Experimental layout (experimental design, number of replications and treatments) is the prime responsibility of the PC/PD through concerned principle investigator. The cultural practices *i.e.* seed rate, date of sowing, row to row and plant to plant spacing, weed, fertilizer and water management etc. shall be strictly followed by the IET centres as per guideline of PC/PD. The plot size of IET is smaller than advanced trials. An IET includes the maximum number of locations across the country to evaluate varietal adaptation and performance. Number of testing locations varies with the crop across the zones. A team of scientists (plant breeder, agronomist, pathologist, entomologists, etc.) will monitor all the trials as per the recommendation of the Project Coordinator. Member of the monitoring team submits their report to the PC based on their observation during trial monitoring. Each crop has different objective and requires different technical requirement for their evaluation. The IET centres will generate information related to days to flowering, physiological maturity, plant height, lodging, recovery percentage, disease and pest tolerance, yield, and different nutritional quality traits. Technical programme is formulated during workshop or annual/ biannual group meet. Project Coordinator will specify the characters on which data shall be recorded. Entries which are superior over the best check in terms of yield and other related traits will be promoted into the Advanced Varietal Trials 1 (AVT-1).
- **Advanced Varietal Trial-I (AVT-I):** Based on the superiority (5–10%) over the best performing check, superior entries will enter into the AVT-I from IET. Number of entries in the AVT-I will be less than IET. The plot size is large in AVT-I as compared to IET, therefore, data generated on yield and other ancillary traits will be more realistic, accurate with minimal chances of error. The number of testing locations should be more as compared to IET in a given zone to generate more realistic data on yield and other economically important traits, varietal adaptation, biotic and abiotic tolerance, quality parameters, etc. National, zonal and local checks (which were used in IET) shall be used for critical analysis along with the entries. During AVT-I, additional data on disease and or insect pest tolerance under artificial epiphytotic condition must be generated by the multi-disciplinary experts (breeder, pathologist, entomologist). Same as IET, monitoring team will be constituted by the PC/PD at different growth stages of the crop and observed data shall be submitted to the concerned PC/PD. Based

on the performance of entry over the best performing check-in the respective zone, the superior entries would enter into the AVT-II.

- **Advanced Varietal Trial-II (AVT-II):** The requirements for laying out of the trial under AVT-II are same as observed for AVT-I. However, few additional data are generated at AVT-II stage *viz.,* response of the entries to different dates of sowing, seed rate, spacing between plant to plant and row to row (population density), response of the entries to different level of the fertilizer and irrigation (as recorded by the sponsored agronomists); response of the diseases and pests (as recorded by the sponsored pathologist and entomologist); important quality parameters (as recorded by the sponsored biochemists). The seed technology centre will develop descriptors to help performing the seed certification process. All the processed and analysed data on yield and other related traits, across the locations/centres (cooperating and volunteer) shall be submitted to the Project Coordinator. On the basis of these data, annual reports are prepared for each crop. All the data of superior entries are comprehensively discussed in the annual workshop/National group meetings by the PC/Project Director. After completion of the AVT-II, the concerned breeders are informed to submit varietal proposal based on the performance of their entries during three years of evaluation.

Monitoring of the trials

All the IET and AVT trials are monitored by a team of scientists deputed by the Project Director/Coordinator. The monitoring teams have the following minimum composition.

- PD/PC/PI/Zonal Coordinator - Team Leader
- Concerned Breeder - Member
- Agronomist - Member
- Pathologist/Entomologist – Member
- Scientist of any other specified discipline - Member

Procedure for varietal identification

Based on three-years performance, the best performing test entries shall be identified in the annual/ biennial group meeting at the institute/university which has been fixed earlier. Superior variety/ varieties are identified by a committee called "Variety Identification Committee" constituted in advance of annual workshop/group meeting with the approval of the Deputy Director General (Crop Sciences/ Horticulture) of Indian Council of Agricultural Research (ICAR).

The breeder should submit a proposal for release as a new variety for approval by the Central sub-committee on Crop standards, Notification and Release of varieties. The proposal for release of new varieties is submitted in a prescribed proforma to variety release committee *viz,* State variety release committee (SVRC) and central variety release committee (CVRC). The variety release proforma contains all the information about the culture *viz.*, parentage, parent's morphology, cultures morphology, key characters of the culture for identification, agronomic practices, pest and disease resistance, quality characters and yield trial results.

These superior entries are presented to the Central Sub-Committee on Crop Standards Release and Notification. The Zonal coordinators and Principal investigators attend the annual group meeting to provide wider aspects of information on the varieties. Principal investigators (PIs) of different disciplines can assist in the process of discussion but they do not have the right to cast vote. Only committee members have the right to cast vote. The Variety Identification Committee (VIC) provides detailed information on the recommended entries to the Central Sub-Committee on crop standards, Notification and release which has the sole right to release and notify the best-performing entry into National wise or zonal wise based on the recommendations of the VIC.

Composition of Variety Identification Committee (VIC)

1. DDG (Crop Science/ Horticulture)/ his or her nominee	Chairman
2. Project Coordinator/Project Director of AICRP	Member Secretary
3. Director of Research of institute/SAUs of that region where the meeting is held	Member
4. Agricultural Commissioner (Department of Agriculture)	Member
5. One nominee of Seed organization (NSC, SSC)	Member
6. One representative of private seed agencies	Member
7. One representative of crop-based industries	Member
8. Project coordinator (seed technology)	Member
9. Two eminent scientists of that institute	Member

Source: Tandon *et al.* (2015)

Eligibility criteria for identification of the variety

- The candidate variety must have a minimum of three years of yield and other ancillary trait data from the multi-location coordinated trials.
- At least two-year data on disease and pest reaction at the hot spot or artificial epiphytotic condition.
- The candidate variety must have at least one-year data on agronomic performance like seed rate, dates of sowing, planting density, irrigation and fertilizer application.
- The concerned breeder must have at least a minimum requirement of nucleus seed so that breeder seed can be generated easily.
- In case of non-availability of seed, in specified quantity, the identification process is postponed to later stage, at the most by one year.

Central sub-committee on crop standards, notification and release of variety

Central sub-committee on crop standards, notification and release of varieties appointed by the Central seed committee under Section 3 of the seed act, 1966 comprised one chairman and 17 members.

Organizational setup of Central sub-committee on seed standards, release and notification of variety

1. Deputy Director General (Crop Sciences/Horticulture), ICAR	Chairman
2. Deputy Commissioner (QC) DAC & FW, GOI	Member Secretary
3. Directors of State seed certification agencies, or their representatives	Member
4. Directors of Departments of Agriculture of all states, or their representatives	Member
5. Project coordinators/Directors of AICCIPs	Member
6. Agricultural Commissioner, GOI	Member
7. Representatives of the seed industry, NSC, State seed corporations, private seed companies	Member
8. Representatives of ICAR, ICAR institutes, NGOs	Member
9. Progressive farmers	Member

Central Sub-Committee releases varieties as per the benefit of the stakeholders and need of the regional, zonal or National importance, and the State seed sub-committee releases varieties beneficial for particular state. Notification of variety is compulsory for regulating the seed quality under the provision of Seed Act, 1966. Notification usually authorizes certified seed production throughout the country, by private or public seed multiplication organizations. Once the Central sub-committee accepts the proposal, the varieties/hybrids will be released for the concerned agro-climatic zone/s (may cover one or more number of states or nationally). Simultaneously, it must be notified for seed certification purpose in the country. During the release, the concerned breeder must have a minimum amount of nucleus seed. Later on, seed multiplication is the responsibility of various seed agencies (NSC, SSC, private seed companies and progressive farmers, etc.).

Table 1: Difference between release and notification of a variety

Release of a variety	Notification of a variety
Release of a variety refers to recommendation of that variety to the farmers for its adoption.	Notification usually authorizes certified seed production throughout the country by private or public seed multiplication organizations.
Release of a variety is aimed to ensure smooth implementation of the Seed Act for large scale cultivation	After release, the variety is notified. Normally, only released varieties are notified but all released varieties are not notified.

Table 2: Critical differences between released and notified varieties

Released variety	Notified variety
It is not a statutory function under the Seed Act, 1966	It is a statutory function and variety will be registered under Section 5 of seed act 1966
It cannot be used for seed certification	Only notified varieties come under seed certification
There is no guarantee for seed quality	It guarantees assured seed quality
Seed law enforcement agencies (seed inspector etc.) cannot draw and test seed samples	Seed law enforcement agencies have the right to draw and test seed samples
These are not the assets of Govt. of India	Notified varieties are assets of Govt. of India
Its main purpose is to make available the information of the cultivar to the public and its area of adoption	The main purpose is regulation of seed quality
It is difficult to trace out the genesis	Notification of the varieties will help to trace out its genesis

Central Seed Committee (CSC)

It is a legal body constituted by the Department of Agriculture, Cooperation and Farmers' Welfare (DAC&FW), Ministry of Agriculture and Farmers' Welfare (MoA & FW), Government of India to advise central and state Government on matters related to the implementation of Seed act, 1966 and other related functions. The core committee includes one chairman and eight members to be nominated by the central Government and one person to be nominated by the Governments of each State. State seed committee (SSC) has a similar role at the state level. The CSC and SSC are empowered to release varieties but only CSC can notify those varieties (Tonapi et al, 2015).

Organizational Setup of Central Seed Committee (CSC)

1. Secretary, DAC&FW, MoA& FW, GOI	Chairman
2. Additional Secretary (In-charge Seeds), MoA & FW, GOI	Member
3. Agricultural Commissioner, MoA & FW, GOI	Member
4. Deputy Director General (Crop Sciences/Horticulture), ICAR	Member
5. Joint Secretary (In-charge Seeds), MoA & FW, GOI	Member
6. Progressive farmers/ seed growers (4) nominated by the Central Government	Member
7. One representative from each State Govt.	Member
8. Director of National Seeds Project, MoA & FW, GOI	Member Secretary

Empowerment of Central Seed Committee

- The CSC has the authority to release varieties (pure lines/hybrids/composites/ synthetics) developed by central research institutes (ICAR/non-ICAR), AICRPs, SAUs, private or corporate sector and other organization as per the scientific data authenticity for zonal basis (which may include more than one state) or at National level.

- The CSC has authority to approve proposals received from the State variety release committees/State seed sub-committees for varieties developed by the State Agricultural University /State research institutes but is considered suitable for areas outside the state (based on their performance).
- The CSC can delimit the regions or tracts for the cultivation of varieties approved for release.
- The CSC can advise the ICAR regarding the manner in which the National register of approved varieties may be maintained, and to suggest the standard description of crop varieties.
- The CSC can notify kinds/varieties for the purpose of the Seeds Act and the areas of their notification.
- The CSC specifies minimum limits of germination percentage and purity for the notified kinds/varieties of seeds as per minimum seed certification standards.
- The CSC specifies the "mark" or "label" in respect of notified kinds/varieties.

The State seed sub-committees are constituted by Central seed committee and are authorized to set up a State seed laboratory, State seed certification agency (SSCA) and an Appellate authority to appoint seed inspectors and seed analysts.

Table 3: Critical differences between central sub-committee and state sub-committee

Central Seed Committee	State Seed Sub-Committees
Authorized by Central seed committee	Authorized by State seed committee
Releases varieties for regional /zonal / National level	Releases varieties only for concerned state/ regions within the state
Statutory body for varietal notification	It cannot notify varieties
Varietal notification is followed by seed certification	Certified seed cannot be produced without notification
Members in the committee are appointed by the Central Government	Members in the committee are appointed by the State Government

Empowerment of state seed sub-committee

There are some rights which have been provided by the Central seed committee for proper functioning of seed chain in respective state in India. These empowerments are-

- The State seed sub-committee will advise the state Government on all matters related to the execution of the Seeds Act, 1966.
- Reviewing the implementation of the Seeds Act in the state and send periodic reports to the state Government and the Central Seed Committee.
- Inspect, analyse and give report on the State Seed Testing Laboratory.
- Advise on educational and promotional measures for proper enforcement and understanding of the Seed Act.
- Planning for different crop varieties to be grown in different regions of the state, and to review the assessment of seed requirements.

- Considering the release of new varieties for the state and recommend their notification to the Central seed committee.
- Monitoring the performance of newly released varieties in the state.

Being agriculture as a state subject in India, centrally released varieties are not directly accepted by all the states for which they have been released. Each state has its own regulatory system which they have to follow for varietal release in the state. They have to pass through all the steps of the concerned state release procedure before they approve for cultivation in the state *viz*., state wise multi-location trials for three years and adaptive trials based on the requirement. The notification requires that the variety must have been tested at least for one year in the AICRP trials and recommended for release in the state by the AICRP varietal identification committee.

Necessity of notification

Since only notified varieties will be under the purview of Seed law enforcement, hence it is necessary to bring the seed of a particular crop variety under notification system. The seed inspector can only draw a sample from the notified variety for analysis and ensure the seed quality (Spielman, and Kennedy, 2016; Kaul *et al.,* 2018). A released variety cannot come under seed chain without notification by the Gazette of India. Therefore, these issues will make the notification as necessary requirement for other things to act on it. The notification is made by the Central Government on the recommendation of the Central seed committee. Thus, notification is prerequisite for production of certified seed which ensures the supply of high quality of seeds to the farmers. After notification, variety becomes the asset of Government of India. The breeder seed can only be produced after the notification of variety and notified varieties enter into seed chain. Notification also helps in the genesis of original variety based on its pedigree and also regulates any kind of infringement in the later stages of varietal promotion.

Denotification of varieties

Released varieties can be denotified if they are not performing well in the area of their adoption or have been in cultivation for more than 15 years or are not much in demand. Denotification can be done based on the recommendation of Central seed committee by the Government of India.

References

Chopra, V.L., 2018. *Plant Breeding: Theory and Practice*, 2nd Revised Edition: NIPA. pp. 110.

Kaul, J., Kumar, R., Nara, U. and Jain, K. 2018. An Overview of Registration of Maize Genetic Resources in India. *Intern. J. Curr. Microbiol.Appl. Sci*., **7**: 933-950

Singh, B.D. 1999. *Plant Breeding: Principles and Methods*, Kalyani publishers. Ludhiana, pp.391-400.

Spielman, D.J. and Kennedy, A. 2016. Towards better metrics and policymaking for seed system development: Insights from Asia's seed industry. *Agril. Syst*., **147**: 111-122.

Tandon, J.P., Sharma, S.P., Sandhu, J.S., Yadava, D.K., Prabhu K.V., and Yadav, O.P. 2015. *Guidelines for Testing Crop Varieties under the All-India Coordinated Crop Improvement Projects*, Indian Council of Agricultural Research, New Delhi, India.

Tonapi, V.A., Bhat, B.V., Kannababu, N., Elangovan, M., Umakanth, A.V., Kulakarni, R., Tonapi, K.V., Rao, K.V.R. and Rao T.G.N. 2015. *Indian regulatory system of plant variety testing, identification, release, and notification* (Part-III). ICAR- Indian Institute of Millet Research, Rajendra Nagar, Hyderabad, Telangana, India, pp. 279-310.

16

Intellectual Property Rights (IPR)

Intellectual Property Rights (IPR)

There are two types of properties *viz*., i) Tangible, physical or material property and ii) Intangible or intellectual property. Intellectual property (IP) pertains to any original creation of the human intellect such as artistic, literary, technical or scientific creation. The tangible or physical properties are protected by the laws of land. In contrast, intellectual property is an idea, a formula, an invention, a design, a variety, etc. which may give rise to a useful product. For obvious reasons the inventors would like to be rewarded and their inventions be protected from imitation. Such a right of an inventor to obtain economic benefit and protection of the invention from imitation is known as Intellectual Property Right (IPR). Thus, Intellectual property rights (IPR) refers to the legal rights given to the inventor or creator to protect his invention or creation for a certain period of time (Singh, 2008).

Why Intellectual Property needs to be protected?

Intellectual property rights (IPR) encourage innovation and protect the rights of the inventors in legalized form. It is an inevitable tool in today's globalized economy. The invention of such things would require intellectual capacity, innovativeness and creative ability of the concerned person(s). These legal rights confer an exclusive right to the inventor/creator or his assignee to fully utilize his/ her invention/creation for a given period of time. It is very well settled that IP play a vital role in the modern economy. It has also been conclusively established that the intellectual labour associated with the innovation should be given due importance so that public good emanates from it. There has been a quantum jump in research and development (R&D) costs with an associated jump in investments required for putting a new technology in the market place. The stakes of the developers of technology have become very high hence, the need to protect the knowledge from unlawful use has become expedient, at least for a period, that would ensure recovery of the R&D and other associated costs and adequate profits for continuous investments in R&D. IPR is a strong tool to protect investments, time, money, effort invested by the inventor/creator of an IP, since it grants the inventor/creator an exclusive right for a certain period of time for use of his invention/creation.

Originally, only patent, trademarks, and industrial designs were protected as 'Industrial Property', but now the term 'Intellectual Property' has a much wider meaning. IPR enhances technology advancement in the following ways (Singh, 2008).

a) It provides a mechanism of handling infringement, piracy and unauthorized use
b) It provides a pool of information to the general public since all forms of IP are published except in case of trade secrets.

Variety of intellectual efforts for which Intellectual Property need to be protected

- Patents.
- Industrial designs related to the features of any shape, configuration, surface pattern, composition of lines and colours applied to an article.
- Trademarks relate to any mark, name or logo under which the trade is conducted for any product or service and by which the manufacturer or the service provider is identified. Trademarks can be bought, sold and licensed. Trademark has no existence apart from the goodwill of the product or service it symbolizes.
- Copyright relates to expression of ideas in material form and includes literary, musical, dramatic, artistic, cinematography work, audio tapes and computer software.
- Geographical indications are indications, which identify as good as originating in the territory of a country or a region or locality in that territory where a given quality, reputation or other characteristic of the goods is essentially attributable to its geographical origin.

Brief historical background of IPR

In earlier days, the local bodies / authorities / Governments formulated various ways to reward the inventors to encourage working with greater effort and devotion toward developing newer and more useful inventions. Such reward was first recorded in the 7th century BC when the discoverer of a food recipe in one Greek colony was given the exclusive right to use this recipe for one year. The first record for law on patent, so far as record reveals, was passed in Venice in 1474 which gave monopoly rights to artisans for their inventions. In 1623, the House of Commons of United Kingdom passed the act of proprietorship. But there was no system of protecting the rights to the knowledge generated within the families/small social groups in India. The Act of protecting the inventions was introduced in India in 1856 which was on the basis of British Patent Law of 1852. In 1872, the Patents and Designs Protection Act was passed. The Protection of Invention Act was consolidated as Inventions and Designs Act in 1888. The Indian patents and design was brought under management of Controller of Patents and Designs on August 15, 1947. Trademark protection has been in force since June 1, 1948 under the Trade Mark Protection Act (1940). Later, this Act was amended as Indian Trade and Merchandise Marks Act (1958) and came in force on November 25, 1959. The Indian Patents Act (1970) was introduced in the parliament in 1965. Later it was modified in 1967 and was passed in 1970. This Act is being currently operated in the country. All types of designs are protected by the Indian Patent and Design Act (1911) and that has been amended in 1978. The copyright laws in India named as The Indian Copyright Act 1957 was amended in 1999. Thereafter, The

Geographical Indications of Goods (Registration & Protection) Act (1999), Protection of Plant Varieties and Farmers' Rights Act (2001), and Biological Diversity Act (2002) have been enforced in the country. National IPR Policy of our country focuses on the vision: 'Intellectual Property stimulates creativity and innovation for the benefit of all'.

Definitions of intellectual properties

1. **Trade Secret:** When an individual or an organization owns an intellectual property and the concerned discoverer does not disclose the property to any one and keeps it as a close secret to support his business interests, it is called trade secret. Trade secret may be related to genetic material, formulae, processes of materials, etc.
2. **Patent:** When an inventor registers his invention with the Government to obtain protection of such invention from imitating, manufacturing, using or selling it by others for commercial use during a specified period by others is called as patenting. Patents are granted for i) an invention (including a product), ii) innovation / improvement in an invention, iii) the process/product of an invention and iv) a concept.
3. **Copyright:** There are certain items that can't be patented and in such cases protection is extended from being imitated by a means known as copy right. The best and common example of such property is authored books. The audio and video CD and DVD are also covered by this protection. The holder of the copy right of a book or a CD and DVD has the right to exclude others from reproducing or imitating the same in any form. The copyright of a CD and DVD of a particular programme is under the custody of the author or the publisher. Presently, the Information Technology Act (2000) has extended copyright to the computer software also.
4. **Designs:** Designs mean only the features of shape, configuration, pattern, ornament or composition of lines or colours applied to any article whether in two dimension or three dimensions or in both forms, by any industrial process or means, whether manual or mechanical or chemical, separate or combined, which is the finished article appeal to and are judged solely by the eye; but it does not include any mode or principle of construction or anything which is in substance a mere mechanical device.
5. **Geographical indications of goods:** 'Geographical indications' (GI) in relation to goods means an indication which identifies such goods as agricultural products, natural goods or manufactured in the particular territory of country or a region or locality in that territory where a given quality, reputation or other characteristic of such products / goods is essentially attributable to its geographical origin. It will also cover the case where either the production or of processing or preparation of the goods concerned take place in such territory region or locality, as the case may be. The GI covers agricultural goods, natural products, goods of handicrafts, manufactured products and even food products.

6. **Plant Variety Protection (PVP):** Systematic breeding programme, devotion and untiring effort of the breeder is required for the development of a variety of a crop plant. Plant variety or animal breed so developed is regarded as intellectual properties of the breeders, who have developed it. Many countries recognize plant varieties as intellectual property and provide them through patent or through suitable form of plant breeders' rights (PBR). Therefore, it may be stated that 'Plant breeders' rights' are the rights granted by the Government to a plant breeder, inventor or possessor of a variety. The protection is given in such a way that no other can produce or make business of the propagating material of the protected variety at least for a period of 15-20 years. However, one person holding PBR for a variety can empower any interested persons or organizations of his desire within limit to produce and sale the propagating material of that variety. A justifiable term for such transfer of PBR label or for sale of the propagating material is required to be framed prior to the authorization, else the Government reserves every right to grant licenses of the titles in the interest of the community.
7. **Farmers' Rights:** Since the beginning of agriculture as early as 11-12 thousand years ago, the genetic resources have been domesticated, managed, selected, developed, tested, used and conserved by the farmers and farming communities of respective gene-rich countries for various crop plants. Almost all the modern high yielding varieties of different crop plants have been developed using those indigenous varieties / land races conserved by the farmers. These indigenous varieties / farmers' varieties / land races are still collected, conserved and are being used as parental materials to breed the modern high yielding varieties of different crops including vegetable crops. Different corporate sectors of seed industry and seed companies are earning enormous profit by selling the seeds and produce of those improved varieties. However, the farmers who not only develop but also conserve the basic genetic materials remain deprived of receiving any benefit of their great contribution. It has been rationally felt that the farmers, who provide the genetic materials for the modern crop varieties should be allowed a share in this profit in recognition of their contribution by way of the development of germplasms of various crops. The FAO has put forward the concept "Farmers' rights" to award recognition of the farmers' contribution towards development of the modern varieties. This include the past, present and future contribution of the farmers in domesticating, conserving, improving and making available plant genetic resources particularly in the centres of origin/ diversity to the institutions or organizations. The FAO has emphatically mentioned that farmers' rights are obligatory and not the privileges. The argument about reward to the farmers still remains pending in certain aspects like to whom the reward is to be paid, what should be the amount of reward and it should go in what manner. However, one suggestion in this respect has been accepted that the rural communities, tribal people and the traditional farming families of the area where the crop is still popular as indicated by the Geographical Indication should be rewarded.

Protection of Plant Varieties and Farmers' Rights Act, 2001

Plants are used in many ways and forms, including food, feed, fibre, medicine and industrial applications. Plant genetic resources are treated as the 'heritage of mankind' and are shared freely among nations, till the concerns for conservation of biological diversity were raised by the Convention on Biological Diversity (CBD) which came into force in 1993. India became the signatory to the Trade Related Aspects of Intellectual Property Rights Agreement (TRIPs) in 1994. The agreement on TRIPs is an international legal agreement between all the member nations of the World Trade Organization. The TRIPs Agreement sets minimum standards of protection for copyrights and related rights, trademarks, geographical indications (GIs), industrial designs, patents, integrated circuit layout designs and undisclosed information. The Article 27.3 (b) of this agreement requires the member countries to provide for protection of plant varieties either by a patent or by an effective *sui generis* system or by any combination thereof. The existing Indian Patent Act, 1970 excluded agriculture and horticultural products and methods of production from patentability. Thus India opted a *sui generis* system to establish "an effective system for the protection of plant varieties, the right of farmers and plant breeders, to encourage the development of new varieties of plants" and accordingly Government of India enacted the Protection of Plant Varieties and Farmers' Rights Act, 2001with the following objectives.

- To recognize and protect the rights of farmers in respect of the contribution made at any time in conserving, improving and making available the plant genetic resources for the development of new plant varieties.
- To accelerate agricultural development in the country, protect plant breeder's rights, stimulate investment for research and development in public / private sector for development of plant variety.
- To facilitate the growth of seed industry which will ensure the availability of high quality seeds and planting material to the farmers.

Chapters and definitions of Plant Varieties and Farmers' Rights Act, 2001

The Act has 11 chapters and is divided in 97 clauses. The first chapter has title, and the definitions used in context of the Act. The last chapter is on miscellaneous clauses. The other nine chapters deal with PPV&FR authority, registration of plant varieties, duration and effect of registration and benefit sharing, surrender and revocation of certificate, farmer's rights, compulsory license, plant varieties protection appellate tribunal, finance, accounts, audit, infringement, offences and penalties, etc. Some of the important definitions in the context of the Act include the following:

Variety

A plant grouping except microorganisms within a single botanical taxon of the lowest known rank, which can be

- defined by the expression of the characteristics resulting from a given genotype of a plant and group of plants;

- distinguished from any other plant group by expression of at least one of the said characteristics;
- considered as a unit with regard to its suitability for being propagated, which remains unchanged after such propagation and includes propagating material of such variety, extant variety, transgenic variety, farmers' variety and essentially derived variety

Extant variety

A variety available in India which is

a) notified under section 5 of Seeds Act, 1966, or
b) farmers' variety, or
c) a variety about which there is common knowledge, or
d) any other variety which is in public domain

Essentially derived variety

A variety shall be said to be essentially derived when it:

(i) is predominantly derived from such initial variety, or from a variety that itself is predominantly derived from such initial variety, while retaining the expression of the essential characteristics that result from the genotype or combination of genotypes of such initial variety;
(ii) is clearly distinguishable from such initial variety, and
(iii) conforms (excepting for the differences which result from the act of derivation) to such initial variety in the expression of the essential characteristics that result from the genotype or combination of genotypes of such initial variety

Farmers' variety

A variety which

(i) has been traditionally cultivated and evolved by the farmers in their fields, or
(ii) is a wild relative or land race of a variety about which the farmers possess common knowledge.

Farmer

Any person who

(i) cultivates crops by cultivating the land himself, or
(ii) cultivates crops by directly supervising the cultivation of land through any other person, or
(iii) conserves and preserves, severally or jointly, with any person any wild species or traditional varieties, or adds value to such wild species or traditional varieties through selection and identification of their useful properties.

Registration

Eligibility

For a variety to be eligible for registration, it must conform to the criteria of novelty, distinctiveness, uniformity and stability (NDUS), as described below [Section 15(1)–(3)]. For the purposes of the Act, a new variety shall be deemed to be:

a) **Novel**: If, at the date of filing of the application for registration for protection, the propagating or harvested material of such a variety has not been sold or otherwise disposed of by or with the consent of its breeder or his successor for the purposes of exploitation of such variety-

   - in India, earlier than one year, or
   - outside India, in the case of trees or vines earlier than six years or in any other case, earlier than four years, before the date of filing such applications.

   Provided that a trial of a new variety which has not been sold or otherwise disposed off shall not affect the right to protection. Provided further the fact that on the date of filing the application for registration, the propagating or harvested material of such variety has become a matter of common knowledge other than through the aforesaid manner shall not affect the criteria of novelty for such variety.

b) **Distinct**: If it is clearly distinguishable by at least one essential characteristic from any other variety whose existence is a matter of common knowledge in any country at the time of filing of the application.

c) **Uniform**: If subject to the variation that may be expected from the particular features of its propagation, it is sufficiently uniform in its essential characteristics.

d) **Stable**: If it's essential characteristics remain unchanged after repeated propagation or, in the case of a particular cycle of propagation, at the end of each such cycle. The variety will be subjected to such distinctiveness, uniformity and stability tests as shall be prescribed.

Crops for Registration

The Act (Section 29.2) provides that the Central Government shall by notification in official Gazettes specify the genera and species for the purpose of registration of varieties. The PPV & FR Authority has developed "Guidelines for the conduct of species specific Distinctiveness, Uniformity and Stability" tests or "Specific Guidelines", for individual crop species. The purpose of these specific guidelines is to provide detailed practical guidance for the harmonized examination of DUS and, in particular, to identify appropriate characteristics for the examination of DUS and production of harmonized variety descriptions. So far, the PPV&FRA Authority has notified 199 crops for registration given below and published 47 specific DUS test guidelines.

Table 1: List of crops notified by PPV&FR Authority for registration

Group	Crop
Cereals	Rice, bread wheat, durum wheat, dicoccum wheat, other *Triticum* species, maize, pearl millet, sorghum, barley, finger millet, foxtail millet, buck wheat (*Fagopyrum tataricum*), grain amaranth, barnyard millet (*Echinochloa esculenta*), kodo millet (*Paspalum scrobiculatum*), little millet (*Panicum sumatrense*) and proso millet (*Panicum miliaceum*), oat (Avena *sativa*)
Pulses and other legume crops	Chickpea, mungbean, urdbean, fieldpea, kidney bean (*Phaseolus vulgaris*), lentil, pigeon pea, horse gram (*Macrotyloma uniflorum*) and moth bean (*Vigna aconitifolia*)
Oilseeds	Indian mustard (*Brassica juncea*), karan rai (*Brassica carinata*), rapeseed (*Brassica napus* var. *napus*), Gobhi sarson (*Brassica napus* L.), groundnut, soyabean, sunflower, safflower, castor, sesame and linseed
Fibre crops	Diploid cotton (*Gossypium arboreum* L., *G. herbaceum* L.), tetraploid Cotton (*G. hirsutum* L., *G.barbadense* L.), tossa Jute (*Corchorus olitorius*) and white Jute (*Corchorus capsularis*).
Sugar crops	Sugarcane
Fruit crops	Mango, apple, pear, pomegranate, grapes, acid lime, lemon, pummelo, sweet orange, mandarin orange, banana, noni (*Morinda citrifolia*), bael, jamun/black plum, papaya, peach, plum, strawberry, cherry, custard apple, apricot, ber, guava, litchi, mulberry (*Morus* spp.), Chironji (*Buchanania lanzan*), tamarind (*Tamarindus indica*), jackfruit, seabuckthorn (*Hippophae rhamnoides*)
Vegetables	Potato, cabbage, cauliflower, garlic, okra, onion, brinjal, tomato, vegetable amaranth, spinach beet/ palak, coriander, chilli, bell pepper and paprika, fenugreek, bitter gourd, bottle gourd, watermelon, muskmelon, pumpkin, ridge gourd, sponge gourd, ash gourd, ivy gourd, snake gourd, pointed gourd, bakla / faba bean, taro, giant swamp taro, elephant foot yam, sweet potato, carrot, radish, garden pea, cowpea, dolichos bean, cluster bean
Flowering plant	Periwinkle, bamboo leaf orchid (*Arundina graminifolia*) or boat orchid, spray orchid or Singapore orchid (*Dendrobium* sp.) and Vanda orchid (*Vanda* spp.), canna, bougainvillea, gladioulus, carnation, tuberose, jasmine, Arabian jasmine or sambac jasmine, rose, damask rose, marigold, gerbera, China aster
Spices	Black pepper, ginger, small cardamom, turmeric, ajwain (*Trachyspermum ammi* L.), anise (*Pimpinella anisum*), celery, cumin, dill, fennel, black cumin, saffron
Medicinal crops	Brahmi (*Bacopa monnieri*), isabgol (*Psyllium* spp.), menthol mint (*Mentha* spp.), periwinkle (*Vinca* spp.), kalmegh (*Andrographis paniculata*), nutmeg (*Myristica fragrans*), amla or Indian gooseberry (*Phyllanthuse mblica*), neem (*Azadirachta indica*), karanj (*Pongamia pinnata*), betel vine (*Piper betle*)
Tree, forest and plantation crops	Almond, walnut, cashew nut, Casurina, eucalyptus, tea, coconut, deodar, Chir pine (*Pinus roxburghii*), arecanut, poplar, greater yam, yam bean, coffee, cocoa, teak, jatropha, willow, date palm, drum stick (*Moringa oleifera*)
Cash crop	Flue-cured Virginia tobacco and bidi tobaccos (*Nicotiana tabacum* L)

Distinctness, Uniformity and Stability (DUS) Testing of new varieties of crops

The application of Intellectual Property Rights (IPRs) to new varieties of crops is a relatively recent phenomenon. The key objective of Plant Variety Protection (PVP), a form of IPR, is to stimulate plant variety innovations. PVP has become well established in developed countries over the last three decades or so. Till the early 1990s, PVP remained almost exclusively a feature of developed countries. While developing countries recognized the importance of varietal improvement for agricultural productivity growth, they generally relied on research by public sector institutions at the National and International level for the development of new varieties. Article 27 (3) (b) of the Agreement on Trade Related Aspects of Intellectual Property Rights (TRIPs) deals with plant variety protection and states that "Members shall provide for the protection of plant varieties either by patents or by an effective *sui generis* system or by any combination thereof". Having research and variety development both in the public and private sectors, traditional communities and farmers conserving landraces/ folk varieties and conserving plant genetic resources for food, fibre and feed, India has adopted a novel *sui generis* system providing the establishment of an effective system for the protection of plant varieties. The Government of India enacted the legislation on "Protection of Plant Varieties and Farmers' Rights (PPV & FR) Act in 2001 and the "Protection of Plant Varieties and Farmers' Rights Rules (2003)" was notified (Anonymous, 2004). Under the Act, a new variety shall be registered if it conforms to the criteria of Novelty, Distinctiveness, Uniformity and Stability (NDUS). The new variety must be clearly distinguishable by one or more essential characteristics from any other variety whose existence is a matter of common knowledge at the time when the protection is applied for. The variety is deemed uniform if subject to the variation that may be expected from the particular features of its propagation, is sufficiently uniform in its relevant characteristics. The variety is deemed to be stable if its relevant characteristics remain unchanged after repeated propagation or, in the case of a particular cycle of propagation, at the end of each such cycle. However, the Act provides for registration of extant/already available variety(s) under extant categories under which varieties of common knowledge; varieties notified under Seeds Act, 1966 and farmers 'varieties can also be protected. Accurate morphological and agronomic descriptions of cultivars and varieties have been the basis of tests for DUS within worldwide PVP systems (Smith and Smith, 1989. Indian guidelines have been developed for DUS testing in different crops (Chakrabarty and Roy Choudhury, 2019). Accordingly, the plant morphological characteristics listed in the crop specific DUS guideline are useful for characterizing varieties. There are no empirical studies to decide many of such issues. There is a need to find out the scientific bases for sole reliance on the UPOV guidelines in our National guidelines.

In order to implement effectively Protection of Plant Varieties and Farmers' Rights Act, 2001 testing of new varieties for Distinctiveness (D), Uniformity (U) and Stability (S) is essential. Based on field and laboratory trials along with the most similar variety a new variety is compared for the characteristics which describe the variety. Distinctiveness assessment of a new variety apparently looks easy but actually it is not so particularly in case of crop species maintained /reproduced through often cross pollination and cross pollination owing to the presence of plants in the population with varied expression in some characteristics. Similarly, uniformity assessment of a variety ensures its genetic purity. The crop specific DUS test guidelines has been developed and followed for plant variety assessment to register a new variety.

DNA Finger printing

DNA fingerprinting is mandatory for registration of newly developed hybrids and varieties with National Bureau of Plant Genetic Resources (NBPGR), New Delhi and Protection of Plant Varieties and Farmers' Rights Authority (PPV&FRA). There are following advantages of DNA-based markers over protein or isozyme based molecular markers.

- **Direct analysis of genome:** In spite of great interest in protein/isozyme marker systems, the major limitations are non-detection of mutation which may occur in non-coded part of genome and will not be translated into protein. But DNA marker could overcome this problem.
- **No tissue and age specificity:** Proteins/isozyme profile varies if taken from different tissues and at different age, which may be due to post translational changes. But this is not the case of DNA-based marker.
- **Independent of environment:** Protein or isozyme profiles vary if taken on normal and stress condition but it is not so in case of DNA- based markers.
- **Unlimited polymorphism:** Proteins/isozymes are functional in nature so polymorphism obtained is less compared DNA-based marker.

DNA Fingerprinting for PPV & FRA

- Microsatellite based fingerprints (simple sequence repeats or SSR) are required for registration at Protection of Plant Varieties & Farmers Right Act Authority. Microsatellites are polymorphic DNA loci comprising repeated sequence motifs of two to five base pairs which are dispersed throughout the genome. Genotyping of microsatellites is a widely accepted tool not only for diagnostic and research purposes such as forensic investigations and parentage testing, but also in clinics, as well as for the agriculture and food industries.
- Next generation sequencing (NGS) based DNA fingerprints analysis offer several advantages over traditional capillary gel electrophoresis methods. NGS allows for high-throughput parallelization of sequencing reactions (deep sequencing) from low DNA concentrations and therefore may, when used in tandem with DNA barcoding, provide greater sensitivity to identify host DNA of low quantity and quality.

Application form

Every application for registration will have to be accompanied with the following information [Section 18 (a–h)]:

a) denomination assigned to such variety by the applicant;
b) an affidavit sworn by the applicant that such variety does not contain any gene or gene sequence involving terminator technology;
c) the application should be in such form as may be specified by regulations;
d) a complete passport data of the parental lines from which the variety has been derived along with the geographical location in India from where the genetic material has been taken and all such information relating to the contribution, if any, of any farmer, village community, institution or organization in breeding, evolving or developing the variety;
e) a statement containing a brief description of the variety, bringing out its characteristics of novelty, distinctiveness, uniformity and stability as required for registration;
f) such fees as may be prescribed;
g) contain a declaration that the genetic material or parental material acquired for breeding, evolving or developing the variety has been lawfully acquired; and
h) such other particulars as may be prescribed.

The conditions stated above (a–h), shall not apply in respect of application for registration of farmers' varieties. The PPV&FRA authority has received 18,296 applications up to 29.03.2024 for registration of plant varieties including open pollinated varieties, hybrids, parental lines and transgenic varieties from different stakeholders such as farmers, public and private sectors including multinational seed companies and 6896 certificates have so far been issued.

Application for registration of plant varieties should be accompanied with the fee of registration prescribed by the Authority. Fee for registration for different types of variety is as mentioned below:

Sl.No	Type of variety	Fees for Registration	
	Essentially derived varieties	Individual	Rs. 7,000/-
		Educational	Rs. 10,000/-
		Commercial	Rs. 50,000/-
2.	Extant variety notified u/s 5 of Seeds Act, 1966		Rs. 2000/-
3.	New variety	Individual	Rs. 7,000/-
		Educational	Rs. 10,000/-
		Commercial	Rs. 50,000/-
4.	Extant Variety about which there is common knowledge (VCK)	Individual	Rs. 7,000/-
		Educational	Rs. 10,000/-
		Commercial	Rs. 50,000/-
5.	Farmers' variety		No fee

Certificate of registration

The certificates of registration are issued for the applications which have fulfilled all requirements and have been finally accepted by the Registrar of the PPV&FR Authority. The number of varieties registered under different crops group up to 29.03.2024 is 6896, which can be grouped into as: Cereals 4780, Fibre crops 575, Vegetables 512, Oilseeds 384, Legumes 369, Fruits 126, Spices 64, Sugar crops 54, Trees, Forest & Plantation crops 17, Flowers 12 and Medicinal & Aromatic plants and cash crop 1. The number of different types of varieties up to 26.12.2023 is classified as: Farmer's variety 3092, Extant VCK 735, Extant notified variety 1527, New Variety 992 and EDV variety 14.

Period of protection

The certificate of registration issued under section 24 or sub-section 98 of section 23 shall be valid for nine years in the case of trees and vines and six years in the case of other crops, and may be reviewed and renewed for the remaining period on payment of such fees as may be fixed by the rules made on this behalf subject to the conditions that the total period of validity shall not exceed

(i) in the case of trees and vines, 18 years from the date of registration of the variety;

(ii) in the case of extant varieties, 15 years from the date of the notification of that variety by the Central Government under Section 5 of the Seed Act, 1996, and

(iii) in the other case, 15 years from the date of registration of the variety.

Payment of annual fee

PPV&FR Authority with the prior approval of the Central Government notified the Annual fee to be paid by every breeder of a variety, agent and licensee thereof registered under the said Act:

Sl.No.	Type of variety	Fees for Registration
1.	New variety	Rs. 2,000/- plus 0.2 per cent of the sales value of the seeds of the registered variety during the previous year plus 1 percent of royalty, if any, received during the previous year from the sale proceed of seeds of a registered variety.
2.	Extant variety notified u/s 5 of the Seeds Act, 1966 (54 of 1996)	Rs. 2,000/- only
3.	Extant variety other than the category specified above	Rs. 2000/- plus 0.1 per cent of the sales value of the seeds of the registered variety during the previous year plus 0.5 percent of royalty, if any, received during the previous year from the sale proceed of seeds of a registered variety
4.	Farmers' Variety	Rs. 10/- only

Annual fee shall be determined on the basis of the declaration given by the registered breeder or agent or licensee regarding the sale value of the seeds of the variety registered under Act during the previous year and royalty, if any, received during the previous year from the sale proceed of seeds of the registered variety and verified by the Authority.

National register of plant varieties

National register of plant varieties has been kept at the head office of the Registry, containing the names of all the registered plant varieties with the name and addresses of their respective breeders, the rights of such breeders in respect of the registered varieties, the particulars of the denomination of each registered variety, its seed or other propagating material along with specification of salient features thereof and such other matters as may be prescribed.

National gene bank of PPV & FRA

Authority has established National gene bank to store the seed material including parental lines submitted by the breeder of the registered variety. The seed lot is stored under low temperature conditions at 5°C for the entire registration period. If necessary, after few years of storage in the National gene bank, the seed lot will be rejuvenated and replenished at the cost of the applicant. The seed stored in the National gene bank will be used for dispute settlement or when an exigency arises for invoking compulsory licensing provision. Such a seed deposition in the National gene bank would dissuade market malpractices or violations as the sample in custody can be drawn to verify the facts. When the period of registration granted lapses, the material automatically moves to public domain.

Rights

Protection of Plant Varieties and Farmers Rights (PPV& FRA) Act, 2001 of Government of India is an Act to provide for the establishment of an effective system for protection of plant varieties, the rights of farmers and plant breeders and to encourage the development of new varieties of plants.

Breeders' rights: Breeders will have exclusive rights to produce, sell, market, distribute, import or export of the protected variety. Breeder can appoint agent / licensee and civil remedy in case of infringement of rights (Section 28 (1)).

Researchers' rights: Researcher can use such variety for conducting experiment or research, use of the variety as an initial source genotype for the purpose of developing another variety however, repeated use needs permission (Section 30)

Farmers' rights: The farmers' rights of the Act define the privilege of farmers and their right to protect varieties developed or conserved by them [Chapter VI]. In the Protection of Plant Varieties and Farmers Rights (PPV& FRA) Act, 2001 of Government of India, farmers are entitled to save, use, sow, re-sow, exchange, share or sell their farm produce, including seed of protected varieties in the same manner as they were entitled to do before the PPV&FR Act has come into force. However, farmers are not entitled to sell branded seed of a variety protected under this Act [Section 39 (1), (i)–(iv)].

Further, the farmers have also been provided protection of innocent infringement when, at the time of infringement, a farmer is not aware of the existence of breeder rights [Section 42 (1)]. A farmer who is engaged in the conservation of genetic resources of landraces and wild relatives of economic plants and their improvement through selection and preservation, shall be entitled in the prescribed manner for recognition and reward from the "Gene fund", provided the material so selected and preserved has been used as donor of genes in varieties registrable under the Act. The expected performance of a variety is to be disclosed to the farmers at the time of sale of seed/ propagating material. A farmer or a group of farmers or an organization of farmers can claim compensation according to the Act, if a variety or the propagating material fails to give the expected performance under given conditions, as claimed by the breeder of the variety.

Communities' rights: The rights of the communities as defined, provide for compensation for the contribution of communities in the evolution of new varieties in quantum to be determined by the PPVFR Authority [Section 41 (1)]

National gene fund

The National gene fund to be constituted under the Act shall be credited thereto:

- the benefit sharing from the breeder
- the annual fee payable to the authority by way of royalties
- by the compensation provided to the communities as defined u/s 41(1)
- contribution from any National and international organization and other sources.

The National gene fund shall be utilized for

- Disbursing shares to benefit claimers, either individuals or organization
- Disbursing compensation to village communities under Section 41(1)
- Supporting the conservation and sustainable use of genetic resources, including *in situ* and *ex situ* collection and for strengthening the capabilities of the panchayat in carrying out such conservation and sustainable use [Section (45)]
- Expenditure of the schemes relating to benefit sharing is framed under Section 46

Benefit sharing

Sharing of benefits accruing to a breeder from a variety developed from indigenously derived plant genetic resources has also been provided in the law [Section 26(1)]. The authority publishes the content of the Certificate for the purpose of inviting claims for benefit sharing under the Act. The advertisement is made in the *Plant Variety Journal of India*. Any person or group of persons, being citizen(s) of India or firm or Governmental or non-Governmental organization formed or established in India shall submit their claims for benefit sharing (u/s 26 (2) of PPV&FR Act, 2001 read with

Rule 41 of PPV&FR Rules, 2003) in Form PV 7 of the First schedule (in triplicate) within a period of six months from the date of publication.

Plant Variety protection appellate tribunal

The Tribunal will be established by a gazette notification by the Government to exercise jurisdiction, powers and authority conferred on it under this Act. The Tribunal will consist of Judicial as well as Technical members.

Generation of awareness about the act

There is a need to create awareness regarding the act among the scientists, policy makers and breeders as well as farmers, village communities and the private seed sector. Wide circulation of the provisions of the Act may be done in all research organizations which are directly or indirectly involved in the development of new crop varieties. For farmers and village communities, awareness generation and information empowerment is a mandatory through vernacular press, radio, television and the internet. A standing committee on awareness generation and information empowerment under the Plant Varieties and Farmers' Rights Protection Authority may be set-up for ensuring the effective spread of credible information concerning the rights of the plant breeders and farmers as cultivators, breeders and conservers. In order to ensure that the farmer as a breeder and conserver secures the recognition and reward provided under the Act, there is need to establish resource centres for farmers' rights and entitlements.

References

Chakrabarty, S.K. and Roy Choudhury, D. 2019. DUS testing for plant variety protection: Some researchable issues. *Indian J. Genet.*, 79 (1) Suppl. : 320-325.

Singh, R. 2008. *Law Relating to Intellectual Property: A Complete Comprehensive Material on Intellectual Property Covering Acts, Rules, Conventions, Treaties, Agreements, Digest of Cases and Much More*. Universal Law Publishing Company, New Delhi.

Smith J.S.C. and Smith O.S. 1989. The description and assessment of distance between inbred lines of maize: I. The use of morphological traits as descriptors. *Maydica*, **34**: 141-150.

17

Agricultural Biotechnology in Vegetable Breeding

Old and Modern biotechnology

Agriculture has evolved since the dawn of human civilization, first as a means to guarantee food supply and then, as a source of family income and improved profitability. Domestication of plants and animals that were found in the wild combined with gradual, long-term changes in their quality and quantity were the first signs of what is now collectively termed "Agriculture". At that time, agriculture replaced the former nomadic habit of food collection for immediate consumption. Domestication of plants and animals, followed by food storage, coincided most probably unintentionally, with the growth of microorganisms. Thus, was born the classical food fermentation, the earliest known application of biotechnology, focusing on the use of microorganisms to produce food products.

Gradual improvements in agricultural techniques, domestication of additional plant and species from the wild, step-by-step selection of better performing and more adapted genotypes, along with intuitive breeding continued at a slow pace. Mendel's discoveries and laws revolutionized genetics, and led at the beginning of the 20th century, to planned and controlled breeding experiments. Breeding or the "old biotechnology" was harnessed as a most successful tool, resulting in better crops and farm animals, and "Green Revolution" of the 1960s was achieved. Many achievements are still being successfully implemented since then. Soon, this "Old Biotechnology" was revolutionized again, when DNA was cloned in 1973. Towards the end of the 1970s and during the early 1980s, recombinant DNA technologies resulted in the first development of transgenic microorganisms and subsequently animals and plants. The novel scientific discoveries and techniques thus paved the way for the "New Agricultural Biotechnology". Biotechnology, in the simplest and broadest sense, is the utilization of living organisms or their components to provide useful products or processes. This definition encompasses essentially all of agriculture, since agriculture is based on the production of plants and animals to provide food, fibre and other products for human use. It also includes familiar uses of microorganisms, such as the yeast used in brewing or baking or the symbiosis of legume plants with nitrogen-fixing *Rhizobiun* bacteria. Thus, humans have been employing biotechnology for at least 10,000 years since the origin of agriculture. Recently, however, the term 'Biotechnology' has become synonymous with 'Genetic Engineering'. The cloning and movement of genes among life forms as diverse as bacteria, tomatoes or sheep is now possible. In this context, biotechnology refers to a wide range of enabling technologies that allow the alteration of heritable traits outside of the living organism and the subsequent re-introduction of the new trait into an organism for specific purposes. These new genotypes are called transgenics because they require the recombination of genetic materials from different organisms to cause the desired change.

Plant tissue culture and *in vitro* propagation

One of the stellar achievements of twentieth century plant biology was the genetic transformation of somatic cells enabling the regeneration of whole plants that were stably transformed and capable of transmitting the inserted genetic material to subsequent generations. This achievement grew out of three independent lines of research initiated early in the twentieth century: plant tissue culture, regeneration of plants from single somatic cells, and the study of crown gall disease. The early discoveries made in these areas represent a combination of basic scientific research and technological innovations which led to the development of genetically transformed crop species expressing traits unobtainable by conventional breeding. Each of these fields can be traced back to a single research publication (Haberlandt, 1902; Smith and Townsend, 1907; White, 1939) that later came to be considered as the foundation of the field.

Plant tissue culture is a collection of techniques used to maintain or grow plant cells, tissues or organs under sterile conditions on a nutrient culture medium of known composition. Plant tissue culture or the aseptic culture of cells, tissues, organs and their components under defined physical and chemical conditions *in vitro* is an important tool in both basic and applied studies as well as in commercial application. It owes its origin to the ideas of the German scientist, Haberlandt (1902) at the beginning of the 20th century. He first proposed that plant cells have the ability to show totipotency, *i.e.*, ability to generate a whole organism from its part in nutrient media, as is found in fertilized egg cell. The first cells cultured by the scientists in artificial media were those of animal rather than plant origin. As regards plant cells, the success came only by the early 1930s.The early studies led to root cultures, embryo cultures and the first true callus/tissue cultures. The period between the 1940s and the 1960s was marked by the development of new techniques and the improvement of those that were already in use. It was the availability of these techniques that led to the application of tissue culture to five broad areas, namely, cell behaviour (including cytology, nutrition, metabolism, morphogenesis, embryogenesis and pathology), plant modification and improvement, pathogen-free plants and germplasm storage, clonal propagation, and product (mainly secondary metabolite) formation, starting in the mid-1960s. The 1990s saw continued expansion in the application of the *in vitro* technologies to an increasing number of plant species. Cell cultures have remained an important tool in the study of basic areas of plant biology and biochemistry and have assumed major significance in studies in molecular biology and agricultural biotechnology. Different applications of in *vitro* culture can be summarized below:

Clonal germplasm

- Quick multiplication of true to the type plants through micro-propagation.
- Isolation of pathogen-free plant materials.
- Production of plants from seeds that otherwise have very low chances of germination, *e.g.*, seeds of orchids and *Nepenthes*.
- Germplasm collection and conservation
- Production of plant metabolites

Crop improvement

- Somatic cell genetics
- Utilization of somaclonal variations
- Production of haploids
- Development of distant hybrids
- Development of somatic hybrids
- DNA-Marker assisted breeding
- Genetic engineering and genomics
- Regeneration of whole plants from plant cells that have been genetically modified

Biological basis of plant tissue culture

Totipotency and Regeneration

Many somatic plant cells, including some fully differentiated types (*e.g.* leaf mesophyll), provided they contain intact nuclear, plastid and mitochondrial genomes, have the capacity to regenerate into whole plants. This phenomenon is called "Totipotency", an amazing developmental plasticity of plant cells. It is the capacity of the cell of higher organism to differentiate into entire organism. Totipotent cell contains all genetic information necessary for complete development.

The first step in expression of regenerative totipotency for mature cells are to re-enter the cell cycle and resume cell division — a process known as "Dedifferentiation".

- **Dedifferentiation**: It is a cellular process by which cells grow in reverse, from a partially or terminally differentiated stage to a less differentiated stage within their own lineage. In general, the phenomenon is manifested by a change in the shape, gene expression pattern, protein expression pattern and function. The non-dividing quiescent cells first undergo certain changes to achieve a meristematic state.

The component cells of the callus have the ability to form a whole plant, a phenomenon described as "Redifferentiation".

- **Redifferentiation**: It is the process by which a dedifferentiated cell attains a particular function and loses its ability to divide again. For example, formation of secondary xylem and phloem from dedifferentiated cells of the cambial ring.

These two phenomenon, "Dedifferentiation" and "Redifferentaition" are inherent in the capacity described as "Cellular Totipotency", a property found only in plant cell and not in animal cells. In other words, while a differentiated plant cell retains its capacity to give rise to whole plant, an animal cell loses its capacity of regeneration after differentiation. Generally, a callus phase is involved before the cell can undergo redifferentaition leading to regeneration of whole plant.

Different pathways of regeneration

Regeneration process of adventitious shoots and roots is referred to as *de novo* shoot organogenesis and *de novo* root organogenesis. Like somatic embryogenesis, *de novo* organogenesis can also be categorized as either a direct or indirect regeneration pathway. Organogenesis and somatic embryogenesis may be direct, originating from the explant and not involving callus formation, or indirect from a callus tissue that is formed first. The difference between direct and indirect regeneration and between the different types of regeneration is important because it relates to genetic stability of the resulting plantlets and to the propagation rate. A callus stage and meristem organization from callus cells usually leads to more genetic aberrations than direct regeneration. The entire micro-propagation process must be carefully controlled and monitored otherwise it may result in generation of plantlets that are not true to the type. Regeneration (both organogenesis and somatic embryogenesis), in contrast to the proliferation of axillary buds, may also be disadvantage for clonal propagation of chimeric varieties. Some micro-propagation stages do not necessarily involve a regeneration process. Culture of shoot tips, for example, may lead to proliferation of already-existing axillary buds, but their further development into plantlets involves a regeneration process (*i.e.*, root formation).

• Axillary bud proliferation

One of the most exciting and important aspects of *in vitro* cell and tissue culture is the capability to regenerate and propagate plants from cultured cells and tissues. The simplest type of *in vitro* plant propagation is the stimulation of axillary bud development. This technique exploits the normal ontogenetic route for branch development by lateral (axillary) meristems. The axillary buds are treated with hormones to break dormancy and produce shoot branches. The shoots are then separated and rooted to produce plants. Alternatively, the shoots are used as propagules for further propagation.

Axillary bud proliferation *in vitro* is usually considered a convenient route for micro-propagation, although the final number of newly regenerated plantlets can sometimes be lower than in other methods. Because axillary bud proliferation does not include a callus stage, it is considered "safer" for the preservation of clonal characteristics. In fact, this pathway does not involve regeneration of new buds, since bud meristems already exist in the axils of leaves. The short stem tip that is used as an explant already contains many axillary buds, at different stages of development, condensed in a small explant. A large number of otherwise quiescent axillary shoot buds grow extensively when the shoot tips or even small apical meristem are excised and cultured in an appropriate medium, usually containing high concentrations of cytokinins. Both the excision of the explant and the cytokinin-rich medium "activate" the buds, leading to massive proliferation of many side shoots. After induction, these shoots are separated for further culture and rooting.

• Organogenesis

Development of adventitious organs or primordia from undifferentiated cell mass in tissue culture by the process of differentiation is called organogenesis. Formation of roots, shoots or flower buds from the cells in culture in manner similar to adventitious root or shoot formation in cuttings is called organogenesis. It involves induction of the vegetative tissue to form organs (shoot or root) which eventually develop into a complete plantlet (small but whole plant).

Formation of organ in callus depends on the constituents of the medium because medium of tissue culture contains growth regulators which determine shoot or root regeneration. For this reason, auxin and cytokinin need to be used in an appropriate ratio which is responsible for callus regeneration. Generally, more concentration of cytokinin compared to auxin generates shoot part in the callus and more quantity of auxin generates roots in the callus. Medium cytokinin/medium auxin concentration in the culture medium regenerate to both shoot and root and medium cytokinin/low auxin content only produce callus.

Once shoot or root meristems have been organized, they start developing, forming small shoots and roots. Further stages of organization, differentiation and growth include the formation of functional vascular connections between the developing shoots and roots, finally giving rise to plantlets *in vitro.* This sequence of events and operations result in new plantlets that are removed from the vials and subjected to *ex vitro* conditions that include acclimatization and cultivation under greenhouse or field conditions. In some cases, organogenesis brings about the formations of modified shoots and roots, usually storage organs such as bulbs and tubers. Here, the shoot or the root meristems forms but instead of further growth and elongation, minibulbs, minitubers or mimcorms (depending on the specific genotype) soon develop. These can grow further or become dormant, depending on the culture conditions. The storage organs can also be developed from axillary buds without true regeneration, especially when basal plates of bulbs or certain corm and tuber tissues are used as primary explants.

• Somatic embryogenesis

Somatic embryogenesis is the process to induce a piece of somatic (vegetative) tissue to develop an embryogenic callus, leading to the formation of a somatic embryo. Embryonic callus induction is followed by the formation of pro-embryonic masses on the surface or within the callus mass, from which single cells or cell clusters develop into somatic embryos (Toonen *et al.*, 1994). Somatic embryos are formed from plant cells that are not normally involved in the development of embryos, *i.e.* ordinary plant tissue. No endosperm or seed coat is formed around a somatic embryo. Somatic embryos are believed to have bipolar structure having both apical and basal meristems which can form shoot and root, respectively.

Somatic embryogenesis is different from organogenesis. In organogenesis, shoots and roots that are regenerated from the callus are monopolar while somatic embryos are bipolar, allowing them to form a whole plant without culturing on multiple media. Cells derived from competent source tissue are cultured to form an undifferentiated mass of cells called a callus. Plant growth regulators in the tissue culture medium can be manipulated to induce callus formation and subsequently changed to induce embryos from the callus. Somatic embryos are mainly produced *in vitro* and for laboratory purposes, using either solid or liquid nutrient media which contain different plant growth regulators (PGRs). The ratio of different plant growth regulators required for inducing callus or embryo formation varies with the type of plant. The main PGRs used in the growing medium are auxins however, it may contain smaller amount of cytokinin .Somatic embryogenesis has served as a model to understand the physiological and biochemical events that occur in plant developmental processes as well as a component to biotechnological advancement. Steward *et al.* (1958) and Reinert (1958) first documented somatic embryogenesis with carrot cell suspension cultures. Applications of this somatic embryogenesis include:

- Clonal propagation of genetically uniform plant material.
- Elimination of viruses.
- Provision of source tissue for genetic transformation.
- Generation of whole plants from protoplasts (single plant cells that have had their cell walls removed by treatment with cellulytic enzymes).
- Development of synthetic seed technology (encapsulating somatic embryos in a protective coating).

Technical elements of plant tissue culture

Tissue culture medium

Different components of plant nutrients, plant growth regulators, gelling agent and others used in the tissue culture medium may be grouped as under:

- **Macro-nutrients:** The essential elements in plant cell or tissue culture media include, besides C, H and O, macro-elements *viz.*, nitrogen (N), phosphorus (P), potassium (K), calcium (Ca), magnesium (Mg) and sulphur (S) for satisfactory growth and morphogenesis. Culture media should contain at least 25-60 mM (millimole i.e., one-thousandth of a mole) of inorganic nitrogen for satisfactory plant cell growth. Potassium is required for cell growth of most plant species. Most media contain K in the form of nitrate or chloride salts of potassium at concentrations ranging between 20 and 30 mM. The optimum concentrations of P, Mg, S and Ca range from 1-3 mM if other requirements for cell growth are provided.
- **Micro-nutrients:** The essential micro-nutrients for plant cell and tissue growth include iron (Fe), manganese (Mn), zinc (Zn), boron (B), copper (Cu) and molybdenum (Mo). Iron is usually the most critical of all the micro-nutrients.

The element is used as either citrate or tartarate salts in culture media, however, there exist some problems with these compounds for their difficulty to dissolve and precipitate after media preparation. This problem was solved by using chelating class of chemical, ethylene diaminete traacetic acid (EDTA)-iron chelate (FeEDTA). A procedure for preparing an iron chelate solution that does not precipitate has also been developed. Cobalt (Co) and iodine (I) may be added to certain media, but their requirements for cell growth has not been precisely established. Sodium (Na) and chlorine (Cl) are also used in some media, in spite of several reports that they are not essential for growth. Copper and cobalt are added to culture media at concentrations of 0.1μM, iron and molybdenum at 1μM, iodine at 5μM, zinc at 5-30 μM, manganese at 20-90 μM and boron at 25-100 μM (micromole).

- **Carbon and energy sources:** Sucrose is often assumed to be the best source of carbon for *in vitro* culture. The levels used are from 2 to 6% and this concentration is to be defined for each species. Besides sucrose, other carbohydrates *viz.*, lactose, galactose, maltose and starch are also used. However, they are less effective than either sucrose or glucose. Glucose is more effective than fructose considering that glucose is utilized by the cells in the beginning, followed by fructose. It was frequently demonstrated that autoclaved sucrose was better for growth than filter sterilized sucrose. Autoclaving seems to hydrolyse sucrose into more efficiently utilizable sugars such as fructose. Sucrose was reported to act as morphogenetic trigger in the formation of auxiliary buds and branching of adventitious roots. It was found that supplements of sugarcane molasses, banana extract and coconut water to basal media can be a good alternative for reducing medium costs. These substrates in addition to sugars are sources of vitamins and inorganic ions required for growth.
- **Vitamins and myo-inositol :** Some vitamins are required for normal growth and development of plants as catalysts in various metabolic processes. They may act as limiting factors for cell growth and differentiation when plant cells and tissues are grown *in vitro*. The vitamins most used in the cell and tissue culture media include thiamine (Vitamin B_1), nicotinic acid and pyridoxine (Vitamin B_6). Thiamine is necessarily required by all cells for growth. Thiamine is used at concentrations ranging from 0.1 to 10 mg per litre. Nicotinic acid and pyridoxine, although not essential for cell growth of many species, are often added to culture media. Nicotinic acid is used at a concentration range 0.1-5 mg/ 1 and pyridoxine is used at 0.1-10 mg/l. Other vitamins such as biotin, folic acid, ascorbic acid, pantothenic acid, tocopherol (vitamin E), riboflavin, p-aminobenzoic acid are used in some cell culture media however they are not growth limiting factors. It is recommended that vitamins should be added to culture media only when the concentration of thiamine is below the desired level or when the cells are required to be grown at low population densities.

  Myo-inositol is a carbocyclic sugar which mediates cell signal transduction in response to a variety of hormones. It is added in small quantities to stimulate cell growth of most plant species. Myo-inositol is believed to play a role in cell division because of its breakdown to ascorbic acid and pectin and incorporation into phosphoinositides (minority phospholipids of all eukaryotic cellular membranes) and phosphatidyl-inositol. It is generally used in plant cell and tissue culture media at concentrations of 50-5000 mg per litre.

- **Amino acids:** The required amino acids for optimal growth are usually synthesized by most plants, however, addition of certain amino acids or amino acid mixtures is particularly important for establishing cultures of cells and protoplasts. Amino acids provide plant cells with a source of nitrogen that is easily assimilated by tissues and cells faster than inorganic nitrogen sources. Amino acid mixtures such as casein hydrolysate, L-glutamine, L-asparagine and adenine are frequently used as sources of organic nitrogen in culture media. Casein hydrolysate is generally used at concentrations between 0.25-1.0 g/l. Amino acids used for enhancement of cell growth in culture media includes glycine at 2 mg/l, glutamine up to 8 mM, asparagine at 100 mg/l, L-arginine and cysteine at 10 mg/l and L-tyrosine at 100 mg per litre.
- **Undefined organic supplements:** A wide variety of organic extracts are commonly added to culture media such as protein hydrolysates, coconut milk, yeast extract, malt extract, ground banana, orange juice and tomato juice to test their effect on growth enhancement. Activated charcoal is sometimes added to culture media where it may have either a beneficial or deleterious effect. Explanation of the mode of action of activated charcoal was based on the adsorption of inhibitory compounds from the medium, adsorption of growth regulators from the culture medium or darkening of the medium. Presence of 1% activated charcoal in the medium largely increases hydrolysis of sucrose during autoclaving which cause acidification of the culture medium.
- **Plant growth regulators:** Plant growth regulators are important in plant tissue culture since they play vital roles in stem elongation, tropism and apical dominance. They are generally classified into five groups; auxins, gibberellins, cytokinins, ethylene and abscisic acid. Moreover, proportion of auxins to cytokinins determines the type and extent of organogenesis in plant cell cultures.

  The common auxins used in plant tissue culture media are indole-3- acetic acid (IAA), indole-3- butyric acid (IBA), 2,4-dichlorophenoxy-acetic acid (2,4-D) and naphthalene-acetic acid (NAA). Other synthetic auxins used in culture media are 4-chlorophenoxy acetic acid or p-chlorophenoxy acetic acid (4-CPA, pCPA), 2,4,5-trichloro-phenoxy acetic acid (2,4,5 T), 3,6- dichloro-2-methoxy-benzoic acid (dicamba) and 4- amino-3,5,6-trichloro-picolinic acid (picloram).

  Cytokinins commonly used in culture media include BAP (6-benzylo aminopurine), 2iP (6-dimethylaminopurine), kinetin (N-2-furanylmethyl-1H-purine-6-amine), zeatin (6-4- hydroxy-3-methyl-trans-2-butenylaminopurine) and TDZ (Thidiazuron (1-phenyl- 3-(1,2,3-thiadiazol-5-yl) urea)

Gibberellins are a large group of phyto-hormones that constitutes around 136 different gibberellin molecules (Hedden and Phillips, 2000), of which GA_3 is the most frequently used gibberellin. These compounds enhance growth of callus and help in elongation of dwarf plantlets. Other growth regulator that sometimes is added to plant tissue culture media is abscisic acid to inhibit or stimulate callus growth, depending upon the species.

- **Gelling agents:** Hardness of the culture medium greatly influences the growth of cultured tissues. A number of gelling agents such as agar, agarose and gellan gum are used. Agar or agar-agar, is a jelly-like substance consisting of polysaccharides obtained from the cell walls of some species of red algae. As found in nature, agar is a mixture of two components, the linear polysaccharide agarose and a heterogeneous mixture of smaller molecules called agaropectin. It is universally used as a gelling agent for preparing semi-solid and solid plant tissue culture media. Agar has several advantages over other gelling agents viz., it is mixed with water easily, it easily melts in a temperature range of 60-100°C and solidifies at approximately 45°C and it forms a gel stable at all feasible incubation temperatures. Agar gels do not react with media constituents and are not digested by plant enzymes. It is commonly used in media at concentrations ranging between 0.8-1.0%.

  Agarose is extracted from agar leaving behind agaropectin and its sulfate groups. It has higher gel strength, highly pure but very expensive. Gellan gum is a water soluble anionic polysaccharide produced by the bacterium, *Sphingomonas elodea.*

- **pH of the medium:** It determines many important aspects of the structure and activity of biological macro-molecules. Nutrient medium pH ranges from 5.0 to 6.0 for suitable *in vitro* growth of explant and pH higher than 7.0 and lower than 4.5 generally stops growth and development.

- **Media preparation:** Appropriate systems for water sterilization or deionization are very important for media preparation. Certain equipment is required for better performance such as a refrigerator, hot plate, stirrer, pH meter, precision balances with different weighing ranges, heater, Bunsen burner in addition to glassware and chemicals. It is well known that mistakes which occur in tissue culture process most frequently originate from inaccurate media preparation and that is why, clean glassware, high quality water, pure chemicals and careful measurement of media components should be ensured.

- **Media selection:** For the establishment of a new protocol for a specific purpose in tissue culture, a suitable medium is better formulated by testing the individual addition of a series of concentrations of a given compound to the universal basal medium for tissue culture such as Murashige and Skoog *medium* (MS), Linsmaier and Skoog Medium (LS) or Gamborg B5 medium *(*B5). The most effective variables in plant tissue culture media are growth regulators, especially auxins and cytokinins. Full strength of salts in media proved good for several species but in some species the reduction of salts level to ½ or ¼ the full concentration gave better results in *in vitro* growth.

- **Media sterilization:** Prevention of contamination of tissue culture media is important for the whole process of plant tissue culture. Contamination of media could be controlled by adding antimicrobial agents, acidification or by filtration through microporous filters. To reduce the possibilities of contamination, it is recommended that sterilization rooms should have the least number of openings. Media preparation and sterilization are preferred to be performed in separate compartments. Sterilization area should also have walls and floor that withstand moisture, heat and steam. Sterilization of media is routinely done by autoclaving at the temperature ranging from 115°-135°C. Advantage of autoclaving is that the method is quick and simple, whereas disadvantages are the changes in media pH, decomposition of some components and loss of effectiveness.
- **Surface-sterilization of the plant material:** Prior good care of stock plants may lessen the level of contamination that is present on the explants. Treatment of stock plants with fungicides and /or antibiotics is sometimes helpful. Covering growing shoots for several days or weeks prior to harvesting the tissue for culture may give cleaner material.

It is administered that explants or plant material should be washed in soapy water and then placed under running water for 1 to 2 hours. Different important surface sterilization chemicals are mentioned below:

- **Sodium hypochlorite:** It is usually diluted to 10-20% of the original concentration, resulting in a final concentration of 0.5-1.0% sodium hypchlorite. Plant material is usually immersed in this solution for 10-20 minutes. A balance between concentration and time must be determined empirically for each type of explant, because this chemical shows phytotoxicity.
- **Ethanol (or Isopropyl alcohol):** It is a powerful sterilizing agent but it is extremely phytotoxic. For this reason, plant material is typically exposed to it for only seconds or minutes. The more tender the tissue, the more it will be damaged by alcohol. Tissues such as dormant buds, seeds or unopened flower buds can be treated for longer periods of time since the tissue that will be explanted or that will develop is actually within the structure that is being surface-sterilized. Generally, 70% ethanol is used prior to treatment with other compounds.
- **Calcium hypochlorite:** It is obtained in a powder form and must be dissolved in water. The concentration that is generally used is 3.25%. The solution must be filtered prior to use since not the entire compound goes into solution. Calcium hypochlorite may be less injurious to plant tissues than sodium hypochlorite.
- **Mercuric chloride:** It is extremely toxic to both plants and humans and must be disposed of with care. Since mercury is very phytotoxic, it is critical that many rinses should be used to remove all traces of the mineral from the plant material.
- **Hydrogen peroxide:** The concentration of hydrogen peroxide used for surface sterilization of plant material is generally 30 per cent.

In vitro culture for clonal germplasm

• Micro-propagation

Multiplication of genetically identical copies of a genetic material / cultivar by asexual reproduction is called clonal propagation. In nature, clonal propagation occurs by apomixis (seed development without meiosis and fertilization) and/or vegetative propagation (regeneration of new plants from vegetative parts). Aseptic method of clonal propagation is called as Micro-propagation and it offer the advantage of producing large number of true-to-type plantlets with relatively short period of time and space from a single individual.

Micro-propagation is the only commercially viable method of clonal propagation of some important horticultural crops like, banana, fern, orchids, etc. Different kinds of explants are used in micro-propagation of orchides. For example, shoot tip (*Anacamptis pyramidalis, Aranthera, Calanthe, Dendrobium*), axillary bud (*Aranda, Brassocattleya, Cattleya, Laelia*), inflorescence segment (*Aranda, Ascofinetia, Neostylis, Vascostylis*), lateral bud (*Cattleya, Rhynocostylis gigantean*), leaf base (*Cattleya*), leaf tip (*Cattleya, Epidendrum*), shoot tip (*Cymbidium, Dendrobium, Odontioda, Odontonia*), nodal segment (*Dendrobium*), flower stalk segment (*Dendrobium, Phalaenopsis*) and root tips (*Neottia, Vanilla*).

Tissue culture in vegetable crops also involves the growth of plant cells, tissues or organs in an artificial nutrient medium under sterile conditions. This method offers numerous advantages, including the production of disease-free, particularly virus free plants.

- **Eradication of diseases:** A major problem in vegetatively propagated crops is transmission of diseases, particular of viruses, through planting material. Several cultivars of potato, sweet potato, garlic, cassava, sweet potato, ginger etc. are sometimes totally contaminated of viruses which drastically reduce the genetic worth of the cultivars. In the absence of genetic resistance within the gene pool, meristem culture has been widely used to transform these cultivars into healthy cultivars, free from diseases. By starting with a small piece of healthy tissue such as a leaf or stem, disease-free plantlet is regenerated from this, which can in turn be propagated *in vitro* or *in vivo* to ensure production of healthy planting material. Its association with disease detection methods such as ELISA or PCR techniques makes it a very powerful tool to ensure propagation of healthy planting material.
- **Rapid multiplication of specific genotypes:** Micro-propagation is particularly useful for rapid multiplication of specific genotypes which are difficult to propagate in pure form through traditional methods. As for example, propagation of self-incompatible line and cytoplasmic male sterile line of the cabbage, cauliflower, broccoli, etc. used for hybrid seed production.

Stages in micro-propagation

Micro-propagation generally involves five stages and each stage has its own requirements.

Stage 0: Preparative stage

This stage involves the preparation of mother plants to provide quality explants for better establishment of aseptic cultures in stage 1. To reduce the contamination problem in the subsequent stages, mother plant should be grown under protective structure (poly house). This will also reduce the need for a harsh sterilization treatment. Stage 0 also includes exposing the stock plants to suitable light, temperature and growth regulator treatments to improve the quality of explants. In the case of photosensitive plants, it may be possible to obtain suitable explants throughout the year by controlling photoperiod in the glasshouse.

Stage 1. Initiation of culture

- **Explant:** The nature of explant to be used for *in vitro* propagation is governed by the method of shoot multiplication.
- **Sterilization:** Special precautions need to be taken when explants are derived from field-grown materials.
- **Browning of medium:** A serious problem with the culture of some plant species is the oxidation of phenolic compounds leached out from the cut surface of the explant. It turns the medium dark brown and is often toxic to the tissues.

Stage 2. Multiplication

This is the most crucial stage since it is the point at which most of failures in micro-propagation occur. Broadly three approaches have been followed to achieve *in vitro* multiplication.

- **Through callusing:** The potentiality of plant cells to multiply indefinitely in cultures and their totipotent nature permit a very rapid multiplication of several plant types. Differentiation of plants from cultured cells may occur *via* shoot-root formation (organogenesis) or somatic embryogenesis. Somatic embryogenesis is most desirable from the commercial angle.
- **Adventitious bud formation:** Buds arising from any place other than leaf axil or the shoot apex are termed adventitious buds. The shoots differentiated from the calli should also be treated as adventitious buds. In many crops, vegetative propagation through adventitious bud formation from root (blackberry, raspberry) and leaf (Begonia, Crassula) cuttings is commercially practiced.
- **Enhanced axillary branching:** In cultures the rate of shoot multiplication by enhanced axillary branching can be substantially enhanced by growing the shoots in a medium containing appropriate concentration of cytokinin with or without auxin. Due to continuous availability of cytokinin, the shoots formed by the bud, present on the explant, develops axillary buds which may grow

directly into shoots. This process may be repeated several times and the initial explant is transformed into a mass of branches.

Stage 3. Rooting of shoots

Somatic embryos carry a pre-formed radicle (embryonic root) and may develop directly into plantlet. However, these embryos often show very poor conversion into plantlets, especially under *in vitro* conditions. They require an additional step of maturation to acquire the capability for normal germination. Adventitious and axillary shoots developed in cultures in the presence of a cytokinin generally lack roots. To obtain full plants, the shoots must be transferred to a rooting medium which is different from the shoot multiplication medium, particularly in its hormonal and salt compositions. For rooting, individual shoots measuring 2 cm in length are excised and transferred to the rooting medium.

Stage 4. Transplantation and acclimatization

The ultimate success of commercial propagation depends on the ability to transfer plants out of culture on a large scale, at low cost and with high survival rates. The plants multiplied *in vitro* are exposed to a unique set of growth conditions (high levels of inorganic and organic nutrients, growth regulators, sucrose as carbon source, high humidity, low light, poor gaseous exchange) which may support rapid growth and multiplication but also induce structural and physiological abnormalities in the plants, rendering them unfit for survival under *in vivo* conditions. Two main deficiencies of *in vitro* grown plants are – poor control of water loss and heterotrophic mode of nutrition. Therefore, gradual acclimatization is necessary for these plants to survive transition from culture to the greenhouse or field condition.

Advantages of Micro-propagation

1. Extremely large number of clonal propagules can be produced within a relatively short time span, reaching an exponential increase of the propagation coefficient. Depending on the multiplication rate, thousands and millions of plants can rapidly be produced *in vitro* from relatively few selected source plants. Plant regeneration has become successful in many vegetable crops, *viz.,* potato, sweet potato, tomato, brinjal, cole crops, cucurbits, carrot, lettuce, taro, onion, etc.
2. Production of disease-free plant material with the possibility of eliminating viral, bacterial, and fungal contamination. Disease causing organisms, particularly virus that are carried over from source plant material are significantly limiting factors in conventional plant propagation. Specific *in vitro* techniques that are designed to control and limit this phenomenon address the issues of contamination and its negative ramifications. Recovery of virus-free plants through shoot-tip culture is based on the assumption that pathogen concentration is not uniform throughout the infected plant. Frequently the growth cone, the apex of the rapidly elongating stem is not invaded by the virus.

In potatoes, through this technique the isolated plant segments like meristem or shoot tips, node, internodes, leaves, etc. are cultured aseptically *in vitro* in some artificial nutrient medium like, MS medium under controlled conditions to produce virus-free planting material. In the same manner, elimination of viruses like, onion yellow dwarf, leek yellow dwarf, pea seed-borne mosaic, etc. and regeneration of virus-free plants of onion, leek and pea, respectively have successfully been achieved.

3. Different breeding materials like, male sterile lines, self-incompatible lines, inbred lines and haploids are difficult to propagate in pure form through sexual reproduction. Isolation of restorer lines for respective male sterile line, bud pollination or CO_2 treatment for maintaining self-incompatible lines are not only cumbersome but also expensive. Regeneration of these breeding lines by tissue culture is of great importance for their convenient maintenance in pure form.
4. The ability to produce and maintain a large stock of true-to-type propagation material, which guarantees a consistently high degree of likeness between the source plant and the product. This issue is of primary importance for commercial propagation.
5. The ability to air-ship large quantities of micro-propagated plants quickly, efficiently and relatively inexpensively. As many as 30,000-50,000 *in vitro* (in closed vials or small containers), 3,000-10,000 *ex vitro* (acclimatized) plants can be packed into a cubic meter of air-shipping space.
6. It provides facility for physiological studies, particularly transition from vegetative to reproductive phases.
7. *In vitro* cultures of different genetic materials can be stored for long time through cryopreservation.
8. Breeding cycle can be shortened.

Disadvantages of micro-propagation

1. Combination of expensive laboratory technology and facilities and labour-intensive nature of tissue culture often render micro-propagation economically unfeasible.
2. Contamination (endogenous and/or exogenous) of plant cultures which cause the greatest economic losses, direct and indirect, in commercial micro-propagation.
3. Poor acclimatization or hyperhydricity (vitrification, glassiness) is the most common physiological disorder associated with *in vitro* cultivation of plants. Hyperhydric (glassy) shoots are characterized by thickened stems, short internodes, translucent, fragile, elongated, and twisted leaves (Polivanova and Bedarev, 2022)
4. Problems in inducing rooting

5. Sometimes somatic variation is higher than acceptable levels causing non-uniformity of plant materials.

In vitro culture for crop improvement

Tissue culture method supplements different conventional breeding methods. It is especially promising as a means for substantially expanding the genetic resources for plant breeding based on the interspecific and intergeneric hybridization, somaclonal variation, haploid plants, preservation and propagation of valuable breeding lines and obtaining virus-free plants.

1. Haploid Production (anther, pollen and ovule culture)

An important aspect of plant breeding is the induction of maximum genetic variability of germplasm sources to secure a wider scope for selection and introduction of better characters in existing crop species. Plant breeders have worked extensively to obtain haploids either *in vitro* or *in vivo*. In nature, haploids arise as a result of parthenogenesis after wide hybridization and these plants rarely produce characters of the male parent. In angiosperms the haploid or gametophytic phase is extremely brief and represented by pollen grains in anther and one cells in the embryo sac of the ovule. Haploid plants can be produced by *in vitro* culture of anther, pollen or ovule.

- **Anther culture:** Anther culture is the technique by which the developing anthers at a precise and critical stage are excised aseptically from unopened flower bud and are cultured on a nutrient medium where the microspore within the cultured anther develop into callus tissue or embryoids that give rise to haploid plantlets either through organogenesis or embryogenesis.
- **Pollen culture:** A typical anther in cross section shows two anther lobes and each lobe possesses two pollen sacs. During micro-sporogenesis, pollen mother cell (PMC) inside the pollen sac form pollen tetrad by meiosis. In each pollen tetrad, four pollens are held together temporarily by their callose wall. Pollens separate as discrete unit by dissolution of callose wall. Each pollen possesses a unique genome where every gene is present as a single copy. Pollen or microspore culture is an *in vitro* technique by which the pollen grains, preferably at the uninucleated stage, are squeezed out aseptically from the intact anther and then cultured on nutrient medium where the microspore, without producing male gametes, develop into haploid embryoids or callus tissues that give rise to haploid plantlets by embryogenesis or organogenesis.
- **Ovule Culture:** Ovule culture is an elegant experimental system by which ovules are aseptically isolated from the ovary and are grown aseptically on chemically defined nutrient medium under controlled conditions.

Exploitation of these unique genetic units and the totipotency of the plant cell is the basis of anther or pollen culture for the production of haploid plants. On the other hand, egg cell produced within the ovule is very difficult to separate from complex tissue integration. In culture, the anther swells and dehisces along its upper margin, lengthwise. This phenomenon helps to expose the pollen grain. Alternatively, huge amount of pollen grains can be isolated manually and can be cultured aseptically very easily. Therefore, pollen is more suitable material than egg cell for the production of haploid.

Development and production of haploid plant *in vitro* is also very important for the study of fundamental and applied aspects of genetics in the higher plants. Production of homozygous diploid by doubling the chromosome number of haploid *in vitro* makes a pure line in single step and such homozygous pure line is of great importance in plant breeding. Haploid plants have been successfully produced in different vegetable crops (*e.g.*, potato, tomato, brinjal, chilli, cauliflower, cabbage, asparagus, cowpea, broad bean, muskmelon, summer squash etc.) and in some other wild relatives.

Production of double haploids

Using *in-vitro* techniques, it is possible to regenerate plants from pollen or ovules. These plants, which contain only one copy of each chromosome, are called haploids. After appropriate colchicine treatment, it is possible to restore the doubling i.e., normal number of chromosomes and to regenerate viable plants. These plants, called double haploids, are homozygous for all their genes. Such plants are of tremendous interests to the breeders, since they allow development of pure line varieties or inbred parental lines much more quickly than through conventional breeding.

Advantages of haploid production

- Haploid production is important for plant breeding because direct plantlets produced from the pollen through embryogenesis are haploids but genetic constitutions of the halpoids produced from heterozygous plants are not alike (gamatoclonal variation) which opens the new direction of crop improvement.
- Simplification in the study of heritability because of only one set of chromosomes in the haploids.
- Easy identification of the recessive mutations.
- Doubling the chromosome number of a haploid to produce doubled haploids result in a completely homozygous plant.
- Use of doubled haploids in breeding programme can greatly reduce the time required for development of improved cultivars.
- The monoploids (haploids) are useful in the studies of genome homology, mutation and gene transfer.

2. Embryo culture

• *Overcoming crossing barriers (Embryo rescue)*

In angiosperms the embryo is the miniature sporophyte resulting from the fertilized egg or zygote. Many unsuccessful crosses result from embryo abortion. Early embryo abortion occurs primarily because the endosperm fails to develop properly. In majority of interspecific, intergeneric crosses and crosses between diploids and tetraploids, the endosperm often develops poorly or not at all developed.

Embryo culture, sometimes called embryo rescue, is an *in vitro* technique to save the hybrid products of fertilization when they might otherwise degenerate. By aseptically culturing the embryo in a nutrient medium, this problem may be overcome. Embryos of some nonviable hybrids may possess the potential for initiating development by avoiding post-zygotic barriers within the mother plant. Several successful cases of embryo rescue technology have been documented with the embryos arising from several interspecific hybrids. *viz*., *Solanum lycopersicum* × *S. peruvianum, Capsicum chinense* × *C. pubescens, C. annuum* × *C. pendulum, Solanum melongena* × *S. torvum, Cucumis sativus* × *C. hystrix, Phaseolus vulgaris* × *P. lanatus, P. vulgaris* × *P. aconitifolius, Cucurbita pepo* × C. *ecuadorensis, Vigna unguiculata* × *V. pubescens, Abelmoschus esculentits* × *A. moschatus,* etc.

Laibach (1925) first described zygotic embryo culture for interspecific hybridization. He observed that seeds from the interspecific cross between *Linum perenne* L. × *Linum austriacum* L. were non-viable however, if embryos were excised early during seed development and cultured *in vitro*, then embryo abortion was overcome. However, contrary to interspecific hybridization, development of transgenics leads to the introduction of solely the target gene, and eliminates the need for several generations of back crossing.

• Mature embryo culture

When dormancy is a problem, culture of mature embryos from ripened seeds is used to eliminate seed germination inhibitors or to shorten the breeding cycle. Dormancy may be caused by the endogenous inhibitors, specific requirements of light, low temperatures, dry storage and embryo immaturity. This culture is easy and only requires a simple nutrient medium with agar, sugar and minerals.

Technique

Embryo culture involves isolating and growing an immature or mature zygotic embryo under sterile conditions on an aseptic nutrient medium with the goal of obtaining a viable plant. The basic premise for this technique is that the integrity of the hybrid genome is retained in a developmentally arrested or an abortive embryo. Its potential to resume normal growth may be realized if supplied with the proper growth substances. Culture of immature embryos is used to rescue embryos that would normally abort or that would not undergo the progressive sequence of ontogeny. This process is difficult due to the tedious dissection necessary and the complex nutrient medium requirements.

Factors affecting the success of embryo culture

Before attempting to elucidate the application of embryo culture method, it is necessary to analyse briefly the factors influencing the embryo culture technique.

- Success in embryo culture technique depends on the genotypes.
- The technique depends on isolating the embryo without injury, formulating a suitable nutrient medium and inducing continued embryogenic growth and seedling formation.

- Success with this type of culture depends strongly on the developmental stage of the embryo when it is isolated. Despite considerable progress in the field of embryo culture, embryo rescuing seems to be difficult where embryo abortion occurs at a very early stage of development because culture of very young embryos is very difficult. To culture very young embryos successfully, the embryo of a particular species is implanted in the endosperm from another seed of the same species. For example, in the cross of *Hordeum* × *Secale* the survival rate with the implantation technique was 30-40 % as compared to 1.0% with traditional method of embryo culture. This technique is termed as embryo-nurse endosperm transplant technique.
- Success with this type of culture depends on the growth conditions of the mother plant.
- Success with embryo culture depends on formulation of suitable nutrient medium. Addition of amino acids and vitamins promote the development of the embryo. Casein hydrolysate, an amino acid complex has been widely used as an additive to the embryo culture media. The natural plant extracts like coconut milk, tomato juice and extracts of banana produce higher recovery of growth and development of embryos. Growth hormones, especially auxins are not used in embryo culture media because of their inhibitory role in embryo growth resulting in structural abnormalities.
- Culture of embryo is promoted with suspensor which is the transitory structure found at the radicle end of the pro-embryo. Mostly embryos cultured without the suspension showed lesser survival and maximum necrosis thus reducing the frequency of plantlet formation. The suspensor is traditionally believed to be a supporting structure during plant embryo development that pushes the embryo proper into the endosperm cavity and connects it to the surrounding maternal and endosperm tissues to facilitate the transfer of nutrients and plant hormones. It promotes growth of the young embryos and degenerate in the later stages of embryo growth, *i.e.*, after the formation of cotyledons.
- Proper light and temperature in the culture room. The optimum temperature depends on plant species, but normally a high range of 25 to 30°C is maintained.

Practical applications of embryo culture

- **Embryo rescue in distant hybrids:** In plants, non-viability of embryo occurs due to many reasons, though there is normal fertilization and development in the early stage. The impairments start subsequently from the endosperm or from the surrounding maternal tissue resulting in the eventual death of embryo. To overcome this post-fertilization barrier for obtaining the hybrids, the embryo culture technique is effective in which the nutritional relationship between the embryo and endosperm is restored by providing the artificial medium to induce and complete growth of the hybrid embryos. The embryo culture technique is not only adopted to produce interspecific hybrids, but also extended to produce viable hybrids between the genera. Intergeneric hybrids have been obtained between *Hordeum* × *Secale*; *Hordeum* × *Hordelymus*, *Triticum* × *Elymus*; *Triticum* × *Secale* and *Tripsacum* × *Zea* through embryo rescue technique.

- **Shortening the breeding cycle:** Embryo culture can shorten the breeding cycle by overcoming dormancy in seeds and eliminating the germination inhibitors.

3. Somaclonal variation

In nature, the genetic diversity and variability within a population are generated *via* recombination events. Factors such as natural selection, mutation, migration and population size influence genetic variability in different ways. Regeneration of whole plants from cultured tissues is employed to obtain genetically uniform plants. However, continuous culture of undifferentiated tissue in the form of callus induces variation, called somaclonal variation.

Steward (1958) first reported this novel, artificially produced source of genetic variability. Cells of higher plant cultured *in vitro* showed a genetic instability that was also the characteristic of regenerated cells. Subsequently, the variability existing in plant tissue and cell cultures received much attention. Larkin and Scowcroft (1981) proposed the term 'somaclone' to refer to the plants derived from any form of cell culture, and the term 'somaclonal variation' was coined to refer to the genetic variation among such plants. Growth of plant cells *in vitro* and their regeneration into whole plants is an asexual process that involves only mitotic division of the cells. In this context, the occurrence of uncontrolled and random spontaneous variation when culturing plant tissue is a major problem. The conditions of *in vitro* culture can be mutagenic and regenerated plants derived from organ cultures, calli, protoplasts and somatic embryos sometimes can show phenotypic and genotypic variation. Some, or all, of the somaclones may be physically different from the stock donor plants. Usually, variability occurs spontaneously and can be a result of temporary or permanent genetic changes in cells or tissue during *in vitro* culture. Temporary changes result from epigenetic or physiological effects and are non-heritable and reversible. In contrast, permanent changes are heritable and often represent expression of pre-existing variation in the source plant or are a result of *de novo* variation. Different studies indicate that somaclonal variation can range in scope from specific trait to the whole plant genome. Somaclonal variation provides a valuable source of genetic variation for the improvement of crops through the selection of novel variants, which may show resistance to disease, improved quality or higher yield. The variation should be spotted early and taken out as soon as possible for stabilization. Changes may be assessed by DNA markers also.

Origin of epigenetic variation in the somaclones

- A mutated cell in the callus and it is not expressed in the whole plant.
- Callus itself is vulnerable for variation because of unorganized state.
- Callus may include nuclei of different size due to deletion of chromosome, ring chromosome, etc.
- Duplication in chromosome due to continuous meristematic state.

- Variation also depends on the tissue used for culture *e.g.*, in embryo culture there is chance of less variation than callus culture.
- Temperature, irradiation (X-ray, UV ray), cycle of light also cause variation.
- Nutrient content (particularly N, P) and hormonal content of medium can also cause variation.

Origin of genetic variation in the somaclones

- Chemical and physical mutagen.
- Gene amplification where one gene may occur in multiple doses (may be due to stresses), which give more expression.
- Change in organelle, though is rare as organelles are more stable than DNA.
- Jumping gene (transposable material) may cause variability by knocking out other gene in the chromosome.
- DNA methylation – methyl group is attached to DNA causing poor transcription.

Utilization of somaclonal variation

- Somaclonal variation can effectively be used for improvement of elite breeding lines without affecting overall agronomic character.
- Desirable traits have been recorded in the somaclonal variations of potato, tomato, garlic, onion, lettuce, asparagus, celery, etc.
- An added advantage of utilizing somaclonal variation is that specific variants can be selected by screening a large population of cells in the laboratory for specific traits like tolerance to salinity, disease resistance, etc.

Table 1: Potential somaclonal variation in different vegetable crops

Crop	Important attributes
Potato	Disease resistance, photoperiodic requirement
Tomato	Resistance to Fusarium wilt, tolerance to tobacco mosaic virus, resistance to bacterial wilt.
Brinjal	Resistance to little leaf disease
Garlic	Increased yield and bulb size
Asparagus	Resistance to Fusarium wilt
Celery	Resistance to Cercospora leaf spot

Problems of somaclonal variation

- There is no control over the variation
- Problem in conditioning the traits

  (a) if changes occur in dominant gene it will be expressed, if recessive then not

  (b) changes in cells or callus are not all the time expressed in complete plant
- Most changes are morphogenetic
- It is difficult to stabilize the genotype

4. Protoplast fusion (Somatic hybridization)

Somatic fusion, also called protoplast fusion, is a type of genetic modification in plants by which isolated protoplasts (*i.e.*, plant cell contents with the plasma membrane but separated enzymatically by cellulase and others from the rigid cellulosic cell wall) of two distinct species of plants are fused together to form a new hybrid plant, a somatic hybrid with the characteristics of both. Somatic hybridization was first introduced by Carlson *et al.* (1972). Protoplast is round shaped because of no cell wall. Cell walls are made of a variety of polysaccharides. Protoplasts can be isolated by degrading cell walls with a mixture of the appropriate polysaccharide-degrading enzymes *viz.*,cellulase, pectinase, xylanase, etc.It should be alive and continue to produce cell wall, which begin to synthesize within 2-4 days. It is mostly isolated from young and fully expanded leaf.

The idea of such somatic hybridization or fusion of protoplasts of somatic cells to produce hybrids is potentially utilized to overcome the natural inherent barriers in producing sexual crosses between closely related as well as unrelated species for breeding different vegetable crops. This method even enables to synthesize new plant forms not existing in nature.

Most of the genetic information in a cell of higher plants is contained in the nucleus, but some of it is also found in cytoplasmic organelles such as, chloroplasts and mitochondria. The genetic novelty of somatic hybrids lay on the combination of nuclear and extra-chromosomal inheritance. Somatic hybrids may show variation in chromosome number because of incomplete elimination of chromosomes of parent.

Methods of protoplasts fusion

Protoplasts fusion is a physical phenomenon. During fusion, two or more protoplasts come in contact and adhere with one another spontaneously or in presence of fusion inducing chemicals. An important aspect has been that incompatibility barriers do not exist during the cell fusion process at interspecific or intergeneric level.

- ***Spontaneous fusion:*** During the enzymatic degradation of cell walls some of the adjacent protoplasts may fuse together to form homocaryons (all the nuclei are genetically identical). These plurinucleate cells sometimes contain 2-3 nuclei, a phenomenon attributed to expansion and subsequent coalescence of the plasmodesmatal connections between the cells. The protoplasts once they are freely isolated do not fuse spontaneously with each other.
- ***Mechanical fusion:*** Protoplast of different species are fused mechanically by pushing together two protoplasts. This fusion does not occur in the presence of fusion inducing agents. However, in this procedure protoplasts are likely to get injury.
- ***Induced fusion:*** Fusion of freely isolated protoplasts from different sources with the help of fusion inducing chemical agents is known as induced fusion. Normally, isolated protoplasts do not fuse with each other because the surface of the isolated protoplasts carries negative charge around the outside of plasma membrane and thus there is a strong tendency for protoplasts to repel one another due to their same electric charges.

- **$NaNO_3$ Treatment:** Equal densities of protoplasts from two different sources are mixed and then centrifuged at 500 rpm for 5 minutes to get a dense pellet. This is followed by addition of 4 ml of 5.5% sodium nitrate in 10.2% sucrose solution to re-suspend the protoplast pellet. The suspended protoplasts are kept in water bath at 35°C for 5 minutes and again centrifuged at 1000 rpm for 5 minutes. The pellet is once again kept in water bath at 30°C for 30 minutes. Fusions of protoplast take place at the time of incubation. Finally, the protoplasts are plated in semisolid culture medium. However, frequency of fusion is not very high in this method.
- **High pH or Ca^{++} Treatment:** Kelier and Melchers (1973) developed a method to effectively induce fusion of tobacco protoplasts at a high temperature (37°C) in media containing high concentration of Ca^{++} ions at a highly alkaline condition (pH 10.5). Equal densities of protoplasts are taken in centrifuge tube and protoplasts are spun at 500 rpm for 5 minutes. The pellet is suspended in 0.5 ml of medium. 4 ml of 0.05 M $CaCl_2$, $2H_2O$ in 0.4 M mannitol at pH 10.5 is mixed to the protoplast suspension. The centrifuge tube containing protoplast at high pH or Ca^{++} is placed in water bath at 30°C for 10 minutes and is spun at 250 rpm for 3 to 4 minutes. This followed by keeping the tubes in water bath (37°C) for 40-50 minutes. About 20-30% protoplast are involved in this fusion process.
- **PEG Treatment:** Polyethylene glycol (PEG), a polyether compound which induces protoplast aggregation and subsequent fusion. But the concentration and molecular weight of PEG are important with respect to fusion. A solution of 37.5% w/v PEG of molecular weight 1500 to 6000 aggregates mesophyll and cultured cell protoplasts during a 45 minutes' incubation at room temperature. Fusion of protoplast takes place during slow elusion of PEG with liquid culture medium. Carrot protoplast can be fused by 28% PEG 1500 and fusion can be promoted by Ca^{++} ion at concentration of 3.5 mM. PEG is more effective when it is mixed with 10-15% dimethyl sulfoxide and addition of concanvalin A (a lectin, carbohydrate-binding protein, originally extracted from the jackbean, *Canavalia ensiformis*) to PEG increases protoplast fusion frequency.
- **Electro fusion:** Mild electrical stimulation is used to fuse protoplast which is known as electrofusion of protoplasts. Two glass capillary microelectrode are placed in contact with the protoplasts. An electric field of low strength ($10kv/m^3$) give rise to dielectrophoretic pole generation within the protoplast suspension which lead to pearl chain arrangement of protoplasts in addition to close contact between membranes. The prerequisite for electrically stimulated cell fusion is dielectric breakdown which leads to changes in the membrane conductance, permeability and probably fluidity. Subsequent application of high intensity electric impulse ($100kv/m^3$) for some micro seconds results in the electric breakdown of membrane and subsequent protoplast fusion occurs.

Table 2: Difference between sexual hybridization and somatic hybridization

Sexual hybridization	Somatic hybridization
In sexual hybridization, fusion occur between two specialized reproductive cells (gametes) which bear the half chromosome number of the somatic cell	In somatic hybridization fusion occur between two somatic cells having diploid chromosome number
In sexual hybridization, the hybrids obtain nuclear genes equally from both the parents	In somatic hybridization also, the hybrids may get nuclear genes equally from both the parents.
In sexual hybridization, the hybrid gets very small cytoplasmic contribution from the male parent that means the extra-chromosomal genes are inherited along the maternal parent	In somatic hybridization, the hybrids have the chance of getting cytoplasmic contribution with the extra-chromosomal genes equally from both the parents.
Sexual hybridization results only two nucleus-cytoplasm combinations	Somatic hybridization results several nucleus-cytoplasm combinations. In fact, utilization of somatic hybridization in plant breeding depends on the isolation of desirable nucleus-cytoplasm combination

Table 3: Achievements of protoplast fusion in vegetable crops

Protoplast fusion between plants	Salient features
Solanum tuberosum (dihaploid) + *Solanum lycopersicum*	The somatic hybrid is sterile and tubers contain more glycoalkaloid content
Solanum tuberosum + *S. chacoense*	The somatic hybrid is close to potato in starch content and to the wild parent in protein content
Solanum melongena + *S. sisymbriifolium*	Resistant to root-knot nematodes (*Meloidogyne incognita*)
Solanum melongena + *S. integrefolium*	Resistant to bacterial wilt
S. melongena + *S. saintwongesi*	Resistant to bacterial wilt
B. oleracea var. *botrytis* + *Sinapis alba*	Resistant to curd and inflorescence blight (*Alternatia brassicicola*)
B. oleracea + *B. campestris*	Disease resistance
B. oleracea + *B. napus*	Atrazine resistance
S. lycopersicum + *S. peruvianum*	Intermediate morphology
Amaranthus tricolor + *Celosia cristata*	Somatic hybrid between C_3 and C_4 species
Allium cepa + *A. ampeloprasum*	Intermediate leaf morphology
S. lycopersicum + *S. peruvianum* + *S. lycopersicoides*	Fertile somatic hybrid

Genetic engineering and recombinant DNA technology

Concept of the genetic material

Genetics is based on the fact that genes carry the information that governs or conditions or specifies the characteristics of the organism from generation to generation. Existence of gene and their effects *i.e.*, expression in terms of phenotype of an organism was demonstrated by Mendel and the history of Genetics elucidating the principles of inheritance began. Genes remained as an abstract in the chromosome until its chemical nature has been discovered. Different historical experiments demonstrated that deoxyribonucleic acid (DNA), mostly present in the nucleus, is the genetic material. The essence of a cell is its ability to grow and divide to produce progeny cells which are likewise capable of generating new cellular molecules and replicating themselves. In the simplest cells of bacteria, DNA that carries the genetic information is located in the inner region of the cell. In these prokaryotes, genetic material is not separated from the cytoplasm by a discrete membrane. On the other hand, in the Eukaryotes, may it be unicellular or multicellular the inner cellular mass is partitioned into a membrane bound spherical body called nucleus and an outer surrounding cytoplasm. In the nucleus of each cell, most of the DNA molecule is packaged into thread-like structures called chromosomes. Each chromosome is made up of DNA tightly coiled many times around proteins called histones that support its structure.

Basic components of the cell and biological molecules

Considering the huge complexity of a living organisms, inorganic components basically used in the formation of organic substances or molecules are too little. Cells containing great variety of organic molecules is composed of restricted set of elements, six of which (C, H, N, O, P, S) make up more than 90% of its weight. Certain combination of atoms such as methyl ($-CH_3$), hydroxyl (–OH), carboxyl (–COOH), sulfhydral (–SH) and amino ($-NH_2$) groups occur repeatedly in organic molecules. The organic molecules in the cell can be grouped into two broad categories.

- **Small organic molecules:** These are compounds with molecular weight in the range of 100-1000 containing up to 30 or so carbon atoms. These kinds of molecules are usually found free in solution in the cytoplasm where they form a pool of intermediates from which large molecules or macromolecules are produced. At least 750 different types of small organic molecules are found in the cell. However, the following three are most important and abundant.
  - **Sugars:** The monosaccharide sugars are molecules composed of carbon, hydrogen and oxygen in the proportion $(CH_2O)_n$ and for this reason, they are called carbohydrates (hydrates of carbon). All sugars contain hydroxyl group (–OH) and either an aldehyde (–CHO) or ketone (–CO) group. The hydroxyl group of one sugar can combine with an aldehyde or ketone group of a second sugar with the elimination of water to form a disaccharide. Addition of more monosaccharides in the same way results in the formation of oligosaccharides and polysaccharides.
  - **Amino acids:** Amino acids are organic acids containing the basic amino group ($-NH_2$) and carboxyl (–COOH) group both linked to a single carbon atom. A general formula for an amino acid is $R-CHNH_2-COOH$. The proteins are basically long linear polymers of amino acids joined head to tail by a peptide bond (–CONH). A peptide bond is a chemical bond, usually

occurs between amino acids. One loses a hydrogen and oxygen from its carboxyl group (COOH) and the other loses hydrogen from its amino group (NH_2). This reaction produces a molecule of water (H_2O) and two amino acids joined by a peptide bond (–CO–NH–). The two joined amino acids are called a dipeptide. There are 20 amino acids in proteins, each with a different side chain attached to the carbon atom. The complete digestion of protein, therefore, yields amino acids.

- **Fatty acids:** The fatty acids are organic acids (carboxylic acid) having the general formula C_XH_Y–COOH with a long aliphatic tail (chain), which is either saturated or unsaturated. Most naturally occurring fatty acids have a chain of an even number of carbon atoms, from 4 to 28. Fatty acids are usually derived from triglycerides or phospholipids. The fatty acids, like carbohydrates, consist of carbon, hydrogen and oxygen, but these elements never occur in the same relative proportions as found in the carbohydrates. There is always less oxygen and relatively more carbon and hydrogen.
- **Macromolecules:** These are polymeric molecules that are formed by joining together specific types of small organic molecules. A large polymeric molecule or more than one polymer associated into a large molecule can be referred to as macromolecule. The most important macromolecules are starch, proteins and nucleic acids. There are no molecules as important as nucleic acids, as they carry within their structure the hereditary information that determines the structure of proteins. A nucleic acid is a polymer of nucleotides. The two nucleic acids found in the cells are DNA (double stranded) and RNA (single stranded).

Structural organization of the genetic material

• *Nucleotides*

Nucleotides are the subunits of DNA (deoxyribonucleic acid) and RNA (ribonudelc acid) meaning that the DNA and RNA consist of the basic monomer units called nucleotides, which are linked by phosphodiester bonds to form the linear polymers (Figure 1). The nucloeotides in turn consist of three simple molecular compounds: a phosphate radical, a pentose ribose sugar and a nitrogenous base. The nitrogenous bases, the ring compounds which can combine with hydrogen atom (H^+) are of two types.

(a) ***Pyrimidine***: These are saturated nitrogen heterocyclic compounds and single derivatives of a six-member ring *e.g.*, cytosine (C), thiamine (T) and uracil (U).

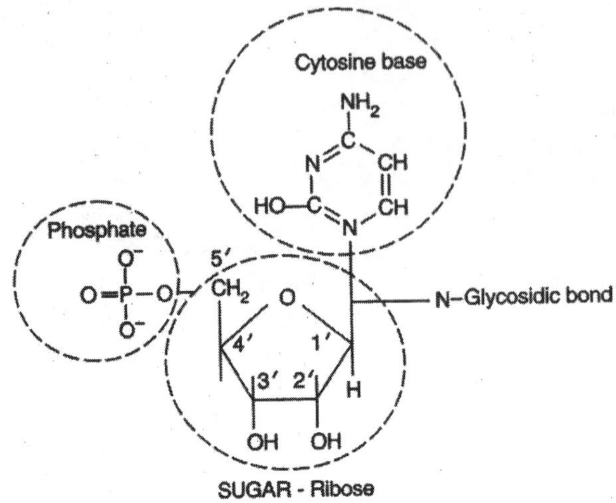

Fig. 1: Structural formula of a nucleotide

(b) ***Purine*:** These saturated nitrogen heterocyclic compounds have a second five members ring fused with the six-member ring *e.g.* adenine (A) and guanine (G).

• *Nucleoside*

When the ribose sugar and nitrogenous base complex is separated from phosphate group, the remaining structure is called nucleoside. It can be expressed as :

Nitrogenous base + Sugar = Nucleoside

Nitrogenous base + Sugar + Phosphate = Nucleotide

The nucleotides may contain several phosphate residues but only monophosphates are present in the DNA and RNA. Monophosphate, diphosphate and triphosphate compounds are denoted by the letters M, D and T, respectively. Nomenclature of the nitrogenous base, nucleoside and nucleotide compounds are given below.

Nitrogen base	Nucleoside	Nucleotide
RNA		
Adenine (A)	Adenosine	Adenylate (AMP)
Guanine (G)	Guanosine	Guanylate (GMP)
Uracil (U)	Uridine	Undylate (UMP)
Cytosine (C)	Cytidine	Cytidylate (CMP)
DNA		
Deoxyadenine (A)	Deoxyadenosine	Deoxyadenylate (dAMP)
Deoxyguanine (G)	Deoxyguanosine	Deoxyguanylate (dGMP)
Thymine (T)	Thymidine	Thymidylate (dTMP)
Deoxycytosine (C)	Deoxycytidine	Deoxycytidylate (dCMP)

Structure of the DNA

Structure of a DNA molecule is built through linking the individual nucleotides together by way of phosphate group to form a polymer, called polynucleotide. The nucleotide monomers are linked together by joining the phosphate group attached to the 5' carbon of one nucleotide to the 3' carbon of the next nucleotide in the chain. Normally a polynucleotide is built up from nucleoside triphosphate subunits. So during polymerization the iPP (β and γ phosphates) are cleaved off. The hydroxyl group attached to the 3' carbon of the second nucleotide is also lost. The linkage between two nucleotides in a polynucleotide is called a phosphodiester bond where, 'phospho' indicate the presence of a phosphorus atom and diester refer to two esters (C–O–P) bonds in each linkage. The relatively rigid and long strands that form the spirals around an axis of the DNA helix are made up of phosphate and pentose sugars and the phosphate-sugar rims are held together firmly by this phosphodiester bond.

The DNA molecules remain invariably in the form of two polynucleotide helix wrapped around one another to form the double helix (Figure 3). Crosswise, these two strands are less rigidly connected by nitrogen bases through hydrogen bonds (H). In other words, the double helix, is held together by hydrogen bond between specific pairs of purine and pyrimidine bases (adenine (A) to thymine (T) or Uracil (U) in RNA and cytosine (C) to guanine (G)). Each strand is complementary to its partner strand with respect to base sequences like, AT/TA or AU/UA and CG/GC. Such 1:1 pairing in relation to the bases do not exist between two purines or two pyrimidines. Moreover, the total amount of purines and pyrimidine bases are equal in the DNA molecule.

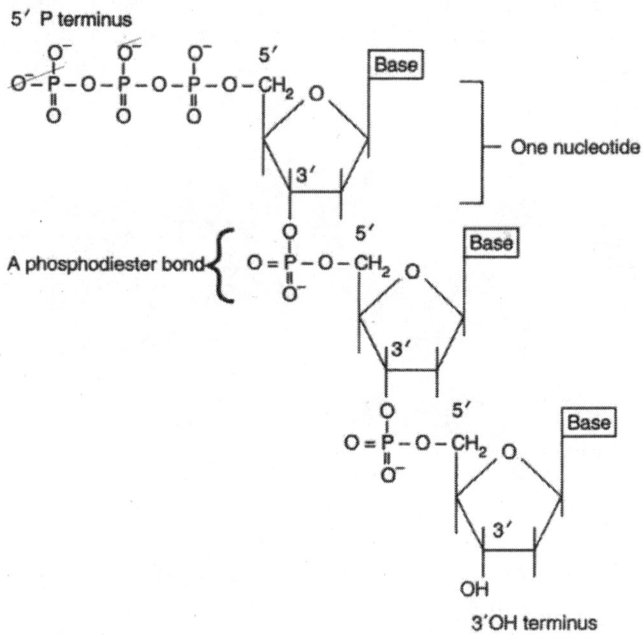

Fig. 2: Single helix of a short DNA molecule

In the polynucleotide helix, the two ends of the molecule are not the same. The top of the polynucleotide ends with a nucleotide in which the triphosphate group attached to the 5' carbon atom has not participated in a phosphodiester bond and α and β phosphates are still in place. This end is called 5' or 5'—P terminus. At the other end of the molecule, the unreacted group is the 3' hydroxyl group. This end is called 3' or 3'—OH terminus. So the polynucleotides or the DNA molecules have a direction which can be looked on as 5' → 3' or 3' → 5'. The two strands of the double helix are antiparallel, with one polynucleotide chain running in 5' → 3' direction and the other is in 3'→ 5' direction thus forming a stable double helix. The double helix structure proposed by Watson and Crick (1953) is hereunder glimpsed.

- DNA double helix comprises of two polynucleotide chains.
- Sugar and phosphate forming the backbone of the molecule remain on the outside of the helix while the nitrogenous bases are stacked inside the helix.
- The two polynucleotide chains are held together by hydrogen bonding.
- The two strands of the double helix are antiparallel.
- The double helix has two different grooves: major and minor and this feature is important in the interaction between the double helix and proteins involved in DNA replication and expression.
- The double helix, 20A° in diameter is right handed and 10 base pairs occur per turn of the helix. The pitch of the helix is 34A° hence, the spacing between the adjacent base pairs is 3.4A°.
- The complementary base pairing in the two polynucleotide chains (AT/TA or AU/UA and GC/CG are of fundamental importance in molecular biology.

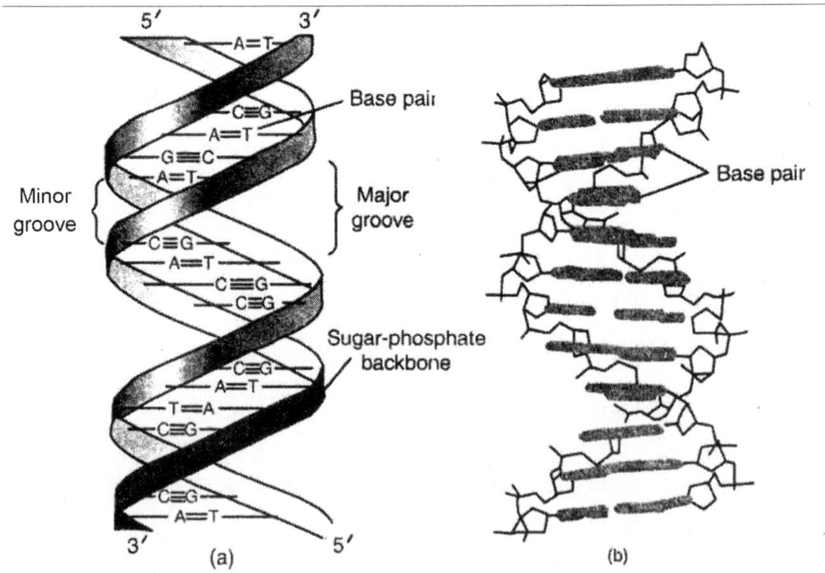

Fig. 3: The Watson-Crick model of DNA double helix

Variability in the genetic material

There are no chemical restrictions on the order in which the nucleotides can join together to form an individual DNA polynucleotide. At any point in the chain, the nucleotide could be A, G, C, or T. In a polynucleotide with just 10 nucleotides in its length it could have any of the $4^{10} = 1048576$ sequences. Different sequence possible for the polynucleotide of 1000 nucleotides is simply unimaginable. It is the variability of the DNA molecule that enables the genetic material to exist in an almost infinite number of forms.

DNA replication

The basic to all development is the replication of nuclear genetic material and its equal distribution between two daughter cells during mitosis. The basis of inheritance lay on faithful copying of the nucleotide sequence of the DNA molecule from the original parental DNA molecule to the two daughter molecules. The clue as to how this duplication is achieved lies in the double helical structure of the DNA molecules. Since complementary DNA strands are held together by comparatively weak hydrogen bonding, their separation can readily be achieved and when separated, each strand serves as a template for the synthesis of a complementary copy of a DNA helix.

Replication of a DNA molecule begins when the two complementary DNA strands are separated. This is usually accomplished by special proteins that unwind the molecule and expose the nucleotide bases. New complementary DNA strands are then synthesized by joining together deoxyribonucleotide triphosphates, one at a time, and with the removal of a di-phosphate. At the end of the process the cell has made to identical copies of all the biological information contained on the original, parental, DNA molecule, but the two new daughter DNA molecules are "Half old" and "Half new". Half the original DNA molecule is saved, or conserved in the daughter molecules. This is why the process is called semi-conservative.

Gene expression

A chromosome is properly defined as a single genetically specific DNA molecule which is attached to large number of proteins that are involved in maintaining chromosome structure and regulating gene expression. In eucaryotes, the DNA molecules are present as chromatin, the nucleoprotein fibre. Nucleoprotein is the conjugated protein consisting of a protein linked to a nucleic acid, either DNA or RNA . The protein combined with DNA is commonly either histone or protamine; the resulting nucleoproteins are found in chromosomes.

Large numbers of DNA binding proteins are likely to play a variety of specific roles in regulating gene expression. While the DNA encodes the genetic information, the proteins are involved in controlling the packaging of DNA and regulating the availability of DNA for transcription. Gene expression depends on many factors related to the DNA molecule.

- **Different forms of DNA:** DNA molecules can have more than one conformation in solution depending on the pH and ionic environment. Besides the conformations, DNA may also form unusual structures that appear to be important in regulating gene expression.

- **DNA supercoiling:** When the DNA helix has the normal number of base pairs per helical turn it is in the relaxed state. Changing this normal amount of twist can be demonstrated by grasping both ends of a short linear model (one to two complete turns) and twisting the ends in opposite directions. If the helix is over-twisted so that it becomes tighter, the edges of the narrow groove move closer together. If the helix is under-twisted, the edges of the narrow groove move further apart. Supercoiling occurs when the molecule relieves the helical stress by twisting around itself. Over-twisting leads to positive supercoiling, while under-twisting leads to negative supercoiling. Supercoiling is important in a number of biological processes, such as compacting DNA. Additionally, certain enzymes such as topoisomerases are able to change DNA topology to facilitate functions such as DNA replication or transcription. Topological characteristics of DNA and specifically DNA supercoiling influence all major DNA transactions in living cells. DNA supercoiling induces the formation of unusual secondary structure by specific DNA repeats which can also affect DNA functioning.
- **DNA content:** All the higher eukaryotes contain significantly more DNA than what appears to be necessary to encode their structures and functions. In human genome, at least a 10-fold excess of DNA exists that is required to produce the human being. The kinetic analysis shows that a large proportion of the DNA is composed of highly repetitive sequences and is unlikely to represent genes coding for proteins hence, are usually not transcribed. Most eukaryotic genomes also contain discrete segments of DNA (DNA sequence) called transposons that can change its position within the genome, sometimes creating or reversing mutations and altering the cell's genome size.
- **Flow of genetic information — the Central dogma:** In the cells of higher organisms all the genes are carried by a small number of chromosomes, each of which contains a single DNA molecule. Each of these DNA molecules carries several hundreds of genes. The majority of genes are spaced more or less randomly along the length of the DNA molecule.

The central dogma of molecular biology is an explanation of the flow of genetic information within a biological system. The central dogma as postulated by Francis Crick (1958) has been described as "DNA makes RNA and RNA makes protein (Figure 4). Such flow of information is unidirectional meaning that proteins cannot direct the synthesis of RNA and RNA cannot direct the synthesis of DNA. It states that such genetic information cannot be transferred back from protein to either protein or nucleic acid.

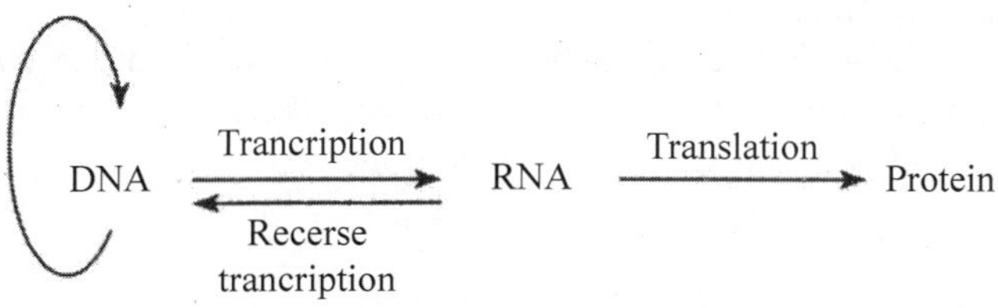

Fig. 4: Representation of central dogma

Later, Baltimore (1970) and Temin and Mizutani (1970) independently discovered that flow of information from DNA to RNA was not unidirectional because certain viruses do transfer biological information from RNA to DNA by an enzyme, reverse transcriptase that synthesizes DNA from an RNA template. Central dogma has been represented in Figure 4.

Organization of genes in DNA molecules

In some cases, genes are grouped into distinct clusters that contain related units of biological information. Examples of such cluster of genes are operon and multigene families.

Cluster of genes

• *Operon*

Operons are fairly common in the organization of genes in bacteria. An operon is a cluster of genes coding for a series of enzymes that work concertedly in an integrated biochemical pathway. The *lac* operon (*lactose* operon) is an operon or a group of genes with a single promoter that encode genes for the transport and metabolism of lactose in *Escherichia coli* and some other enteric bacteria. The *lac* operon consists of three structural genes and a promoter. The three structural genes are: *lacZ*, *lacY*, and *lacA*.

- *LacZ* encodes β-galactosidase , an intracellular enzyme that cleaves the disaccharide lactose into glucose and galactose.
- *LacY* encodes lactose permease , a transmembrane symporter that pumps β-galactosides into the cell using a proton gradient in the same direction.
- *Lac A* encodes galactoside O-acetyltransferase , an enzyme that transfers an acetyl group from acetyl-CoA to β-galactosides.

Only *lacZ* and *lacY* appear to be necessary for lactose catabolism. Gene regulation of the *lac* operon was the first genetic regulatory mechanism to be understood clearly. Hence, it has become a foremost example of prokaryotic gene regulation.

• *Multigene family*

A gene family is a set of several similar genes, formed by duplication of a single original gene, and generally with similar biochemical functions that encode proteins with similar sequences either over their full lengths or limited to a specific domain.

Examples of multigene families include those that encode the actins (family of globular multi-functional proteins), hemoglobins, immunoglobulins, tubulins, interferons, histones, etc. In human haemoglobin gene family, ten genes are there in two clusters on different chromosomes, called the α-globin and β-globin loci.

Structure of plant gene

The genes carrying the biological information are split into distinct units separated by intervening segments of DNA. The sections containing the coding sequence (that are expressed and translated into proteins) are called exons and the intervening sequences are called as introns (Figure 5). These introns are transcribed but not represented in mature mRNA and hence, are not translated into proteins. Introns always have two distinct nucleotides at either end. The 5′ cap protects the nascent mRNA from degradation and assists in ribosome binding during translation. A poly (A) tail is added to the 3′ end of the pre-mRNA once elongation is complete.These nucleotides are part of the splicing sites. Conservation of the two dinucleotides at the intron/exon junctions in all eukaryotic genes suggests that similar RNA splicing mechanisms are involved.

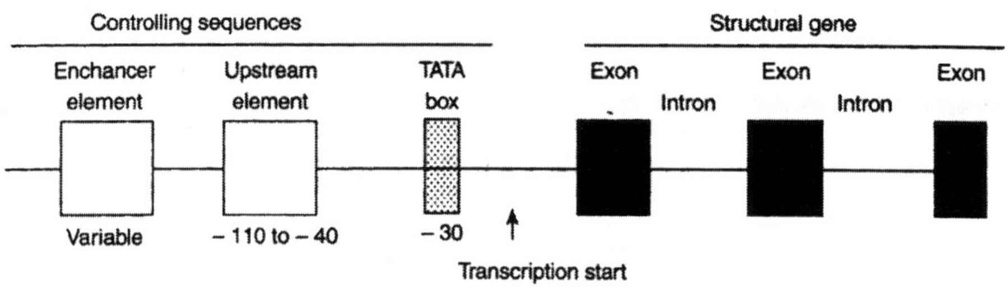

Fig. 5: Diagrammatic representation of a typical nuclear gene

Control of gene expression

Initiation of transcription occurs when RNA polymerase binds to a specific DNA sequence called the promoter at the beginning of an operon that occurs just upstream of genes. Promoter is a region of DNA that initiates transcription of a particular gene. Promoters are located near the transcription start sites of genes, on the same strand and upstream on the DNA (towards the 5' region of the sense strand). Promoters can be about 100–1000 base pairs long.

The maximum rate of transcription from a particular segment of DNA depends on the sequence of bases in its promoter. The actual DNA sequence at a promoter site varies from gene to gene, but all promoters have a consensus sequence, which in *E. coli* has two distinct components, approximately –10 and –35 nucleotides upstream from the transcription start site. The negative numbers refer to the position of each box to the site at which transcription starts, which are designated as +1. The +1 position therefore, corresponds to the site at which the first nucleotide is polymerized into the encoded RNA molecule.

- 35 box 5' TTGACA3'
- 10 box 5' TATAAT 3'

TATA box, the promoter region

The TATA box has the core DNA sequence 5'-TATAAA-3' or a variant, which is usually followed by three or more adenine bases. It is usually located 25 base pairs upstream of the transcription start site. The sequence is believed to have remained consistent throughout much of the evolutionary process, possibly originating in an ancient eukaryotic organism.

Gene expression

• Transcription

The first stage of gene expression, where the template strand of the gene directs the synthesis of a complementary RNA molecule is known as transcription (Figure 6). The RNA molecule that is synthesized is called a transcript. The RNA molecules produced by transcription can be grouped into three major classes according to function. These are messenger or mRNA, transfer or tRNA and ribosomal or rRNA. Transfer and ribosomal RNAs are the end products of gene expression and perform their roles in the cell as RNA molecules. Messenger RNA undergoes the second stage of gene expression, acting as an intermediate between a gene and its final expression product, a polypeptide translation product. Ribosomal and transfer RNAs are often referred to as stable RNAs, indicating these molecules are long lived in the cell, in contrast to mRNA, which has a relatively rapid turnover rate.

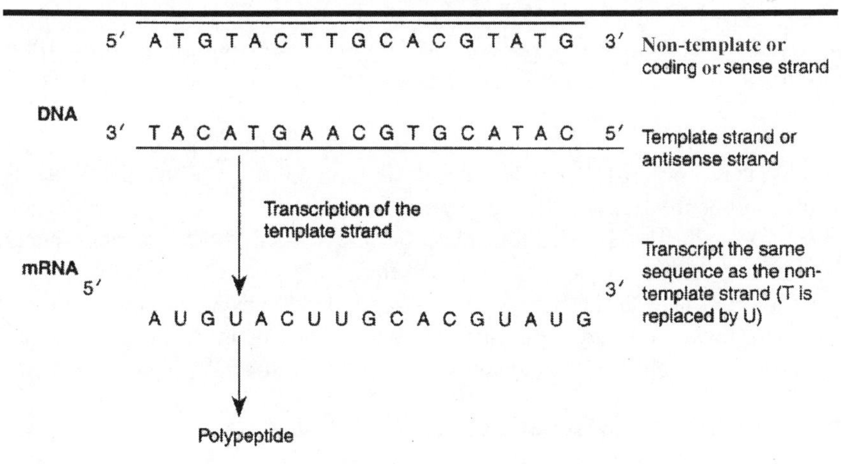

Fig. 6: Transcription of messenger RNA

- **Messenger RNA:** In the prokaryotes, the mRNA molecules are the direct copies of the gene. However, in eukaryotes, most mRNAs undergo a fairly series of modifications and processing events before translation occurs. The DNA strand that bears the nucleotide sequence identical to the mRNA (except for possessing T instead of U) is called the coding strand or sense strand or non-template strand. The other strand of DNA which directs the synthesis of mRNA *via* complementary base pairing is called template strand or antisense strand. The antisense strand has 3' → 5' direction which directs the synthesis of mRNA in the 5' → 3' direction. Synthesis of mRNA occurs by polymerization of nucleotide subunits in the presence of a DNA template which directs the order in which the individual nucleotide is polymerized into RNA. During polymerization, the 3' OH group of one nucleotide reacts with the 5' P group of the second nucleotide to form a phosphodiester bond. The transcript is therefore, synthesized in 5' → 3' direction. The enzyme that catalyzes the reaction is called DNA dependent RNA polymerase.

- **Ribosomal RNA:** Ribosomal RNA molecules are components of ribosomes that act as factories for protein synthesis. Ribosomes are made up of rRNA molecules and proteins and are numerous in most cells. The rRNA comprises about 80% of the total cell RNA and about 10% of the total protein. During translation, ribosomes attach to mRNA molecules and migrate along them, synthesizing polypeptides.
- **Transfer RNA:** The tRNAs are transcribed initially as precursor tRNA which are subsequently processed to release the mature molecules. Transfer RNAs are the adapter molecules that read the nucleotide sequences of mRNA transcript and convert it into a sequence of amino acids. Each cell contains many different types of tRNA molecules. The tRNA molecules are relatively small, mostly between 74 to 95 nucleotides and almost every tRNA molecule can be folded into a base paired structure referred to as clover leaf. The anticodon arm of the tRNA plays the central role in decoding the biological information carried by the mRNA. During protein synthesis, an amino acid is attached to the acceptor arm of the tRNA.

• Translation

Translation is the process by which the information in the mRNA molecule is decoded into polypeptide chain. This polymerization of the amino acids is carried out by ribosomes and tRNAs. The sequence of amino acids in the polypeptide being synthesized in the ribosome is specified by the sequence of nucleotides in the mRNA molecule being translated. Information in mRNA is indeed translated in units of three nucleotides, and these units are called triplet codons. The assignment of particular codons to particular amino acids is called Genetic code (Figure 7).

Important features of the Genetic code

1. The genetic code is the set of rules by which information encoded within genetic material (DNA or mRNA sequences) is translated into proteins by living cells. Biological decoding is accomplished by the ribosome, which links amino acids in an order specified by mRNA, using transfer RNA (tRNA) molecules to carry amino acids and to read the mRNA three nucleotides at a time. The genetic code is highly similar among all organisms and can be expressed in a simple table with

64 entries. Each code word or codon is a triplet of nucleotides in the mRNA that code for a single amino acid. Since there are only four different nucleotide bases in DNA (A, G, C, T) and correspondingly in mRNA (A, G, C, U), a one base code could only specify four different amino acids and similarly a two base code could specify 16 amino acids (4 × 4). A three base code is capable of specifying all 20 amino acids since there are 64 possible combinations (4 × 4 × 4).

2. Of the 64 possible codons, 61 specify a particular amino acid; with the remaining 3 codons (UAA, UAG, UGA) being the stop signals which indicate to the translation machinery that protein synthesis is complete.
3. The codon AUG (which specifies methionine) is also known as start codon since in almost all cases the start site for translation is determined by an AUG in the mRNA. Thus most newly synthesized polypeptides will have methionine at the amino terminus, though it may be removed subsequently by post translational processing of protein.
4. All amino acids except methionine and triptophan have more than one codon.
5. Most synonymous codons are grouped into families (GGA, GGU, GGG, GGC), all code for glycine.
6. The stop or termination codons, UAA, UAG and UGA do not code for amino acid if present in the middle of a heteropolymer cause protein synthesis to stop. One of the three codons always occurs at the end of a gene at the point where translation must stop.
7. This genetic code is followed by virtually in all nuclear genes.

UUU	Phe	UCU	Ser	UAU	Tyr	UGU	Cys
UUC	Phe	UCC	Ser	UAC	Tyr	UGC	Cys
UUA	Leu	UCA	Ser	UAA	Stop	UGA	Stop
UUG	Leu	UCG	Ser	UAG	Stop	UGG	Trp
CUU	Leu	CCU	Pro	CAU	Hits	CGU	Arg
CUC	Leu	CCC	Pro	CAC	Hits	CGC	Arg
CUA	Leu	CCA	Pro	CAA	Gln	CGA	Arg
CUG	Ile	CCG	Pro	CAG	Gln	CGG	Arg
AUU	Ile	ACU	Thr	AAU	Asn	AGU	Ser
AUC	Ile	ACC	Thr	AAC	Asn	AGC	Ser
AUA	Met	ACA	Thr	AAA	Lys	AGA	Arg
AUG	Vat	ACG	Thr	AAG	Lys	AGG	Arg
GUU	Vat	GCU	Ala	GAU	Asp	GGU	Gly
GUC	Vat	GCC	Ala	GAC	Asp	GGC	Gly
GUA	Vat	GCA	Ala	GAA	Glu	GGA	Gly
GUG	Vat	GCG	Ala	GAG	Glu	GGG	Gly

Key to amino acid abbreviations: Ala: alanine; Arg: arginine; Asn: asparagine; Asp: aspartic acid; Cys: cysteine; Gln: glutamine; Glu: glutamic acid; Gly: glycine; Hits: histidine; Ile: isolecucine; Leu: leucine; Lys: lysine; Met: methionine; Phe: phenylalanine; Pro: proline; Ser: serine; Thr: threonine; Trp: tryptophan; Tyr: tyrosine; Val: valine.

Fig. 7: Genetic code

Cloning of genes or Recombinant DNA technology or Genetic engineering

In gene cloning, a fragment of DNA that code for the required gene product is isolated from the host organism and inserted to a vector molecule (plasmid, cosmid, bacteriophage Lambda, etc.) to construct the recombinant DNA molecule where it can be replicated. The construction of such composite or artificial DNA molecules is also termed as genetic engineering or gene manipulation because of the potential for creating novel genetic combinations by molecular biology. The recombinant DNA is then incorporated into the host genome in which they do not naturally occur but in which they are capable of continued propagation. Fundamental steps in gene cloning are presented hereunder.

- Identification and isolation of the gene of interest.
- Incorporation of the gene of interest into a small replicating DNA molecule called vector by cutting and joining DNA molecules at specific sites through enzyme system. The vector with an incorporated gene is called a recombinant vector.
- Introduction of the recombinant vector into a suitable host cell.
- Monitoring and selection of the host cell which have acquired recombinant DNA molecule.
- Culture of the plant cell or tissues having the recombinant DNA sequence *in vitro* to develop the transgenic plant.

Prime requirements of genetic engineering

Vectors

One of the most important elements in recombinant DNA (rDNA) technology is the vector, the cloning vehicle. Different types of cloning vectors are plasmid, cosmid, Bacteriophage, Yeast artificial chromosome, etc. A vector should have the following features.

- It must contain a replicon (A replicon is a DNA molecule or RNA molecule, or a region of DNA or RNA that replicates from single origin of replication) that enables it to replicate in the host cell.
- It should have several marker genes which will help to identify the transformed cells containing the recombinant DNA.
- It should have a unique cleavage site within one of the marker genes so that insertion of foreign DNA into the marker gene leads to its inactivation and identification of recombinant DNA molecule.
- The vector DNA should contain suitable control elements such as promoters, terminators, ribosome binding sites, etc. for the expression of the cloned gene.

Different types of cloning vectors

- **Plasmids:** A plasmid is a small DNA molecule within a cell as extra-chromosomal units that is physically separated from a chromosomal DNA and can replicate independently. They are most commonly found in bacteria as small, circular, double-stranded DNA molecules; however, plasmids are sometimes present in archaea and eukaryotic organisms. Most cloning vectors for plants are based on the *Ti* plasmid of *Agrobacterium tumefaciens* and *Ri* plasmid of *Agrobacterium rhizogenes.*

- **Cosmids:** A cosmid is a type of hybrid plasmid that contains a Lambda phage cos sequence. Cosmids' DNA sequences are originally from the lambda phage. Cosmids can contain 37 to 52 (normally 45) kb of DNA. They can replicate as plasmids if they have a suitable origin of replication: for example "SV40 ori" in mammalian cells, "ColE1 ori" for double-stranded DNA replication or "f1 ori" for single-stranded DNA replication in prokaryotes. Cosmids are developed to overcome the technical problems of introducing large pieces of DNA into *E. coli e.g.,* pJB8 cosmid.
- **Bacteriophage:** A bacteriophage is a virus that infects and replicates within a bacterium. Phages replicate within the bacterium following the injection of their genome into its cytoplasm. During infection a phage attaches to a bacterium and inserts its genetic material into the cell. The bacteriophages used for cloning are the phage λ and M13 phage. There is an upper limit on the amount of DNA that can be packed into a phage (a maximum of 53 kb), therefore to allow foreign DNA to be inserted into phage DNA, phage cloning vectors need to have some non-essential genes deleted, for example the genes for lysogeny (Lysogeny is a condition in which the genome of a bacteriophage remains inside a host cell in a repressed condition) in phage λ.
- **Yeast artificial chromosome:** Yeast artificial chromosomes (YAC) are genetically engineered chromosomes derived from the DNA of the yeast, *Saccharomyces cerevisiae*, which is then ligated into a bacterial plasmid. By inserting large fragments of DNA, from 100–1000 kb, the inserted sequences can be cloned and physically mapped using a process called chromosome walking.
- **Bacterial artificial chromosome:** A bacterial artificial chromosome (BAC) is a DNA construct, based on a functional fertility plasmid (or F-plasmid), used for transforming and cloning in bacteria, usually *E. coli.* F-plasmids play a crucial role because they contain partition genes that promote the even distribution of plasmids after bacterial cell division. The bacterial artificial chromosome's usual insert size is 150-350 kbp.

Enzymes for cutting and joining of DNA molecule

• Enzymes for cutting the DNA

Restriction endonucleases are enzymes that cut the DNA at precisely defined sequences, thus act as molecular scissors. Smith and Wilcox (1970) first isolated restriction endonucleases enzymes from *Haemophilus influenza.* Now, many types of restriction endonuclease enzymes have been isolated from a wide variety of bacteria. The relevant endonucleases for gene cloning have the following characteristics.

1. These enzymes recognize specific nucleotide sequence and cut a DNA molecule at this sequence and nowhere else.
2. These enzymes are very stable and only require Mg^{++} as cofactors.
3. The majority of the known recognition sites have palindromic sequences (sequences have an axis of rotational symmetry). For instance, the recognition sequence of *EcoRI* is:

   5'-GAATTC-3'

   3'-CTTAAG-5'
4. Many restriction enzymes make a simple double stranded cut in the middle of the recognition sequence resulting a blunt end *e.g.*, *PvuII, AluI, HaeIII,* etc.

   HaeIII 5'- GG↓CC-3' 5'- GG_{OH} 3'

   3'- CC↓GG-5' 3'- CC_P 5'
5. Some restriction enzymes do not cut the two DNA strands at exactly the same position. Instead, the cleavage is staggered usually by 2 or 4 nucleotides so that the resulting DNA fragments have short single stranded overhangs at each end called sticky ends. For example, *EcoRI, BamHI, HindIII, PslI, TaqI*

   EcoRI 5'-G↓AATTC-3' 5'-G_{OH}5'

   3'-CTTAA↓G-5' 3'-$CTTAA_P$5'

   PstI 5'-CTGCA↓G-3' 5'-$CTGCA_{OH}$3'

   3'-G↓ACGTC-5' 3'-Gp5'
6. The DNA ends of one molecule generated by a given restriction enzyme with a specified recognition sequence can form complementary base pairs with other DNA molecule, provided they possess the same ends.

   Cleavage and complementary base pairing *of EcoRI*

   5'G3' + AATTC – 3' G-5'

   5'-GAATTC-3'

   3'-CTTAAG-5'

• Enzymes for joining the DNA

E. coli and bacteriophage T4 encode an enzyme which joins DNA fragments are called DNA ligase or molecular glue. DNA ligase is a specific type of enzyme that facilitates the joining of DNA strands together by catalyzing the formation of a phosphodiester bond. It plays a role in repairing single-strand breaks in duplex DNA in living organisms, but some forms (such as DNA ligase IV) may specifically repair double-strand breaks (*i.e.* a break in both complementary strands of DNA). Single-strand breaks are repaired by DNA ligase using the complementary strand of the double helix as a template, with DNA ligase creating the final phosphodiester bond to fully repair the DNA. When the vector DNA and foreign DNA (gene of interest) are both cut with the same restriction enzyme, the overlapping ends of the vector and the foreign DNA are compatible and complementary. These DNA fragments and vector molecules are mixed together, which form complementary base pairs between overlapping terminal single stranded DNA sequences. Ligases act on DNA substrates with 5' terminal phosphate groups and forms the phosphodiester bond between the two DNA sequences (vector molecule and the DNA to be cloned) to join them together. This process, called ligation is the final step in the construction of a recombinant DNA molecule.

• Linkers and Adaptors:

If suitable enzyme sites are not available for manipulation on the DNA, the cleavage sites can be added with linker or adaptor molecules.

- ***Linkers*:** These are synthetic single stranded oligonucleotides which self-associate to form symmetrical double stranded molecules containing recognition sequence for a restriction enzyme. For example, the oligonucleotide dCCGAATTCGG will self-associate to give a duplex structure containing the *EcoRl* recognition sequence.

  5'CCG ↓ AATTCGG3' 3'

  GGCTTAA ↓ GCC 5'

- ***Adaptor*:** Adaptors, the synthesized DNA molecules are required when the target site for the restriction enzyme used is present within the DNA sequence to be cloned. The adaptor, used in the 5' hydroxyl form is ligated to the foreign DNA to prevent self-polymerization.

• Marker genes

The marker genes are introduced into the vector along with the target gene at the time of construction of recombinant DNA to know whether the desired DNA has been successfully transferred into the recipient cells. Marker genes are grouped in two categories.

i) **Reporter gene:** A reporter gene is a gene that is attached to a regulatory sequence of another gene of interest in bacteria, cell culture, animals or plants. Certain genes are chosen as reporters because the characteristics they confer on organisms expressing them are easily identified and measured. Reporter genes are often used as an indication of whether a certain gene has been taken up by or expressed in the cell or organism population. Different reporter gene and gene product is mentioned below.

Gene name	Gene product
LacZ	β galactosidase
gus	β glucuronidase
Lux F2	Bacterial luciferase
cat	Chloramphenicol acetyltransferase
neo	Neomycin phosphotransferase
nos	Nopaline synthase
ocs	Octopine synthase
gfp	Green fluorescent protein
rfp	Red fluorescent protein

iii) **Selectable markers:** A selectable marker is a gene introduced into a cell, especially a bacterium or to cells in culture, that confers a trait suitable for artificial selection. They are a type of reporter gene used in laboratory microbiology, molecular biology and genetic engineering to indicate the success of a transfection or other procedure meant to introduce foreign DNA into a cell. Selectable markers are often antibiotic resistance genes. Bacteria that have been subjected to a procedure to introduce foreign DNA are grown on a medium containing an antibiotic, and those bacterial colonies that can grow have successfully taken up and expressed the introduced genetic material. The selectable marker genes enable the transformed cells to survive on media containing toxic levels of selection agents (*e.g.*, kanamycin, neomycin, gentamycin, streptomycin, etc.). Examples of different selectable marker genes are mentioned below.

- *β lactamase* are enzymes produced by bacteria that provide multi-resistance to *β lactam* antibiotics such as penicillins, cephalosporins, etc. to bacterial hosts.
- *Neomycin phosphotransferase II* (*nptII*) conferring resistance to neomycin, kanamycin, gentimycin in the bacteria.
- *Gentamycin acetyl transferase* conferring resistance to gentamycin in the bacteria.
- *Streptomycin phosphotransferase* conferring resistance to streptomycin in the bacteria.

• Identification and analysis of the cloned genes

The cloned DNA (gene) is analyzed mostly by hybridization probe method before transfer to the host cells. In molecular biology, a hybridization probe is a fragment of DNA or RNA of variable length (usually 100-1000 bases long) which is used in DNA or RNA samples to detect the presence of nucleotide sequences (the DNA target) that are complementary to the sequence in the probe. The probe thereby hybridizes to single-stranded nucleic acid (DNA or RNA) whose base sequence allows probe-target base pairing due to complementarity between the probe and target. The labelled probe is first denatured (by heating or under alkaline conditions such as exposure to sodium hydroxide) into single stranded DNA (ssDNA) and then hybridized to the target ssDNA (Southern blotting) or RNA (Northern blotting) immobilized on a membrane or *in situ*. Hybridization probing is based on the following principles.

- Any two single stranded nucleic acids have the potential to base pair with one another.
- If the polynucleotides are complementary, the extensive base pairing will occur to form a stable double stranded molecule.
- The base pairing can occur between single stranded DNA molecules to form a double helix, between single stranded RNA molecules, and also between combination of one DNA and one RNA strand.

- If the two strands of nucleic acids are not exactly complementary to each other, they will form an unstable hybrid structure because only a small number of individual inter-stand bonds will be formed.

Methods of gene transfer

Gene transfer or DNA uptake refers to *in vitro* laboratory technique that moves a specific piece of recombinant DNA (desired DNA segment ligated to the vector) into the host cells. The directed desirable gene transfers from one organism to other and the subsequent stable integration and expression of a foreign gene into the genome is referred to as genetic transformation. The transferred gene is known as transgene and the organisms that are developed after a successful gene transfer are known as transgenics. Transgenic plants carry integrated foreign DNA in its gnome. Gene transfer techniques can be grouped in two main categories: i) vector mediated gene transfer and ii) direct gene transfer.

Vector mediated gene transfer

The term plant gene vector refers to the potential vectors both for the transfer of genetic information between plants and the transfer of genetic information from other organisms (bacteria, fungi, etc.) to plants. Plant gene vectors being exploited for the transfer of genes are plasmids of *Agrobacterium,* viruses and transposable elements.

a) ***Agrobacterium*-mediated gene transfer:** *Agrobacterium* species, the soil bacteria have been known as plant pathogens since the beginning of the 20th century. *A. tumefaciens* causes crown gall disease and *A. rhizogenes* causes hairy root disease. The ability of *Agrobacterium*, the "natural genetic engineer" to transfer DNA to plant cells has been harnessed for the purposes of plant genetic engineering since early 1980s. The molecular basis of genetic transformation of plant cells by *Agrobacterium* is transfer from the bacterium and integration into the plant nuclear genome of a region of a large tumor-inducing (Ti) or rhizogenic (Ri) plasmid resident in *Agrobacterium.* The bacterial T-DNA is about 24,000 base pairs long and contains genes that code for enzymes synthesizing opines and phytohormones. Opines are produced from common plant metabolites like sugars, amino acids and α-keto acids and are used by the bacterium as an important energy, carbon and nitrogen source. By transferring the T-DNA into the plant genome, the bacterium essentially reprograms the plant cells to grow into a tumor and produce a unique food source for the bacteria. The synthesis of the plant hormones auxin and cytokinin enables the plant cell to grow uncontrollably, thus forming the crown gall tumors. *Agrobacterium*-mediated T-DNA transfer is widely used as a tool in biotechnology. In genetic engineering, the tumor-promoting and opine-synthesis genes are removed from the T-DNA and replaced with a gene of interest and/or a selection marker, which is required to establish which plants have been successfully transformed. Examples of selection markers include neomycin phosphotransferase, hygromycin B phosphotransferase (both phosphorylate antibiotics) and phosphinothricin acetyltransferase (which acetylates and deactivates phosphinothricin, a potent inhibitor of glutamine synthetase, an enzyme that plays an essential role in the metabolism of nitrogen by catalyzing the condensation of glutamate). *Agrobacterium* is then used as a vector to transfer the engineered T-DNA into the plant cells where it integrates into the plant genome. Transfer of the plasmid of the bacteria with the recombinant DNA inserted in it into the host cell begins naturally with the invasion of the bacteria into the wound of plant parts. This method is widely used to generate transgenic plants carrying a foreign gene.

b) **Viruses-mediated gene transfer:** High efficiency of gene transfer can be obtained by infection and the amplification of transferred genes that occurs *via* viral genome replication. Virus-based vectors require the ability to modify plant viruses to express foreign genes. It is potentially very useful for transient and whole-plant expression studies, and for obtaining high levels of expression in plants. Host-range limitations can be overcome by virus-mediated gene transfer technology. However, it is not practical for producing stably transformed plants. Hence, for specialized studies (production of foreign proteins in plants, expression of potentially lethal genes, etc.), virus-based vectors and systems may be superior. Three groups of viruses generally employed for gene transfer are Caulimo viruses, Gemini viruses and RNA viruses.

Direct DNA transfer

In the direct gene transfer methods, the foreign gene of interest is delivered into the host plant cell without the help of a vector. The methods used for direct gene transfer in plants are:

(a) **Chemical mediated gene transfer:** Chemical gene transfer method involves treatment of plant protoplast with different chemicals like, polyethylene glycol, dextran sulphate, calcium phosphate, etc. that creates reversible channels in the plasma membrane. Polyethylene glycol is mostly used in this method. The recombinant DNA diffuses passively through these channels into the cytoplasm, a small portion enters the nucleus and integrates into the plant genome.

(b) **Microinjection:** In this method, the DNA is directly injected into plant protoplasts or cells (specifically into the nucleus or cytoplasm) using fine tipped (0.5-1.0 micrometer (μm) diameter) glass needle or micropipette. This method of gene transfer is used to introduce DNA into large cells such as oocytes, eggs, and the cells of early embryo.

(c) **Electroporation:** In this method, a pulse of high voltage is applied to protoplasts/cells/tissues to make transient (temporary) pores in the plasma membrane which facilitates the uptake of foreign DNA. The cells are placed in a solution containing DNA and subjected to electrical shocks to cause holes in the membranes. The foreign DNA fragments enter through the holes into the cytoplasm and then to nucleus.

(d) **Particle gun/Particle bombardment**: In this method, the foreign DNA containing the genes to be transferred is coated onto the surface of minute gold or tungsten particles (1-3 micrometers, μm) and bombarded (helium driven) onto the target tissue or cells using a particle gun (also called as gene gun/shot gun/micro-projectile gun). The micro-projectile bombardment method was initially named as biolistics by its inventor Sanford (1988). Two types of plant tissue are commonly used for particle bombardment *viz*., primary explants and the proliferating embryonic tissues.

(e) **Transformation**: This method is used for introducing foreign DNA into bacterial cells *e.g. E. coli*. The transformation frequency (the fraction of cell population that can be transferred) is very good in this method. The uptake of plasmid DNA by *E. coli* is carried out in ice cold $CaCl_2$ (0-5°C) followed by heat shock treatment at 37-45°C for about 90 second. The $CaCl_2$ breaks the cell wall at certain regions and binds the DNA to the cell surface.

(f) **Conjugation**: It is the process by which one bacterium transfers genetic material to another through direct contact. It is a natural microbial recombination process and is used as a method for gene transfer. In conjugation, two live bacteria come together and the single stranded DNA is transferred *via* cytoplasmic bridges from the donor bacteria to the recipient bacteria.

(g) **Liposome mediated gene transfer or Lipofection:** Liposomes are circular lipid molecules with an aqueous interior that can carry nucleic acids. Liposomes encapsulate the DNA fragments and then adhere to the cell membranes and fuse with them to transfer DNA fragments. Thus, the DNA enters the cell and then to the nucleus. Lipofection is a very efficient technique used to transfer genes in bacterial, animal and plant cells.

Selection of the transformed cells

Identification of specific recombinant clone from among the population of recombinant clones and at the same time, in the background of non-recombinant clones is the most crucial step in genetic engineering. Selection for resistance to antibiotics, herbicides, etc. and growth of plant cells on selective media provide initial phenotypic evidence for transformation. However, molecular evidences (Northern, Southern, Western blot, PCR) are essential to confirm integration of the transferred genes in the host cell genome.

Development of the transgenic plant

The plant cells or tissue having recombinant DNA sequence are cultured in artificial nutrient media under aseptic condition in the laboratory to develop the transgenic (transformed) plant.

Status and expression of the transferred genes in transgenic plant

Generally, the inserted DNA integrates at a single locus. The transgenic plant is self-fertilized and seeds are harvested. The seeds are grown and simple inheritance pattern of 3: 1 of the trait conferred by the transgene indicates a single site of insertion which is confirmed by Southern blotting technique of molecular evidence. The inserted DNA is usually transmitted as a simple Mendelian trait and expressed in the transgenic plant.

However, sometimes the transgene may not show its expression, called gene silencing, due to various causes including hypermethylation, suppression by antisense genes, etc.

Backcross breeding is the final step in producing genetically engineered crops. This is done by crossing the transgenic plant with elite lines using conventional plant breeding methods. This enables the combination of the desired traits of the elite parents and the transgenic into a single line. The offspring are repeatedly crossed back to the elite line to obtain a transgenic line possessing high yield and other desirable characters.

The length of time in developing transgenic plant depends upon the gene, crop species, available resources and regulatory approval. It varies from 6 to 15 years before a new transgenic hybrid is ready for commercial release.

Cisgenic

Cisgenesis is one term for organisms that have been engineered using a process in which genes are artificially transferred between organisms that could otherwise be conventionally bred. Cisgenic modification is useful for plants that are difficult to cross breed by conventional means and that plants in the cisgenic category should not require the same regulatory scrutiny as transgenic.

Markers and marker assisted selection

Concept of marker is based on the genetic polymorphism which does mean the existence of two or more clearly different phenotypes in the same population of a species. Co-inheritance of the characters was first identified with the classical phenomenon, linkage. Linkage is a familiar concept in genetics referring that combination of genes tends to inherit together because they are close to each other in the same chromosome. So, inheritance of the particular character serves as a marker for the existence of the gene of interest say, the resistance gene in the genotype. No further confirmation is needed to prove its existence in the genotype. A marker must be polymorphic *i.e.*, it must exist in different genotypes so that chromosome carrying the mutant gene can be distinguished from the chromosome with normal gene by the form of marker it carries. Polymorphism in marker can be detected at four levels; phenotypic difference (morphological marker), difference on the segments of the chromosome (cytological marker), difference in proteins *i.e.*, gene products (biochemical marker) and difference in nucleotide sequence of DNA in the chromosome (molecular marker). Genetic markers are classified into three different types i.e., morphological markers, biochemical markers, and molecular markers (Nadeem *et al.*, 2018).

• Morphological marker

Visual indications of phenotypically distinct traits, such as colour, size, and form of flowers, seeds and leaves; plant growth, inflorescences or root systems; pigmentation; and habit are known as morphological markers (Chesnokov *et al.*, 2020). These markers are often detectable by simple visual inspection. Morphological markers generally refer to different monogenic or oligogenic qualitative traits under either dominant or recessive genetic control. As for example, enhancement of fruit carotenoid content and colour of tomato by different mutant genes *viz.*, *high pigment-1* (*hp-1*), *old gold crimson* (*og^c^*), *dark green* (*dg*) and *Beta-carotene* (*B*) and anthocyanin expression in tomato fruit by *Aft* gene. In well-characterized crops like tomato, pea, hot and sweet pepper, carrot, etc. tens or hundreds of genes that determine morphological traits have been mapped to specific chromosome locations.

Use of morphological markers doesn't require specialized equipment and is simple to detect. They are not dependent on complex biochemical and molecular methods. Such markers have been successfully employed by breeders in breeding programmes for a variety of crops. Inspite of the fact that these biomarkers are simpler to assess than other markers, they also have drawbacks that restrict their use in breeding programmes (Sepahvand *et al.* 2021). The main drawbacks of morphological markers include their scarcity, sensitivity to different environmental conditions, and dependence on plant growth phases. Since the dawn of time, people have effectively investigated variation for use in plant breeding using a variety of morphological markers. Some of these markers can be utilized as indirect selection criterion in breeding because they are connected to other agronomic features. The dwarfism governed by the gene *Rht10* in Norin 10 is a classic example for successful use of morphological marker in wheat breeding. Similarly, dwarfing gene *sd1* of rice from Dee-geo-woo-gen has been extensively utilized by rice breeders all over the world based on morphological markers. These might be viewed as a successful utilization of morphological markers in contemporary breeding (Jiang, 2013). Novel breeding line "Purple tomato" has been developed through combining the lycopene enhancing gene *dg* or *hp-2*^{dg} present in chromosome 1 and anthocyanin expressing gene, *Aft* present in chromosome 10 in one genotype based on morphological marker characters conditioned by two genes, dark green fruit of *dgdg* and purple colour for anthocyanin pigment of *Aft*- gene (Hazra *et al.*, 2018).

Description of such morphological marker trait has significance only when accompanied by properly documented pedigree information and environmental condition. Such factors may modify a gene's expression of phenotype which limits its usefulness as a genetic marker. The constraints of using the morphological markers in plant breeding are mentioned.

- Morphological markers sometimes put forth so huge and undesirable effects on the phenotype that their uses are avoided in the breeding programme.
- Most of the morphological markers mask the effects of linked polygenes, making it nearly impossible to identify the desirable linkages to be used in selection.
- Morphological markers are sometimes highly influenced by the environment, making it irreproducible over a range of environments.

• Cytological marker

Cytological markers are those that correspond to differences in number, banding pattern, size, shape, order, and location of the chromosomes. Cytological markers rely on the segments of the chromosome, which can preferentially be stained to reveal the presence or absence of the certain tracks of the genome, for example, G banding. Such marker techniques are time consuming and inefficient to be used routinely in a plant breeding programme. However, development of genomic *in situ* hybridization (GISH) technique has offered great potential as a marker system. Such *in situ* hybridization based markers may be utilized in the disease resistance breeding programme utilizing wild species as a donor parent to analyse interspecific backcross progenies for the amount and chromosome sites introgressed in the desired line from the wild species.

• Biochemical markers

▪ lysozyme

Among the biochemical markers, lsozyme markers have been utilized widely in plant breeding. Isozymes refer to the different molecular forms of the same enzyme that catalyze the same reaction. Isozymes are differently charged protein molecules that can be separated by electrophoresis. Since enzymes catalyze specific biochemical reactions, it is possible to visualize the location of a particular enzyme on a gel by supplying the appropriate substrate and cofactors, and involving the product of the enzyme reaction in a colour producing reaction. The coloured product becomes deposited on the gel, forming a visible band where a particular enzyme has been electrophoretically localized. Bands visualized from specific enzymes represent protein products, have a genetic basis and thus provide genetic information as codominant markers. However, their utility is restricted because of

i) the paucity of lsozyme loci,
ii) the fact that they are subject to post-translational modifications, and
iii) the change in fingerprinting with the growth stages of the plant.

An example of the use of biochemical markers in vegetable breeding is the introduction of the root knot nematode resistance gene (*Mi*) into commercial tomato varieties. This gene was originally isolated in a wild relative of tomato (*Solanum peruvianum*) and was transferred into cultivated tomatoes by conventional sexual crosses. However, the wild tomato parent contributed many additional traits that were not desirable, such as small fruit, poor colour, and toxic compounds. After many generations (more than 10 years) of selection, most of the wild type chromosomes had been eliminated except for a short section containing the *Mi* gene. However, this section also contained a distinct form (or isozyme) of a common enzyme, acid phosphatase, that was directly adjacent (linked) to the *Mi* gene. Once this linkage was discovered, breeders could determine whether an individual plant had inherited the *Mi* gene simply by testing which form of the acid phosphatase enzyme was present. This made further breeding and selection of nematode resistant plants much more convenient, as the isozyme tests can be conducted on very small seedlings, while tests for nematode resistance require inoculating test soil with the pest and then growing the plants to full size. Unfortunately, since the acid phosphatase gene is not identical with the *Mi* gene, plants were eventually found which contained the isozyme marker but did not have nematode resistance, so a linked marker such as this is not a guarantee that the desired gene has been inherited.

▪ Storage protein

Polymorphism in storage protein can also be used as a marker. However, the protein may be present in only small amounts and therefore, be difficult to detect, or it may be affected by other genes or by the environment in which the plant is growing. In addition, in many vegetables including lettuce or tomato, relatively little variation is found among enzymes or proteins to be used as markers to identify different lines.

▪ Inducible chemical compounds

Different inducible chemical compounds associated with the disease resistance/ susceptibility like, glucosinolate with resistance to downy mildew in cauliflower, phenols with resistance to tomato leaf curl virus in tomato, etc. can also be used as markers. However, precision in the level of inoculation and concentration of the chemicals after specific days of inoculation is required for its utilization in resistance breeding programme.

• Molecular (DNA) markers

Limitation of biochemical or protein markers can be minimized or eliminated using DNA-based markers, one of the new tools of biotechnology. Molecular markers are defined as the fragments of DNA sequence associated with a genome, which are used to identify a particular DNA sequence. These molecular markers aid in the identification of a particular DNA sequence at a specific position, whether it be in a pool of unknown DNA or throughout a whole chromosome.

Molecular markers are 'tags' that can be used to identify specific genes and locate them on the chromosomes. In the case of the *Mi* gene, DNA sequences have been isolated that overlap with parts of the desired gene and therefore, are more certain to be inherited with it. In other cases, the gene of interest has been isolated, or cloned, and copies are available for use as the marker. An advantage of these types of markers is that the DNA of the gene itself is identified, rather than the expression of enzymes or proteins. Molecular marker thus refers to a polymorphic tag which gives the clue for the existence as well as location of a specific gene in the chromosome and can be compared with the book mark to identify a specific page in a book. A simple way to appreciate this powerful tool for selection is to visualize the UPC bar-code found on most products in the supermarket. There are a number of different kinds of molecular markers that can be used to confirm the presence of a gene or even locate its relative position on the chromosome.

So, molecular marker is a DNA sequence which is readily detected and whose inheritance can be monitored. Naturally occurring DNA polymorphism (polymorphism in nucleotide sequence) forms the basis of the use of molecular markers. Virtually inexhaustible number of DNA-based markers can be generated to distinguish among even closely related varieties. The clear cut advantages of the DNA based molecular marker techniques over the other marker system are mentioned below.

- Highly reproducible fingerprint pattern.
- Fingerprinting can be done at any stage of plant growth.

- Only small amount of plant tissue is needed.
- It allows reliable tracking of beneficial traits during variety selection.
- It is the only practical technique for ensuring the presence of multiple beneficial genes in a single variety.
- DNA-based marker systems are more readily automated to increase cost effectiveness.
- It is the best currently available technology to establish differentiation for patent protection and Plant Variety Protection Act Certification.

A molecular marker should have some desirable properties although no single molecular marker meets all these requirements.

- It must be polymorphic *i.e.*, the nucleotide sequence for the mutant gene can be distinguished from the nucleotide sequence for the normal gene of the wild genotype.
- It should show codominant inheritance to allow discrimination of homo and heterozygote.
- The marker should be evenly and frequently distributed throughout the genome.
- It should be reproducible, easy, fast and cheap to detect.

A wide range of molecular techniques are available which detect polymorphism at the DNA level which can be grouped into two categories according to basic strategy for utilization.

▪ Non-PCR based approaches

i) **Restriction fragment length polymorphism (RFLP):** RFLPs involves fragmenting a sample of DNA by a restriction enzyme, which can recognize and cut DNA wherever a specific short sequence occurs. A RFLP occurs when the length of a detected fragment varies between individuals and can be used in genetic analysis. It is the most reliable technique, which is being used for accurate scoring of the genotypes. The DNA polymorphism is detected by the use of restriction enzymes that cut genomic DNA molecules at specific nucleotide sequences (restriction sites), thereby yielding variable size of DNA fragments. Identification of genomic DNA fragments is made by Southern blotting, a procedure whereby DNA fragments, separated by electrophoreses, are transferred to nitrocellulose or nylon filter. Filter immobilized DNA is allowed to hybridize to radioactively labeled probe DNA. Probes are usually small (500 to 3000 base pair) cloned DNA segments (*e.g.*, genomic or cDNA). The filter is placed against photographic film, where radioactive disintegration from the probe results in visible bands. Such bands are visualizations of RFLPs, which are codominant markers.

▪ PCR based techniques

These marker techniques are based on in *vitro* method for amplification of DNA fragments referred to as polymerase chain reaction (PCR) through semi conservative

DNA replication carried out by DNA polymerase in prokaryotic and eukaryotic cells. The amplification is achieved by repetitive series of cycles involving three steps

a) Denaturation of a template DNA duplex by heating at 94°C,

b) Annealing of oligonucleotide primers to the target sequence of separated DNA strands at 35-65°C,

c) DNA synthesis from the 3'- OH end of each primer by DNA polymerase.

The PCR has been used to develop several DNA marker systems.

i) **Amplified fragment length polymorphism (AFLP):** AFLP is a technique based on the detection of genomic restriction fragments by PCR amplification and can be used for DNAs of any origin or complexity. The fingerprints are produced without any prior knowledge of sequence, using a limited set of generic primers. The number of fragments detected in a single reaction can be 'tuned' by selection of specific primer sets. AFLP technique is reliable since stringent reaction conditions are used for primer annealing. This technique thus shows an ingenious combination of RFLP and PCR techniques and is extremely useful in detection of polymorphism between closely related genotypes.

ii. **Random amplification of polymorphic DNA (RAPD):** RAPD is a type of PCR reaction, but the segments of DNA that are amplified are random. This technique is fast, relatively inexpensive and highly variable. However, the main disadvantages are: markers are dominant, presence of a band could mean the individual is either homozygous or heterozygous for the sequence and data analysis is more complicated.

iii. **Simple sequences repeat:** This technique is based on the microsatellites or simple sequence repeats (SSRs) which are repeating sequences of 1-6 base pairs of DNA. The main advantages of this technique are highly variable, fast evolving and co dominant in nature (If the genetic pattern of homozygotes can be distinguished from that of heterozygotes, then a marker is said to be co-dominant).

iv. **Other PCR based techniques:** These techniques include VNTR (Variable number tandem repeat), SNP (Single nucleotide polymorphism), STR (Short tandem repeat), SFP (Single feature polymorphism), DArT (Diversity Arrays Technology), RAD markers (Restriction site associated DNA markers)

Marker assisted selection (MAS), the new tool for vegetable breeding

Genetic variation among the genotypes is of utmost importance for the effective selection of superior ones. Marker assisted selection (MAS) implies a breeding strategy that involves the use of DNA markers associated with or closely linked to desirable traits controlled by genes of interest (Singh, 2009). Traditional breeding involves phenotypic selection approaches, which are labour-intensive and time-consuming as well. In recent days, use of some novel genetics and genomics tools, techniques and approaches can enhance the efficiency of plant breeding programmes. Some examples of these tools and techniques are molecular markers, whole genome sequencing, genetic linkage maps, marker-assisted selection (MAS), marker-assisted recurrent selection (MARS), marker assisted backcrossing (MABC), genomic selection (GS),etc.These precision breeding approaches can be utilized at different developmental stages of crops depending on the complexity of genetics of the species. These tools and techniques are frequently applied to some major crops like rice, wheat, maize, etc. (Simko 2015). These tools and techniques give ample scope to gradually find their way into vegetable breeding (Simko *et al.*, 2021).

Marker assisted selection has become a promising and potent approach for integrating biotechnology with conventional breeding approach. In this technique, linkages are sought between the DNA markers and economically important traits such as resistance to the pathogens (fungus, bacteria, and virus), insects and nematodes, tolerance to abiotic stresses, different quality traits and other economic characters. In this breeding scheme, instead of selecting the genotypes with the targeted character based on phenotype, the easily detectable markers in the genotypes are identified. Having multiple markers or codes is particularly useful if the breeder wishes to simultaneously follow the inheritance of several genes. For example, there are at least 15 genes that can confer resistance to various strains of lettuce downy mildew. In order to develop cultivars which possess several of these genes to give broad spectrum resistance, the rare individuals who inherit all of the desired genes from both the parents must be identified. Rather than screening each seedling against all of the strains of the pathogen, which is difficult to accomplish, the seedlings can be screened using molecular markers specific for each gene. Only those possessing all of the desired genes are grown for the next generation. Similarly, for traits that are governed by number of genes, such as soluble solids content of tomatoes, multiple markers closely associated with desirable genes can be screened and specific individuals can be selected. This technology has the advantage of being suitable for automation to process thousands of samples. This markedly reduces the time required to develop horticulturally acceptable cultivars containing genes from their distant relatives. One can think of the desired gene as the 'needle in the [genetic] haystack'. Traditional methods essentially require sorting through the entire stack to find the needle, while molecular markers can be thought of as providing a metal detector to easily find the specific part of the haystack where the needle is located. These tools allow breeders to narrow their search for desirable traits to much smaller plant populations, greatly streamlining the development and evaluation of new cultivars. Essential requirements for MAS in any plant breeding programme are: the efficient means of screening large populations for the molecular marker and cost effective screening technique with very high reproducibility. The following broad steps are involved during MAS for any character.

- Identification of the tightly linked markers with the desired character so that the identified DNA marker co-segregates or be closely linked with the desired trait.
- Testing validity of the linked markers in the breeding materials.
- Marker based indirect selection for plants with the desired character in segregating and non-segregating generations.

Strategies for the marker assisted selection (MAS)

Strategies for the MAS are based on the screening of large number of random, unmapped molecular markers (RFLP, RAPD, AFLP, SSR, etc.) in a relatively short time and to select just those few markers that reside in the vicinity of the target gene. The methods of MAS rely on two principles.

- To generate huge number of potentially polymorphic DNA segments from single preparations of DNA.
- Development of mapping population for molecular marker studies. Near isogenic line (NIL) and recombinant inbred line (RIL) are the two primary types of mapping populations are used for molecular mapping.

• Near isogenic line (NIL)

Near isogenic lines (NILs) are developed through repeated backcrossing of genetically distinct parental lines. This mapping population is developed utilizing inbred lines that differ at the targeted locus. Nearly isogenic lines are created through crossing the donor line (P_1) to the recipient line (P_2), backcrossing of the F_1 to the P_2 (BC_1), selection of single individual containing the dominant allele of the targeted gene from the P_1 on the basis of phenotype and repeated backcrossings with the P_2 to develop BC_7 generation. In this generation, most if not all of the genome will be derived from P_2 except for a small chromosome segment containing the selected dominant allele, which is derived from the donor parent P_1. The lines homozygous for the target gene can be selected from the selfed generation of BC_7 ($BC_7 F_2$). NILs are most useful when the gene or one particular locus of interest is available. This will allow measuring the effect of allelic variation at that locus only, while eliminating background genetic variation. NILs are more powerful than RILs but can only be used to study one locus at a time.

• Recombinant inbred lines (RIL)

A recombinant inbred line is an organism with chromosomes that incorporate an essentially permanent set of recombination events between chromosomes inherited from two or more inbred strains. Families of recombinant inbred strains numbering from 25 to 5000 are often used to map the locations of DNA sequence differences (quantitative trait loci) that contributed to differences in phenotype in model organisms.

The F_1 hybrids are developed by crossing two parents that show significant polymorphism for whichever type of loci that has been targeted to score and the F_2 population is developed by selfing the F_1 individuals or intermating among the F_1 individuals. Recombinant inbred lines are developed by single-seed selections from individual plants of an F_2 population. Because of this procedure, these lines are also called F_2-derived lines. Single-seed descent is repeated for several generations. At this point, all of the seed from an individual plant is bulked. For example, in a $F_{3:4}$ generations, the RI population underwent single-seed descent through the F_3 generation, and was bulked to develop the F_4. This population of seed can then be grown to obtain a large quantity of seed of each individual line. Importantly, each of the lines is fixed for many recombination events.

RILs are useful for preliminary mapping of any trait that differs between the parental strains used to generate the population. The great thing about RILs is that the same mapping population can be maintained and used over and over again to map all kinds of different traits. They can also reveal multiple loci contributing to any trait of interest. The downside is that they are less statistically powerful for analysing effects of any one particular locus, because each RIL also harbours potentially confounding background genetic variation.

Table 3: Molecular markers linked to major resistant genes in vegetable crops

Crop	Pathogen	Gene	Marker(s)
Pepper	Tomato spotted wilt virus	*Tsw*	RAPD
	Tomato spotted wilt virus	*Tsw*	CAP
	Xanthomonas vesicatoria	*Bs2*	AFLP
	Leveillula taurica	*Lv*	RFLP
Bean	Common bean mosaic virus	*I*	*RAPD*
	Uromyces appendiculatus	*Up2*	RAPD
Pea	Pea common mosaic virus	*MO*	RFLP
	Erysiphe polygone	*Er*	RAPD
	Fusarium oxysporum	*Fw*	µSAT
Cucumber	*Fusarium oxysporum* f sp. *melonis*	*Fom2*	SSP
Melon	*Fusarium oxysporum* f sp. *melonis*	*Fom2*	RAPD
Tomato	*Meloidogyne incognita*	*Mi*	RAPD
	Meloidogyne incognita/javanica	*Mi3*	RAPD, RFLP
	Cladosporium fulvum	*Cf2*	RFLP
	Phytophthora infestans	*Ph2*	RFLP
	Verticillium dahliae	*Ve*	RAPD
	Verticillium dahliae	*Ve*	SCAR
	Verticillium dahliae	*Ve*	RFLP
	F. oxysporum f. sp. *radicis-lycopersici*	*Fr2*	RAPD
	Cucumber mosaic virus	*Cmr*	RFLP
	Yellow leaf curl virus	*Ty 1*	RFLP
	Yellow leaf curl virus	*Ty 2*	RFLP
	Tomato spotted wilt virus	*Sw-5.*	RFLP
	Tomato spotted wilt virus	*Sw-5*	RAPD
	Tomato mosaic virus Tomato mosaic virus	*Tm2nv* *Tm2a*	RAPD RAPD
	Tomato mosaic virus	*Tm 1*	RAPD/ SCAR
	Tomato mosaic virus	*Tm2(2)*	SCAR
	Tomato mosaic virus	*Tm2*	SCAR
	Pyrenochaeta lycopersici	*Py*	RAPD

CAPS: Cleaved amplified polymorphism sequence; **RAPD:** Random amplified polymorphic DNA; **RFLP:** Restriction fragment length polymorphism; **µSAT:** Microsatelites; **SCAR:** Sequence characterized amplified region.

• Genomics

Genomics is a discipline in genetics that applies recombinant DNA, DNA sequencing methods, and bioinformatics to sequence, assemble, and analyse the function and structure of genomes (the complete set of DNA within a single cell of an organism). Advances in genomics have triggered a revolution in discovery-based research to understand even the most complex biological systems such as brain. The field includes efforts to determine the entire DNA sequence of organisms and fine-scale genetic mapping. The field also includes studies of intra-genomic phenomena such as heterosis, epistasis, pleiotropy and other interactions between loci and alleles within the genome. In contrast, the investigation of the roles and functions of single genes is a primary focus of molecular biology or genetics and is a common topic of modern medical and biological research.

The genome projects involve three components: the sequencing of DNA, the assembly of that sequence to create a representation of the original chromosome, and the annotation and analysis of that representation. Whole exome or whole genome sequencing is typically used when single gene or panel testing has not provided a diagnosis, or when the suspected condition or genetic cause is unclear.

Achivement of genetic engineering in important of vetetable crops

• Genome editing

Genome on gene editing is a group of technologies to change an organism's DNA. Several approaches to allow genetic material to be added, removed or altered at particular locations in the genome include CRISPR-Cas9, TALEN nucleases and Zinc-finger nucleases. CRISPR-Cas9 is well known which is short for clustere regularly interspaced short palindromic repeats and CRISPR is associated to protein 9. This technology was adapted from a naturally occurring genome editing system that bacteria use as an immune defence against infection of virus. The bacteria use Cas 9 or a similar enzyme to cut the DNA apart, which disables the virus.

• Insect resistance genes

Bacillus thuringiensis (Bt) crystal proteins have largely been used to develop insect resistant transgenic varieties of different vegetable crops. *Bacillus thuringiensis* is a gram-positive bacterium commonly found in soils. Different strains ofthis bacterium produce one ormore proteins referred to as d-endotoxins. These crystal proteins or d-endotoxins are characterized by their toxicity to the larvae of distinct range of host insects. Based onhost range, the d-endotoxins are classified into 4 groups:

- Cry I: Lepidopteran specific
- Cry II: Lepidopteran and Dipteran specific
- Cry III: Coleopteran specific
- Cry IV: Dipteran specific

***Bt cry* proteins used in transgenic vegetable crops**

Bt *Cry* protein	Vegetable crop (s)
Cry 1Aa	Rutabaga
Cry 1Ab	Potato, tomato, brinjal, potato, corn
Cry 1B	Brinjal
Cry 1Ac	Broccoli, cabbage, tomato
Cry 1C	Broccoli, Chinese cabbage
Cry 3Aa	Brinjal, potato

A durable and eco-friendly alternative for insect pest management in vegetables is to genetically transform elite genotypes and breeding lines of vegetable crops using genes that encode insecticidal proteins which can be safely expressed in transgenic plants for insect control (Table 3).

Table 3: Different insecticidal proteins showing resistance to insects in vegetable crops

Insecticidal protein	Vegetable Crops
Proteinase inhibitors	
Soybean serine- proteinase inhibitor	Potato
Cowpea trypsin inhibitor	Potato, sweet potato, tomato
Tomato proteinase Tomato inhibitor I	Tomato
Tomato proteinase Tomato inhibitor II	Tomato
Amylase inhibitors	
Amylase inhibitor of common bean	Garden pea, potato
Lectins	
Snowdrop lectin	Potato, sweet potato, tomato
Pea lectin	Potato
Miscellaneous	
Bean chitinase	Potato
Tobacco anionic peroxidase	Tomato

- **Disease resistance genes:** There are several approaches to develop transgenic plants resistant to virus, *viz.*, introducing genes encoding viral-coat protein to interfere viral replication and expression, engineering satellite RNA and using antisense virus RNA to govern transgenic resistance. Genetic engineering of the plants resistant to fungal disease has been achieved by using chitinase gene of plant and/or bacterial origin. Chitinase enzyme degrades chitin rich cell walls of fungi and delays symptom development. Genes responsible for detoxification of pathogenic toxins isolated from bacteria or other organisms have been used to generate transgenic plants resistant to bacteria.
- **Resistance to herbicide:** Generation of transgenic plants resistant to herbicides like glyphosate, sulfonylureas, imidazolinones, etc. have been accomplished either by transferring a gene enabling plants to synthesize an enzyme which is not inhibited by the herbicide to catalyze the essential metabolic activities or by adding a gene that directs the synthesis of an enzyme that detoxifies the herbicide, thus preventing it from inhibiting its target.

- **Detection of diseases:** The commonly used methods for the detection of disease causing microorganisms are: a) ELISA (Enzyme Linked Immunosorbent Assay) which is widely used technique based on an antibody-antigen reaction, and used for diagnostic determination of the presence of specific molecules in a sample and b) PCR (Polymerase Chain Reaction) which is a technical procedure that amplifies and makes it possible to detect a specific piece of DNA of the disease causing microorganisms.
- **Improving shelf-life and fruit quality:** Post-harvest losses take a heavy toll in vegetable crops. Different biotechnological tools help to remove the plant components that cause early deterioration after harvesting. This technique involves engineering plants with an antisense gene or RNA interference technology so that production of enzymes responsible for fruit ripening (ACC synthase, ACC Oxidase) and subsequent softening of fruits (Polygalacturonase, pectin methyl esterase and β-galactosidase) is significantly reduced. Improved storage and shelf-life of genetically modified vegetables with quality traits would increase the availability and improve the possibilities of balanced diet.

  Target genes for quality improvement are against the enzymes mainly polygalacturonase (delay in fruit spoilage), polyphenoloxidase (discolouration in fruit vegetables), starch synthases (reduction of starch content) and chalcone synthase (modification of flower colour).
- **Improving processing quality:** Genetic engineering of starch biosynthesis in potato and cassava may lead to modification of starch granule shape, amylase to amylopectin ratio, amylopectin chain length, crystallinity, gelling properties, phosphorylation, lipid content etc. which alter significantly the technological properties of starch for use in both food and non-food industry.

  At low temperatures, potatoes accumulate reducing sugars making them unfit for processing. This is because of vascular invertase activity. Invertase is an enzyme that catalyzes the hydrolysis of the table sugar sucrose into fructose and glucose. An invertase inhibitor (INH) gene (accession AY145781) from tobacco has been cloned and integrated in TDNA of pBinAR. This construct is used for transformation of both table and processing potatoes to reduce sugar accumulation during storage.
- **Enhancing quality protein:** Nutritionally improved transgenic potatoes have been developed by transferring the amaranth seed albumin gene *(AmA1)* from *Amaranthus hypochondriacus* into potato. The transgenic potatoes show a significant increase in all essential amino acids, including lysine, tryptophan, tyrosine and sulphur containing amino acids, such as cysteine and methionine. Interestingly, transgenic plants also show enhanced total protein in tubers as compared with control plants.
- **Plant vaccines**: Utilization of plants as 'Biofactories' for the production of vaccines has been examined in many countries. This is attainable by use of transgenic plants for the production of adequate levels of desired vaccines

and facilitates distribution to target population, which may otherwise not have access to such vaccines.

- **Parthenocarpy:** Parthenocarpic fruit development in tomato and brinjal has been achieved by using *DefH9-iaaM* gene, which consists of the coding region of the *iaaM* gene from *Pseudomonas syringae* pr. *savastanoi* under the control of the placental and ovule specific *DefH9* promoter from *Anthirrhinum majus*. The *DefH9-iaaM gene* fusion which is expressed specifically in placenta/ovules promotes auxin synthesis and induces parthenocarpic fruit.
- **Male sterility:** Genetically engineering male sterility induction system is based on the suppression of pollen development by expressing ribonuclease gene under tapetum specific TA 29 promoter, which removes the tapetum that supports pollen development. Transgenic male sterile plants have been generated using this sterility restorer system.
- **Control of varietal identity and purity:** Distict, Unique and Stable phenotypic characters (DUS characters), which are expression of genes are used to describe particular varieties. These characteristics may be either morphological or physiological. Different markers have successfully been used to have a fingerprint for identity of many vegetable crops. DNA-based fingerprinting provides the best available technology to establish differentiation for patent protection and Plant Variety Protection Act Certification.

Table 4: Successful genetic engineering to develop transgenic vegetable crops

Purpose of genetic engineering	Transgene product	Origin of transgene	Transformed plant
A. Resistance against virus			
Tomato mosaic virus	Viral-coat protein	ToMV	Tomato
Cucumber mosaic virus (CMV)	-do-	CMV	Cucumber
Potato virus X (PVX)	-do-	PVX	Potato
Potato virus Y (PVY)	-do-	PVY	Potato
Tomato mosaic virus (ToMV)	Antisense RNA	ToMV	Tomato
Cucumber mosaic virus (CMV)	Satellite RNA	CMV	Cucumber
Tomato spotted wilt virus (TSWV)	N-gene	TSWV	Tomato
B. Resistance against bacteria			
Erwinia carotovora	Lysozyme	T4 bacterophage	Potato
E. carotovora	Tachyplesin	Horseshoe crab	Potato
E. carotovora	Pectate lyase	*E. carotovora*	Potato
E. carotovora	Glucose oxidase	*Aspergillus niger*	Potato
Pseudomonas syringae pv. phaseolicola	Phaseolotoxin	*P. syringae pv. phaseolicola*	Bean
C. Resistance against insect pests			
Lepidopteran pest	Cry IAB	*Bacillus thuringiensis*	Tomato
-do-	Cry I H	-do-	Corn

Purpose of genetic engineering	Transgene product	Origin of transgene	Transformed plant
-do-	Cry 9C	-do-	Corn
Coleopteran pest	Cry 3A	-do-	Brinjal
-do-	CpT I (Cowpea trypsininhibitor)	Cowpea	Tomato
Lepidopteran pest	Tomato proteinase inhibitor I	Tomato	Tomato'
Lepidopteran pest	Tomato proteinaseinhibitor II	Tomato	Tomato
Lepidopteran pest	Bovine pancreatic trypsin inhibitor (BPT-1)	Animal	Lettuce
D. Improving post-harvest traits			
Chips with less fat	Starch production gene	*Escherichia coli*	Potato
Improved shelf-life	Antisense poly-galacturonase	Tomato	Tomato
Improved ripening	Antisense ACC oxidase	Tomato	Tomato
Improved fruit pigmentation	Phytoene synthase gene	Tomato	Tomato
Slow ripening	Antisense RNA for ACC synthase	Tomato	Tomato
E. Induction of parthenocarpic development of fruits			
Parthenocarpic fruits	*rolB* gene	*Agrobacterium rhizogenes*	Tomato
-do-	*iaaM* gene	*Pseudontonas syringae* pv. *savastanoi*	Brinjal

References

Baltimore, D. 1970. RNA-dependent DNA polymerase in virions of RNA tumour viruses, *Nature,* **226**: 1209–1211.

Carlson, P.S., Smith, H.H., Dearing, R.D.1972. Parasexual inter-specifIc plant hybridization. *Proc. Nat. Acad. Sci.* (Wash.), **69**: 2292-2294

Chesnokov, Y.V., Kosolapov,V.M. and Savchenko, I.V. 2020. Morphological genetic markers in plants. *Russ. J. Genet.*, **56**:1406–1415

Crick, F.H.C. 1958. On protein synthesis. *Symp. Soc. Exp. Biol.*, 12:138–163.

Haberlandt, G. 1902. Culturversuche mit isolierten Pflanzenzellen. Sitz-Ber. *Mat. Nat. Kl. Kais. Akad. Wiss. Wien.*, **111**: 69–92.

Hazra, P., Longjam, M. and Chattopadhyay, A. 2018. Stacking of mutant genes in the development of "purple tomato" rich in both lycopene and anthocyanin contents. *Sci. Hort.*, **239**: 253-258.

Hedden, P. and Phillips, A.L. 2000. Gibberellin metabolism: new insights revealed by the genes. *Trends Plant Sci.*, **5**: 523-30 doi: 10.1016/s1360-1385(00)01790-8.

Jiang, G.L. 2013. Molecular markers and marker-assisted breeding in plants, In: *Plant Breeding from Laboratories to Felds* (Ed. Andersen S.B.). InTech, Rijeka, pp. 45–83, DOI. 10.5772/3362.

Keller, W.A. and Melchers, G. 1973. The effect of high pH and calcium on tobacco leaf protoplast fusion. *Z. Naturforsch C.*, **28**: 737-41.

Laibach, F. 1925. Das Taubwerden von Bastardsamen und die künstliche Aufzucht früh absterbender Bastardembryonen. *Z. Bot.*, **17**: 417–459

Larkin, P.J. and Scowcroft, W.R. 1981. Somaclonal variation - a novel source of variability from cell cultures for plant improvement. *Theor. Appl. Genet.*, **60**:197-214.

Nadeem, M.A., Nawaz, M.A., Shahid, M.Q., Dogan,Y., Comertpay,G., Yıldız, M., Hatipoglu, R., Ahmad, F., Alsaleh, A., Labhance, N., Ozkan, H., Chung, G. and Baloch, F.S. 2018. DNA molecular markers in plant breeding. *Cent. Asian J. Plant Sci. Innov.*, **1**: 192–200.

Polivanova, O.B. and Bedarev, V.A. 2022. Hyper hydricity in Plant Tissue Culture. *Plants*, **11**(23): 3313. doi: 10.3390/plants11233313

Reinert, J. 1958. Morphogenese und ihre Kontrolle an Gewebekulturen aus Carotten. *Die Naturwissenschaften*, **45**: 344–345.

Sanford, J.C. 1988. The biolistic process. *Trends Biotechnol.*, **6**: 299-302

Sepahvand, D., Matinizadeh, M., Etemad, V. and Shirvany, A. 2021. Changes in morphological and biochemical properties of *Celtis caucasica* L. mycorrhizal fungi-inoculated under drought stress condition. *Cent. Asian J. Environ. Sci. Technol. Innov.*, **2**:142–155

Simko, I. 2015. Analysis of bibliometric indicators to determine citation bias. *Palgrave Commun.*,1:15011.https://doi.org/10.1057/palcomms.2015.11

Simko, I., Jia, M., Venkatesh, J., Kang, B.C., Weng, Y., Barcaccia, G., Lanteri, S., Bhattarai, G. and Foolad, M.R. 2021. Genomics and marker-assisted improvement of vegetable crops. *Crit. Rev. Plant Sci.*, **40**:303–365.

Singh, B.D. 2009. *Plant Breeding: Principles & Methods*, 11th edn. Kalyani Publishers, New Delhi, 722p.

Smith H.O. and Wilcox, K.W. 1970. A restriction enzyme from *Hemophilus influenzae*. I. Purification and general properties. *J. Mol. Biol.*, **51**: 379–391

Smith, E.F., and Townsend, C.O. 1907. A plant-tumour of bacterial origin. *Science*, **25**: 671–673.

Steward, F. C., Mapes, M. O., and Mears, K. 1958. Growth and organized development of cultured cells II: Organization in cultures grown from freely suspended cells. *Amer. J. Bot.*, **45**: 705–708.

Steward, F.C. 1958. Growth and organized development of cultured cells. III. Interpretations of the growth from free cell to carrot plant. *Amer. J. Bot.*, **45**: 709–713.

Temin, H.M and Mizutani, S. 1970. RNA-dependent DNA polymerase in virions of *Rous sarcoma* virus, *Nature,* **226**: 1211–1213.

Toonen, M. A. J., Hendriks, T., Schmidt, E.D.L., Verhoeven, H. A., Vankammen, A., and Devries, S.C. 1994. Description of somatic-embryo-forming single cells in carrot suspension-cultures employing video cell tracking. *Planta*, **194**: 565–572.

Watson, J. D. and Crick, F.H.C. 1953. Molecular structure of nucleic acids. *Nature*, **171**: 737–738.

White, P.R. 1939. Potentially unlimited growth of excised plant callus in an artificial nutrient. *Amer. J. Bot.*, **26**: 59–64.

18

Tomato

Tomato, *Solanum lycopersicum* L under the family Solanaceae (Chromosome No.: 2n = 2x = 24) is one of the major vegetable crops for fresh consumption as well as for processing purpose in India and the world as a whole. Tomato is the world's most popular and widely cultivated vegetable crop and grown in all kinds of climate *viz.,* temperate, subtropical and tropical. Tomatoes are the major dietary source of the antioxidant lycopene, which has been linked to many health benefits, including reduced risk of heart disease and cancer. They are also a great source of vitamin C, potassium, folate, and vitamin K.

Origin, domestication and adaptation in India

Tomato was classified under *Lycopersicon* for several years although, initially Linnaeus (1753) classified it in the genus *Solanum.* Recently, molecular studies provided new data on the relationships of tomato-potato and based on these results, tomato has been again assigned to the genus *Solanum. Solanum* section *Lycopersicon* includes the cultivated tomato (*Solanum lycopersicum*) and its wild relatives, *Solanum lycopersicum* being the only domesticated species. The wild cherry tomato, *Solanum lycopersicum* var *cerasiforme*, the ancestor of the modern cultivated tomato has been originated in Peru-Ecuador-Bolivia region of the Andes. The Veracruz-Puebla region of Mexico is the centre of domestication of the cultivated tomato. From Mexico, tomato was taken to Italy, Spain, Portugal and other European countries, Africa and Middle East by the explorers in the 16th century. It was taken to Philippines after its discovery by Magellan in 1521, and from there it was taken to other Asian countries. Tomatoes did not have an easy passage even in Europe. The fruit got tagged as the "poison apple" in Britain, France, Spain and other European countries. Tomato moved to the USA from northern Europe around 1781 where it was first grown by Thomas Jefferson in Virginia.

Tomatoes made their way to India through Portuguese explorers in the early 16th century. However, tomato cultivation got high momentum during the British period through the introduction of several varieties, particularly from England and South Africa. The Agri Horticultural Society of India was founded in 1820 by Rev. William Carey at Calcutta with the main objectives to introduce useful plants into India. Several cultivars of tomato were introduced in India during mid-19th century through this society at Calcutta. Thriving in warm and sunny conditions, tomatoes adapted well to Indian soils. At present, India is the second largest producer of tomatoes globally, following China.

Crop botany and floral biology

- Two types of growth habits *viz.,* determinate and indeterminate are basically found in different genotypes of tomato; determinate tomatoes grow to a certain height, then flower and set their fruit within a short time; indeterminate tomatoes continue to grow and produce flowers and fruits until senesce naturally; leaves are more or less hairy, strongly odorous and pinnately compound in nature.
- Inflorescence is extra-axillary raceme cymes with dichotomous or polychotomous branching arising from the stem opposite to the leaf; in cyme inflorescence, the shoot apex differentiates into a flower and subsequent growth occurs due to activity in an axillary branch which will eventually terminate in a flower; the cyme may be simple or branched depending on the genotype; the primary axis ends in a flower and one lateral branch is produced ending in a flower although different modifications of uniparous cyme as well as multiparous cyme i.e., primary axis terminates in a flower and produces behind the apex more than two lateral secondary axes in a whorled manner; flowering comes in tomato without any definite peak.
- The five-petaled flowers are yellow and rarely orange in colour, generally 2 cm across, pendant, and cluster of 5 to 12 flowers are borne in small forked raceme cyme; ovary is superior (above the attachment of the petals, sepals and stamens, and is free from the receptacle) and with 2-9 compartments (locules of the fruit).
- In case of determinate and self-topping bushy growth habit, the growing point is ceased, the lateral axis terminates in a blossom and there are less than three leaves between the flower clusters.
- In case of indeterminate growth habit, the terminal bud and lateral buds always give vegetative growth and the internodes produce inflorescence after three leaves.
- Calyx tube is short and bears 5-6 persistent and enlarged lobes; calyx persists until fruit matures.
- The yellow petals are united in a short tube with five or more lobes which are often recurved with a short tube with 5-6 wrinkled lobes.
- Stamens 5-6 in number, inserted on the throat of corolla tube, filament bright yellow; the long anthers are partly united in the form of a cone surrounding the pistil.
- Ovary is hypogynous, carpels are 5-6 in number, united and the basal ovary is typically 6 celled with a central fleshy placenta.
- The pistil consists of a long slender style reaching the tip or projecting from the staminal cone as much as 2 mm with a capitate, single, narrow or somewhat bulbous stigma; style is short in the early stages and later elongated with the age of the bud.
- Fruit is berry, locular with fleshy mesocarp.

Reproductive behaviour

Tomato is basically a self-pollinated crop, but very low level cross-pollination may occur. Certain degree of cross-pollination is accomplished mainly by different bees when the stigma protrudes beyond the anther cone. Anthesis starts from 6.00 am, continues up to 10.30 am and anther dehiscence occurs during 7.00 to 10.00 am. Expansion of endothecial layer and subsequent drying are required for dehiscence. The endothecium tissue is responsible for the tensions that lead to splitting of the anther. This tissue is usually one to several layers thick, with cell walls of uneven thickness due to uneven lignification. Dehiscence of anther is longitudinal which generally coincides with anthesis although dehiscence of some anthers in the same flower may occur even 1-2 days after anthesis. Generally opening of flowers coincides with self pollination (chasmogamy). The optimum temperature for pollination is around 21°C. The stigma receptivity is maximum at anthesis although it become receptive 15-20 hours prior to anthesis, and the receptivity continues even up to five days after anthesis at the optimum temperature condition of 18 –25° C. Emasculated flowers are generally cross-pollinated in the morning hours next day or two days later depending on the bud maturity. Pollen remains viable for one day at moderate room temperature however, dried pollen can be stored in a sealed vial and kept at 0°C for a month. Gametophytic system of self-incompatibility is found in different wild relatives of tomato.

Table 1: Reproductive features of cultivated tomato and different wild relatives

Species	Common name	Reproductive features
Solanum lycopersicum L.	Cultivated tomato	Self pollinated
Solanum lycopersicum var. *cerasiforme*	Ancestor of cultivated tomato	Self pollinated
S. pimpinellifolium (Jusl.)	Wild relative of tomato	Self pollinated
S. cheesmaniae	Wild relative of tomato	Self pollinated
S. chmielewskii	Wild relative of tomato	Cross pollinated
S. chilense	Wild relative of tomato	Self-incompatible
S. *parviflorum*	Wild relative of tomato	Self pollinated
S. peruvianum	Wild relative of tomato	Self-incompatible
S. *habrochaites*	Wild relative of tomato	Self-fertile & Self-incompatible
S. *pennellii*	Wild relative of tomato	Self-incompatible

Many useful genes, especially for disease resistance present in different wild relatives of tomato could be incorporated to the cultivated tomato although different types of breeding barrier are encountered in interspecific hybridization. An inter-specific crossing programme was carried out employing 3 widely divergent variety / breeding lines of tomato (Berika, BCT-115 *dg* and Alisha Craig) and 4 wild relatives *viz*., *Solanum lycopersicum* var. *cerasiformae, S. pimpinellifolium, S. chilense* and *S. peruvianum*. The number of fruits set and developed up to maturity, which contained viable seeds, determined crossability. In most of the cases, incompatible crosses resulted drop of the flowers, which were cross-pollinated. *Solanum lycopersicum* x *S. chilense* and *Solanum lycopersicum* x *S. peruvianum* crosses were not successful. In few crosses of *Solanum lycopersicum* x *S. chilense* and *Solanum lycopersicum* x *S. peruvianum*, some under-sized fruits containing no seed developed up to the maturity, which might have happened due to stimulative parthenocarpy. The F_1 seeds of the inter-specific hybrids of *Solanum lycopersicum* x *Solanum lycopersicum* var. *cerasiformae* and *Solanum lycopersicum* x *S. pimpinellifolium* showed complete viability (Pradeepkumar *et al*, 2019).

Table 2: Breeding barriers between tomato and its wild relatives

Female/Male	***S. lycopersicum***	***S. pimpinellifolium***	***S. habrochaites***	***S. chilense***	***S. peruvianum***
S. lycopersicum	No barrier	No barrier	No barrier	Embryo abortion	Embryo abortion
S. pimpinellifolium	No barrier	No barrier	No barrier	Embryo abortion	Embryo abortion
S. *habrochaites*	Unilateral incompatability	Unilateral incompatibility	Self-incompatibility	Embryo abortion	Embryo abortion
S. chilense	Embryo abortion	Embryo abortion	Embryo abortion	Self-incompatibility	Embryo abortion
S. peruvianum	Unilateral incompatability	Unilateral incompatability	Unilateral incompatability	Embryo abortion	Self incompatibiliy

Genetics of important characters

The basic and haploid chromosome numbers of tomato and allied *Solanum* species are same, x = n =12 (2n = 24). There are 12 linkage groups in cultivated tomato. The number of gene loci mapped has increased to more than 1500, using classical genetics, isozyme and different DNA based markers viz., RFLP, RAPD, AFLP, SSR, etc. A few important genes of great interest to plant breeders and their locations on different chromosomes are mentioned.

Table 3: Some important monogenic characters of tomato

Gene symbol	Phenotype of plant character
a	Anthocyaninless; absence of anthocyanin in all plant parts
aa	Anthocyanin absent; absence of anthocyanin in all plant parts
ae	Entirely anthocyaninless
aw	Anthocyanin without; absence of anthocyanin in all plant parts
Aft	Anthocyanin fruit
alc	Alcobaca, delay in fruit ripening
Ad	Resistance to *Alternaria solani*
Asc	Resistance to *Alternaria alternate f.sp. lycopersici*
B	High β carotene pigment in fruit
bls	Baby lea syndrome; absence of anthocyanin in all plant parts
Br	Resistance to buckeye rot
br	Brachtic; shortened internode
bk	Beaked; pointed styler end of the fruit
bs	Brown seed; brown endosperm colour
bs-2	Brown seed-2; brown endosperm colour
c	Potato leaf; reduced leaf segments
Cf, Cf-2, Cf-3, Cf-4, Cf-11	Resistance to specific race of *Cladosporium falvum*
cm	Leaves curly mottled, sensitive to environment
cl-2	Cleistogamous flower; functional male sterility, petals fail to open
Cor-1/2	Resistance to bacterial canker, *Clavibactor michiganensis* ssp. *michiganensis*
d	Dwarf; all plant part shortened, leaf dark green and rugose
dg	Dark green fruit and exaggerated photoresponsiveness
dl	Functional male sterility; growth of epidermal hair on anther surface is prevented
ex	Excerted stigma; pollen viable but stigma protrudes from anther cone
f	Fasciated fruit with many locules
Fri	Resistance to Fusarium crown and root rot caused by *Fusarium oxysporum* f.sp. *radilis lycopersici*
gf	Green flesh fruit, persistent chlorophyll
Hr	Hirsute; long hair on upper leaf surface
h	Hair absent; incompletely dominant

hp, hp-1, hp-2, hp-3, hp-4, hp-5	High pigment; high chlorophyll, carotenoids and ascorbic acid in fruits
I	Immunity; resistant to *Fusarium oxysporum* f. *lycopersici*
I-2	Immunity-2; resistant to *Fusarium oxysporum* f. *lycopersici*
I-3	Immunity-3; resistant to *Fusarium oxysporum* f. *lycopersici*
j	Jointless pedicel
j-2	Jointless pedicel-2
lg-5	Foliage uniformly light green
Lv	Resistance to powdery mildew, *Leveillula taurica*
Meu-1	Resistance to potato aphid, *Macrossiphum euphobia*
Mi	Resistance to root knot nematode, *Meloidogyne* spp.
ms-1 to ms-55	Male sterility; aborted pollen, sporogenous sterility
nor	Non-ripening; very retarded fruit ripening process
Nr	Never ripe; fruits ripen to dirty-orange in colour
o	Ovate; fruit ovate or pyriform
og^c	Old gold crimson; high lycopene but low β carotene contents
pat, pat-2	Parthenocarpic fruit
Ph, Ph-2	Resistance to race T_1 of *Phytophthora infestans*
ph-1	Resistance to *Pyrenochaeta lycopersici*
ps	Positional sterile mutant; connate petal, non-dehiscent anther, viable pollen
ps-2	Positional sterile-2 mutant; viable pollen, non-dehiscent anther due to structural alterations in the zone of anther dehiscence
pst	Persistent style resulting beak in fruit
r	Yellow flesh of the fruit
rin	Ripening inhibitor; delayed ripening, bright yellow fruit at maturity
s	Compound inflorescence
Se	Resistance to *Septoria lycopersici*
sh	Sherry
sl	Stamenless; male sterility
sp	Self-pruning; determinate growth habit
Sm	Resistance to *Stemphylium solani*
Sw-5	Resistance to tomato spotted wilt virus
suer	Sugar (sucrose) accumulator gene
t	Tangerine fruit colour
Tm	Resistance to tobacco mosaic virus
Tv-1/2	Resistance to greenhouse white fly, *Trialeurodes vaporariorum*
Ty-1, Ty-3, Ty-4, Ty-6	Resistance to tomato yellow leaf curl virus (*from S. chilense*)
Ty-2	Resistance to tomato yellow leaf curl virus (from *S. habrochaites*)
u	Uniform ripening of fruit
Ve	Resistance to *Verticillium albo-atrum, V. dahlia*
y	Unpigmented fruit epidermis

Source: Swarup (2006), Attanasova and Hazra (2012)

Table 4: Location of some important genes on different chromosomes

Chromosome No.	Genes present on the chromosome
Ch-1	*br, bs, y, ms-32, Cf, Cf-4, ms-6, Nr-2, dg (hp-2)*
Ch-2	*s, bk, ms-10, aa, o, ms-15, ms-2, ms-5, ms-10,ms-26, ms-35*
Ch-3	*Ms-9, r, sucr*
Ch-4	*cm,sl*
Ch-5	*rin*
Ch-6	*Mi, ms-16, ms-33, Cf-2,c, sp, B, cl-2, Ty-1*
Ch-7	*pst, lg-5, bs-2, Ph, I-3*
Ch-8	*ms-8,gf,Hr,ae,ms-4, ms-17*
Ch-9	*TM-2, Nr*
Ch-10	*nor, u, h, t, ms-31, Ty-6, Aft*
Ch-11	*j-2, Cf-3, j, ms-3, ms-7, ms-12, ms-14, ms-42, s, I-2, f, I, Ty-2*
Ch-12	*Cf-11*

Source: Swarup (2006), Attanasova and Hazra (2012)

Different quantitative characters *viz*., plant height, number of branches/ plant, earliness, total yield/plant, fruit number/plant, fruit weight, pericarp thickness, locule number and fruit shape are controlled by polygenes. Similarly, different fruit quality characters *viz*., total soluble solids content, acidity, ascorbic acid, lycopene, β carotene, reducing sugar and total sugar contents are also governed by polygenes.

Breeding objectives

In the early period, basic goal of tomato breeding has primarily been to increase yield followed by disease resistance to maintain production in challenging environment. Prior to 1930, tomato improvement depended largely upon selection of chance variants that originated as a result of mutation, spontaneous out-crossing or recombination of pre-existing genetic variation. Such selections succeeded in developing cultivars adapted to widely different environments in different parts of the world. Recent breeding trend has been towards development of cultivars to meet the specific uses rather than multipurpose cultivars to meet several needs.

Some of the important traits that have been the main focus for different purpose of tomato are outlined below (Tigchelaar, 1986; Foolad, 2007a; Gascuel *et al*., 2017).

Universal tomato breeding objectives

- Growth habit (both determinate and indeterminate)
- Sufficient leaf cover
- Earliness in flowering, fruiting and maturity
- High early and total fruit yield
- Appearance of fruit (fruit size and weight according to preference, generally 80-90 g fruit with flattish-round, round or oblong shape is preferred), uniform fruit colour, ribbed surface for fresh and smooth surface for processing tomatoes)
- Bushy plant frame and concentrated fruiting suitable for machine harvest

- Resistance to important diseases *viz*., tomato leaf curl virus, tomato yellow leaf curl virus tomato spotted wilt virus, early blight (*Alternaria solani*), late blight (*Phytophthora infestans*), buck eye rot (*Phytophthora parasitica*), Fusarium wilt (*Fusarium oxysporum* f. sp. *lycopersici*), bacterial wilt (*Ralstonia solanacearum*), etc.
- Resistance to insect pests *viz*., fruit borer (*Helicoverpa armigera),* white fly *(Bemisia tabaci)* and nematodes (*Meloidogyne incognita*)
- Tolerance to high and low temperature and soil salinity
- Wide adaptability and fruit set in both high and low temperature conditions
- Resistance to rain induced cracking of fruits

Breeding objectives for fresh market tomatoes

- Large, round and flattish-round fruits for different forms of use like salad and cooking
- Firmness and longer shelf life is desirable trait for transportability; thick pericarp and low locule number (2-3) are associated with firm fruits
- Minimum handling and shipping damage of fruits
- Slow fruit ripening character
- Uniformity in size and colour development
- Fruit shape (ribbed surface is preferred) and colour
- Attractive red fruit colour without green shoulder
- Freedom from external blemishes
- Good fruit quality (high lycopene, TSS, appreciable acidity (above 0.5%) contents and good flavour.
- Good taste and flavour (Fruit flavour depends on sugar and acid contents; pericarp tissue contains more sugar than the locular tissue; acid content in the locular tissue has predominant influence on fruit flavour)

Breeding objectives for processing tomatoes

- Compact and determinate plant habit
- Indeterminate growth habit with good foliage cover
- Concentrated flowering
- Suited for mechanical harvest
- Ease of fruit separation from vine
- High total soluble solids contents, preferably 5.5 or higher °Brix
- Tritrable acidity should be 0.4 to 0.5 percent and pH between 4.2 to 4.3
- Solid: acid ratio of 15 and sugar: acid ratio of 8.5
- High viscosity of fruit pulp (viscosity is correlated with alcohol insoluble solids content of pulp)

- Deep red fruit colour is preferred (deep red colour is determined by lycopene content and it should be around 8-9 mg/ 100g fresh)
- Good flavour (most important volatile compound for good flavour is 2-isobutylthiazole, besides methyl salicylate and eugenol)

Breeding methods and development of variety

Tomato breeding started in India with the introduction of exotic varieties. Some of them have been released directly for cultivation. Tomato adapted fast in the tropical conditions of India. Few indigenous selections from the early introduced varieties in different names like, Annaji of Tamil Nadu, Meeruti of Uttar Pradesh, Patharkutchi of West Bengal and some others are still being cultivated commercially. These adaptable selections have also been used in hybridization and selection method of breeding programmes. New varieties have also been developed through individual plant selection or pure line selection from the introduced varieties or lines. However, most of the varieties have been developed by hybridization and selection methods through mostly inter-varietal hybridization and in some cases inter-specific hybridization followed by pedigree, backcross or some modified methods of handling the segregating populations. Mutation breeding was also employed in developing some varieties in India. Breeding methods like pedigree methods, single descent method in combination with the molecular approaches have been found to be a useful method for breeding tomato (Fentik, 2017).

The varieties developed in India can be grouped as per their growth habit *viz*., determinate (terminal bud ends in a floral bud and further vegetative growth is ceased), semi-determinate (intermediate between determinate and indeterminate growth habit) and indeterminate (terminal bud ends in a vegetative bud and the main axis continues to grow).

Determinate variety

Pusa Early Dwarf, HS 101, HS 102, Hisar Arun, Punjab Chhuhara, Sonali, Roma, Pusa Sheetal, Sweet 72, Puss Gaurav, S-12, CO-1, CO-3 (Maruthan), PunujabKeshri, La Bonita, Azad T-2 (KS-2), Pusa Red Plum, Hisar Lalima, VC 48-1, Narendra Tomato 2, Sel-32, Hisar Anmol (H-24), LE 79-5, Arka Alok, Paiyur-1, Arka Ashish, Pusa Sadabahar, Pusa Rohini, Pant Bahar, PunjabNR-7, Vasundhara, Utkal Pallavi, Mukthi, Kashi Vishesh, Kashi Amrit (DVRT-1), Kashi Anupam (DVRT-2), Kashi Hemant, Kashi Aman, Kashi Adarsh, Kashi Amul, Swarna Naveen, Swarna Lalima, Mani Leima, Gujarat Tomato-2 (GT-2), Anand Tomato-3, Utkal Pragyan, Narendra Tomato-3, Azad T-6 , NDT8-2001-3, Punjab Varkha Bahar-4

Indeterminate / Semi-determinate variety

HS 110, Solan Vajr, Hisar Lalit, Pant T3, Parbhani Yashashri, Pant Poly House Tomato-1, Arka Vikas, Arka Saurabh, Pusa 120, Arka Ahuti, Arka Abha, Punjab Tropic, Pusa Ruby, Sioux, Marglobe, Best of All, Keck-Ruth, CO-2, Pant T-1, PKM-l, Arka Meghali, Utkal Kumari (BT-10), Utkal Urbashi (BT-12), Pusa Uphar, Bhagyashree,

Dhanashree, Narendra Tomato 1, Solan Gola, Kalianpur Angoorlata, Junagadh Ruby, Kashi Sharad, DT-10, Narendra Tomato-4, Utkal Raja, LE-415, BT-136, Phule Raja, Azad T-5, Punjab Ratta, Punjab Upma, Punjab Varkha Bahar 1, Punjab Varkha Bahar 2, Punjab Swarna, Punjab Gaurav, Punjab Sartaj

Germplasm resources and plant introduction

The Tomato Genetics Stock Centre (TGSC) at the University of California, Davis, USA maintains large collections of genetic stocks of mutants, wild relatives and Latin American landraces, and important old and modern cultivars. Large germplasm assemblage of tomato and wild relatives are also maintained at Asian Vegetable Research and Development Center (The World Vegetable Center, WorldVeg), Taiwan, United States Department of Agriculture (USDA, USA), Vavilov Institute of Plant Industry (VIR, Russia), Institute for Horticultural Plant Breeding (IVT, Netherlands), Departamento de Horticultura de la Universidad Nacional Agraria (DHUNA, Peru) and The National Institute of Agrobiological Sciences (NIAS, Japan).

The National Bureau of Plant genetic resources (NBPGR), New Delhi has a collection of more than 3000 germplasm from 47 countries. Tomato breeding started in India with the introduction of exotic varieties. Some of them have been released directly for cultivation. Many early introductions viz., Sioux, Roma, Marglobe , Rutgers, Best of All, La Bonita, Floradade, etc. were either used directly for large scale cultivation or utilized in the hybridization programme for development of new and better varieties.

Table 5: Exotic variety directly released for cultivation

Name of the variety	Country from where introduced
Roma (EC 13513)	USA
Red Cloud	USA
La Bonita	USA
Sioux	USA
Marglobe	USA (parentage: Marvel × Globe)
Best of All	USA
Keck-Ruth	USA
Money Maker	USA
Rutgers	USA
Floradade	USA

Source: Hazra and Som (2006); Hazra *et al* (2022)

Variability and selection

Phenotypic characteristics have been used to study genetic differences and for genetic diversity analysis and cultivar development. Assessment of diversity in commercial cultivars provides the status of a crop in relation to its wild and domesticated relatives and assists in creating a breeding scheme to bring in the lost alleles due to extreme selection and bottlenecks during domestication or introduction (Ranc *et al*., 2008). In the absence of genetic variation, breeding efforts in the crops remain without any gain which may lead to severe crop failure with the onset of new biotic or abiotic stress. (Tanksley and Mccouch, 1997). Most of the disease resistant genes in modern varieties are not only the result of selection from these varieties but also originated from the wild species (Rick and Chetelat, 1995). Selection has played an important role in tomato improvement in India. From different studies of correlation and path coefficient analyses, characters *viz.*, plant height, number of flower clusters/plant, number of fruits/cluster, fruit weight, fruit diameter, number of fruits/plant, number of locules/fruit and number of seeds per fruit have been found most important selection indices of tomato. Large number of varieties have been developed in India mainly through single plant selection or pure line selection mostly from heterogenous exotic stock and a few indigenously adapted genotype which were introduced long back in India. Individual plant selection in heterogenous population of a locally adapted cultivar in Uttar Pradesh named "Meeruti" gave a new variety "Improved Meeruti". Similarly, PKM-1 was selected from a locally adapted cultivar in Tamil Nadu named "Annaji". Later, the variety "Kalyanpur Angoorlata" was selected from a locally adapted material in Uttar Pradesh. Several improved varieties have been developed through selection from heterogeneous populations of exotic varieties or breeding lines such as, Arka Vikas from Tip Top, Narendra Tomato 1 from EC 130041, Arka Ashish from UC83B, Arka Abha from VC8-1-2-1, Pant Bahar from AC-238, etc.

Table 6: Varieties developed through selection

Name of variety	Indigenous / Exotic material
Improved Meeruti	Locally adapted variety named "Meeruti"
PKM-1	Locally adapted variety named Annaji
Kalianpur Angoorlata	Locally adapted genotype
CO-1	Reselection from a Kalyanpur selection
Sonali	Exotic
VC 48-1	Introduced material from Taiwan
Narendra Tomato 1	Exotic line EC 130041
Narendra Tomato 2	Exotic line EC 119201
LE 79-5	Exotic
Arka Alok	Exotic line (CL-114-5-1-0) from AVRDC, Taiwan
Arka Ashish	Exotic line UC83B from California
Pant Bahar	Exotic line AC-238
Utkal Pallavi	Exotic
Mukthi	Exotic
Swarna Naveen	Exotic
Swarna Lalima	Exotic
Utkal Pragyan	Exotic
Punjab Varkha Bahar-4	Exotic
HS 110	Exotic
Pant T3	Exotic
Arka Vikas	Exotic line Tip Top from USA
Arka Saurabh	Exotic line V 685 from Canada
Pusa 120	Exotic line from Hawai

Name of variety	Indigenous / Exotic material
Arka Ahuti	Exotic line Ottawa-60 from Canada
Arka Abha	VC8-1-2-1 from AVRDC, Taiwan
CO-2	Exotic line from Russia
Pant T-1	Exotic
Bhagyashree	Exotic
Utkal Raja	Exotic
LE-415	Exotic
Punjab Red cherry	Exotic

Source: Hazra and Som (2006); Hazra *et al* (2022)

Hybridization and selection

Inter-varietal hybridization with the involvement of exotic variety and already developed variety followed by pedigree method of selection in segregating generations has been followed for the development of most of the varieties in India *viz*., Pusa Ruby (Sioux × Improved Meeruti), Punjab Chuhara (EC 55055 x Punjab Tropic), Pusa Gaurav (Glamour (exotic) × Watch (exotic), etc. Single seed descent method using single seed from each of the selected plant to advance the generation from F_2 to F_5 or F_6 followed by pedigree selection is effective in tomato breeding. Double seed descent method utilizing two seeds from the selected lines is also useful in isolating promising progenies in the advanced generation of the hybrid. Back cross method has been successfully employed in breeding varieties resistant to diseases. Double back cross method has been used to combine two negatively associated characters like, earliness and fruit size.

Table 7: Varieties bred through hybridization followed by selection

Variety	Parentage
Pusa Ruby	Sioux × Improved Meeruti
Pusa Early Dwarf	Improved Meeruti × Red Cloud
Marglobe	Marvel × Globe
Pusa Red Plum	*S. lycopersicum* × *S.pimpinellifolium*
Punjab Chuhara	EC 55055 x Punjab Tropic
Punjab Kesri	Punjab Tropic × EC 55055
HS-101	Sel. 2-3 × Exotic cultivar
HS-102	Sel. 12 × Pusa Early Dwarf
Hisar Arun (Sel. 7)	Pusa Early Dwarf × K2
Hisar Lalima (Sel. 18)	Pusa Early Dwarf × HS-101
Hisar Lalit	Bangalore × HS-101
Hisar Anmol (H-24)	Hisar Arun × L. *hirsutum* f. *glabratum*
Pusa Sheetal	Balkan (Bulgaria) × Jemnorrosnej (Russia)
Pusa Gaurav	Glamour (exotic) × Watch (exotic)
Sweet-72	Pusa Red Plum × Sioux
Arka Meghali	Arka Vikas × IIHR 554
Utkal Kumari	Selection 22 × BT-2

Variety	Parentage
Paiyur-1	Pusa Ruby × CO-3
Azad T-2 (KS-2)	HB-5 × Kuber
Punjab NR-7	S-12 × NMR-1
Kashi Hemant	Sel-18 x Flora Dade
Gujarat Tomato-2	Angurlata x Punjab Chhuhara
Narendra Tomato-3	Resistant Bangalore x Kalyanpur Kuber
NDT8-2001-3	H-24 x NDT-120
Utkal Urbashi	Punjab Chhuhara × LE-79
Dhanashree	407 DPDBK × LA-124
Junagadh Ruby	Big Silari × Pusa Ruby
Narendra Tomato-4	Pusa Ruby x Kalyanpur Kuber

Source: Hazra and Som (2006); Hazra *et al* (2022)

Crop wild relatives and inter-specific hybridization

Several research findings have revealed that resistance against nearly all serious pests and diseases are available amongst wild *Solanum* species. The genus *Solanum* can be divided into 2 groups: the esculentum-complex and the peruvianum-complex. The *esculentum*-complex includes 6 species, 3 of which *(S. lycopersicum, S. cheesmanii* and *S. pimpinellifolium*) are able to synthesize colour carotenoids in their fruits and 3 others (*S. hirsutum* or *S. habrochiates, S. parviflorum* and *S. pennellii*) are green-fruited. All representatives of esculentum - complex group are easily crossable with tomato cultivars. The peruvianum-complex includes 2 green-fruited species (*S. peruvianum* and *S. chilense*) which carry a number of economically important traits, arc sexually incompatible with the cultivated tomato.

Several methods have been employed to overcome the crossing barrier. Different techniques to raise some viable hyrbids of *S. lycopersicum* x *S. peruvianum* are i) embryo culture, ii) application of growth substances, iii) grafting, iv) application of mixed pollens and (v) crossing tetraploid tomato with diploid *peruvianum* (Choudhury, 1959). This sexual incompatibility can also be overcome using bridge lines or by some tissue culture methods such as embryo-rescue and somatic hybridization (Lebeda and Mieslerova, 2000). Initial interspecific hybridization mainly targeted major qualitative traits like disease and pest resistance governed by major genes, thus often neglecting the agronomic traits controlled by polygenes (Passam *et al.*, 2007). Seven promising segregates could be isolated based on important yield attributes from three interspecific hybrids (Alisa craig *Aft* × *Solanum pimpinellifolium*, Beri ka × *Solanum lycopersicum* var. *cerasiforme* and BCT 115 *dg.* × *Solanum lycopersicum* var. *cerasiformae*) in F_5 generation (Acharya *et al.*, 2024)

Seven improved varieties possessing resistance to tomato leaf curl virus have been developed by inter-specific hybridization between *Solanum lycopersicum* x *Solanum habrochiates* (for introgression of *Ty2* gene) and between *Solanum lycopersicum* x *Solanum chilense* (for introgression of *Ty3* gene) followed by pedigree or back cross or backcross-pedigree method breeding. Generally, back cross method has been successfully employed in breeding varieties resistant to diseases. One variety, Pusa Red Plum having enhanced sugar and vitamin C contents in the fruits has been developed by inter-specific hybridization between *Solanum lycopersicum* x *Solanum pimpinellifolium* followed by pedigree method of breeding.

Table 8: Varieties developed through inter-specific hybridization

Variety	Inter-specific hybridization
Pusa Red Plum	*Solanum lycopersicum* x *Solanum pimpinellifolium* (High sugar and vitamin C content)
Hisar Anmol (H-24)	Hisar Arun × L. *hirsutum* f. *glabratum* (*Solanum habrochiates*) (TLCV resistant variety)
Kashi Vishesh	*Solanum lycopersicum* × *L. hirsutum f. glabratum* B-6013 (*Solanum habrochaites*) (TLCV resistant variety)
Kashi Amrit (DVRT-I)	Hisar Arun × L. *hirsutum* f. *glabratum* (*Solanum habrochiates*) (TLCV resistant variety)
Kashi Anupam (DVRT-2)	Hisar Arun × L. *hirsutum* f. *glabratum* (*Solanum habrochiates*) (TLCV resistant variety)
Kashi Aman	Introgression of *Ty3* gene from *Solanum chilense* (TLCV resistant variety)
Kashi Adarsh	Introgression of *Ty3* gene from *Solanum chilense* (TLCV resistant variety)
Kashi Amul	Introgression of *Ty3* gene from *Solanum chilense* (TLCV resistant variety)

Source: Hazra and Som (2006); Hazra *et al* (2022)

Utilization of mutants and mutation breeding

Significant number of induced and spontaneous mutations are described and based on their phenotypes they are classified into 26 classes however, majority of these genes belong mainly to the classes concerning plant habit and size, fruit colour and flavour, ripening modification, fruit form and surface texture and disease resistance (Chetelat, 2002). Although, the number of mutant genes used in tomato genetic improvement is rather limited, if compared to the total number of mutant genes described. However, their substitution in new cultivars contributed significantly to breeding tomato varieties possessing useful new traits and adapted for different purposes (Balacheva *et al.*, 2008).

Gene *sp* (*self prunning*) appeared as spontaneous mutation in Florida in 1914 (Rick, 1978) which controls compact, determinate habit in contrast to the usual unlimited indeterminate growth of the tomato plant. One of the greatest steps in the genetic improvement of tomato was made in the 1960's by incorporating this gene in the cultivar VF 145 which facilitated harvesting, particularly by machine, and the fruiting concentrated in a shorter season. It was probably the most spectacular advance in the history of the tomato genetic improvement at this period. Genes *j* (*jointless*) and *j 2* (*jointless 2*) control the formation of flower pedicels which lack an abscission zone. This character is of value in the development of mechanically harvestable varieties. The two mutations are genetically mapped to chromosome 11 and chromosome 12 respectively (Rick and Yoder, 1988).

Fruit firmness and longevity are traits of crucial importance for processing and fresh market tomatoes, and are important for growers, processors and marketers. One way for developing hybrid varieties possessing fruits characterized by prolonged shelf life is by incorporating in one of the parent lines a gene that controls delayed fruit ripening, such as: *rin* (*ripening inhibitor*), *alc* (*alcobaca*), *nor* (*non-ripening*). Mutations *ripening inhibitor* (*rin*) and *non-ripening* (*nor*) result in non-ripening fruit deficient in climacteric respiration and the associated burst in ethylene biosynthesis. Both genes are isolated through positional cloning strategies and it was shown that ripening is regulated, in part, by a MADS-box transcription factor at the *rin* locus (Moore *et al.*, 2002). The *rin* gene in heterozygous condition improved shelf-life although colour development varied within the hybrids (Kitagawa *et al.* 2005). The *alc* mutation affects the ripening and storability of tomato fruit and its heterozygous condition caused no detrimental effect on fruit colour, firmness or size (Mutschler *et al.*, 1992)

A few genes which result in enhanced carotenoid accumulation have been identified. The *high pigment* (*hp-1* and *hp 2*) mutations result in increased accumulation of both lycopene and β-carotene during fruit development. The *hp-1*and *hp-2* are separate loci with *hp-1* located on chromosome 2 and *hp-2* located on chromosome 1 (Van Tuinenen *et al.* 1997). Mutant plants possessed dark-green immature fruits due to the overproduction of chlorophyll pigments and elevated levels of lycopene and ascorbic acid were characterized by their exaggerated photoresponsiveness. The third mutant *dark green* (*dg*), characterized by exaggerated photo-responsiveness, as well, was allelic to *hp 2* (Levin *et al.*, 2003). The *old gold crimson* (*og^c^*) gene increases lycopene content in tomato fruit by about 50% while reducing β-carotene by the same amount (O'Hare *et al.,* 2005). Introgression of og^c and *hp* genes in one tomato genotype results in production of fruits with significantly higher lycopene level (three-to four times) than that of conventional red fruited tomatoes (Stommel, 2007). It was found that lycopene content in the fruits of two anthocyaninless lines Alisha Craig *ah* and Alisha Craig *aw* was significantly higher that of the wild type line exceeding it by 28-29%, (Karceva *et al.*, 2008) which suggests that the introgression of these genes in tomato genotypes might be one more resource for enhancing lycopene content in tomato fruit.

The dominant *B* gene is used for developing tomato varieties possessing high β carotene content however, the effect of gene *B* was influenced by modifier gene Mo_B and these two genes are linked on chromosome 6 (Zhang and Stommel, 2000). Tomato fruit carotenoid content and colour might be influenced by other genes, different from *hp-1, dg,* og^c, and *B*, such as *tangerine* (*t*), *yellow flesh* (*r*), *sherry* (*sh*) (Stommel, 2007). Gene *y* controls colorless fruit skin and is included in breeding programs aiming at the development of cultivars with pink fruits which are typically grown in the Southern Italian area, Bulgaria and in a number of countries in Asia (Acciarri *et al.*, 2007). Gene *Aft* (*Anthocyanin fruit*), described for the first time as *Af* (Georgiev, 1972) controls elevated anthocyanin expression in the hypocotyls of seedlings and anthocyanin in the skin and pericarp tissues of the fruit (Jones *et al.*, 2003).

Genes *pat* and *pat 2* (parthenocarpic fruit) entail ovary development into a fruit without fertilization and seed formation (Mazzuccato *et al.,* 1998) Parthenocarpy was manifested through up-regulation of both auxin and gibberellins in the ovary concomitant with precocious ovary development and impaired fertilization (Hazra *et al.*, 2010). Expression of parthenocarpy in three segregating populations indicated the involvement of an independent gene designated as "ii" which influenced the gene *pat-2* having major effect for the control of parthenocarpy and no linkage was detected between parthenocarpy and two marker genes, stem pigmentation (aw+/aw) and leaf character (c+/c) (Hazra and Dutta, 2011)

The Gene *u* (*uniform ripening*) prevents formation of green shoulders and is largely used in breeding programs. Unripe fruits of the genotypes carrying this gene are of uniform light green colour.

The use of male sterility in breeding and hybrid seed production in tomato has become applicable in practice during the last two decades and there are three types of sterility used in breeding and hybrid seed production *viz*., *ms 10 aa*, *ps* and *ps 2* (Atanassova and Georgiev, 2007).

A number of morphological markers such as *potato leaf* (*c*), *anthocyaninless of Hoffmann* (*ah*), *anthocyanin without* (*aw*) etc. were introduced in seed parents of a number of commercial hybrids to evaluate the purity of the hybrid seed produced (Atanassova *et al.* 2001). It permitted testing the hybridity of the seed at germination or seedling stage. Moreover, three antocyaninless genes *viz*., *ah, aw* and *bls* (*baby lea syndrome*) enhanced germination under optimal and a large scale of stress conditions (Balacheva *et al.*, 2008)

Different lycopene enhancing mutants *viz*., *dg, hp-1* and og^C ; anthocyanin producing mutant *viz*., *Aft* and different fruit ripening manipulator mutants *viz*., *rin, nor* and *Nr* of tomato have been fully characterized at both homozygous and heterozygous form (Biswas *et al*., 2016), at the level of bioactive compound (Siddiqui *et al.,* 2014), at the level of metabolic behaviour (Pal *et al.,* 2019) and postharvest physico-chemical changes in the fruits of those mutants (Siddiqui *et al*., 2014; Siddiqui *et al.,* 2015). These spontaneous mutants were utilized in developing commercially exploitable hybrids rich in both lycopene and anthocyanin content BCT-115 *dg/dg* x Alisa Craig *Aft/Aft* and high yielding with enhanced shelf life BCT-115 dg/dg x BCT *rin/rin* (Biswas *et al*., 2016).

Exposure of dry seeds to 50-150 Gy dose gamma radiation was the most efficient in inducing desirable macro-mutation in tomato (Sikder *et al*., 2013). The macro-mutants could be isolated on the basis of wide array of distinct, unique and stable characters *viz*., plant stature, leaf morphology, inflorescence type, floral morphology, pedicel character, fruit morphology and chlorophyll content of leaf and fruit. Two mutants *viz*.,"Multiparus cyme mutant" from Berika, the line bred variety of Bulgaria and "Dark green fruit mutant" from the cultivar, Patharkutchi emerged as most promising for further use in breeding programme (Sikder *et al*., 2013; Das *et al*., 2019; Das *et al*., 2020).

Three varieties, CO3, S-12 and PKM-1 has been developed through induced mutation.

Table 9: Varieties/ mutants developed through applied mutagenesis

Variety/ Mutant	Applied mutagenesis / Remark
S-12	X-ray mutation of Sioux
CO-3 (Maruthan)	EMS mutant of CO-1
Multiparus cyme mutant	50Gy gamma radiation of EC-620177 (This mutant manifests a unique character "multiparous cyme" where the primary axis of the inflorescence "cyme" produce lateral branches with increased flower as well as fruit number per plant)
Dark green fruit mutant	150 Gy gamma radiation of Patharkutchi (This mutant bearing dark green leaf and fruit, dark red ripe fruits with high lycopene content (7.49 mg/100 g fresh)

Source: Hazra *et al* (2022) ; Sikder *et al.*, 2013; Das *et al.*, 2020

Heterosis breeding

Hedrick and Booth (1907) first reported heterosis in tomato. Since then, heterosis breeding is the most commonly used breeding method in tomato all over the world to develop superior F_1 hybrids characterized by earliness, high yield, extended fruiting duration, better fruit quality attributes, resistance to diseases and abiotic stresses particularly, high temperature stress. Different F_1 hybrids have replaced line-bred varieties in tomato which has remarkably increased productivity to a great extent in the last 50 years in different countries including India. At present, more than 90% of tomato area in the country is under different hybrids, mostly developed and marketed by different private or corporate sector although, 17 hybrids have so far been developed in the public sector (ICAR Institutes, State Agricultural Universities, etc.). Research and developmental works in tomato and other vegetable crops received ample fillip with the liberalization of import of parental materials of the hybrids and even hybrid seeds. At present, almost all the hybrid seeds of tomato are produced indigenously by manual emasculation and pollination method utilizing indigenously developed or exotic parental lines.

Hybridization technique

Emasculation

Emasculation generally begins about 50-55 days after transplanting. Flowers from the first cluster are removed. Flower buds, from the 2nd to 4th cluster on each branch are selected. Flower buds, which will open in 1-2 days are chosen for emasculation. The bud but not yet opened, and the corolla colour is slightly yellow or even paler. The forceps, scissors and hands should be sterilized with alcohol before emasculation is started. The selected buds are force opened with sharp-pointed forceps holding gently at the base. The anther cone is split open and then pulled out of the bud carefully leaving the calyx, corolla and pistil. The anther cone along with the corolla can also be removed by single upward pull with the forceps. Care should be taken to avoid any injury to the style and stigma. Skilled persons particularly women can emasculate even by hand without the help of any forceps. In this technique, the selected flower bud is held by left hand, base of the flower is gently pressed by the fore finger and thumb of the right hand and finally the anthers along with the corolla are pulled off by one upward pull. Two sepals of the bud are cut to identify the hybrid fruits. No protection from out crossing is needed because of very low natural cross - pollination.

Pollen collection

The best time for pollen collection is during the early morning before the anther dehiscence. Anther cones are removed from the opened flowers and kept on glass petridish or cellophane papers under sunlight for an hour or so to facilitate dehiscence of the anthers. Gentle crushing or shaking of the dried anthers helps easy liberation of the pollens. The pollens after screening are collected in the small glass vial or petridish or pollinating ring for onward pollination. Placing the undehisced anther cones 30 cm below a 60 to100-watt lamp for 20- 24 hours can effectively dry large quantity anther cones. The lamp creates a drying temperature of about 30°C. Next day, the dried anther cones are put in plastic container. It is covered with a fine mesh (200-300 mesh) cloth or net. The container with dried anther cones is then sealed with a tight fitting lid. The screened pollens are collected in the lid of the container by shaking it 10-20 times.

Pollen storage

Fresh pollen is the best for good fruit set. It can be kept for one day at moderate room temperature. The pollen stored under room conditions (mean maximum temperature 26°C and relative humidity 47%) retained high viability even after five days of storage whereas under refrigerated condition (constant temperature of 9-10°C) they showed no significant drop in viability even after 7 days of storage. Dried pollen can be stored in a sealed vial and kept at 0°C for about a mouth. The pollens should be taken from the refrigerator and kept closed until the container warms to room temperature to prevent the pollens from getting wet due to condensation.

Pollination

Emasculated flowers are generally pollinated in the morning hours between 8.00 to 11.00 a.m. next day or two days later depending on bud maturity. Pollination done 2 days after emasculation increased fruit set. The stigma is dipped into the pollen mass in the pollen container (small plastic spoon, petridish, pollinating ring, etc). Successful pollination and subsequent fertilization can easily be recognized within a week by the enlargement of the ovary. Pollination is usually done three times weekly over a three (for determinate female) to five (for indeterminate female) weeks period. In general, pollination period is restricted to 25 days only after flowering.

Use of male sterility in development of hybrids

The sporogenous male sterile mutant *ms 10* deviates from the wild type genotype in the absence of fertile pollen. The gene *ms 10* is closely linked to gene *anthocyanin absent* (*aa*) and by developing *ms 10 aa* genotypes, the anthocyaninless sterile plants can be easily distinguished in early developmental stages (Philouze, 1974). The sporogenous male-sterile line *ms 45* was a good combiner for fruit weight and total yield (Kaur *et al*, 2004). In mild winter condition of West Bengal, the male sterility gene *ms-45* in the background VFN-8 emerged as the most stable male sterile line while, in other male sterile lines, *ms-2*, *ms-16* and *ms10*^{36} pollen sterility did not persist in the advanced stages of some plants (Roy *et al.*, 2011).

The genes *positional sterile* (*ps*) (Larson and Paur, 1948) and *positional sterile - 2*, (*ps 2*) (Tronickova, 1962) control functional male sterility. The sterility controlled by the gene *ps* is characterized by petals showing coalescence of the corolla nearly to their extremity and the gene *ps 2* controls presence of fertile pollen and indehiscent anthers. Positional male-sterility (male sterility is caused due to non-dehiscence of anthers) or exserted stigma sterility (stigma protrudes beyond the anther cone at the bud stage due to high length of style) is convenient to utilize because of easy maintenance by forced selfing. Only manual cross-pollination is required for producing hybrid seeds. Based on the *positional sterile 2* lines (*ps-2*) a number of tomato hybrids were developed and released in Bulgaria, Poland, Czech Republic and Moldova (Atanassova 1999; Staniaszek *et al.* ,2000; Atanassova and Georgiev, 2002). However, in mild winter condition of West Bengal, selfed generation of both *ps* and *ps-2* breeding lines under open field condition showed inconsistent functional male sterility. However, considering 75% plants with stable anther non-dehiscence character along with very low of 5.6 % selfing it appeared that the *ps-2* male sterile line EC 620175 hold ample promise for utilization in commercial hybrid development in this condition (Hazra *et al.*, 2009).

Two stamenless mutants, *viz.*, stamenless-1 (*sl-1*) and stamcnless-2 (*sl-2*) produce flowers without stamen when grown at higher temperature, while those grown at low temperature produce flowers with abnormal stamens and often with viable pollen (Sawhney,1997). However, these male sterile lines have not yet been used in hybrid seed production.

Breeding for resistance to biotic stresses

Tomato is highly prone to different biotic stresses *viz.*, diseases, insects and nematodes which are responsible for significant yield reduction.

Breeding for resistance to diseases

Tomato is susceptible to many diseases caused by pathogenic fungi, bacteria, viruses, or nematodes. Breeding for disease resistance remains a major goal of most public and private sector tomato breeding programmes as new diseases achieve significance or new races of existing pathogens become established. In fact, the greatest contribution of modern tomato breeding programme has been development of varieties with improved disease resistance. Resistance has been identified, and in many cases characterized, for more than 30 major tomato diseases (Foolad, 2007a). The major resistance genes that have been identified include fusarium wilt, verticillium wilt, root-knot nematode, *Alternaria* stem canker, gray leaf spot, and some bacterial and viral diseases particularly tomato yellow leaf curl virus. However, horizontal or polygenic resistance has also been reported for several tomato diseases, where major genes for resistance to a particular pathogen or race are not found, such as early blight, powdery mildew, bacterial canker, and bacterial wilt. Resistance resources have been identified in most wild relatives of tomato particularly, *S. pimpinellifolium*, *S. peruvianum*, *S. hirsutum* (*S. habrochaites*) and *S. chilense*. For some tomato diseases *viz*., late blight (caused by the fungus *Phytophthora infestans*) and powdery mildew (caused by the fungus *Oidium lycopersicum*), both vertical and horizontal resistances have been identified.

Table 10: Sources of resistance to important diseases of tomato

Diseases	Sources
Fungal disease	
Anthracnose	PI272636
Brown root rot	*L. pimpinellifolium*
Collar rot	*L. pimpinellifolium*
Corky root	*L. peruvianum*, PI 262906, PI198601
Foot and stem rot	*L. hirsutum, L. hirsutum* 66087 (IVT61292)
Early blight	*L. hirsutum* f. *glabratum* CI 943, *L. peruvianum* var. *dentatum*, *L. pimpinellifolium* 'Pan American', *L. peruvianum* B 6002177, NCEBR-1 and NCEBR-2
Fusarium wilt Race1.	*L. pimpinellifolium, L. hirsutum,* Walter, Columbia, Roma, HS 110, Homestead, Flora-Dade, Sel. 22 and Sel. 30. PI126915, Rustrel.
Race 2.	*L. pimpinellifolium,* CR6
Race 3.	*L. pimpinellifolium*, Manapal, Manalucie
Grey leaf spot	Campbell 28, Rotella, Caraiba, Ont. 7620 and Flora-Dade
Late blight	*L. pimpinellifolium, L. esculentum* var. *cerasiforme,* PT 251305, PI 126445, LA 1255
Leaf mould	*L. pimpinellifolium,* *L. hirsutum* f. *glabratum, L. peruvianum,* *L. pennellii*, Vegabond, Vinequeen, V548, Vantage, Wetham.
Leaf spot	*L. hirsutum, L. pimpinellifolium,* *l. esculentum* var. *cerasiforme*, Targinnie Red
Bacterial disease	
Bacterial wilt	Louisiana Pink, PI 127805A Venus, Saturn, *L. pimpinellifolium*
Bacterial canker	*L. hirsutum* PI251305, *L. pimpinellifolium* lines Utah 737, Uttah 20
Bacterial spot	*L. esculentum* var. *cerasiforme*, Hawaii 7998
Bacterial speck	*L. pimpinellifolium* PI 112215, *L. hirsutum* f. *glabratum* Pl 129157

Viral disease	
Tomato yellow leaf curl virus	*L. pimpinellifolium* A 1921, *L. chilense* *L. hirsutum* f. *glabratum* B 6013, *L. hirsutum* f. *typicum* A 1904, *L. peruvianum*
Tomato mosaic virus	*L. peruvianum* PI 128650, *L. hirsutum*
Spotted wilt virus	*L. pimpinellifolium*, Pearl Harbor

Source: Kalloo (1988)

Resistance to most of the diseases is governed by single dominant gene while resistance to bacterial wilt and bacterial spot is polygenically controlled and bacterial canker is controlled by three dominant and one recessive gene (Lukyanenko, 1991). Tomato yellow leaf curl virus (TYLCV) is one of the most harmful afflictions in the world that affects tomato growth and production. Six regular antagonistic genes (*Ty-1*, *Ty-2*, *Ty-3*, *Ty-4*, *ty-5*, and *Ty-6*) have been transferred from wild germplasms to commercial cultivars as TYLCV protections. *Ty-1, Ty-3, Ty-4*, and *Ty-6* genes were all introduced from *S. chilense* accessions and are located on chromosomes 6, 6, 3, and 10, respectively (Gill *et al*., 2019). The *ty-5* and *Ty-2* were introgressed from *S. peruvianum* and *S. habrochaites*, respectively (Hanson *et al*., 2006). The *Ty-1* and *Ty-3* genes were introgressed from *S. chilense Ty-1*, *Ty-2*, and *Ty-3* displayed substantial levels of opposition against few strains. It has been possible to clone three TYLCV opposition genes (*Ty-1*/*Ty-3*, *Ty-2*, and *ty-5*) that target three antiviral safety mechanisms. Several varieties resistant to tomato leaf curl virus or tomato yellow leaf curl virus across the world has been developed through introgression of these resistant or antagonistic genes.

Table 11: Inheritance of resistance to disease, insect pests, and environmental stresses

Nature of resistance	Diseases/insect pests/environmental stress
A. Genetic factors	
i) Monogenic dominant	Fusarium wilt, leaf mould, Rhizoctonia root rot, leaf spot, Verticillium wilt, spotted wilt virus, tobacco mosaic virus, tomato leaf curl virus and root knot nematodes
ii) Monogenic recessive	Fruit borer, root knot nematodes and fruit set at low temperature
iii) Oligogenic factors	Bacterial canker, early bight, rhizoctonia root rot and spotted wilt virus
iv) Polygenic factors	Bacterial wilt, bacterial canker, late blight and fruit set at low and high temperatures
B. Plant biochemical factors	
i) Antibiosis	Fruit borer
ii) Antibiosis and plant morphological factors	Aphids, white flies and spider mites

Source: Kalloo (1988)

Backcross method is commonly followed to transfer resistant genes from donor parents to the desired variety. It is also useful in transgression of the desired genes from the wild relatives of tomato into the cultivated species. However, introgression of genes from *S. chilense*, *S. peruvianum* and *S. habrochaites* need biotechnological interventions like, embryo rescue technique to develop viable and utilizable F_1 hybrids. In many cases selection, pedigree method, combination of backcross and pedigree methods have been employed to develop disease resistant variety. The commercial varieties developed with the resistant genes introgressed from the wild species has also been utilized as donor parent to develop another disease resistant variety. The leaf curl virus resistant line, NDT8-2001-3 has been developed through hybridization of H-24 (leaf curl virus resistant) x NDT-120 followed by selection. In this case pedigree and combination of backcross and pedigree methods are employed to develop disease resistant variety.

Breeding for resistance to insect pests

Tomato is susceptible to number of insects which cause huge loss to the crop in most parts of the world. Various species of whiteflies, aphids, mites, Lepidopteran insects (e.g., tomato fruit borer (*Helicoverpa armigera*), Coleopteran insects (e.g., Colorado potato beetle and tobacco flea beetle), Dipteran insects (e.g., leaf miners and fruit fly), thrips and cutworms, attack tomatoes and many of them are capable of causing devastating losses. Though chemical control is very common practice, use of resistant varieties is the most economical measure to check such pests through reducing the use of pesticides.

Resistance to major insect pests of tomato has been identified within the related wild species, particularly, *S. habrochiates* and *S. pennellii* (Muigai *et al.* 2003). Resistance to at least nine insect species has been reported in *S. pennellii*, including greenhouse whitefly, carmine and two spotted spider mites, and the potato aphid (Muigai *et al.*, 2003).

Table 12. Sources of resistance to insects and mites in *Lycopersicon* (*Solanum*) spp.

Common name	Source of resistance
Fruit borer	*L.hirsutum, L.hirsutum* f. *glabratum*
Coloradopotato beetle	*L.hirsutum* (*S. habrochiates*)
Whitefly	*L.hirsutum* f. *glabratum*, Red cloud, Strain B., *S. pennellii*
Leaf miner	*L. hirsutum, L.hirsutum* f. *glabratum*, PearsonVF6L, *S. pennellii, S. cheesmanii*
Tobacco hornworm	*L. hirsutum* f. *glabratum* PI134417, PI LA 407
Potato aphid	*S. peruvianum, S. pennellii*
Tobacco flea beetle	*S. hirsutum* f. *glabratum*, PI126449, *S. hirsutum*
Pinworm	*S. cheesmanii* f. *minor*. *L. glandulosum*, *S..hirsutum*
Greenhouse whitefly	*S. hirsutum, L.hirsutum*f.*glabratum*, *S. pennellii*
Carmine spider mite	*S. hirsutum*, PI126449 *L. hirsutum* f. *glabratum, L. pennellii*
Two-spotted spider mite	*L. hirsutum* f. *glabratum, L. pennellii*

Source: Kalloo (1988), Swarup (2006)

Breeding for resistance to root knot nematodes

Root knot nematodes are responsible for marked reduction in yield and sometimes for complete crop failures in many parts of the world. A number of *Meloidogyne* species are responsible for inflicting crop loss. Among these, *M. incognita* is the most common followed by *M. javanica, M. hapla* and *M. arevaria*. Chemical control measures are most common but it is too expensive and has adverse effects on environment. This necessitates exploitation of non-chemical methods like crop rotation, fallow, trap crops, exposure of soil to summer heat, flooding, biological control and development of resistant varieties.

Bailey (1941) first reported high level of resistance in *S. peruvianum* to *Meloidgyne* spp. which was later supported by other workers. Many reports suggest that single dominant or partially dominant gene controls the resistance. With the use of embryo culture technique in interspecific hybridization, it has been possible to combine the resistance gene of *S. peruvianum* in cultivated cultivar by the backcross method

Table 13: Host resistance to root knot nematodes in tomato

Root knot nematode	Resistant genotype
M. incognita	Anahu, Nemared, Ronita, Piernita, Nemacross BBF1, Nemared F_1, VFN 14,Monita,Rossol,Santa Cruz,Venus,Saturno,VFN 8, Atkinson, Monalucie66 N,569 N – 10, *S. peruvianum*
M. hapla	Rutgers, Ace, Pi270435.
M. javanica	Ronita, Rossol,SantaCruz,Venus, Saturno,V14, V 275, NC 802, Healani,Anahu, Nematex, Bonus VFN, Atkinson, Manalucie66 N_1, 569N-10.
M. arenari	Ronita, Rossol, Santa Cruz, Venus, Saturno, Atkinson, Manalucie 66 N1,569 N-10

Source: Kalloo (1988), Swarup (2006)

Breeding for resistance to abiotic stresses

Tomato is highly sensitive to different abiotic stress namely, high temperature, salinity and moisture, etc. which markedly reduce the yield and quality.

Table 14: Source of resistance to various abiotic stresses

Traits	Sources of resistance
Cold tolerance (germination)	PI 120256, PI 174263, PI 342988 *L. hirsutum* f. *glabratum* B 6013
Fruits set	Precoce, Apedice, Montfavet 63-4 Super marmande, Severianin
Chilling tolerance	*L. hirsutum, L. hirsutum*f *.glabratum.*
Heat tolerance (germination at 40°C)	Nema 1200, P 28693, UC 82 L.
Fruits set	Saladette, PS Mini Rose (84-58), PI 205009, *L. pimpinellifolium* (Pan American), EC 130042, *L. cheesmanii*
Drought resistance	*L. pennellii*, EC 130042, Sel.28, PI 205009,
Salt tolerance	*L. pimpinellifolium, L. cheesmanii.*
Ozone tolerance	*L.cheesmanii, L. pennellii.* Heinz 1439, New Yorker, Tatiner, Pierette, Kharkov
Flood tolerance	PI 406966 (*L. esculentum*)

Source: Kalloo (1988)

Breeding for resistance to high /cold temperature stress

Night temperature is more important for fruit setting than day temperature and normal fruit set takes place at night temperature range of 15-20°C. When the temperature goes beyond these two extremes, fruit set is drastically reduced in open field condition. At higher average temperature (>29°C), fruit set is seriously affected because of style elongation, lack of pollination, reduced pollen germination and slow pollen tube growth. Main factor responsible for low fruit set at low temperatures is poor anther dehiscence. Different wild relatives are tolerant to environmental stresses *viz*., *S. cheesmanii* (salt, heat, cold and drought tolerance), *S. pimpinellifolium* (tolerance to heat, drought and water logging), *S. chilense* (cold and drought tolerance), *S. habrochaites* (tolerance to cold, chilling and salt), *S. pennellii* (tolerance to drought, moisture stress and salt), *S. lycopersicum var. cerasiforme* (tolerance to water logging and moisture stress. Wild relatives like *S. habrochaites* and *S. chilense* have the ability to resist temperature fluctuation of both low and high temperature. It has been identified recently that Sucrose synthase gene *SUS3* could enhance cold tolerance in tomato (Li *et al*, 2024).

Great variation in tomato accessions including wild relatives regarding fruit set at high temperature has been documented through several research works.

Table 15: Genotypes setting fruits at high temperature

Genotypes
HS-102, Sparton Red 8, OK-7-2, OK-7-3, Saladette, HS-101, PI 262934, BL 6807,S 6916,CIA 5161, VF 36, Punjab Tropic, Marzano, P-4, Avalanche, ChikoIII, Merit, C 28,Red Rock, Narcarlang, Breeding Line 165, Red Cherry, KS-1, KS-2,Walter, Sweet 72, Sel.58, Sel.84, CL 9-0-0-1, Palican, Tropic, Caraibe, Severianin, L 569, L567, L 573, L 571, L 125, L 226, L 232, L 2972, L 3690, L 246, L 2991, Burgess, *L. pimpinellifolium,* CLN-2413R, CLN 2116B and COML CR-7

Source: Kalloo (1988), Akhtar *et al.* (2012)

Different studies on conventional breeding strategy based on fruit vegetable crops indicated that single plant selection in the F_2 for heat tolerance will not be effective. Hence, advancement of generation of the cross involving heat tolerant genotype and selection for some important characters *viz*., flower/ plant, flower drop %, pollen germination, fruits / plant, fruit weight and fruit yield/ plant based on replicated family testing in F_3 and later generations may be effective for improving heat tolerance in different fruit vegetable crops, particularly tomato (Hazra and Ansary, 2008).

The tomato variety, Pusa Sheetal developed through hybridization of two varieties from Bulgaria and Russia, Balkan (Bulgaria) × Jemnorrosnej (Russia) followed by pedigree selection can tolerate low night temperature up to 8°C for fruit set. The hybrid, Pusa Hybrid-1 sets fruits at high average temperature of 28°C and a dwarf and determinate variety, Pusa Sadabahar set fruit at both low (6-8°C) and high average temperature of 30°C.

Breeding for resistance to moisture stress and flood stress

Tomato production is highly restricted under excess moisture conditions. Even short periods of flooding due to rain may cause wilting. Kuo *et al.* (1982) conducted an experiment on screening for flood tolerance in the genus *Solanum* in which a total of 4630 accessions were included. They observed that seven *Solanum lycopersicum* accessions (L-123, L-125, L-973, L-3072, L-3091, L-4313 and L-4360) and one accession of *S. pimpinellifolium* (L-4422) exhibited some degree of tolerance to flood for a short period at high temperature condition. The variety Arka Vikas, developed by single plant selection from the exotic line "Tip Top" introduced from USA is tolerant to moisture stress. The wild relative, *S. pennellii* has been reported to be drought resistance on the basis of plant growth.

Breeding for resistance to salt stress

Soil salinity poses a very serious problem especially in arid and semi arid parts of the world and thus limits the production of tomato. Development of salt tolerant types is the only economic approach to grow a profitable crop under such conditions. Almost all the cultivated varieties are sensitive to salinity while wild species like *S. peruvianum, S. cheesmani* spp. *minor* and *S. pennellii* exhibit some degrees of tolerance to sodium chloride.

The usefulness of marker-assisted selection (MAS) to develop salt-tolerant breeding lines from a F_2 derived from *S. lycopersicum* × *S. pimpinellifolium* was studied by Monforte *et al.* (1996) by interval mapping methodology of quantitative trait locus (QTL) analysis to locate more precisely the previously detected salt tolerance QTLs. A new QTL for total fruit weight under salinity (TW) near TG24 was detected. Most of the detected QTLs [(3for TW, 5 for fruit number (FN) and 4 for fruit weight (FW)] had low R^2 values, except the FW QTL in the TG180-TG48 interval, which explains 36.6% of the total variance. Dominant and over dominant effects were detected at the QTLs for TW, whereas gene effects at the QTLs for FN and FW ranged from additive to partial dominance. Phenotypic selection of F_2 families and marker-assisted selection of F_3families were carried out. Yield under salinity decreased in the F_2 generation. F_3 means were similar to those of the F_1 as a consequence of phenotypic selection.

Breeding for quality

Quality is a component that adds value to crops and their products. To cater the needs of processing industry, variety / hybrid should be high yielding, uniform maturity, rich in colour, vitamin C, TSS and acidity. Besides, there is an urgent need to develop variety / hybrid suitable for long distance transportation for internal market as well as for export.

Fruit colour

Development of proper colour is an important quality parameter both for table purpose and for processing varieties from the point of view of consumer preference. Red and yellow are the two major colours of tomato which are controlled by the carotene content of the fruits. Red fruits have mostly lycopene and only a small quantity of β carotene. In normal red fruit the ratio of lycopene and β carotene is generally 12 to 15: 1. Besides, small amounts of phytoene, phytofluene, ζ-carotene, γ-carotene, neorosporine, and lutein are also present.

The genetics of colour in tomato is well documented and several lycopene enhancing mutant (*hp-1*, *hp2*/ *dg*, *og*^{C}) and β carotene enhancing mutant (*B*) have been utilized in tomato breeding programme for increased lycopene and β carotene content. The basic breeding strategy for development of purple tomato has been framed to combine the lycopene enhancing gene *dg* or *hp-2*^{dg} present in chromosome 1 and anthocyanin expressing gene, *Aft* present in chromosome 10 in one genotype taking advantage of independent assortment in the ratio of 1:15 segregation in the F_2 population. This breeding scheme took only one crossing Alisa Craig *Aft/Aft* x BCT-115 *dg/dg* and two selfings to select the desirable double homozygous genotype *AftAft dg/dg* based on morphological marker characters conditioned by two genes, dark green fruit of *dgdg* and purple colour for anthocyanin pigment of *Aft*- gene. The novel breeding line "Purple tomato" contains both high lycopene (6.13 mg/100 g fresh) and anthocyanin (20.73 mg/100 g fresh) pigments in the fruits (Hazra *et al.*, 2018).

Ascorbic acid

In the quality breeding programme of tomato, *S. pimpinellifolium* and *S. peruvianum* have been reported to contribute genes for high ascorbic acid content. There existed a relationship between ascorbic acid content and fruit size, shape and extent of fruit exposed to light (Tigchelaar, 1986). Small fruited genotypes are richer in ascorbic acid content and firm fruited cultivars have exhibited lower content.

Acidity

It is yet another important quality parameter which not only contributes to sourness but is also major factors in flavour intensity of tomato. Acidity influences the storability of processed products by inhibiting the spore germination of thermophilic organisms. There is negative correlation between size/weight of fruit and acidity. Locular tissues have more acidity than the pericarp tissues. So for high acidity, a variety with more locular area should be preferred. Titratable acidity value greater than 0.35% is desirable for processing tomato.

Sugars and solids

The total solids of the cultivated tomato comprise 4 – 7.5% of its fresh weight, though this percentage can be much higher in some wild species (Fridman *et al.*, 2000). The total solids are composed of all fruit components except water and volatiles. In the cultivated tomato, the soluble and insoluble solids account for about 75% and 25%, respectively, of the total solids. Reducing sugars glucose and fructose are the major components of the soluble solids. Sucrose is also present but in very small quantities, although some wild species of tomato, including *S. chmielewskii* and *S. habrochaites* have higher concentration of sucrose (Chetelat *et al.*, 1995). The remaining soluble solids are composed of organic acids, lipids, minerals, and pigments. The insoluble solids include proteins, cellulose, hemicellulose, pectins, and polysaccharides, which determine fruit viscosity (Hewitt and Garvey, 1987). Viscosity of fruit pulp determines the quality of the processed products. The increase in total solids of tomatoes has often been found to enhance the yield of processed products. The taste and flavour are also improved by higher solids content (Foolad, 2007a). Soluble solids in tomato have been found to be up to 4.6% in fresh market tomatoes and up to 6.3% in most processing varieties. However, fruits of some wild relatives *viz*., *S. pimpinellifolium*, *S. chmielewskii* and *S. cheesmanii* have much higher concentrations (9–15%) of soluble solids (Hewitt and Garvey, 1987). However, the complex and quantitative nature of the trait and the negative relationship between yield and percentage of soluble solids have limited success of getting genotypes with high soluble solids and high yield (Foolad, 2007a).

Sugar content is another important qualitative attribute which deserves due attention in the quality breeding programme. Both reducing and non-reducing sugars are found in tomato. Sugars contribute to sweetness and may lead to increase in flavour. It is reported that at the pink stage, there is a higher content of sugars than at other stages. A single incomplete-dominant gene (*Fgr*) has been identified in *S. hirsutum* that increases the proportion of fructose over glucose, thus contributing to fruit sweetness (Levin *et al*., 2000).

Postharvest shelf-life

Besides nutritive and processing qualities, certain post-harvest qualitative traits like slow ripening leading to long shelf-life, thick skin facilitating long distance transport are also equally important in the quality improvement programme. The most effective method of breeding synchronous cultivars is the combination of high synchrony of fruit maturity with their slower ripening. Tomato being a climacteric fruit shows a dramatic increase in respiration at the onset of ripening accompanied by increased synthesis of the phytohormone, ethylene. In tomato, three single gene mutants namely *rin* (*ripening inhibitor*), *nor* (*non ripening*) and *Nr* (*never ripe*) which alter ripening process have been identified which are located on the chromosomes 5, 10 and 9 respectively (Foolad, 2007b). Enough scope is there to utilize these mutants in breeding tomato for enhanced shelf-life.

References

Acciarri, N., Rotino, G. L., Tamietti, G., Valentino, D., Voltattorni, S. and Sabatini, E. 2007. Molecular markers for *Ve1* and *Ve2 Verticillium* resistance genes from Italian tomato germplasm. *Plant Breed.*, **126**: 617-621.

Acharya, B., Ghorai, A. K., Dutta, S., Maurya, P. K., Dutta, S., Mandal, A. K. and Hazra, P. 2018. Genetics of tolerance to bacterial wilt disease in tomato (*Solanum lycopersicum* L.). *Austral. Plant Path.*, **47**: 591-600.

Acharya, B., Pradeep Kumar, P. Hazra, S., Dutta, S., Saha, S., Roy, S. Maji, A., Chakraborty, I. Chattopadhyay, A. and Hazra, P. 2024. Genetic control of improtant yield attributing character prodicted through mechanie learing in segragating generatinons of interspecific crosses of tomato (*Solanum lycopersicum* L.) *Acta Physiol Plant.* **46**: 78, doi.org10.1007/S 11738-024-03702-3.

Akhtar, S., Ansary, S. H., Dutta, A. K., Karak, C. and Hazra, P. 2012. Crucial reproductive char Crucial reproductive characters as screening indices for tomato (*Lycopersicon esculentum* Mill.) under high temperature stress. *J. Crop Weed*, **8**: 114–117.

Atanassova, B. 1999. Functional male sterility (*ps 2*) in tomato (*Lycopersicon esculentum* Mill.) and its application in breeding and hybrid seed production. *Euphytica.* **107** : 13-21.

Atanassova, B. and Georgiev, Hr. 2007. Expression of Heterosis by Hybridization. In: *Genetic improvement of Solanaceous crops. Vol.2. Tomato.* (Ed. Razdan, M. and Mattoo, A.K.). *Science Publishers*, USA, pp. 113-151.

Atanassova, B. and Hazra, P. 2012. Tomato. In: *Hand book of Vegetables* (Eds. K. V. Peter and P. Hazra), Studium Press LLC, Houston, Texas, USA, pp. 533-593.

Atanassova, B., Daskalov, S., Shtereva, L. and Balatcheva, E. 2001. Anthocyaninless mutations improving tomato and pepper tolerance to adverse climatic conditions. *Euphytica*, **120**:357-365.

Bailey, D.M, 1941. The seedling test method for root-knot nematode resistance. *Proc. Amer. Soc. Hort. Sci.,* 38: 573–575.

Balatcheva, E., Atanassova, B. and Hazra, P. 2008. Application of contemporary achievements in plant genetic improvement for tomato - the basic tool for sustainable development of agriculture in the coastal area. *J. Indian Soc. Coastal Agric. Res.*, **26**: 22-27.

Biswas, P., Chattopadhyay, A.and Hazra, P. 2016. Genetic control of fruit yield and quality characters in tomato genotypes possessing *hp-1*, *og^c^*, *dg*, *Aft* and *rin*genes. *Indian J. Genet.*, **76**: 326-332.

Chattopadhyay, A., Dutta, S., Dutta, P. and Hazra, P. 2011. Studies on heterobeltiosis, combining ability and gene action in tomato (*Solanum lycopersicum*). *Intern.J. Plant Breed.*, **5**: 88-93.

Chetelat, R.T. 2002. Revised list of monogenic stocks. *Rep. Tomato Genet Coop.*, **52**:41-62.

Chetelat, R. T., DeVerna, J. W. and Bennett, A. B. 1995. Introgression into tomato (*Lycopersicon esculentum*) of the *L. chmielewskii* sucrose accumulator gene (*sucr*) controlling fruit sugar composition. *Theor.Appl. Genet.*, **91**: 327-333.

Choudhury, B. 1959. Development of Seed and its Parts-II of Hybrid Seed of *Lycopersicon esculentum* x *Lycopersicon peruvianvm. Indian J.Hort.*, **16**: 23-30.

Das, I., Hazra, P., Longjam, M., Bhattacharjee, T., Maurya, P.K., Banerjee, S. and Chattopadhyay, A. 2020. Genetic control of reproductive and fruit quality traits in crosses involving cultivars and induced mutants of tomato (*Solanum lycopersicum* L.). *J. Genet.*, **99**: 56 https://doi.org/10.1007/s12041-020-01209-7

Das, I., Hazra, P., Longjam, M., Bhattacharjee,T., Maurya, P.K., Banerjee, S. and Chattopadhyay, A. 2019. Characterization of induced mutants and their hybrids of tomato (*Solanum lycopersicum* L.) for growth, yield and fruit quality traits to explore the feasibility in future breeding. *Genet. Resour. Crop Evol.*, **66**:1421-1441.

Fentik, D.A. 2017. Review on genetics and breeding of tomato (*Lycopersicon esculentum* Mill.). *Adv.Crop Sci.Technol.,* 5: 306 , DOI: 10.4172/2329-8863.1000306

Foolad, M. R. 2007a. Genome mapping and molecular breeding of tomato. *Intern. J. plant Genom.*, **10**: 1-52. doi: 10.1155/2007/64358

Foolad, M. R. (2007b). Tolerance to abiotic stresses. In: *Genetic Improvement of Solanaceous Crops: Tomato* (Ed. Rajdan, M.K.). CRC Press, Boca Raton, USA, **12:** 521-590.

Gascuel, Q., Diretto, G., Monforte, A. J., Fortes, A. M. and Granell, A. 2017. Use of natural diversity and biotechnology to increase the quality and nutritional content of tomato and grape. *Front. Plant Sci.*, **8**: doi: 10.3389/fpls.2017.00652.

Georgiev, C. 1972. Anthocyanin fruit (*Af*). *Rep. Tomato Genet. Coop.*, **22**:10.

Gill, U., Scott, J. W., Shekasteband, R., Ogundiwin, E., Schuit, C. and Francis, D. M. 2019. Ty-6, a major begomovirus resistance gene on chromosome 10, is effective against tomato yellow leaf curl virus and tomato mottle virus. *Theor. Appl. Genet.*, **132**: 1543–1554.

Hanson, P., Green, S., and Kuo, G. 2006. *Ty-2*, a gene on chromosome 11 conditioning geminivirus resistance in tomato. *Tomato Genet. Coop Rep.* **56**: 17–18.

Hazra, P. and Ansary, S.H. 2008. Genetics of heat tolerance for floral and fruit set to high temperature stress in tomato (*Lycopersicon esculentum* Mill.). *SABRAO J. Breeding Genet.,* **40**: 117- 125.

Hazra P., Roy, M., Akhtar, S., Balacheva, E. and Atanassova, B. 2009. Expressivity of two genes controlling functional male sterility (*positional sterile* (*ps*) and *positional sterile-2* (*ps-2*) in tomato under the climatic conditions of Eastern India. *Genetics and Breeding*, Bulgaria **38**: 10-16.

Hazra, P. and Som, M.G. 2006. *Vegetable Science*. Kalyani Publishers, New Delhi, Ludhiana, pp.37-108

Hazra, P., Banerjee, M.K. and Chattopadhyay, A. 2022. *Varieties of Vegetable crops in India*, Kalyani Publishers, New Delhi, Ludhiana, 204p.

Hazra, P. and Dutta, A.K. 2011. Parthenocarpy in tomato (*Solanum lycopersicum*), its inheritance pattern and association with marker characters. *Indian J. Genet.Plant Breed.*, 71 : 248-253.

Hazra, P., Dutta, A.K. and Chatterjee, P. 2010. Altered gibberellin and auxin levels in the ovaries in the manifestation of genetic parthenocarpy in tomato (*Solanum lycopersicum*). *Curr. Sci.*, 99: 1439 – 1443.

Hazra, P., Longjam, M. and Chattopadhyay, A. 2018. Stacking of mutant genes in the development of "purple tomato" rich in both lycopene and anthocyanin contents. *Sci. Hort.*,**239**: 253-258.

Hedrick, U. P. and Booth, N. O. 1907. Mendelian characters in tomatoes. *Proc.* Amer. *Soc. Hort. Sci.*, **5**:19-24.

Hewitt, J. D. and Garvey, T. C. 1987. Wild sources of high soluble solids in tomato. In: *Plant Biology*, vol 4: *Tomato Biotechnology* (Eds. Nevins, D.J. and Jones, R.A.), Liss,New York, USA, pp. 45–54.

Jones, C.M., Mes, P. and Myers, R. 2003. Characteriazation and inheritance of the Anthocyanin fruit (*Aft*) tomato. *J. Hered.*, **94**: 449-456.

Kalloo, G. 1988. *Vegetable Breeding*, Vol I, II, III, CRC Press, Boca Raton , USA.

Kartseva, T., Balacheva, E., Molle, E. and Atanassova, B. 2008. Effect of anthocyaninless mutants genes on the content of compounds related to nutritive quality in tomato (*Solanum lycopersicum*) fruits. *Acta Hort.*, **830**: 285-289.

Kitagawa, M., Ito, H., Shiina, T., Nakamura, N., Inakuma, T., Kazumi, T., Ishiguro, Y., Yabe, K. and Ito, Y. 2005. Characterization of tomato fruit ripening and analysis of gene expression in F_1 hybrids of the ripening inhibitor (*rin*) mutant. *Physiol. Plant.,* **123**:331-338.

Kuo, C. G., Tsay, J. S., Chen, B. W. and Lin, P. Y. 1982. Screening for flooding tolerance in the genus *Lycopersicon. Hort Sci.*, **17**: 76-78.

Larson, R.E. and Paur, S. 1948. The description and the inheritance of a functional sterile flower mutant in tomato and its probable value in hybrid tomato seed production. *J. Am. Soc. Hort. Sci.*, **52**: 355-364.

Lebeda, A. and Mieslerova, B. 2000. Case study of host-pathogen interaction: Tomato (*Lycopersicon* spp.) - tomato powdery mildew (*Oidium lycopersici*). *Plant Protect. Sci.* Prague, **36**: 156-162.

Levin, I., Frankel, P., Gilboa, N., Tanny, S. and Lalazar, A. 2003. The tomato dark green mutation is a novel allel of the tomato homolog of DEETIOLATED1 gene. *Theor. Appl. Genet.*, **106**: 454-460.

Levin, I., Gilboa, N., Yeselson, E., Shen, S. and Schaffer, A. A. 2000. *Fgr*, a major locus that modulates the fructose to glucose ratio in mature tomato fruits. *Theor.Appl. Genet.*, **100**: 256-262.

Li, S., Wang, Y., Liu, Y., Liu, C., Xu, W., Lu, Y. and Ye, Z. 2024.Sucrose synthase gene *SUS3* could enhance cold tolerance in tomato. *Front. Plant Sci.*, 14-2023 | https://doi.org/10.3389/fpls.2023.1324401

Lukyanenko, A.N. 1991. Disease Resistance in Tomato. In: *Genetic Improvement of Tomato*, Vol. 14 (Ed. Kalloo, G.), Monographs on Theoretical and Applied Genetics, Springer, Berlin, pp. 99-119.

Mazzuccato, A., Taddei, A.R. and Soressi, G.P. 1998. The parthenocarpic fruit (*pat*) mutant of tomato (*Lycopersicon esculentum* Mill.) sets seedless fruits and has aberrant anther and ovule development. *Development*, **125**: 107-114.

Monforte, A. J., Asins, M. J. and Carbonell, E. A. 1996. Salt tolerance in Lycopersicon species. IV. Efficiency of marker-assisted selection for salt tolerance improvement. *Theor.Appl. Genet.*, **93**: 765-772.

Moore, S., Vrebalov, J., Payton, P. and Giovanini, J. 2002. Use of genomics tools to isolate key ripening genes and analyse fruit maturation in tomato. *J. Exp. Bot.,* **377**: 2023-2030.

Muigai, S. G., Bassett, M. J., Schuster, D. J. and Scott, J. W. 2003. Greenhouse and field screening of wild Lycopersicon germplasm for resistance to the whitefly *Bemisia argentifolii*. *Phytoparasitica*, **31**: 27-38.

Mutschler, M., Guttieri, M., Kinzer, S., Grierson, D. and Tucker, G. 1988. Changes in ripening-related processes in tomato conditioned by the *alc* mutant. *Theor. Appl. Genet.,* **76**:285-292.

O'Hare, T.J., Wong, L.S. and McGrath, D.J. 2005. Increasing visual appeal and health benefit of fresh-cut tomato slices using high lycopene varieties. *Acta Hort.*, **687**:395-397.

Pal, H., Kundu, A., Sahu, R., Sethi, A., Hazra, P. and Chatterjee, S. 2019. Unravelling the metabolic behavior in Tomato high pigment mutants (*hp-1, hp-2dg, og^c*) and non ripening mutant (*rin*) during fruit ripening. *Sci. Hort.*, **246**: 652-663.

Passam, H. C., Karapanos, I. C., Bebeli, P. J. and Savvas, D. 2007. A review of recent research on tomato nutrition, breeding and post-harvest technology with reference to fruit quality. *The European J.Plant Sci. Biotechnol.*, **1**: 1-21.

Philouze, J. 1974. Gènes marqueur liés aux gènes de sterilité mâle *ms-32* et *ms 35* chez la tomate. *Ann. Amelior. Plantes*., **24**:77-82.

Pradeep Kumar, P., Acharya, B., Dutta, S., Saha, S. and Hazra, P. 2019. Variation among the genotypes of tomato for crossability to its wild relatives. *Intern. J. Chem. Stud.*, **7**: 3423-3426.

Ranc, N., Muños, S., Santoni, S. and Causse, M. 2008. A clarified position for *Solanum lycopersicum* var. *cerasiforme* in the evolutionary history of tomatoes (Solanaceae). *BMC Plant Biol.*, **8**:1-16.

Roy, T., Bhattacharya, S., Dutta, A. K. Akhtar, S., Karak, C. and Hazra, P. 2011. Characterization of four genic male sterile lines of tomato (*Solanum lycopersicum*). *The Hort. J.*, **24** : 47-52.

Rick, C. M. and Chetelat, R. T. 1995. Utilization of related wild species for tomato improvement. *Acta Hort.*, **412**: 21-38.

Rick, C.M. 1978. The tomato. *Scientific American*, **239**: 77-87

Rick, C.M. and Yoder, J.I. 1988. Classical and molecular genetics of tomato: highlights and perspectives. *Ann. Rev. Genet.,* **22**:281-300.

Sawhney, V.K. 1997. Genic male sterility. In: *Pollen Biotechnology for Crop Production and Improvement* (Eds. Shivanna, K.R. and Sawhney, V.K.), Cambridge Univ. Press, pp. 183-198.

Siddiqui, Wasim, Md., Chakraborty, I., Mishra, P. and Hazra, P. 2014. Bioactive attributes of tomatoes involving *dg*, *og*^{c}, and *rin* genes. *Food &Function* **5**: 936-943.

Siddiqui, Wasim, Md., Chakraborty,I., Mishra,P., Hazra, P. and Ayala-Zavalad, J.F. 2015. Postharvest physico-chemical changes in mutant (*dg, og*^{c}, and *rin*) and non-mutant tomatoes. *Acta Physiol. Plant.*, **37**:1-11.

Siddiqui, Wasim,Md.,Chakraborty,I.,Hazra,P.and Ayala-Zavalad,J.F. 2014.Characterization of quality indices on storage of puree of mutant (*dg* and *og*^{c}) and normal tomatoes..*Acta Alim.*, **43:** 426-436.

Sikder, S., Biswas, P., Hazra, P., Chattopadhyay, A., Badigannavar, A. M. and D'Souza, S.F.2013. Induction of mutation in tomato (*Solanum lycopersicum* L.) by gamma irradiation, EMS and their combinations. *Indian J. Genet.*, **73**:392-399.

Staniaszek, M., Marczewski,W., Habdas, H and Potaczek, H. 2000. Identification of RAPD markers linked to the *ps* gene and their usefulness for purity determination of breeding lines and F_1 tomato hybrids. *Acta Physiol. Plant.*, **22**:303-306.

Stommel, J. R. 2007. Genetic enhancement of tomato fruit nutritive value. In: *Genetic Improvement of Solanaceous crops.* Vol.2. *Tomato.* (Razdan, M.K.. and Mattoo, A.K.). *Science Publishers*, USA. pp.193-238.

Tanksley, S.D. and McCouch, S.R. 1997. Seed banks and molecular maps: unlocking genetic potential from the wild. *Science*, **277**: 1063-1066.

Tigchelaar, E. C. 1986. Tomato breeding In: *Breeding Vegetable Crops* (Ed. Basset, M.J.), AVI Publishing Co, pp. 135-171.

Tronickova, E. 1962. New type of functional male sterility in tomato. *Ved. Prace Vysk. Ust. Rostl. Vyr. Praha-Ruzine*, **6** : 29-39 (in Czech)

Van Tuinen, A., Cordonnier-Pratt, M., Pratt, L.H., Verkerk, R., Koorneef, M and Zabel, P. 1997. The mapping of phytochrome genes and photomorphogenic mutants of tomato. *Theor Appl Genet*, **94**:115-122.

Zhang, Y. and Stommel, J.R. 2000. RAPD and AFLP tagging and mapping of *Beta* (*B*) and *Beta* modifier (Mo_B), two genes which influence β-carotene accumulation in fruit of tomato (*Lycopersicon esculentum* Mill.). *Theor. Appl. Genet.*, **100**:368-375.

19

Brinjal

Brinjal is the fifth most economically important Solanaceous crop after potato, tomato, chilli, and tobacco. Apart from the well-known brinjal species (*Solanum melongena* L.), two other under-utilized species, the scarlet eggplant (*S. aethiopicum* L.) and the gboma eggplant *S. macrocarpon* are also cultivated. Brinjal, also known as aubergine or eggplant, under the family Solanaceae (Chromosome No.: 2n = 2x =24) is one of the important vegetables of the tropical countries particularly India, Bangladesh, Pakistan, China, Japan, Indonesia, Thiland, Malaysia and the Philippines. Brinjal is the most common, popular and principal vegetable crop grown widely in India for its various shape, size and colour of fruits. Brinjals are rich sources of potential antioxidants: anthocyanin and phenolic compounds.

Origin, domestication and adaptation in India

Brinjal (*Solanum melongena*) is believed to have been originated in Indo-Burma region and perhaps China is the secondary centre of origin of this crop. Various data indicate, that from the several species that evolved in Africa, one (*S. insanum*), gave rise to a distinct species which spread to South-East Asia as the wild ancestor of *S. melongena* (Lester 1998; Ranil *et al.*, 2017). This wild progenitor of the cultivated species has wide distribution in at least 10 habitats in India which amply suggests India being the centre of origin of brinjal.

Primitive characters of brinjal are tall plants with large, spiny leaves, flowering in clusters. Their fruits are small, green, and bitter in taste, with thick skin and hard flesh. Domestication, mutation, natural hybridization and human selection brought extensive genetic diversity of brinjal cultivars grown all over the world. Cultivar differences concern mainly the colour, shape and length of fruits, but chemical composition of fruits, the earliness of fruiting, yield, environmental requirements, etc. are also taken into consideration.

The first record of brinjal in India was during 300 BC to 300 AD. At least 33 Sanskrit names for brinjal have been mentioned in the ancient literatures of India and the most common being Varttaka, Bhantaki and Vattingan. It was cultivated in Africa around 9th century A.D. It was known in Italy at the end of the 14th century and introduced into southern Europe in the 15th century and the name" eggplant" was probably derived from the white egg like fruits.

Crop botany and floral biology

- Brinjal is a ligneous, tropical perennial plants often cultivated as a tender or half-hardy annual vegetable crop; stem is often spiny. It grows 40 to 150 cm tall, with large, coarsely lobed leaves that are 10 to 20 cm long and 5 to 10 cm broad; semi-wild types can grow much larger, to 225 cm with large leaves over 30 cm long and 15 cm broad.
- Inflorescence is cymose and helicoids; flowering pattern of brinjal may be solitary, cluster or mixed (both solitary and cyme)
- At the level of each flower, there is dichotomous branching that grows more or less regularly, depending on the species and variety; the sympodia generally consist of two leaves and the axillary bud of the leaf below each flower frequently gives rise to a new branch. Growth and flowering are continuous throughout the life of the plant.
- Flowers are pentamerous, hermaphrodite and regular; calyx persistent, 5 -6 and densely hairy, prickles may or may not present. In most varieties the perfect flowers are borne singly and opposite the leaves.
- Corolla 5-6 with triangular lobes, large and generally light purple-blue outside deeper colour within, although white, whitish and whitish-purple flowers are also found in some cultivars; stamens 5 -6, adnate, erect and yellow.
- Carpels are united, ovary superior and styles are in variable length; hypogynus gynoecium is syncarp located obliquely in relation to the median.
- Heterostyly is a common feature in brinjal; four distinct types of flowers according to length of style and size of ovary *viz*., long styled with big size ovary; medium styled with medium sized ovary; pseudo-short styled with rudimentary ovary and true short styled with rudimentary ovary; percentage of long and medium styled flowers is a varietal character.
- Only long and medium styled flowers produce fruits, where as pseudo-short and true short -styled flowers do not set fruits.
- The fruit is botanically classified as a fleshy berry; white flesh with a meaty texture; fruits are borne singly or in clusters.
- Fruit length varies between 4-45 cm, and thickness 2-35 cm, at different shapes and weight ranging between 15-1500 g. The fruits are set as single or in clusters; physiologically ripe fruits become brown, red or yellow in colour
- The fruit contains numerous small, soft seeds that, though edible, taste bitter because they contain nicotinoidalkaloids like that is found in tobacco belonging to the same family.

Reproductive behaviour

Brinjal is basically self- pollinated crop but some degree of cross pollination (generally 8-10 % but it may go up to 30% in some cases) makes it some times as often cross pollinated crop (Hazra *et al.*, 2003). The extent of cross-pollination may go up in predominantly long styled flowers when population of pollinators is high. Heterostyly present in brinjal cultivars has also helped in widening the genetic diversity through partial cross pollination occurring in nature. Major pollinators are bumble bees (*Bombus* species) and honey bees (*Apis* species). Anthesis generally starts from 7.30-8.00 am which continues up to 11.00 – 11.30 am with peak time between 8.30 to 10.30 am however, it depends on the prevailing temperature and humidity condition. Anther dehiscence occurs at least half an hour after anthesis. Anther dehiscence occurs through the development of pores at the tip of the anthers. The stigma receptivity is maximum at the time of opening of the flower although it become effectively receptive a day prior to anthesis, and the receptivity sometimes continues even up to four days after anthesis. The flowers emasculated one to two days earlier should have completely blossomed under the paper bag and are ready for cross pollination in the morning hours of the next day. Pollen viability is retained for 8-10 days at moderate temperature of 20-22°C with 50-55% relative humidity. Dried pollens can be stored in sealed vial at 0°C for 1-2 months.

Interspecific compatibility of brinjal (*Solanum melongena*) with wild relatives

The cultivated species, *Solanum melongena* is crossable and produce fertile hybrids with *S. indicum, S. incanum, S. insanum, S. aethiopicum, S. surattense* and *S. macrocarpon*. However, few seeds are generally produced. In cross between *S. melongena* × *S. hispidum*, the hybrid had partial fertility. *Solanum aethiopicum* cv.gr. *gilo* (EC 769658) was crossable with cultivated brinjal but the F_1 generation was sterile which restrict the advancement of the inter-specific hybrids (Chattopadhyay *et al*, 2009). Amphidiploid fertile hybrids were obtained in the crosses of S. *melongena* × *S.torvum* and *S. melongena*×*S. integrifolium*. It was found easier to cross *S. integrifolium* with *S. melongena* as a female parent. Somatic hybrids were produced between *S .melongena* + *S.sisymbrifolium* and *S.melongena* + *S.torvum*.

Table 1: Crossability between cultivated brinjal (*Solanum melongena*) and wild species

Parents for Cross	**Status of hybrid**
S.melongena×*S.indicum*	F_4 plants obtained; Partially fertile plants
S.melongena x S.sodomeum	Fertile F_1 plants
S.melongena x S.macrocarpon	Fertile F_1 and F_2 plants
S.melongena x S. khasianum	Successful F_1 hybrids after embryo rescue
S.melongena x S. aethiopicum	F_1 hybrids obtained after embryo rescue; Fertile hybrids
S.melongena xS.insanum	F_1hybrids obtained after embryo rescue; Fertile hybrids
S.melongena x S. gilo	F_1 hybrids obtained; Sterile F_1 hybrids Partially fertile hybrids
S.melongena x S.hispdum	Sterile hybrids
S.melongena x S.torvum	F_1 hybrids after embryo rescue but the plants had very low fertility (1.73%); Unsuccessful cross
S.melongena x S.integriifolium	Sterile F_1 plants; Partially fertile hybrids

Parents for Cross	Status of hybrid
S.melongena x S.sisymbriifolium	Embryo rescued hybrid plantlets did not survive; Embryo rescued hybrid plantlets were sterile
S.melongena x S.xanthocarpum	F_1 hybrids were sterile; Partially fertile hybrids
S.melongena x S.grandifolium	Partially fertile hybrids
S.melongena x S.cumingii	Fertile hybrids
S.melongena x S.zuccanianum	Sterile hybrids
S.melongena x S.surattense	Sterile hybrids
S.melongena x S.indicum	Sterile hybrids
S.melongena x S.capsicoides	No germination of seeds
S.melongena x S.stramonifolium	No fruit set

Source: Daunay and Hazra (2012); Singh *et al.* (2015)

Genetics of important characters

Cultivated brinjal (*Solanum melongena*) is a diploid species with basic and haploid chromosome number x = n =12 (2n = 2x = 24). In fact, basic and haploid chromosome numbers of all the allied *Solanum* species including tomato are same. In brinjal, chromosome aberrations like, chromatid bridges, laggards, fragments and elimination of univalent have been found. High frequency of multivalents indicated the presence of translocation interchanges. It has been suggested that hybridization and structural chromosome changes might have influenced the evolution of brinjal.

Qualitative characters

The breeding method in any crop depends upon its breeding system, genetic architecture and the pattern of inheritance of characters. The inheritance of some qualitative characters in brinjal may be simple and governed by few genes only *viz.*, hypocotyl colour, growth habit, colour of stem, petiole, leaf and leaf vein, spines, corolla, fruit surface and fruit flesh, fruit shape, fruit bearing habit, male sterility, etc. Several reports on the genetics of different plant characters have been documented and presented.

Table 2: Inheritance pattern of different qualitative characters of brinjal

Characters	Inheritance pattern / Gene symbol
Hypocotyl colour	Purple dominant over green
Stern colour/pigmentation	Single gene, Purple (*Pst*) dominant over green (*pst*)
Purple pigmentation of stem, petiole	Pigmentation in all three plant parts controlled by
and veins	Duplicate genes designated *Pst1* and *Pst2*
	Pigmentation in all plant parts governed by three or
Leaf vein colour, purple shade on	more non-allelic genes.
young leaves, corolla and calyx colour	Monogenic, Purple colour dominant over green
Stem and leaf colour	leaf and stem
	Monogenic, Purple dominant over white or light
Corolla colour	purple
	Purple dominant over white (interaction of 3 or
Spinyiness	more non-allelic genes
Spinyiness and purple pigmentation	Monogenic dominant over spineless
Pigmentation on leaves and veins	Digenic dominant over spineless and green colour
	Governed by two complementary genes, C_1a_1,
Spines on stem, leaf and petiole	C_1b_1, $P_1v_1$and P_1v_2 respectively
Spines of leaves and stem	Single dominant gene
	Governed by three complementary genes of
Spines on calyx	which any two must be dominant
Erect versus spreading habit of plant	Two complementary genes and an inhibitor
	Controlled by two genes (*Egr*$_1$*and Egr*$_2$) with
Plant height	erectness being dominant
Fruit shape	Monogenic, tall being dominant over dwarf
	Controlled by three partially dominant genes (*Ofa,*
Fruit colour	*Ojb*$_1$*, Ojb*$_2$*, and Ojb*$_3$)
	Purple dominant over green and cream (Complex
	inheritance); monogenic, digenic, trigenic: three
	complimentary genes, *Pfa, Pfb1, Pfb2*
	Variegated dominant over white and green.
	Green (*Gm*) is incompletely dominant over white
	(*gm*).
	Green streaked purple is dominant over greenish
Fruit end shape	white
	Elongated dominant over round
Fruit bearing habit	Round dominant over oval
	Cluster dominant over solitary
	Controlled by single gene
Male sterility (Functional male	Complex and governed by 3-4 genes
sterility due to non-dehiscent anther)	Controlled single recessive allele (*fms*) which is
	linked to purple colour of fruit
	Single recessive allele (*fms*) along with two
Male sterility (Sporogenous sterility)	modifier genes
	Under cytoplasmic control

Source: Kalda (1998), Ram (1998), Hazra *et al* (2008); Daunay and Hazra (2012)

Quantitative characters

Several quantitative characters *viz*., plant height, leaf area, number of branches/plant, plant spread, days to flowering, number of flowers per cluster, number of fruits per plant, number of fruits per cluster, fruit length, girth and weight, early and total yield, etc. are polygenically control. Resistance to some disease and pests is also sometimes under polygenic control. Unlike the qualitative characters which have a simple inheritance, the polygenic characters are greatly influenced by environment and hence, in these cases a phenotype is often not a true indicator of its genotype. In a quantitative character, the cumulative effects of the individual genes and their interactions, both allelic or dominance and non-allelic or epistasis, determines the genotypic value. Estimation of various types of gene effects would provide useful information for future crop improvement programme in brinjal.

Several research reports suggest the predominance of additive gene action for number of fruits per plant, fruit length, fruit weight, number of flowers per inflorescence, number of long-styled flowers, total sugar content and fruit borer infestation. A breeding method that increases the frequency of favourable alleles and identifies the superior combinations by repeated crossing and selection could be the best method to exploit the additive gene effects. Several other studies suggested the importance of non-additive gene action for the conditioning of days to first flowering, days to 50 % flowering, plant height, fruit length, phenol content, marketable fruit yield per plant, fruit borer infestation, total sugar, number of marketable fruits per plant and average fruit weight. Overwhelming importance of non-additive gene action for the control of these characters suggests their slow genetic improvement through selection. In such case, heterosis breeding would be the best possible option for improving these characters. Different studies indicated importance of both additive as well as non-additive gene effects have also been noted for fruit yield per plant, fruits per plant, fruit girth, total fruit phenol content and plant height. Use of diallel selective mating or inter-mating the most desirable segregates followed by selection might be some useful breeding strategies for the improvement of the characters governed by both additive and non-additive type of gene action.

Breeding objectives

Some of the important traits that have been the main focus for different purpose of brinjal are outlined below.

- Erect and sturdy growth habit with compact branching habit
- Earliness in flowering and fruiting
- High early and total yield of marketable fruits
- Fruit shape, size and colour as per regional preference
- Good fruit quality (thickness, texture, flavour, less seediness, soft flesh, high fruit colour retention in summer, low solanine content, appreciable chlorogenic acid content and good shelf-life)

- Resistance to different diseases *viz.*, bacterial wilt (*Ralstonia solanacearum*), phomopsis blight (*Phomopsis vexans*), little leaf (Phytoplasma), mosaic virus disease, etc.
- Tolerance to insect pests *viz.*, shoot and fruit borer (*Leucinodes orbonalis*), jassids (*Amrasca biguttula biguttula*), epilachna beetle (*Henosepilachna vigintioctopunctata*) and root knot nematodes (*Meloidogyne incognita*)
- Tolerance to abiotic stress (drought, low or high temperatures, soil salinity).

Breeding methods and development of variety

Both natural and planned hybridization and continuous human selection have played important role in the evolution of the widely diverse cultivars of brinjal. Prevailing climatic condition and highly differential selection pressures according to the regional preference for fruit characters have also resulted in the development of widely divergent cultivars in different regions of the country. In India, this huge genetic variability has been utilized either for direct selection from the variable population or for combination breeding programme as per specific breeding objectives. Some of the popular farmers' varieties in different names are Muktakeshi, Bhangar, Patakata, Elokeski, Nababganj Local, Begor, Makra, Soyla, Kanta Makra, Nurki, etc. of West Bengal; Singhnath, Bholanath of Tripura; Ratan Mani, Burupada, Borigumma, Kutiguda, Dhenkanal, Batparsi, Banmalipur Local, Narayanpur Local, Nisan Local, Radha, Green Rocket, Chitrakonda, etc. of Odisha; Hydarpur, Improved Muktakeshi, Sharanpur Long, Kathvaia, etc. of Bihar; Muktakeshi, Ramnagar Giant, Batia, Sheonath, Bundelkhand, Dudhya baingan, Jafrabadi, etc. of Uttar Pradesh and north-central region; Manjari Gota, Bourad Local, Borgona Local, Khandesh, Khed Shivpur, Malkapuri, Shivpur Khandala, Chikhalgaon Local, Hingana, Dorla, etc. of Maharashtra; Junagadh Bhattu, Surati Gota, Gulabi Dorla, etc. of Gujarat; Coorg Local, Irangeri, West Coast Green Round, Udupi Gulla, etc.of Karnataka; Nurki Round, Nurki Long, Banjari, etc. of Madhya Pradesh; Wynad Giant, Pattabiram, Annamalai, etc. of Tamil Nadu; Gola Baingan, Ludhiana Local, Batia, etc. of Punjab.

Wide variability in brinjal can be characterized according to fruit shape and size and fruit bearing habit. Fruit shape is a very stable genotype-dependent character although different skin colours are found in the genotypes having a particular fruit shape. Most common fruit shapes are long and slender, round/oval, oblong and longish-oblong/ cylindrical. The cluster fruited varieties generally bear small fruits of 20-45 g weight. Some varieties bearing 50-80 g fruits of round or oblong shape stand as a distinct group "small fruited variety" which has regional preference (preferred particularly in South India and Maharashtra). In brinjal, pattern of flowering and fruiting is same *viz.*, solitary, cluster and mixed. Fruiting habit not only depends on flowering pattern but also on number of fertile flowers (long/medium styled flower) in the cyme. Most of the genotypes show solitary fruiting habit and some of them show occasional clustered habit in combination (mixed fruiting habit). In this context, cluster fruited varieties show marked distinction.

Brinjal breeding started in India through selection according to regional preference from the huge and diverse indigenous germplasm and maximum number of varieties have been developed by single plant selection. However, few varieties have also been developed by pure line selection. Good number of the varieties have been developed by inter-varietal hybridization followed by pedigree, backcross or some modified methods of handling the segregating populations. Keeping in view of low narrow sense heritability for some important characters *viz*., fruit weight, fruit yield per plant and tolerance to fruit and shoot brown selection should be based on replicated family testing in F_3 and later generations (Mondal *et al*., 2021). Breeding methods like pedigree methods, single descent method in combination with the molecular approaches have been found to be a useful method for breeding brinjal. Unlike tomato, exotic cultivars did not play any significant role in improvement of brinjal in India.

The varieties developed in India can conveniently be categorized as per fruit shape, fruit size (especially small fruit) and bearing habit (especially cluster bearing).

Long / medium long/ cylindrical fruited variety

Pusa Purple Long, Hisar Pragati, Swetha (SM-6-6), Phule Harit (RHRB-16), Pusa Bhairav, H -4, KS-331, JB-15, Green Long, Punjab Barsati, PH-4 (Punjab Haryana Brinjal-4), Azad Kranti, BB-64, Swarna Abhilamb, Azad B-1, Azad B-3, Arka Sheel, Arka Shirish, Junagadh Long, Kashi Taru, GBL-1, Punjab Sadabahar, JB-15, NDB-25, Punjab Chamkila, Narendra Brinjal-2, Pusa Shyamala, Pusa Kaushal, JC-I, Arka Keshav, Bidhan Supreme, Arka Unnathi, Arka Harshitha, Arka Avinash, Kashi Himani, HABR-21, Utkal Madhuri, Utkal Keshari, Kashi Prakash, Pusa Kranti, Pusa Anupam

Round/ oval /oblong fruited variety

MDU-1, BR-112, Punjab Bahar, Surya (SM6-7), CHBR-I, Swarna Shyamali, Swarna Shree, DBR-8, Pusa Uttam, Pusa Upkar, Pusa Purple Round, Pant Rituraj, Azad B-1, Azad B-2, Azad B-4, Kalyanpur T-3, Swarna Pratibha, Punjab Jamuni Gola, Punjab Neelam, Hisar Shyamal, Black Beauty, CO-1, Annamalai, Bidhan Super, JC-2, Junagadh Oblong, CO-2, Gujarat Oblong Brinjal-1, Bhagyamati, Utkal Tarini, Hisar Jamuni, Pusa Safed Baingan 1, Pusa Hara Baingan 1, Arka Neelanchal Shyama, Phule Harit, Phule Arjun, VRM-1, Swarna Mani, Swarna Shobha

Small fruited variety

PKM-1, Shyamala, Aruna, JB 64-1-2, Pusa Ankur, Pusa Bindu, Jawahar Brinjal-64, Anand Brinjal Small Round-2, Manjari Gota

Cluster bearing variety

Pant Samrat, PLR-1, Arka Neelkanth, Utkal Jyoti, KKM-1, ARU-1, Arka Nidhi, Utkal Jyoti, Utkal Anushree, Gulabi, Vaishali, Pragati, Pusa Purple Cluster, ARU-2C, Arka Kushumkar, Punjab Barsati, Punjab Bharpoor

Germplasm resources and Plant introduction

The genus *Solanum* holds 15,910 eggplant accessions, including the three cultivated species and some crop wild relatives (Solberg *et al*., 2022). The largest collections are held in 1) The National Bureau of Plant Genetic Resources, India (4236 accessions); 2) World Vegetable Centre, Taiwan (3,200 accessions); 3) The NARO Gene bank, Japan (1501 accessions); 4) The USDA collection in Griffin, Georgia, USA (870 accessions); and 5) The Centre for Genetic Resources, the Netherlands (516 accessions).

Asian Vegetable Research and Development Center (The World Vegetable Center), Taiwan holds a large public germplasm collection of brinjal, which includes the three cultivated species and more than 30 eggplant wild relatives, with more than 3,200 accessions collected from 90 countries. The French National Research Institute for Agriculture, Food and the Environment (INRAE) conserve around 2500 germplasm, collected from geographically diverse origins. More than 3500 germplasm is conserved at DISAFA – Plant Genetics, University of Turin, Grugliasco, Torino, Italy.

National Bureau of Plant Genetic Resources (NBPGR), New Delhi maintains huge brinjal germplasm collected from more than 50 countries. Rich diversity in brinjal exists in different parts of India notably, Eastern Ghats, north-eastern region, central India, northern India and eastern India particularly in Odisha and West Bengal. There are more than 2500 farmers' varieties and land races are available in India.

In India indigenous diversity of brinjal is very wide hence, exotic cultivars did not play any significant role in improvement of brinjal in India.

Table 3: Exotic variety directly released for cultivation

Name of the variety	Country from where introduced
Black Beauty	USA

Variability and selection

As in many other crops, in brinjal too, the major efforts for the crop improvement have been directed towards stepping up the yield. Yield is very complex and polygenic in nature. The knowledge of the association among the yield components is the touch stone to a breeder to evolve potentially a better plant type through selection either from the existing genotypes or from the segregating generation of a cross. Hence the genetic potentialities of yield contributing characters and their inter-relationship should be properly assessed for improving the crop. It is desirable to know the existing variability present for various growth, flower and fruit characters in different genotypes of brinjal. This type of information can be of immense utility for further crop improvement programme.

Maximum genotypic coefficient of variation and phenotypic coefficient of variation for number of flowers per inflorescence, number of fruits per picking, fruit girth, fruit weight, fruit yield per plant and shoot and fruit borer incidence indicated that selection can be predicted to improve the brinjal genotypes for these characters. High heritability and high genetic gain have also been observed for different characters like number of flowers per cluster, number of fruits per cluster, fruit length, fruit diameter, fruit weight, long and medium styled flowers per plant, number of fruits per plant, yield per plant, anthocyanin in peel, total phenols and DPPH free radical scavenging capacity. These characters are suggested to be governed mostly by additive gene action and hence, simple selection based on phenotypic performance of these traits would be effective.

Yield per plant has strong positive association with number of fruits per plant, plant height, leaves per plant, primary branches per plant, number of pickings, fruit volume, fruit weight, fruit length, fruit girth, harvest index, total soluble solids and total phenols at both phenotypic and genotypic level. Path analysis indicated that fruit weight, fruit diameter, number of fruits per plant, leaves per plant, number of long styled flowers per plant, number of short styled flowers per plant, days to first harvesting and shoot and fruit borer incidence imparted the highest positive direct effect on yield. These factors emerged as the most important fruit yield contributing characters which may be used as important selection parameters because of their probable conditioning by additive gene action. Selection methods for self-pollinated crops *viz.*, single plant selection and pure line selection are applicable for brinjal however, most of the varieties have been developed by single plant selection. Maximum number of varieties have been developed in India through selection from the variability of indigenous germplasm.

Table 4: Varieties developed through selection

Name of variety	**Indigenous / Exotic material**
Pusa Purple Long	Indigenous (Batia from Punjab)
Swetha (SM-6-6)	Indigenous (Bacterial wilt resistant genotype)
Phule Harit	Indigenous
KS-331	Indigenous
JB-15	Indigenous
Green Long	Indigenous
BB-64	Indigenous (Bacterial wilt resistant genotype)
Swarna Abhilamb	Indigenous (Dhubri, Assam)
Azad B-3	Indigenous
Arka Sheel	Indigenous (Coorg, Karnataka)
Arka Shirish	Indigenous ("Irangeri" of Karnataka)
Kashi Taru	Indigenous
GBL-1	Indigenous
JB-15	Indigenous
NDB-25	Indigenous
Punjab Chamkila	Indigenous
Pusa Shyamala	Indigenous
Pusa Kaushal	Indigenous
JC-I	Indigenous
Bidhan Supreme	Indigenous (Bacterial wilt tolerant)
Kashi Himani	Indigenous
HABR-21	Indigenous
MDU-1	Indigenous ("Kallampatti" from Madurai)
BR-112	Indigenous
Punjab Bahar	Indigenous
Surya (SM6-7)	Indigenous (Bacterial wilt tolerant)
CHBR-I	Indigenous
Swarna Shyamali	Indigenous (Bacterial wilt tolerant)

Name of variety	Indigenous / Exotic material
Swarna Shree	Indigenous
DBR-8	Indigenous
Pusa Purple Round	Indigenous
Kalyanpur T-3	Indigenous
Swarna Pratibha	Indigenous
Punjab Jamuni Gola	Indigenous
CO-1	Indigenous
Annamalai	Indigenous
Bidhan Super	Indigenous (from Murshidabad)
JC-2	Indigenous
Junagadh Oblong	Indigenous
CO-2	Indigenous
Gujarat Oblong Brinjal-1	Indigenous
Bhagyamati	Indigenous
Pusa Safed Baingan 1	Indigenous
Pusa Hara Baingan 1	Indigenous
Arka Neelanchal Shyama	Indigenous
Phule Harit	Indigenous
Phule Arjun	Indigenous
VRM-1	Indigenous
Swarna Mani	Indigenous
Swarna Shobha	Indigenous
PKM-1	Indigenous
Shyamala	Indigenous
Aruna	Indigenous
Pusa Ankur	Indigenous
Pusa Bindu	Indigenous
Jawahar Brinjal-64	Indigenous
Anand Brinjal Small Round-2	Indigenous
Manjari Gota	Indigenous
Pant Samrat	Indigenous
PLR-I	Indigenous
KKM-1	Indigenous
ARU-1	Indigenous
Gulabi	Indigenous
Vaishali	Indigenous
Pragati	Indigenous
Pusa Purple Cluster	Indigenous
ARU-2C	Indigenous
Arka Kushumkar	Indigenous
Punjab Barsati	Indigenous
Punjab Bharpoor	Indigenous

Hybridization and selection

It is suggested that while selecting parents for hybridization purpose, their genetic divergence and specific characters should be considered. A number of studies have been carried out on genetic divergence studies in brinjal in order to identify genetically diverse parents which will have a good combining ability. Combination of such diverse parental lines may result in high heterosis for different characters and fruit yield and at the same time, release superior segregates in the advanced generations. In many studies, biological diversity was not related to genetic diversity and the clustering pattern of the germplasm which indicated that either a common character constellation was manifested simultaneously in the genotypes or mutual balancing in character expression was operative in the genotypes with no character contributing overwhelmingly towards divergence. Inter-varietal hybridization followed by pedigree method of selection in segregating generations have been followed for the development of good number of varieties in India (Hazra *et al*, 2022).

Table 5: Varieties bred through hybridization followed by selection

Variety	Parentage
Hisar Pragati	R-34 × Sel. 26
Pusa Bhairav	Pusa Purple Long × 11 a-12-2-1 (Phomopsis blight resistant)
H -4	Hyderpur x Pusa Purple Long
Punjab Barsati	Pusa Purple Cluster × PH-4
PH-4	Hyderpur × Pusa Purple Long
Azad Kranti	Banarasi Gaint Long × Pusa Purple Long
Junagadh Long	Doli-5 x Panch Mahal
Punjab Sadabahar	Japanese Long × R-34
Narendra Brinjal-2	NDB-25 x NDB-2
Arka Keshav	Dingrass Multiple Purple x Arka Sheel
Arka Neelkanth	Dingrass Multiple Purple x Arka Sheel
Arka Nidhi	Dingrass Multiple Purple x Arka Sheel
Bidhan Suphala	Bidhan Supreme × Punjab Sadabahar (Bacterial wilt tolerant)
Arka Unnathi	Arka Kusumakar × IIHR-3
Arka Harshitha	Arka Kusumakar × IIHR-3
Arka Avinash	Arka Kusumakar × IIHR-3
Utkal Madhuri	PBR 125-5 × Pipili-4 (Bacterial wilt resistant)
Utkal Keshari	BB-11 × KJ 3-1 (Fusarium wilt resistant)
Pusa Kranti	(Pusa Purple Long × Hyderpur) x Wynad Gaint
Pusa Anupam	Pusa Kranti × Pusa Purple Cluster
Pusa Uttam	GR × Pant Rituraj
Pusa Upkar	Gr × PB 91-1
Pant Rituraj	T -3 × Pusa Purple Cluster
Azad B-1	Banarasi Giant Round x 7103
Azad B-2	Pusa Purple Round × Arka Sheel

Variety	Parentage
Punjab Neelam	Jamuni Gola × Pant Rituraj
Hisar Shyamal	Aushey × BR-112
Utkal Tarini	Pusa Kranti × Gopa Local (Bacterial wilt resistant)
Hisar Jamuni	Aushey × R 34
Utkal Jyoti	KT-4 x BB-11

Heterosis breeding

Hybrid vigour in brinjal was first recorded in Japan by Nagai and Kida (1926). Since then, hybrid brinjal has gained popularity in many countries. Several studies suggested the importance of non-additive gene action for the conditioning of days to first flowering, days to 50 % flowering, plant height, fruit length, phenol content, marketable fruit yield per plant, fruit borer infestation, total sugar, number of marketable fruits per plant and average fruit weight. Overwhelming importance of non-additive gene action for the control of these characters suggests that heterosis breeding would be the best possible option for improving these characters.

Among the different vegetable crops grown in India, highest area goes to brinjal however, most of the area is covered with wide array of very good quality and adaptable farmers' varieties. For this reason, brinjal area under hybrids is comparatively much low of 20 percent. Hybrid seeds of all brinjal hybrids grown in India are exclusively produced within the country utilizing mostly the indigenously developed parental lines by hand emasculation and pollination method. Hazra *et al* (2010) demonstrated positive relationship between genetic distance of the parents and both relative heterosis and heterobeltiosis for fruit yield/plant, however, the relationship was not strong enough to confidently predict the level of heterosis based on a given value of genetic distance between the parents. It indicated that there might be optimum level of genetic divergence between parents to obtain heterosis in the F_1 generation.

Hybridization technique

Emasculaton

The flowers are borne solitarily or in cluster. Of the four types of flower, fruit set occurs only in long and medium styled flowers (stigma is either above or on the same level as the stamen) and short or pseudo-short styled flower generally do not set fruits. Most of the solitary flowers and basal flowers of cluster habit are generally long styled and borne on thick pedicles. These flower buds should be selected for emasculation. However, length of style should be glanced during emasculation. Before starting emasculation, all opened flowers and developing fruits are removed along with undesirable flowers buds. Flower buds about one to two days away from opening should be chosen for emasculation. At this stage, corolla is whitish with slight violet tinge. Emasculating very young buds may lead to injury to the style and ovary, while emasculating large buds whose petals have turned violet colour are not recommended since the likelihood of self pollination is high. The unopened bud is opened carefully from the side with sharp forceps, length of the style is glanced and the anthers are removed leaving only petals, ovary and style. All the five anthers should be removed intact carefully without having any puncture from the forceps so that chances of adhering pollens on the stigma become remote. The emasculated flower buds are covered with small paper bags because extent of natural out crossing is about 10 per cent (Hazra *et al*, 2003).

Pollen collection

Opened flowers of any type are collected from the intended male parent in early morning hours before anther dehiscence. The anthers are removed from the flowers, kept in a petridish or cellophane paper and dried in sun in the morning hours. Gentle crushing or shaking of the dried anthers helps easy liberation of the pollens. The pollens after straining with fine mesh cloth are collected in the small glass vial or petridish or pollinating ring for onward pollination. Placing the undehisced anthers 30 cm below a 60 to 100-watt lamp for 20-24 hours can effectively dry large number of anthers. Next day, the dried anthers are put in plastic container. It is covered with a fine mesh (200-300 mesh) cloth or net. The container with dried anthers is then sealed with a tight fitting lid. The screened pollens are collected in the lid of the container by shaking it 10-20 times.

Pollen storage

Fresh pollen is the best for fruit set. Pollen viability is retained for 8-10 days at moderate temperature of 20-22°C with 50- 55% relative humidity. Best results are obtained when pollen is used within four days after collection. Dried pollens can be stored in sealed vial and kept in freezer at 0°C for 1-2 months.

Pollination

The flowers emasculated one or two days earlier should have completely blossomed under the paper bag and are ready for cross-pollination. The stigma is dipped into pollen mass kept in suitable pollen container in the morning hours. Hand pollination of emasculated flowers is done on the next day between 9 a.m. and 12.30 p.m. in winter and 7 am and 11 am in summer season. Pollination may also be done using brush. The pollinated flowers are marked with either strings or rubber bands around the peduncles. The pollinated buds should again be protected by paper bag because of prolonged stigma receptivity in brinjal, and it should be removed 2-3 days after pollination. Any non-crossed old flowers on the female parent should be removed. The pollination period ranges between 25-35 days depending on the flowering and fruiting behaviour of the female parent. Successful pollination and subsequent fertilization can be recognized within 8-10 days by enlargement of the ovary.

Use of male sterility in development of hybrids

Cost of hybrid seed production of brinjal is not high as compared to other vegetables because each hand pollination results good number of hybrid seeds. However, cost of production can further be reduced with the use of male sterile lines. A naturally occurring genic male sterile line UGA 1-MS was derived from a male sterile spontaneous mutant of "Florida Highbush". This functional male sterility with non-dehiscent anther of UGA1-MS is governed by a single recessive allele, for which the symbol *fms* was proposed (Phatak *et al*, 1991). However, Hazra *et al* (2008) suggested two modifier genes in recessive condition along with the major recessive gene symbolized as *fms/fms +modl/modl + mod2/mod2* for the control of this functional male sterility.

Fang *et al.* (1985) developed two male sterile lines by three back crosses to *S. melongena* after crossing *S. gilo* with *S. melongena* lines 9334A and 2578A and suggested that the male sterility is under cytoplasmic control. CMS systems in brinjal is classified into two types namely, the anther indehiscent type in which anther contains normal pollen but does not open to release and the pollen non-formation type in which anther of the male sterile lines is completely devoid of pollen. Furthermore, in pollen non-formation CMS systems, two independent dominant fertility restorer (*Rf*) genes were discovered. Interspecific hybrid *S. gilo* x *S. melongena* displays high frequency of univalent at meiosis which is responsible for the high degree of sterility (Behra and Singh, 2002). Absence of anther dehiscence together with very low pollen fertility was also obtained in the backcross progenies of *S. virginianum* x *S. melongena* (Khan and Isshiki, 2008). However, male sterility has not yet been employed in commercial hybrid seed production of brinjal. Several research reports suggested large number of indigenous germplasm (improved variety, breeding line, farmers' variety, etc.) as potential parental lines for the development of promising hybrids.

Breeding for resistance to biotic stresses

Breeding for resistant to diseases

Most of the cultivars of brinjal are susceptible to a number of diseases. Screening of cultivars and wild relatives of brinjal for resistance to diseases and pests is of great significance for evolving resistant cultivars. Resistance to different diseases and nematodes is available in a number of *Solanum* species. Crosses involving *S. gilo* as one of the parents has shown high level of resistance against *Phomopsis* blight and successful gene transfer from these wild relatives was possible through back cross. However, support of biotechnological tools was required to develop viable *S. khasianum* x *S. melongena* hybrid.

Resistance reactions and mechanism of resistance against bacterial wilt (*Ralstonia solanacearum*) in brinjal are location- and environment-specific. This indicates that brinjal germplasm should be evaluated at a particular place against specific isolates (Biovar) of the pathogen of that area. *S. toxicarium* was found immune to 4 and 5 strains of *Pseudomonas solanacearum*. Namisy *et al.* (2019) screened eggplant germplasm including cultivated and wild against two strains of *R. solanacearum* Pss97 and Pss2016 and the results indicated that one accession of *S. sisymbriifolium* (SIS1) and two accessions of *S. torvum* (TOR2 and TOR3) were resistant or moderately resistant to the strain Pss97. Three resistance QTLs (*EBWR2, EBWR14*, and *EBWR9*) control strains belonging to one or more phylotypes. By combining three resistance QTLs, breeders could be able to develop varieties with a large spectrum of bacterial wilt resistance (Salgon *et al.,* 2017). Breeding for stable resistance is a challenge due to the fact that resistance in brinjal to bacterial wilt can be location-specific, and incorporation of resistance into large-fruited varieties has also been a major challenge because of probable linkage between small fruit size and bacterial wilt resistance (Lalramhlimi *et al.*, 2019). Selection of cultivar with higher inner area of large vessel in both stem and roots and higher number of large over small vessel per square mm of stem vascular area may be an important selection criterion in the breeding programme of brinjal resistance / tolerance to bacterial wilt disease (Chattopadhyay *et al.*, 2012; Ghosh *et al.*, 2015).

Genetic behavior of bacterial wilt can be depicted from segregating populations. The dominant and recessive major gene or multiple genes determine a breeding strategy to develop resistant varieties or hybrids as well as suitability of donor sources and recurrent parents. The inheritance pattern for bacterial wilt tolerance in eggplant has been reported ranging from monogenic dominant/recessive to polygenic. Available sources of bacterial wilt resistance resistance are mostly under polygenic control. Bidhan Suphala, a medium-sized variety tolerant to bacterial wilt under field condition has been developed through planned hybridization Bidhan Supreme × Punjab Sadabahar (Bacterial wilt tolerant), followed by pedigree method of selection in the advanced generations.

Accessions of *S. torvum* collected from Java Island are valuable source of resistance against both *Ralstonia* and *Fusarium* wilts and can be utilized for the resistance breeding programme.

High resistance to *Verticillium dahlia* was shown by *S. abutiloidea*, *S. torvum*, *S. caripense*, *S. persicum* and *S. scabrum*. Interspecific hybridization between *S.torvum* and *S.melongena* for resistance to *Verticillium* wilt clearly indicated that *S. torvum* was a good donor of resistant genes to eggplant cultivars.

Table 6. *Solanum* species as source of resistance to diseases, insect-pests and nematode

Species	**Insect-pests/Nematode**	**Diseases**
Solanum integrifolium	Red spider mite	Little leaf, *Fusarium* wilt and *Verticillium* wilt
S. gilo	Shoot and fruit borer	Little leaf, *Phomopsis* fruit rot
S. indicum	-do-	*Fusarium* wilt, *Phomopsis* fruit rot
S. incanum	-do-	*Fusarium* wilt
S. sisymbrifolium	Root knot nematode	*Fusarium* wilt, *Phomopsis* fruit rot, *Verticillium* wilt
S. torvum	-do-	*Verticillium* wilt, *Phomopsis* fruit rot
S. aculeatissimum	-do-	*Verticillium* wilt
S. sodamaem	-do-	*Verticillium* wilt
S. xanthocarpum	-do-	*Phomopsis* fruit rot
S. khasianum	-do-	*Phomopsis* fruit rot
S. nigrum	-do-	*Phomopsis* fruit rot
S. testiculatum	-do-	*Phomopsis* fruit rot

Source: Kalda and Kumar (1999)

Several improved varieties have been developed which are resistant / tolerant to different diseases *viz.*, bacterial wilt, *Fusarium* wilt, *Phomopsis* fruit rot, *Cercospora* leaf spot, little leaf and root knot nematode. In most of cases, resistance source was indigenous cultivar or landraces *viz.*, Pusa Purple Cluster, Improved Muktakeshi, Swarna Shyamali, Utkal Madhuri, Utkal Tarini, etc. which shows resistance to bacterial wilt. The variety, Pusa Bhairav, resistant to phomopsis blight has been bred through planned hybridization between Pusa Purple Long × 11 a-12-2-1 (Phomopsis blight resistant line) followed by bulk method of selection because resistance to phomopsis blight is governed by multiple genes.

Table 7: Brinjal cultivars showing resistance / tolerance to different diseases

Disease	Cultivars
Bacterial wilt (*Ralstonia solanacearum*)	Pusa Purple Cluster, Improved Muktakeshi, Black Beauty, Dingrass Multiple Purple, BWR-12, SM6-1, SM6-7, BB-7, BB-44, BB-49, SM1-5, SM1-10, Punjab Barsati, Annamalai, Arka Nidhi, Arka Keshab, Arka Neelkantha, Bhagyamati, Surya, Swetha, BB-64, Haritha, Bidhan Supreme, Bidhan Suphala, Swarna Shyamali, Utkal Madhuri, Utkal Tarini etc.
Fusarium wilt (*Fusarium oxysporum*)	K-61, K-72, Utkal Keshari, Ghana Local, Kopek, Black Beauty, Nihon Nassu
Vertcillium wilt (*Verticillium dahlia*)	Florida market, Haris Hybrid, Harris 468 Special
Phomopsis blight and fruit rot (*Phomopsis vexans*)	Pusa Bhairav, Pusa Purple Cluster, Florida Beauty, Florida Market
Cercospora leaf spot (*Cercospora solani*)	Improved Muktakeshi, Udupi Gulla
Little leaf (Phytoplasma like organism)	Pusa Purple Cluster, H-8, H-9, BB-2, Hisar Shyamal, Pant Rituraj
Root Knot Nematode (*Meloidogyne incognita*)	Vijay, Black Beauty, MM-392, MM-450, Gulla

Source: Kalda and Kumar (1999), Chattopadhyay *et al* (2020)

Solanum melongena var. *insanum* carries a single dominant gene for resistance to bacterial wilt. Three lines SM6-1SP, SM6-1M and MS-7SP were found resistant to *Ralstonia solanacearum*. Three elite brinjal genotypes 16/BRBW RES-2, 2012/BRBW RES-5 and 2012/BRBW RES-6, which had recorded high bacterial wilt resistance coupled with diverse horticultural traits could be utilized as donor parents either for developing hybrids or to isolate promising lines in segregating generation.

Breeding for resistance to insect pest and nematodes

Brinjal is also very susceptible to the attack of different insect-pests and root-knot nematodes. *S. sisymbrifolium* and Pusa Purple Cluster are the least susceptible wild *Solanum* species and variety, respectively to the attack of brinjal shoot and fruit borer. *Solanum viarum*, closely related and cross compatible to brinjal was found to be resistant to shoot and fruit borer. Mandal *et al.* (2019) identified some genotypes BCB-50, BCB-11, BCB-30, BCB-40, Garia and Punjab Sadabahar as potential donors that could be utilized not only for resistant breeding against fruit and shoot borer but also for improvement in yield and better fruit quality.

Similarly, *S. torvum and S. sisymbrifolium* are reported to be highly resistant to epilachna beetle with lowest infestation percentage and lowest number of grubs/adults per plant. Two elite varieties, Vijaya and Black Beauty are reported to be tolerant against root-knot nematodes (*M. incognita*).

Table 8: Cultivars and wild relatives resistant to different insect pests

Insects	Resistant/ Tolerant source
Shoot and Fruit borer (*Leucinodes orbonalis*)	*S.xanthocarpum, S.khasianum S.integrifoilium, S.sisymbriifolium* Arka Mahima, Arka Sanjivani, Punjab Sadabahar, Arka Kusumkar, Kashi Taru, SM-6- 6, Arka Nidhi, Kali Rawai, Arka Nilkanth, Singnath, IC 136299, BCB-50, BCB-11, BCB-30, BCB-40 and Garia
Aphid (*Aphis gossypii)*	*S. sisymbriifolium, S.macrocarpon*
Jassids (*Amrasca biguttula biguttula*)	Dorli Jumbli Malayalum, Manjari Gota
Spider mite (*Tetranychus* spp.)	*S. sisymbriifolium, S.dasyphyllum, S. torvum* Rampur Giant, Swarna Pratibha, Kashi Taru

Breeding for resistance to abiotic stresses

Brinjal shows wide variation in morphological and physiological characteristics, therefore, it is necessary to explore the germplasm for tolerance to abiotic stresses (drought, heat, cold, salinity, heavy metal concentration in soils). Sufficient water availability at the first stage of plant growth is very important. Limited water supply may lead to drought stress. The long period of drought stress may disturb the metabolic process and lead the irreversible damage of the plant tissues, so the plant cannot recover. This abiotic stress has been significantly proved to be one of factors decreasing the plant growth and productivity. As brinjal is one of the most important vegetables for the Asian and Mediterranean regions, and climate change will enforce the frequency as well as the intensity of combined stresses, it is of great importance to address the issues of climate-resilient vegetables. Several cultivated brinjal and its wild relatives are reported to be resistant to different abiotic stresses. Different cultivars show promise in breeding for resistance to abiotic stresses *viz*., Arka Shirish, Arka Neelkanth, Pusa Shaymala, Pusa Purple Long, Annamalai, Jawahar Brinjal-69, Jawahar Brinjal-64, Pusa Upkar for drought tolerance; BL205, BL215 and BR123, R-34 for high temperature tolerance; SL-8-PB-1-3-1-4 , KBSR-343-1, IC136296, IC136041, IC136096 and IC136461 for soil salinity, etc.

Table 9: Cultivated brinjal and wild relatives as a source of tolerance to abiotic stress

Stress	Source of tolerance
Frost tolerance	*S. incanum*
Drought tolerance	*Solanum macrocarpon*
	BMG-1, Arka Shirish, Arka Neelkanth
	Pusa Shaymala, Pusa Purple Long, Annamalai, Jawahar Brinjal-69, Jawahar Brinjal-64, Pusa Upkar
High temperature	R-34
	BL205, BL215 and BR123
Soil salinity	SL-8-PB-1-3-1-4 (Small Long) KBSR-343-1 (Small round) BL-215 (Long)
	IC136296, IC136041, IC136096 and IC136461
Flooding	Arka Nidhi and Arka Keshav

Breeding for quality

Brinjal has a very low caloric value and is considered among the healthiest vegetables for its high content of protein, various vitamins, minerals like, iron, calcium, potassium, magnesium, dietary fibre and different bio-active phyto-chemicals (Docimo *et al.* 2016). The bioactive properties of brinjal are mostly associated with high content in phenolic compounds having scavenging activities which are mostly phenolic acids, particularly chlorogenic and caffeic acid in the fruit flesh (Stommel *et al.* 2015) and anthocyanins (flavonoids) in the fruit skin, mainly nasunin and delphinidin- 3-(coumaroyl-rutinoside)- 5-glucoside (Cassidy *et al.* 2013). Wide genetic variation was recorded in total phenol content in cultivated brinjal and its wild relatives. According to the study conducted by Kaushik (2020), the highest content of chlorogenic acid around 4.5 mg g^{-1} of fruit dry weight was present in the species *S. linnaeanum* and *S. torvum* whereas, the highest content of total fruit phenolics were found in the secondary gene pool species *S. linnaeanum*

References

Behra, T.K. and Singh, N. 2002. Inter-specific crosses between eggplant (*Solanum melongena*) and related *Solanum* species. *Sci. Hort.*, **95**: 165-172.

Cassidy, A., Mukamal, K.J., Liu, L., Franz, M., Eliassen, A.H. snd Rimm, E.B. 2013. High anthocyanin intake is associated with a reduced risk of myocardial infarction in young and middle-aged women. *Circulation,* **127**: 188-196.

Chattopadhyay, A., Dutta, S., Mandal, A.K., Maurya, P.K., Banerjee, S., Bhattacharjee, T. and Hazra, P. 2020. Bidhan Supreme and Bidhan Super: New brinjal varieties. *Indian Horticulture*, March-April, 2020 issue, pp. 16-17.

Chattopadhyay, A., Seth, T., Dutta, S., Ghosh, P.P., Chattopadhyay, S.B., Majumder, D. and Hazra, P. 2012. Breeding eggplant for higher productivity and bacterial wilt tolerance. *Inern. J. Veg. Sci.,* **18**: 376-392.

Chattopadhyay, S.B., Sarkar, A. and Hazra, P. 2009. Study on the crossability of three species of *Solanum* in Gangetic plains of West Bengal. *J. Crop Weed,* **5**: 53-56

Docimo, T., Francese, G., Ruggiero, A., Batelli, G., De Palma, M. and Bassolino L. 2016. Phenylpropanoids accumulation in eggplant fruit: characterization of biosynthetic genes and regulation by a MYB transcription factor. *Front. Plant Sci.*, **6**:1233. doi: 10.3389/fpls.2015.01233.

Duanay, M.C. and Hazra, P. 2012. Eggplant. In: *Hand book of Vegetables* (Eds. K. V. Peter and P. Hazra), Studium Press LLC, Houston, Texas, USA, pp. 257-322.

Fang, M.R., Mao, R.C. and Xie, W.M. 1985. Breeding cytoplasmically male sterile lines of eggplant. *Acta. Hort. Sinica.*, **12**: 261-266.

Ghosh, P.P., Dutta, S., Seth, T. and Chattopadhyay, A. 2015. Influence of stem and root vascular anatomy of eggplant germplasm on severity of vascular bacterial wilt disease. *Indian Phytopath.*, **68**: 363-367.

Hazra, P., Banerjee, M. K. and Chattopadhyay, A. 2022. *Varieties of Vegetable crops in India* (Third edition), Kalyani Publishers, Ludhiana, New Delhi, 199p.

Hazra, P., Mandal, J and Mukhopadhyay, T.P. 2003. Pollination behaviour and natural hybridization in *Solanum melongena* L., and utilization of the functional male sterile line in hybrid seed production. *Capsicum and Eggplant Newsl.*, **22**: 143-146.

Hazra, P., Roy, T. and Choudhury, J. 2008. Characterization of genic functional male sterility in brinjal and its expression in different genetic background. *Indian J. Genet.*, **68**: 47-51.

Hazra, P., Sahu, P. K., Roy, U., Dutta, R., Roy, T. and Chattopadhyay, A. 2010. Heterosis in relation to multivariate genetic divergence in brinjal (*Solanum melongena*). *Indian J. Agril. Sci.*, **80**: 119 – 124.

Kalda, T.S. 1998. Advanced technology in improvement of vegetable crops including cole crops, In: Compendium of lectures of Summer School, held at ICAR-IARI, New Delhi, 24th May-6th June, 1998, pp. 132-140.

Kalda, T.S. and Kumar, R. 1999. Advances in breeding of temperate and tropical vegetables for biotic and abiotic stresses In: *Compendium of lectures of Summer School*, held at ICAR-IARI, New Delhi,17th May-6th June, 1999, pp. 102-107.

Kaushik, P. 2020. Characterization of cultivated eggplant and its wild relatives based on important fruit biochemical traits. *Pak J. Biol. Sci.*, **23**:1220-1226.

Khan, M.M.R. and Isshiki, S. 2008. Development of male sterile eggplant by utilizing the cytoplasm of *Solanum virginianum* and a biparental transmission of chroplastDNA in backcrossing. *Sci. Hort.*, 117: 316-320.

Lalramhlimi, B., Dutta, T., Jamir, I., Maurya, P.K., Bhattacharjee, T., Banerjee, S., Chatterjee, S., Mandal, A.K. and Chattopadhyay, A. 2019. Choosing of brinjal parents for future breeding aiming at bacterial wilt disease tolerance. *Int. J. Curr. Microbiol. Appl. Sci*, **8**: 2370-2390.

Lester, R.N., 1998. Genetic resources of capsicum and eggplants. *Proc. Xth EUCARPIA Meeting on Genetic and Breeding of Capsicum and Eggplant* (Eds. Palloix, A.and Daunay, M.C.),7-11 September, *1998*, Avignon, France, pp. 25-30.

Mondal, R., Maurya, P.K., Bhattacharjee, T., Banerjee, S., Chattopadhyay, A. and Hazra, P. 2019. Breeding potential of brinjal genotypes for fruit and shoot borer resistance using multivariate analysis. *J. Entomol. Zool. Stud.*, **7**: 1340-1347.

Mondal, R., Hazra, P. Hazra, S. and Chattopadhyay, A. 2021. Combining ability studies for yield components, quality characters and relative susceptibility to fruit and shoot borer in brinjal *Veg. Sci.*, **48**: 86-94.

Nagai, K. and Kida, S. 1926. Experiments on hybridization of various strains of *Solanum melongena*. *Japan. J. Genet*, **4:** 10-30.

Namisy, A., Chen, J.R., Prohens, J., Metwally, E., Elmahrouk, M. and Rakha, M. 2019. Screening cultivated eggplant and wild relatives for resistance to bacterial wilt (*Ralstonia solanacearum*). *Agric.,* **9**: 157.

Phatak, S.C., Liu, J., Jaworski, C.A. and Sultanbawa, A.F. 1991. Functional male sterility in eggplant: inheritance and linkage to the purple fruit colour gene. *J. Hered.*, 82: 81-83.

Ram, H.H. 1998. *Vegetable breeding: Principles and Practices*, Kalyani Publishers, Ludhiana, pp. 188-194.

Ranil, R.H.G., Prohens, J., Aubriot, X., Niran, H.M.L., Plazas, M., Fonseka, R.M., Vilanova, S., Fonseka, H.H., Gramazio, P. and Knapp, S. 2017. *Solanum insanum* L.(subgenus Leptostemonum Bitter, Solanaceae), the neglected wild progenitor of eggplant (*S. melongena* L.): a review of taxonomy, characteristics and uses aimed at its enhancement for improved eggplant breeding. *Genet. Resour. Crop.Evol.,* **64**:1707-1722.

Salgon, S., Jourda, C., Sauvage, C., Dauna, M.C., Reynaud, B., Wicker, E. and Dintinger, J. 2017. Eggplant resistance to the *Ralstonia solanacearum* species complex involves both broad-spectrum and strain-specific quantitative trait loci. *Front. Plant Sci.,* **8**: 828. doi: 10.3389/fpls.2017.00828.

Singh, T.H., Sadashiva, A.T. and Reddy, K.M. 2015. Advanced breeding strategies for biotic and abiotic stress tolerance in brinjal (*Solanum melongena* L.), In: *ICAR-Winter School on Advanced Breeding Strategies for Biotic and Abiotic Stress Tolerance in vegetable Crops*, held on Oct 8-28, 2015 at ICAR-IIHR, Bengaluru, India.

Solberg, S., van Zonneveld, M., Rakha, M.T., Taher, D.I., Prohens, J., Jarret, R., van Dooijeweert, W., and Giovannini, P. 2022. *Global strategy for the conservation and use of eggplants.* Global Crop Diversity Trust. Bonn, Germany. 67p.

Stommel, J. R., Whitaker, B. D., Haynes, K. G. and Prohens, J. 2015. Genotype × environment interactions in eggplant for fruit phenolic acid content. *Euphytica*, **205**: 823–836.

20

Chilli

Chilli or hot pepper, the *Capsicum* species (principally *Capsicum annuum*) under the family Solanaceae (Chromosome No. 2n =2x =24) is a tropical vegetable-cum spice crop commonly used throughout the world as spice for its pungency and red colour of ripe dried fruits and also for its green fruits for pungency and flavour.

Origin, domestication and adaptation in India

It is generally accepted that the *Capsicum* genus belonging to the family Solanaceae was originated in Bolivia which consists of 25–30 species. Five of these *Capsicum* species *viz*., *C. annuum*, *C. baccatum*, *C. chinense*, *C. frutescens*, and *C. pubescens* were domesticated. The largest group of varieties is found under *C. annuum* which is grown worldwide. Both hot pepper and sweet pepper belonging to the same species *Capsicum annuum* originated from the wild and weedy *Capsicum annuum* var. *minimum* distributed from the southern United States to the northern South America. The primary centre of origin of chilli is in southern and central America and secondary centre in Guatemala.

Excavations at Tehuacan in Mexico have indicated the existence of *C. annuum* around 7000 B.C. Columbus on his return from Americas, brought chilli, *Capsicum annuum* to Spain in 1493. It spread to other parts of the Mediterranean region. Later, it went to north in England in 1548 and moved to central Europe in the 16th century.

Another cultivated species, *Capsicum baccatum* var. *pendulum* was originated from its wild progenitor, *Capsicum baccatum* var. *baccatum* in South America. The species *Capsicum frutescens* is wide spread as a wild, weedy or semi-domesticated species in low land tropical America including Mexico, central South America and lowland South America. Perhaps, *Capsicum chinense* has been originated from the wild type of *Capsicum frutescens*. No ancestral type could be identified for the other cultivated species *Capsicum pubescens*.

Domestication, human selection and breeding have produced hundreds of varieties in the chilli growing areas of the of world from the great natural diversity of the *Capsicum* species differing greatly in fruit size and shape, pungency and ranging in colours from orange, red, yellow, violet to green. Of the five domesticated *Capsicum* species, *C. annuum* is the most important in the world and obviously in India. Although land races of other species are cultivated in a limited scale, improvement of chilli in India is based on *C. annuum* L.

Mixed stock of *Capsicum* sp. was introduced in India by the Portuguese during early 16th century. Christian missionaries also introduced *Capsicum* sp. in the North-Eastern states separately. Adaptation to Indian condition particularly South Indian condition had resulted in greater diversification in hot or pungent varieties. The farmers' selection pressure in all the chilli growing regions was in favour of red colour both in fresh and after drying which suited for long distance transport. During this process, sweet or partially sweet types, if at all have been introduced in India as mixed stock, possibly got eliminated.

Table 1: Chief characters of the domesticated *Capsicum* species

Species	Anther	Corolla	Calyx	Fruits at node	Seed colour
C. annuum	Blue	Milky white, rarely purple	Inconspicuous calyx teeth	One, occasionally two	Straw coloured
C. baccatum	Yellow, brown	Cream, yellow, tan or brown spots	Prominent calyx teeth	One or two	Straw coloured
C. frutescens	Blue	Pale-greenish or yellowish-white	No calyx teeth	Two or more	Straw coloured
C. chinense	Blue	White, milky-greenish or yellowish-white	A constriction on calyx	Two or more	Straw coloured
C. pubescens	Purple	Purple	Conspicuous calyx teeth	One	Black

Source: Bosland and Votava (1999)

Crop botany and floral biology

- *Capsicum annuum* encompassing the varieties of chilli or hot pepper, sweet pepper and paprika peppers is a perennial herbaceous plants but often cultivated as an annual; it is a many branched plant, growing up to 75-80 cm in cultivated varieties and often shrubby in appearance; the leaves are simple and alternate, elliptical to lanceolate, with smooth margins (entire); some *Capsicum* species synthesize and accumulate anthocyanins in different plant parts including fruits.
- The small flowers (around 1.5 cm diameter) are borne singly or, rarely, in pairs in the axils tending to be leaf opposed; they are white or occasionally purple, campanulate (bell-shaped), often with 5 lobes, and contain 5 bluish stamens.
- Flower bracteolate, pedicellate, bisexual, radially symmetrical, hypogynous, pentamerous and heterochlamydeous; sometimes flowers appear in cluster; flowers are borne one, two or more at a node depending on the species of *Capsicum*.
- Calyx campanulate, sepals usally five (sometimes six), gamosepalous, ranging from tubular to deeply cleft, persistent in fruits.
- Calyx margin vary *viz*., inconspicuous calyx teeth, prominent calyx teeth, no calyx teeth (tubular) and constriction on the calyx

- Corolla rotate, petals five (sometimes six), gamosepalous, valvate; flower colour may be white, whitish-green, yellowish-white, purple, yellow-tan or brown spots depending on the species of *Capsicum*; anther colour is bluish, yellow, brown or purple depending on the *Capsicum* species.
- Stamens are five (sometimes six), distinct, alternating with the lobes of corolla and adnate to the corolla tube or perigynous zone.
- Anthers are ditheceous, introse, non-convenient in cone, not longer than filament, dehisce longitudinally by lateral slits.
- Ovary superior, bicarpellary, syncarpous, bilocular with numerous ovules in each locule, axile placentation.
- Style is terminal, linear, white or purple; stigma is sub-capitate and faintly bifid.
- The fruits are many-seeded berries and pod-like, but no sutures are there, pericarp leathery or succulent; shape of the fruit varies from a flattened oblate to long slender and tapering; fruits vary considerably in size and shape, ripening to green, yellow, orange, red, or purple.
- Seeds within the fruits mature as the fruit ripens; the seeds are borne in a compact formation on the plancentae and usually at the basal end of the fruit.

Reproductive behaviour

Chilli is basically a self- pollinated crop but there is wide variation in natural cross-pollination ranging from 7 to 37% which makes it most of the times as often cross pollinated crop. High degree of cross pollination even up to 68% may happen in some condition. With an active group of insect pollinators, a number of different cultivars grown in the area, and concentrated production of chilli in most of the growing areas, there is ample opportunity for seed fields to become contaminated with foreign pollen. Hence, rouging of off types is considered as the most important operation for the maintenance of parental lines for the purpose of breeding and seed production. Major pollinators are honey bees (*Apis* species), ants and wind. Anthesis generally starts from 5.30 am and majority of the flowers open by 9.00 a.m. however, it depends on the prevailing temperature and humidity condition. Anther dehiscence occurs at least one hour after anthesis. The pollen grains in the anther become fertile a day before anthesis with maximum fertility on the day of anthesis. The stigma receptivity is the maximum at the time of opening of the flower although it becomes effectively receptive a day prior to anthesis, and the receptivity continues upto two days after anthesis. At high temperature condition, fair crossing success can be achieved if hybridization is taken up either early in the morning or late afternoon. Pollen viability is retained for a day in ambient condition. Pollens can be kept in refrigerator under airtight condition for 2-3 days. Dried pollens can be stored in sealed vial at - 5.0°C for 10 days.

Table 2: Crossability among different *Capsicum* species

Inter-specific cross	**Crossing behaviour**
C. annuum × *C. frutescens*	In general, about 2 per cent of the F_1 seeds are viable when C. *frutescens* was used as female parent.
C. annuum × *C. chinense*	Crosses can be made in either direction but crossing success is higher when *C. annuum* is used as the female parent.
C. annuum × *C. pendulum*	No viable F_1 seed had ever been obtained using *C. pendulum* as female parent but viable F_1 seeds could be obtained through embryo rescue technique.
C. annuum × *C. pubescens*	These species are completely cross-sterile. Crosses rarely resulted in fruits and those obtained are without seed.
C. frutescens × *C. chinense*	Viable crosses could be obtained in either direction although with considerable difficulty particularly when *C. chinense* is used as the female parent. The F_1 plants are completely to moderately self-sterile. Backcrosses to both parents can be obtained.
C. frutescens × *C. pendulum*	Crosses can be made in either direction, but more readily when *C. pendulum* is used as the female parent. The F_1 plants are highly sterile and viable F_2 seeds are very rarely formed. Back-crosses to each parent resulted in only occasional viable seed.
C. frutescens × *C. pubescens*	No fruit had ever been obtained from repeated pollinations in both directions.
C. chinense × *C. pendulum*	Crosses can be made with some difficulty with *C. pendulum* as the female parent. A few F_2 seeds can be obtained but none are viable. A limited number of back-crosses are successful.
C. chinense × *C. pubescens*	Fruits are formed when *C. chinense* is used as the seed parent. The fruit at maturity contained many seeds with fully developed embryos but without endosperm. Such embryos could be grown only on artificial growing media. The F_1 plants are completely sterile and could not be back-crossed to either parent.
C. pendulum × *C. pubescens*	No successful crosses could be obtained.
C. annuum × *C. chacoense*	F_1 hybrids are partially fertile by using *C. chacoense* as female parent. F_1 plants formed mostly bivalents at meiosis with a few multivalents and univalents suggesting substantial homology between the genomes of the two species.

Source: Bosland and Votava (1999)

Interspecific hybrid performance and meiotic chromosome behaviour of the F_1 hybrids were studied by Tong and Bosland (1999) to elucidate genetic relationship between *C. tovarii* and other *Capsicum* species. *C. tovarii* was hybridized, as a female and a male parent, to *C. annuum, C. chinense, C. frutescens, C. chacoense, C. galapogense, C. baccatum, C. praetermissum, C. cardenasii, C. eximium and C. pubescens*. When the hybridization of *C. baccatum* × *C. tovarii* was performed, F_1, F_2 and backcross progenies were successfully obtained. In addition, a successful hybridization of *C.*

praetermisum × *C. tovarii* could also obtained. A cytological investigation of the F_1 hybrids of *C. baccatum* × *C. tovarii* revealed that most meiotic chromosomes paired as bivalents. However, multivalents, chromosome bridges and chromosome lags were also observed. These results suggest that *C. baccatum* differs from *C. tovarii* by at least a chromosomal reciprocal translocation. Crosses of *C. tovarii* x *C. chinense* and *C. tovarii* x *C. frutescens* produced plump seeds, but none of the seeds germinated. Hybridization of *C. tovarii* to *C. pubescens, C. eximium and C. cardenasii* did not produce seed. These hybridization results indicate that *C. tovarii* is genetically more closely related to the *C. baccatum* complex than to the *C. annuum* complex or the *C. pubescens* complex.

Genetics of important characters

All the species of *Capsicum*, both cultivated and wild are diploids, 2n = 2x = 24. The chief characteristics of five cultivated *Capsicum* species (Table1) indicated that *C. frutescens* and *C. chinense* are alike except the latter has a conspicuous constriction below the calyx. The species, *C. pubescens* has dark or black seeds while other cultivated species have straw coloured seeds.

The qualitative traits having monogenic inheritance in chillies include anthocyanin pigmentation in leaves, stems, flowers and fruits, carotenoid content, fruit colour, plant growth habit, fruit bearing pattern, hairiness of stem, fruit shape, fruit apex, fruit orientation, male sterility, pungency, etc. Several quantitative characters *viz.*, plant height, earliness, fruit yield, fruit length, fruit number/plant, fruit weight, etc. are governed by polygenes. Resistance to diseases has simple inheritance in many cases except the resistance to *Cercospora* leaf spot and anthracnose or fruit rot (*Colletotrichum capsici*) which are polygenic in control.

Table 3: Genes for different qualitative characters

Gene symbol	Phenotype of plant character
A	Anthocyanin; purple coloration of leaves, stem, flower, and immature fruits; incompletely dominant over green
B	β carotene in fruits; interacts with the gene '*t*' for higher levels
bl	Branchless; recessive
C	Capsaicin, pungent fruit; modifier increase or decrease capsaicin content
cl1, cl2	Carotenoid pigment inhibitor; recessive
cf-1, cf-1	Closed flower; recessive
ct	Compact habit with numerious shoots; recessive
dw_1 *to* dw_9	Dwarf plant habit; tall dominant over dwarf
ef	Early flowering; recessive
fa	Fasciculate; fruits borne in clusters, recessive character
fr	Frilly leaf margin; recessive character
fms	Functional male sterility, degenerated corolla
H_1	Hairless, smooth stem, dominant over hairy stem
lf	Late flowering

Gene symbol	Phenotype of plant character
ira	Single recessive gene; inheritance of foliage colour and fruit colour transition phenotype
Mf-1, Mf-2, Mf-3	Multiple flowers per node; recessive homozygosity at any two loci is epistatic to the dominant allele present in the third locus.
ms1 to ms15	Genic male sterility
O	Oblate or round fruit; dominant over elongate
P	Pointed fruit apex; incompletely dominant over blunt apex
pun-1	Non-pungency; absence of capsaicinoids in fruits
rl	Round leaf; recessive character
ru-1, ru-2	Rugose leaf
sl-1, sl-2	Styleless flower
Smsms	Cytoplasmic male sterility; occurrence of dominant allele in both the loci is necessary to restore fertility
t	High β carotene; complementary with 'B'
up	Upright pedicel and fruit; intermediate expression in some genotypes
ya	Yellow anther

Source: Wang and Bosland (2006)

The inheritance of mature fruit colour in peppers including chilli is controlled by several genes. However, the inheritance of the transition of colours the fruit undergo during ripening has not been described extensively. The jalapeno culivar NuMex Pinata (NP) is unique in the transition of colours and fruit undergo as they mature (light green, through yellow and orange, and finally red). The inheritance of a unique gene which affects foliage colour and fruit colour transition was studied in reciprocal crosses between Early Jalapeno (EJ) and NP. Foliage and fruit colour transition phenotypes of the entire F_1 population were that of EJ, regardless of whether EJ was the female or male parent, indicating that the NP phenotype is recessive and displays no maternal effects. Data from the segregating F_2 population indicated that the inheritance of foliage colour and fruit colour transition phenotype of NP is due to single homozygous recessive gene. The gene responsible is designated *ira* (Votava *et al.*, 2000).

Breeding objectives

Some of the important characters that have been the main focus for breeding chilli are outlined below.

- Preferably erect growth habit with good branching, short internodes and good canopy and early maturity
- High early and total of both green and dry chilli depending upon the variety and usage.
- Rejuvenation ability of the plant after winter under north Indian condition
- Good quality for fresh market, dried spice, fresh processing and oleoresin extraction

a) Fresh market: Good fruit shape, size, colour, no wrinkleness, pungency, pleasant flavour, endocarp: seed ratio, carotenoid and vitamin C content

b) Dry chilli (whole fruit and powder): Fruit colour of ripe, dried fruit, pungency, retention of fruit colour during storage, flavour, dry weight, endocarp: seed ratio.

c) Fresh processing (sauce, paste, pickling): Fruit colour, pungency, flavour, pericarp thickness, endocarp: seed ratio.

d) Oleoresin extraction: High oleoresin content, good colour and pungency

- Resistance to diseases particularly, leaf-curl virus, anthracnose or fruit rot (*Colletotrichum capsici*), cercospora leaf spot (*Cercospora capsici*), powdery mildew (*Leveillula taurica*), , bacterial leaf spot (*Xanthomonas campestris* pv. *vesicatoria*) , etc.
- Resistance/tolerance to insect-pests particularly, thrips (*Scirtothrips dorsalis*), mite (*Polyphagotarsonemus latus*), aphid (*Aphis gossypii*), and root knot nematode (*Meloidogyne incognita*)
- Resistance/ tolerance to abiotic stresses particularly, high temperature, water stress, soil salinity

Breeding methods and development of variety

Highly variable land races of the hot pepper or chilli are available in different states of the country. These land races are cultivated in different local names. A good number of the improved varieties have been developed from these indigenous cultivars through selection. Some varieties have also been developed through selection from introduced variety or line. However, majority of the varieties have been developed by pedigree breeding following inter-varietal hybridization. Interspecific hybridization and selection in the advanced generations by pedigree or backcross-pedigree method was also employed to develop some improved varieties. However, choice of the best method or combination of them depends mainly on the type of inheritance (monogenic, digenic, or polygenic) of traits to be improved (Lee *et al*., 2013).

It is difficult to classify the chilli varieties based on fruit characters. Some region specific classifications are in vogue. As for example, in Southern India, chillies are conventionally recognized in two groups: "Samba type" referring to the fruits which are long and pendant and "Gundu type" referring to the fruits that are short and somewhat round, oblong or conical in shape. The "Gundu" types are generally more pungent than Samba types. Although, most of the improved varieties belong to the so-called "Samba type" there exists huge variation among the varieties for pungency of fruits. Hence, the varieties can also be grouped irrespective of fruit shape, size and orientation in two categories according to pungency of the fruit *i.e.*, capsaicin content in it *viz*., pungent to highly pungent (capsaicin content more than 0.55 mg/g fruit) and mild to medium pungent (capsaicin content 0.20 to 0.55 mg/g fruit).

- **Pungent variety**

K-l (Kovilpatti-1), K-2 (Kovilpatti-2), Bhagyalakshmi (G-4), Pusa Jwala, CO-3 (CA 586), Jwala Mukhi, Jwala Sakhi, Ujwala, RHRC Pendant Cluster, Andhra Jyothi (G-5), PMK-1, PLR-1, Muslawadi, CA 960, NP 46 A, Prakash (LCA-206), Sabour Arun, Agnirekha, Phule Jyoti, Phule Suryamukhi, Arka Neelanchal Prabha, LCA-305 (LAM-305), Jayanti (AKC-86-39), Kashmir Long-1, Arka Suphal (PMR-57-88-K), Gujarat Vegetable Chilli-101, Kashi Anmol (KA-2), Prashanth (LCA-334), Anugraha, Gujarat Vegetable Chilli-111, Gujarat Vegetable Chilli-121, RCH-1, LCA-353, CO-1, Pant C-1, Pant C-2, PKM-1, Arka Lohit, Pusa Sadabahar, Jawahar Mirch 218 (J-218), Punjab Lal, Hisar Vijay (H-28), Hisar Shakti (H-44), CO-2, CO-4, Bhaskar (X-235), G-2, G-3, MDU-1, Sindhur, X-197, Phule Sai, Phule Mukta, Jawahar Mirch 283, Hot Portugal (CA-960), DH-76-6, Chanchal, Punjab Guchhedar, Punjab Surkh, Punjab Sindhuri, Punjab Tej, Jawahar-283, Utkal Ava (BC-14-2), Azad Mirch-1, Bidhan Chilli-4, Kashi Gaurav, Kashi Abha (VR-339), Swarna Praphulya

- **Mildly pungent / pickling type variety**

JCA 154, KTPL-19, Arka Abir, Kashi Sinduri

Germplasm resources and Plant introduction

Asian Vegetable Research and Development Centre (The World Vegetable Center, WorldVeg), Taiwan holds more than 6000 germplasm collection of chilli under different *Capsicum* species. Wide germplasm of chilli is also maintained at the Centre for Genetic Resources, the Netherlands (CGN) through International Plant Genetic Resources Institute (PGRI), National Agricultural Research Centre (NARC) Islamabad, Pakistan and other organizations / countries.

In India, chilli germplasm including cultivars belonging to different *Capsicum* species is maintained at National Bureau of Plant genetic resources (NBPGR), New Delhi, ICAR-IIVR, Varanasi, ICAR-IIHR, Bengaluru and many All India Coordinated Research Centres for Vegetable crops *viz*., Lam, Guntur (Andhra Pradesh), Kovilpatti (Tamil Nadu), Kalyanpur (Uttar Pradesh), etc.

In India although indigenous diversity of chilli is very wide yet some exotic genetic materials has been employed for selection of desirable genotypes. Some exotic varieties / breeding lines have also been employed in hybridization programme to develop new variety. However, no exotic variety has so far been directly released as variety.

Variability and Selection

Rich diversity in chilli exists in different parts of India notably, Kerala (Cochin), northern part of Karnataka (Belgaum, Bellary, Dharwad), Andhra Pradesh (Guntur), Tamil Nadu, Uttar Pradesh, Goa, West Bengal (Murshidabad, South 24 Parganas), Odisha, North-eastern region (Tripura, Assam, Sikkim).

High GCV, heritability and genetic advance has been reported for plant height, number of primary branches, fruit length, fruit width, days to first flower, seed number per fruit, pericarp thickness, number of fruits per plant, fruit yield per plant, ascorbic acid content. These characters show prospect as selection indices in the selection programme for improving yield. From the study of correlation and path coefficient analyses, yield of green chilli is positively associated with dry weight of fruit, leaf area index, leaf number, plant height, number of primary branches, length of fruits, girth of fruit, pericarp thickness, seed content and number of fruits, whereas earliness register negative association with fruit yield. Yield of dry chilli is positively correlated with weight of dry fruit and the number of fruits per plant. Capsaicin content is negatively correlated with fresh fruit weight, fruit diameter, length and thickness and carotene content.

Chilli is basically self-pollinated crop but better termed as often-cross pollinated crop due to considerably wide and high of 7 to 37% natural cross-pollination. Selection methods for self-pollinated crops *viz*., single plant selection and pure line selection are applicable for chilli. However, rouging of off types should meticulously be done to maintain the germplasm to be used for the purpose of breeding. Maximum number of varieties has been developed in India through selection mainly from the heterozygous or heterogeneous land races or local cultivars *viz*., Kovilpatti-1, Bhagyalakshmi, Andhra Jyothi, Sabour Arun, Punjab Surkh, Muslawadi, etc. A few varieties have been developed through selection from the exotic materials *viz*., Bhagyalakshmi from "Thohian chilli" of Sri Lanka, Hot Portugal from the material of Portugal, CO-3 and Kashi Anmol from the materials of Sri Lanka, Punjab Guchhedar from the material of Indonesia, etc. (Hazra *et al*, 2022).

Table 4: Varieties developed through selection

Name of variety	Indigenous / Exotic material
K-l (Kovilpatti-1)	Indigenous (B72A from Assam)
Bhagyalakshmi (G-4)	Exotic ("Thohian chilli" from Sri Lanka)
CO-3 (CA 586)	Exotic (from Sri Lanka)
Ujwala	Indigenous
RHRC Pendent Cluster	Indigenous
Andhra Jyothi (G-5)	Indigenous (from Sabour, Bihar)
PLR-I	Indigenous (CA(P)8) called "Kandangadu type"
Muslawadi	Indigenous
Hot Portugal (CA 960)	Exotic (from Portugal)
Sabour Arun	Indigenous
Phule Jyoti	Indigenous
Phule Suryamukhi	Indigenous
Arka Neelanchal Prabha	Indigenous
LCA-305	Indigenous
Jayanti	Indigenous
Kashmir Long-1	Indigenous
Gujarat Vegetable Chilli-101	Indigenous
Gujarat Vegetable Chilli-111	Indigenous
Gujarat Vegetable Chilli-121	Indigenous
Kashi Anmol	Exotic (from Sri Lanka)

Prashanth	Indigenous
Anugraha	Indigenous
RCH-1	Indigenous
CO-1	Indigenous (Sattur Samba (CA(P)247)
Arka Lohit	Indigenous
Hisar Vijay	Indigenous
Hisar Shakti	Indigenous (tolerant to TMV, CMV, PVY and LCV)
CO-2	Indigenous (Nambiyur Local (CA(P) 63)
CO 4	Indigenous
G-2	Indigenous (re-selection from NP464)
G-3	Indigenous (re-selection from NP464)
Sindhur	Exotic (CA960)
Phule Mukta	Indigenous (tolerant to Fusarium wilt , powdery mildew)
Jawahar Mirch 283	Indigenous
Chanchal	Indigenous
Punjab Guchhedar	Exotic (Indonesia)
Punjab Surkh	Indigenous
Punjab Sindhuri	Indigenous
Punjab Tej	Indigenous
Jawahar-283	Indigenous
Utkal Ava	Indigenous
Azad Mirch-1	Indigenous
Bidhan Chilli-4	Indigenous (from Coochbehar district, West Bengal)
Kashi Gaurav	Indigenous
Kashi Abha	Indigenous
Swarna Praphulya	Indigenous
JCA 154	Indigenous
KTPL-19	Indigenous
Arka Abir	Indigenous
Kashi Sinduri	Indigenous

Hybridization and selection

Development of high yielding chilli varieties having high pungency, antioxidants, vitamins and appreciable tolerance against leaf curl virus, important pests and high temperature stress is the present-day need in major part of the tropics. Choice of parents is the most important consideration in any sound breeding programme based on hybridization and selection. Inter-varietal hybridization followed by pedigree method of selection in segregating generations is generally followed for the development of good number of varieties in India. In most the cases, adapted and indigenous germplasm and already developed variety have been utilized in the hybridization programme for the development of variety *viz*., Kovilpatti-2 (K-1 × Sattur Samba), Pusa Jwala (NP 46A × Puri Red), NP 46 A (Local × Puri Red), Jawahar Mirch 218 (Kalipeeth × Pusa

Jwala), Agnirekha (Dondaicha × Jwala), Phule Sai (Pant C-1 × Kamandalow), etc. In few cases, exotic genotypes were included in hybridization programme *viz.*, Prakash (G-3 × Huntaka (a Japanese line).

Table 5: Varieties bred through hybridization and selection

Variety	Parentage
K-2 (Kovilpatti-2)	K-1 × Sattur Samba
Pusa Jwala	NP 46A × Puri Red
Jwala Mukhi	Vellanotchi × Pusa Jwala
Jwala Sakhi	Vellanotchi × Pusa Jwala
PMK-I	CO-2 × Ramanathanpuram Gundu
NP 46 A	Local × Puri Red
Prakash (LCA-206)	G-3 × Huntaka (a Japanese line)
Agnirekha	Dondaicha × Jwala
Arka Suphal	Pant C-1 x IIHR-517
LCA-353	LCA-305 x Sel.-116
PKM-1	CO2 × Ramanathapuram Gundu
Jawahar Mirch 218	Kalipeeth × Pusa Jawala
Bhaskar	G-4 × Yellow Anther mutant
X-197	G-3 x Assam chilli
Phule Sai	Pant C-1 × Kamandalow
DH-76-6	Puri Red x Byadgi Local

Utilization of mutants and Mutation breeding

A constraint of narrow genetic variability of chilli warrants the mutation breeding programme as worth-pursuing in increasing the genetic variability. Research findings showed that LD_{50} points were in the range of 422.64-629.68 Gy (Gaswanto *et al*, 2016). However, only one variety, MDU-1 has so far been developed through applied mutagenesis by Gamma irradiation of the variety, Kovilpatti-1 (K-1).

Table 6: Varieties developed through applied mutagenesis

Variety	Applied mutagenesis
MDU-1	Gamma radiation mutation of K-1

Heterosis breeding

Disease resistant and pungent-fruited hybrids have replaced widely divergent line-bred varieties in chilli which has remarkably increased productivity to a great extent in the last 30 years in India. Heterosis in chilli was first mentioned by Deshpande (1933). Since then, chilli hybrids have shown considerable promise and at present, more than 35% of the chilli area in the country is under different hybrids. Many studies suggested that heterosis in chilli is manifested for uniformity, greater vigour, faster growth and development, earliness in maturity, increased productivity, better quality attributes and higher levels of resistance to biotic stresses. Chilli flowers are small and delicate, causing high flower-drop after manual emasculation. Moreover, hand pollination may result set of maximum 20-25 seeds per fruit which substantially increases the cost of hybrid seed production.

Use of male sterility in development of hybrids

Commercially, hybrid seed production is done using either genic or cytoplasmic-genic male-sterile lines in open pollination with the provision of supplementary hand pollination for better hybrid fruit set.

Use of genic male-sterile line

The induced male-sterile gene in France, *ms-509* (Pochard, 1970) was renamed *ms-10* by Daskalov and Poulos (1994). It is allelic to *msk* allele, isolated spontaneously in Korea (Woong Yu, 1985). The *ms-2* line identified by Shifriss and Rylski (1972) is non-allelic to *ms-1* isolated by Shifriss and Frankel (1969). The *ms-509* line in bell pepper back ground was introduced in India at PAU, Ludhiana and the male sterile gene was introgressed in three chilli genotypes, *viz. ms-12*, *ms-13* and *ms-41* (Singh and Kaur, 1986). The *ms-12* (*ms-10*) line has been utilized to develop CH-1 and CH-3 hybrids of hot pepper in Punjab. Meiotic analyses of the male-fertile and male-sterile plants revealed that *MS-12* was a sporogenous male-sterile line, in which microspores degenerated immediately after formation of the tetrads (Dash *et al*, 2001).

Seeds of the back cross generation (heterozygous seed stock) are sown in the nursery in line with adequate spacing. In the population, 50% plant will be homozygous male sterile (*msms*) and 50% plants are heterozygous male fertile (*Msms*). The male sterile plants are generally recognized only by seeing the flower. Male sterile plants could be identified easily if male sterility is linked with recessive marker gene. However, genic male sterility linked with marker character has not been reported so far in chilli.

Use of genic-cytoplasmic male sterile line

In chilli, stable cytoplasmic-genic male-sterile (CGMS) lines are available which has been utilized as female parent in hybrid seed production (Ledo *et al*, 1992). Wang *et al* (2006) described the cytoplasmic male sterile (CMS) lines of the hot pepper genotypes 1A, Yan A and 8A, their maintainer lines 1B, Yan B and 8B and the fertility restoring lines Xiangzi and LS. In the CGMS male-sterile line 1A, microspore abortion occurred from the tetrad stage to the uninucleate pollen grain stage and during this period, tapetum cells were hypertrophied so that tetrads were extruded to crack and break down so that they could not form normal uninucleate pollen grains. Scanning electron microscope observation indicated that the maintainer 1B had structurally complete pollen grains, which had obvious germination furrows. The pollen grains of the male-sterile line 1A were irregular in shape, hollow and wrinkled, and some broke down and completely aborted which indicated that the line was of high value in pepper breeding (Lu *et al*, 2006).

However, to develop hybrids with high resistance it is necessary to develop the parents with high resistance, good combining abilities and having effective restorer system (Qian *et al*, 1998). Leveraging phenotypic and marker-assisted selection techniques, IIHR developed cytoplasmic-genetic male sterility (CGMS) lines to create F_1 hybrids using LCV-resistant male parents. In a stable nuclear-cytoplasmic line of chilli CCA-4261, non-dehiscence of anther as a major mechanism and irregular meiosis at telophase II as an additional mechanism that provides support to the former in relation to the complete expression of male-sterility (Kumar *et al*, 2001). This stable nuclear-cytoplasmic male sterile line (CCA-4261) and a strong restorer line (Pant C-1) were used to tag the male sterility (*ms*) and restorer (*Rf*) genes in chilli through RAPD protocol. The primer OPP13 generated a unique fragment of about 1400 bp that was present in the DNA bulk of restorer plants and absent in the pooled DNA bulks of maintainer plants and it was concluded that the 1400 bp fragment may be linked to the *Rf* gene in Pant C-1 (Kumar *et al*, 2002). Kumar *et al* (2007) also studied the genetics of fertility restoration and examined distribution of RAPD markers (OPW 19800 and OPP 131400) linked with fertility restoration gene (*Rf*) in pepper inbreds in relation to the CGMS line CCA-4261. The results revealed that most of the hot pepper lines possessed *Rf* gene. The *Rf* gene associated two markers, *viz.,* OPW 19800 and OPP 131400 were not frequently distributed in the restorer inbred lines because presence of marker bands often does not coincide with the presence of *Rf* gene identified in many restorer inbreds.

The male-sterile line (A line) is the female parent having the genetic constitution "*Smsms*" which produce the hybrid seed. The maintainer line (B line) having genetic constitution "*Nmsms*" is the male fertile counterpart for the maintenance of male-sterility of "A" line. The crossing of 'A' line x 'B' line will produce all male-sterile plants because sterile cytoplasm is inherited through the female (A line) and thus the cms-based male-sterility is maintained. For the development of male sterile 'cms' line (A line) in new genetic background, the already available 'A' line is backcrossed to the selected maintainer line (B line) for 6-7 generations. This generates a pair of almost identical 'A' and 'B' lines in the new genetic background. Restorer allele (*Rf*) is either introgressed into identified male parent, 'C' line (restorer breeding) or male parent (C line) is directly used for hybrid seed production along with the CMS based male-sterile 'A' line, if the restorer (*Rf*) gene in homozygous state is already available in male parent (most of the chilli and sweet pepper genotypes possess *Rf* and *rf* genes, respectively). Similarly, if maintainer (*rf*) gene is not available in the line, which is to be developed as 'A' line, then maintainer breeding would be required to transfer recessive gene (*rf*) in the desirable line.

Multiplication of maintainer (B) line, male-sterile (A) line and C line

This male fertile maintainer (B) line should be grown with adequate isolation, at least 200 meter away from other chilli or sweet pepper lines and by covering the plants with 16 mesh nylon nets to keep out bees. Rigorous roguing should be done till harvesting of fruit to maintain genetic purity.

The male-sterile line (A line) is maintained by hybridization, in isolation with the B line. All the plants will be male sterile. Tall barrier crops like maize or sorghum should be planted at least 60 days before planting the 'A' and 'B' line to restrict the movement of bees into and out of the plot. It was more efficient to propagate and maintain the CMS lines and produce hybrid seeds by artificial supplementary pollination or natural pollination in a plastic tunnel than by artificial supplementary pollination or natural pollination in an open isolated plot (Wang *et al*, 2006).

Seeds of the male line, the restorer line (C line) should be multiplied in same isolation and restriction as done for multiplying B line.

Hybrid seed production

Commercial hybrid seeds are produced through hybridizing the male sterile line (A line) and the specified male fertile line, the fertility restorer line (C line) in open and isolated field at least 200 m away from any chilli and sweet pepper lines. It is better to grow tall barrier crops like, maize or sorghum at least 60 days before planting the seedlings of "A" and "C" lines.

The nursery beds of both the parents (A and C lines) should be prepared separately. Since chilli is attacked by several pathogens, adequate plant protection measures should be taken in the nursery to get healthy seedlings of both the parents. Plants of the 'A' line (female line) and 'C' line (male line) are planted together in intermingling manner at 50 cm spacing both ways at the planting ratio of 3-4 female: 1 male.

Bee activity and supplementary hand pollination

Bee activity should be ensured in the seed production plot for effective open cross-pollination. Cross-pollination principally by bees should be supplemented by hand pollination during the periods of peak flowering.

In the early morning, fresh flowers should be plucked from intended male parent. The pollen is dry and powdery and easily dislodged from the anthers with a gentle shake. Pollens can be kept in refrigerator under airtight condition for 2-3 days. Dried pollens can be stored in sealed vial at - 5.0°C for 10 days. Hand pollination is done between 8.00 to 12.00 a.m. by dipping the stigma into the pollen mass contained in the pollinating ring or small glass vial.

Breeding resistance to biotic stresses

Breeding resistance to diseases

Many chilli cultivars are resistant to many diseases and insect pests. Some cultivars have multiple resistance. These cultivars may be used as a resistant variety as such or utilized as donor parent in resistance breeding programme.

Table 7: Cultivars showing resistance to different diseases

Cultivar	Resistance to diseases
Perennial, PSP- 11	CMV, TMV, Chilli leaf curl virus, Potato virus X
Punjab Lal	CMV, CMMV, Bacterial wilt
Pusa Sadabahar, Pusa Jwala	CMV, TMV, Chilli leaf curl virus
Bidhan Chilli-4, JCA 248, JCA 218, JCA 196	Chilli leaf curl virus
Pant C-1	CMV, TMV, Chilli leaf curl virus, Bacterial wilt, Phytophthora root rot, root knot nematode

White Kandhari	Bacterial wilt, Phytophthora root rot, root knot nematode
Tambeli 1, Tambeli 2	Tocacco etch virus, TMV, Potato virus X
Myliddy 1, Myliddy 2	Chilli leaf curl virus, pepper mottle virus
KAU Cluster	Bacterial wilt
Ujjwala, Anugraha	Bacterial wilt
Phule Mukta	Powdery mildew
Cipanas , Indragiri	Root knot nematode

Source: Munshi (1999), Chattopadhyay *et al* (2020)

Note: CMV: Cucumber mosaic virus, TMV: Tobacco mosaic virus; CMMV: Cucumber mottle mosaic virus

Many *Capsicum* species have been found to be resistant to several diseases, insect pests and abiotic stresses. However, most of the *Capsicum* species having useful resistance characters cannot easily be crossed with the major cultivated species *Capsicum annuum*. For this reason, these domesticated as well as wild species have not been utilized in breeding varieties resistant to different biotic and abiotic stresses. Only, *C. frutescens* could be utilized in wide hybridization programme followed by selection to develop four chilli leaf curl virus tolerant varieties.

Table 8: Different species of *Capsicum* showing resistant to biotic and abiotic stresses

Species	**Resistance to biotic and abiotic stresses**
C. chinense	TMV, poty virus, root rot, tomato spotted wilt virus, potato virus X, drought, aphids, thrips
C. chacoense	TMV, bacterial spot, poty virus, anthracnose, aphids, thrips
C. frutescens	Chilli leaf curl virus
C. baccatum var. pendulum	CMV, potato virus Y, powdery mildew, drought, high temperature
C. ciliaum	Root rot, water logging.
C. eximium	Potato virus Y, leaf curl, drought, high temperature.
C. flexuosum	CMV, potato virus X, soil salinity.
C. cadenasii	Aphids, thrips, low temperature.
C. pubescens	Powdery mildew
C. microcarpum	Powdery mildew

Munshi (1999)

Table 9: Inheritance pattern of resistance to diseases and root knot nematode

Disease	**Resistance / genetic control**
Damping off	Digenic control
Root rot, *Phytopthora capsici*	Monogenic recessive (*C. chinense*, *C. annuum*) *Pfo*, *Pfr*: Monogenic dominant (*C. chinense*, *C. annuum*) Polygenic control (*C. annuum*)
Bacterial wilt, *Ralstonia solanacearum*	Monogenic dominant

Bacterial leaf spot, *Xanthomonas vesicatoria*	Monogenic dominant, BS_1, BS_2, BS_3, BS_4, BS_5, BS_6 Monogenic *recessive, gds*
Cercospora leaf spot	Recessive; three complementary genes, additive and dominance gene effect
Cucumber mosaic virus (CMV)	Monogenic dominant; polygenic (*C. frutescens*)
Tobacco mosaic virus (TMV)	Monogenic dominant Monogenic partially dominant (*C. chinense*)
Leaf curl virus	Polygenic
Potato virus X	Monogenic recessive
Potato virus Y	Pvr-4: Monogenic dominant
Anthracnose or fruit rot, *Colletotrichum capsici*	*Anr-1* to *Anr-5*; polygenic, recessive; additive , dominant and epistatic effects
Root knot nematode, *Meloidogyne incognita*	*N* ; Monogenic dominant *Me1* to *Me 7*: monogenic dominanat

Source: Swarup (2006); Wang and Bosland (2006)

Resistance has simple inheritance in many diseases *viz*., root rot (*Phytopthora capsici*), bacterial wilt (*Ralstonia solanacearum),* bacterial leaf spot (*Xanthomonas vesicatoria*), cucumber mosaic virus, tobacco mosaic virus and root knot nematode (*Meloidogyne incognita*). In these cases, backcross method of breeding is effective in transferring the resistance character to the otherwise good variety. Resistance to *Cercospora* leaf spot (*Cercospora capsici*), anthracnose or fruit rot (*Colletotrichum capsici*) and leaf curl virus are under polygenic control. In these cases, pedigree or backcross-pedigree method of breeding will be effective to develop resistant variety.

Inter-specific hybridization for resistance breeding

Different *Capsicum* species have been found to be resistant to several diseases, insect pests and abiotic stresses. Introgression of resistance gene against chilli leaf curl virus disease from *C. frutescens* through wide hybridization followed by pedigree or backcross-pedigree selection have been employed to develop four chilli leaf curl virus tolerant varieties *viz*., Punjab Lal, Pusa Sadabahar, Pant C-1 and Pant C-2.

Table 10: Varieties developed through inter-specific hybridization and selection

Variety	**Inter-specific hybridization**
Pant C-1	NP46-4 × Kandhari (local cultivar under *C. frutescens*) (tolerant to chilli leaf curl virus)
Pant C-2	NP46-4 × Kandhari (local cultivar under *C. frutescens*) (tolerant to chilli leaf curl virus)
Pusa Sadabahar	Pusa Jwala × IC 81339 (*C. frutescens*) (tolerant to chilli leaf curl virus)
Punjab Lal	Perennial (*C. frutescens*) × Long Red

Mutation breeding is a viable alternative for resistance breeding of *Begomovirus* disease in chilli. In a resistance breeding programme utilizing gamma irradiation, there were some chilli genotypes in the population of M_2 that had high genetic coefficient of variance genetic and broad sense heritability value for disease incubation time which could be used as a resistance parameter to *Begomovirus* (Gaswanto *et al*, 2016).

Breeding resistance to insect-pests

Several cultivars have tolerance to insect-pests like, thrips, mites and aphids. The varieties Pant C-1, Myliddy 1, Myliddy 2 are tolerant to all these three important insect-pests. The variety Pusa Jwala has tolerance to thrips and aphids and Punjab Lal and Kalyanpur Red show tolerance to mites and aphids. Some of these cultivars have also been utilized in hybridization programme to develop new variety.

Table 11: Cultivars showing resistance to insect pests

Insect pests	Cultivars showing tolerance
Thrips	Chamatkar, NP46A, Pusa Jwala, Pant C-1, Pusa Sadabahar, Myliddy 1, Myliddy 2
Mites	LEC-1, Kalyanpur Red, Goli Kalyanpur, Punjab Lal, Pant C-1, Myliddy 1, Myliddy 2
Aphids	LEC-28, LEC-30, Kalyanpur Red, Pusa Jwala, Pant C-1, Punjab Lal, Myliddy 1, Myliddy 2
Leaf eating caterpillar (*Spodoptera litura*)	Punjab Lal

Source: Munshi (1999)

Breeding for quality

Most important quality characters of chilli include red colour due to the carotenoid pigment, "capsanthin" and biting pungency and spicy ingredient attributed to "capsaicin". Long staying colour of dry chilli is also an important quality character. Oleoresin, the concentrated extract from chilli is also important quality compound of chilli. It is different from essential oils because oleoresins comprise, in addition to volatile compounds, non-volatile compounds (pigments, pungency), and making them more complex extracts (Sowbhagya, 2019) .The other quality parameters of chilli are length, width and skin thickness of fruit. All these quality characters should be taken into consideration during chilli breeding programme for yield and resistance to both biotic and abiotic stresses.

References

Bosland, P.W. and Votava, E.J. 1999. *Peppers: Vegetable and Spice Capsicums*. CABI Publishing, Massachusetts, USA.

Chattopadhyay, A., Karmakar, K., Dutta, S., Mandal, A.K. , Maurya, P. K. ,Banerjee, S., Bhattacharjee, T. and Hazra, P. 2020. Bidhan Chilli-4: Dual purpose chilli variety. *Indian Hort.*, November-December, 2020 issue, pp.10-11.

Dash, S.S., Kumar, S. and Singh, J.N. 2001. Cytomorphological characterization of a nuclear male sterile line of chilli pepper (*Capsicum annuum* L.). *Cytologia*, **66**: 365-371.

Daskalov, H. and Poulos, J.M. 1994. Updated *Capsicum* gene list. *Capsicum and Eggplant Newsl.*, **13** : 15-26.

Deshpande, R.B. 1933. The inheritance of some characters in *Capsicum annuum* L. *Indian J. Agric.Sci.*, **3** : 219-300.

Gaswanto, R., Syukur, M., Purwoko, B.S. and Hidayat, S. 2016.Induced mutation by gamma rays irradiation to increase chilli resistance to begomovirus. *J. Agril. Sci.*,38: DOI:10.17503/agrivita.v38i1.581

Hazra, P., Banerjee, M.K. and Chattopadhyay, A. 2022. *Varieties of Vegetable crops in India*, Kalyani Publishers, New Delhi, Ludhiana, 204p

Kumar, S., Rai, S.K., Banerjee, M.K. and Kalloo, G. 2001. Cytological mechanisms of male sterility in a nuclear-cytoplasmic line of chilli pepper (*Capsicum annuum* L.). *Capsicum and Eggplant Newsl.*, **20**: 64-67.

Kumar, S., Singh, V., Kumar, S., Singh, M., Rai, M. and Kalloo, G. 2002. RAPD protocol for tagging of fertility restorer and male sterility genes in chilli (*Capsicum annuum* L.). *Veg. Sci.*, **29**: 101-105.

Kumar, S., Singh, V., Singh, M., Rai, S., Kumar, S., Rai, S. K. and Rai , M. 2007. Genetics and distribution of fertility restoration associated RAPD markers in inbreds of pepper (*Capsicum annuum*L.). *Scientia Hort.*, **111**: 197-202.

Ledo, H.D., Anikeenko, V. S. and Voronina, M.V. 1992. Examination of five CMS lines in *Capsicum annuum* L. in Hungary. *Capsicum Newsl.* , Special issue, pp. 66- 68.

Lee, J., Park, S.J., Do, J.W., Han, J.H., Choi, D., and Yoon, J. B. 2013. Development of a genetic map of chili pepper using single nucleotide polymorphism markers generated from next generation resequencing of parents. *Kor. J. Hort. Sci. Tech.*, **31** : 473-482.

Lu, H. D., Gong, Z. H., Wang, X. M., Huang,W. and Lu, M. H. 2006. Cytomorphological observation of the microsporogenesis of male sterility materials in pepper (*Capsicum annuum* l.). *Acta Bot. Boreali Occidentalia Sin.*, **26**: 1842-1845.

Munshi, A.D. 1999. *Compendium of Lectures of Summer School on Advances in Breeding of Temperate and Tropical Vegetables for Biotic and Abiotic Stresses*, ICAR, New Delhi,17th May to 6th June, 1999, pp. 133-140.

Pochard, E. 1970. Obtaining three new male sterile mutants of pepper (*C. annum)* through application of mutagenes on monoploid material. *Eucarpia,* Versailles, France, pp. 93-95.

Qian, Z.L., Sun, J.B., Ding, L.P. and Yuan, C.Y. 1998. The genetic effects of three male sterile F_1 lines of chilli. *Jiangsu Agril. Sci.,* **4**: 54-57.

Shifriss, C. and Frankel, R. 1969. A new male sterility gene in *Capsicum annum. J. Amer. Soc. Hort. Sci.,* **94**: 385-387.

Shifriss, C. and Rylski, I. 1972. Comparative performance of F_1 hybrids and open pollinated bell pepper varieties (*C. annum* L.) under sub-optimal temperature-regimes. *Euphytica,* **22** : 530- 534.

Singh, J. and Kaur, S.1986. Present status of hot pepper breeding for multiple resistance in Punjab. *Proc. VIIth Meet.Breed. Capsicum and Eggplant,* Zaragoza, Spain, pp. 111-114.

Sowbhagya, H.B. 2019. Value-added processing of by-products from spice industry. *Food Quality and Safety* 3(2):73-80

Swarup, V. *Vegetable Science and Technology in India*. Kalyani Publishers, Ludhiana, 656p.

Tong, N. and Bosland, PW. 1999. *Capsicum tovarii*, a new member of the *Capsicum baccatum* complex. *Euphytica*, **109** : 71-77.

Votava, E.G., Balok, C., Coon, D. and Bosland, P.W. 2000. Inheritance of unique fruit and foliage color mutation in NuMex Piñata. *J. Hered.*, **91** : 60-61.

Wang, D. and Bosland, P. W. 2006. The genes of *Capsicum*. *Hort Sci*., 41: 1169-1187.

Wang, S. B., Liu, J. B. and Pan, B. G . 2006. Genetic effects of the male sterile cytoplasmic gene in hot pepper. *Acta Agric. Shanghai*, **22**: 14-17.

Woong Yu, L. 1985. Inheritance of Cytoplasmic male sterlity in pepper (*Capsicum annuum* L.). M.Sc. thesis, Kyung Hee University, South Korea, 43p.

21

Sweet Pepper

Sweet pepper or bell pepper, a cultivar group of the species *Capsicum annuum* under the family Solanaceae (Chromosome No. 2n =2x =24) is grown worldwide as green vegetable for its thick and fleshy fruits having delicate taste, pleasant flavour and colour. Sweet pepper requires slightly cooler climate than hot chillies. The fruits are generally non-pungent hence, it is commonly known as sweet pepper. It is also popular as bell pepper because of blocky fruit shape. China is the largest producer of sweet pepper in the world.

Origin, domestication and adaptation in India

Hot chillies and sweet pepper have originated from the wild and weedy species *Capsicum annuum* var. *minimum* whose domestication occurred in Mexico and distributed from the southern United states to northern south America. Domestication, cultivation and human selection brought about changes in fruit size and shape and fruit position from erect to pendant. It also resulted in more self-pollinating character due to inserted stigma unlike the exserted stigma in the wild types.

In India, mixed stock of *Capsicum annuum* including both hot and sweet peppers was introduced in Kerala by the Portuguese from Brazil in 1584. Christian missionaries also introduced *Capsicum* species in the North-eastern states long ago separately. However, chillies or hot pepper spread much more widely and rapidly than sweet pepper in India. High adaptation in the tropical Indian condition, wide spread, extensive course of cultivation almost throughout the country and farmers' selections particularly for pungent and red types have resulted in the development of huge numbers of local cultivars of chilli under different names in India. In contrast, very few local cultivars have been developed in sweet pepper from the early introductions suggesting that Indian farmers were in favour of pungent, red and hot pepper. Hence, most of the non-pungent or mildly pungent types, if at all have been introduced in India as mixed stock, possibly got eliminated. Very few local cultivars of sweet pepper having mildly pungent or non-pungent, small, roundish fruits are still being grown by the farmers in Karnataka, Goa and coastal areas of Kerala.

Crop botany and floral biology

- Plants are tender annuals or perennials; although the plants are technically perennials; they are not worth keeping after fruiting once.

- Two growth habits are found in sweet pepper: determinate and indeterminate; cultivars of determinate growth habit grow to a certain size, produce fruit and stop growing and eventually dies off; cultivars of indeterminate growth habit are mostly grown under polyhouse in which the plants continually develop and grow from new meristems that produce new stems, leaves, flowers and fruit.
- The leaves are oval and taper to a point; leaves are usually bright to dark green, but can also be mottled; size of the leaf corresponds somewhat to the size of the fruit.
- Inflorescence is solitary tending to be leaf opposed and axillary.
- Flowers are star-shaped, in the axils of the leaves, bracteolate, pedicellate, bisexual, actinomorphic or slighltly zygomorphic, larger than chilli, hypogynous, pentamerous and heterochlamydeous.
- Calyx campanulate, sepals are usally five (sometimes six), gamosepalous, enclosing the base of the fruits and persistent in fruits.
- Corolla rotate, campanulate, deeply 5- 6 partite, white.
- Stamens are five (sometimes six), distinct, alternating with the lobes of corolla and adnate to the corolla tube or perigynous zone.
- Anthers are bluish, ditheceous, introse, non-convenient in cone, not longer than filament, dehisce longitudinally by lateral slits.
- Ovary superior, bicarpellary, syncarpous, bilocular with numerous ovules in each locule, axile placentation.
- Style is terminal, linear, white or purple.
- Stigma sub-capitate and faintly bifid.
- Fruit is berry with an accrescent calyx (growing larger after flowering) of variable shapes, pericarp leathery or succulent.
- Fruit is an indehiscent, many seeded berry with no sutures, large, with basal depression, inflated, 3-4 lobed, thick flesh, skins generally quite tough, shiny and mildly pungent.
- Fruit shape varies from "large, heart shaped, thick walled"; "large, blocky, thick walled"; "long and thick"; "small, cone-shaped", etc.
- All the fruits are green at the early stages; fruits of some variety remain green at harvest maturity stage while others change to different colours including red, yellow, orange, chocolate, brown, purple, vanilla and even white; most common colours at matured stage of different varieties are green, red, yellow and orange.
- Seeds are flat, kidney-shaped and whitish-brown in colour.

Reproductive behaviour

Sweet pepper bears hermaphrodite flowers and is basically a self-pollinated crop, but considered as often cross pollinated crop due to appreciable degree of cross pollination (even 16% in some condition) due to open nature of floral morphology, very fine dry pollen and frequent visit of many insect pollinators. Major pollinators are honey bees (*Apis* species), ants and wind. Anthesis generally starts from 5.30 am and majority of the flowers open by 9.00 a.m. however, it depends on the prevailing temperature and humidity condition. Anther dehiscence occurs after anthesis. Pollen grains in the anther become fertile a day before anthesis with maximum fertility on the day of anthesis. Stigma receptivity is the maximum at anthesis and receptivity continues up to two days after. At comparatively high temperature condition, emasculation should be done 2 days prior to anthesis and cross pollination is carried out on the day of anthesis. At high temperature condition, fair crossing success can be achieved if hybridization is taken up early in the morning. Pollen viability is retained for a day in ambient condition. Pollens can be kept in refrigerator under airtight condition for 2-3 days. Dried pollens can be stored in sealed vial at - 5.0°C for 10 days.

Classification of cultivars of *Capsicum annuum*

The species, *Capsicum annuum* is a source of both popular sweet peppers and hot chillis, with numerous varieties cultivated all around the world. Capsaicinoids is the class of pungent polyphenol derivative compound and are plant secondary metabolites. The main pungent capsaicinoid is capsaicin. Capsaicin is an active component of red chili peppers that is used for flavouring, colouring, and preserving food, as well as for medical purposes. Content of the pungent principle widely varies among the cultivar groups of *Capsicum annuum*.

Cultivar group	**Major characters**
Ancho	Large, heart-shaped, thin-walled fruit, mildly pungent
Bell	Large fruits, blocky with a blunt end of 3-4 lobes, thick-walled, sweet, very mild pungency; fruit colour green, red, yellow, orange, purple, etc.
Cayenne	Fruits are tapered, slender, wrinkled with thin wall, highly pungent
Cheese	Mostly globose-shaped fruits, small to medium- sized, almost non-pungent, yellow or green maturing to red in colour
Cherry	Spherical, small, slightly flattened fruits, thin-walled, sweet and pungent
Chiltepin	Very small, egg-shaped fruits, thin-walled, highly pungent
Cuban	Fruits are of irregular and blunt-shaped, mildly pungent, thin-walled
Jalapeno	Fruit are small, almost cylindrical with round ends, thick-walled, fruit skin russeted in some varieties, highly pungent
Long Wax	Fruits are long, tapered but in some cases blunt, mildly pungent
New Mexican	Long, slender with a tapered end, thin-walled, moderately pungent however, some varieties have sweet fruits
Pimiento (Pimento)	Fruits large, cone or heart-shaped, thick-walled, non-pungent
Serrano	Short, elongated, blunt-end fruits, thin-walled and very pungent
Squash	Fruits are flattened and scallop-shaped, medium to thick-walled, pungent

The cultivar group "Bell", "Cheese", "Long Wax", "Pimiento", etc. are generally considered as sweet pepper.

Crossability among different *Capsicum* species

Chillies and sweet pepper can be crossed easily and viable seeds are obtained in F_1 and F_2 generations. Crossing behaviour or interspecific hybridizations allow transfer of genes of interest between different species, especially those involved in disease resistance, allowing breeders to develop genetically superior genotypes (Bosland and Votava, 2000). However, for the success of such a transfer, the species must be genetically close, minimizing problems of incompatibility and thus enabling hybridization. Moreover, it is imperative to know about the direction of the cross, since, in some species, the interspecific cross is possible in one direction only.

Incompatibility barriers in interspecific crosses may be expressed before fertilization in prezygotic barriers, and/or after fertilization, in postzygotic barriers. Among the former, non-germination of pollen grains on the stigma of the receptor plant and a slow pollen tube growth towards the ovule may be cited and among the latter, embryo death caused by endosperm degeneration, a lack of vigour and plant sterility of the hybrid plant (Pickersgill, 1997).

As a general rule, the closer the species involved in the cross are genetically the easier it will be to develop hybrids and the more fertile the progenies. Thus, the existence of barriers in interspecific crosses indicates the genetic distance of the species. The genus *Capsicum* comprises three gene complexes, formed on the basis of crossability between species: 1) Complex *C. annuum* (varieties of *C. annuum* and *C. annuum* L. var. *glabriusculum*): *C. frutescens*, *C. chinense*, *C. chacoense* and *C. galapagoense* ; 2) Complex *C. baccatum*: *C. baccatum* (*C. baccatum* var. *baccatum,* var. *pendulum* and var. *praetermissum*) and *C. tovarii* and 3) Complex *C. pubescens*: *C. cardenasii*, *C. eximium* and *C. pubescens* (Bosland and Votava, 2000).

Martins *et al.* (2015) reported crossability among *C. annuum*, *C. chinense*, *C. frutescens*, *C. baccatum* var. *pendulum*, and *C. baccatum* var. *baccatum.* The resulting hybrids were evaluated for their pollen viability and in the crosses that produced no hybrids. The *in vivo* germination of pollen grains was assessed. Some crosses generated hybrids, while others produced no fruit or, despite generating fruits, no plants grew from them. The crossability rate between the species of the complex *C. annuum* and *C. baccatum* ranged from 2.2 % to 3.7 %, and was 14.6 % between species of the complex *C. annuum*. A pre-fertilization barrier was observed in the non-germination of pollen grains, as well as post-fertilization barriers consisting of embryo death, lack of vigour and hybrid sterility.

Manzur *et al.* (2015) studied crossability of four cultivated species of *Capsicum* including *C. annuum* , *C. baccatum* , *C. chinense* , and *C. frutescens* . *C. chinense* performed as a good bridge species between *C. annuum* and *C. baccatum*, with the best results being obtained with the cross combination [*C. baccatum* (♀) × *C. chinense* (♂)] (♀) × *C. annuum* (♂), while *C. frutescens* gave poor results as bridge species due to strong prezygotic and postzygotic barriers. Virus-like-syndrome or dwarfism was observed in the F_1 hybrids when both *C. chinense* and *C. frutescens* were used as female parents. Regarding the embryo rescue (ER) strategy, the best response was found in *C. annuum* (♀) × *C. baccatum* (♂) crosses. First backcrosses to *C. annuum* (BC_1s) were obtained according to the crossing scheme [*C. annuum* (♀) × *C. baccatum* (♂)] (♀) × *C. annuum* (♂) using ER. These results provided the breeders with useful practical information for the regular utilization of the *C. baccatum* gene pool in *C. annuum* breeding. Wei *et al.* (2019) produced eight fertile and reciprocal F_1 hybrids between *Capsicum annuum* x *C. frutescens*. Interspecific hybrids were intermediate, although they were closer to *C. frutescens* in most of the agronomic traits.

Genetics of important characters

All the species of *Capsicum*, both cultivated and wild are diploids, 2n = 2x = 24. The qualitative traits having simple inheritance in *Capsicum* sp. include fruit shape, fruit apex, fruit surface, carotenoid content of fruit, fruit colour, plant growth habit, male sterility, pungency, etc.

Fruit colour and shape are deciding factors affecting market value in bell pepper. The plant has undergone selection during domestication, which resulted in the fruit colour at physiological maturity of: red, yellow, orange, yellow-orange, cream, white, chocolate, purple, and shapes like, blocky, almost round, elongate and triangular. Since diverse colours and shapes can be found in different sweet pepper varieties, the plant has become a good model system for studying the genetic control of fruit colour change and shape. Fruits of the colored bell pepper genotypes attain specific colour of specific physiological maturity. Fruit colour of bell pepper is due to presence of colour pigments. Fruit colour at physiological maturity emerges due to degradation of chlorophyll and accumulation of anthocyanin and carotenoid pigments. The colour found in fruit mainly depends on variable accumulation of pigments capsanthin and capsorubin, both of which are involved in the carotenoid biosynthetic pathway.

Red bell peppers accumulate higher levels of total carotenoid during ripening. Expression levels of carotenoid biosynthesis-related genes are directly linked to high total carotenoid accumulation in bell pepper. The expression levels of some carotenoid biosynthetic gene, including phytoene synthase (*Psy*), phytoene desaturase (*Pds*) and capsanthin-capsorubin synthase (*Ccs*) are relatively high in red peppers; some of these genes are not expressed in peppers with lower levels of total carotenoid (Ha *et al.*, 2007). The capsanthin-capsorubin synthase (*Ccs*) gene is responsible for capsanthin synthesis controlling orange and red colour of bell pepper by changing the carotenoid composition. The orange colour is due to lack of expression of *Ccs* gene. There is a hypothesis for involvement of the *Ccs* gene in segregation of yellow and red color (Popovsky and Paran, 2000). Control of colour development in orange sweet pepper involves a more complex process than presence or deletion in a structural gene for a step in pigment biosyntheses; orange colour may involve transcriptional regulation of *Psy* and / or *Ccs* expression (Rodriguez-Uribe *et al.*, 2012). Yellow sweet pepper fruit phenotype, which is recessive to red, is controlled by the *y* locus. Linkage analysis indicated the candidate gene of the *y* locus is the *Ccs* gene (Ha *et al.*, 2007).

McArdle and Bouwkamp (1983) studied the genetics of different fruit characteristics and they found that fruit length and fruit apex configuration were simply inherited and linked. Average locule number and fruit width were found to be quantitatively inherited. Fruit curvature and fruit base configuration probably are heavily influenced by environmental factors. Kumar *et al.* (2009) studied the inheritance pattern of mature fruit colour in sweet pepper. Red colour was found dominant over others whereas yellow-orange was recessive to all others. From the analysis of a series of F_2 generations and test cross of all possible F_1 progeny it was established that when the *c* locus is silent (i.e., homozygous recessive) the *y* locus controls the appearance of orange colour (*Y-cc*) *vs*. yellow orange (*yycc*). On the other hand, when the *y* locus is homozygous recessive, the *c* locus controls the expression of yellow (*yyC-*) vs. yellow-orange (*yycc*) colour. Expression of fruit colour and shape at physiological maturity stage in the F_1 generation indicated the dominance of red coloured group over other fruit colour groups and dominance of blocky fruit shape over elongate and round fruit shape (Banerjee *et al.*, 2024).

Vilarinho *et al.* (2015) studied inheritance of fruit-related traits in a population generated by crossing two heirloom pepper cultivars *viz.*, "Round of Hungary" and "Bulgarian Carrot". Inheritance of corrugated pericarp phenotype of 'Round of Hungary' behaved as a recessive trait controlled by two genes while round fruit shape behaved as a single gene. Srivastava *et al.* (2019) studied inheritance pattern for different fruits traits. They found that fruit bearing habit, its colour at unripe stage and fruit apex are simply inherited traits with or without epistatic interactions while fruit length is a quantitative character. Banerjee *et al.*, (2022) studied inheritance pattern of fruit colour and shape at physiological maturity in six genetic populations generated from four crosses involving parents bearing different fruit colour and shape. They found duplicate recessive epistasis gene action in the expression of fruit colour in Red × Yellow cross. The segregation pattern of the crosses Red × Purple and Red × White indicated involvement of dominant and recessive epistasis, respectively. Fruit shape segregation of the cross Blocky × Elongate indicated recessive epistasis gene interaction. The inheritance pattern indicated that fruit colour and shape in bell pepper were conditioned by two genes with either dominant or recessive epistatic action. The segregation pattern of the test crosses revealed involvement of two loci and there is no linkage between the genes and their two alternative alleles on both the loci.

Several quantitative characters *viz.*, plant height, earliness, fruit length, fruit number/ plant, fruit weight, fruit yield, etc. are governed by polygenes. Resistance to diseases has simple inheritance in many cases except the resistance to *Cercospora* leaf spot and anthracnose or fruit rot (*Colletotrichum capsici*) which are polygenic in control.

Breeding objectives

Some of the important characters that have been the main focus for breeding sweet pepper are outlined below.

- Preferably erect growth habit with good branching and good leaf canopy for the variety having determinate growth habit

- Good growth with adequate leaf canopy which response to pruning and training on a regular basis in order to ensure a balanced growth for maximum fruit production for the variety having indeterminate growth habit
- High early and total yield depending upon the type and variety.
- Good quality for fresh market and fresh processing purpose
  - e) Fresh market: Desirable fruit shape (preferably blocky) and good size with preferably 3-lobes, thick exocarp, mild or non-pungent, pleasant flavour, enhanced shelf-life, high carotenoid and vitamin C content
  - f) Fruits of different colour at harvest maturity stage *viz*., green, red, yellow, orange, purple or white according to the variety and consumers' preference (consumers mostly prefer green and red fruited-varieties)
  - g) Fresh processing (frozen product): Fruit must be free of physiological defects such as calcium deficiency (blossom end rot), bruising and pitting
- Resistance to diseases particularly, leaf-curl virus, tobacco mosaic virus (TMV), anthracnose or fruit rot (*Colletotrichum capsici*), cercospora leaf spot (*Cercospora capsici*), fruit rot (*Phytophthora capsici*), powdery mildew (*Leveillula taurica*), bacterial leaf spot (*Xanthomonas campestris* pv. *vesicatoria*) , etc.
- Resistance/tolerance to insect-pests particularly, fruit borer (*Helicoverpa armigera*), thrips (*Scirtothrips dorsalis*), mite (*Polyphagotarsonemus latus*), aphid (*Aphis gossypii*), and root knot nematode (*Meloidogyne incognita*)
- Resistance/ tolerance to abiotic stresses particularly, high temperature, water stress, soil salinity

Breeding methods and development of variety

Sweet pepper varieties were introduced in India during the British periods and their cultivation took momentum with the increase in demand basically from the urban areas. Introductions played major role for the development of varieties in India. However, some varieties have also been developed through single plant or pure line selection from the introduced varieties or breeding lines. It is difficult to classify the sweet pepper varieties based on fruit characters because all the varieties grown and developed in India with preference towards green blocky types than to elephant trunk types. The varieties so far developed in India are mentioned below.

California Wonder, Bull Nose, Chinese Giant, Yolo Wonder, World Beater, Arka Mohini, Arka Gaurav, Arka Basant, Nishat-1, Solan Bharpur, VL Shimla Mirch 2, VL Shimla Mirch 3.

Germplasm resources and plant introduction

Asian Vegetable Research and Development Centre (The World Vegetable Center, WorldVeg), Taiwan holds more than 6000 germplasm collection of sweet pepper, chilli and different *Capsicum* species. Wide germplasm of sweet pepper is also maintained at USA, Spain, Bulgaria, Mexico, Peru, Bolivia, Ethiopia, Nigeria, Ghana, China, India, Indonesia and other countries.

In India germplasm of sweet pepper is maintained at NBPGR, New Delhi, ICAR-IARI Regional Vegetable Research Station, Katrain, Kullu Valley, Himachal Pradesh, Himachal Pradesh Agricultural University, Palampur, Himachal Pradesh, Y.S. Parmar University of Horticulture and Forestry, Solan, Himachal Pradesh, Sher-e-Kashmir University of Agricultural Sciences and Technology, Kashmir and other places.

Most of the varieties grown in India are introduced varieties mainly from USA.

Table 1: Exotic variety directly released for cultivation

Name of the variety	Country from where introduced
California Wonder	USA
Bull Nose	USA
Chinese Giant	Heirloom variety from the USA
Yolo Wonder	USA
World Beater	USA

Variability and selection

Estimation of genetic variability parameters is a pre-requisite for success in any breeding programme. Assessment of genetic variability in the germplasm by estimating various parameters of variability helps to identify the potential genotypes for their use either directly as varieties or as parents in future crop improvement programmes. The phenotypic coefficient of variation (PCV) and genotypic coefficient of variation (GCV) were high for fruit yield per plant and ascorbic acid content indicating that these traits would respond better to selection. Yield is a complex character, determined by the interaction of several factors, including genetic, physiological and environmental factors. High heritability and high genetic advance were recorded for average fruit weight, fruit yield per plant, fruit diameter, number of lobes per fruit, days to first harvest, leaf area and ascorbic acid, total chlorophyll, TSS, capsanthin, and total carotenoids contents indicating the role of additive gene action for the genetic control of these traits. Maximum positive correlation of fruit yield was found with number of fruits per plant followed by fruit size, number of lobes per fruit, number of pickings, plant height, number of laterals, pericarp thickness and harvest duration (Thakur *et al.*, 2019; Banerjee *et al.*, 2023). Hence, these characters can be used as reliable selection indices for improvement of yield and quality of sweet pepper.

Sweet pepper is basically self-pollinated crop but better termed as often-cross pollinated crop due to appreciable degree of cross pollination (even 16% in some condition). Selection methods for self-pollinated crops *viz*., single plant selection and pure line selection are applicable for sweet pepper. However, rouging of off types should meticulously be done to maintain the germplasm to be used for the purpose of breeding. All the varieties have so far been developed in India through selection mainly from the heterozygous or heterogeneous introduced varieties or lines.

Table 2: Varieties developed through selection

Name of variety	Indigenous / Exotic material
Arka Mohini	Exotic ("Titan" from USA)
Arka Gaurav	Exotic ("Golden Calwonder" from USA)
Arka Basant	Exotic ("Soroksari" from Hungary)
Nishat-1	Exotic
Solan Bharpur	Exotic
VL Shimla Mirch 2	Exotic
VL Shimla Mirch 3	Exotic

Heterosis breeding

Exploitation of heterosis is one of the potential means for improvement of sweet pepper. Preponderance of non-additive gene action for most of the characters suggested the usefulness of exploitation of hybrid vigour (Banerjee *et al.*, 2024). Sweet pepper hybrids are popular among the farmers due to blocky shape, desirable fruit size, attractive fruit colour, high early and total yield, enhanced cropping duration, resistance to some major disease and adapted to tropical and subtropical climate.

Some promising hybrids identified through different studies are Arya × Baby Bell, 8/4 × Baby Bell and C/4 × 8/4 (Banerjee *et al.*, 2024), Solan Bharpur × Yolo Wonder, Yolo Wonder × Nishath-1, California Wonder × Solan Bharpur, California Wonder × UHFBP-3 and UHFBP-3 × KC-12 (Sharma *et al.*, 2020), DARL-10 × California Wonder, DARL-01 × California Wonder, LC × Solan Bharpur and Kandaghat Sel-9 × Solan Bharpur (Devi and Pathania,2018), etc.

Assessment of genetic divergence through Mahalanobis D^2 statistics (Sharma *et al.*, 2017) and molecular marker, AFLP (Hussein and Mahdi, 2017) helps in identification of breeding lines with broad spectrum of genetic variability thus providing better scope to select superior parental lines for the development of hybrids.

Hybrid seed production

Hybrid seeds are commonly produced by hand emasculation and hand pollination because reasonably good number of hybrid seeds are produced in hand pollination and less quantity of seeds are required for its cultivation. However, genetic and cytoplasmic male-sterility is also being used but on a limited scale. Sweet pepper grows best between 21 and 24°C. When temperature fall below 18°C or exceed 27°C for extended periods, growth and yield are usually decreased. It can tolerate day temperature over 30°C, as long as night temperatures remain within 20-22°C. It is better to produce hybrid seeds under polyhouse because sweet pepper is highly susceptible to different viral diseases. Plants of the male parent should be planted 10-15 days earlier. Generally, one male parent is sufficient for supplying the pollens to pollinate every 3- 4 female plants. Seedlings should be raised in seedling tray filled with sterile growing medium under insect proof net so that insect pests cannot transmit the virus disease early.

Emasculation

Sweet pepper flowers are sufficiently large for easy emasculation and pollination. However, selection of appropriate bud size is important for crossing success and getting maximum hybrid seed production. Pollens are abundant and sufficient seeds (about 200) are obtained per pollination. First 2-3 flowers buds are pinched off to encourage vegetative growth. The flowers are white with blue to purple anthers, toothed calyx and typically single fruited (rarely two fruits /node). The bulging buds with corolla turning white are selected a day prior to anthesis. The petals open on slight pressure at the top. Then the anthers alternate with petals are pulled off intact with sharp forceps holding at the base. Care should be taken so that the stigma should not get mechanical injury because stigma remains at the level of anther. The emasculated buds are protected from pollen contamination by small paper bagging or thin cotton stretched over it. All open un-emasculated flowers should be removed.

Pollen collection

Matured anthers from open flowers, which are about to dehisce, should be collected in the early morning hours and kept in petridish or cellophane paper under sunlight. The dried anthers are gently crushed to collect the pollens. The pollens can also be collected by shaking in a container and strained through a 200- 300 mesh cloth.

Pollen storage

Use of fresh pollen within 24 hours gives higher set of fruits and seeds. For this reason, pollens are collected from dried and dehisced anthers in the morning on the day of pollination. Pollens can be kept in refrigerator under airtight condition for 2-3 days. However, it can be stored in sealed vial at -5°C for 10 days.

Pollination

Pollination is done between 8.00 to 11.00 a.m. by dipping the stigma into the pollen mass kept in the pollen ring or glass vial. It should be ensured that maximum pollens are deposited on the stigma to get maximum seeds per fruit. After pollination, paper bagging again protects the buds. The paper bags should be removed 2-3 days after pollination. A thread is tied around the peduncle to identify the crossed fruits. Spraying the plants with 10 ppm NAA, an auxinic growth substance, one at flowering and next seven days later increases hybrid fruit set. Enlargement of the ovary can visualize development of the crossed fruit 7-8 days after pollination.

Use of male sterility in development of hybrids

Both genic or cytoplasmic-genic male-sterile lines are available in sweet pepper. However, unlike chilli, both these male sterility systems are used in the development of hybrids on a limited scale.

Use of genic male-sterile line

This male sterile line controlled by single recessive gene (*msms*) is maintained through backcross generation with its heterozygous male fertile sister line (*Msms*) every year.

In the population, 50% plant will be homozygous male-sterile (*msms*) and 50% plants are heterozygous male fertile (*Msms*). The male sterile plants are generally recognized only by seeing the flower because marker characters are not available.

Use of genic-cytoplasmic male-sterile line

Cytoplasmic male sterility interacting with the nuclear gene (CGMS) was first discovered by Peterson (1958) in *Capsicum annuum* PI-164835, a stock from India. Stable cytoplasmic-nuclear male-sterile (cms) lines can be utilized as female parent in hybrid seed production. The male sterile line (A line) is the female parent having the genetic constitution "*Smsms*" which produce the hybrid seed. The maintainer line (B line) having genetic constitution "*Nmsms*" is the male fertile counterpart for the maintenance of male sterility of "A" line. The crossing of 'A' line x 'B' line will produce all male-sterile plants because sterile cytoplasm is inherited through the female (A line) and thus the CMS-based male-sterility is maintained. Crossing the 'A' line with a good combiner line (male or 'C' line) produce the commercial hybrid. Hybrid seeds are produced by hand pollination. However, in chilli hybrid seeds using the male sterility system is produced under open condition by insect pollinators as well as supplementary hand pollination.

Breeding for disease resistance

Sweet pepper suffers from many diseases particularly, leaf-curl virus, tobacco mosaic virus (TMV), anthracnose or fruit rot (*Colletotrichum capsici*), cercospora leaf spot (*Cercospora capsici*), fruit rot (*Phytophthora capsici*), powdery mildew (*Leveillula taurica*), bacterial leaf spot (*Xanthomonas campestris* pv. *vesicatoria*) , etc. Heavy application of pesticides causes environmental pollution, increases the cost of cultivation as well as residual toxicity of the chemicals in the produce causes health hazards upon consumption.

According to Sharma *et al.* (1989) bell pepper suffers considerable losses from a strain of potato virus Y (PVYo-sbp). Crosses were attempted between two resistant lines Perennial and S41-1 and two highly susceptible bell pepper commercial cultivars California Wonder and Yolo Wonder. Studies of F_1's, F_2's, back crosses and F_3's indicated that Perennial and S41-1 carry a recessive gene imparting resistance to potato virus Y.

Inheritance of resistance to bacterial spot caused by *Xanthomonas axonopodis* pv. *vesicatoria* was investigated by Riva *et al.* (2004) with P_1, P_2, F_1, F_2, BC_1, and BC_2 generations obtained from the cross 'Hercules' (P_1) × UENF 1381 (P_2). These plants were inoculated with the isolate ENA 4135 (10^3 cells mL^{-1} per 1.0 cm^2 mesophile) to assess their reaction to bacterial spot. Broad-and narrow-sense heritability were 82.54 % and 50.17%, respectively. The average dominance degree indicated presence of over-dominance and at least three recessive genes control resistance.

Depending upon the source of resistance and progenies studied in *Capsicum annuum*, inheritance of bacterial wilt resistance was reported as digenic recessive, two to five genes with additive effects (Lafortune *et al.*, 2005), duplicate epistasis (Tran and

ByungSoo, 2010) and incomplete dominance to complete dominance (Sharma *et al.*, 2013). However, according to Devi *et al.* (2015), bacterial wilt resistance in bell paper was governed by a single dominant gene in the cross EC 464107 × Sweet Happy I, whereas two genes with dominant and recessive epistasis conditioned resistance in the crosses EC 464107 × Kandaghat Selection and EC 464115 × Kandaghat Selection, respectively. Backcrossing with commercial cultivars and recombinant breeding accompanied by selection would also help to develop bacterial wilt resistant, high yielding and horticulturally desirable pure line varieties of sweet pepper.

Kumari and Sharma (2014) screened 15 genotypes of bell pepper against *Phytophthora* leaf blight and fruit rot disease under natural epiphytic conditions and the genotype Feroz and LC-1 emerged as moderately resistant which could be utilized the resistance breeding programme.

Many *Capsicum* species have been found to be resistant to several diseases, insect pests and abiotic stresses. However, most of the *Capsicum* species having useful resistance characters cannot easily be crossed with the major cultivated species *Capsicum annuum*. For this reason, these domesticated as well as wild species have not been utilized in breeding varieties resistant to different biotic and abiotic stresses.

Table 3: Different species of *Capsicum* showing resistant to biotic and abiotic stresses

Species	**Resistance to biotic and abiotic stresses**
C. chinense	TMV, poty virus, root rot, tomato spotted wilt virus, potato virus X, drought, aphids, thrips
C. chacoense	TMV, bacterial spot, poty virus, anthracnose, aphids, Thrips
C. frutescens	Chilli leaf curl virus
C. baccatum var. pendulum	CMV, potato virus Y, powdery mildew, drought, high temperature
C. ciliaum	Root rot, water logging.
C. eximium	Potato virus Y, leaf curl, drought, high temperature.
C. flexuosum	CMV, potato virus X, soil salinity.
C. cadenasii	Aphids, thrips, low temperature.
C. pubescens	Powdery mildew
C. microcarpum	Powdery mildew

Source: Munshi (1999); Negi *et al.* (2018)

Many sweet pepper cultivars are tolerant to many diseases and insect pests. Some cultivars have multiple resistance. These cultivars may be used as a tolerant variety as such or utilized as donor parent in resistance breeding programme.

Table 4: Cultivars of sweet pepper showing tolerance to different diseases

Disease	**Tolerant cultivars**
Fruit rot	Hungarian Yellow Wax, Spartan, Chinese Giant, Yolo Y
Cercospora Leaf Spot	California Wonder, Hungarian Wax
Bacterial Leaf Spot	PI-163192, 260435, Aristotle, Revolution, King Arthur, Lafayette
Root Rot	Santaka, Charleston Belle, Charleston Hot, Carolina Wonder, Carolina Cayenne, and Mississippi Nemaheart
TMV	Sonnette, Keystone Resistant Giant, Yolo Y, Aladdin
Phytophthora blight	Paladin
PYV	Aristotle, Lafayette

Source: Negi *et al.* (2018)

Breeding for quality

Sweet pepper, as a vegetable crop is used to season or flavour food. The fruits stimulate appetite by increasing the flow of gastric juice. Basic quality characters are desirable fruit shape (preferably blocky) and good size with preferably 3-lobes, thick exocarp, mild or non-pungent, pleasant flavour, enhanced shelf-life, high carotenoid and vitamin C content. Sweet peppers are packed with vitamin C. In addition to its biosynthetic and antioxidant functions, vitamin C plays an important role in immune function and improves the absorption of non-heme iron. It is also considered high antioxidant containing vegetable crop. Different studies have shown that there are significant differences in carbohydrate, protein, fat and oil, carotene, mineral and moisture contents of pepper species notably, *Capsicum annuum* and *Capsicum frutescens* L. (Ilodibia *et al.*, 2015) which open ample scope for development of variety with enhanced nutrient content.

References

Banerjee, S., Bhattacharjee, T., Maurya, P.K., Mukherjee, D., Islam, S.M., Chattopadhyay, A., Ghosh, D.K. and Hazra, P. 2022. Genetic control of qualitative and quantitative traits in bell pepper crosses involving varie*d fruit co*lors and shapes. Intern. J. Veg. Sci., **28**: 477-492.

Banerjee, S., Bhattacharjee, T., Maurya, P.K., Pramanik, S., Ghosh, T., Kundu, S., Parvin, N., Baul, D., Mallick, R.G., Ghosh, S. and Ghosh, D.K. 2023. Breeding potential of sweet pepper genotypes involving d*ifferent fruit colou*rs and shapes using multivariate analysis. Indian J. Ecol., **50**: 1425-1439.

Banerjee, S., Pramanik, S., Bhattacharjee, T., Maurya, P.K., Islam, S.M., Ghosh, D.K., Chattopadhyay, A. and Hazra, P. 2024. Breeding sweet pepper for improvement in yield components and fruit quality traits unde*r low cost protec*ted structure. J. Agric. Sci. Technol., **26**: 343-357.

Bosland, P.W. and Votava, E.J. 2000. Peppers: vegetable and spice Capsicums Crop Production Science in Horticulture 12. CAB International, Wallingford, p. 204.

Devi, J., Sood, S., Vidyasagar and Singh, Y. *2015. Inheritance of bacterial* wilt resistance and performance of horticultu*ral traits in bell pepper (C*a*psicum annuum* var. grossum). *Indian J. Agric. Sci.,* **85**: 126-131.

Devi, M.B. and Pathania, N.K. 2018. Heterosis studies in bell pepper (Capsicum annuum L. var. grossum Sendt) under modified naturally ventilated polyhouse in mid hills of Himala*yas. Veg. Sci.,* **45**: 3*1-37.*

Ha, S.H., Kim, J.B., Park, J.S., Lee, S.W. and Cho, K.J. 2007. A comparison of the carotenoid accumulation in *Capsicum variet*ies that show different ripening colours: deletion of the capsanthin-capsorubin synthase gene is not a prerequisite for the formation of a yellow pepper. J. Exp. Bot., **58**: 3135-3144.

Hussein, I.A. and Mahdi, S.S. 2017. Genetic variation in Bell peppers (Capsicum annuum L.) 8th International Conference on Agricultural, Environment, Biology and Medical Sciences, 21-22 December, 2017, Dubai (UAE), *p.*123.

Ilodibia, C.V. , Ugwoke, 1C.E., Egboka, T.P., Akachukwu, E.E. , Chukwuma, U.M. and Aziagba, B.*O. 2015. Breed*ing P*epper for enhanced food nutrients. Asian J.Crop Sci.,* **7**: *214-218*

Kumar, P.R., Lal, S.K., Thakur, P.C. and Sharma, S.R. 2009. Inheritance of mature fruit colour in capsicum. Indian J. Plant Genet. Resour., **22**: 36-40.

Kumari, S. and Sharma, M. 2014. Screening bell pepper germplasm *for resistance* against Phytophthora leaf blight and fruit rot disease. Int. J. Farm Sci., **4**: 76-80.

Lafortune, D., Beramis, M., Daubeze, *A. M., Boissot, N. and Palloi*x, A. 2005. Partial resistance of pepper to bacterial wilt is oligogenic and stable under tropical conditions. Plant Dis., **89**: 501-506.

Manzur, J.P., *Fita, A., Prohe*ns, J., Rodríguez-Burruezo, A. 2015. Successful wide hybridization and introgression breeding in a diverse set of common peppers (Capsicum annuum) using different cultivated Ají (C. baccatum*) access*ions **a**s donor parents. PLoS ONE, **10**: e0144142.

Martins, K.C., Pereira, T.N.S., Souza, S.A.M., Rodrigues, R. and Junior, A.T.A. 2015. Crossability and evaluation of incompatibil*ity barriers in* crosses between Capsicum species. *Crop Breed.* Appl. Biotechnol., **15**: 139-14*5.*

McArdle, R.N. and Bouwkamp, J.C. 1983. Inheritance of several fruit characters in Capsicum annuum L. J. Hered., **74**: 125-127.

Munshi, A.D. 1999. Compendium of Lectures of Summer Schoo*l* on Advances in Breeding of Temperate and Tropical Vegetables for Biotic and Abiotic Stresses, ICAR, New Delhi,17th May to 6th June, 1999, pp. 133-140.

Negi, R., Thakur, S. and Sharma, P. 2018. Advances in *the breeding of bell pepper - a review. Inernt. J. Curr. Microbiol. Appl. Sci.,* **7:** *2272-2281.*

Peterson, P.A. 1958. *Cytoplasmically* inherited male steri[lit]y in Cap[sic]um. Amer. Nat., **92**: 111-119.

Pickersgill, B. 1997. Genetic resources and breeding of Capsicum spp. Euphytica., 96: 129-*133.*

Popovsky, S. and Paran, I. 2000. Molecular genetics of the Y locus in pepper: its relation to capsanthin - capsoru*bin synt*ha*se and fr*uit color. Theor. Appl. Genet., **101**: 80-89.

Riva, E.M., Rodrigues, R., Pereira, M.G., Sudré, C.P. and Karasawa, M. 2004. Inheritance of bacterial spot disease in Capsicum annuum L. Crop Breed. Appl. Biotechnol., **4**: 490-494.

Rodriguez-Uribe, L., Guzman, I., Rajapakse, W., Richins, R.D. and O'Connell, M.A. 2012. Carotenoid accumulation in orange-pigmented Capsicum annuum fruit, regulated at multiple level*s. J. Exp. Bot.*, **63**: 517-526.

Sharma, A., Kanwar, H.S., Singh, P.S. and Singh, S. 2020. Heterotic potential, potence ratio, combining ability and genetic control of quality and yield traits in b*ell pepper unde*r net-house conditions of NW Himalayas. *Agric. Res*., **9**: 526-535.

Sharma, A., Kumar, M. and Kumar, N. 2017. Genetic divergence for crop improvement in bell pepper. *J. Hill Agric.*, **8**: 31-34.

Sharma, O.P., Sharma, P.P. and Chowfla, S.C. 1989. Inheritance of resistance to potato virus Y in garden pepper (*Capsicum annuum* L.). *Euphytica.*, **42**: 31-33.

Sharma, V. K., Punetha, S. and Sharma, B.B. 2013. Heterosis studies for earliness, fruit yield and yield attributing traits in bell pepper. *African J. Agric. Res.*, **8**: 4088-4098.

Srivastava, A., Sharma, V. R., Dishri, M., Dikshit, A., Mangal, M. and Kalia, P. 2019. Inheritance of fruit attributes in chilli pepper. *Indian J. Hort.*, **76**: 86-93.

Thakur, M., Sood, S. and Gupta, N. 2019. Genetic analysis of quantitative and quality traits in bacterial wilt resistant genotypes of bell pepper under sub-temperate conditions of north-western Himalayas. *J. Pharmacog. Phytochem.*, SP**1** : 66-73.

Tran, N. H. and Byungsoo, K. 2010. Inheritance of resistance to bacterial wilt (*Ralstonia solanacearum*) in pepper (*Capsicum annuum* L.). *Hortic. Environ. Biotechnol.*, **51**: 431-439.

Vilarinho, L.B.O., Silva, D.J.H., Greene, A., Salazar, K.D., Alves, C., Eveleth, M., Nichols, B., Tehseen, S., Jr. Khoury, J.K., Johnson, J.V., Sargent, S.A. and Rathinasabapathi, B. 2015. Inheritance of fruit traits in *Capsicum annuum*: Heirloom cultivars as sources of quality parameters relating to pericarp shape, color, thickness, and total soluble solids. *J. Am. Soc. Hortic. Sci.*, **140**: 597–604.

Wei, J., Zheng, J., Yu, J., Zhao, D., Cheng, Y., Ruan, M., Ye, Q., Yao, Z., Wang, R., Zhou, G., Yang, Y., Li, Z., and Wan, H. 2019. Production and identification of interspecific hybrids between pepper (*Capsicum annuum* L.) and the wild relative (*Capsicum frutescens* L.). *J. Agr. Sci. Tech.*, **21**: 761-769.

22

Potato

Potato, *Solanum tuberosum* L. under the family Solanaceae (Chromosome No. 2n = 4x = 48) is the world's most important vegetable crop and the fourth most important food crop after rice, wheat and maize. They contain a wealth of health benefits that make them essential as a staple dietary item for much of the world's population. It is valuable both as a fresh product and as a processed product, and the current trend is for rapid growth in processed forms, mainly chips and French fries. It is now grown in nearly 150 countries. It is extensively cultivated in China, Russia, Ukarine, Poland, Ireland, Great Britain, Germany, Netherlands, France, Spain, South America, India and USA.

Origin, domestication and adaptation in India

Potato is a native of tropical South American region where it grows wild in nature and presents the widest diversity of forms like, tuber shape, size, colour, taste, etc. Cultivated diploid potatoes are believed to have originated in the high plateau region of Peru and Bolivia in South America, more specifically the Lake Titicaca basin in the Peru-Bolivian region (Simmonds, 1976). Cultivated tetraploid potatoes are also found in this region. Wild potato tubers are all bitter in taste and contain potentially toxic steroidal alkaloids. So, the first step in the evolution of the crop is the emergence of alkaloid free diploids which might have happened in the period 2000 to 5000 B.C. This New World crop was originally domesticated about 8000 years ago by the communities of hunters and gatherers in the Andes mountain range of South America. The Spaniards found cultivated potatoes in Peru at the time of their conquest beginning in 1536 (Hawkes, 1990).

Wild potato species grew in 16 countries, from the south-western United States to central coastal Chile and adjacent Argentina. High species richness occurs in northern Argentina, central Bolivia, central Ecuador, central Mexico, and south and north-central Peru.

Potatoes were domesticated there approximately 7,000–10,000 years ago, from a species in the *Solanum brevicaule* complex. In the Andes region of South America, where the species is indigenous, some close relatives of the potato are cultivated. Potatoes were introduced to Europe from the Americas in the second half of the 16th century by the Spanish. Today they are staple food in many parts of the world and an integral part of much of the world's food supply.

Potato was introduced in India from Europe in early 17th century probably by the Portuguese, about 40 years after its introduction in Europe. By the end of 19th century, potato cultivation spread throughout India. Since then, potato is grown widely almost throughout India. Potato cultivation in India started with the introduced varieties from Europe namely, Up-to Date, Ackersegan, Magnum Bonum, Royal Kidney, Great Scot, Darjeeling Red Round, etc. These long day adapted and long duration varieties suitable for European conditions did not perform well under short day and short growing seasons of the tropical and sub-tropical plains of India where high temperature prevails both at the time of planting and harvesting.

Crop botany and floral biology

- The subsection Potato is distinguished from all other subsections within the genus *Solanum* by the tubers borne on underground stolons, which are true stems, not roots.
- The series tuberosa is characterized by "imparipinnate or simple leaves, forked peduncle, rotate to pentagonal corolla and round berries".
- The species *S. tuberosum* is characterized by pedicel articulation placed in the middle third, short calyx lobes arranged regularly, leaves often slightly arched, leaflets always ovate to lanceolate, about twice as long as broad, tubers with well marked dormancy period.
- Potato is an annual herbaceous plant and growth habit varies between and within species.
- The plant has a rosette or semi-rosette habit; stem is erect in the early stage but becomes spreading and prostrate later on.
- Stems can be green to purple or mottled mixtures of these colours, and can have "wings" (ridges or stem tissue growing outward from the stem), or lack wings; compound leaves arise spirally on the aerial stems. In the angle between the stems and leaves (the leaf axil) branches or flower clusters (inflorescences) are produced; typically, at the base of the leaves, flaps of tissue called "pseudostipular leaflets" are produced. These are either sometimes not produced or fall off.
- The leaves are compound and alternate, irregularly odd pinnate; leaves of the cultivated and wild potatoes vary tremendously in relative degree of dissection and shape.
- The tuber is an enlarged portion of an underground stem or stolon; tuber eyes are the buds from which next season's growth will emerge; eyes are concentrated near the apical end of the tuber, with fewer near the stolon or basal end; eye number and distribution are characteristic of the variety.
- Tubers of various species vary considerably in skin and flesh colour, shape and size; in modern cultivars skin colours vary from white to yellow to red to tan, but most have white tubers (some cream to yellow); tuber shapes vary from round to elongate, and tuber skin surface varies from smooth to netted (russets).

- Potato is mainly propagated vegetatively by means of tubers; hybridization and selection, the major breeding method requires production of botanical seeds, i.e., "True Potato seeds".

True potato seed

True potato seed (TPS) refers to the matured ovule developed in the fruit (berry) of the potato plant as a result of sexual reproduction. In general, sexual reproduction is mainly employed for developing new varieties of potato by "Hybridization and selection" method of breeding. Depending on the genotype and environment, a single potato plant may bears 15 to 30 fruits and having 50-500 seeds in each fruit and 1.0 g TPS may contain 1500-1750 seeds. These TPS are used to raise plants for onward production of tuber in them for necessary evaluation.

- The potato plant developed from a seed tuber consists of one or more above ground shoots; each stem is terminated by an inflorescence but stem growth may continue from the lateral buds and the new branches will again terminate with an inflorescence and this process continues for several cycles.
- The inflorescence is single or compound cymes arising from extra-axillary position; number of flowers per inflorescence and per cyme depends on genotype, the environment and the position of inflorescence in the shoot system; inflorescence at higher position is characterized by fewer flowers than at lower positions.
- *Solanum tuberosum* ssp. *tuberosum* (common white potato) usually does not flower under short day condition; it flowers in comparatively low temperature regime under long day condition.
- Extending the 12-hour photoperiod with 4-hours incandescent light during winter promote flower production in the plain.
- Flowers are hermaphrodite, regular, pedicelled and bracteates; calyx is gamosepalous, five-lobed and tubular; corolla is gamopetalous, five lobed and rotate; corolla may be white, yellow, blue, purple or striped according to the variety.
- Colours and shapes of corolla also vary with the species; a group of species showing characteristics of the probable earliest ancestors of potato have corollas that are star shaped (stellate) and generally white, or white tinged with blue or purple; corolla colours are more typically blue to purple, and in the cultivated species sometimes pink; pink corolla colour is rarely seen in the wild species; corolla shapes also vary from pentagonal to more circular or rotate in outline.
- Stamens are five in number, epipetalous, alternate with corolla lobes and borne on the corolla tube; anthers erect and longer than filaments with apical dehiscence.
- Pistil consists of two carpels that form two-loculed ovary with single style and stigma; flowers produce no nectar ; style simple ending in capitate stigma.

- Ovary superior, bicarpellary, axile placentation; ovules are many.
- Fruits of most wild and cultivated species are globose, but some are conical, with intermediate shapes that are ovoid; fruits can be green or purple or almost white, or with mottled or dotted combinations of these colours; 1.2 to 1.9 cm in diameter; it has two compartments and contains numerous small seeds ranging in number from 50 to 500.

Reproductive behaviour

The crop is generally propagated vegetatively by means of tubers except for genetical and breeding purposes. Potato is a long-day crop and flowers and set fruits sparingly in certain favourable climates: cool temperature, moderate humidity and long day conditions of more than 15 hours light period which is found in the high hill situation in India during spring-summer season. For this reason, Central Potato Research Institute was shifted to high hill situation of Shimla, Himachal Pradesh in 1956 to start systematic breeding of potato under suitable climatic condition for induction of flowering and subsequent breeding of potato. Long days hasten potato flower primordial initiation and development. In some potato genotypes, extending 12 hours photoperiod with 4-hours incandescent light (1000 lux) sparingly promote flower production and berry set in the prolonged winter season in some plain areas of India.

Potato is basically self- pollinated crop, but some degree of cross-pollination occurs due to protruding stigma and self-incompatibility. Main obstacles to seed formation of *Solanum tuberosum* genotypes are high degree of male-sterility, gametophytic self-incompatibility, low pollen production and dropping of buds and flowers either before or after fertilization. Generally, all diploid species are gametophytic self-incompatible while tetraploids and hexaploids are comparatively self-compatible. However, the tetraploids and hexaploids produce very low number of viable pollens. Controlled selfing has been possible with the identification of a dominant gene, named *S*-locus inhibitor gene (*Sli*), which is sporophytically expressed to inhibit the function of *S*-alleles in pollen, and alters self-incompatible plants to self-compatible ones. The *Sli* gene was introduced into cultivated diploid potatoes by crossing (Birhman and Hosaka 2000) and continuous selfing produced two series of highly homozygous inbred lines, inbred series A and B (Phumichai *et al.* 2005).

Major pollinators are bumble bees (*Bombus* spp.) and honey bees (*Apis* species). Anthesis generally starts from 5.00 am which continues up to 8.00 a.m. Anther dehiscence occurs through the development of pores at the tip of the anthers just after anthesis. The stigma receptivity is at its maximum during anthesis although it becomes effectively receptive a day prior to anthesis. The flowers emasculated one to two days earlier should have completely blossomed under the paper bag and are ready for cross-pollination in the morning hours. Pollens remains viable from the day of anthesis to the next morning. Dried pollens can be stored in sealed vial at 0°C for 1-2 months. True potato seeds (TPS) are collected from ripe berries which can be stored safely at ordinary room temperature and under low humidity for 2-3 years.

Genetics of important characters

Wild potatoes grow from sea level to 4300 m, but typically occur between 2000 and 4000 m altitudes (Hijmans and Spooner, 2001). They grow in an amazing variety of habitats from very seasonal wet/dry climates (e.g., *S. bulbocastanum*) to high altitude grasslands (e.g., *S. colombianum*), to sea-level (e.g., *S. tuberosum*), while others grow in upland rainforests (e.g., *S. candolleanum*). The basic chromosome number (x) of *Solanum* is 12. Their ploidy level varies from 2x to 6x with 73% of these being diploids, 4% triploids, 15% tetraploids, 2% pentaploids and 6% hexaploids (Hawkes 1990). The main cultivated potato species *S. tuberosum* is a tetraploid (2n=4x=48). It has two subspecies, namely ssp. *tuberosum* adapted to long days and ssp. *andigena* Hawkes, adapted to short days. *Solanum. tuberosum* ssp. a*ndigena* is widely distributed in the Andean regions of Venezuela and northern Argentina, whereas *S. tuberosum* ssp. *tuberosum* occurs naturally in southern Chile only.

Table 1: Primitive cultivated species of potato

Species	Ploidy level	Distribution
S. phureja Juz.et. Buk	2n=2*x*=24	Venezuela to northern Bolivia between 2,000 to 3,500 m
S. chaucha Juz. et Buk	2n =3*x* =36	Northern Peru and central Bolivia between 3,000 to 3,900 m
S. stenotomum Juz. et Buk.	2n=2*x*=24	Northern Peru and central Bolivia between 3,000 to 3,900 m
S. goniocalyx Juz. et Buk.	2n=2*x*=24	Northern Peru and central Bolivia between 3,000 to 3,900 m
S. curtilobum Juz. et Buk.	2n =5*x* =60	Central Peru and northern Argentina at altitudes ranging from 3,800 to 4,200 m.
S. juzepczukii Buk.	2n =3*x* =36	Central Peru and northern Argentina at altitudes ranging from 3,800 to 4,200 m.
S. ajanhuiri Juz. et Buk.	2n=2*x*=24	Central Peru and northern Argentina at altitudes ranging from 3,800 to 4,200 m.

Source: Hawkes (1990)

However, this nine cultivated species taxonomy is not universally accepted rather, all the primitive and cultivated species of potatoes as landrace populations has been recognized as a single species, *S. tuberosum*, with the eight cultivar-groups: Ajanhuiri Group, Andigenum Group, Chaucha Group, Chilotanum Group, Curtilobum Group, Juzepczukii Group, Phureja Group, and Stenotomum Group (Huamán and Spooner, 2002).

Cultivated potato *S. tuberosum* L. is a tetraploid, displays tetrasomic inheritance and comprises highly heterozygous individuals, which suffer inbreeding depression on selfing. Hence, in potato genetics, it has not been possible to achieve the same degree of sophistication as in the genetical analysis of crosses between true-breeding inbred lines that display disomic inheritance (as found in tomato, brinjal, chilli of Solanaceae family). Genetics of some important characters is presented below.

Table 2: Genetic control of some important qualitative characters of potato

Character	Genetic control and gene symbol
Tuber shape	Continuous variation for the length to breadth ratio One major gene controls the round (compressed plus round) and long (oblong and long) shapes with round dominant over the long
Anthocyanin in the tuber skin	The gene *R* linked to locus I, controls the expression of anthocyanin in the tuber skin. The dominant allele *P* appeared epistatic to *R* in the tubers, but not in the flowers. There appeared to be three alleles at the *R* locus, *R*, R^{pw} and *r*, with R^{pw} giving pink tubers and white flowers. White: *R-P-ii, rr pp I-* Pink: $R^{pw}r$ *pp I-* Red: *R- pp I-* Purple: *rr P- I-, R- P- I-*
Anthocyanin in the flower	The gene *R* linked to locus I, controls the expression of anthocyanin in the tuber skin. The dominant allele *P* appeared epistatic to *R* in the tubers, but not in the flowers. White: *R- P- ff, rr pp F-*, $R^{pw}r$ *pp F-* Red: *R- pp F-* Blue: *rr P- F-* Purple: *R- P- F-*

Source: Gopal and Govindakrishnan (2012)

Most of the quantitative characters *viz*., tuber number per plant, tuber length, breadth and length to breadth ratio, TSS, starch content, tuber yield per plant, etc. are under polygenic control. Information on general and specific combining ability is usually specific to the particular population of genotypes under investigation and hence, must be sought afresh when new material is introduced into the breeding programme. Combining ability and regression analyses have both pointed to high heritability and the importance of general combining ability for crop maturity. Dormancy is the obligate period of non-sprouting after harvest, even under conditions favourable to sprouting. Simmonds (1964) showed that true seed and tuber dormancies were highly correlated in cultivated diploids and that the former was under polygenic control. The diploid species, *Solanum phureja* contributed dominant alleles for short dormancy (Freyre and Douches, 1994).

Breeding objectives

Modern cultivars of potato have been bred for relative uniformity in shape and colours with good processing quality. The followings are the main objectives of potato breeding.

- Breeding objectives vary according to usage like boiling, frying, mashing, cooking and processing in the forms of canned, frozen, dehydrated, chips and French fries.
- Wide adaptability to climatic conditions according to photoperiod and temperature, cultural practices, maturity period, etc.
- Tubers should be uniform and preferably be round to elongate in shape and tuber skin surface should be smooth.

- Skin colours may vary from white to yellow to red to tan however, white or cream-skinned tubers are mostly preferred.
- In processing type varieties, high proportion of tubers should be large (>85mm), white skinned, oblong with shallow eyes, white- or cream-fleshed and free from internal defects.
- In processing type varieties, tubers should be low in reducing sugars, phenols and glycoalkaloid content of less than 20 mg per 100 g fresh weight and high in dry matter content of more than 21% at harvest.
- Resistance to disease *viz.*, late blight (*Phytophthora infestans*), early blight (*Alternaria solani*), scab (*Streptomyces scabies*), charcoal rot (*Macrophomina phaseolina*), wart (*Synchytrium endobioticum*), bactretial wilt (*Ralstonia solanacearum*) and viruses (potato virus X, potato virus Y, etc.)
- Resistance to insect pest and nematodes *viz.*, potato tuber worm (*Pthorimaea operculella*), aphids (*Myzus persicae*), root knot nematode (*Meloidogyne incognita*)
- Resistance to abiotic stresses like, high temperature, frost, soil salinity, drought, etc.

Breeding methods and development of variety

Central Potato Research Institute was established at Patna, Bihar during August, 1949. The institute later shifted to Shimla, Himachal Pradesh in 1956 to start systematic breeding of potato suitable for Indian condition. The varieties bred at CPRI always have the prefix 'Kufri' denoting the name of the regional station of CPRI at Himachal Pradesh where segregating populations are produced by hybridization involving different parental lines. Earlier, some varieties were developed by clonal selection either from introduced varieties (e.g. Kufri Red from Darjeeling Red Round) or from developed line (e.g., Phulwa was developed by selecting seedling segregates from an exotic cultivar and subsequently Kufri Safed was developed from Phulwa by clonal selection). Thereafter, all the varieties have been bred through hybridization involving different exotic or indigenously bred varieties, breeding lines and different species particularly, *S. tuberosum* ssp. *andigena, Solanum demissum, Solanum curtilobum*, etc. mainly for introgression of resistance attributes. Hybrids which are generated by single cross, backcross or complex multi-cross show segregations as the cultivated tetraploid varieties or lines of *S. tuberosum* are highly heterozygous in nature. Clonal selection is practiced in F_1 and subsequent generations. Valuable elite plants are selected at any stage of breeding works and their properties are reinforced by subsequent vegetative propagation.

The varieties developed in India can conveniently be grouped according to the maturity period and their processing qualities.

- **Early maturing varieties (Economic yield within 80-100 days after planting):** Kufri Kuber, Kufri Chandramukhi, Kufri Alankar, Kufri Lauvkar, Kufri Jawahar, Kufri Ashoka, Kufri Khyati

- **Medium maturing varieties (Economic yield within 90-110 days after planting):** Kufri Kundan, Kufri Red, Kufri Jyoti, Kufri Sheetman, Kufri Muthu, Kufri Badshah, Kufri Bahar, Kufri Lalima, Kufri Sherpa , Kufri Megha, Kufri Sutlez, Kufri Pukhraj, Kufri Giriraj, Kufri Anand, Kufri Kanchan, Kufri Arun, Kufri Pushkar, Kufri Sailaja, Kufri Himalini, Kufri Sadabahar, Kufri Girdhari, Kufri Garima, Kufri Gaurav, Kufri Neelima, Kufri Lalit
- **Late Maturing Varieties (Economic yield within 110-140 days after planting):** Kufri Kumar, Kufri Safed, Kufri Neela, Kufri Sindhuri, Kufri Jeevan, Kufri Khasigaro, Kufri Naveen, Kufri Chamatkar, Kufri Dewa, Kufri Swarna
- **Processing quality varieties:** Kufri Chipsona 1, Kufri Chipsona 2, Kufri Chipsona-3, Kufri Chipsona-4, Kufri Surya, Kufri Himsona, Kufri Frysona

Germplasm resources and plant introduction

The gene bank at the International Potato Center (CIP), Lima, Peru maintains clonal and seed collections of potato, sweet potato, and Andean roots and tubers (ARTC's). The gene bank is maintained as a global public good under the International Treaty on Plant Genetic Resources for Food and Agriculture (ITPGRFA).

In India, the ICAR-CPRI (Central Potato Research Institute), Shimla, Himachal Pradesh presently holds a good collection of over 4,500 accessions, comprising of cultivated species (subsp. *tuberosum* and *andigena*), parental lines and wild/semi-wild tuber bearing *Solanum* species.

Potato cultivation in India started with the introduced varieties from Europe namely, Up-to Date, Ackersegan, Magnum Bonum, Royal Kidney, Great Scot, Darjeeling Red Round, etc. These long day adapted and long duration varieties suitable for European conditions did not perform well under short day and short growing seasons of the tropical and sub-tropical plains of India where high temperature prevails both at the time of planting and harvesting.

Variability and clonal selection

Earlier, some varieties were developed by clonal selection either from introduced varieties (e.g. Kufri Red from Darjeeling Red Round) or from developed line (e.g., Phulwa was developed by selecting seedling segregates from an exotic cultivar and subsequently Kufri Safed was developed from Phulwa by clonal selection). Very few old varieties have been developed by sole clonal selection.

Table 3: Varieties developed through clonal selection

Name of variety	Indigenous / Exotic material
Kufri Red	Clonal selection from Darjeeling Red Round
Kufri Safed	Clonal selection from Phulwa

Hybridization and clonal selection

Although potato reproduces sexually yet, in commercial cultivation it is propagated vegetatively through tubers. All the present-day varieties have been bred through hybridization involving different exotic or indigenously bred varieties, breeding lines and different species particularly, *S. tuberosum* ssp. *andigena*, *Solanum demissum*, *Solanum curtilobum*, etc. mainly for introgression of resistance attributes. Hybrids which are generated by single cross, backcross or complex multi-cross show segregations as the cultivated tetraploid varieties or lines of *S. tuberosum* are highly heterozygous in nature. Clonal selection is practiced in F_1 and subsequent generations. By this means, any genotype can be fixed and all its intra- and inter-locus interactions responsible for phenotypic expression can be maintained indefinitely by asexual propagation. Valuable elite plants are selected at any stage of breeding works and their properties are reinforced by subsequent vegetative propagation. The desirable clone can be multiplied for commercial cultivation even though initially it may be present as a single plant.

Exploitation of positive heterosis is by far the most satisfactory method of potato breeding aimed to enhance tuber yield. For improving tuber yield, there is a scope to exploit heterosis in crosses of Tuberosum with Neotuberosum, Andigena and Phureja/Stenotomum. Under long days, Tuberosum x long-day Andigena / Neotuberosum hybrids have shown yield advantages of 13-17 % over Tuberosum x Tuberosum hybrids. Under short day condition of sub-tropical environment, Tuberosum x Andigena hybrids had the yield advantage of 19-56 percent over Tuberosum x Tuberosum hybrids (Gopal *et al.* 2000). To determine whether in potatoes the tetraploid level is preferable to the diploid level, the diploid (2n = 2x = 24) *Solanum phureja* x dihaploid *Solanum tuberosum* hybrid parents and their vegetatively doubled, tetraploid (2n = 2x = 48) counterparts are intermated which resulted in two F_1 hybrid families at both levels of ploidy (Mendiburu and Peloquin, 1977). It appears that the extent of heterotic advantage is dependent on day length. The tetraploid offspring from Tuberosum (4*x*) x Tuberosum/Phureja hybrids (2*x*) have shown yield advantages of 27 % over their 4*x* parents and controls.

Selection of parents

Improved potato varieties are developed by inter-mating desired parental lines and selecting superior clones from the progeny. The success of any potato-breeding programme thus depends on the selection of superior parents and crosses, and selection of superior clones from the progeny. Generally, regardless of breeding objectives, choice of parent is most often based on its *de facto* phenotypic expression of the characters of interest. Normally breeders try to pick up pairs of clones/varieties for inter-mating based on complementary sets of characters with desirable level of expression. Breeders generally try to use improved clones or varieties as parents because these are taken to be of acceptable “horticultural types”. However, the choice of parents based merely on phenotypic expression of desired characters is often ineffective particularly when only non-additive gene action is significant for the trait of interest. In such a situation, it is essential to select the parents based on the performance of their progeny. However, where both additive and non-additive gene actions are important, which is generally the case for most of the characters, general combining ability can be a useful parameter for selecting parents. However, estimating the general combining ability of parents is labour intensive, time consuming and tedious due to problems of large scale sterility and cross-incompatibility in potatoes.

Hybridization

- **Flowering:** Potato is a long-day crop and requires cool climate, moderate humidity and long day conditions of more than 15 hours' light period for flowering and seed production. For this reason, Central Potato Research Institute was shifted to high hill condition of Shimla, Himachal Pradeshin in 1956 to start systematic breeding of potato under suitable climatic condition for induction of flowering and seed production of potato. In the prolong winter condition of some plain areas of India, more than 15 hours' day length through arrangement of extra illumination for 4-5 hour induces flowering and seed set. Artificial illumination is provided with 150-watt sodium vapour lamp for every 100 sq.m block area. Additional photoperiod is provided from 20 days after planting and continued till the completion of flowering. The plants, regenerated from the tuberlets flower after 35 to 40 days of planting.
- **Emasculation:** Potato flowers are hermaphrodite in sex form and therefore, emasculation is done in the flowers of the selected female parent mostly in the afternoon hours. The five stamens forming a cone surrounding the protruding stigma are removed carefully from the flower buds that will open next day. The emasculated buds are then bagged with butter paper bags ensuring no pollination from foreign pollens.
- **Flower collection for pollen extraction:** Freshly opened flowers are collected early in the morning from the plants of the male line. Anther lobes are separated from the flowers and are allowed to dry under room condition. Flowers from the selected fertile male parent can also be collected a day in advance, shade dried and pollens are extracted next day in the morning.
- **Pollen extraction:** Pollen grains are extracted by shaking the dried anthers through a nylon net sieve. Immediately after extraction, the pollens are taken in small plastic vial for doing cross-pollination.
- **Cross-pollination:** Cross-pollination is done by dipping the protruded and receptive stigma of the emasculated flowers of the female line into the pollen mass collected in the small plastic vial.
- **Harvesting of fruits:** The number of flowers per inflorescence, berry set, berry weight, number of seeds per berry and seed weight generally decrease from primary inflorescences to tertiary inflorescences and inflorescences on lateral stems. The berries are ready for harvest in about 45 – 50 days after pollination. The well matured berries are harvested and collected in plastic baskets. The berries are allowed to ripe fully at room temperature for about 10 – 12 days.
- **Seed extraction:** The ripe berries are crushed with the help of screw type juice extractor and allowed to ferment overnight. After fermentation, the pulp mixture is washed in running water through 20 mesh wire sieve. Water suspension along

with the mucilaginous substances is decanted. The precipitated seed lot at the bottom of the plastic tub is washed separately.

- **Seed treatment and storage:** The seeds are then dried at room temperature and subsequently in the sun. The seeds are packed in moisture proof containers when seed moisture is lowered to 4-5%, which enables the retention of viability in the seeds for longer duration at room temperature and low relative humidity. At 4°C TPS can be stored for many years. The dried seeds are stored in air tight container using silica gel at the rate of 100 g per kg of seed and kept at room temperature.
- **TPS dormancy:** The true potato seeds of the parental lines and hybrids have a dormancy period of 4-6 months depending upon the progenies. Dormancy in fresh TPS can be broken by immersing the seeds in the solution of 1,500 ppm gibberellic acid for 24 hours or by storing fresh TPS (with seed moisture between 4-5% and packed in airtight containers) at 30° C for 3-4 months depending on the genotypes.

Selection of crosses

It is important to decide in what combinations the selected parents are inter-mated so as to have good chance of finding superior clones among their progenies. Crosses may be selected based on the following considerations.

- ***Mid-parent and mid-self-value:*** If parental general combining ability effects are highly correlated with their phenotypes, the mid-parental values should provide a good prediction of the mean performance of the progeny. Mid-self-values (in analogy to mid-parent values) measured as mean performance of the self-progenies of parents involved in a cross could be another method of cross prediction. By evaluating self-progenies of "n" parents (one offspring population per parent), the value of n(n-1) cross combinations can be predicted. However, this can be used only for self-fertile parents. Mid-self-values are, in general, slightly better than mid-parent values for cross prediction. This may be because mid-self-values reflect the potential of the parents based on the performance of new genotypic combinations produced after selfing, whereas, mid-parent values are based on the phenotypic performance of the parents, which may or may not be reflected in their progeny.
- ***Specific combining ability:*** The departure of a progeny mean value from that expected on the basis of the general combining ability of parents is called specific combining ability. Low predictive values of mid-parent and mid-self-values suggest that specific combining ability effects are more important than general combining ability effects for characters like tuber yield and its components. In such situations, it is desirable to select the specific cross combinations based on their high specific combining ability for the traits of interest. However, this requires crossing the parents in a specific pattern as for estimating the general combining ability. When this is not possible progeny tests can be used.

- ***Progeny test:*** Number of efforts have been made to develop progeny tests which can be used to identify superior cross combinations from unplanned crosses. The objective is to identify the superior progenies based on a few genotypes per cross (may be 40-50) at the earliest opportunity so that larger populations of the identified crosses are raised to go for individual clone selection in the segregating progenies. The overriding criterion is that progeny means in initial generations must adequately reflect the subsequent performance of the clones when evaluated in advanced generations. Identification of superior crosses based on progeny means in seedling and/or early clonal generations have been found to be effective for a number of characters including resistance to potato cyst nematodes, late blight in foliage and in tubers, powdery scab and even for tuber yield and its components. Selection of superior crosses based on only progeny mean can be as effective as the one based on both progeny mean and within progeny standard deviation.
- ***Diploid progenies and use of 4x × 2x test crosses:*** The 4x × 2x test crosses as predictors of progeny performance provides good estimates for the actual values of the progenies.
- ***Heterosis in relation to genetic divergence of parents:*** Genetic divergence based on morphological data is of low to moderate effectiveness in predicting tuber yield and heterosis for tuber yield (Gopal and Mincoha, 1997a). This reflects the importance of genotype x environment interaction on the use of apparent genetic diversity determined on phenotypic basis of parents for predicting the progeny's performance.

Though all the above criteria can be used for identification of superior parents and crosses with varying degrees of effectiveness, the ideal strategy would be to identify the superior parents based on their general combining ability and inter-mating them in all feasible combinations to select a few top combinations based on progeny test. Further, the effectiveness of various genetic parameters in identification of superior parents and crosses vary with the genotypes, and also by the specific conditions under which these parameters are estimated. Thus, it is important to use them judiciously taking into consideration the extent of their relationship with each other and more importantly with the performance of the progenies generated from parents selected based on these parameters.

Selection of superior clones from the progeny

• *Selection in early generations*

In the Netherlands, on an average one potato variety is developed out of 1, 80,000 seedlings (Neele *et al.* 1988). Wenzel *et al.* (1983) calculated that 500, 000 to 2,000,000 seedlings must be evaluated in order to select a single new cultivar. This estimate is still consistent with the practices used for most modern breeding programmes. These large populations must quickly be reduced to a manageable number before statistically accurate and consistent tests are performed. Major problem in early generations of potato breeding programmes is the limitation in the availability of number of tubers per clone. Assessment is based on non-replicated plots of just a single plant per genotype in the seedling (F_1) generation and a few hills per genotype (3-12) in first (F_1C_1) and second clonal (F_1C_2) generations. The tubers used for planting various clones also often vary in size. These factors in addition to normal year and location effects make the identification of potentially desirable genotypes in early generations difficult. Clones are selected based on visual assessment of their commercial worth. This appraisal, known as general impression or breeder's preference, involves a complex series of visible traits that may be genetically unrelated. There is a general consensus that selection of individual clones in early generations on the basis of general impression results in a random reduction in number of genotypes. The worst problem appeared to be the inadvertent elimination of superior genotypes when selection was based on overall appearance. In most of the studies aimed at estimating repeatability of clones' performance in early generations, seedlings were raised in pots in greenhouses. In the plains of sub-tropics where assured irrigation is available, seedlings can be grown very well in the fields, and this is being followed in all potato breeding programmes in India. The following selection procedure has been suggested for an early and safe elimination of unproductive genotypes for tuber yield and its components, including tuber colour, shape, eye depth, etc.

- Seedlings of poor vigour may be rejected prior to transplanting in the field as they are likely to have low yields in later generations.
- Clones with undesirable tuber colour, tuber shape, eye depth and cracking may be rejected from the seedling stage onward, as these characters have a high repeatability over generations.
- Rejection should not be done on the basis of tuber yield, average tuber weight or number of tubers in the seedling generation.
- Negative selection (rejection of the poor phenotype) for tuber yield and tuber weight can be initiated from the first clonal generation onwards, whereas number of tubers can be considered for the rejection of undesirable types from the second clonal generation onward.
- In the subsequent generation, for simultaneous improvement in tuber yield and its components, average tuber weight should be used as a selection parameter after fixing a standard for the minimum number of tubers required in the selected types.

In order to further improve the efficiency of selection in early generations, the possibilities of using *in vitro* micro-propagation technique has been explored. *In vitro* selection was found to be highly effective for tuber colour, number of eyes, and moderately effective for average tuber weight, plant vigour and foliage senescence under specific conditions (Gopal *et al.*, 2004). Production of micro-tubers in sufficient numbers, and the subsequent production of normal seed tubers from them, will result in earlier large-scale trials thereby substantially reducing the period taken to develop a variety.

• *Selection in later generations*

In these generations, since enough tubers are available, clones can be evaluated in replicated trials of adequate plot size. Plot size and number of replications can be increased as material is multiplied. In these generations, depending on the size of plot, parts of the same plot may be harvested at different dates after planting to judge the potential of clones at different crop duration. Such trials cover the obvious areas of Distinctness, Uniformity and Stability (DUS) of the clone.

Table 4: Varieties developed through hybridization and clonal selection

Name of variety	Parentage
Kufri Kuber	(*S. curtilobum* × *S. tuberosum*) × *S. andigenum*
Kufri Chandramukhi	4485 × Kufri Kuber
Kufri Alankar	ON 2090 (Ekishirazu × Majestic) × Kennebec
Kufri Lauvkar	Sarkov × Adina
Kufri Jawahar	Kufri Neelamani × Kufri Jyoti
Kufri Ashoka	EM/C-1020 × Allerfrüheste Gelbe
Kufri Khyati	MS/82-638 × Kufri Pukhraj
Kufri Kundan	Ekishirazu × Katahdin
Kufri Jyoti	Cross 3069d(4) × 2814a (1) which was supplied by Dr. W. Black of the Scottish Plant Breeding Station, Edinburgh, UK
Kufri Sheetman	Craigs Defiance × Phulwa
Kufri Muthu	3046(1) × M109-3 which was supplied by Dr. W. Black of the Scottish Plant Breeding Station, Edinburgh, U.K
Kufri Badshah	Kufri Jyoti × Kufri Alankar
Kufri Bahar	Kufri Red × Gineke
Kufri Lalima	Kufri Red × AG 14 (Wis x 37)
Kufri Sherpa	Ultimus × Adina
Kufri Megha	SLB/K-37× SLB/Z-73
Kufri Sutlez	Kufri Bahar × Kufri Alankar
Kufri Pukhraj	Craigs Defiance × JEX/B-687
Kufri Giriraj	SLB/J132 × EX/A-680-16
Kufri Anand	Kufri Ashoka × PH/F-1430
Kufri Kanchan	SLB/Z-405(a) × Pimpernel
Kufri Arun	Kufri Lalima × MS/82–797
Kufri Pushkar	QB/A 9-120 × Spatz
Kufri Sailaja	Kufri Jyoti × EX/A 680-16
Kufri Himalini	I-1062 × Tollocan
Kufri Sadabahar	MS/81–145 and PH/F-1545
Kufri Girdhari	Kufri Megha × Bulk pollen of 10 genotypes
Kufri Garima	PH/F 1045 × MS/82-638
Kufri Gaurav	JE 812 × Kufri Jyoti
Kufri Neelima	E/79-15 × E/79-42
Kufri Lalit	85-P-670 × CP 3192
Kufri Kumar	Lumbri × Katahdin

Name of variety	Parentage
Kufri Neela	Kathadin × Shamrock
Kufri Sindhuri	Kufri Red × Kufri Kundan (Ekishirazu × Kathadin)
Kufri Jeevan	M 109-3 × Seedling 698-D
Kufri Khasigaro	Taborky × Seedling 698-D
Kufri Naveen	3070d (4) × Seedling 692-D
Kufri Chamatkar	Phulwa × Ekishirazu
Kufri Dewa	Craigs Defiance × Phulwa
Kufri Swarna	Kufri Jyoti × (VTn)2 62.33.3
Kufri Chipsona 1	MEX.750826 × MS/78-79
Kufri Chipsona 2	F-6 × QB/B 92-4
Kufri Chipsona-3	MP/91-86 × Kufri Chipsona-2
Kufri Chipsona-4	Atlantic × MP/92-35
Kufri Surya	Kufri Lauvkar × LT-1
Kufri Himsona	MP/92-35 × Kufri Chipsona-2
Kufri Frysona	MP/92-30 × MP/90-94
Kufri Sangam	Kufri Himsona x Kufri Pukhraj
Kufri Neelkanth	MS/89-1095 x CP3290
Kufri Lohit	Kufri Kanchan x Kufri Khyati

Multiplication of the variety

Seed plot technique is a method for the production of quality seed tubers through autumn planting in the subtropical plains of India. It has been proved that in such a seed-crop, the annual virus infiltration rate is less than 1% as against 20-25% observed in spring potatoes. The certified seed tuber must be raised from fresh nucleus / foundation seed of recommended variety from reliable sources. The basic principles of quality seed production entail following steps particularly for north-western/north eastern plains of India mainly because of high virus vector population during the main potato season in these areas.

- Clonal selection for development of healthy, virus-tested nucleus seed stocks for further multiplication and distribution as foundation/certified seed through National/State Govt. Agencies and start with disease free seed stocks.
- Selection of aphid (vector) free locations and/or periods of cropping and adoption of 'hygienic' cultural practices.
- Timely planting, good crop management: proper crop rotation, isolation and roguing.
- Proper plant-protection measures for the virus vector aphids and other important diseases especially late blight.
- Use of pre-sprouted tubers and adoption of relatively close spacing
- Judicious nutrient management to raise the crop
- Restricted irrigation after adequate bulking of tuber and withholding irrigation 10-15 days before haulms are cut.

- De-haulming the crop as soon as aphid vectors attain critical limits to prevent deposition of viruses in the plant.

Breeding for resistance to diseases

Potato is affected by many pathogens, among which *Phytophthora infestans*, causing late blight, is of the most economic importance. Different wild species, elite and natural accessions have also been used for late blight resistance breeding. Introgression of major resistance genes (*R*) from its wild species is the most common method of resistance breeding to late blight. Using late blight resistance genes targeting conservative effectors of *Phytophthora infestans* and the constructing gene pyramids may lead to durable, broad-spectrum resistance, which could be accelerated through genetic engineering. Therefore, new potato cultivars have been bred using resistance genes against *Phytophthora infestans* (*Rpi* genes) that originate from wild relatives of potato *viz*., *Solanum demissum*, *S. bulbocastanum*, *S. stoloniferum*, *S. chacoense*, etc. (Paluchowska *et al*, 2022).

However, *Rpi* gene introgression from wild relatives of the potato into commercial cultivars through crossing is time-consuming, especially in the case of species separated from potato with crossing barriers such as different endosperm balance numbers (EBNs). For example, the introgression of a single *Rpi* gene (*Rpi-blb2*) from the wild species *Solanum bulbocastanum* to potato cultivars Bionica and Toluca necessitated more than 45years (Haverkort *et al*. 2016). Compared with conventional breeding, genetic engineering techniques such as cisgenesis facilitate a faster introduction of *Rpi* genes into commercial cultivars (Ghislain *et al*. 2019). However, to avoid a rapid overcoming of resistance in newly engineered cultivars, introducing not one but several *Rpi* genes at a time has been proposed (Haverkort *et al*. 2016).

The varieties which show resistance to late blight are Kufri Jyoti, Kufri Muthu, Kufri Badshah, Kufri Sherpa, Kufri Megha, Kufri Giriraj, Kufri Swarna, Kufri Chipsona 1, Kufri Chipsona 2 and moderately resistant are Kufri Jawahar, Kufri Khyati, Kufri Kundan, Kufri Sheetman, Kufri Anand, Kufri Kanchan, Kufri Sailaja, Kufri Himalini, Kufri Sadabahar, Kufri Kumar, Kufri Neela, Kufri Khasigaro, Kufri Pukhraj, etc.

The varieties resistant or moderately resistant to different diseases include, Kufri Badshah (resistant to late blight, early blight and potato virus X and tolerant to frost); Kufri Jeevan (moderately resistant to late blight, immune to wart and moderately resistant to early blight and *Cercospora* leaf blight); Kufri Sindhuri (resistance to early blight and fairly resistant to frost); Kufri Kuber (immune to potato virus Y); Kufri Lalima (moderately resistant to early blight); Kufri Swarna (highly resistant to cyst nematode); Kufri Sherpa (moderately resistant to early blight, late blight, virus diseases and immune to wart); Kufri Naveen (resistant to late blight and immune to wart); etc.

Breeding for resistance to abiotic stress

The varieties which shows resistance or moderate resistance to frost include, Kufri Dewa, Kufri Sheetman and Kufri Sindhuri. The variety Kufri Sheetman is also resistant to drought.

Breeding for processing quality

Processing of potatoes in the factory involves several steps, including cleaning, sorting, peeling, slicing, blanching, frying and mashing. These steps are carefully designed to create different kinds of final products such as potato chips, French fries, and mashed potatoes. Different processing qualities includes high proportion of large size (>85mm) tubers of high specific gravity, white skinned, oblong tuber with shallow eyes, white- or cream coloured flesh and free from internal defects. The tubers should be low in reducing sugars, phenols and glycoalkaloid content of less than 20 mg per 100 g fresh weight and high in starch and dry matter content of more than 21% at harvest.

Starch is the primary storage carbohydrate in tubers accounting up to 70% of tuber dry matter. It is an important industrial raw material for more than 500 different commercial products. It is used in the food and beverage industries, as a thickener in sauces, custards, desserts, etc. The starch gives the potato its characteristic fluffiness.

Reducing sugars participate in 'Malliard reaction' during frying, thus producing dark coloured finished product. The 'Maillard reaction' is a chemical reaction between amino acids and reducing sugars to create melanoidins, the compounds which give browned food its distinctive flavour. For this reason, while potatoes having high reducing sugar content are processed or cooked at high temperature, the chemical reaction between amino acids and reducing sugars generate different flavours and brown colour making the product inferior in quality. Hence, suitability of potato tubers for development of processed products such as chips, French fries etc. is mainly determined by the low reducing sugars or glucose content in potato tubers.

The varieties, Kufri Chipsona 1, Kufri Chipsona 2, Kufri Chipsona-3, Kufri Chipsona-4 and Kufri Frysona having high dry matter and low reducing sugar contents in tuber are excellent processing quality varieties because of less browning of the processed products like, potato chips and French fries.

Breeding for nutrient and phytochemical content

Kufri Neelkanth is a purple skin-coloured, table purpose, medium maturing and biofortified potato variety developed at ICAR- Central Potato Research Institute, Shimla, Himachal Pradesh through hybridization (MS/89-1095 x CP3290) followed by clonal selection. It is rich in antioxidants (anthocyanins > 100µg/100g fresh weight and carotenoids~200 µg/100g fresh wt.) and the average yield of the hybrid is 35-38 t/ha. However, the first ever purple-fleshed variety is Kufri Jamunia, developed at developed at ICAR-CPRI. High level of anthocyanin content which are powerful antioxidants is found in its vibrant purple flesh. It is a high-yielding variety, with oblong tuber, shallow eyes and an average yield of 32-35 tons per hectare.

References

Birhman, R.K. and Hosaka, K. 2000. Production of inbred progenies of diploid potatoes using an S-locus inhibitor (*Sli*) gene, and their characterization. Genome · July 2000 DOI: 10.1139/gen-43-3-495 · Genome · July 2000 DOI: 10.1139/gen-43-3-495 · Source: PubMed

Freyre, R., and Douches, D.S. 1994. Development of a model for marker-assisted selection of specific gravity in diploid potato across environments. *Crop Sci.,* **34**: 1361-1368.

Ghislain, M., Byarugaba, A.A., Magembe, E., Njoroge, A., Rivera, C., Román, M.L., Tovar, J.C., Gamboa, S., Forbes, G.A., Kreuze, J.F., Barekye, A. and Kiggundu, A. 2019. Stacking three late blight resistance genes from wild species directly into African highland potato varieties confers complete field resistance to local blight races. *Plant Biotechnol. J.*, **17**:1119–1129.

Gopal, J and Minocha, J.L. 1997. Genetic divergence for cross prediction in potato. *Euphytica*, **103**: 269-275.

Gopal, J, Chahal, G.S. and Minocha, J.L. 2000. Progeny mean, heterosis and heterobeltiosis in *Solanum tuberosum* x *tuberosum* and *Solanum tuberosum* x *andigena* families under a short day sub – tropic environment. *Potato Res.,* **43**: 61-70.

Gopal, J, Chamail, A. and Sarkar, D. 2004. *In vitro* production of microtubers for conservation of potato germplasm: Effect of genotype, abscisic acid and sucrose. *In vitro Cellular Dev. Biol.-Plant*, **40:** 485-490.

Gopal, J. and Govindakrishnan, P.M. 2012. Potato. In: *Hand book of Vegetables* (Eds. Peter, K. V. and Hazra, P.), Studium Press LLC, Houston, Texas, USA, pp. 441-489

Haverkort, A.J, Boonekamp, P.M., Hutten, R., Jacobsen, E., Lotz, L.A.P., Kessel, G.J.T., Vossen, J.H. and Visser, R.G.F. 2016. Durable late blight resistance in potato through dynamic varieties obtained by cisgenesis: scientific and societal advances in the DuRPh project. *Potato Res.*, **59**:35–66.

Hawkes, J.G. 1990. *The Potato: Evolution, Biodiversity and genetic* Resources. Belhaen Press, Washington, DC, USA.

Hijmans, R. and Spooner, D.M. 2001. Geographic distribution of wild potato species. *Amer. J. Bot.,* **88**: 2101-2112.

Huaman, Z. and Spooner, D.M. 2002. Reclassifiction of landrace populatons of cultiated potato (*Solanum* section *Petota*). *Amer. J. Bot.,* **89**: 947-965.

Mendibrane, A.O. and Peloquin, S.G. 1977. The significance of 2N gametes in potato breeding *Theor. Appl. Genet.*, **49:** 53-61.

Neele, A.E.F., Barten, J.H.M. and Louwes, K.M. 1988. Effects of plot size and selection intensity on efficiency of selection in the first clonal generation of potato. *Euphytica* **39**: 27-35.

Paluchowska , P., Śliwka , J. Yin , Z. and 2022. Late blight resistance genes in potato breeding. *Planta* 255:127 https: // doi.org / 10.1007 / s00425-022-03910-6

Phumichai, C., Mori, M., Kobayashi, A., Kamijima, O., and Hosak, K. 2005. Toward the development of highly homozygous diploid potato lines using the self-compatibility controlling Sli gene. *Genome*· January 2006 DOI: 10.1139/g05-066.

Simmonds, N.W. 1964. The genetics of seed and tuber dormancy in the cultivated potatoes. *Heredity*, **19**: 489-504.

Simmonds, N.W. 1976. Potatoes. In: *Evolution of Crop Plants* (ed. Simmonds, N.W.). Longman, London, 339p.

Wenzel, G, Bapat, V.A. and Uhrig, H. 1983. New strategy to tackle breeding problems of potato. In: *Plant Cell Culture in Crop Improvement* (eds. Sen, S.K. and Giles, K.L.), Plenum Press, New York, pp.337-349.

23

Sweet Potato

Sweet potato, *Ipomoea batatas* under the family Convolvulaceae (Chromosome No. $2n = 6x = 90$) is an important tuber crop grown mostly in tropical and subtropical regions of Asia, Latin America, the Pacific islands, Papua New Guinea and Africa. India and China are the leading sweet potato growing countries in the world. In India, it is largely grown in Uttar Pradesh, Bihar, Odisha, Madhya Pradesh, Maharashtra, Tamil Nadu, Kerala, Karnataka and Andhra Pradesh. Its large, swollen, starchy, sweet-tasting, adventitious roots are used as tuberous vegetable. Orange-fleshed sweet potato, in particular, with high β carotene content has become a prominent example of the effectiveness of biofortified staple crops to combat vitamin A deficiency. Sweet potatoes are rich in dietary fibers and are also known for improving gut health and digestion. It also contains a high amount of phytosterol that has a protective effect on the digestive system. Sometimes, tender leaves and shoots are consumed as a green leafy vegetable. Sweet potato tubers are used in agro-industry for the preparation of starch, alcohol and animal feed.

Origin, domestication and adaptation in India

Sweet potato has probably been originated between the Yucatán Peninsula of Mexico and the mouth of the Orinoco River in Venezuela. Domestication of sweet potato occurred in either Central or South America. In Central America, domesticated sweet potatoes were present at least 5,000 years ago. It was grown in Peru in ancient times as evidenced from the excavations of caves. It was domesticated in tropical America and Polynesia. Perhaps, *Ipomoea triloba* and *Ipomoea trifida* are the two close wild relatives of sweet potato.

Sweet potato has been introduced in India through Portuguese and British trade and colonization.

Crop botany and floral biology

- Sweet potato is an herbaceous and perennial plant; however, it is grown as an annual plant by vegetative propagation using either storage roots or stem cuttings.
- Growth habit is predominantly prostrate with vine that expands rapidly and horizontally on the ground; types of growth habits are erect, semi erect, spreading and very spreading; stem is cylindrical and its length, like that of the internodes, depends on the growth habit of the cultivar and of the availability of

water in the soil; the erect cultivars are approximately 1 m long, while the very spreading vine can reach more than 5 m long; some cultivars have stems with twining characteristics.

- The internode length can vary from short to very long; plant can be thin or very thick depending on stem diameter; stem colour varies with the cultivar from green to totally pigmented with anthocyanins (red-purple color).
- The hairiness in the apical shoots, and in some cultivars also in the stems, varies from glabrous (without hairs) to very pubescent (with hairs).
- Leaves are simple and spirally arranged alternately on the stem in a pattern known as 2/5 phyllotaxy (5 leaves spirally arranged in 2 circles around the stem for any two leaves be located in the same vertical plane on the stem).
- Depending on the cultivar, the edge of the lamina can be entire, toothed or lobed; base of the lamina generally has two lobes that can be almost straight or rounded.
- Shape of the leaves can be rounded, reniform (kidney-shaped), cordate (heart-shaped), triangular, hastate (trilobular and spear-shaped with the two basal loves divergent), lobed and almost divided; lobed leaves differ in the degree of the cut, ranging from superficial to deeply lobed; number of lobes generally range from 3 to 7 and can be easily determined by counting the veins that go from the junction of the petiole up to the edge of the lamina; however, toothed leaves have minute lobes called teeth which could number from 1 to more than 9.
- Some cultivars show variation in leaf shape on the same plant; leaf colour can be green-yellowish, green or can have purple pigmentation in part or the entire leaf blade; some cultivars show purple young leaves and green mature leaves.
- Leaf size and the degree of hairiness vary according to the cultivar and environmental conditions.; hairs are glandular and generally are more numerous in the lower surface of the leaf. Leaf characters are very useful to differentiate cultivars; length of the petiole ranges from very short to very long.
- Petioles can be green or with purple pigmentation at the junction with the lamina and/or with the stem or throughout the petiole.
- Root system consists of fibrous roots that absorb nutrients and water and anchor the plant, and storage roots that are lateral roots which store photosynthetic products; the root system in plants produced by vegetative propagation starts with adventitious roots that develop into primary fibrous roots, which are branched into lateral roots.
- As the plant matures, pencil- thick roots that have some lignification are produced; other roots that have no lignification are fleshy and thicken a lot, are called storage roots; plants grown from true seed form a typical root with a central axle with lateral branches and later on, the central axle functions as a storage root.

- A tuberous root or storage root is a modified lateral root, enlarged to function as a storage organ; the enlarged area of the root-tuber or storage root can be produced at the end or middle of a root or involve the entire root; it is thus different in origin but similar in function and appearance to a stem tuber.
- Root tubers are perennating organs, which stores enough nutrients to sustain the organism during the unfavorable season and it develops into one or more new plants the following year; thickened roots that store nutrients over periods when the plant cannot actively grow, thus permitting survival from one year to the next ; massive enlargement of the secondary roots have the internal and external cell and tissue structures of a normal root; they produce adventitious roots and stems which again produce adventitious roots.
- In root-tubers, there are no nodes and internodes or reduced leaves (as found in stem tuber of potato); root tubers have one end called the proximal end, which is the end that was attached to the old plant; this end has crown tissue that produces buds which grow into new stems and foliage; the other end of the root tuber is called the distal end, and it normally produces unmodified roots; in stem tubers the order is reversed, with the distal end producing stems.
- A transverse section of the storage roots shows the protective periderm or skin, the cortex or cortical parenchyma that, depending on the cultivar, varies from very thin to very thick, the cambium ring where the latex vessels are found, and the medulla or central parenchyma.
- The amount of the latex formed depends on the maturity of the storage root, the cultivar, and the soil moisture during the growing period; the latex drops are produced when the storage roots are cut and they darken very quickly due to oxidation.
- The storage root skin colour maybe whitish, cream, yellow, orange, brown-orange, pink, red-purple, and very dark purple; the intensity of the colour depends on the environmental conditions where the plant is grown; the flesh colour can be white, cream, yellow or orange; however, some cultivars show red-purple pigmentation in the flesh in very few scattered spots, pigmented rings or, in some cases, throughout the entire flesh of the root.
- Sweet potato cultivars differ in their ability of producing flower; under normal conditions in the field, some cultivars do not flower, others produce very few flowers, and others flower profusely.
- The inflorescence is generally a cyme in which the peduncle is divided in two axillary peduncles; each one is further divided in two after the flower is produced (biparous cyme).
- In general, buds of first, second, and third order are developed; however, single flowers are also formed; the flower buds are joined to the pedundle through a very short stalk called pedicel.
- The colour of the flower bud pedicel, and peduncle varies from green to completely purple pigmented; flower is bisexual; besides the calyx and corolla, they contain androecium and gynoecium.

- The calyx consists of 5 sepals, 2 outer and 3 inner, that stay attached to the floral axle after the petals dry up and fall.
- The corolla consists of 5 petals that are fused forming a funnel, generally with lilac or pale purple and with reddish to purple throat (the inside of the tube); some cultivars produce white flowers.
- The androecium consists of five stamens with filaments that are covered with glandular hairs and that are partly fused to the corolla; length of the filaments is variable in relation to the position of the stigma; anthers are whitish, yellow or pink, with a longitudinal dehiscence.
- Pollen grains are spherical with the surface covered with very small glandular hairs; gynoecium consists of a pistil with a superior ovary, two carpels, and two locules that contain one or two ovules.
- Style is relatively short and ends in a broad stigma that is divided into two lobes that are covered with glandular hairs; at the base of the ovary there are basal yellow glands that contain insect-attracting nectar.
- The stigma is receptive early in the morning and the pollination is done mainly by bees.
- The fruit is a capsule, more or less spherical with a terminal tip, and can be pubescent or glabrous; the capsule turns brown when mature; each capsule contains one to four seeds that are slightly flattened on one side and convex on the other.
- Seed shape maybe irregular, slightly angular or rounded; the colour ranges from brown to black; and the size is approximately 3 mm; embryo and endosperm are protected by a thick, very hard and impermeable testa.
- Seed germination is difficult and requires scarification by mechanical abrasion or chemical treatment; seeds do not have dormancy period but maintain their viability for many years.

Reproductive behavior

Flowering itself is a major constraint in this crop. The crop does bloom occasionally and a number of elite clones do not flower. Generally, flowering occur best under short day conditions and begins about 1.5 months after the vines are planted/established and continues for ~3-5 months. The flowers open soon after daybreak and wither depending on prevailing environmental conditions (Jones, 1980). Flowers open longer on cool and cloudy days compared to hot and sunny days.

Flowers are bisexual but self-fertilization is rare because sweet potato possesses a homomorphic, sporophytic type of self-incompatibility that is not affected by environment, chemical treatment, and cannot be overcome using bud pollination. This system is likely controlled by a single *S*-locus with multiple alleles with a dominant-recessive relationship. Heterostyly also occurs in sweet potato; however, it does not appear to affect fertility. Besides self-incompatibility, sterility of different intensities is also found because of hexaploidy and accumulation of recessive sub-lethal mutations through long clonal propagation.

Cross incompatibility among different varieties can limit recombination and seed production and hinders targeted breeding especially when the parents with desirable traits of interest, such as disease resistance, drought tolerance, enhanced levels of protein, vitamins, macro- and micro-nutrients and dry matter are closely related and belong to the same incompatibility group. For this reason, large populations that contain non-related accessions with complementary traits need to be maintained.

Three types of cross compatibility exist depending on the success or failure of reciprocal crosses *viz.*, reciprocal fertility when fertility occurs in both the directions, reciprocal incompatibility when incompatibility occurs in both directions, and unilateral fertility/incompatibility when fertility occurs only when a genotype is used as the female but not when used as a male and *vice versa*. Higher fertilization success results for flowers pollinated in early than late hours of the day and additionally, insect pollination produce more seeds compared to hand pollination (Gasura *et al.*, 2010).

Table 1: Types of cross-compatibilities in sweet potato

Type of cross-compatibility	Expected outcome; A and B are different genotypes
Reciprocal fertility	A x B and B x A both produce seed
Reciprocal incompatibility	A x B and B x A both do not produce seeds
Unilateral fertility/incompatibility	A x B produces seed but B x A does not or B x A produces seed but A x B does not

Source: Barb *et al.* (2023)

Seeds mature 4 to 6 weeks after pollination. The capsule and pedicel both turn brown and dry and begin to shrivel when the seed is ready to harvest. Capsules that are left too long will dehisce (split open) so care must be taken to harvest the seeds before they are lost. Seeds can be hand-scarified by scratching a small notch in each seed coat with a sharp needle or a small, 3-cornered file or acid scarified with concentrated sulfuric acid for 40 minutes followed by a 5 to10-minute rinse under running water. It is highly heterozygous due to cross-pollination. Hence, the sexual seedlings are highly variable in nature. Sexual reproduction is only encouraged for breeding purpose.

Genetics of important characters

Of the approximately 50 genera and more than 1,000 species under the family, Convolvulaceae, *Ipomoea batatas* is the only crop plant of major importance. It is the only known natural hexaploid occurring in this family. It is natural hexaploid (2n = 6x = 90). The basic chromosome number is x = 15 in *Ipomoea* species and there are also diploids (2n = 2x = 30) and tetraploids (2n = 4x = 60).

The related species, water spinach or "kang kong" (*Ipomoea aquatica*) consumed as leafy green is diploid (2n = 2x = 30).

Genetic improvement of sweet potato is complicated by a number of factors *viz.*, self- and cross-incompatibility, highly heterozygous nature of individual clones and large number of chromosomes (2n = 6x = 90). It is a hexaploid with 2 non-homologous genomes (B1B1B2B2B2B2) with tetradisomic inheritance so the genetics of simple traits is more complex with up to six alleles per locus (Lebot, 2010). Cultivars of sweet potato are phenotypically homogenous (because they are clonally reproduced) and genetically heterozygous (because they are self-incompatible out-crossers). For this reason, in sweet potato genetics, it has not been possible to achieve the same degree of sophistication as in the genetical analysis of crosses between true-breeding inbred lines that display disomic inheritance of the diploid species. Furthermore, because it is a hexaploid, heterozygous genotypes occur in much larger frequencies making heterosis more important for quantitatively inherited traits (e.g., yield, yield stability, vigor after planting, etc.). However, it has been found that most of the characters in sweet potato are quantitatively in nature with high heritability for several characters.

Breeding objectives

Sweet potato is grown for human consumption (cooked vegetable), processed for starch, bioethanol, colourants /dyes, and for foliage for human and animal consumption. Important breeding objectives are mentioned below.

- High vine survival and vigour after planting especially during the periods of drought and vine with dense foliage
- Improved and stable yield, good size, shape, and uniformity of root tuber
- Yield stability and high dry matter content
- Orange-fleshed varieties with high β carotene content and improved nutrient, mineral and phytochemical composition *viz.*, starch, cellulose, sugars, protein, zinc, iron, anthocyanins, etc.
- Extended harvest for subsistence cropping, drought tolerance, with improved palatability and digestibility of tuber
- Improved storage, resistance to skinning, and lower acrylamide potential.
- Resistance to different diseases *viz.*, leaf petiole and stem blight (*Alternaria bataticola*), black rot (*Ceratocystis fimbriata*), Fusarium rot (*Fusarium solani*), bacterial rot (*Erwinia chrysanthemi*), sweet potato virus disease (SPVD) caused by sweet potato chlorotic stunt virus and sweet potato feathery mottle virus.
- Resistance to different insect-pests particularly sweet potato weevil (*Cylas formicarius*) and nematodes (*Meloidogyne incognita, M. javanica, etc.)*

Breeding methods and development of variety

Sweet potato has traditionally been viewed as a "poor person's crop" and it has attracted limited attention compared to other staple crops. However, during the last three decades, this perception has changed, and it is widely acknowledged that sweet potato has great potential to contribute to the alleviation of malnutrition and hunger in the developing world.

Presence of self-incompatibility and out-crossing behavior generate lot of natural variability. It provides ample scope for selection even among the collections of local cultivars. Wide variation exists among the cultivars for different morphological characters like shape and size, colour of skin and flesh of the tubers, flower colour, flesh texture (firm and soft flesh), depth and period of rooting, etc. This vegetatively propagated, root tuber crop is highly heterozygous in nature so wide variation can be created in the seedlings from the seeds of even natural settings. Some varieties have been developed in India through selection of elite chance seedlings and their subsequent clonal propagation. Intervarietal hybridization and selection of desirable clones in F_1 generation itself and further maintaining them clonally is the best breeding method of sweet potato. Wide array of F_1 population is raised through hybridization by different methods like, (a) open pollination for random intermating among various parental lines (b) polycross for random intermating among the tested male and female parents and (c) biparental cross for specific cross combination. A number of varieties have been developed by this method.

The cultivars grown in India can conveniently be grouped according to tuber skin colour *viz*, white skinned, brown skinned and red or pink skinned.

- **White/cream skinned Variety:** Pusa Safed, Rajendra Shakarkand 5, Sree Nandini, Sree Varun, Goutam
- **Brown skinned variety:** Cross-4, Samrat, V 35, Kalmegh, Sree Kanaka, Pusa Sunheri
- **Red/pink skinned varieties:** H-41, H-635, H-620, CO-1, CO-2, Pusa Lal, H-42, H-268, Sree Vardhini, Sree Bhadra, Sree Ratna, CO-3, Konkan Ashwini, Gouri, Varsha, Sankar, Sree Arun, Kalinga, Sourin, Kishan, Bidhan Jagannath

Germplasm resources and Plant introduction

Huge collections are maintained at International Potato Center in Peru (CIP), Crops Research Institute (CSIR) in Ghana, Mozambique Institute of Agricultural Research (IIAM), National Crops Resources Research Institute (NaCRRI) in Uganda, North Carolina State University (NCSU) in the USA, Louisiana State University (LSU) in the USA, Agricultural Research Council (ARC) in South Africa, Kenya Agricultural Research Institute (KARI), Agricultural Research Institute (ARI) in Tanzania, Zambia Agricultural Research Institute (ZARI), Department for Agricultural Research Services (DARS) in Malawi, Rwanda Agriculture Board (RAB), National Root Crops Research Institute (NRCRI) in Nigeria, Ethiopian Institute of Agricultural Research (EIAR) and Environment and Agricultural Research Institute (INERA) in Burkina Faso.

Huge exotic collections mainly from Japan, China, Taiwan, Nigeria, USA, Argentina, Sudan and Peru enriched the germplasm resource of India. ICAR-CTCRI (Central Tuber Crops Research Institute), Thiruvananthapuram, Kerala at present holds a good collection of sweet potato genotypes.

However, only one exotic variety, V 35 was directly released for cultivation in India.

Table 2: Exotic variety directly released for cultivation

Name of the variety	Country from where introduced
V 35	USA

Variability and clonal selection

Sweet potato is an open-pollinated and clonally propagated crop. Characteristics common to clonally propagated crops include:

- Strong and positive relationship between productivity/vigour and level of heterozygosity
- Selfing reduces vigour due to inbreeding depression
- Vigour/heterozygosity can be fixed and maintained for the life of the clone
- Polyploids /aneuploids can be maintained *via* clonal propagation
- Difficult to create a large quantity of clones from one plant
- Clonal propagules are typically bulky and difficult to store

Presence of self-incompatibility and out-crossing behaviour generate lot of natural variability. It provides ample scope for selection even among the collections of local cultivars. Wide variation exists among the cultivars for different morphological characters like shape and size, colour of skin and flesh of the tubers, flower colour, flesh texture (firm and soft flesh), depth and period of rooting, etc. A number of varieties *viz*., Pusa Safed , Pusa Sunheri , Kalmegh , Pusa Lal , etc. have been developed by clonal selection from both indigenous material and exotic germplasm resource collected from Japan, China, Taiwan, Nigeria, USA, Argentina, Sudan and Peru.

Table 3: Varieties developed through clonal selection

Name of variety	Indigenous / Exotic material
Pusa Safed	Exotic , China ((FA-17 -white)
Samrat	Indigenous
Kalmegh	Indigenous
Pusa Sunheri	Exotic (USA) This variety was originally bred through clonal selection from a double hybrid (Puerto Blanco × Wannop) × Australian Canner
CO-1	Indigenous
Pusa Lal	Exotic, Japanese variety, Norin
Sree Bhadra	Exotic, IITA, Nigeria
CO-3	Indigenous
Konkan Ashwini	Indigenous

Selection of seedling segregates

Cultivars of sweet potato are phenotypically homogenous (because they are clonally reproduced) and genetically heterozygous (because they are self-incompatible out-crossers). Hence, wide variation can be created in the open pollinated progeny from the seeds of even natural settings. Enhanced vigour, yield and other qualities of the elite chance seedlings from the open pollinated progenies can be fixed and maintained for the life through clonal propagation. Good number varieties *viz*., Sree Nandini , Sree Varun, Sree Vardhini, Bidhan Jagannath, etc. have been developed in India through this method of breeding.

Table 4: Varieties developed through selection of seedling segregates

Name of variety	Indigenous / Exotic material
Sree Nandini	Exotic, open pollinated progenies of the mother parent S-32, an introduced cultivar from Sudan
Sree Varun	Exotic, seedling selection from seeds from CIP, Peru
CO-2	Indigenous, Clonal selection from the seedlings of natural settings
Sree Vardhini	Open pollinated progeny of the mother parent "8-13"
Kalinga	Selection from open pollinated progeny
Bidhan Jagannath	Seedling segregate selection from the open population of an indigenous collection of Odisha

Hybridization and clonal selection

Sweet potato, the vegetatively propagated tuber crop is highly heterozygous in nature hence, the hybrids resulting from either biparental mating or random intermating show wide variability. Intervarietal hybridization and selection of desirable clones in the F_1 generation itself and further maintaining them clonally is the best breeding method of sweet potato. The basic breeding procedure by hybridization and selection includes the following steps:

- Framing of specific breeding objectives *viz*., yield stability, adaptation, taste pest and disease resistance, etc.
- Assemblage of germplasm (i.e., local varieties, wild species, cultivars developed in other parts of the world, etc.) to establish a breeding nursery
- Development of segregating populations *via* hybridization through biparental cross, polycross nursery or diallel crosses and/or induced mutagenesis
- Evaluation and selection of superior clones

Wide array of F_1 population is raised through hybridization by different methods like, (a) open pollination for random intermating among various parental lines (b) polycross for random intermating among the tested male and female parents and (c) biparental cross for specific cross combination.

Recurrent selection is a common method for improving a breeding population that captures additive variance and is most effective for traits with moderate to high heritability. The basic scheme consists of three phases:

- Selection of genotypes and their hybridization in an insect-pollinated polycross nursery or using controlled crosses
- Evaluation of progeny
- Selection of superior progeny and the creation of a new polycross nursery or controlled crosses with or without the best parents from the initial polycross nursery in Step 1.

Recurrent selection is a proven method for increasing the frequency of desired alleles and for creating a genetically broad-based population especially when new germplasm (i.e., new parents/genotypes and/or superior progeny from previous cycles) is added to the polycross nursery/controlled crossing block after each cycle.

Recurrent selection allows rapid increase in the percentage of minor and recessive alleles, but it requires accurate screening techniques to be successful. When considering which parents to choose for a polycross nursery or controlled crosses it is also important to consider the combining ability of individual clones. The combining ability of a clone can be estimated by comparing the performance of a clone versus the average performance of open-pollinated progeny harvested from that clone when it is grown in a polycross nursery or other population of mixed genotypes (i.e., seedling nursery, yield trial, etc.). If the average performance of the progeny is greater than the performance of the maternal clone, then it is assumed that a particular clone has good combining ability when crossed with other clones.

Often a combination of open-pollination (i.e., polycross nursery) and hand-pollination (i.e., controlled crosses) is used to ensure that specific cross combinations are made. Some of the advantages and disadvantages of using a polycross nursery *vs*. controlled crosses is mentioned below.

Table 5: Advantages and disadvantages of using an insect-pollinated polycross nursery *vs.* controlled crosses to generate progeny

Polycross nursery		**Controlled crosses**	
Advantages	**Disadvantages**	**Advantages**	**Disadvantages**
Requires less labour	Only the female parent is known so genetic advancement is based on only ½ of the genetic variation that is available	Requires more labour to make the crosses by hand	Both the parents are known so genetic advancement is based on all of the genetic variation that is available
Makes more seed/more cross combinations	Unbalanced contribution from some clones (poor seed set, low pollen production, limited and/or asynchronous flowering)	Less seed/less combination are created	Superior combinations are tracked and can be recreated

Source: Barb *et al* (2023)

Furthermore, because it is a hexaploid, heterozygous genotypes occur in much larger frequencies making heterosis more important for quantitatively inherited traits (e.g., yield, yield stability, vigor after planting, etc.). Heterosis manifested in the F_1 generation can be fixed indefinitely by vegetative propagation. At the same time, the hexaploid nature and self-incompatibility makes it extremely difficult to fix recessively inherited traits like resistance to certain viruses and some quality traits e.g., orange flesh.

Maximum number of varieties *viz*., Rajendra Shakarkand 5, Goutam, Cross-4, Sree Ratna, Sree Arun, etc. have been developed in India through selection of desirable seedlings from the widely variable hybrid seedlings of the biparental cross or polycross involving exotic or indigenous material in the F_1 generation. Enhanced vigour, yield and other qualities of the seedlings from the hybrid progenies are fixed and maintained for the life through clonal propagation.

Table 6: Varieties developed through hybridization followed by clonal selection

Name of variety	Parentage
Rajendra Shakarkand 5	Cross 4 × H 5 followed by clonal selections
Goutam	Polycross material with Dhenkanal local, a popular land race being the female parent
Cross-4	Controlled hybridization followed by clonal selection
Sree Kanaka	S-187 x H-633 followed by clonal selections
H-41	Norin (introduction from Japan) × a local type followed by clonal selection from the F_1 population
H-635	Controlled cross followed by clonal selection
H-620	Controlled cross followed by clonal selection
H-42	Vella Damph (indigenous cultivar) × Triumph (American variety)
H-268	H-74 × H-42 (cross between two hybrids)
Sree Ratna	S187 × Sree Vardhini followed by clonal selections
Gouri	H-219 x H-42 followed by clonal selections
Varsha	Clonal selections from a hybrid
Sankar	H-219 x S-73 followed by clonal selections
Sree Arun	Polycross followed by recurrent selection
Sourin	Polycross with the female parent "accession no. 1162" followed by clonal selection
Kishan	Polycross with the female parent "accession no. 1016" followed by clonal selection

Breeding for disease resistance

Sweet potato is attacked by number of fungal and bacterial diseases *viz.*, leaf petiole and stem blight (*Alternaria bataticola*), black rot (*Ceratocystis fimbriata*), Fusarium rot (*Fusarium solani*), bacterial rot (*Erwinia chrysanthemi*), etc. However, breeding for resistance against Sweet potato virus disease (SPVD) caused by the dual infection and synergistic interaction of sweet potato chlorotic stunt virus and sweet potato feathery mottle virus remains the most important component to improve yield and reduce the impact of SPVD. Breeding for resistant cultivars is indispensable to control the disease for ascertained food security and incomes of rural and marginalized communities depending on this subsector. However, reduced flowering and fertility, self-or cross-incompatibility are the major challenges of conventional breeding in sweet potato breeding. Conventional breeding in combination with non-conventional techniques such as biotechnology, mutation breeding and genetic engineering have significant role in developing new sweet potato cultivars that are high yielding and resistant to SPVD.

Breeding for resistance to insect pests

Production of sweet potato is highly affected by sweet potato weevil (*Cylas formicarius*) which tunnel and feed on the vines and storage roots thereby reducing the quality and yield of the crop. Sweet potato genotypes differed significantly in resistance determined by the percentage of roots damaged by the sweet potato weevil. Moderate to high levels of resistance to this soil insect pests exhibited by many of these traditional and heirloom cultivars may provide useful sources of germplasm for sweet potato breeding programmes (Jackson and Harrison Jr., 2013). Several sweet potato weevil tolerant varieties identified in India are H-42, CO-1, Rajendra Shakarkand 5, etc.

Breeding for quality

Quality can be improved by exploiting heterosis through creating two genetically distinct germplasm pools and selecting and recombining within each pool based on the general combining ability of clones that are crossed with the other germplasm pool. Work at the international Potato Center (CIP), Lima, Peru is focused on creating high yielding, orange-fleshed varieties using this approach. For this experiment, two populations were created *viz.*, "Jewel" population and the "Zapallo" population. The "Jewel" population consists of orange-fleshed clones released prior to 2004 and includes mostly North American varieties including Jewel and Resisto. The "Zapallo" population was created using a factorial crossing design using 8 male parents (Jonathan, Zapallo, Huambachero, Tanzania, Yurimaguas, Wagabolige, Xushu18, and Ninshu1) and 200 orange fleshed females. More than 200 families were then produced from cross combinations involving 49 clones from the Jewel population crossed with 31 clones from the Zapallo population. Some cross combination showing high yield and high β carotene content could be developed. Several orange to deep orange–fleshed varieties having 4.5-5.5 mg β carotene content /100g fresh *viz.*, Varsha, Gouri, Sree Ratna, Pusa Sunheri, Sree Kanaka have been developed in India by clonal selection and hybridization followed by clonal selection.

References

Barb, J., Singh, A.K., Singh, A. and Mahama, A. A. 2023. Sweet potato Breeding. In: *Crop Improvement* (Eds. Suza, W. P. and Lamkey, K. R.), Iowa State University, USA.

Gasura, E., Mashingaidze, A.B. and Mukasa, S.B. 2010. Genetic variability for tuber yield, quality, and virus disease complex traits in Uganda sweet potato germplasm. *Afr. Crop Sci. J.*, 16:147-160.

Jackson, D.M. and Harrison Jr., H.F. 2013. Insect resistance in traditional and heirloom sweet potato varieties. *J. Econ. Entomol.*, **106**:1456-62

Jones, A. 1980. Sweet potato. *Amer. Soc. Agron.*, 46:645-655.

Lebot, V. 2010. Sweet potato. In: *Root and tuber crops* (Ed. Bradshaw, J.E.) Vol 7. Springer Science & Business Media.

24

Cauliflower and Cabbage

Cauliflower

Cauliflower, *Brassica oleracea* L var. *botrytis* under the family Brassicaceae (Cruciferae) having chromosome No. 2n = 2x =18 is the most popular vegetable among the cole crops, particularly in Asia. It is grown in all continents of the world, of which Asia is the leading one followed by Europe and covers small acerages in North America, South America, Africa, Australia and New Zealand. India and China together have about 60 % of total cauliflower area of the world. Cauliflower is used in different curry and fry preparations with potato, garden pea, sweet pepper and other vegetables. It contains high quality proteins and good in stability of vitamin C after cooking. Frozen products are also prepared from cauliflower. It is rich in minerals such as potassium, sodium, iron, phosphorus, calcium, magnesium, etc.

Origin, domestication and adaptation in India

The cole crops have originated from the wild diploid species *Brassica oleracea* var. *sylvestris* mainly by mutation, human selection and adaptation. This wild species, a leafy kale-like plant which occurs in the northern shore of the Mediterranean Sea and along the lime cliffs in north-eastern England and North Sea is the progenitor of the cultivated kale and leafy non-heading cabbage. Probably kale was the first cole crop to have selected and adapted by man. From this leafy kale-like plant (*Brassica oleracea* var. *sylvestris*), other cole crops *viz*., kale, collard, broccoli, cauliflower and cabbage have been descended.

Brassica cretica ssp. nivea is the primitive form of broccoli from which cultivated broccoli evolved several thousand years ago. Natural crossing between the primitive form of broccoli and the biennial wild types / species have resulted evolution of many forms of cultivated *Brassica oleracea viz*., cultivated broccoli and its various forms and cauliflower.

Cauliflower is comparatively of a later origin, about 500 years ago, probably by introgression within broccoli gene pool having close genetic affinity. In the Middle Ages, early forms of cauliflower were associated with the island of Cyprus, with the 12th- and 13th-century Arab botanists Ibn al-'Awwam and Ibn al-Baitar claiming its origin to be Cyprus. It was probably introduced in Italy by the Genoese from Levant (large area in the Eastern Mediterranean region of West Asia and core territory of the political term Middle East) or Cyprus around 1490. There was extensive trade of cauliflower seeds in Western Europe from Cyprus. Italy is the centre of genetic diversity of cauliflower and broccoli where several landraces of these two crops occur even today. At the beginning of 17th century, cauliflower reached Germany, France and England.

Cauliflower was introduced in India from England in 1822 by Dr. Jemson, a botanist from Royal Botanic garden, Kew, UK who was in-charge of the Company Bagh, Saharanpur, United Provinces (now Uttar Pradesh). The Royal Agri-Horticultural Society, Calcutta also introduced cauliflower, tomato and other English vegetables in 1824 from South Africa. The Cornish group of cauliflower of England was the first to have been introduced, which got adapted to higher temperature and humid conditions.

During a period of more than 100 years, introduced cauliflower varieties underwent involuntary selection by the local growers when seed production was attempted for early maturity and adaptability to hot and humid weather which has resulted in the development of a separate group of cauliflower: Indian or Tropical cauliflower. Besides being heat tolerant, the Indian cauliflowers have genes for resistance to several diseases *viz*., downy mildew (caused by the fungus *Peronospora parasitica*), black rot (caused by the bacteria *Xanthomonas campestris* pv. *campestris*) and curd and inflorescence blight (caused by the fungus *Alternaria brassicicola*).

Crop botany and floral biology

- The pattern of development has been modified in various ways during the domestication of different cole vegetable crops; the leaves on the main stem have a different shape in both kales (wrinkled and curly) and cabbages (packed into a head).
- In Brussels sprouts, the many secondary shoots arising along the main stem generate small heads of leaves; in knolkhol or kohlrabi, the region of the main stem that produces leaves is itself grossly swollen.
- In both cauliflower and broccoli, the flowering stem is modified; in broccoli, the stems are much shorter, and many buds are densely packed.
- In cauliflowers, flower buds are not usually produced at all; instead, the inflorescence meristem continuously generates replicas of itself in a spiral on its flanks; each new meristem can in turn produce more, and so on until the tenth order of branching and beyond may be generated; this results in a closely packed, geometric cluster of undifferentiated inflorescence meristems (called 'curd') that has a novel form and texture not present in wildtype plant, *Brassica oleracea.*
- The edible part of cauliflower is 'curd' which is botanically the pre-floral fleshy apical meristem; morphologically, the curd is made up of numerous divided hypertrophic branches which terminate the main stem of the plant; it is highly suppressed and extremely ramified and hypertrophied flower stalk.
- Late and mid-late (group III and IV) cauliflowers require vernalization for induction of flowering and hilly regions of India (Kulu valley, Kashmir vally

for group III and IV varieties and Nilgiri hills for group III varieties) are suitable for breeding and seed production of those cauliflower types.

- Early and mid season cauliflower varieties (group I and II) does not require vernalization hence, breeding and seed production of these varieties can be done in the northern and eastern plains of India.
- The inflorescence develops from the cauliflower curd; the inflorescence is a terminal raceme which develop on the main stem and corymbose towards the top.
- The inflorescence is dwarfer and more umbrella-shaped than that of cabbage; there is absence of central main stem above the point where branching begins.
- Flowers are hermaphrodite, singular and complete; sepals are 4 and free.
- Corolla bright yellow in colour with four petals, standing opposite to each other in a square cross called cruciform corolla.
- Stamens are six, tetradynamous *i.e.* two are short having nectarines at their base and four are long.
- Carpels are two, syncarpous which form a superior ovary with a false septum and two rows of campylotropous ovules with parietal placentation.
- Fruits are special type of bicarpellary pod called the siliqua, which are glabrous, 4-5 mm wide and sometimes over 10 cm long, with two rows of seeds lying along the edges of the septum; the two valves dehisce when the pod is fully mature; a siliqua contains 8-15 seeds depending on the genotype and effectiveness of insect pollination.
- Seeds are small, globular, smooth and dark brown in colour, non-endospermic with large embryo.
- Cotyledons are folded; average 100 seed weight is 0.28 gram.

Reproductive behaviour

The tropical or Indian cauliflower types (Group I and II) behave like annual and bolt in response to cool temperature followed by long day condition. They produce flowers and set seeds freely in the plains of India. Breeding and seed production of this cauliflower group is done in Punjab, Haryana, Uttarakhand, Uttar Pradesh and Bihar where temperature my drop down to even 1-2°C in some period of winter. However, breeding and seed production of the earliest group of cauliflower (Group IA, Kanwari group) can even be done in the western part of West Bengal where winter is comparatively prolonged. The temperate or Snowball types (group IV) behave like biennial crop which require 6-8 weeks of vernalization at mean temperature range of 4° to 8°C after they have passed the initial curding phase for the production of seed stalks. Breeding and seed production of these types of cauliflower can only be done in the high hill situations of Himachal Pradesh and Kashmir.

Cauliflower bear hermaphrodite flowers but varying levels of cross fertilization occurs even upto 70 % in different open pollinated varieties due to manifestation of self-incompatibility of different intensities. Self-incompatibilty system in cauliflower is sporophytic in nature and governed by multiallelic single gene (*S* gene) with generally dominant in effect. Intensities of self-incompatibility depend on the presence of *S* allele in the genotype. In the early bud stage, 2-3 days before anthesis when the stigma has already become receptive due to protogyny, self-fertilization can be done by self-pollination with the pollens from other opened flowers of the same plant. The *S* alleles are active in the stigma in flowering stage but remain inactive in the bud stage, i.e., 2-4 days before anthesis. Heat tolerant tropical cauliflowers show significantly high level of self-incompatibility. The European annual or Snowball types show very weak self-compatibility. In cauliflower, cross pollination is generally accomplished by bees, more particularly honey bees. Bees and other insect pollinators frequently visit the flowers of cauliflower because of the presence of nectar in the flowers.

Genic male sterility controlled by single recessive gene and duplicate dominant genes with cumulative effect have been found in cauliflower however, such sterility is not generally utilized in hybrid seed production. The male sterility transgene *barnase* from *Bacillus amyloliquifaciens* which destroys the tapetum layer in the anther sac has been utilized to develop stable cytoplasmic male sterile line in cauliflower. Hybrids of cauliflower have been developed mainly in Japan, Korea, Taiwan and USA utilizing self-incompatibility and cytoplasmic male sterility mechanisms.

Anthesis starts early in the morning and continues up to 11.30 am depending on the prevailing weather conditions. Dehiscence of anther is longitudinal and it generally takes place few hours after anthesis. The stigma becomes receptive at least two days before anthesis although, receptivity is the maximum at anthesis and it continues at least one day after.

Genetics of important characters

The cauliflower *Brassica oleracea* L var. *botrytis* is an important species of the genus Brasica (tribe Brassiceae). *Brassica oleracea* is diploid, 2n = 2x =18, with *C* genome. The molecular genetic maps constructed for *Brassica* indicated that within a species there is almost complete collinearity, except for small inversions that differ among some *Brassica oleracea* genotypes.

Several plant characters in cauliflower have simple inheritance *viz.*, plant type, leaf characters, stalk length, curd colour, flower colour, riciness of curd.

Table 1: Important simple inherited characters in cauliflower

Plant character	**Inheritance pattern**
Leaf characters	
a) Petiolate	Single gene; petiolate dominant (*PET*)
b) Leaf apex	Single gene; round apex dominant (*RO*)
c) Leaf margin	Single gene; wavy dominant (*WY*) dominant
Leaf arrangement around curd	Two genes; semi-blanched dominant (Bl_1, Bl_2)
Glossy leaf	Single gene; recessive (*gl*)
Stalk length	Single gene; long stalk dominant

Plant character	Inheritance pattern
Curd colour	
a) Orange	Single gene, dominant
b) Green	Two genes
Retentive white	Three genes control peroxidase activity and thereby bleaching
Flower colour (white, yellow, cream)	Two independent dominant genes showing epistatic interaction with few modifiers

Source: Swarup and Chatterjee (1998)

Genic male sterility

Genic male sterility controlled by a single recessive gene, *ms* and duplicate dominant genes with cumulative effect have been found in cauliflower. In broccoli, both single recessive and dominant gene governs male sterility.

Transgenic male sterility

The transgenes, *barnese*, *bar* and *barstar* have been utilized to develop transgenic male sterility in cauliflower. The male sterility transgene *barnese* protein comes from *Bacillus amyloliquefaciens*, in which ribonucleases destroy the tapetum cell layer of the anther and thus produce stable male sterile plant. The male sterility gene is linked with the herbicide-resistance gene, *bar*. The restorer line for the *barnese* male sterile lines can be obtained by expressing the gene coding for *barstar*.

Cytoplasmic male sterility

The *Ogura* sterile cytoplasm of *Raphanus* is the most important source of sterile cytoplasm for cauliflower, broccoli and cabbage. Protoplasm fusion technique has been employed to transfer the sterile cytoplasm from *Ogura* radish. The CMS male sterility is being used in hybrid seed production of cauliflower.

Self-incompatibility

Most of heat tolerant tropical Indian cauliflower varieties express self-incompatibility and comparatively high inbreeding depression. The late or Group IV or Snaoball cauliflower which are basically European annual types show very low self-incompatibility and comparatively low inbreeding depression. In cauliflower, self-incompatibility system is sporophytic and governed by single multiallelic gene (*S* gene) with mostly dominance relationships.

Quantitative characters

In cauliflower, important quantitative characters governed by polygenes include curd diameter, curd compactness, curd maturity, curd weight, curd depth, curd size index, curd shape and curd yield. Loose curds and bracted curds which are greatly influenced by environment are perhaps governed by polygenes.

Table 2: Associated genes/QTLs and nature of gene action for some important characters in cauliflower

Trait	Genes/QTLs	Inheritance	Reference (s)
Curd weight	Polygenic	Additive and dominance gene action	Jyoti and Vashistha (1986)
Curd compactness	Polygenic	Dominance and additive gene action	Lal *et al.* (1977)
Earliness	Polygenic	Additive and dominant gene action	Chand (1980)
Purple curd	*Pr*	Single semi-dominant gene	Singh *et al.* (2020)
Large size and biomass	*CDAG1* (curd development associated gene)	-	Li *et al.* (2017)
Stalk length	*(qSL.C6–1, qSL. C6–2*	Quantitative	Zhao *et al.* (2020)
Curd solidity	*qCS.C6–1 and qCS.C6–2*	Quantitative	Zhao *et al.* (2020)
Reproductive development	*BoFLC2*	-	Ridge *et al.* (2015)
Riceyness	*Bo4g024850*	Monogenic	Zhao *et al.* (2020)
Orange color curd mutation	*Or*	Single dominant	Li *et al.* (2006)
Arrest of curding, control leafy floral differentiation	*BoCAL-a and BoAP-1*	-	Labate *et al.* (2006)
Black rot	*Xca1bo*	Monogenic dominant	Saha *et al.* (2014)

Breeding objectives

The following characters are considered important in the improvement programme of cauliflower.

- Plant type including frame (erect, semi-erect and spreading); in Indian cauliflower, semi-erect plant frame with medium stalk length is desirable as in this plant type, the uppermost leaves cover the curd which helps in blanching and whiteness of the curd.
- Curd shape (hemispherical shape is mostly preferred)
- High nutrient content of curd (particularly β carotene, vitamin C and anthocyanin content (in purple couliflower)
- Curd of good diameter, depth, compactness and weight
- High curd yield per unit area
- Curd quality (colour, maturity, compactness, weight, flavour, taste, uniformity)
- Curd favour is an important quality character to be taken into consideration; two volatile chemical compounds *viz.*, nitriles and isothiocyanates produced on

hydrolysis of glucosinolates by the enzyme myrosinase, impart flavour to the curd.

- Desired level of glucosinolates in the curd because it shows anticarcinogenic property and varietal difference exists for this sulphur-containing chemical.
- Resistance to common diseases *viz.*, downy mildew (*Peronospora parasitica*), black rot (*Xanthomonas campestris* pv. *campestris*) and curd and inflorescence blight (*Alternaria brassicicola*).
- Tolerance to high temperature for initiation and development of curd (particularly for tropical or Indian cauliflower group)

Breeding methods and development of variety

At present two separate groups of cauliflower are grown in India *viz.*, Indian or tropical cauliflower and Temperate annual or Snowball group. The Indian or tropical cauliflowers are adaptive to hot weather and high rainfall condition. They generally form curd and mature at average daily temperature around 20°C and above. The typical Indian cauliflowers were developed as the result of inter-crossings between the introduced cultivars of Cornish and other European types (Roscoff, Italian, Angiers, etc.) and subsequent selections by local growers of North Indian plains for early maturity and wider adaptability to hot and humid weather conditions during long period of more than 100 years. General characters of Indian cauliflowers are long stalked plant with broadly wavy and loosely arranged leaves producing highly flavoured, flat to semi-spherical, uneven and some what loose and yellow to cream coloured curds. Most of the curd and leaf characters are typical of Cornish, while some leaf and curd characters can also be traced to Roscoff and Italian group of European cauliflower. Cultivars of maturity group I and II belong to typical Indian cauliflower group.

The Temperate annual or Snowball group of cauliflower are not adaptive to hot weather and high temperature condition. They generally form curd within the temperature range of 10°C and 16°C. This temperate annual cauliflower group show similarity with Erfurt and Alpha types of cauliflower originated in Germany and the Netherlands (European summer group) and Italian types (autumn group). Different varieties under this group have been developed in India utilizing mainly introduced cultivars. General characters of snowball group are dwarf plant with dark green, wavy, short, erect (sometimes spreading) leaves producing solid, very deep, knobbly, bright white, less flavoured and high quality curds which are generally less sensitive to riceyness. Cultivars of maturity group IV (late) come under this group in strict sense. Cultivars of maturity group III (mid-late) though are inclined to this group yet do not show all the group characters.

From 1940 onwards several open pollinated varieties under Indian cauliflower group have been developed utilizing different locally adapted cultivars employing different population improvement programmes *viz.*, mass selection, line breeding, family breeding, inbreeding, selfing and massing, recurrent selection (simple recurrent selection and reciprocal recurrent selection), disruptive selection, development of synthetics, heterosis breeding, etc.

Cauliflower varieties grown in India are described under four maturity groups according to their temperature requirement and time of curd maturity in North Indian condition.

1. **Early group (Maturity group I):** Curd maturity September-November; temperature requirement for curd initiation and proper curd development 20-27°C. Several locally adaptable cultivars of Uttar Pradesh and Bihar in particular which are all farmers' selections over the years are available for cultivation in the name of Early Patna, Hajipur Local, Kunwari, Katki, etc.
   a) **Earliest group (Maturity group IA):** Curd maturity mid-September to mid October; Temperature requirement for curd initiation and proper curd development 20-27°C.

      Early Kanwari, Pusa Early Synthetic 78-1, Pusa Early Synthetic, Pant Gobi-3, Kashi Kunwari, Pusa Meghna

   b) **Early group (Maturity group IB):** Curd maturity mid-October to mid-November; Temperature requirement for curd initiation and proper curd development is 20-25°C.

      Pusa Kartiki, Pusa Deepali, Pusa Ashwini, Pusa Shubhra, Pant Gobi-2, Pusa Kartik Sankar

2. **Mid-early group (Maturity group II):** Curd maturity mid-November to mid-December; Temperature requirement for curd initiation and development is 16°-20°C.

   Improved Japanese, Pant Shubra, Pant Gobi-4, Pusa Sharad, Pusa Betakesari-1, Pusa Hybrid-2, Pusa Hybrid-3

3. **Mid-late group (Maturity group III):** Curd maturity mid-December to mid-January; Temperature requirement for curd initiation and development is 12°-16°C. Several adaptable local cultivars are available in the name of Poosi or Pousi depicting the month of December and January when curds of these cultivars are matured.

   D-96, Pusa Synthetic, Pusa Himjyoti, Punjab Giant-35, Punjab Giant-26, Hisar-1, Pusa Shukti, Pusa Paushja, Pusa Hybrid-301

4. **Late group (Maturity group IV):** Curd matures in January-February; temperature requirement for curd initiation and development is 10-16°C.

   Dania, Dania Kalimpong, Pusa Snowball, Pusa Snowball-1, Pusa Snowball-2, Snowball-16, Pusa Snowball K-1, Pusa Snowball K-53, Pusa Purple Cauliflower-1, Ooty-1, Pusa Snowball Hybrid-2

Germplasm resources and Plant introduction

Wide range of germplasm is maintained at the National Vegetable Research Station, Wellesbourne, England. The greatest genetic diversity of cauliflower, broccoli and kale is available in the Mediterranean gene centre including Greece, Syria, Sicily, Cyprus, Italy, Spain and Portugal.

In India, wide range of germplasm of tropical or Indian group is at NBPGR, New Delhi, Indian Agricultural Research Institute, New Delhi, Punjal Agricultural University, Ludhiana, Choudhury Charan Singh Haryana Agricultural University, Hisar, G.B. Pant University of Agriculture and Technology, Pantnagar and other places. Wide range of germplasm of Group IV cauliflower (Snowball group) is maintained at ICAR-IARI Regional Vegetable Research Station, Katrain, Kullu valley, Himachal Pradesh.

Few early introduced varieties belonging to Group III (mid-late) have been released directly for cultivation in India. Howevr, these varieties have been replaced by other varieties developed later in India.

Table 3: Exotic variety directly released for cultivation

Name of the variety	Country from where introduced
Improved Japanese	Israel
D-96	Israel
Dania	Exotic

Variability and population improvement programme

In India, breeding of tropical or Indian group of cauliflower is done at IARI, New Delhi, Punjal Agricultural University, Ludhiana, Choudhury Charan Singh Haryana Agricultural University, G.B. Pant University of Agriculture and Technology, Pantnagar, Uttarakhand and other places because these group of cauliflower flower and set seeds freely in the plains of northern India. Breeding of late or Group IV cauliflower has been undertaken only at ICAR-IARI Regional Vegetble Research Station, Katrain, Kullu Valley, Himachal Pradesh because it requires 6-8 weeks of vernalization at mean temperature range of 4° to 8°C after they have passed the juvenile phase for induction of flowering. Breeding of late cauliflower was also started at Y.S. Parmar University of Horticulture and Forestry, Solan, Himachal Pradesh, Himachal Pradesh Agricultural University, Palampur, Himachal Predesh, Uttarakhand University of Horticulture and Forestry, Ranichauri, Uttarakhand.

Cauliflower bear hermaphrodite flowers but varying levels of cross fertilization occurs even upto 70 % in different open pollinated varieties due to manifestation of self-incompatibility of different intensities. Usually, the cross-pollinated crops are highly heterozygous to such a degree that so many new gene combinations emerge in the course of continuous cross-pollination. Hence. systematic selection combined with controlled pollination always makes it possible to develop a new variety with new properties. Different selection methods or population improvement schemes for cross-pollinated vegetable crops are applicable for cauliflower *viz*., mass selection, recurrent selection, line breeding, family breeding, etc.

Line breeding is an important population improvement scheme for improvement of cauliflower. In this method, the best plants are selected from the base population. Progeny of the selected plants, raised by controlled pollination, are grown separately in rows for evaluation. After several cycles of selection, the seeds of some superior and almost similar progeny lines based on progeny tests are composited and treated as a new variety.

Recurrent is designed to concentrate favourable genes scattered among number of individuals in the base population. This method is based on selection of superior plants followed by controlled mating to produce the new base population. This method is effective in improving yield and other quantitative characters by accumulation of additive genetic variance in the population. The source population for recurrent selection may be an open pollinated variety, F_1 hybrid, synthetic variety, composite variety or inter-crossed progenies of selected inbred lines. Period for selection can be reduced by employing suitable molecular marker. The new and improved genetic base after recurrent selection can be used in developing hybrids, synthetics, or composites.

Inbred development is an important breeding method of developing uniform genetic material from the highly heterozygous and heterogeneous and open-pollinated source population. In cauliflower varying intensity of depression is manifested due to inbreeding. Late cauliflower varieties grown in India show comparatively less inbreeding depression than the early or tropical cauliflower group. In cauliflower, self-pollination by bud pollination can be done up to 3-4 generations without much inbreeding depression to develop inbred lines. These inbred lines can be developed as a new variety or may be used for the development of synthetics or hybrids.

Most of the varieties, particularly under Group I &II (tropical Indian cauliflower group) have been developed through different population improvement programmes from widely variable indigenous genetic materials *viz*., Early Kanwari, Pusa Early Synthetic 78-1, Pusa Early Synthetic, Pant Gobi-3, Kashi Kunwari, Pusa Meghna, Pusa Kartiki, Pusa Deepali, Pusa Ashwini, Pusa Shubhra, Pant Gobi-2, Pant Shubra, Pant Gobi-4, Pusa Sharad, Pusa Betakesari-1, etc.

Some varieties of Group III have been developed through different population improvement programmes from widely variable indigenous genetic materials *viz*., Pusa Himjyoti, Punjab Giant-35, Punjab Giant-26, Hisar-1, Pusa Shukti, Pusa Paushja, etc. and few varieties of Group IV have been developed through different population improvement programmes from exotic genetic materials *viz*., Pusa Snowball-2, Snowball-16, Pusa Snowball K-1, etc.

Table 4: Varieties developed through population improvement programme

Name of variety	Indigenous / Exotic material
Early Kanwari	Indigenous (from Punjab); mass selection
Kashi Kunwari	Indigenous; mass selection
Pusa Meghna	Indigenous; mass selection
Pusa Kartiki	Indigenous; mass selection
Pusa Deepali	Indigenous (from Punjab); mass selection
Pusa Ashwini	Indigenous; mass selection
Plant Gobi-2	Indigenous; mass selection
Pant Shubra	Indigenous; mass selection
Pant Gobi-4	Indigenous; recurrent selection
Pusa Sharad	Indigenous; line breeding
Pusa Betakesari-1	Exotic (β carotene rich (800-1000 µg/100 g fresh) variety)

Name of variety	Indigenous / Exotic material
Pusa Purple Cauliflower-1	Exotic (anthocyanin concentration 43.7 mg/100 g); line breeding
Pusa Himjyoti	Indigenous; mass selection
Punjab Giant-35	Indigenous; mass selection
Punjab Giant-26	Indigenous; mass selection
Ooty-1	Indigenous (from Nilgiri hills) ; mass selection
Hisar-1	Indigenous; mass selection
Pusa Shukti	Indigenous; mass selection
Pusa Paushja	Indigenous; mass selection
Dania Kalimpong	Selection from Dania; line breeding
Pusa Snowball-2	Exotic; mass selection
Snowball-16	Exotic; mass selection
Pusa Snowball K-1	Exotic; mass selection

Source: Hazra *et al.* (2022)

Development of synthetic variety

In cauliflower considerable heterosis is manifested but commercial production of F_1 hybrid seeds needs utilization of different inherent pollination control mechanisms like self-incompatibility and male sterility. In cauliflower, synthetic variety has been developed with advantage by intercrossing a number of genotypes *viz.*, inbred lines, mass selected varieties and other lines, commercial variety of known superior combining ability i.e., genotypes that are known to give superior hybrid performance when crossed in all combinations. By contrast, a variety developed by mass selection is made up of genotypes bulked together without having undergone preliminary testing to determine their performance in hybrid combination. Synthetic varieties are known for manifesting hybrid vigour and for their ability to produce usable seed for succeeding seasons. Because of these advantages, synthetic varieties have become increasingly favoured in production of many cross-pollinated vegetable crops including cauliflower in which expense prohibits the development or use of hybrids. In this population improvement scheme, additive genetic variance is basically exploited in developing a synthetic variety. Different synthetic varieties of cauliflower developed through synthesis of 6-8 inbred lines are Pusa Early Synthetic 78-1, Pusa Early Synthetic, Pant Gobi-3, Pusa Synthetic, etc.

Table 5: Development of synthetic variety

Name of variety	Indigenous / Exotic material
Pusa Early Synthetic 78-1	Synthesis of 6 inbred lines
Pusa Early Synthetic	Synthesis of 6 inbred lines
Pant Gobi-3	Synthesis of 8 inbred lines
Pusa Synthetic	Synthesis of 7 inbred lines

Hybridization and population improvement programme

Intervarietal hybridization or hybridization with the genotype (s) having specific character (as for example, disease resistance) is useful in recombinant breeding for incorporating desired genes from two or more genetic sources within the genepool of particular cauliflower group. As a consequence of such hybridization programme huge variation is created in the segregating population, each being to some extent different from the other ones in that population. Different population improvement approaches *viz*., mass selection, recurrent selection, line breeding, family breeding, etc. are followed to handle the segregating generations for development of the varieties like, Pusa Shubhra, Pusa Snowball, Pusa Snowball-1, Pusa Snowball K-53, etc.

Table 6: Varieties developed through hybridization and population improvement

Name of variety	Parentage / Remarks
Pusa Shubhra	Triple cross, MG 2-3 (a line of Indian cauliflower highly resistant to black rot and curd blight), 15-1-1(a line having acceptable curd quality) and D-96
Pusa Snowball	EC-12013 × EC-12012
Pusa Snowball-1	EC 12012 × EC12013
Pusa Snowball K-53	Lawyana Old × 13-85

Cabbage

Cabbage, also known as white cabbage, *Brassica oleracea* L var. *capitata* under the family Brassicaceae (Cruciferae) having chromosome No. 2n = 2x =18 is one of the most popular vegetables in the world. It is widely grown in China, India, South Korea, Japan, Russia, Ukraine, Poland, USA and many other countries. China produces nearly half of the world's cabbage while Russia consumes the most per person. It is a biennial plant, grown as an annual vegetable crop for its dense-leaved heads. Cabbage heads generally range from 0.5 to 4.0 kg in weight, and can be green, purple and greenish-white. It can be processed to many products *viz*., pickled, fermented product (such as sauerkraut, kimchi). It can be consumed as steamed, stewed, roasted, sautéed and braised, or eaten raw. Raw cabbage is a rich source of vitamin K, vitamin C, and dietary fibre.

Origin, domestication and adaptation in India

All cole crops including cabbage have been originated in the Mediterrnean region. The wild diploid species, *Brassica oleracea* var. *sylvestris*, a leafy kale like plant which occurs in the northern shore of Mediterranian Sea and on the lime cliffs in north-eastern England and along the North Sea is the progenitor of the cultivated kale and leafy non-heading cabbage. The hard headed white cabbage selected in North Europe descended from wild, leafy, non-heading cabbage originating in eastern Mediterranean region and Asia Minor which was probably brought into Western Europe by the Celts. The Celts were the collection of tribes with origins in central Europe (Ireland, Scotland, Wales, Cornwall and Brittany) that shared a similar language, religious beliefs, traditions and culture. The oldest record of cabbage was in 2000 to 2500 BC. The modern-day white cabbage was probably selected in Germany around 1150 AD. which later moved to England in the 14th century. However, savoy cabbages (*Brassica oleracea* var. *sabauda*) were not developed until the 16th century AD in Italy. It was probably developed from Portuguese kale (*Brassica oleracea* var. *costata*). By the middle ages, cabbage had become a prominent part of European cuisine.

Cabbage was introduced in India much earlier than cauliflower probably by the Portuguese. It is known to be cultivated during the Mughal period also. However, its popularity grew during the British regime in the country. At that time, seeds were imported from Europe. First organized seed production started in the country in early part of 1940 and cabbage breeding started from 1956 at ICAR-IARI Regional station, Katrain, Kullu, Himachal Pradesh. The character of wide adaptability to fluctuating temperature conditions has not been exploited in cabbage to the same extent as done in cauliflower in India, although the potentialities for adaptation was there in cabbage also. Adaptation potential of cabbage to tropical conditions has been fully exploited by the breeders of Japan, Korea, Taiwan and Vietnam through the development of tropical cabbage varieties. The classical 'KK' and 'KY' crosses form compact heads even under day temperature condition of 30-35°C.

Crop botany and floral biology

- The plants have fibrous and shallow root systems and about 90 percent of the root mass is confined in the upper 20–30 cm of soil.
- The edible part of cabbage is 'head', which is formed by thickening of apical bud with thick tightly packed overlapping leaves; smooth-leafed and firm-headed green cabbages are the most common; some varieties produce smooth and purple-leafed head while, crinkle-leafed "Savoy" cabbages of both green and purple colour are also found.
- Most cabbages have thick, alternating leaves, with margins that range from wavy or lobed to highly dissected; some varieties have a waxy bloom on the leaves.
- Different varieties produce heads of different shapes (oblate, round and pointed), colour and leaf texture.
- Cabbage is a biennial and temperate crop which requires vernalization for induction of flowering hence, high hilly regions of India (Kulu valley, Kashmir vally) are suitable for its breeding and seed production.
- The inflorescence is a terminal raceme which is developed on the main stem and corymbose towards the top.
- Flowers are hermaphrodite, singular and complete, sepals are four and free; corolla is bright yellow in colour with four petals, standing opposite to each other in a square cross, called cruciform corolla
- Stamens are six in number, tetradynamous *i.e.* two short having nectaries at their base and four long.

- Carpels are two in number, syncarpous, which form a superior ovary with a false septum and two rows of campylotropous ovules with parietal placentation.
- Fruits are special type of bicarpellary pod called the siliqua, which are glabrous, 4-5 mm wide and sometimes over 10 cm long, with two rows of seeds lying along the edges of the septum; the two valves dehisce when the pod is fully mature; a siliqua contains 8-15 seeds.
- Seeds are small, globular, smooth and dark brown in colour, non-endospermic with large embryo.
- Cotyledons are folded; average 100 seed weight is 0.33 gram.

Reproductive behaviour

Cabbage is biennial in nature for flowering and seed production. It requires 6-8 weeks of vernalization at mean temperature range of 4° to 8°C after they have passed the juvenile phase (at least 8-10 weeks after transplanting) for the production of seed stalks. In India, breeding of cabbage is done in the high hill situations of Himachal Pradesh and Kashmir. Cabbage though bears hermaphrodite flowers yet varying levels of cross-fertilization takes place as high as 75 % due to manifestation of self-incompatibility of different intensities. In cabbage, self-incompatibilty system is sporophytic in nature and governed by multiallelic single gene (*S* gene) with generally dominant in effect. Intensities of self-incompatibility depend on the presence of *S* allele in the genotype. The *S* alleles are active in the stigma in flowering stage but remain inactive in the bud stage, i.e., 2-4 days before anthesis. The stigma becomes receptive five days before anthesis due to protogyny and remains so 3-4 days after anthesis. Due to protogynous nature, self-fertilization can be done by self-pollination with the pollens from other opened flowers of the same plant. In cabbage, cross-pollination is generally done by bees, more particularly honeybees and flies. Bees and other insect pollinators frequently visit the flowers of cabbage because of the presence of nectar in the flowers. Genic male sterility though identified in cabbage yet is not utilized in hybrid seed production. Cytoplasmic male sterility system is generally utilized in hybrid seed production. The cytoplasm of the *Ogura* male sterile radish has been successfully transferred in developing cytoplasmic male sterility in cabbage. Anthesis starts early in the morning between 7.30 and 8.00 am and continues up to 11.30 am depending on the prevailing weather conditions. Dehiscence of anther is longitudinal and it generally occurs few hours after anthesis.

Genetics of important characters

The cabbage *Brassica oleracea* L var. *capitata* is an important species of the genus Brasica (tribe Brassiceae). *Brassica oleracea* is diploid, 2n = 2x =18, with *C* genome. Cabbage is a secondary polyploidy with a basic chromosome number, x = 6. Three basic chromosomes are present in duplicate and the remainders are single with the genome, ABBCCDEEF. The molecular genetic maps constructed for *Brassica* indicated that within a species there is almost complete collinearity, except for small inversions that differ among some *Brassica oleracea* genotypes. The diploid species *Brassica rapa* (2n= 2x = 20, A genome) and *Brassica oleracea* (2n = 2x =18, with *C* genome) have close evolutionary relationship between them but indicates that deletions and insertions may have occurred after divergence of these two species.

Plant habit, leaf characters (colour, morphology and heading habit), flower colour and morphology, genic male sterility, self-incompatibility and resistance to diseases (black rot, club root, etc.) have simple inheritance.

Table 1: Important simple inherited characters in cabbage

Gene symbol	Inheritance pattern / Phenotype
T	Tall (*T*) dominant over short (*t*)
dw	Dwarf, short internode, round leaves; recessive character
Ax	Axil sprout, axillary heading habit; dominant character
A	Basic anthocyanin production gene
A^{rc}	Coloured lamina in red cabbage, dominant
B	Light red mid rib; dominant, dominant
c_1, c_2	Inhibits anthocyanin production; epistatic to *A*
M	Magenta (reddish purple) plant colour, dominant
En	Entire leaf margin, dominant
fc	Fused cotyledon
GL	Glossy green foliage; dominant over waxy
Hr-1	Hairy first leaf; hairs on margin
Sm	Smooth leaves; with *wr* gene
Pet	Petiolate leaves; dominant over sessile (*pet*)
W_1, W_2, W_3, W_4	Series of genes govern frilled leaves
W	Wide leaf ; dominant over narrow leaf
Wr	Recessive gene *wr* with *sm* govern smooth leaves
K	Control heading; dominant
Wh	White petal; dominant over yellow, *wh*
cp	Crinkly petal; recessive character
cr	Cream coloured petal; recessive character
ap	Male sterility; aborted pollen
ms	Nuclear gene for male sterility with *R* radish cytoplasm
S	Self-incompatibility, multiple alleles

Source: Chiang *et al.* (1993), Ram (1998), Swarup (2006)

The characters inherited quantitatively are head size, head weight, head maturity and inner core length which are under additive genetic control and these characters show high heritability. The other quantitative characters controlled by partial dominance, overdominance and epistasis include dry matter content, plant height, frame size, head splitting, number of wrapper leaves, head diameter and axillary heading. Complete dominance was recorded in case of net head weight and maturity and overdominance was predominant for head yield. All the characters had non-allelic complementary interaction. Inheritance of head shape is controlled by additive gene action.

Table 2: Associated genes/QTLs and nature of gene action for some important characters in cabbage

Trait	**Genes/QTLs**	**Inheritance**	**Reference (s)**
Earliness of head formation	Polygenic	Additive and average dominant effects	Tanaka and Niikura (2006)
Head shape	*Htd 7.1 and Hsi 7.2*	Quantitative	Zhang *et al.* (2016)
Head splitting resistance	Three major QTLs (*Hsr 3.2, 4.2, 9.2*)	additive-epistatic effects of three major gene pairs combined with those of polygenes	Su *et al.* (2015)
Wax deposition in cabbage leaves	CER_3	Monogenic dominant for high-wax deposition	Laila *et al.* (2017)
Wax-less gene	*Bo-GL3*	Single dominant	Dong *et al.* (2019)
Wax biosynthesis locus	*BoWax1*	Monogenic recessive for glossy green trait	Liu *et al.* (2018)
Glossy trait	*Cgl1*	Single recessive	Liu *et al.* (2018)
Fusarium wilt resistance	*FOC1*	Single dominant	Lv *et al.* (2014)
Black rot resistance	*QTL-1,2,3*	Quantitative	Kifuji *et al.* (2012)
GMS	*Ms-cd1*	Single dominant	Fang *et al.* (2004)
GMS	*ms-3*	Single recessive	Han *et al.* (2018)
Glossy green trait	*BoGL1*	Single dominant	Liu *et al.* (2018)
Yellow green leaf	*ygl1*	Single recessive	Liu *et al.* (2018)
Club root	*Rcr7*	Polygenic recessive	Dakouri *et al.* (2018)
Club root	*QTL, pb-Bo(Anju)1, pb-Bo(GC)1*	Quantitative	Nagaoka *et al.* (2010)

Genic male sterility

Earlier, genic male sterility controlled by a single recessive gene, *ms* was reported in cabbage. Later, male sterility governed by a single dominant gene (*Ms*) has been found. However, both these forms of male starility are not being utilized for development of hybrids commercially.

Cytoplasmic male sterility

The *Ogura* sterile cytoplasm of *Raphanus* is the most important source of sterile cytoplasm for cauliflower, broccoli and cabbage. Ogura cytoplasmic male sterility (OguCMS), controlled by the mitochondrial gene *orf138*, has been extensively used for cabbage hybrid production with complete and stable male sterility. Light and transmission electron (TE) microscopy revealed fat abnormal tapetum rather than vacuolization and degradation at the tetrad and microspore stages of the OguCMS line. Protoplasm fusion technique has been employed to transfer *Ogura* sterile cytoplasm from broccoli to cabbage. The CMS male sterility is being used in hybrid seed production of cabbage.

The two other sources of CMS were from *Brassica nigra* sterile cytoplasm and sterile "Anand" cytoplasm from *Brassica rapa* in which it was originally derived from the wild species, *Brassica tournefortii.* In CMS system of hybrid seed production, three lines used are male sterile "A" line (*Smsms*), male sterility maintainer "B" line (*Nmsms*) and fertility restorer pollen parent "C" line (*NMsMs* /*NMsms*).

Self-incompatibility

Self-incompatibility (*SI*) promotes outcrossing in *Brassica* species and has been exploited for producing F_1 hybrids in cole crops including cabbage. The *SI* locus (*S*-locus) is very polymorphic and cross-compatibility between plants depending on the allelic forms (*S*-haplotypes) of genes at the *S*-locus. About 50-70 *S*-alleles have been identified in *Brassica olercea.* In cabbage, self-incompatibility system is sporophytic and governed by single multiallelic gene (*S* allele) with mostly dominance relationships. However, in cabbage sensitivity of the self-incompatibility reaction to environmental factors, mainly temperature is influenced by modifier genes. Most of the cabbage varieties express self-incompatibility and low inbreeding depression as compared to tropical or Indian cauliflower.

Breeding objectives

The following characters are considered important in the improvement programme of cabbage

- Plant with short frame and stout stem, early head maturity, uniform heading and less number of non-wrapper leaves
- Leaf colour green in white cabbage and red in red cabbage, leaf size medium, soft textured, leaf margin entire, leaf surface glossy or waxy
- Head shape preferably spherical in round shaped variety with polar and equatorial diameter ratio 0.8 to 1.0; flat shaped variety in drum headed variety with polar and equatorial diameter ratio 0.5 to 0.6; conical head in pointed headed variety with polar and equatorial diameter ratio of more than 1.0.
- Head size or weight preferably medium (1.0 to 1.5 Kg), compact head, short inner core length and narrow and soft core
- Field holding capacity of head should be 15 days or more
- High nutrient content of head (particularly β carotene, vitamin C and anthocyanin content (in red cabbage)
- Desired level of glucosinolates in the head because it shows anticarcinogenic property and varietal difference exists for this sulphur-containing chemical.
- Resistance to common diseases *viz.*, downy mildew (*Peronospora parasitica*), black rot (*Xanthomonas campestris* pv. *campestris*) and cabbage yellow (*Fusarium oxysporum* f.sp. *conglutinans*).

- Resistance to important insect pests *viz.*, diamondback moth (*Plutella xylostella*), aphids (*Brevicoryne brassicae*), etc.
- Tolerance to high temperature for initiation and development of head (particularly for tropical hybrids)

Breeding methods and development of variety

Most of the cabbage cultivars grown in India are either open-pollinated old European introductions or tropical hybrids imported from different countries mainly, Japan, Korea, Taiwan and Vietnam. Some open pollinated varieties have also been developed utilizing the exotic genetic stock through different population improvement schemes *viz.*, mass selection, line breeding, family breeding, inbred development and recurrent selection.

Intervarietal hybridization and hybridization with the genotype (s) having specific character (as for example, disease resistance) followed by population improvement, development of synthetics and hybrids have also been employed for the cabbage improvement programme.

The commonly grown cabbage cultivars belong to three cultivar groups of cabbage.

1. **White Cabbage** (*Brassica oleracea* var. *capitata* f. *alba*). Cultivars of this group are the only commercial cabbage grown in India. Leaves are generally smooth and light green in colour. Inside colour of the head is creamy-white. These cabbages are available in three different head shapes namely, round, flat or drum head and pointed. Round-headed cultivars are most popular in India.
2. **Savoy Cabbage** (*Brassica oleracea* var. *capitata* f. *sabauda*). Characteristic features of this cabbage group are blistered and puckered leaves of light green colour. Varieties of this group also form heads of three different shapes namely, round, flat and pointed. This group is not popular in India.
3. **Red Cabbage** (*Brassica oleracea* var. *capitata* f. *rubra*). Leaves of this cabbage group are smooth, red and having distinct coating of wax. Varieties of this group form only round head. This group is also not popular in India.

White cabbage group

• Round and Flat or Drum-headed Varieties

All the open pollinated varieties in this category are either traditional European cultivars or developed from the genetic stock of these types, which grow and form heads best in the temperature range of 15° to 20°C. All the tropical hybrids available in India that form head in the temperature range of 20° to 30° C are imported mainly from Japan, Korea, Taiwan and Vietnam. Many tropical cabbage hybrids were introduced in India e.g. Golden Cross, KK Cross, OS Cross, Resistlaka, Green Cornet, Autumn Queen, Green Ball, Green Boy, Green Express, Herculis, Stone Head, Regalia. Japanese seed companies have also undertaken seed multiplication of their self-incompatible lines in South Asian Countries.

Open pollinated varieties of this category of white cabbage fall under three maturity groups *viz*., early, medium and late on the basis of days required for maturity of heads after transplanting.

a) **Early group:** Copenhagen Market, Golden Acre, Pride of India, Pusa Ageti, Pusa Mukta, Pusa Synthetic, Early Drumhead, Palampur Green, Pusa Cabbage Hybrid-81
b) **Mid-season group:** Glory of Enkhuizen, Kalimpong English Ball, Pusa Cabbage-1
c) **Late Season Group:** Pusa Drumhead, Late Large Drumhead, September, Kalimpong Eclipse Drmhead

Pointed headed varieties

Jersey Wakefield, Express

Savoy cabbage group

Perfection

Red cabbage group

Red Acre, Kinner Red

Germplasm resources and Plant introduction

Wide range of germplasm of cabbage is maintained at the National Vegetable Research Station, Wellesbourne, England. Wide genetic diversity of cabbage including tropical lines are also maintained in Turkey, Portugal, Italy, Korea, Japan and many other countries.

In India, appreciably wide germplasm of cabbage is maintained at ICAR-IARI Regional Vegetable Research Station, Katrain, Kullu valley, Himachal Pradesh. The major countries from where the temperate vegetables including cabbage and Group III and IV (mid-late , late) cauliflower have been introduced are Russia, Canada, USA, Australia, Brazil, France, U.K., Netherlands, Germany, Italy, Polland, Hungary, Denmark, Turkey, Yugoslavia, Egypt, Zambia, Israel, Korea, Japan, Taiwan, etc (Verma and Prakash, 2006).

Several open pollinated varieties *viz*., Copenhagen Market, Glory of Enkhuizen, September, Red Acre, etc. imported from European countries have been released directly for cultivation in India. However, these varieties have been replaced by different tropical hybrids of white cabbage group imported mainly from Japan, Korea, Taiwan, Vietnum, etc.

Table 3: Exotic variety directly released for cultivation

Name of the variety	Country from where introduced
Copenhagen Market	Denmark
Glory of Enkhuizen	Netherlands
Late Large Drumhead	Europe
September	Germany

Jaersey Wakefield	USA (Pointed headed cabbage)
Express	Europe (Pointed headed cabbage)
Perfection	Europe (Savoy cabbage)
Red Acre	Europe (Red cabbage)

Variability and population improvement programme

Breeding of cabbage in India has been undertaken only at ICAR-IARI Regional Vegetble Research Station, Katrain, Kullu Valley, Himachal Pradesh because it requires 6-8 weeks of vernalization at mean temperature range of 4° to 8°C after they have passed the juvenile phase for induction of flowering. Later, cabbage breeding was also started at Himachal Pradesh Agricultural University, Palampur, Y.S. Parmar University of Horticulture and Forestry, Solan, Uttarakhand University of Horticulture and Forestry, Ranichauri, Uttarakhand. Cabbage bear hermaphrodite flowers but but mainly cross-pollinating due to manifestation of self-incompatibility of different intensities which results heterozygous population upon open pollination. Heterozygosity and heterogenity due to many new gene combinations occurs to such a degree that systematic selection combined with controlled pollination always makes it possible to develop a new variety with new properties.

Different selection methods or population improvement schemes for cross-pollinated vegetable crops are applicable for cabbage *viz.*, mass selection, recurrent selection, line breeding, family breeding, etc. Line breeding is an important population improvement scheme for cabbage. In this method, the best plants are selected from the base population. Progeny of the selected plants, raised by controlled pollination, are grown separately in rows for evaluation. After several cycles of selection, the seeds of some superior and almost similar progeny lines based on progeny tests are composited and treated as a new variety.

Recurrent is designed to concentrate favourable genes scattered among number of individuals in the base population. This method is based on selection of superior plants followed by controlled mating to produce the new base population. This method is effective in improving yield and other quantitative characters by accumulation of additive genetic variance in the population. The source population for recurrent selection may be an open pollinated variety, F_1 hybrid, synthetic variety, composite variety or inter-crossed progenies of selected inbred lines. Period for selection can be reduced by employing suitable molecular marker. The new and improved genetic base after recurrent selection can be used in developing hybrids, synthetics, or composites.

Inbred development is an important breeding method of developing uniform genetic material from the highly heterozygous and heterogeneous and open-pollinated source population. Cabbage generally show comparatively less inbreeding depression than the early or tropical cauliflower group. In cabbage, self-pollination by bud pollination can be done up to 3-4 generations without much inbreeding depression to develop inbred lines. These inbred lines can be developed as a new variety or may be used for the development of synthetics or hybrids.

All the varieties *viz.*, Golden Acre, Palampur Green, Pusa Drumhead, Kinner Red, etc have been developed through improvement of different exotic varieties or lines (Hazra *et al.*, 2022) following different population improvement schemes.

Table 4: Varieties developed through population improvement schemes

Name of variety	Indigenous / Exotic material
Golden Acre	Exotic (from Copenhagen Market); line breeding
Pride of India	Exotic (from Copenhagen Market); line breeding
Early Drumhead	Exotic; mass selection
Palampur Green	Exotic; mass selection
Kalimpong English Ball	Exotic; mass selection
Pusa Drumhead	Exotic; mass selection
Kinner Red	Exotic; mass selection (Red cabbage)

Development of synthetic variety

In cabbage considerable heterosis is manifested but commercial production of the F_1 hybrid seeds needs utilization of inherent pollination control mechanisms like self-incompatibility and male sterility. Synthetic variety can be developed with advantage by intercrossing a number of genotypes *viz.*, inbred lines, mass selected varieties and other lines, commercial variety of known superior combining ability i.e., genotypes that are known to give superior hybrid performance when crossed in all combinations. Synthetic varieties are known for manifesting hybrid vigour and for their ability to produce usable seed for succeeding seasons. In this population improvement scheme, additive genetic variance is basically exploited in developing a synthetic variety. One synthetic variety, Pusa Synthetic has been developed through random mating (synthesis) of 6 inbred lines having good general combining ability in isolation.

Table 5: Development of synthetic variety

Name of variety	Indigenous / Exotic material
Pusa Synthetic	Synthesis of 6 inbred lines having good general combining ability

Hybridization and population improvement programme

Intervarietal hybridization or hybridization with the genotype (s) having specific character (as for example, disease resistance) is useful in recombinant breeding for incorporating desired genes from two or more genetic sources within the genepool of particular cultivar group. As a consequence of such hybridization programme huge variation is created in the segregating population, each being to some extent different from the other ones in that population. Different population improvement approaches *viz.*, mass selection, recurrent selection, line breeding, family breeding, etc. are followed to handle the segregating generations for developing a new variety. Pusa Ageti, the first variety seeds of which can be produced in subtropical climate has been developed from the F_1 hybrid (TKCBH-28) procured from Taiwan and selections were made F_1 generation onwards by employing simple recurrent selection. Pusa Mukta developed through selection from the F_1 (EC24855 x ECIO109) and subsequent generations has been identified as resistant variety against black rot disease. Another line, BRH-5 selected from the F_1 (Pusa Mukta x AC 204) and subsequent generations showed promise as a black rot resistant variety.

Table 6: Varieties developed through hybridization and population improvement

Name of variety	Parentage / Remarks
Pusa Ageti	F_1 hybrid (TKCBH-28); simple recurrent selections
Pusa Mukta	EC 24855 × EC 10109; line breeding (black rot resistant)
BRH-5	Pusa Mukta x AC 204; line breeding

Heterosis breeding in cole crops (cauliflower and cabbage)

Unlike synthetics and composites, heterosis breeding utilizes mainly dominance variance. Both cabbage and cauliflower show pronounced heterosis in earliness, net weight of head or curd, yield per unit area, resistance to diseases, hardiness, uniformity, etc. In most of the advanced countries, the F_1 hybrids in cole crops are being preferred for uniformity of harvest and resistance to diseases. The yield may or may not be of much value since the level of yield otherwise is also very high. This has become more popular, possibly due to comparatively easier method of F_1 seed production with the help of self-incompatibility, cytoplasmic male sterility system and comparatively cheaper cost of seed production.

At present CMS system is most commonly used in developing F_1 hybrids of cauliflower and cabbage (Dey *et al.*, 2011; 2014). The CMS line Ogu12A emerged as good general combiner. Singh *et al.* (2018) studied the heterosis and combining ability for antioxidant compounds in cauliflower using CMS and doubled haploid lines. However, availability of strong S allele with good stability in crop like cabbage also offesrs good opportunity to develop F_1 hynrids.

It is important to select suitable parents in developing heterotic hybrids for different yield related and quality traits. Both morphological diversity and molecular markers uniformly distributed throughout the genome can be employed for this purpose. Singh *et al.* (2019) reported the suitability of genomic- SSRs and EST-SSRs based genetic distance (GD) and agronomic trait based phenotypic distance (PD) for predicting heterosis in F_1 hybrids using CMS and DH based parents.

Genetic male sterility

Most of sterile mutants originated spontaneously having homozygous alleles (*msms*) have been documented in about 175 plant species including *Brassica* vegetables. Among the *Brassica* species, the first report of genic male sterility (GMS) was documented in sprouting broccoli (*Brassica oleracea* var. *italica* L.) (Anstey and Moore, 1954) and during the past 70 years, other cases of GMS have been reported in most of the cultivated vegetable Brassicas.

Utilization of GMS in hybrid breeding

The cost effective and simplified hybrid development *via* natural genetic mechanisms of male sterility have encouraged the researchers to pay more attention to look for genetic resources of male sterility and their exploitation in heterosis breeding of *Brassica* crops. Both dominant (DGMS, DNMS) and recessive (RGMS) types of nuclear (genic) male sterility have been reported in *Brassica* crops and can be effectively utilized in hybrid breeding (Ji *et al.*, 2017). Fang *et al.* (1997) reported the two step breeding programme (production, multiplication of male sterile line and subsequent F_1 hybrid seed development) with use of DGMS line in cabbage (*Ms-cdl* gene) for the production of hybrid seeds. The propagation of homozygous male sterile plants (*Ms-cdl*/*Ms-cdl*) was carried out with tissue culture approach and then *via* natural pollination of DGMS plants (*Ms-cdl*/*Ms-cdl*) with a sister inbred line, obtained the male sterile line. Subsequently, interplanting of male sterile line (as female parent) with male fertile inbred line (as male), having desirable features, in the ratio of 3:1 in panmictic plot was done and the F_1 hybrid seeds were collected from male sterile line (Fang *et al.*, 1997). In China, this two step method derived from DGMS line in cabbage has been utilized commercially in cabbage hybrid breeding. On the other hand, RGMS has not yet been commercially utilized in hybrid breeding owing to its limitation of not isolating 100 % male sterile plants. The homozygous recessive (RGMS) male sterile (*msms*) line is maintained by repeated backcrossing of it with heterozygous isogenic B line (also called as maintainer line) (*Msms*). Hence, in the hybrid seed production block, 50 % of male fertile segregants (*Msms*) should be identified and removed before pollen shedding. In some crops, the tight linkage of recessive male sterile gene '*ms*' with recessive morphological markers, especially at seedling stage, is of prime significance in identification of male sterile and male fertile plants in female parent e.g. in broccoli (bright green hypocotyl). Hence, the time and labour needed for identification and removal of male fertile plants in female parent, in hybrid seed production block, is largely avoided.

Cytoplasmic male sterility

Most reliable system of F_1 seed production of cole crops is based on cytoplasmic male sterility (CMS). Various types of CMS have been developed and adopted in practice to breed *Brassica* oil seed and vegetable crops. CMS is a maternally inherited trait encoded in the mitochondrial genome, and the male sterile phenotype arises as a result of interaction of a mitochondrial CMS gene and a nuclear fertility restoring (*Rf*) gene. Therefore, CMS has been intensively investigated for gaining basic insights into molecular aspects of nuclear-mitochondrial genome interactions and for practical applications in plant breeding. Several CMS genes have been identified by molecular genetic studies, including Ogura CMS from Japanese radish, which is the most extensively studied and most widely used.

Bannerot *et al.* (1977) reported successful introduction of cabbage nucleus into the cytoplasm of the male sterile Japanese radish through repeated backcrossings. They observed that the level of 18 chromosomes was achieved after fourth backcross followed by *in vitro* culture of the immature embryos. All the *B. oleracea* genotypes that were introduced into this radish cytoplasm exhibited severe yellowing of the young leaves, when the young plants were grown at temperature below 12°C. It seemed that the nucleus of *oleracea* did not function properly in the radish cytoplasm. The degree of yellowing differed with recurrent male parents used. The most severe effect was observed with cauliflower while Brussels sprouts and savoy cabbage F_1 hybrids were greenish though yellowing remained. Bannerot *et al.* (1974) intended to replace the chloroplast set of radish origin by a set of cabbage chloroplast *via* protoplast fusion but it was not successful. Pelletier *et al.* (1994) gave information on transfer of cytoplasmic male sterility (CMS) through protoplast fusion in *Brassica oleracea.*

Development of CMS lines in *B. oleracea* in India

Systematic development of CMS lines in *Brassica* vegetable like cabbage and cauliflower was started at ICAR-Indian Agricultural Research Institute, Regional Station, Katrain, Kullu, Himachal Pradesh, India since 1995. The *Ogura* based cytoplasm was introgressed into a large number of cabbage and cauliflower through back-cross introgression. Three CMS lines in cauliflower, *Ogu1A*, *Ogu2A* amd *Ogu3A* were developed by Dey *et al.* (2011) with good floral traits and seed setting capacity. These lines were evaluated and compared with their respective fertile maintainer (B) line for various agronomic traits like curd yield and size, floral like size of petals and style, shape of style and seed setting traits like number of seeds per pod and seed yield per plant. Nuclear-cytoplasmic interaction played an important role in determining various agronomic, floral and seed setting related traits. All the three CMS lines were similar with fertile maintainer lines for agronomic traits like days to curd maturity. Introgression of alien *Ogura* cytoplasm in the background of snowball cauliflower significantly reduced the petal size, filament length, style and stamen length. Number of pods per plant was significantly lower in all three CMS lines, when compared with their respective B lines. Number of seeds per pod and seed yield per plant in Ogu3A and its respective B line was *at par* but reduced significantly in Ogu1A and Ogu2A after introgression of *Ogura* cytoplasm. These Three CMS lines are being used widely in hybrid development in cauliflower. Two CMS lines Ogu3A (Snowball/ late group) and Ogu33A (mid-late groups) were developed and registered with NBPGR by IARI, Regional Station, Katrai, Kullu. They are the first of CMS lines in cole crops in India which are available in public domain and can be accessed by any breeders working on cauliflower improvement. Another set of 7 CMS lines were characterised by Dey *et al.* (2014) and found that they were suitable for use in hybrid breeding. However, introgression of alien cytoplasm resulted into reduced flower size in each line because of nuclear-cytoplasmic interaction. Later, Dey *et al.* (2017) reported a large number (25) of cauliflower CMS lines developed after 9 generations of back-crossing in different maturity groups and characterised them for all important agronomic and floral traits. Besides, these lines, were also grouped using a large number of SSR markers. These lines were also evaluated for combining ability to reveal their breeding potential. Two CMS lines, Ogu402-6A and Ogu76- 4A were distantly placed from rest of the CMS genotypes based on morphological characterization. Molecular analysis through SSR primers also revealed the genetic distance of the CMS lines Ogu402-6A from rest of the CMS pool. Three genotypes (Ogu402-6A, Ogu76-4A and Ogu119-2A) with early maturity were identified for their use in the development of short duration hybrids. Similarly, 3 CMS lines, Ogu118-2A, OguHL-3A and Ogu126- 1A were identified with yield potential of more than 60 t/ha. Based on 13 polymorphic SSR markers, 25 CMS lines were grouped into 4 major clusters. Effects of improved cytoplasmic lines in agronomic and floral traits were studied in detail by Dey *et al.* (2013), and it was concluded that introgressed cytoplasm mainly altered the floral and reproductive traits and did not have any influence for different agronomic traits of the CMS lines and F_1 hybrids developed using these CMS lines. In cabbage, one CMS line, CMS-1 was developed indigenously by Parkash *et al.* (2015) and was was found suitable for use in commercial scle based on the performance of the F_1 hybrids developed using this line.

Development of F_1 hybrids using CMS lines

Cytoplasmic male sterile lines are most preferred pollination control mechanisms in cole crops used in development of F_1 hybrids worldwide. Vegetative parts are the economically important edible portions in the cole crops hence, there is no need of successful fertilization for proper development of edible part unlike fruit vegetable. Therefore, restoration of male fertility is not necessary in the F_1 hybrids. In crops like *B. juncea* and *B. napus* which need successful development of seeds from the F_1 hybrids, restoration of male fertility is necessary in hybrid breeding and they need the restorer line with *Rf* gene used as pollen/ male aprent in hybrid development. However, in case of cole crops there is no need of any restorers with *Rf* gene to be used as male parent. Any inbred which is good combiner with the male sterile female parent and gives superior hybrid with higher yield and quality is suitable as male parent.

Development of CMS lines have three major steps:

i. Identification of source of male sterile cytoplasm without any adverse effect when introgressed into the nuclear background of recurring parent.

ii. Back-cross substitution of the nuclear genome of the souce line (non-recurreng parent) through 6-7 generations of back-crossing using the source parent only once as female parent.

iii. Selection of the male sterile plants with higher resemblance to recurrent parent for different morphological traits and better floral traits and normal development of style and stigma.

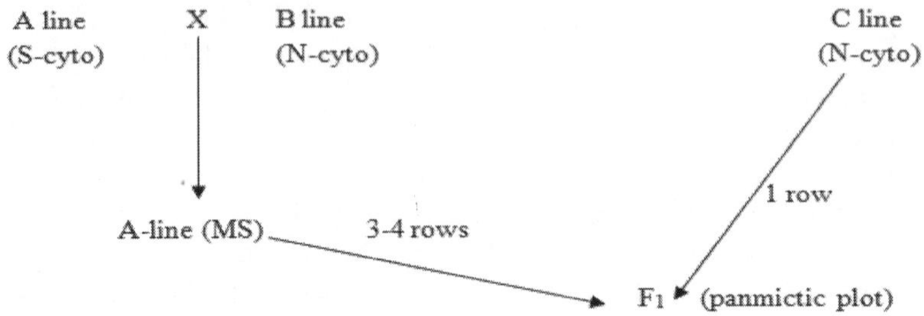

At the end of 6-8 generations of back-crossing the male sterile sterile lines developed will be the A line and the recurring male fertile parent will the maintainer of the male sterile line (B line). While introgressing the male sterile cytoplasm it is evident that introgression of alien cytoplasm results into reduced flower size. However, utmost care should be taken while selecting the plants after each generation of back-crossing. The back cross-progenies with normal flower structure and development of nectaries should be selected. The cole crops are highly heterozygous with varied extent of self-incompatibility and expresses inbreeding depression while selfing or sibbing with limited number of plants. Therefore, development of strategy to select suitable number of plants used for back-crossing is very much essential. Unlike the other field crops, growing of large numner of plants in each generation of back-crossing is not possible. Moreover, while introgressing the male sterile cytoplasm into a large number of lines subsequently, it may not be possible to raise and pollinate a large number of plants.

Dey *et al.* (2018) developed a small population based breeding strategy for introgression of male sterile cytoplasm into a large number of genotypes. Three methods of back-crossings were adopted after 3 generations of single plant selection.

i) Twelve plants both from recurrent and non-recurrent parents till BC9 (M1)

ii) Six plants both from recurrent and non-recurrent parents till BC9 (M2)

iii) Single plant selection from recurrent and non-recurrent parent till BC9 (M3).

Later on, two generations of bulking were practiced for their use in hybrid breeding. The agronomic and floral performance was tested at different stages. Inbreeding depression was determined for various traits at different stage of development. Performance of the hybrids using the CMS lines was evaluated to find out the suitable method and ideal population structure. Among the different methods of backcrossing the M2 (6 plants from BC4 onwards) was found to be most effective in ensuring uniformity and complete recovery of recurrent genotype nuclear genome. However, in evaluating the performance of the hybrids M2 was *at par* with the M1. Inbreeding depression for various traits was highest at M3 followed by M2. However, the inbreeding depression was reversed when the plants generated through M2 undergone two generations of bulking. The plants generated through M3 did not recover even after bulking indicating the permanent loss in vigor because of several generations of inbreeding using single plant. Therefore, it was recommemded to adopt single plant selection for three generation of back-crossing followed by back-crossing using 6 plants in each generation. This model can be widely used in development of CMS lines in different cole crops. This innovative model is summerised as follows:

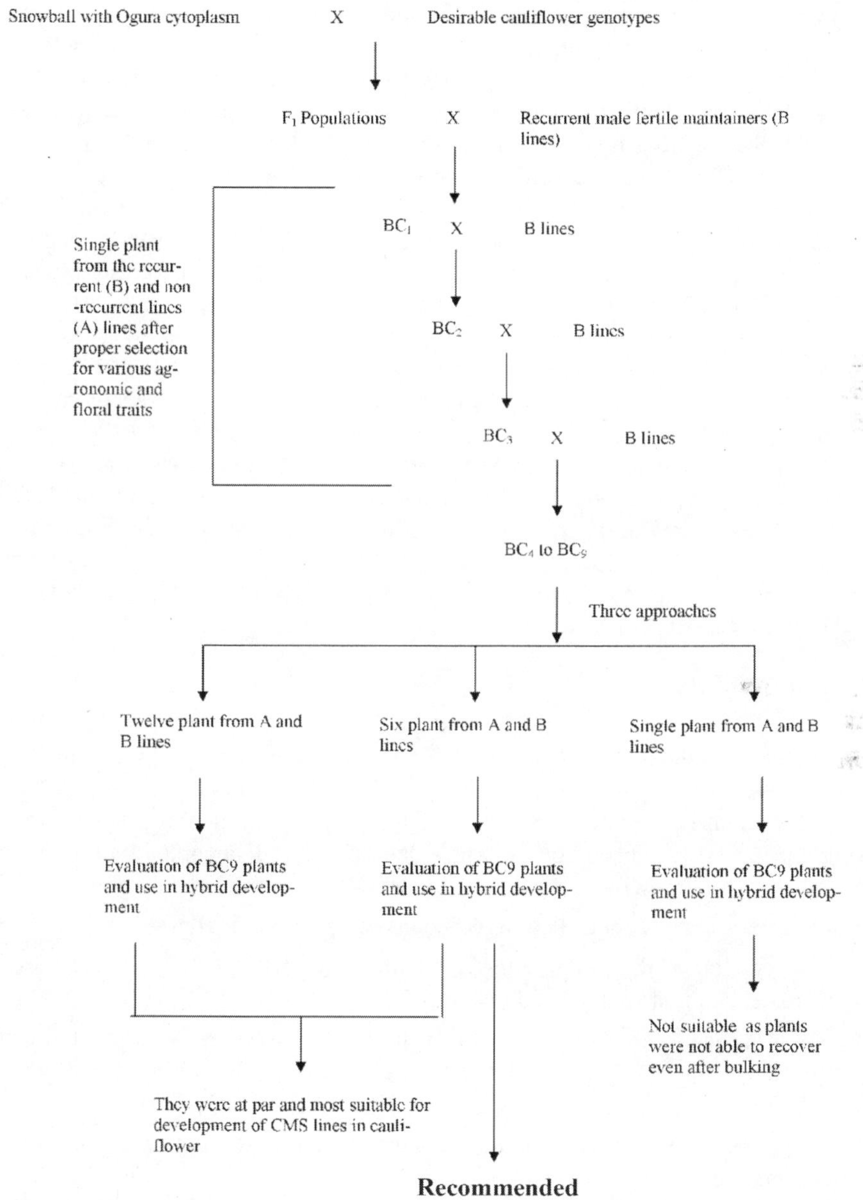

Fig. 2: The model backcross breeding for introgression of wild male sterile cytoplasm into the nuclear background of *Brassica oleracea* (Dey *et al.*, 2018)

Problems of using CMS lines

The main problem being faced is the lower quantity of hybrid seed due to preferential foraging of honey-bees on the male fertile C-line. This could be overcome to a certain extent with appropriate flower morphology and manipulating the ratio of plants of A and C lines, spacing and planting design. A unique honey bee mediated selection strategy was devised by Dey *et al.* (2019). Honey bees were used as selection agent in addition to manual selection to improve the flower structure and seed setting ability of the CMS lines. This study adopted an alternative selection strategy using honey bee (*Apis cerena indica*) as the selection agent in five selection cycles in three CMS lines of cauliflower (Ogu1A, Ogu2A and Ogu3A). Selected plants with the highest seed yield, determined through honey bee foraging, were bulked to raise the plants for next generation. The reproductive structures of the CMS plants improved significantly after five cycles of honey bee foraging-mediated selection. Seed yield/plant, nectar quantity, size of nectary and flower size improved significantly in all three CMS lines at the end of the selection cycles. Quantity of nectar increased from 0.45 to 1.45 µl and from 0.37 to 1.49 µl per flower in CMS lines Ogu1A and Ogu3A, respectively, after five cycles of selection, quantity of seeds per plant improved from 13.48 to 20.71 g and from 12.77 to 17.57 g in the same CMS lines, respectively. The extent of improvement in nectar quantity and seed yield was comparatively lower in CMS line Ogu2A. The occurrence of different floral deformities was also reduced in all three CMS lines after five cycles of honey bee foraging-mediated selection. Once CMS lines with superior flower structure and normal nectar is developed after complete nuclear substitution, they can be used successfully in development of F_1 hybrids.

Self-incompatibility

Self-incompatibility is the term used to describe the phenomenon of inability of plants having fuctional gametes to set seeds when either self-pollinated or crossed with some of their genetic relative (Frankel and Galum, 1977). Self-incompatibility system in all species of the family Cruciferae is of sporophytic type. The main genetic feature of the single locus multi-alleleic homomorphic sporophytic self-incompatibility may be summarized in the following (Frankel and Gatun, 1977):

(i) Incompatibility is controlled by 1 *S*-locus having several alleles.

(ii) The reaction of the pollen is determined by the genotype of the sporophytic tissue in which it was formed and thus it is controlled by 2 S-alleles.

(iii) Resulting from the above type of control, all the pollen grains of a plant have the same incompatibility reaction regardless of the S-allele received during meiosis.

(iv) The 2 S-alleles may react independently or they may interact by one being dominant over the other.

(v) Active S-allele identity in both pollen and pistil leads to incompatibility.

(vi) The dominant/independence relationships of the S-alleles in the pollen and in the pistil may differ.

An individual plant has only 1 or 2 *S*-factors so the number of *S*-factors or alleles in the line derived from it never exceed two. Where only 1 *S*-factor occurs, all plants are homozygous for it and the line is self-incompatible and sib-incompatible. If the line in the progeny of a plant is heterozygous for the S-factor, the line is also heterozygous for the S-factor. The first generation of an S_xS_y plant consists of S_xS_x, S_xS_y and S_yS_y genotype.

Constraints in exploiting SI lines in delopment of hybrids

- Continuous inbreeding in many *Brassica* crops may lead to complete loss of inbred lines. Therefore, it is more desirable to use tissue culture techniques for propagation of self-incompatible inbred lines.
- Pseudo-incompatibility may lead to self seed instead of hybrid seed as found in cauliflower.
- Maintenance of S-allele lines is costly (hybrid seed will be expensive).
- Effect of environmental factors on the level of self-incompatibility (high temperature and relative humidity) may weaken S-allele action.
- Synchronization of flowering among the SI lines is difficult as found cauliflower.
- Difficulties to produce homozygous inbred lines.

Table 1: Development of cauliflower hybrids utilizing different pollination control mechanisms

Name of hybrid	Pollination control mechanisms/ Parentage / Remarks
Pusa Hybrid-2	Utilization of self-incompatible lines
Pusa Hybrid-3	Utilization of self-incompatible lines
Pusa Kartik Sankar	Utilization of self-incompatible lines
Pusa Hybrid-301	Utilization of cytoplasmic male sterility as female line
Pusa Snowball Hybrid-1	Utilization of cytoplasmic male sterility as female line
Pusa Snowball Hybrid-2	CMS lines as female parent and one doubled haploid line as male aprent

Table 2. Development of cabbage hybrids utilizing different pollination control mechanisms

Name of hybrid	Pollination control mechanisms/ Parentage / Remarks
Pusa Cabbage Hybrid-81	Utilization of cytoplasmic male sterility as female parent
Pusa Cabbage Hybrid-822	Utilization of cytoplasmic male sterility as female parent
Pusa Cabbage-1	Utilization of self-incompatible lines

Breeding for resistance to disease in cole crops (cauliflower and cabbage)

Development of disease resistant varieties is the priority in breeding programmes of cole crops. The evaluation and screening of new sources of resistance to the most common pests and pathogens, the study of the genetic basis of resistance, and the identification of genes for resistance are important goals of breeding programs in cabbage, cauliflower and other *Brassica* vegetables. Genetic basis of multiple disease resistance is complicated due to environmental interactions with pathogens and correlated gene expression of different resistance traits. Notable advances have been made in the use of molecular methods for locating the position of resistance genes on genetic maps of *Brassica* vegetables.

Black Rot

In cauliflower, Sn 445, MGS 2-3, Puakea, EC162587 and RBS-l possess resistance to black rot. Sel-12 evolved through inter-varietal hybridization between Sn 445 and EC12012, has been identified as a source of resistance. Kt-9 (Sel-12 x PSB-l) also showed field resistance. In cabbages, Greenland Hammer, EC10I09 and Spitzkool have been identified. Pusa Mukta (EC24855 x EClO109) has been identified as resistant cultivar. BRH-5 (Pusa Mukta x AC 204) is another combination showing resistance.

Sclerotinia rot

Heading broccoli EC I03576, EC131592, Janavon, Early Winter Adam's White Head, EC177283 and Kn-81 possess resistance to *Sclerotinia* rot. The cauliflower line, Kt-25 identified as Pusa Snowball K-25 has been developed as resistant variety.

Other diseases

MR-l, an introduction from USA, in cabbage carries multiple resistance to soft rot, wire stem, black rot and *Sclerotinia* rot. AC 238, Spitzkool and EC93559 have been found resistant against cabbage yellows. BR-2, MGS 2-3, 3-5-1-1, CC-12, EWAWH in cauliflower and Spitzkool in cabbage possess resistance to downy mildew. MGS 2-3, Puakea, 1-6-1-4, 6-1-2- 1 and Pusa Subhra in cauliflower have resistance against black spot disease. Selection-1301 has high degree of tolerance to diseases like *Sclerotinia* rot, downy mildew and black rot coupled with desirable horticultural traits.

Breeding for resistance to insect pests

In cabbage, cultivars Large Blood Red and Red Pickling (all red types) were identified to possess preferential resistance against caterpillars of cabbage butterfly (*Pieris brassicae*). The white cabbage varieties All Season and Round Sure Head showed resistance to these pests also. Cabbage genotypes IRCH-4 to 6 and KK Cross have shown tolerance to aphids.

Breeding for phytonutrient contents in cole crops

The *Brassica* vegetables are often entitled as "super foods" on account of their health-promoting properties due to the presence of potentially protective phytochemical and nutraceutical compounds. Ascribed to these functional food properties, the intake of cole vegetables is linked with the reduction of several types of cancer, chronic and cardiovascular ailments. Particularly, the antioxidant and anticancer properties of bioactive compounds present in *Brassica* vegetables has gained much attention, and has been investigated extensively over the last few years. Carotenoids (β carotenes and xanthophylls) are lipid soluble antioxidants imparting yellow, orange and red pigments in fruits and vegetables. Some of the carotenoids are precursors of vitamin A (β-carotene, β-cryptoxanthin and γ-carotene). The anticancer activity of carotenoids is suggested to be due to different mechanisms such as reduction in cell proliferation, antioxidant activity on mutagenesis and genotoxicity, immunomodulatory efffects, enhancement of gap-junction cell communication, and increasing apoptosis of cancerous cells. The most abundant forms of carotenoids in *Brassica* vegetables are lutein and β-carotene. The highest content of lutein + zeaxanthin is recorded in kale (3.04-39.55 mg/100g), moderate in broccoli and brussels sprouts (0.78-3.50 mg/100g). In case of orange cauliflower, the single semi-dominant '*Or*' gene mutation confers β-carotene accumulation in the curd tissue of cauliflower. The orange colour in cauliflower exhibiting novel *'Or'* gene mutation is due to massive and highly ordered sheets doing carotenoid (β-carotene) sequestration in plastids, and classified as membranous chromoplasts. The advent of molecular markers has given the impetus to molecular breeding techniques and overcome the locks of conventional breeding. However, through the conventional breeding approach, a β-carotene rich orange cauliflower hybrid 'Jinyu 60' was developed involving CMS line as female parent and homozygous orange inbred 93-4 as donor parent. During the recent years, significant progress has been made in the molecular mapping and identifiation of carotenoid biosynthetic genes in plant species.

Anthocyanins are water soluble flavonoid antioxidants imparting red, blue, purple colour in many flowers, fruits and vegetables. They perform the biological functions in plant defense mechanism against various biotic and abiotic stresses. Numerous regulatory genes including R2R3 MYB transcription factors (TF), basic helix-loop-helix (bHLH) transcription factors, and WD40 proteins have been cloned from different plant species, such as Arabidopsis, maize, petunia, *B. rapa,* cauliflower, red cabbage, purple chinese cabbage, purple kale and other *Brasssica species* and other species. In purple cauliflower a spontaneous mutant, single semi-dominant *Pr* gene encoding MYB transcription factor is responsible for giving tissue specific expression of anthocyanin accumulation in different plant parts. The major anthocyanin compound responsible for anthocyanin pigmentation in purple cauliflower is cyanidin, 3, 5-diglucoside. The R2R3MYB transcription factor, *BoPAP1* gene and transcriptional activation of *BoPAP2* and *BoTT8* accounting for upregulation of anthocyanin biosynthetic structural genes, is responsible for anthocyanin accumulation in purple kale and purple kohlrabi, respectively.

There is currently an increasing interest in the alteration of levels of specific glucosinolates in crop plants as certain glucosinolates have desirable properties in cancer prevention and crop protection. The study of inheritance of genes controlling glucosinolate biosynthesis will allow for manipulation of these genes and the development of lines of cabbage and kale with specific glucosinolate profiles. The biology and biochemistry of glucosinolates has been widely reviewed and significant progress has been made in understanding the genetics of glucosinolate biosynthesis. Recent studies have identified key genes responsible for structure regulation and glucosinolate content in different *Brassica* species. The cloning of some key genes provides the opportunity to engineer *Brassica* crops with specific glucosinolate content by modifying their glucosinolate side-chains. This manipulation can have several possible applications. For example, down regulation of *BoGSL-ALK*, which controls side-chain desaturation and cosegregates with *BoGSL-OH*, responsible for side-chain hydroxylation, could produce brassica varieties lacking the antinutrient progoitrin and would simultaneously produce plants accumulating glucoraphanin as a source of anticarcinogens. Although considerable progress is being made in that field, presently, it is not possible to identify the genes responsible for the variations in glucosinolate content. Most of the genes controlling glucosinolate breakdown are still unknown. In the near future, the isolation of genes of glucosinolate metabolism could facilitate the development of plants with modified glucosinolate profiles. The glucosinolate metabolism will allow the study of the potential of glucosinolates to improve human nutrition and to improve the pest resistance.

Table 5: Genes controlling anthocyanin concentration in different *Brassica* sopecies

Brassica species	**Transcription factor/genes/enzymes**
Brassica oleracea var. *botrytis*	*Pr* gene mutation encoding R2R3 MYB transcription factor
Red cabbage (*B. oleracea* var. *capitata*)	*BoMYB2* with *bHLH* transcription factor *BoTT8*
Purple heading chinese cabbage (*B. rapa* L. *ssp. pekinensis*)	*BrPur*
Ornamental kale (*B. oleracea* L. var. *acephala*)	*BoPr*
B. rapa L. ssp. *chinensis var. purpurea*	*BrEGL3.1* and *BrEGL3.2* encoding *bHLH* transcription factor
Purple kale (*B. oleracea* var. *acephala* f. *tricolor*)	*BoPAP 1*
Brassica rapa	*Anp*
Brassica napus	*BnaA.PL1*
Brassica rapa var. *chinensis*	*BrTT8*
Brassica oleracea var. *gongylodes*	*BoPAP2* and *BoTT8*

References

Bannerot, H., Boulidard, L. and Cauderon, Y. 1977. Unexpected difficulties met with the radish cytoplasm. *Eucarpia Cruciferae Newsl.* **2**: 16

Bannerot, H., Boulidard, L., Cauderon, Y. and Temp, J. 1974. Transfer of cytoplasmic male sterility from *Raphanus sativus* to *Brassica oleracea.* Eucarpia Meeting Cruciferae, Dundee, Scotland, pp. 52–54

Chand, J. 1980. *Studies on selection for yield and quality in cauliflower. Ph.D. Thesis*, Indian Agricultural Research Institute, New Delhi.

Chiang, M.S., Chong, C., Landry, B.S. and Grete, R. 1993. Cabbage, *Brassica oleracea* subsp. *capitata* L. In: *Genetic Improvement of Vegetable crops* (Eds. Kalloo, G. and Bergh, B.O.), Pergamon Press, Oxford, New York, Deoul, Tokyo, pp. 113-155.

Dakouri, A., Zhang, X., Peng, G., Falk, K.C., Gossen, B.D., Strelkov, S.E. and Yu, F. 2018. Analysis of genome-wide variants through bulked segregant RNA sequencing reveals a major gene for resistance to *Plasmodiophora brassicae* in *Brassica oleracea. Sci. Rep.*, **8**: 17657.

Dey, S.S., Bhatia, R., Parkash, C., Sharma, S., Dabral, M., Mishra, V., Bhardwaj, I., Sharma, K., Sharma, V.K. and Kumar, R. 2017. Alteration in important quality traits and antioxidant activities in *Brassica oleracea* with *Ogura* cybrid cytoplasm. *Plant Breed.*, **136**: 400-409.

Dey, S.S., Bhatia, R., Sharma, K., Bhardwaj, I., Pandey, C.D., Parkash, C. and Kumar, R. 2019. Frequent introgression of European cauliflowers in the present day cultivated Indian cauliflowers and role of Indian genotypes in the evolution of tropical cauliflower. *Euphytica*, **215**: 1-17.

Dey, S.S., Bhatia, R., Sharma, S.R., Sharma, K., Parkash, C. and Kumar, R. 2018. Population dynamics in introgression of wild male sterile cytoplasm into *Brassica oleracea*: a small population based breeding modelr. *Sci. Hort.*, **232**: 231-239.

Dey, S.S., Sharma, S.R., Bhatia, R., Parkash, C. and Barwal, R.N. 2011. Superior CMS (Ogura) lines with better combining ability improve yield and maturity in cauliflower (*Brassica oleracea* var. *botrytis*). *Euphytica*, **182**: 187-197.

Dey, S.S., Singh, N., Bhatia, R., Parkash, C. and Chandel, C. 2014. Genetic combining ability and heterosis for important vitamins and antioxidant pigments in cauliflower (*Brassica oleracea* var. *botrytis* L.). *Euphytica,* **195**: 169-181.

Dong, X., Ji, J. and Yang, L. 2019. Fine-mapping and transcriptome analysis of BoGL-3, a wax-less gene in cabbage (*Brassica oleracea* L. var. *capitata*). *Mol. Genet. Genomics*, **294**: 1231-1239.

Fang, Z.Y., Liu, Y.M., Yang, L.M., Wang, X.W., Zhuang, M., Zhang, Y.Y. and Sun, P.T. 2004. Breeding and seed production technique of dominant genic male sterile (DGMs) line and cytoplasmic male sterile (CMS) line in cabbage. *Sci. Agric. Sin.*, **37**: 717-723.

Fang, Z.Y., Sun, PT., Liu, Y.M., Yang, L.M., Wang, X.W., Hou, A.F. and Bian, X.S. 1997. A male sterile line with dominant gene (Ms) in cabbage (*Brassica oleracea* var. *capitata*) and its utilization for hybrid seed production. *Euphytica*, **97**: 265-268.

Frankel, R. and Galun, E. 1977. Pollination mechanisms, reproduction and plant breeding. Monographs on Theoretical and Applied Genetics , Vol 2. Springer, Heidelberg. Berlin,

Han, F., Zhang, X. and Liu, X. 2017. Comparative analysis of genome wide DNA methylation profiles for the genic male sterile cabbage line 01-20S and its maintainer line. *Genes*, **8**: 159.

Hazra, P., Banerjee, M.K. and Chattopadhyay, A. 2022. *Varieties of Vegetable crops in India*, Kalyani Publishers, New Delhi, Ludhiana, 204p.

Ji, J.L., Yang, L.M., Fang, Z.Y., Zhuang, M., Zhang, Y.Y., Lv, H.H., Liu, Y.M. and Li, Z.S. 2017. Recessive male sterility in cabbage (*Brassica oleracea* var. *capitata*) caused by loss of function of BoCYP704B1 due to the insertion of a LTR-retrotransposon. *Theor. Appl. Genet.*, **130**: 1441-1451.

Jyoti, S. and Vashistha, R.N. 1986. Gene effects studies of curd weight in mid-season cauliflower (*Brassica oleracea* var. *botrytis* L.). *Haryana J. Hort. Sci.*, **15**: 263-266.

Kifuji, K., Hanzawa, H., Terasawa, Y., Ashutosh and Nishio, T. 2013. QTL analysis of black rot resistance in cabbage using newly developed EST-SNP markers. *Euphytica,* **190**: 289-295.

Labate, J.A., Robertson, L.D, and Baldo, A.M. 2006. Inflorescence identity gene alleles are poor predictors of inflorescence type in broccoli and cauliflower. *J. Am. Soc. Hortic. Sci.,* **131**: 667-673.

Laila, R., Robin, A.H.K., Yang, K., Park, J.I., Suh, M.C., Kim, J. and Nou, I.S. 2017. Developmental and genotypic variation in leaf wax content and composition, and in expression of wax biosynthetic genes in *Brassica oleracea* var. *capitata*. *Front. Plant Sci.,* **7**: 1972. doi.org/10.3389/fpls.2016.01972

Lal, G., Swarup, V., and Chatterjee, S. S. 1977. Combining ability in early Indian cauliflower. *J.Agril. Sci.*, **89**: 169-175.

Li, H., Liu, Q., Zhang, Q., Qin, E., Jin, C., Wang, Y., Wu, M., Shen, G., Chen, C., Song, W and Wang, C. 2017. Curd development associated gene (CDAG1) in cauliflower (*Brassica oleracea* L. var. *botrytis*) could result in enlarged organ size and increased biomass. *Plant Sci.*, . 2017 , 4:82-94. doi: 10.1016/j.plantsci.2016.10.009

Li, L., Lu, S., Cosman, K.M., Earle, E.D., Garvin, D.F., and O'Neill, J. 2006. β-Carotene accumulation induced by the cauliflower *Or* gene is not due to an increased capacity of biosynthesis. *Phytochem.*, **67**: 1177–1184.

Liu, D., Dong, X., Liu, Z., Tang, J., Zhuang, M., Zhang, Y., Lv, H., Liu, Y., Li, Z., Fang, Z., and Yang, L. 2018. Fine mapping and candidate gene identification for wax biosynthesis locus, BoWax1 in *Brassica oleracea* L. var. *capitata. Front. Plant Sci.*, **9**: 309. doi.org/10.3389/fpls.2018.00309

Lv, H., Fang, Z., Yang, L., Zhang, Y., Wang, Q., Liu, Y., Zhuang, M., Yang, Y., Xie, B., Liu, B. and Liu, J. 2014. Mapping and analysis of a novel candidate Fusarium wilt resistance gene FOC1 in *Brassica oleracea. BMC Genom.*, **15**: 1-10

Nagaoka, T., Doullah, M. A. U., Matsumoto, S., Kawasaki, S., Ishikawa, T., Hori, H. and Okazaki, K. 2010. Identification of QTLs that control clubroot resistance in Brassica oleracea and comparative analysis of clubroot resistance genes between *B. rapa* and *B. oleracea. Theor. Appl. Genet.,* **120**: 1335–1346.

Parkash, C., Dey, S.S., Bhatia, R. and Dhiman, M.R. 2015. Indigenously developed SI and CMS lines in hybrid breeding of cabbage. *Indian J. Hortic.*, **72:** 212-217.

Pelletier, G., Ferault, M., Lancelin, D., Boulidard, L., Dore, C., Bonhomme, S., Grelon, M. and Budar, F. 1994. Engineering of cytoplasmic male sterility in vegetables by protoplast fusion. *Acta Hort.*, **392**: 11-18

Ram, H.H. 1998. *Vegetable Breeding: Principles and Practices*, Kalyani Publishers, Ludhiana.

Ridge, S., Brown, P.H., Hecht, V., Driessen, R.G. and Weller, J. L. 2015. The role of BoFLC2 in cauliflower (*Brassica oleracea* var. *botrytis* L.) reproductive development. *J. Exp. Bot.*, **66**: 125-135.

Saha, P., Kalia, P., Sonah, H. and Sharma, T.R. 2014. Molecular mapping of black rot resistance locus X ca1bo on chromosome 3 in Indian cauliflower (*Brassica oleracea* var. *botrytis* L.). *Plant Breed.*, **133**: 268-274.

Singh, S., Bhatia, R., Kumar, R., Sharma, K., Dash, S. and Dey, S.S. 2018. Cytoplasmic male sterile and doubled haploid lines with desirable combining ability enhances the concentration of important antioxidant attributes in Brassica oleracea. *Euphytica*, **214**: 1-19.

Singh, S., Kalia, P., Meena, R.K., Mangal, M., Islam, S., Saha, S., Tomar, B.S. 2020. Genetics and expression analysis of anthocyanin accumulation in curd portion of Sicilian purple to facilitate biofortification of Indian cauliflower. *Front. Plant Sci.*, **10**: 1766. doi.org/10.3389/fpls.2019.01766.

Su, Y., Liu, Y., Li, Z., Fang, Z., Yang, L. and Zhuang, M. 2015. QTL analysis of head splitting resistance in cabbage (*Brassica oleracea* L. var. *capitata*) using SSR and InDel makers based on whole-genome re-sequencing. *PLoS One,* **10**: e0138073.

Swarup, V. 2006. *Vegetable Science and Technology*, Kalyani Publishers, Ludhians, 656p.

Swarup, V. and Chatterjee, S.S. 1998. Improvement of cauliflower (*Brassica oleraea* L.con var. *botrytis* (L.) Alef. *botrytis* L. In: *Crop Productivity and Sustainability. Shaping the future* (Eds. Chopra, V.L., Singh, R.B. and Varma, A.), *Proc. 2nd Intern. Crop Sci. Congr.*, Oxford and IBH Publishing Co. Pvt. Ltd., New Delhi, Calcutta.

Tanaka, N. and Niikura, S. 2006. Genetic analysis of the developmental characteristics related to the earliness of head formation in cabbage (*Brassica oleracea* L.). *Breed. Sci.* **56**: 147-153.

Verma, T.S. and Chander Parkash. 2006. Temperate Vegetables: Germplasm Resources, their Utilization and Achievements in India. *Indian J. Plant Genet. Resour.*, **19**: 397-403

Zhang, J., Liu, Z., Liu, X., Dong, J., Pang, H. and Yu, C. 2016. Proteomic alteration of a thermo-sensitive male sterility SP 2 S in rapeseed (*Brassica napus*) in response to mild temperature stress. *Plant Breed.*, **135**: 191-199.

Zhao, Z., Sheng, X., Yu, H., Wang, J., Shen, Y. and Gu, H. 2020. Identification of candidate genes involved in curd riceyness in cauliflower. *Int. J. Mol. Sci.*, **21**(6): https://doi.org/10.3390/ijms21061999

25

Onion

Onion, *Allium cepa* under the family Alliaceae (Amarylidaceae) having chromosome No. = $2n \times 2x = 16$ is an important vegetable cum spice crop grown in almost all parts of the world. The four major onion growing countries in the world are China, India, USA and Turkey. Other important onion growing countries are Iran, Italy, Egypt, The Netherlands, Russia, Thailand, Indonesia, Korea, Japan and Brazil. Onions are the fundamental ingredient in many dishes of the Asian countries. In India, important onion growing states are Maharashtra, Karnataka, Tamil Nadu, Andhra Pradesh, Gujarat, Punjab, Haryana, Rajasthan, Uttar Pradesh, Bihar and Madhya Pradesh. In India, onions are cooked for the preparation of stews, curries and as a base for rice pilafs or pilau or used fresh as salads. Onions are widely used in American cuisine, from burgers and sandwiches to soups and stews. Fried onion rings are a popular side dish. Onions are healthy whether consumed as raw or cooked however, raw onions have higher levels of organic sulfur compounds that provide many health benefits.

Origin, domestication and adaptation in India

The geographic origin of the onion is uncertain because the wild onion is extinct and ancient records of using onions span western and eastern Asia. However, it is presumed that domestication of onion likely took place in Southwest or Central Asia comprising Iran, Afganistan, western Pakistan Tadzhikistan and Uzbekistan. Its secondary centre of origin is in the Mediterranean region. Its cultivation has spread in Western Asia from Palestine to India. It entered northern Europe through Italy and north Africa during 500 AD. It was introduced in the USA by the early Spanish explorers around 1625 AD.

Onion is an outstanding example of crop adaptation carried through from the very early times before the Christian era. It was mentioned in the Bible, Koran and early Greek literature. The use of onions as food, medicine and religious object dates back 3200 B.C. in Egypt. The Greeks and Romans were using onion in 400-300 B.C. Onion dates back to antiquity in India. It was cultivated in India about 600 B.C. It was mentioned in Apastamba Dharma Sutra-1 during 860- 300 B.C. and Carak Samhita around 6th century A.D. However, popularity of onion in India might have been promoted by the Mughals.

Onion basically shows perennial/biennial habit and long day bulbing character because of its origin in temperate region. In North India, "Rabi" (winter) onion production has been established with such biennial types. Concentration of onion growing in Peninsular and Westem India tropicalized the crop with the evolution of tropical and short day cultivars. In Western Maharashtra and Gujarat, two crops, one in kharif and other in rabi season, are regularly raised. Adaptation of the genotypes to high temperature, high rainfall and short photoperiod condition has resulted in the development of wide diversity of tropicalized cultivars in Western and Peninsular India.

Crop botany and floral biology

- The leaves are yellowish-to bluish-green and grow alternately in a flattened, fan-shaped swathe; they are fleshy, hollow, and cylindrical with one flattened side; they are at their broadest about a quarter of the way up, beyond which they taper towards a blunt tip; the base of each leaf is a flattened, usually white sheath that grows out of a basal disc.
- From the underside of the disc, a bundle of fibrous roots extends for a short way into the soil; as the onion matures, food reserves begin to accumulate in the leaf bases and the bulb of the onion begins to swell at the advent of certain day-length.
- The bulbs are composed of shortened, compressed, underground stems surrounded by fleshy modified scale (leaves) that envelop a central bud at the tip of the stem; in the autumn (or in spring, in the case of over-wintering onions), the foliage dies down and the outer layers of the bulb become dry and brittle (scale leaves).
- The crop is normally harvested when the leaves die back and the outer scales of the bulb become dry and brittle; if left in the soil over winter, the growing point in the middle of the bulb begins to develop in the spring; new leaves appear and a long, stout, hollow stem expands, topped by a bract protecting a developing inflorescence.
- Onion is biennial, foetid scapigerous and bulbous herb; inflorescence is a terminal cymose umbel carried on a long peduncle called scape, which is hollow, when mature, round in cross section and somewhat swollen at the middle or near the base; the inflorescence is at first enclosed in a spathe of three united membranous bracts, which split open, by the pressure of developing flower buds.
- The common onion has one or more leafless flower stalks that reach a height of 75–180 cm, terminating in a spherical cluster of small greenish white flowers; in an inflorescence, flower opening in the umbel is quite irregular and generally takes a period of 2-3 weeks.
- The flowers are small, white to greenish-white in colour, and number of flowers per umbel may range between 50-2000 depending on the variety, time of planting and bulb characters; the flowers are trimerous, bisexual, actinomorphic, hypogynous and bracteates; the flowers are characterized by the presence of five whorls of three plant parts each, starting at the centre: 3 carpels (united in a single pistil), 3 inner stamens, 3 outer stamens, 3 inner perianth and 3 outer perianth segments.

- Perianth lobes are six, free, imbricate aestivation and are arranged in two whorls as three inner perianth segments and three outer perianth segments.
- The stamens are free, six in number in two whorls as three inner and three outer stamens; androecium hypogynous, filaments exerted, the inner ones toothed at the base; anthers are oblong, bilocular (dithecous); the anthers of the inner whorl dehisce first and dehiscence occurs over a long span between 9.00 a.m. and 5.00 p.m.; all the pollens are shed over a period of 2-3 days.
- The pistil has three-celled (tricarpellary), syncarpous, superior ovary with two ovules in each with axile placentation; the style is filiform, approximately 1 mm long when flower first opens and it elongates to the length of about 5mm, its full length after 1-2 days when all the pollens have been shed.
- Protandry is operative and stigma becomes receptive 1-2 days after the dehiscence of anthers; the fruit is a capsule, loculicidal.
- Seeds are small, somewhat semicircular or oval, compressed with black testa; irregularly shaped and short lived.
- The greater portion of the seed is endosperm; the spirally twisted cylindrical embryo, which is embedded in the endosperm, is made up mostly of the cotyledon; average 100 seed weight is 0.36 gram.

Reproductive behaviour

Onion bears hermaphrodite flowers but it is predominantly cross-pollinated crop due to protandry and wide span of flower opening in the umbel. The flowers are protandrous in nature and stigma becomes receptive one to two days after anthers have all dehisced. Such delay in receptivity of stigma reduces self-fertilization of single flower. The anthers of inner whorl dehisce first and it occurs between 9 a.m and 5 p.m. An inflorescence may continue opening for 2-3 weeks and a plant may be in bloom for more than 30 days. It is pollinated chiefly by honeybees (*Apis* sp.) through transferring the pollens among the flowers within an umbel and between different plants. Blowflies are also excellent pollinators as they go from flower to flower.

Genetics of important characters

The bulb onion, *Allium cepa* L. is diploid species (2n = 2x = 16). Chromosome number of different *Allium* species is mentioned below.

Chromosome number	***Allium*** **species**
2n =16	*Allium cepa* L., *Allium fistulosum, Allium sativum, Allium cepa* var. *ascalonicum* (*Allium ascalonicum*), *Allium x proliferum, Allium galanthum, Allium roylei*
2n = 24	*Allium schoenoprasum*
2n =32	*Allium schoenoprasum, Allium ampeloprasum var. porrum, Allium tuberosum*

Common bulb onion, *Allium cepa* is crossable with *Allium fistulosum.* Genomic evidence has conclusively shown that tree onion (*Allium* × *proliferum*) is a diploid hybrid of the shallot (*Allium cepa* var. *ascalonicum* or *Allium ascalonicum*) and the Welsh onion (*Allium fistulosum*). The wild species *Allium galanthum* and *Allium pskemense* cross with *Allium cepa* but fertility level of the hybrids is low. The interspecific hybrid between *Allium oschaninii* x *Allium cepa* is completely sterile. In onion breeding, *Allium roylei* Stearn has been frequently used as a source of resistance to downy mildew for cultivars of bulb onion, *Allium cepa* L. However, genome dominance, the phenomenon in wide hybrids when one of the parental genomes becomes "dominant," while the other genome turns to be "submissive" is manifested in this inter-specific hybrid. Kopecký *et al.* (2022) demonstrated that in *A. cepa* × *A. roylei* hybrids, chromosomes of *A. cepa* are frequently substituted by those of *A. roylei* and in just one generation, the genomic constitution shifts from 8 *A. cepa* + 8 *A. roylei* chromosomes in the F_1 generation to the average of 6.7 *A. cepa* + 9.3 *A. roylei* chromosomes in the F_2 generation. Screening of the backcross generation *A. cepa* × (*A. cepa* × *A. roylei*) revealed that this shift does not appear during male meiosis, which is perfectly regular and results with balanced segregation of parental chromosomes, which are equally transmitted to the next generation. This indicates that female meiotic drive is the key factor underlying *A. roylei* genome dominance. This dominance may express itself in several ways including homoeologous gene expression bias and modified epigenetic regulation. This may hamper employment of such wide hybridization in practical breeding due to the potential elimination of introgressed segments from progeny.

Keller *et al.* (1996) reported interspecific crosses of onion with distant *Allium* species and characterization of the presumed hybrids by means of flow cytometry, karyotype analysis and genomic *in situ* hybridization. The viable hybrid of the most distant cross resulted from crossing *A. cepa* with *A. sphaerocephalon. Allium fistulosum* possesses a number of traits which would be desirable in *A. cepa*. Peffley and Hou (2000) first demonstrated bulb-type onion introgressants possessing *Allium fistulosum* L. genes recovered from interspecific hybrid backcrosses *between A. cepa and A. fistulosum.*

Colour of anthers, foliage, seed coat, dwarf scape, restoration of cytoplasmic male sterility and resistance to different diseases like, pink root (caused by the fungus *Phoma terrestris*), basal rot (caused by the fungus *Fusarium oxysporum* f.sp. *cepae*) are different qualitative and simply inherited characters of onion. Yellow seedling character has genetic linkage with glossy foliage. Onion bulb colour may be white, red, brown and yellow. Five major genes, I, C, R, G and L control the inheritance of bulb colour. There are several genes which intensify bulb colour. Two genes control wax production in foliage and scapes; recessive alleles at one locus determines glossy foliage and the second conditions glossy scapes. However, glossy scape is only expressed in presence of glossy foliage.

Bulb yield is positively correlated with bulb diameter, bulb weight, number of leaves per plant and plant height but negatively correlated with total soluble solids content. Total soluble solids content is negatively correlated with bulb size suggesting that comparatively low yielding genotype with small bulb size contain high total soluble solids.

Table 1: Associated genes/QTLs and nature of gene action for some important characters in onion

Trait	Genes/QTLs
Albino seedlings	*a/a*
Yellow seedling linked with glossy	y^1/y^1
Yellow seedling not linked with glossy	y^2/y^2
Pale green seedling	*pg / pg*
Virescent seedling	*v/v*
Glossy foliage	*gl/gl*
Exposed anther	*ea/ea*
Yellow anther	*ya/ya*
Pink root resistance	*pr/pr*
Male sterility	*ms/ms*
Bulb colour	*I-C-R-*
Red, homozygous	*ii-CC-RR*
Red, heterozygous	*ii-Cc-RR, ii-CC-Rr, ii-Cc-Rr*
Yellow, homozygous	*ii-CC-rr*
Yellow, heterozygous	*ii-Cc-rr*
White, homozygous recessive	*ii-cc-RR, ii-cc-rr*
White, heterozygous recessive	*ii-cc-Rr*
White, homozygous dominant	*II- -*
White, homozygous recessive (buff colour)	*Ii- -*

Source: Ram (1998)

The quantitative characters controlled by polygenes include number of leaves per plant, number of rings per bulb, ring thickness, bulb firmness, bulb diameter, bulb weight, bulb shape index (bulb height over diameter), bulb shape (round, pyriform flallish-round), ring per bulb, bulb dormancy, days to maturity of bulb, bulb yield, total soluble solids content, dry matter content, pungency in terms of pyruvic acid concentration, storability of bulb.

Bulb size, bulb maturity and dry matter content of bulb are controlled by additive gene action. Inheritance of number of leaves per plant, bulb yield and total soluble acid content is governed by both additive and non-additive gene action. Predominance of non-additive gene action control bulb shape.

Male sterility

Cytoplasmic-genic male-sterility in onion was first reported from the variety Italian Red 13-53 (Jones and Emsweller, 1936). Male sterility is determined by the interaction of a cytoplasmic factor (*N* for normal cytoplasm and *S* for sterile cytoplasm) and a nuclear gene (*msms* for sterility and *MsMs* /*Msms* for fertility). The male sterile line (CMS) line has the genetic constitution *Smsms* and male fertile lines are *Nmsms*, *NMsMs*, *NMsms* and *SMsms*, *SMsMs*. The male sterility restorer line is *Nmsms* because the hybrid *Smsms* x *Nmsms* will be male sterile.

Breeding objectives

The following characters are considered important in the improvement programme of onion. However, foliage color in green onion and shape and colour of onion bulb are most important characters for choosing the product in the market (Solanki *et al.*, 2015).

- Varieties/ hybrids for kharif season should be high yielding and tolerant to onion smudge (*Colletotrichum circinans*) and purple blotch (*Alternaria porri*) and longer storage life of bulb.
- Varieties/hybrids for rangda season (late kharif) should be high yielding, with low bolting and twins with better keeping quality and resistant to *Stemphylium* blight (*Stemphylium vesicarium*) and thrips (*Thrips tabaci).*
- Varieties/hybrids for rabi season should be high yielding and having dark red as well as light red colour with very good storabiity and resistant to *Stemphylium* blight (*Stemphylium vesicarium*) and thrips (*Thrips tabaci*).
- Yellow onion varieties/hybrids for export in European countries should be high yielding.
- White onion varieties/hybrids with good bulb quality (size, shape, colour, pungency, firmness, dormancy and high TSS content (> 20%) for dehydration.
- Red or light red onion varieties/hybrids with good bulb quality (size, shape, colour, pungency, firmness) and longer storage life of bulb.
- Pungency should be determined indirectly by measuring enzymatically produced pyruvic acid in individual bulbs; pyruvic acid concentration lower than 3 μmol/g fresh weight corresponds to sweet onion, 3-7 μmol/g fresh weight to moderately pungent and > 7 μmol/g fresh weight to pungent onion.
- Resistance to important diseases *viz*., purple blotch (*Alternaria porri*), basal rot (*Fusarium oxysporum* and *Fusarium proliferatum*), stemphylium blight (*Stemphylium vesicarium*), bacterial storage rot (*Pseudomonas*, *Erwinia, etc.*)
- Resistance to insect pests particularly thrips (*Thrips tabaci).*
- Varieties/hybrids resistant to abiotic stresses (moisture stress, high temperature, salinity, alkalinity) and resistant to bolting.

Breeding methods and development of variety

Onion is biennial in nature for flowering and seed production but grown as an annual crop for bulb production. In the bulb production, seedlings are transplanted in the field for the production of bulb however, for seed production, the bulbs after harvesting are stored for at least 6 months and then replanted in winter for the initiation of flowering and seed set in the long day condition with the advent of spring. It is a cross-pollinating crop due to protandry and insect pollinated mainly by bees. Onion shows pronounced inbreeding depression and generally inbreeding is not done for more than two generations. Loss of vigour can be restored through crossing among the inbred lines.

The commonly practiced population improvement schemes include mass selection, line breeding and selfing and massing however, mass selection is the predominant method of selection. It is a biennial crop hence, each cycle of selection takes two seasons covering two years to complete. Hybridization and bulk or other population improvement schemes are followed to develop some varieties. Heterosis breeding utilizing cytoplasmic male sterile or female line (*Smsms*) or "A", sterility maintainer line (*Nmsms*) or "B" line and pollen parent or male line (*NMsMs*) or "C" line is the most important and widely practiced onion improvement programme in most of the developed countries.

Huge genetic diversity available in the rabi onion of the Northem and Eastern India have been utilized mainly by mass selection and other population improvement schemes to develop many varieties. Wide arrays of germplasm of short-day kharif onion have also been improved mainly by mass selection to develop varieties suitable for kharif season only or for both rabi and kharif seasons. Vegetatively propagated multiplier onions have been developed through selection of germplasm from Tamil Nadu. Commercial varieties of onion generally fall under *Allium cepa* L. (Common onion) and *A. cepa* var. *aggregatum* (Multiplier onion).

Winter or Rabi onion

These are basically long day cultivars or cultivars requiring intermediate photoperiod and bulb formation in them is not initiated unless their longer critical value of photoperiod for bulb formation is reached. Rabi onion can broadly be classified according to skin colour *viz*., red skinned varieties and white or silver skinned varieties (Hazra *et al*, 2022).

Common onion

- **Deep red-skinned varieties:** Pusa Red, Pusa Ratnar, Hisar Onion 4, Udaipur-101, Udaipur-103, Nasik Red (N-2-4-1), Arka Pragati, Arka Bindu, Punjab Naroya, Punjab Red Round, Rajasthan Onion-1(RO-1), Aprita (RO-59), Line-28, Pusa Riddhi, Arka Bheem, Arka Akshay, Arka Vishwas, Bhima Shakti, NHRDF Red, NHRDF Red 4, PRO 6, PRO 7, Gujarat Junagadh Red Onion 11, Palam Lohit
- **Light-red or Bronze-red skinned varieties:** Bhima Light Red, Punjab Selection, Arka Niketan, Pusa Madhvi, Hisar 2, Hisar Onion 3, Agrifound Light Red, Kalyanpur Red Round, Patna Red, VL Piaz 3, HOS-1, Bhima Kiran, NHRDF Red 2, NHRDF Red 3
- **White or Silver-skinned varieties:** N-257-9-1, Pusa White Round, Pusa White Flat, Punjab 48 (S-48), Udaipur-102, Punjab White, Akola Safed, Arka Yojith, Arka Swadista, Agrifound White, N-257-9-1, Phule Safed

  Akola Safed, PWO-2, Gujarat White Onion 1, Gujarat Junagadh White Onion 3, Gujarat Anand White Onion 2 (GAWO 2)
- **Yellow-skinned varieties:** Early Grano, Brown Spanish, Bermuda Yellow, Arka Pitambar, Phule Suvarna, Arka Sona, PYO 1

Kharif onion

These are so-called short day cultivars in which bulb formation is initiated when shorter critical photoperiodic value limit for bulb formation is passed, and bulb development is continued under short day condition. A number of varieties in India can be cultivated in *Kharif* season with transplanting in mid July to beginning of August, particularly in Maharashtra condition and harvesting during November to December. Karak and Hazra (2014) recorded that bulbing in all the 14 varieties occurred at 11-hour photoperiod which classified these materials under short day category. Planting of seedlings in the first week of October and harvesting of bulbs during last week of February has been found to be the best option for growing early-rabi onion under New Allubial zone of West Bengal.

• Deep or light red-skinned varieties

Niphad 53 (N 53), Arka Kalyan, Agrifound Dark Red, Baswant 780, Phule Samarth, Bhima Super, Bhima Shubra, Bhima Safed, Bhima Red, Bhima Raj, Bhima Dark Red

Rose onion

Rose onion varieties have bulbs with flat base and are spherical. The other characteristics which make them unique are their deep scarlet red colour, high anthocyanin, phenols and high pungency. They also contain higher levels of protein, phosphorus, iron and carotene. Rose onions are grown in the districts of Bangalore Urban, Bangalore Rural, Chikkaballapur and Kolar. They were also grown in Tumkur, Hassan, Davanagere, Dharwad and Bagalkot of Karnataka. Rose onions are not cultivated in any other place in India. Few varieties of Rose onion have been developed in India.

Agrifound Rose, Arka Vishwas, Bangalore Rose

Multiplier onion

Multiplier onion (*Allium cepa* var. *aggregatum*) is also known as Aggregatum group of onion or Potato onion. It forms closely packed clusters of small sized bulblets underground, rather than on the surface like the common onion. Multiplier onion sets produce abundant offsets from each bulb planted. In Thailand, Indonesia, the Philippines, Sri Lanka and India, multiplier onions are grown on a commercial scale for export and for internal consumption. These are commonly hardy and early maturing types. Commercial cultivation of multiplier onion is limited. Varieties under this onion group so far developed have red skin colour.

• Red-skinned Varieties

CO-1, CO-2, CO-3, CO-4, CO-5, MDU-1, Agrifound Red, Arka Ujjwal

Bunching onion

Bunching onion (*Allium fistulosum*) is commonly called Welsh onion, long green onion, Japanese bunching onion and spring onion. It is a perennial species and does not develop bulbs, and possesses hollow leaves and scapes. The species is very similar in taste and odor to the common onion. *A. fistulosum* can be multiplied by perennial evergreen clumps.

Onion breeding in India

Organized and systematic onion breeding in India was initiated during 1960 at Nashik and consecutively at IARI, New Delhi, IIHR, Bengaluru and National Horticultural Research and Development Foundation (NHRDF), Nashik in 1977 (Lawande *et al.*, 2009). NHRDF played vital role by developing high yielding varieties and production technology. During early nineties, State Agricultural Universities (SAUs) and All-India Coordinated Research Project (AICRP) strengthened onion research in the country through development of new varieties for various agro-climatic zones with refined agronomical practices. To boost productivity with specific focus on onion research, ICAR established National Research Center on Onion and Garlic (NRCOG) at Nashik in 1994 that was shifted to Rajgurunagar in 1998 and upgraded as Directorate of Onion and Garlic Research in 2009. ICAR institutes along with few SAUs such as MPKV Rahuri, PAU Ludhiana and CCSHAU Hisar have contributed towards onion research. In India, it was well emphasized to develop F_1 hybrids in short-day tropical onion during late nineties, but not much success has been achieved and is still an imperative and attention seeking aspect of onion research (Lawande *et al.*, 2009). More than 65 onion varieties, including 2 F_1 hybrids and 6 varieties of multiplier type have so far been developed and released from public sectors for different locations and seasons.

Germplasm resources and Plant introduction

Germplasm repositories play an integral role in the collection, preservation, and distribution of germplasm. Numerous well-established seed banks and germplasm repositories, spanning both national and international levels, maintain extensive collections of onions and other *Allium* genetic resources include Asian Vegetable Research and Development Center (AVRDC), Taiwan, Nordic Genetic Resources Center (NordGen) representing a genebank for the Nordic countries (including Denmark, Finland, Iceland, Norway, and Sweden), Brazilian Agricultural Research Corporation, New Zealand Institute of Crop and Food Research, National Plant Germplasm System, USDA, USA, Vegetable Germplasm Bank of Zaragoza, Spain, etc.

In India, wide range of germplasm of long day cultivars or cultivars requiring intermediate photoperiod (Rabi onion) and short day cultivars (Kharif onion) are maintained at National Bureau of Plant Genetic Resources (NBPGR), New Delhi; ICAR- Directorate of Onion and Garlic Research, Rajgurunagar, Pune; National Horticultural Research and Development Foundation (NHRDF), Nashik; different ICAR institutes and few State Agricultural Universities *viz*., ICAR-IARI, New Delhi; ICAR-IIHR, Bengaluru; MPKV Rahuri, Maharashtra; PAU, Ludhiana, Punjab, CCSHAU, Hisar, Haryana, etc.

Yellow or brown-skinned onion variety have a greenish-white, light yellow, or white inside, very mild and its layers of papery skin have a yellow-brown or pale golden colour. This group of onion cultivars are mostly cultivated in Eurpean countries and USA. Several varieties of this group *viz.*, Early Grano, Granex, Brown Spanish, etc. introduced from USA, Spain, South Africa and Phillippines have been released directly for cultivation mainly in the hills (1000 m and above).

Table 2: Exotic variety directly released for cultivation

Name of the variety	Country from where introduced
Early Grano	USA (yellow-skinned variety)
Granex	USA (yellow-skinned variety developed at Texus through crossing 1015 x Bermuda
Red Creole	South Africa
Brown Spanish	Spain (yellow-skinned variety)
Bermuda Yellow	Phillipines (yellow-skinned variety)

Variability and population improvement programme

Onion is essentially a cross-pollinated crop due to protandry and wide span of flower opening in the umbel. It influences greatly the breeding methodology adopted for improvement. Onion also shows abrupt inbreeding depression, especially after two generations of selfing, which makes it impossible to go in for pure line selection or pedigree method of breeding. Usually, the cross-pollinated crops are highly heterozygous to such a degree and so many new gene combinations emerge in the course of continuous cross-pollination that systematic selection combined with controlled pollination always makes it possible to develop a new variety with new properties. Different selection methods or population improvement schemes for cross-pollinated vegetable crops are applicable for onion *viz.*, mass selection, recurrent selection, line breeding, family breeding, etc. However, mass selection and bulb-to-row are the common method of population improvement. Even in seed production, mass selection of bulbs is necessary, since in seed production without bulb selection, rapid deterioration of the purity of the cultivars usually occurs.

In order to minimize the adverse effects of inbreeding, Jones and Mann (1963) recommended alternate selfing and sibmating. The selected individual bulbs are planted in isolation and selfing is done in each plant. In the second year, progeny rows are grown and considering the typical characters of the cultivar or characters under selection, the progeny rows are selected strictly to the conformity of the character under selection. Further selection of bulbs is done within the progeny. The selected bulbs are planted in isolation for sibmating among the selected progeny. The next cycles of selection are repeated till the uniformity in the characters is obtained. Even in this method, a certain amount of loss in vigour in noticed. Only one variety, Pusa Madhvi has so far been developed by “Selfing and massing” method of population improvement.

The bulb shape index (bulb height over diameter) is influenced by environment, particularly time taken for bulb diameter to increase. Bulb size, bulb diameter, bulb yield, bolting percentage, dry matter content, seed stalk per plant and sprouting percentage show high heritability hence, these characters respond positively during selection. Bulb yield is positively correlated with bulb diameter, bulb weight, number of leaves per plant and plant height hence, these characters deserve due attention as important selection indices for bulb yielb. Negative correlation of bulb size and bulb yield with total soluble solids content indicate that there will be less likelihood of selecting the genotype having big bulb size with high total soluble solids content.

Mass selection and bulb-to-row method of population improvement has been adopted for the development of most of the Rabi and Kharif varieties in India utilizing huge indigenous diversity *viz.*, Pusa Red, Nasik Red, Arka Pragati, Bhima Shakti, Arka Niketan, Agrifound Dark Red, Baswant 780, Agrifound Rose, etc. Few yellow-skinned variety *viz.*, Arka Sona, PYO 1, etc. have been developed through mass selection from exotic genetic materials.

Cultivars of Multiplier onion (*Allium cepa* var. *aggregatum*) group generally do not produce seeds and thus propagated vegetatively by bulblets. However, some types produce sparse seeds. The plants can also remain in the ground as perennial clumps of bulbs year after year. Several varieties of multiplier onion *viz.*, CO-1, CO-2, CO-3, CO-4, CO-5, MDU-1, Agrifound Red, etc. have been developed by clonal selection from the indigenous clonal variability.

Table 3: Varieties developed through population improvement programme

Name of variety	Indigenous / Exotic material
Pusa Red	Indigenous, Mass selection
Hisar Onion 4	Indigenous, Mass selection
Udaipur-101	Indigenous, Mass selection
Udaipur-103	Indigenous, Mass selection
Nasik Red	Indigenous, Mass selection
Arka Pragati	Indigenous, Mass selection
Arka Bindu	Indigenous, Mass selection
Punjab Naroya	Indigenous, Mass selection
Punjab Red Round	Indigenous, Mass selection
Rajasthan Onion-1	Indigenous, Mass selection
Arpita	Indigenous, Mass selection
Pusa Riddhi	Indigenous, Mass selection
Arka Vishwas	Indigenous, Mass selection
Bhima Shakti	Indigenous, Mass selection
NHRDF Red	Indigenous, Mass selection
NHRDF Red 4	Indigenous, Mass selection
PRO 6	Indigenous, Mass selection
PRO 7	Indigenous, Mass selection
Gujarat Junagadh Red Onion 11	Indigenous, Mass selection
Palam Lohit	Indigenous, Mass selection
Bhima Light Red	Indigenous, Mass selection
Punjab Selection	Indigenous, Mass selection

Name of variety	**Indigenous / Exotic material**
Arka Niketan	Indigenous, Mass selection
Pusa Madhvi	Indigenous, Selfing and massing
Hisar 2	Indigenous, Mass selection
Hisar Onion 3	Indigenous, Mass selection
Agrifound Light Red	Indigenous, Mass selection
Kalyanpur Red Round	Indigenous, Mass selection
Patna Red	Indigenous, Mass selection
HOS-1	Indigenous, Mass selection
Bhima Kiran	Indigenous, Mass selection
NHRDF Red 2	Indigenous, Mass selection
NHRDF Red 3	Indigenous, Mass selection
N-257-9-1	Indigenous, Mass selection
Pusa White Round	Indigenous, Mass selection
Pusa White Flat	Indigenous, Mass selection
Punjab 48	Indigenous, Mass selection
Udaipur-102	Indigenous, Mass selection
Punjab White	Indigenous, Mass selection
Akola Safed	Indigenous, Mass selection
Arka Yojith	Indigenous, Mass selection
Arka Swadista	Indigenous, Mass selection
Bhima Shweta	Indigenous, Mass selection
Agrifound White	Indigenous (Nimad area of Madhya Pradesh), Mass selection
N-257-9-1	Indigenous (Maharashtra), Mass selection
Phule Safed	Indigenous (Kagal , Maharashtra), Mass selection
Akola Safed	Indigenous, Mass selection
PWO-2	Indigenous, Mass selection
Gujarat White Onion 1	Indigenous, Mass selection
Gujarat Junagadh White Onion 3	Indigenous, Mass selection
Gujarat Anand White Onion 2	Indigenous, Mass selection
Arka Sona	Exotic, Mass selection; yellow-skinned onion
PYO 1	Exotic, Mass selection; yellow-skinned onion
Niphad 53	Indigenous, Mass selection; Kharif onion
Arka Kalyan	Indigenous (Ralawana taluk , Maharashtra), Mass selection; Kharif onion
Agrifound Dark Red	Indigenous (Nasik , Maharashtra), Mass selection; Kharif onion
Baswant 780	Indigenous (Maharashtra), Mass selection; Kharif onion
Phule Samarth	Indigenous (Sangamner , Maharashtra), Mass selection; Kharif onion
Bhima Super	Indigenous, Bulb-to-row selection; Kharif onion
Bhima Shubra	Indigenous, Mass selection; Kharif onion

Name of variety	Indigenous / Exotic material
Bhima Safed	Indigenous, Mass selection; Kharif onion
Bhima Red	Indigenous, Bulb-to-row selection; Kharif onion
Bhima Raj	Indigenous, Bulb-to-row selection; Kharif onion
Bhima Dark Red	Indigenous, Bulb-to-row selection; Kharif onion
Agrifound Rose	Indigenous, Mass selection; Rose onion
Arka Vishwas	Indigenous, Mass selection; Rose onion
Bangalore Rose	Indigenous ("Gulabi Eerulli"), Mass selection; Rose onion
Pusa Soumya	Indigenous, Mass selection; Bunching onion
Pusa Ratnar	Exotic 'Granex' variety introduced from USA

Table 4: Varieties of multiplier onion developed through clonal selection

CO-1	Indigenous, Clonal selection; Multiplier onion
CO-2	Indigenous, Clonal selection; Multiplier onion
CO-3	Indigenous, Clonal selection; Multiplier onion
CO-4	Indigenous, Clonal selection; Multiplier onion
CO-5	Indigenous, Clonal selection; Multiplier onion
MDU-1	Indigenous, Clonal selection; Multiplier onion
Agrifound Red	Indigenous, Clonal selection; Multiplier onion

Development of synthetic variety

In onion considerable heterosis is manifested but commercial production of F_1 hybrid seeds needs utilization of inherent pollination control mechanisms like cytoplasmic-genic male sterility. In onion, synthetic variety can be developed with advantage by intercrossing a number of genotypes of known superior combining ability i.e., genotypes that are known to give superior hybrid performance when crossed in all combinations. Synthetic varieties are known for manifesting hybrid vigour and for their ability to produce usable seed for succeeding seasons. In this population improvement scheme, additive genetic variance is basically exploited in developing a synthetic variety. In India, two synthetic varieties *viz*., Arka Bheem and Arka Akshay have been developed through tri-parental open pollination in isolated field.

Table 5: Development of synthetic variety

Name of variety	Indigenous / Exotic material
Arka Bheem	Tri-parental synthetic variety
Arka Akshay	Tri-parental synthetic variety

Hybridization and population improvement programme

Intervarietal hybridization or hybridization between the genotypes having specific character (as for example, bulb colour) is useful in recombinant breeding. Different population improvement approaches particularly, mass selection is adopted to handle the segregating generations for developing a new variety. A yellow-skinned variety, Arka Pitambar has been developed through hybridization of white-skinned line, UD 102 x IHR 396, a red-skinned line followed by mass selection. Another yellow-skinned variety, Phule Suvarna has been developed through hybridization of RHR-200 x RHR-155 followed by mass selection.

Table 6: Varieties developed through hybridization and population improvement

Name of variety	Parentage / Remarks
VL Piaz 3	BYG 2207 (an exotic line) x Indigenous line followed by 3 generation of mass selection
Arka Pitambar	UD 102 (White) × IHR 396 (red); yellow-skinned variety
Phule Suvarna	RHR-200 x RHR-155; yellow-skinned variety

Utilization of mutants and Mutation breeding

Applied mutagenesis has become one of the important ways for modern crop breeding because it may shorten breeding cycle and break the linkage of disadvantageous characters. In onion, mutation breeding was attempted with a view to improving some specific characters. Low doses of some mutagens induced stimulatory effects on seedling height and bulb weight in onion. However, the increasing doses of mutagens decreased germination, seedling height, survival, bulb weight and seed setting. The damages due to mutagens were not similar to all the onion however, 125 gray (Gy) gamma-radiation increased the induced mutations rate to 14 times and different morphological mutations affecting foliage, bulb formation, flowering, seed and growth habit. Both chemical and physical mutagens (ethyl-methane-sulphonate, N-methyl-N-nitrosourea and ethyl-imine along with gamma radiation increased TSS content from 12.12 to 14.01% in Pusa Red and 10.88 to 13.07% in White Warangal (Kataria and Singh, 1990).

Kim *et al.* (2004a) identified a new locus (Pink; P) responsible for a pink trait in onions resulting from natural mutations of anthocyanidin synthase. This unusual pink onion was isolated in haploid populations induced from an F_1 hybrid between yellow and dark red parents and in F_3 populations originating from the same cross. Segregation ratios of red to pink in F_2, backcross and F_3 populations indicated that this pink trait is determined by a single recessive locus. The transcript level of anthocyanidin synthase (ANS) was significantly reduced in the pink line. Reduced transcription of the ANS gene caused by mutations in a cis-acting element is likely to result in the pink trait in onions. Unlike the five loci determining qualitative colour changes, this locus is the first reported gene whose mutant phenotype results in reduced intensity of a color in onions. Kim *et al.* (2004b) also observed a natural mutation that yield an unusual gold-colored onion from a F_3 family originating from a cross between US type yellow and Brazilian yellow onions. The gold onions contained a significantly reduced amount of quercetin. They identified a premature stop codon and a subsequent single base-pair addition causing a frameshift in the coding region of the chalcone into flavanone (CHI) gene in the gold onions.

Inactivation of the gene (DFR-A) coding for dihydroflavonol 4-reductase (DFR) involved in the anthocyanin biosynthesis pathway results in a yellow bulb color in onion. Phylogenetic analysis of DFR-A alleles showed that all inactive alleles were independently derived from four different active alleles. In addition, the close relatedness and diversity of DFR-A mutants implied that all these mutations might have occurred after domestication of onions and had probably been maintained by artificial selection. Song *et al.* (2014) suggested that at least nine independent natural mutations of the DFR-A gene are responsible for appearance of yellow onions from red progenitors.

A laser induced new red onion variety" Chang Ji 99-3 was developed (Li, 2004). A new yellow-peel onion variety "Changji 07-9" with high yield, high quality, multi-resistance and stronger adaptability was bred from the variation progenies of Japanese yellow-peel onion "96203", irradiated by laser. Yang *et al.* (2017) isolated the glossy, bright green welsh onion (*Allium fistulosum* L.) (GLBG) which was derived from a natural variation in the field, and it was tentatively demonstrated that the GLBG or glossy phenotype is conditioned by a single recessive gene.

Heterosis breeding

Being a cross-pollinated crop, heterosis breeding in onion can be taken up profitably, both to avoid ill effects of inbreeding depression and to bring about uniformity in all the economic characters, especially in maturity. Importance of heterosis in onion has attracted plant breeders ever since the discovery of pollination control mechanisms like cytoplasmic male sterility. Different developed countries like USA, UK, Netherlands, Japan, Korea exploited heterosis extensively to develop onion hybrids. Exploitation of heterosis in onion was possible commercially in these countries due to the isolation, maintenance and easier production of cytoplasmic male sterility.

Production of hybrid onions became possible with the discovery of male sterile cytoplasm in the onion cultivar 'Italian red' by Jones and Emsweller (1936). They propagated this male sterile bulb clonally and used it in crosses with different pollen parents and demonstrated that these F_1 showed heterosis for yield. In their classical paper on inheritance of male sterility, Jones and Clarke (1943) showed that sterility was determined by the combination of a cytoplasmic factor '*S*' together with a recessive nuclear gene in its homozygous form *ms/ms*. This hypothesis enabled them to devise practical methods for the commercial production of F_1 hybrid seeds using cytoplasmic male sterile lines and for the regeneration of the male sterile lines by pollination with a maintainer line of constitution *Nmsms*. Later on, another form of cytoplasm conferring male sterility *i.e.*, CMS-(*T*) was discovered in the onion cultivar 'Jaune paille des Vertus' (Berninger, 1965). It was reported that one independent and two complementary genes (three recessive genes *a*, *b* and *c*) were involved in fertility restoration of the male sterility and in very low temperature condition, male fertility sometimes is restored. However, of the two systems, CMS-S system is most widely used because of its stability in various environments (Havey, 2000). Modern onion breeding is mainly based on hybrid seed production by means of CMS induced by the interaction between S or T cytoplasm and one or three nuclear genes.

Interspecific hybrids between *Allium cepa* and *A. ampeloprasum* were generated as a first step for the introduction of *S*-cytoplasm from onion into leek. Pre-zygotic barriers of crossability were observed after the arrival of pollen tubes at the end of the style when entering the cavity. Nevertheless, micropyle penetration of pollen tubes and the formation of hybrid embryos were also observed. After accomplishing *in vitro* culture of ovaries and ovules successively, triploid hyrbid plants with 24 chromosomes were obtained. Their hyrbid nature was confirmed by RAPD analysis, genomic *in situ* hybridization, and morphological analysis. Southern hybridization with a cytoplasmic probe indicated the transfer of unaltered *S*-cytoplasm into the hybrid plants.

Normal (*N*) fertile and (*S*) sterile cytoplasms are distinguishable by restriction-enzyme (RE) analysis of the chloroplast genome. RE analysis of the chloroplast DNA established that *S* cytoplasm has been introgressed into OP onion cultivars since its discovery. A new source of sterility has been backcrossed into onions from *A. galanthum* (Havey, 1998).

Heterosis breeding in India

India, being the major onion producing country, still lag behind to exploit heterosis in onion crop. That is the primary reason for lesser productivity in India compared to other onion growing countries. The open-pollinated improved varieties, at present in cultivation, are by the large, highly heterozygous with at least 15-20 per cent off-types in respect of maturity, bulb size and shape, etc. Further, in the short period available for vegetative growth and bulb development, especially in the *rabi* onion crop of North India, which follows an early winter potato crop, the F_1 hybrids would likely to prove more advantageous for quicker bulb development and maturity.

To develop heterosis breeding programme in Indian scenario, some of the critical breeding objectives should be:

- Production of suitable and stable inbred lines
- Identification of male sterile lines and their maintainers from local and commercial varieties grown in India
- Exploitation of heterosis for enhancing productivity and wider adaptability
- Resistance/tolerance against various biotic and abiotic stresses
- Development of hybrids having desired quality and high storability with great potential to export.

The first hybrid in the name VL 67 was released for Uttarakhand Hills in 1973 (Khar and Saini, 2016). After the initial success, few attempts have been made by Indian scientists for identification and isolation of male sterile plants from indigenously grown cultivars. Male-sterile plants were identified from the local cultivars like Niphad-2-4-1, Nasik White Globe, IIHR-20 and Pusa Red. Systematic work on hybrid onion was started at IIHR, Bengaluru. It has been found that only two varieties namely Arka Pragati (Indian variety) and Red Coral (exotic variety) possessed S cytoplasm which was the initial report on S-cytoplasm in Indian short-day onion having *S* cytoplasm. Pathak and Gowda (1993) tried to exploit exotic male sterile lines in their heterosis breeding program at IIHR, but they failed to get success owing to short photoperiodic conditions. Furthermore, Pathak (1997) obtained male sterile hybrids through crossing between a male-sterile line and fertile lines. He observed that all the test crosses were sterile and opined that either all the fertile lines used in the crossing possess *ms* recessive loci or there is no control of nuclear genes on sterile cytoplasm. Based on his findings, he hypothesized that cytoplasm of Indian onion was different from *S* and *T* cytoplasm and was indeed a new form of male sterile cytoplasm.

Further, he observed that flowers having light green anthers were male sterile and dark green anthers were male fertile. Khar and Saini (2016) recorded that male sterile flowers do possess light green, dark green or yellow anthers. Saini *et al.* (2015) also didn't observe any correlation among the male sterility and anther colour, but they observed irregular production of fertile pollen in male-sterile lines that might be because of higher temperature condition which induce pollen production. Later, IIHR Bengaluru released two F_1 onion hybrids namely, Arka Kirthiman and Arka Lalima however, these hybrids did not get commercial popularity and adoption among the onion growers at National level (Khar and Saini, 2016). Manjunathagowda and Anjanappa (2020) identified male sterile lines from Indian onion population. They observed that male sterile flowers had longer style, narrow stigmatic knob, produced functionally non-viable pollens and perianths did not open fully. Light green to slightly yellowish translucent, shrivelled, seldom empty anther without pollen grains with fused anther sacs were observed.

Sheemar and Dhatt (2015) used a mitochondrial DNA based marker cytochrome b (cob), which was used to identify *S* and *N* cytoplasms in onion, to cytoplasmically characterize Indian onion populations. The oligonucleotides amplified 414 bp DNA for *S*-cytotype and 180 bp for *N*-cytotype in open-pollinated populations, namely, Punjab Naroya, Punjab Selection, Punjab White, PRO-6, PBR-4 and P-266. Out of the 200 seedlings screened in each population, 19 in Punjab Naroya, 15 in Punjab Selection and 5 in Punjab White showed amplification of 414 bp revealing *S*-cytoplasmic proportion of 9.5, 7.5 and 2.5%, respectively. While PRO-6, PBR-4 and P-266 showed amplification of 180 bp only, indicating the absence of *S*-cytoplasm. The results suggest that Punjab Naroya, Punjab Selection and Punjab White can be exploited for development of male sterile (A) and maintainer (B) lines, while PRO-6, PBR-4 and P-266 can be utilized to isolate maintainer and restorer (R) lines. The study established a rapid method of cytoplasm identification in onion, which is the pre-requisite to develop CMS lines. In contrast, 4–8 years are taken by conventional approach. This technique identifies cytoplasm within 15 days of sowing and proves its practical efficacy for future onion F_1 breeding in the country. Malik *et al.* (2017) tried marker assisted selection (MAS) using mitochondrial DNA based marker cytochrome b (cob) integrated with phenotypic evaluation to isolate male sterile and maintainer lines from open-pollinated onion varieties adapted to North Indian agro-climatic region. They determined the cytotype (N/S) by cob marker followed by morphological and microscopic study of pollen and discovered male sterile plants (*Smsms*) at frequencies of 0.015 in Punjab Naroya, 0.020 in Punjab Selection, and 0.006 in Punjab White. They reported that the progeny scoring of test-crosses between male sterile and N-cytoplasmic plants isolated the maintainers (*Nmsms*) at frequencies of 0.133 in Punjab Naroya, 0.231 in Punjab Selection and 0.182 in Punjab White. As a novel approach, they used Trait Recovery Programme to reduce the population size required to recover a male sterile plant by 91.08% in Punjab Naroya, 92.99% in Punjab Selection and 97.66% in Punjab White. For recovering a maintainer, they calculated 10% reduction in Punjab Naroya and 9.10% in Punjab Selection. It validated that in a randomly mating onion population, frequency of recessive *ms* allele squared is equal to the frequency of male sterile plants among S-cytotype and frequency of maintainers among N-cytotype (*fms2* = *fSmsms/fS* = *fNmsms/fN*). Emphasis should be given for the development of short day onion hybrids by using male sterile lines for higher yield and economization of cost of hybrid seed production.

Induction of doubled haploid (DH) for inbred development

For heterosis breeding programme of any crop, development of complete homozygous inbred lines plays a vital role. It is very easy to develop inbred lines in self-pollinated crop as compared to cross-pollinated crops such as onion. This bulbous crop exhibits greater heterozygosity, outbreeding nature, biennial growth habit, protandry and massive genome size. High inbreeding depression in onion led to decline of yield due to reduction of bulb size after 2-3 generations of selfing. To overcome above-mentioned issues, double haploid (DH) development is an important biotechnological tool to speed up the inbred development. In onion, use of anthers has not been ideal, but *in-vitro* gynogenesis has been preferred for haploid induction (Khar *et al.,* 2019). For *in-vitro* culture of flower buds, the perfect stage is around 3-5 mm size that corresponds to 3 to 5 days earlier to anthesis (Alan *et al.,* 2004). An efficient and simplified single step protocol consisting of entire flower bud culture on *in-vitro* induction medium until embryo stage developed is widely adopted by the onion breeders across the world. Gamborg's B5 medium fortified with 2 mg/l 2,4-D and 2 mg/l BA was found to be the best media for short day Indian onion for the haploid induction (Khar *et al.*, 2018).

Hybrid seed production

Hybrid seed production of onion involves male-sterile line as a female parent with a genetic constitution of sterile (S) cytoplasm and homozygous recessive *Ms* locus (*msms*) and a male fertile line known as maintainer line or B lines having normal (N) cytoplasm and a homozygous recessive *Ms* locus (*msms*). Hence, the sterile parent (A line) will have a genetic constitution of *Smsms* and a maintainer line (B line) will have a genetic constitution of *Nmsms*. The B line is used to maintain the sterile line. For hybrid seed production, 'C' line, a male fertile line, with normal (N) cytoplasm and any of the genetic constitution at *Ms* locus (*MsMs/Msms/msms*) can be used. Both A lines and B lines are isogenic and are developed by backcrossing for 4-5 generations and C lines are developed by inbreeding of superior male fertile lines for at least two generations and then sibmating for two generations to achieve uniformity. In onion, the main part to be used as vegetable is bulb, it is immaterial whether the C line is maintainer or restorer line. In case of other crops like rice, where seed is the economical part, a restorer as "C" line is needed. In hybrid seed production of onion, it is very important to select the 'C' line which should be diverse from the A line so that heterosis in terms of yield and productivity is achieved.

Table 7. Development of onion hybrids in India utilizing cytoplasmic-genic male sterility

Name of hybrid	**Remarks**
Arka Kirthiman	High yielding F_1 hybrid suitable for paste making. TSS -12.4^{0}Brix & dry matter-13.5%. Yield 37t/ha in 140 days
Arka Lalima	High yielding F_1 hybrid suitable for paste making. TSS -11.8^{0}Brix & dry matter-12%. Yield 40t/ha in 140 days

Breeding for resistance to biotic stresses

In onion the major diseases are purple blotch (caused by the fungus *Alternaria porri*), *Stemphylium* rot (caused by the fungus *Stemphylium vesicarium*) and anthracnose (*Colletotrichum gloeosporioides*) in the field and black mould (caused by the fungus *Aspergillus niger*) and *Penicillium* rot in the storage. Some reports are available on the partial tolerance to these diseases. Sources of resistance to stemphylium leaf blight was identified in Welsh onion (*A. fistulosum*) and this trait was transfered to onion lines (Pathak *et al.*, 2001).

Insufficient disease resistance in onions prompted the researchers to explore wild-related species for source of resistance. Downy mildew resistance governed by single dominant gene originating from *A. roylei* could be introgressed into *A. cepa* through marker-assisted breeding (Scholten *et al.*, 2007). Kim *et al.* (2021) developed the SCAR-OPAN1 and SNP-3 HRM markers that could be utilized in a breeding programme in order to reduce the damage caused by the grey mold disease. Different viruses *viz.*, potyviruses (e.g. onion yellow-dwarf virus, leek yellow-stripe virus), carlaviruses (e.g. garlic latent virus, shallot latent virus) , and garlic and shallot virus X are the most devastating *Allium* viruses. While *in vitro* elimination is possible (Fletcher *et al.*, 1998), inbuilt resistance would provide a simpler solution. Researchers have isolated and sequenced coat-protein gene sequences from *Allium* carla virus and and potyvirus types (Tsuneyoshi *et al.*, 1998). Screening for resistance to the virus was undertaken with 30 Welsh onion (*Allium fistulosum*) varieties under artificial inoculation and more than 80 varieties under natural field infection. It revealed that varieties with dark coloured leaves or with a thicker leaf wax layer were considered most resistant.

Hamilton *et al.* (1999) suggested that greater genetic gains for thrips resistance in onion can be achieved by selection on a family basis rather than using single plant selection.

Different Indian varieties and breeding lines *viz.*, Bellary Red, L-819 and Arka Kalyan during *kharif* and NHRDF Red3, Bellary Red, Bheema Red and Bheema Shubra during *rabi* were moderately resistant to thrips. Other set of genotypes *viz.*, Bhima Super, NRCRO-4, NRCWO-3 and Arka Niketan showed tolerance to both onion thrips and purple blotch disease.

Table 8: Sources of resistance for different biotic stresses

Biotic stresses and causal organism	**Source of resistance**	**Genes/QTLs**
Purple blotch (*Alternaria porri* (Ellis) Ciferri)	Arka Kalyan is source for ApR1 gene Other resistant cultivars: IIHR 56-1, VL-67, Pusa Red, IIHR 25, N-2-4-1, Punjab Selection-3, and Pusa Ratnar	A single dominant gene, *ApR1,* determine resistant to purple blotch, Seven putatively linked ISSR and SSR markers flank the ApR1 gene, which may be useful in MAS.
Stemphyllium blight *(Stemphylium vesicarium)*	Five lines of welsh onion (*A. fistulosum*) showed resistance to *Stemphyllium* blight and was reported to be under dominant gene control (Pathak *et al.*, 2001) Other resistant cultivars: IIHR IC-32176, 42900, 47954, 48001, 48575, 49121 and 48503, RO-1, Pusa Soumya, Red Creole 2	-
Leaf blight (*Botrytis squamosa*)	*A. roylei*	A single dominant gene, *Bs 1 resistant to leaf blight* QTL for resistance identified on the chromosome 6
Downey mildew (*Peronospora destructor* [Berk.] Casp.)	*A. roylei* Resistant inbred line No. 136	Two genes *Pd* 1 and *Pd* 2 determine downey mildew resistance
Fusarium basal rot	*A. fistulosum;* other resistant cultivars are (Poona Red, Patna Red, N-257-9-1. IIHR-141, Telagi Red, White Large, Bellari Red, IIHR-506 and Udaipur 1037	Steroidal saponins compound, alliospiroside A showed antagostic report
Anthracnose (*Colletotrichum* spp*)*	Moderate resistance: red onion DOGR-1666 and DOGR Hy-50	
Thrips (*Thrips tabaci*) transmit Iris yellow spot virus	N-2-4-1, Sel-104, Pusa Ratnar, Kalyanpur Red Round, Udaipur-103, White Persian and BDR	Cuticlar and epicuticular wax depositions on onion leaves, Two loci on Chro 2 and 5 for thrips resistance

Source: Mahajan and Gupta (2023)

Breeding for resistance to abiotic stresses

Onion crop is highly sensitive to moisture deficit stress due to its shallow root system, which restricts its water absorption capacity zone to only about 18-25 cm of the top soil. Water deficit stress results into split, double centric bulbs, sometime skin cracking of bulb causing tremendous yield losses.

Bulb yield of onion is not severely affected under 10-30% water deficit stress however water deficit stress beyond 30% during any growth phase reduces the bulb yield by 15-30%. Water deficit stress during bulb enlargement stage resulted into significant yield losses with poor size bulbs and post harvest losses are also higher with 50% water deficit stress at this stage. Similarly, in onion seed crop, the reproductive stage is highly sensitive to water deficit stress limiting the seed production and quality. Water deficit decrease the amount of sugary nector produced by the florets that reduce the number of pollinator visit. Water deficit stress also reduces the amount of viable pollen grains per flower, increase floret abortion and finally contributes to the lower seed yield. Bolting and anthesis stage are the most sensitive for water deficit stress in seed crop.

Identification of water deficit tolerance genotypes is the prerequisite in stress breeding programme. In addition, stay-green characters like high chlorophyll content with better photosynthesis efficiency and less leaf senescence rate are considered valuable traits for water deficit stress toletance. Likewise, induced ROS scavenging mechanism with increased antioxidant enzyme activity for maintaining the osmoregulation within the plant that confer the drought tolerance further reduces the risk of extreme water deficit situation. ICAR-DOGR identified onion genotypes W-397 (white bulb colour) and Accession 1656 (red bulb colour) which performed better under limited moisture condition and Bhima Shubhra has moderate water deficit tolerance ability.

Onion is a relatively salt sensitive crop. It is postulated that short root hairs of onion contribute to a lower salt tolerance, whereas long root hairs of other crop may show salt tolerance due to enhanced water uptake from saline soils (Schleiff, 2008). Soil salinity influence bulbing and the quality of the bulbs harvested. Salts cause alterations in key physiological processes, due to the salt-induced osmotic stress and the specific ionic effect, which in turn result in water deficit, ionic toxicity and plant nutritional imbalances (García *et al.*, 2020). Cultivar variation was recorded for tolerance to soil salinity and identification of onion cultivars which are able to maintain productivity at low or moderate levels of salt stress may provide a cost-effective solution to this situation. García *et al.* (2020) suggested that the lower sensitivity to salt in some onion genotypes could be related to a better performance of the antioxidant machinery under salinity conditions. They elaborated that as salt stress being first perceived by the root system, the root apoplastic antioxidant defenses were enhanced in the salt-resistant genotype compared to the salt-sensitive genotype. El-baky *et al.* (2003) compared the degree of lipid peroxidation, CAT, SOD and POX enzymes behaviour and glutathione content in leaves of three onion varieties under saline conditions and they found that the most tolerant cultivar showed an increased antioxidant capacity. Sudha and Riazunnisa (2015) identified Agrifound white as tolerant one and LINE - 28 as susceptible to salt stress.

In India, heavy rainfall or soil flooding is a serious environmental consequence that severely limits onion production during *kharif* season. The overall damage in onion bulb yield due to heavy rainfall in standing crop ranges from 50-80% and sometime prolonged waterlogged condition causes the total crop failure. Onion being a shallow rooted crop is extremely sensitive to the stress due to excess moisture condition. Leaf yellowing due to premature leaf senescence and wilting are the visual morphological symptoms of excess moisture stress. Flooded soils also suffer from low light that affect photosynthesis. Additionally, it limits the other gas exchange processes like transpiration and respiration in both leaves and roots. Excess soil moisture causes the negative impact on plant growth and development by reducing the plant inorganic nutrient absorption efficiency. Heavy rainfall and high atmospheric moisture during *kharif* season favours out break of various diseases both in nursery and standing crop. Identification of excess moisture stress and waterlogging tolerant genotypes is required to address this problem. W-208 (white onion) and KH-M-2 (red onion) were identified as promising lines which can withstand frequent to prolonged excess moisture condition. Additionally, Bhima Shweta, Bhima Shubhra and Bhima Dark Red have moderate tolerance to waterlogging.

Table 9: Sources of tolerance to different abiotic stresses

Abiotic stresses	Source of tolerance
Moisture deficit stress	Acc. 1656, Acc. 1658, W-009, and W-085, Breeding lines U12, U47, U49 and K52
Soil salinity	Ac Bog 409, Ac Bog 414, Ac Bog 424, Ac Bog 430, Ac Bog 417, Ac Bog 419, Ac Bog 420, Ac Bog 422, and Ac Bog 425, POS35, NHRDF Red (L-28), GWO 1, POS36, NHRDF Red-4 (L-744), POS37, POS38, U12, U47, U49, K52
Water logging	Acc. 1666, Acc. 1622, W-355, W-208, KH-M-2 and RGP-5 DOGR-1666 and DOGR Hy-50

Source: Mahajan and Gupta (2023)

Breeding for quality

Pungency, sweet taste, texture, total soluble solids contents are important quality characters of onion. The pyruvate scale measures pungency in onion and garlic with units of μmol/g fresh weight (micromoles per gram fresh weight). It is named after pyruvic acid, the alpha-keto acid co-product created in the biochemical pathway that forms syn-propanethial-S-oxide, the main lachrymatory agent in onions. It is a member of a class of organosulfur compounds known as thiocarbonyl S-oxides. It is a volatile liquid that acts as a lachrymatory agent (triggers tearing and stinging on contact with the eyes). Pyruvic acid concentration lower than 3 μmol/g fresh weight corresponds to sweet onion, 3-7 μmol/g fresh weight to moderately pungent and > 7 μmol/g fresh weight to pungent onion.

The most important characters for the processing type cultivars for dehydration are uniform white colour, globe shape for convenient handling, high soluble solids (more than 20° brix) and insoluble solids preferably more of non-reducing sugar, and ratio of reducing to non-reducing sugar should be low to lessen discolouration and browning during drying, high degree of pungency (pyruvic acid 0.50-0.70%) and its better retention after dehydration (Mitra *et al*., 2012). Other important characters include better recovery, small root zone and free from doubles, etc.

Pal and Singh (1987) reported that pyruvic acid content which is positively associated with storage and drying quality is controlled by both additive and dominance gene effects, the former being more important. While breeding, attention should be given on good sized bulbs having high solids, since these characters are usually negatively correlated. Development of non-lachrymatory onion can be possible by suppressing the lachrymatory factor synthase gene while increasing the yield of thiosulphinate (Imai *et al.*, 2002).

Lachrymatory factor (LF) is the cause of the pungent aroma associated with cut onions and the cause of heat and pungency flavour notes. Disulfide and dihydrothiophene compounds are associated with the sweeter aroma of cooked or fried onions. Trans- and cis-zwiebelanes (any of a series of 1,3-dithietane compounds) are associated with sweet raw onion sulfur tastes and sweet brown saute flavour notes, respectively. Many countries prefer sweet and less- or non-pungent salad type onions. The "tearless" onion cultivars (e.g. 'Vidalia') could be developed through deficient uptake and partitioning of sulfur and/or growth in sulfur-deficient soils. Yoo *et al.* (2020) studied the recurrent selective breeding programme to produce low-pungency lines in short-day type onions. They tested two parent populations, 'T81079' (n = 754) and 'T81082' (n = 347), with mean pungencies of 4.1 and 3.5 mM pyruvic acid, respectively, for pungency. Genetic shifts towards the selection differential were observed. The fourth cycle of selection from a population with a mean pungency of 1.5 mM pyruvic acid was recorded. The genes controlling pungency became fairly homogeneous after four generations. The recurrent selection of low pungency shifted the pungency distribution toward lower pungency levels, and therefore developing low-pungency (1.5 mM pyruvic acid) lines will be feasible in about 10 years in short-day onions. A selection pressure of 1–3 % is necessary for fast genetic progress.

However, Eady *et al.* (2008) genetically manipulated the sulfur secondary metabolite pathway of onion using RNAi. They successfully decreased endogenous LFS (lachrymatory factor synthase) production through a single simple transformation event. This unambiguously lowered LFS activity in the plant, and the consequence of this was a significant reduction in the production of the irritant LF (lachrymatory factor). This reduced lachrymatory synthase activity by up to 1,544-fold, so that when wounded the onions produced significantly reduced levels of tear-inducing lachrymatory factor. They noted that this silencing had shifted the trans-S-1-propenyl-L-cysteine sulfoxide breakdown pathway so that more 1-propenyl sulfenic acid was converted into di-1-propenyl thiosulfinate. A consequence of this raised thiosulfinate level was a marked increase in the downstream production of a non-enzymatically produced zwiebelane isomer and other volatile sulfur compounds, di-1-propenyl disulfide and 2-mercapto-3,4-dimethyl-2,3-dihydrothiophene, which had previously been reported in trace amounts or had not been detected in onion.

Onion cultivars not only vary in pungency and sweetness, but also in flavour profile which is reflected by differences in flavour precursor ratios and the type of soluble sugar accumulated (Randle and Galmarini, 1997). Barbieri *et al.* (1998) evaluated volatile components contents in onion cultivars and found that cv. Density had more aroma-giving volatile and sulphurous compounds. Kandoliya *et al.* (2015) studied the nutritional quality along with various parameters contributing antioxidant activity from onion of different red (AGFL Red, Pillipati, JDRO-07-13, Talaja Red) and white (JWO-05-07, GWO-1, PWF-131) type local varieties. They recorded 58.14 to 77.67 % DPPH value, comparable amount of flavanoids (0.422 to 1.232 mg/g) and anthocyanin content along with total phenol (8.96-18.23 mg/100g), pyruvic acid (1.09 to 1.33 mg/g), ascorbic acid (1.18 to 3.89 mg/100g) , protein (0.79 to 1.27%) and titrable acidity (0.34 0.75%). The red variety, JDRO-07-13 and white variety GWO-1 was found nutritionally better due to higher antioxidant property, proteins, carbohydrates and reducing sugar. Aggarwal *et al.* (2016) analyzed the bioactive compounds in seven genotypes of onions (Punjab Naroya, Punjab White, PRO-6, P-305, POH-1, POH-2 and CLX-18021) and PRO-6 contained maximum anthocyanin (2.32 mg/100g), ascorbic acid (11.40 mg/100g) and total phenols (283.57 mg/100g). P-305 contained maximum total carotenoids (86.43 μg/100g) and β-carotene content (13.72 μg/100g) and PRO-6 had maximum antioxidant activity (13.63%). Dangi *et al.* (2018) recorded significant variability in different morphological and biochemical traits of 58 onion genotypes. Pusa White Flat was the highest yielder and 106BS2 had maximum TSS content (15.30 ± 3.03°B). Arka Kalyan had highest dry matter (15.70 ± 0.57%), Red Creole 3 had highest pyruvic acid (6.32 ± 0.17 μmol/ml) and total phenolic content was recorded highest (36.96 ± 2.00 mg/g fresh weight) in Juni. They also identified genotypes with waxy leaves and firm skin. Highly significant correlation emerged between TSS and dry matter (r = 0.45), pyruvic acid and dry matter (r = 0.35) and significant and negative correlation between TSS and TPC (total phenolic content) (r =-0.32).

Quercetin is a trait of interest in onion breeding programmes due to its potential health benefit. Smith *et al.* (2003) estimated total flavonol (TF) content in the short-day onion 'Lady Raider' (red and grano shaped) which registered 79 to 431 mg/kg TF in the parent population. They grouped the bulbs by TF concentration into high (>232 mg/kg), medium (203-223 mg/kg), or low (<203 mg/kg). These three populations were sib-pollinated which formed three Sib-one (S1) populations which designated as Sib-one high (S1H), Sib-one medium (S1M), and Sib-one low (S1L) based on the parent population from which it was generated. They noted different TF ranges in the sibs. TF values of the S1 populations ranged from 228 - 675 mg/kg. S1H had the highest TF flavonol mean (456 mg/kg), S1M had medium-range (329 mg/kg), and S1L had the lowest mean (286 mg/kg). It was concluded that flavonol content in the S1 generation segregated into classes similar to those of parental populations and quercetin content was highly heritable and selection for TFs for increased levels could be achieved.

Breeding for wide adaptation

Bulbing in onion is a photoperiodically driven process. Besides day length requirement, temperature is yet another pivotal factor which governs bulbing and flowering in onion. Both of these physiologically antagonistic phenomena are driven by the same group of genes. Economic production of bulb requires adaptation to photoperiod and temperature such that a bulb is formed in the first year followed by emergence of inflorescence (umbel) in the second year. 'Bolting', or premature flowering before bulb maturation, is an undesirable trait which is strongly bred out during adaptation of germplasm.

In short day onion, very low (<10°C) temperature during bulb development leads to bolting of onion. On the other hand, excessive high temperatures (>42°C) at the time of maturity during April-May leads to reduction in bulb sizes. Many varieties developed worldwide are restricted for cultivation to specific season or climate and sensitive to climate change. In India short day varieties (Kharif or late-*kharif* onion) have been developed but these varieties cannot be cultivated in *rabi* season and *vice-versa.* Efforts were initiated to develop varieties which can sustain environmental vagaries and the variety 'Bhima Super', 'Bhima Raj', and 'Bhima Red' were identified which can be grown in all the three seasons *viz. kharif* (May to October), late-*kharif* (August to February) and *rabi* (October to April). In white onion group, the varieties Bhima Shubhra and Shweta are less influenced by photoperiod and temperature and can sustain up to greater extent in changing climate (Mahajan and Gupta, 2016; 2023).

Baldwin *et al.* (2014) conducted linkage mapping and population genetic analyses of candidate genes, and QTL analysis of bolting using a low-density linkage map. They performed tagged amplicon sequencing of ten candidate genes, including the FT-like gene family, in eight diverse populations to identify polymorphisms and sought evidence of differentiation. They observed low nucleotide diversity for most of the genes, which were consistent with purifying selection. Significant population differentiation was observed only in AcFT2 and AcSOC1. Genotyping in a large 'Nasik Red × CUDH2150' F_2 family revealed genome regions on chromosomes 1, 3 and 6 associated with bolting. Two F_2 families grown in two environments confirmed that a QTL on chromosome 1, designated as AcBlt1, consistently conditioned bolting susceptibility in this cross.

Breeding for increased shelf-life

The bulbing and yield of dry onion vary among the cultivars however, they depend on the contents of high dry matter and non-structural carbohybrates (fructo-oligosaccharides) which contribute to storability. Different study suggested positive and significant association of bulb yield with TSS and number of rings per bulb, and neck thickness was significantly correlated with rotting and total storage loss.

References

Aggarwal, P., Rajput, H., Dhatt, A.S. and Kaur, A. 2016. Bioactive compounds and antioxidant activity in different genotypes of onion. *Agril. Res. J.*, **53** : 97-100.

Alan, A. R., Brants, A., Cobb, E. D., Goldschmied, P. A., Mutschler, M. A. and Earle, E. D. 2004. Fecund gynogenic lines from onion (*Allium cepa* L.) breeding materials. *Plant Sci.*, 167: 1055-1066.

Baldwin, S., Revanna, R., Pither-Joyce, M., Shaw, M., Wright, K., Thomson, S., Moya, L., Lee, R., Macknight, R. and McCallum, J. 2014. Genetic analyses of bolting in bulb onion (*Allium cepa* L.). *Theor. Appl. Genet.*, **127** : 535-547.

Barbieri, G., Bolzoni, L., Siviero, P. and Macchiavelli, L. 1998. Valutazione sul contenuto di componenti volatili nella cipolla. *Informatore Agrario*, **54** : 43-46.

Berninger, E. 1965. Contribution to the study of male sterility in the onion (*Allium cepa* L.). *Annales de l'Amélioration des Plantes*, **15**: 183–199.

Eady, C.C., Kamoi, T., Kato, M., Porter, N.G., Davis, S., Shaw, M., Kamoi, A. and Imai, S. 2008. Silencing onion lachrymatory factor synthase causes a significant change in the sulfur secondary metabolite profile. *Plant Physiol.*, **147** : 2096-2106.

El-baky, A., Hanaa, H., Hussein, M.M. and Amal, M.A. 2003. Influence of salinity on lipid peroxidation, antioxidant enzymes and electrophoretic patterns of protein and isoenzymes in leaves of some onion cultivars. *Asian J. Plant Sci.*, **2** : 633-638.

Fletcher, P.J., Fletcher, J.D. and Lewthwaite, S.L. 1998. *In vitro* elimination of onion yellow dwarf and shallot latent viruses in shallots (*Allium cepa* var. *ascalonicum* L.). *New Zealand J. Crop Hort. Sci.*, **26** : 23-26.

García, G., Clemente-Moreno, M.J., Díaz-Vivancos, P., García, M. and Hernández, J.A. 2020. The apoplastic and symplastic antioxidant system in onion: Response to long-term salt stress. *Antioxidants (Basel),* **67** : 1-17.

Hamilton, B.K., Pike, L.M., Sparks, A.N., Bender, D.A., Jones, R.W., Candeia, J. and Franca, G. de 1999. Heritability of thrips resistance in the 'IPA-3'onion cultivar in South Texas. *Euphytica*, **109** : 117-122.

Havey, M.J. 1998. Male sterile cytoplasms used commercially to produce hybrid-onion seed. In: *Proc. Nat. Onion (and Other Allium) Res. Conf.,* 10–12 December,1998 University of California, Davis, California, USA, pp. 35–38.

Hazra, P., Banerjee, M.K. and Chattopadhyay, A. 2022. *Varieties of Vegetable crops in India.* Kalyani publishers, Ludhiana, 215p.

Imai, S., Tsuge, N., Tomotake, M., Nagatome, Y., Sawada, H., Nagata, T. and Kumagai, H. 2002. An onion enzyme that makes the eyes water. *Nature*, **419** : 685.

Jones, H.A. and Clark, A.E. 1943. Inheritance of male-sterility in the onion and the production of hybrid seed. *Proc. Amer. Soc. Hort. Sci.*, **43** : 189-194.

Jones, H.A. and Emsweller, S.L. 1936. A male-sterile onion. *Proc. Amer. Soc. Hort. Sci.*, **34** : 582-585.

Jones, H.A. and Mann, L.K. 1963. *Onion and their Allies, Botany, Cultivation and Utilization,* World Crops Books, Leonard Hill (Books) Ltd., London, 169p.

Kandoliya, U.K., Bodar, N.P., Bajaniya, V.K., Bhadja, N.V. and Golakiya, B.A. 2015. Determination of nutritional value and antioxidant from bulbs of different onion (*Allium cepa*) variety: a comparative study. *Int. J. Curr. Microbiol. App. Sci,* **4** : 635-641.

Karak, C. and Hazra, P. 2014. Influence of environment on growth and bulbing of onion (*Allium cepa* L.) under Gangetic alluvial plain of West Bengal. *Indian J. Agril. Sci.* ,84: 492-497.

Kataria, A. S. and Singh, N. 1990. Mutation studies in onion (*Allium cepa* L.). IV-effect of mutagens on total soluble solid contents. *Indian J. Hort.*, 47: 216-219.

Keller, E.R.J., Schubert, I., Fuchs, J. and Meister, A. 1996. Interspecific crosses of onion with distant *Allium* species and characterization of the presumed hybrids by means of flow cytometry, karyotype analysis and genomic *in situ* hybridization. *Theor. Appl. Genet.,* **92** : 417–424.

Khar, A., and Saini, N. 2016. Limitations of PCR-based molecular markers to identify male-sterile and maintainer plants from Indian onion (*Allium cepa* L.) populations. *Plant Breed.*, 135: 519-524.

Khar, A., Islam, S., Kalia, P., Bhatia, R. and Kumar, A. 2019. Present status of haploidy research in onion (*Allium cepa*) –A review. *Indian J. Agril. Sci.,* **89** : 396-405.

Khar, A., Kumar, A., Islam, S., Kumar, A. and Agarwal, A. 2018. Genotypic response towards haploid induction in short day tropical Indian onion (*Allium cepa* L.). *Indian J. Agril. Sci.* **88**: 709–713.

Kim, S. J., Park, J. S., Park, T., Lee, H. M., Choi, J. R., and Park, Y. D. 2021. Development of molecular markers associated with resistance to gray mold disease in onion (*Allium cepa* L.) through RAPD-PCR and transcriptome analysis. *Horticulturae*, **7**: 436. DOI:10.3390/horticulturae7110436

Kim, S., Binzel, M.L., Yoo, K.S., Park, S. and Pike, L.M. 2004a. Pink (P), a new locus responsible for a pink trait in onions (*Allium cepa*) resulting from natural mutations of anthocyanidin synthase. *Mol. Genet. Genom.*, **272** : 18-27.

Kim, S., Jones, R., Yoo, K.S.and Pike, L.M. 2004b. Gold color in onions (*Allium cepa*): a natural mutation of the chalcone isomerase gene resulting in a premature stop codon. *Mol. Genet. Genom.*, **272** : 411–419.

Kopecký , D., Scholten , O., Majka , J., Burger-Meijer , K., Duchoslav , M. and Bartoš , J. 2022. Genome Dominance in *Allium* Hybrids (*A. cepa* × *A. roylei*). *Front Plant Sci.*, 2022; **13**: 854127. doi: 10.3389/fpls.2022.854127

Lawande, K.E., Khar, A. and Mahajan, V. 2009. Onion and Garlic research in India. *J. Hort. Sci.*, **4:** 91-119

Li, C.Z. 2004. A laser induced new red onion variety" Chang Ji 99-3". *Acta Laser Biol. Sin.* **13**: 258-261.

Mahajan, V. and Gupta, A. J. 2016. Development of photo and thermo-insensitive cultivars of short day onion in India. *Acta Hort.*, **1143**: 61-68.

Mahajan, V. and Gupta, A. J. 2023. Onion: Breeding and Genomics. *Veg. Sci.*, **50** (Spl.): 244-260.

Malik, G., Dhatt, A.S. and Malik, A. 2017. Isolation of Male Sterile and Maintainer Lines from North-Indian Onion (*Allium cepa* L.) Populations with the Aid of PCR-Based Molecular Marker. *Intern. J. Plant Res.*, 30(2):94 DOI:10.5958/2229-4473.2017.00142.2

Manjunathagowda, D. C., and Anjanappa, M. 2020. Identification and development of male sterile and their maintainer lines in short-day onion (*Allium cepa* L.) genotypes. *Genet. Res. Crop Evol.*, **67**: 357-365.

Mitra, J., Shrivastava, S. L. and Rao, P. S. 2012. Onion dehydration: a review. *J. food Sci. Tech.*, 49: 267-277.

Pal, N. and Singh, N. 1987. Analysis of genetic architecture for pungency (pyruvic acid) in onion (*Allium cepa* L.). *Curr. Sci.*, **56** : 719-720.

Pathak C S and Gowda R. 1993. Breeding for the development of onion hybrids in India: problems and prospects. *Acta Hort.*, 358: 239-42.

Pathak C S. 1997. A possible new source of male sterility in onion. *Acta Hort.*, 433: 313–316

Pathak, C.S., Black, L.L., Cherng, S.J., Wang, T.C.. Ko, S.S. and Armstrong, J. 2001. Breeding onions for stemphylium leaf blight resistance. *Acta Hort.*, **555** : 77-81.

Peffley, E.B. and Hou, A. 2000. Bulb-type onion introgressants possessing *Allium fistulosum* L. genes recovered from interspecific hybrid backcrosses between *A. cepa* L. and *A. fistulosum* L. *Theor. Appl. Genet.,* **100** : 528-534.

Ram, H.H. 1998. *Vegetable Breeding: Principles and Practices*, Kalyani Publishers, Ludhiana

Randle, W.M. and Galmarini, C.R. 1997. Genetic and environmental effects influencing flavor in onion. *Acta Hort.*, **433** : 299-311.

Saini, N., Hedau, N.K., Khar, A., Yadav, S., Bhatt, J.C. and Agrawal, P.K. 2015. Successful deployment of marker assisted selection (MAS) for inbred and hybrid development in long-day onion (*Allium cepa* L.). *Indian J. Genet.*, **75** : 93-98.

Schleiff, U. 2008. Analysis of water supply of plants under saline soil conditions and conclusions for research on crop salt tolerance. *J. Agron. Crop Sci.*, **194** : 1-8.

Scholten, O. E., Van Heusden, A. W., Khrustaleva, L. I., Burger-Meijer, K., Mank, R. A., Antonise, R. G. C. and Kik, C. 2007. The long and winding road leading to the successful introgression of downy mildew resistance into onion. *Euphytica*, 156: 345-353.

Sheemar, G. and Dhatt, A.S. 2015. PCR-based rapid identification of Indian onion populations possessing S-cytoplasm for isolation of CMS lines. *Research on Crops*, **16** : 133-138.

Smith, C., Lombard, K.A., Peffley, E.B. and Liu, W. 2003. Genetic analysis of quercetin in onion (*Allium cepa* L.)'Lady Raider'. *The Texas J. Agril. Nat. Resour.*, **16** : 24-28.

Solanki, P., Jain, P.K., Prajapati, S., Raghuwanshi, N., Khandait, R.N. and Patel, S. 2015. Genetic analysis and character association in different genotypes of onion (*Allium cepa* L.). *Intern. J. Agric., Environ. Biotech.*, **8** : 783-793.

Song, S., Kim, C.W., Moon, J.S. and Kim, S. 2014. At least nine independent natural mutations of the DFR-A gene are responsible for appearance of yellow onions (*Allium cepa* L.) from red progenitors. *Mol. Breed.*, **33** : 173-186.

Sudha, G.S. and Riazunnisa, K. 2015. Effect of salt stress (NaCl) on morphological parameters of onion (*Allium cepa* L.) seedlings. *Inter. J. Plant, Anim. Environ. Sci.*, **5** : 125-128.

Tsuneyoshi, T., Matsumi, T., Natsuaki, K.T. and Sumi, S. 1998. Nucleotide sequence analysis of virus isolates indicates the presence of three potyvirus species in *Allium* plants. *Arch. Virol.,* **143**: 97-113.

Yang, L., Liu, Q., Wang, Y. and Liu, L. 2017. Identification and characterization of a glossy mutant in Welsh onion (*Allium fistulosum* L.). *Sci. Hort.*, **225** : 122-127.

Yoo, K.S., Pike, L.M., Patil, B.S. and Lee, E.J. 2020. Developing sweet onions by recurrent selection in a short-day onion breeding program. *Sci. Hort.*, **266** : 109269. DOI:10.1016

26

Carrot

Carrot, *Daucus carota* L. under the family Apiaceae (Umbelliferae) having chromosome numbers 2n = 2x = 18 is a root vegetable, grown in spring, summer and autumn in temperate regions and during winter in tropical and sub-tropical regions all over the world. It is important root crop of the European countries, USA, China, Japan, Korea and India. It is grown in almost all parts of India, both in plains and hills. Most of the carrots are typically orange in colour, though purple, black, red, white and yellow cultivars are also available. It is consumed fresh as salad or cooked as vegetable. It is also used for fresh juice, pickles and for the preparation of sweets. Leaves are used as fodder. Essential oil from the seeds is used for flavouring liquors and foods. It contains highest amount of β carotene (pro-vitamin A) among the common vegetables. It is also a good source of vitamin K, and vitamin B6.

Origin, domestication and adaptation in India

Carrot probably originated in Afghanistanin, the region where the Himalaya and Hindukush mountains are confluent. The secondary centres of diversity are probably Ehiopia and north America. Its domestication occurred in Afganistan and adjacent regions of Russia, Iran (previously called Persia), India, Pakistan and Anatolia (Van der Vossen and Sambas, 1993). From its centre of origin in Afghanistan, it was taken by the Arabs to eastern Mediterranean and Spain in 12th century. It was introduced in China in 13th or 14th century, in England during 15th century, in Japan in early 17th century and in the USA in 1620 AD.

All the carrot genotypes showing different form and root colour are the domesticated forms of the wild carrot, *Daucus carota* ssp. *carota*, commonly known as "Queen Ann's lace, which is still found in Afganistan, south-western Asia, southern Europe, north Africa and north America. In India, this wild species is also found in Punjab, Kashmir and Himachal Pradesh which amply suggests India as the place of its domestication. It was originally cultivated for its leaves and seeds. The cultivated species *Daucus carota* L. has been developed from the wild species by spontaneous mutation and human selection.The domestic carrot has been selectively bred for its greatly enlarged, more palatable, less woody-textured taproot. The purple and yellow coloured carrots were the first to have been cultivated in the Mediterranean region and Europe as early as 11th century. The white and orange forms developed from yellow forms through mutation. The orange carrots were first cultivated in the Netherlands and adjacent areas in the 17th century. Reddish-orange carrots are cultivated in Germany, the Netherlands and England by mid of 18th century.

Carrot was introduced into India early in its history. The red or purple types might have been introduced from the Central Asian region of its origin-probably Afghanistan and Persia. These types with low carotenoid content were mostly annual, juicy and forming roots under comparatively higher temperature condition. These types showing wide variation in colour, shape and size of roots form the distinct tropical or oriental type. Adapted cultivars of this type have been utilized in combination breeding programmes to develop Oriental or Tropical varieties in India. The orange coloured carrots having higher carotenoid content and believed to have originated in Europe was introduced in India much later with the European traders. Most of the varieties under this group grown at present in India are exotic varieties.

Crop botany and floral biology

- Edible portion of carrot is the hypocotyl together with the taproot; the tap root system develops from the hypocotyls and the secondary lateral branching arises from the xylem.
- Taproots are typically long and conical, although cylindrical and nearly-spherical rooted cultivars are available.
- Flower development begins when the flat meristem changes from producing leaves to an uplifted, conical meristem capable of producing stem elongation and a cluster of flowers.
- Inflorescence is a compound umbel and each umbel contains several smaller umbels (umbellets); the first (primary) umbel occurs at the end of the main floral stem; smaller secondary umbels grow from the main branch, and these further branches into third, fourth, and even later-flowering umbels.
- Compound umbel is the king umbel (primary umbel) or the umbel of first order that flowers first; the secondary, tertiary and other orders of umbels flower at interval of 8-12 days from each other. Flowers come in the peripheral umbellets first followed by inner umbellets.
- The large, primary umbel can contain up to 50 umbellets, each of which may have as many as 50 flowers; subsequent umbels have fewer flowers.
- Individual flowers are small and white, sometimes with a light green or yellow tint; they consist of five petals, five stamens, and an entire calyx.
- Sepals are five in number, gamosepalous; corolla is white, petals are five in number
- Stamens are 5, free, alternipetalous, filaments are equal in length.
- Carpel two, syncarpous, ovary inferior and bilocular, placentation marginal and partital
- The stamens usually split and fall off before the stigma becomes receptive to receive pollen; stamens of the brown, male, sterile flowers degenerate and shrivel before theflower fully opens; in the other type of male sterile flower, the stamens are replaced by petals, and these petals do not fall off.

- A nectar-containing disc is present on the upper surface of the carpels.
- The stamens release their pollen before the stigma of the same flower is receptive; the arrangement is centripetal, meaning the oldest flowers are near the edge and the youngest flowers are in the centre; flowers usually first open at the outer edge of the primary umbel, followed about a week later on the secondary umbels, and then in subsequent weeks in higher-order umbels.
- The usual flowering period of individual umbels is 7 to 10 days, so a plant can be in the process of flowering for 30–50 days.
- The distinctive umbels and floral nectaries attract pollinating insects.
- The fruit that develops is a schizocarp consisting of two mericarps; each mericarp is a true seed; the paired mericarps are easily separated when they are dry producing one hairy seed.
- Seeds are flat, ribbed, and spiny which vary greatly in size; seeds are very light and average 100 seed weight is 0.08gram.

Reproductive behaviour

Carrot is a long day crop and produce inflorescence and seeds after winter months with the advent of spring. The tropical cultivars are annual crop, which do not require specific low temperature exposure for flowering and set seeds freely in the plains after the winter months with the advent of long day condition in the spring. The European types behave like biennial crop, which require 6-8 weeks of vernalization at mean temperature range of 4° to 8°C after they have passed the full root development phase for the production of inflorescence. Hence, breeding and seeds production of these types can only be done in the high hill situations of Himachal Pradesh and Kashmir. Carrot bear hermaphrodite flowers but varying levels of cross-fertilization takes place even up to 80% in different open pollinated varieties due to protandry, high foraging of insects for nectars and strong smell and prolonged receptivity of stigma. Honeybees and wild honeybees generally accomplish cross-pollination. It shows very high inbreeding depression on selfing. Anthesis starts in the morning at about 7.00 a.m. and continues up to 11.30 a.m. depending on the prevailing weather conditions. Anthesis in a single umbel is completed in 7-9 days. Anther deshiscence occurs shortly after anthesis. The stigma becomes receptive 2-3 days before anthesis and it remains so for 6-8 days. Better fruit set occur with effective pollination 5 – 6 days after anthesis. Pollen becomes viable 3-4 days before stigma becomes receptive. For experimental hybridization purpose, the umbel having maximum buds at the time of emasculation are selected and opened flowers along with young buds are removed. Maximum number of buds are emasculated and then bagged. The emasculated umbel is pollinated by enclosing an umbel from the intended male parent or by rubbing the umbel from the male parent on the emasculated umbel of the female parent. However, commercially hybrid seed production of carrot is done utilizing the cytoplasmic-genic male sterile lines.

Genetics of important characters

There are 80 species of *Daucus* but only *Daucus carota* L. is cultivated as root crop. All the 80 species of *Daucus* are diploid with basic chromosome number ranging between n = 7 to n =11. The cultivated carrot, *Daucus carota* L. and its wild relative *Daucus carota* ssp. *carota* are diploid having basic chromosome number, n = 9; 2n = 18. Cultivated carrot crosses freely with its wild relative. Besides the wild carrot, *Daucus carota* ssp. *carota*, other wild species are *D. communalis, D. maritimus, D. hispanicus, D. gummifer, D. fontanesii, D. bocconei, D. capillifolius and D. major.* Two wild species, *Daucus capillifolius* and *Daucus gummifer* also have the basic chromosome number, n = 9; 2n = 18 and they are also cross-compatible with the cultivated carrot.

Carrot cultivars grown in India fall under two broad groups namely, "Temperate or European" and "Tropical or Asiatic" type according to their temperature sensitivity, root characters and seed production behaviour.

Table 1. Important characters of different cultivar groups of carrot

Group of carrot	Chief characters
Temperate or European type	This cultivar group behaves like true biennial crop which require 6-8 weeks of vernalization at mean temperature range of 4° to 8°C after they have passed the full root development phase for the production of seed stalks. Most of the cultivars under temperate type are introduced from Europe and America. Most of the temperate cultivars are hardy, firm textured with less moisture content, sweet, highly flavoured, yellowish-orange to strongly orange in colour due to higher concentration of β-carotene content, and acclimatized to cool temperature condition. This cultivar group does not produce seeds in the plains of India. Seeds of these types of carrot can only be produced in the high hill situations of Himachal Pradesh and Kashmir from autumn sown crop. Hence, breeding works for this group of carrot can be done in high hill situations through induction of flowering and seed set.
Tropical or Asiatic type	This cultivar group is annual and adapted to high temperature condition. Cultivars of this type do not require low temperature exposure (vernalization) for flowering. Most of the cultivars of this type are slightly softer in texture with more moisture content, less sweet, low flavoured, bolt easily, adapted to warm temperatures, and often have scarlet or reddish-orange coloured roots due to anthocyanin pigments. They produce seeds freely in the plains of India at the advent of spring from autumn-winter sown crop. Hence, breeding works for this group of carrot can be done in the plains.

The simply inherited qualitative characters of carrot include cracking of roots, green petiole, glabrous seed stalk, intense orange xylem, lycopene synthesis, male sterility, orange xylem, purple root, reducing sugars in root, yellow xylem, differential xylem / phloem carotene levels, etc. Inheritance of colour, sugar and dry matter content was studied by several workers. Timin and Vasilevsky (1997) investigated inheritance of a number of traits in carrot. They found that the petaloid form of cytoplasmic male sterility (CMS) is determined by a sterility factor in the cytoplasm (*Sp*) and 3 dominant nuclear genes (*Ms 3, Ms 4* and *Ms 5*). Inheritance of root colour is determined by the interaction of 5 genes, while flower colour is governed by two complementary genes. Inheritance of glabrousness of the seed stalk and purple colour of the surface roots and the leaf stalk are dependent on single dominant genes (*Gls, Pr1* and *Pr*, respectively).

The genetic control of β-carotene synthesis in carrot root tissue indicating the presence of orange pigment (i.e., orange roots) is recessive to white or non-pigmented roots. The *Y* allele, which conditions lack of pigmentation (or white roots) was dominant to orange which was conditioned by *yy*. The presence of *Y* and Y_2 always resulted in white roots demonstrating the dominance of lack of pigmentation. Genetic constitution of orange carrots is yyy_2y_2, while that of white carrots is YYY_2Y_2, with yellow and pale orange colour in other genotypes.

Purple carrots bring a different category of phytochemical to the consumers, the anthocyanins. A single dominant gene, P_1, conditions purple root colour, while the second gene, P_2, is hypostatic to P_1 and conditions purple pigmentations in aerial parts (Simon, 1996).

A single gene controls sugar type (sucrose *vs.* reducing sugar) in carrots (Freeman and Simon, 1983). When the dominant allele (*Rs*) is present, the carrot root will accumulate the reducing sugars, fructose and glucose; when both the recessive alleles (*rsrs*) are present, the carrot root accumulates primarily sucrose.

Table 2: Some important monogenic characters of carrot

Gene symbol	**Phenotype of plant character**
A	α-carotene synthesis (may be identical to *Io* or *O*)
Io	Intense orange xylem
L	Lycopene synthesis
Ms-1, Ms-2, Ms-3, Ms-4, Ms-5	Maintenance of male sterility
O	Orange xylem
P_1	Purple root
P_2	Purple node
P_3, P_4	Purple root
rp	Reduced carotenoid pigmentation
rs	Reducing sugar in root
y	Yellow xylem
Y_1, Y_2	Differential xylem/phloem carotene levels
Ce	*Cercospora* leaf spot resistance
Eh	Powdery mildew (*Erysip heheraclei*) resistance
mh-1, mh-2	*Meloidogyne hapla* resistance
Mj-1	*Meloidogyne javanica*, resistance
Cr	*Cracking* roots (dominant to non-cracking)
g	Green petiole (*g*) recessive to purple petiole (*G*)
gls	*Glabrous* seed stalk
sp-1, sp-2	Seed spine formation

Source: Simon (2000), Sureja and Kalia (2012)

The quantitative characters *viz.*, root length and girth, top to root ratio, root yield, carotenoid content and isocoumarin content responsible for bitter flavour are controlled by polygenes.

Cytoplasmic-genic male sterility

The male-sterility found in carrot is due to interaction of a cytoplasmic factor and a number of nuclear genes. Maternally inherited defects in the formation of male flower parts leading to cytoplasmic malesterility (CMS) indicate the involvement of mitochondrial genes in the control of flower formation. In the 'carpeloid' CMS type of carrot, stamens are replaced by carpels. Two type of sterile cytoplasm have been reported: (i) petaloid and (ii) brown anther (Welch and Grimball, 1947; Morelock, 1974). In the 'brown anther type' the anthers are shrivelled and brown without any viable pollen and in the 'petalloid type' the anthers are modified into petalloid structure. The cytoplasm donors of these male sterile lines were thought to be wild carrots (Nakajima, 1997). Seeds of the plants with cytoplasmic male sterility of the brown and petalloid types contained more mineral elements but less starch than seeds of fertile plants. The leaves and inflorescences of the male-sterile forms also had lower carbohydrate content, suggesting disturbances in carbohydrate metabolism (Litvinova *et al*, 1986). In USA, majority of the hybrids are produced from one cytoplasm i.e., Cornell cytoplasm. Considering the risk of disease vulnerability of the hybrids due to monolistic use of Cornell sterile cytoplasm, USDA has released a petaloid typed new CMS line derived from sterile cytoplasm of 'Winconsin' (Morelock *et al.* 1996). Three potential novel CMS sources with different phenotypes were identified, based on the cytoplasms of *Daucus carota* subspp. *gummifer*, *maritimus* and *gadacei* (Nothnagel *et al*, 2000).

Breeding objectives

Main breeding objectives of carrot are mentioned below.

- Early and high root yield
- Desirable, uniform root size, shape (cylindrical), dark orange external and internal colour, i.e., uniform colour in xylem and phloem,
- High top-root ratio, i.e., small tops and smooth heavy tender roots; broad shouldered, uniform tapering or stump-rooted, thin and self-coloured, slow bolting character
- Improvement in quality, especially β carotene, lycopene and anthocyanin content, flavour, texture, high sugar and dry matter in roots and low in harsh flavour
- Resistance to cracking and breaking of the root during harvesting and post harvest handling
- Tolerance to environmental stresses and wider adaptability
- Resistance to different defects such as excessive secondary root development, splitting, secondary growth and cavity spot

- Development of widely adaptable F_1 hybrids using cytoplasmic male sterility
- Resistance to important diseases *viz*., alternaria leaf blight (*Alternaria dauci*) and cercospora blight (*Cercospora carotae*)
- Resistance to insect pest particularly carrot fly (*Psila rosae*) and root knot nematode (*Meloidogyne incognita*)

Breeding methods and development of variety

The black and red types have been introduced in India from the Central Asian region long back. These types with low carotenoid content are mostly annual, juicy and forming roots under comparatively higher temperature condition. These types showing wide variation in colour, shape and size of roots form the distinct "Tropical or Asiatic" type.

Carrot is a cross –pollinated, annual and biennial and insect pollinated mainly by bees. It shows very high inbreeding depression on selfing hence, recurrent selection, mass selection and other population improvement methods are adopted utilizing wide indigenous variability of germplasm of both temperate and tropical types. Few temperate type varieties have been developed through population improvement of indigenous temperate types. However, wide variability of indigenous tropical types has mostly been utilized to develop many tropical type varieties. Hybridization and selection method of breeding has been the most important breeding methods. Adapted cultivars have been utilized in hybridization programme between tropical type x temperate type followed by different population improvement schemes to develop many tropical type varieties in India. The varieties which have been developed by combining temperate and tropical types behave like tropical types as far as seed production is concerned.

The orange coloured 'Temperate' types having higher carotenoid content and believed to have originated in Europe was introduced in India much later with the European traders. Most of the varieties under this group grown in India are exotic introductions. However, different population improvement schemes particularly recurrent selection is adopted to develop few varieties of this temperate group. Few varieties of this group have also been developed through hybridization between Temperate x Temperate types followed by different population improvement schemes.

Carrot cultivars grown in India consisting of indigenous, improved and introduced types are described under two broad groups namely, "Temperate or European" and "Tropical or Asiatic" type according to their temperature sensitivity, root characters and seed production behaviour.

- **Temperate or European Type:** Nantes, Early Nantes, Nantes Half Long, Chanteny, Danvers, Imperator, Pusa Yamdagni, Zeno, Chaman, Ooty-1, Shalimar Carrot-1, Solan Rachna
- **Tropical Asiatic Type:** Hisar Selection-1, Selection 233, Pusa Kesar, Pusa Meghali, Hisar Gairic, Sel-21, Pusa Rudhira, Pusa Asita, Pusa Vrishti , Punjab Carrot Red, Punjab Black Beauty, Kashi Krishna, Arka Suraj, PC 161

Germplasm resources and Plant introduction

Carrot germplasms are maintained through the U.S. National Plant Germplasm System (NPGS) at the North Central Regional Plant Introduction Station (NCRPIS) in Ames, Iowa. These cultivated carrots represent global carrot germplasm, collected over multiple plant exploration trips between 1947 and 2015 from 60 countries, with over 80% of accessions originating from the Eurasian supercontinent: 53% from Asia, 34% from Europe and the Caucasus, and 13% (in descending order) collected from the Americas, Africa, Australia, and New Zealand.

In India, tropical carrot germplasm, in particular is maintained at NBPGR, New Delhi, ICAR-IIVR, Varanasi, ICAR-IARI, New Delhi, PAU, Ludhiana, Punjab, CCSHAU, Hisar, Haryana, GBPUAT, Pantnagar, Uttarakhand, HPKV, Palampur, Himachal Pradesh and other places.

Most of the cultivars under temperate type were introduced from Europe and America. Most of the temperate cultivars are hardy, firm textured with less moisture content, sweet, highly flavoured, yellowish-orange to strongly orange in colour due to higher concentration of β-carotene content and acclimatized to cool temperature condition.

Table 3: Exotic variety directly released for cultivation

Name of the variety	Country from where introduced
Nantes	France
Early Nantes	Europe
Nantes Half Long	Europe
Chanteny	France
Danvers	USA
Zeno	Germany

Variability and population improvement

Initial selection is based upon individual root performance in the field as observed for horticultural characteristics deemed important by the breeder. Field performance and quality attributes such as flavour and β carotene content can be evaluated before seed production begins. Only those plants making satisfactory field and quality performance should be selected for seed production. Systematic carrot breeding work with a view to developing high yielding cultivars with good adaptability has started much late. The method previously employed was mass selection or combined mass-pedigree selection. In the early 17th century 'Long Orange' cultivar was developed in the Netherlands. From this cultivar, three orange coloured cultivars differing in earliness and size were developed *viz.*, Late Half-Long (the biggest), Early Half-Long and Early Scarlet-Horn (the smallest). All present day cultivars of the western carotene carrots have been developed from these four closely related cultivars, either by simple selection or by intercrossing different types.

Carrot is a cross –pollinated, annual and biennial and insect pollinated mainly by bees. Tropical or Asiatic or desi types having low carotenoid content are juicy and form roots under comparatively higher temperature condition. These types behave like annual and sets seeds freely in the plains of northern India. Hence, breeding works of this carrot group is mostly done in different Agricultural Universities / ICAR- Institutes around Delhi, Punjab, Haryana, Uttarakhand, and Uttar Pradesh. Wide variation in colour, shape and size of roots of these types are available in India.

The orange-coloured, temperate or European types are biennial and the root requires vernalization for 45 -60 days for bolting and therefore, it produces seeds only in the high hill situation in India. Hence, breeding works this carrot group is only done in the high altitude areas in different Agricultural Universities / ICAR- Institutes in Kulu valley of Himachal Pradesh, Kashmir valley and valley of Nilgiri hills. Available variation in temperate types is not as wide as found in the tropical types in India.

Carrot shows very high inbreeding depression on selfing. For this reason, selfing is generally done only up to 2 to 3 generations. Generally, different population improvement schemes *viz.*, recurrent selection, mass selection, etc. were adopted utilizing indigenous variability of germplasm of temperate types to develop few temperate type varieties *viz.*, Chaman, Ooty-1, Shalimar Carrot-1, etc. Most of the tropical type varieties *viz.*, Hisar Selection-1, Hisar Gairic, Pusa Rudhira, Pusa Asita, Pusa Vrishti, etc. have been developed through population improvement from wide indigenous variability of tropical types in India (Hazra *et al*, 2022). Exotic introductions have also been utilized in the population improvement programme to develop very few "Temperate" type variety *viz.*, Solan Rachna.

Table 4: Varieties developed through population improvement schemes

Name of variety	**Indigenous / Exotic material**
Chaman	Indigenous
Ooty-1	Indigenous, selection from half-sib progeny of local type DC 3
Shalimar Carrot-1	Indigenous, Mass selection from an indigenous population
Solan Rachna	Exotic (from Netherlands)
Hisar Selection-1	Indigenous (mass selection)
HisarGairic	Indigenous (mass selection)
Sel-21	Indigenous (from Punjab)
PusaRudhira	Indigenous (recurrent selection)
PusaAsita	Indigenous (recurrent selection)
PusaVrishti	Indigenous (recurrent selection)
Punjab Carrot Red	Indigenous (recurrent selection)
Punjab Black Beauty	Indigenous (recurrent selection)
Kashi Krishna	Indigenous (recurrent selection)
Arka Suraj	Indigenous (recurrent selection)
PC 161	Indigenous (recurrent selection)

Hybridization and population improvement

Line selection involving only self-pollination is most effective in fixing traits. Carrot shows very high inbreeding depression on selfing hence, selfing is generally done only up to 2 to 3 generations. Because of high inbreeding depression, a broader genetic base for carrot inbred development is required to achieve uniformity and maintain vigour. In carrots, starting at the F_3 or even the F_2 generation, several individual roots from a given family are self-pollinated, whereas others from the same family are sib-mated in two to five plant masses. Seed produced from self-pollination and sib-mating are grown in adjacent rows the next year, and if vigour is adequate, the best performing self-pollinating plants are selected to continue the line. If severe inbreeding depression is observed in the form of small plants, then plants from the sib-mated sib-population are selected. This process is repeated two to five generations beyond the F_2 generation until plant characteristics are satisfactory and uniform. This constitutes a carrot 'inbred' which is used as a pollinator in a crossing cage to create hybrid combinations for evaluating combining ability and hybrid development.

Adapted cultivars of tropical Asiatic type have also been utilized in hybridization programme between tropical x temperate type followed by different population improvement schemes to develop many tropical type varieties in India. The varieties developed by combining temperate and tropical types behave like tropical types as far as seed production is concerned *viz*., Selection 233, Pusa Kesar, Pusa Meghali, etc. (Hazra *et al*, 2022)

Temperate types are biennial in nature and produce seeds only in high hill situation of India. Most of the exotic cultivars under temperate type are introduced from Europe and America. Most of the temperate cultivars are hardy, firm textured with less moisture content, sweet, highly flavoured, yellowish orange to strongly orange in colour due to higher concentration of β-carotene content, and acclimatized to cool temperature condition. However, very few temperate type varieties have also been bred in India through temperate x temperate cross involving the exotic varieties in the crossing programme followed by population improvement *viz*., Pusa Yamdagni.

Table 5: Varieties bred through hybridization and population improvement

Name of variety	Parentage
Imperator (Temperate)	Nantes × Chanteny
PusaYamdagni (Temperate)	EC 9981 (having quick growing habit) × Nantes Half Long
Selection 233 (Tropical)	Nantes × No. 29
Pusa Kesar (Tropical)	Local Red × Nantes Half Long
Pusa Meghali (Tropical)	Pusa Kesar × Nantes

Heterosis breeding

Poole (1937) first reported heterosis in carrot where the inbred lines with average 12.8 and 24.9g root produced hybrids with roots weighing 80.5 g. Commercially, hybrid seed production of carrot is done utilizing the cytoplasmic-genic male sterile lines. The male sterility is conditioned by the interaction between a cytoplasmic factor (*N* for normal cytoplasm and *S* for sterile cytoplasm) and a nuclear gene (recessive or dominant). Cytoplasmic-genic male sterility in carrot occurs in two morphologically distinct phenotypes *viz*., "brown anther" type and "petaloid type" male sterility. The brown anther male sterility is conditioned by sterile cytoplasm and homozygous recessive locus *ms5* and dominant allele would restore the fertility. Petaloid sterility, which has arisen due to homeotic mutation, is commercially used for hybrid seed production in the world. This male sterility is manifested as the replacement of stamens with petals (white petaloidy) or both stamens and petals with green bract like structures (green petaloidy). This male sterility is stable across a wide range of environments which is conditioned by sterile cytoplasm and homozygous dominant gene. The male-sterile line has the genetic constitution *SMsMs* or *SMsms* and the male fertiles are *NMsMs* or *NMsms*.

Maintenance of male-sterile line

An excellent uniformity of hybrids demands a high degree of homozygosity in the male-sterile and maintainer lines. The male-sterile line is maintained by crossing it with male-fertile sister line. This male-fertile sister plants are homozygous dominant for one or two *Ms* loci, free from epistasism and possess normal cytoplasm (*NMsMs or NMsms*). The male sterile plants are crossed with the normal male fertile plants of the same cultivars/line. If all the F_1 plants of such crossings are male-sterile, it is presumed that the male fertile plants have the genetic constitution *NMsMs*, as the normal male fertile plants generally possess normal cytoplasm (N).

Development of pollen parent

Homozygous inbred pollinator lines having desirable characters are developed from good adapted cultivars. Since inbreeding depression is pronounced, it is difficult to isolate inbred lines with enough vigour. Alternate generation of selfing and sibmating in strict isolation helps in overcoming the effect of inbreeding depression. Sibmating is done under the net cage. A population of 8-honey bees/sq m is sufficient to get good seed set. For getting small quantity of selfed seeds, hand pollination is effective. In this case, king or main umbel is enclosed in muslin bag and a second order umbel of the same plant is also put in the same bag or bagged separately. For a couple of days these two umbels are rubbed gently against each other by hand daily to ensure self-pollination. Sometimes, it is convenient to enclose two or three umbels of the sample plant in a muslin bag before anthesis. After anthesis, the umbels are rubbed against each other daily for a few days.

Production of hybrid seeds

Male-sterile lines with high seed producing ability and stable expression of male-sterility are essential for the development of superior F_1 hybrids of carrot. Male-sterility of carrot is highly stable in temperate region. Frequent fluctuation in temperature in different seasons in hilly areas breakdown male sterility and thus render male-fertility. So, specific locations in the hilly areas need to be identified to ensure stability of the male sterile lines. Success in the hybrid seed production largely depends on the location where hybrid seed production programmes are taken. In hybrid seed production programme, three lines are used, namely CMS-based male-sterile line (female line), male-fertile sister line (maintainer line), and pollinator line (male line) which is male-fertile and has a good combining ability with male-sterile line. The male-sterile and pollinator lines are inbred for several generations for attaining uniformity even though there may be decline in vigour of these two lines. In European type of carrots, roots of both seed and pollen parents are vernalized for 6-7 weeks at 3° to 5°C temperature before planting. Stecklings of 30 g weight which gives higher seed yield are prepared from the vernalized roots of both seed and pollen parents. These roots are then planted in isolated field in the ratio of 4-5: 2 (seed parent: pollen parent). Under cool condition (16/22°C night/day temperature), seed set is better. Seeds collected from male- sterile seed parents are hybrid seeds. Hybrid seeds are produced through cross pollination of male-sterile line by the pollens of the male-fertile inbred lines (cytoplasmic male-sterility, petaloid form or brownish anthers type) in insect-proof enclosures, each containing a small hive (3 frames) of honeybees (*Apis mellifera*) because bees are the major pollinating agents. The degree of pollination by bees depends on the distance between the last rows of the maternal line (male-sterile line) and the pollen source (male- fertile line) and the distance no greater than 3 m is the best for complete cross-pollination in hybrid seed production.

Pusa Nayanjyoti is the first "Temperate" type, orange coloured hybrid developed by Public Sector in India at IARI, New Delhi using CMS system. Roots are orange, smooth, uniform, cylindrical, stumpy with small indistinct, self-coloured core. Roots possess high β-carotene content (7.55 mg/100g fresh weight). PusaVasuda is the first public sector "Tropical" type hybrid developed at IARI, New Delhi using CMS system. Roots are smooth, attractive, vigorous, self-coloured, red, juicy, rich in total carotenoids, lycopene, TSS and minerals.

Breeding for resistance to diseases

Alternaria leaf blight caused by *Alternaria dauci* is one of the most important foliar diseases, considerably reducing the carrot yield around the world. Asian and Brazilian cultivars have been used as the source of resistance genes for *Alternaria* leaf blight. The genetics of resistance is conditioned by several genes with varying levels of dominance (Simon and Strandberg, 1998). Different wild relatives *viz*., *D. carota* subsp. *capillifolius* and *D. carota* subsp. *maximus* and *D. crinitus* are the sources of resistance to *Alternaria* leaf blight (Arbizu *et al.*,2017). Resistance to *Cercospora* leaf spot (*Cercospora carotae*) is controlled by a dominant gene (*Ce*) and resistance to powdery mildew caused by *Erysiphe heracle* is also controlled a dominant gene (*Eh*) (Lebeda *et al.*, 1988).

Cavity spot caused by *Pythium* sp. is among the most destructive post-harvest storage diseases of carrots. Resistance to cavity spot in carrot is generally quantitative in nature based on relatively minor differences in severity among the cultivars. Partial resistance to cavity spot has been reported in several varieties *viz*., Six Pak, SR-481, Eagle, Red Core, Panther, Caropride, Fannia, Navajo, CS 732 and CS 736 (McDonald *et al.*, 2017).

Carrot cultivars 'Panther' and 'Caropak' have been found to be relatively resistant to black rot disease caused by *Alternaria radicina* (Pryor *et al.*, 2000). Grzebelus *et al.* (2013) isolated somaclonal variants within protoplast cultures that were challenged by fungal culture filtrates and obtained regenerated plants with greater tolerance.

Aster yellows (AY), an aster leafhopper transmitted disease caused by a mycoplasma-like organisms (MLO), is a destructive disease of carrot in the U.S.A, Canada and other parts of the temperate world. Germplasm derived from University of Wisconsin inbreds and four European open-pollinated cultivars provided significant protection to aster yellows (Gabelman *et al.*, 1994).

Breeding for resistance to insect-pests

Carrot fly (*Psila rosae*) is the most important insect pest of carrot in Europe and Canada. Michalik and Wiech (2000) reported five breeding lines (HRI 93171, HRI 96732, HRI 93104, HRI 96382 and AR 694) and cultivar Puma as tolerant genotypes. *Daucus capillifolius* (Gilli), a wild relative of carrot found in Northern Africa, is resistant to carrot fly.

Several species of nematodes attack carrots. Northern root-knot nematode, *Meloidogyne hapla* is prevalent in cool temperate areas of northern Europe and North and South America, while the tropical and Southern root-knot nematodes *M. javanica* and *M. incognita* are widespread in warmer carrot production areas of southern Europe, southern U.S.A, northern South America, and Asia. A two-gene model for *M. hapla* resistance has been reported by Wang and Goldman (1996) whereas a single gene confers resistance to *M. javanica* and *M. incognita* (Simon *et al.*, 2000). Yunhee *et al.* (2014) also reported that resistance to *M. incognita* may be governed by one or few genes. Yunhee *et al.* (2014) screened Korean carrot lines and reported 61 lines resistant to *M. incognita* race 1. A genotype from north India, 'DR-333' showed tolerance to *M. incognita* (Siddiqui *et al.*, 2011).

Breeding for quality

Carrots are a natural target for carotenoid biofortification due to their high levels of consumption, affordability to consumers, widespread cultivation worldwide, and potential for increased carotene biosynthesis. Early varieties of orange carrots during 1960s in the USA had β carotene contents in the range of 70 ppm, but breeding efforts, through phenotypic recurrent selection of roots with a darker orange colour, increased that value to 90 ppm by the 1970s. Modern spectrophotometric detection of high levels of root carotenes has allowed breeders to elevate the concentration above 130 ppm in typical orange and dark orange carrots and high-carotene mass can reach 500 ppm (Simon *et al.*, 1989).

Visual appearance traits are important targets in carrot breeding in Japan, and the 'best bright orange color' is selected based on a comparison of minute color differences. High-carotene transgressive segregates were observed in crosses among European cultivars and also in European x Asian crosses. The elevated carotene content in one European x Asian cross was especially interesting, since the F_2 mean for β carotene content was greater than the higher parental stock (Simon *et al.*, 1989). Derivatives of this population have been used as a source of high carotene content in the development of new breeding stocks and hybrids.

Purple carrots bring a different category of phytochemical to consumers, the anthocyanins. A single dominant gene, P_1, conditions purple root colour, while a second gene, P_2, is hypostatic to P_1 and conditions purple pigmentations in aerial parts (Simon, 1996). The wide range of genetic variation for purple pigment type is found in tropical Asiatic carrot type.

Four high anthocyanin containing (anthocyanin content: 182-520 mg/100g fresh) tropical type varieties *viz.*, Pusa Rudhira, Pusa Asita, Punjab Carrot Red and Punjab Black Beauty have been developed through phenotypic recurrent selection from the indigenous variability. With the use of divergent population and efficient estimation method, anthocyanin level can be improved to the higher level.

Nitrates have anti-nutritional value, especially for carrots used to make baby food. The inheritance of nitrate content in carrot is complex within complete dominance so that low-nitrate parents are necessary to obtain low-nitrate hybrids. Progress in genetic selection of low-nitrate carrots has been successful in Poland.

References

Arbizu, C.I., Tas, P.M., Simon, P.W. and Spooner, D.M. 2017. Phylogenetic Prediction of Alternaria Leaf Blight Resistance in Wild and Cultivated Species of Carrots. *Crop Sci.*, **57**: 2645–2653.

Freeman, R.E. and Simon, P.W. 1983. Evidence for Simple Genetic Control of Sugar Typein Carrot (*Daucus carota* L.). *J. Amer. Soc. Hort. Sci.,* **108**: 50-54.

Gabelman, W.H., Goldman, I.L., and Breitbach, D.W. 1994. Evaluation and selection for resistance to aster yellows in carrot (*Daucus carota* L.). *J. Amer. Soc. Hort. Sci.*,**119:** 1293-1297.

Grzebelus, E., Kruk, M., Macko-Podgorni, A. and Grzebelus, D. 2013. Response of carrot protoplasts and protoplast-derived aggre-gates to selection using a fungal culture filtrate of *Alternaria radicina. Plant Cell Tiss. Org. Cult.*, **115**: 209–222.

Hazra, P., Banerjee, M.K. and Chattopadhyay, A. 2022. *Varieties of Vegetable crops in India*, Kalyani Publishers, New Delhi, Ludhiana, 204p.

Lebeda, A., Coufal, J. and Kvasnicka, P. 1988. Evaluation of susceptibility of *Daucus carota* varieties to natural infection with *Erysiphe heracleid. Euphytica,* **39**: 285-288.

Litvinova, M.K., Lesnevich, L.A. and Frolova, N.F. 1986. Physiological and biochemical features of the seeds in male-sterile and fertile forms of carrot. *Selektsiya ovoshchnykh kul'tur*, **1**: 29-38.

McDonald, M.R., van der Kooi, K. and Simon, P. 2017. *Muck Crops Research Station Annual Report*, University of Guelph, Ontario, p.6. https://www.uoguelph.ca/muckcrop/pdfs/ Muck.

Michalik, B. and Wiech, K. 2000. Differences in the resistance of carrot lines and cultivars to carrot fly (*Psila rosae* (Fabr.)) attack. *Folia Hort.*, **12**: 43-51.

Morelock, T.E. 1974. *Influence of protoplasm source expression of male sterility in carrot (D. carota)*. Ph. D. Thesis, Wisconsin University, USA.

Morelock, T.E., Simon, P.W. and Peterson, C.E. 1996. Wisconsin Wild: Another petaloid male sterility cytoplasm for carrot. *Hort Sci.*, **31** : 887-88.

Nakajima, Y., Yamamoto, T. and Oeda, K. 1997. Genetic variation of mitochondrial and nuclear genomes in carrots revealed by random amplified polymorphic DNA (RAPD). *Euphytica*, **95**: 259-267.

Nothnagel, T., Straka, P. and Linke, B. 2000. Male sterility in populations of Daucus and the development of alloplasmic male-sterile carrot lines. *Plant Breed.*, **119**: 145-152.

Poole, C.F. 1937. *Improving the root vegetables*. U.S. Dept. Agr. Yearbook, pp.300-325.

Pryor, B., Davis, R.M. and Gilbertson, R.L. 2000. A toothpick inoculation method for evaluation of carrot cultivars for resistance to *Alternaria radicina. Hort Sci.*, **35**: 1099–1102.

Siddiqui, Z.A., Nesha, R. and Varshney, A. 2011. Response of Carrot cultivars to *Meloidogyne incognita* and *Pectobacterium carotovorum* subsp. *carotovorum. J. Plant Pathol.*, **93**: 503–506.

Simon, P.W. 1996. Inheritance and expression of purple and yellow storage root color in carrot. *J. Hered.,* **87**: 63-66.

Simon, P.W. 2000. Domestication, Historical development and modern breeding of Carrot. *Plant Breed. Rev.,* **19**: 157-190.

Simon, P.W. and Strandberg, J.O. 1998. Diallel analysis of resistance in carrot to alternaria leaf blight. *J. Amer. Soc. Hort. Sci.*, **123**: 412-415.

Simon, P.W., Wolff, X.Y., Peterson, C.E., Kammerlohr, D.S., Rubatzky, V.E., Strandberg, J.O., Bassett, M.J. and White, J.M. 1989. High carotene mass carrot population. *Hort Sci.,* **24**: 174.

Sureja, A.K. and Kalia, P. 2012. Carrot. In: *Hand book of Vegetables* (Eds. Peter, K. V. and Hazra, P.), Studium Press LLC, Houston, Texas, USA, pp. 115-148.

Timin, N.I. and Vasilevsky, V.A. 1997. Genetic peculiarities of carrot (*Daucus carota* L.). *J. Appl. Genet.*, **38A**: 232-236.

Van der Vossen, H.A.M. and Sambas, E.N. 1993. *Daucus carota* L. In: *Plant Resources of South-east Asia* (Eds. Siemonsma, J.S. and Piluek, K.), No.9, Vegetables, Pudoc Scientific Publishers, Wageningen.

Wang, M. and Goldman, I.L. 1996. Resistance to root knot nematode (*Meloidogyne hapla* Chitwood) in carrot is controlled by two recessive genes. *J. Hered.*, **87**: 119–123.

Welch, J.E. and Grimball, Jr. E.L. 1947. Male sterility in carrot. *Science*, **106** : 594.

Yunhee, S., Park, J., Kim, Y.S., Park, Y. and Kim, Y.H. 2014. Screening and histopathological characterization of Korean carrot lines for resistance to the root-knot nematode *Meloidogyne incognita. Plant Pathol. J.,* **30**: 75–81.

27

Radish

Radish, *Raphanus sativus* under the family Brassicaceae (Crucifereae) having chromosome No. 2n = 2x =18 is grown mainly for fleshy taproot roots in both tropical and temperate regions throughout the world. Radish is cultivated widely in China, Japan, Korea, India, Europe and USA. It is grown throughout India and particularly in Uttar Pradesh, Bihar, West Bengal, Punjab, Haryana, Himachal Pradesh, Assam and Gujarat. Radish is mostly eaten raw as a crunchy salad or cooked as a vegetable. Leaves are sometimes eaten as cooked leafy vegetable. Immature pods of rat-tail radish, *Raphanus sativus* var. *caudatus* are also consumed as vegetable.

Origin, domestication and adaptation in India

Radish, one of the ancient root vegetables was domesticated in Asia prior to Roman times. Varieties of radish are now broadly distributed around the world, but almost no archeological records are available to help determine their early history and domestication. However, different research works have tentatively located the origin of *Raphanus sativus* in Southeast Asia, as this is the only region where truly wild forms have been discovered. This radish might be Oriental or Asiatic radish. India, Central China, and Central Asia appear to have been the secondary centers where different forms were developed. However, according to other hypothesis, eastern Mediterranean region, China and middle Asia are considered to be the origin of radish where wild radish still exists.

Mediterranean region has been suggested as the origin of European radish because of the presence of maximum diversity including different wild species *viz.*, *Raphanus raphanistrum, R. maritimus, R. landra, R. microcarpus* and *R. rostratus* in this region extending eastward to the Caspian Sea. The small-rooted European types most likely evolved from more than one of these wild species. However, origin of cultivated radish is obscure because its wild progenitor could not be identified.

Radishes enter in the historical record in the third century BC. Greek and Roman agriculturalists of the first century AD gave details of small, large, round, long, mild and sharp varieties of radish. Radish seems to be one of the first European crops introduced to the Americas. The cultivated radish, *Raphanus sativus* was introduced in China from Eastern Mediterranean region through silk route more than 2400 years ago. It was then introduced in Japan more than 1250 years ago. It was known in Egypt during 2700 – 2000 BC. The Chinese radishes are large, mild and white East Asian form which was developed in China are native to western region of Central Asia and southwest China.

The Japanese radish has close relationship with *R. sativus* var, *raphanistroides* known as "daikon" which are very big in size. This species occurs wild in the coastal regions of Japan, Korea and South China. The Indian group of radishes including the rat-tail radish evolved in north-western region but their ancestors do not exist.

Introduction of radish in India has not been chronicled however, mentioning of radish in different ancient literatures points its introduction before Christian era. Radish was mentioned in the Ramayana, c.1450 BC. Development of large range of tropicalized cultivars suitable for growing in subtropical and tropical condition with heat tolerance and capable of forming roots even under day temperature of 35°C is a significant feature of adaptation of Asiatic races in India.

The temperate or European radish has been originated in Europe and is basically temperate in nature. The European radish maintains its distinctness as temperate race. Some semi-domesticate type like that of "Black Radish" possessing temperate habit is found in the North Indian hills.

Crop botany and floral biology

- Radish is the crop of tropical, subtropical and temperate climate; the tropical, high temperature tolerant types mainly developed in India are more pungent than both Oriental and European types; the European radishes are early, juicy and mild type and crispy.
- True Oriental radishes are also mildly pungent and crispy in nature.
- Radishes are annual or biennial crops grown for their swollen taproots, which can be globular, tapering or cylindrical.
- Roots vary widely in size, flavour, colour, and length of time they take to mature; smaller types which are normally eaten raw in salads have a few leaves about 13 cm long with round roots up to 2.5 cm in diameter or more in long and slender roots up to 7 cm long.
- Long–rooted radish including oriental radishes, daikon and winter radishes, grows up to 60 cm long with foliage about 60 cm high with a spread of 45 cm.
- The root skin colour ranges from white throughout, pink, red, purple, yellow and green to black, but the flesh is usually white; pink or black skin colour is due to anthocyanin pigments.
- Leaves are arranged in a rosette; lyrate shape, meaning they are divided pinnately with an enlarged terminal lobe and smaller lateral lobes.
- The flesh of radishes harvested timely is crisp and sweet, but becomes bitter and tough if the root is left in the ground for too long time after full development stage.
- The inflorescence is a typical terminal raceme characteristic of the Brassicaceae family.
- The flowers are bisexual, small, white, rose or lilac in colour with purple veins; sepals are four in number, polysepalous and erect.

- Petals are also four in number and polypetalous, regular, each petal differentiates into claw; four sepals and 4 petals standing opposite to each other in the form of a cross.
- It has six stamens having nectarines at their base in two whorls; two outer stamens are short, while four interior stamens are long.
- Two carpels form a superior ovary with a false septum and two rows of campylotropous ovules.
- Fruit is a special type of indehiscent, bicarpellary pod called the siliqua, 3.5 to 3.7 cm long and upto 1.5 cm in diameter with 6-12 seeds with two rows of seeds lying along the edges of the replum; pods of *R. caudatus* (rat-tail radish) are usually 20-30cm in length but may be even 100cm in length in some varieties; small pods can be eaten when young.
- Radish seeds are exalbuminous and globose in shape, about 3 mm in diameter; embryo curved, attached to septum and 100 seeds make 1.0 to 1.2 gram; mature seed colour is yellowish but turning reddish-brown with age and remain viable for 5 years in moisture proof condition.

Reproductive behaviour

The tropicalized Asiatic types, mainly developed in India are high temperature tolerant and behave like annual and set seeds freely in the plains with the advent of long day condition in spring. The Oriental (Chinese and Japanese) radishes are not true biennial but require low temperature for bolting and seed set. The European types behave like true biennial crop which require 6-8 weeks of vernalization at mean temperature range of 4° to 8°C after they have passed the full root development phase for the production of seed stalks. Seeds of the European types can only be produced in the high hill situations of Himachal Pradesh and Kashmir. Radish bear hermaphrodite flowers but varying levels of cross-fertilization takes place even upto 80 % in different open pollinated varieties due to manifestation of self-incompatibility of different intensities. It shows considerable inbreeding depression on selfing. In the early bud stage, 2-3 days before anthesis when the stigma has already become receptive due to protogyny, self-fertilization can be accomplished by self-pollination with the pollens from other opened flowers of the same plant. Experimental hybrids can be developed by emasculation of the hermaphrodite flower in the afternoon hours of the day before anthesis followed by protection from pollen contamination by keeping the inflorescence inside a paper bag. Cross pollination with the pollens from the intended male parent is done in the next day morning hours followed by enveloping the inflorescence again for another two days. However, hybrid seeds of radish are commercially produced utilizing sporophytic self-incompatibility and cytoplasmic male sterility system.

Genetics of important characters

Radish, *Raphanus sativus* L. and its allied species are diploid and have basic chromosome number n = 9, 2n= 18. Chromosome are very small (1.5 to 3.5 μm) in size. There are several reports on cytogenetical studies and phylogenetic relationship in intergeneric and interspecific hybrids involving the genera, *Brassica*, *Eruca* and *Sinapis*. Such relationship indicated that the species of *Raphanus* and *Brassica* evolved from a common ancestral form with chromosome number 2n = 6 by chromosome duplication.

Radish cultivars grown in India fall under two broad groups namely, "Tropical or Asiatic" and "Temperate or European" type according to their temperature sensitivity, root characters and seed production behaviour. Characters of the "Oriental" type cultivars are inclined towards "Temperate or European" type.

Table 1: Important characters of different cultivar groups of radish

Group of carrot	Chief characters
Tropical or Asiatic Type	The "Asiatic" cultivar group, mainly developed in India, is true annual type and adapted to tropical and subtropical growing condition. Some cultivars under this group are heat tolerant and capable of forming roots even under day temperature of 35°C which shows the significant feature of adaptation of the Asiatic races in India. A good number of indigenous cultivars are grown under different names particularly in Northern and Eastern India. These types can tolerate high temperature and can be grown almost throughout the year under mild temperature condition. Radish is essentially a crop of cool or moderate climate. Naturally, performance of Asiatic radish will be better in autumn-winter than in spring-summer season. These types are more pungent than European types and produce seeds freely in the plains of India. Hence, breeding works for this group of radish can be done in the plains.
Oriental type	The Oriental cultivar group of radish, mainly developed in China and Japan, is mildly pungent and crispy in nature. These types cannot tolerate high temperature and are best grown under cool temperature condition. Although these types are not true biennial type yet, it requires low temperature exposure for some period for bolting and seed set. Hence, seed production and breeding works of these types is done in the hill regions of India through induction of flowering and fruit set.
Temperate or European type	The temperate or European radish has been originated in Mediterranean region and is basically temperate in nature. The European radish maintains its distinctness as temperate race. These types are cool season, generally early and having crispy root with mild pungency. European radish is not very popular in India. They do not set seeds in the plains of India and from seed production point of view, they are either i) winter biennial type (require vernalization at mean temperature range of 4° to 8°C for 6-8 weeks after they have passed the full root development phase for the production of seed stalks) or ii) summer type of temperate region. Most of the varieties grown in India (e.g., White Icicle, Rapid Red White Tipped, Scarlet Globe, Chinese Pink, etc.) under this group are summer types of the temperate region. They behave more or less like winter biennial type and require vernalization for seed production. So their seed production and breeding works is done in the hill regions of India through induction of flowering and fruit set.

Several reports are available regarding genetical studies in radish. In the study of genetic variability for different quantitative characters, it has been found that most of the characters *viz.*, leaf width, leaf length, leaves per plant, top fresh weight, root weight, root length, diameter and root yield show high heritability and are less influenced by environment. In another study, late bolting showed dominance over early bolting. Inheritance of root yield per plot and important yield components *viz.*, root weight, length and diameter, plant height, number of leaves and root: top ratio was governed by over-dominance gene action. Genetic control of different simply inherited characters is presented below.

Table 2: Genetic control of different simply inherited characters of radish

Character	Phenotype / inheritance pattern/ Gene symbol
Pale green neck or hypocotyl	Monogenic dominant, *G*
Hypocotyl colour	Single gene; violet colour "*C*" incompletely dominant over white "*c*"
Leaf colour	Single gene; green dominant "*Yg*" over yellow-green "*yg*"
Green fleck on leaves	Two duplicate recessive genes, *gf1* and *gf2*
Leaf shape	Single gene; lyrate shape dominant
Root colour	Single gene; red, purple and yellow dominant over white Gene "*R*" for red colour; epistatic over "*Y*", for yellow colour gene *Rp* for purple colour, epistatic over the gene "*Y*"
Root shape	Long root dominant over round root; conditioned by two unlinked dominant genes
Root texture	Corky epidermal layer dominant over smooth epidermis
Hairy leaf	Single gene; hairy dominant over glabrous leaf
Plant colour	Single gene; pink being dominant "*Pi*"
Seed coat colour	Single gene; brown seed coat being dominant
Petal colour	Single gene; white petal being dominant
Pollen colour	Single gene; normal bright yellow pollen dominant (*Cp*) over cream coloured pollen (*cp*)
Hypocotyl length	Polygenically controlled
Earliness / Maturity	Polygenically controlled
Cytoplasmic male sterility	Cytoplasmic factor and one recessive gene
Genic male sterility	Single recessive gene Two recessive genes acting independently with the involvement of minor genes
Self-incompatibility	Single multiallelic gene "*S*"; sporophytic system with six alleles (*S1* to *S6*)

Source: Akhtar (2015)

Self-incompatibility

Radish exhibits sporophytic self-incompatibility with multiple *S*-alleles located at a single locus and 37 *S*-alleles, designated as S201 to S237 were identified (Niikura and Matsuura, 2001).

Cytoplasmic male-sterility

On the basis of floral morphology, radish male-sterile (MS) plants can be classified into 3 types: i) degenerative corolla, ii) shrivelled stamen and iii) abortive pollen (Zhou and Zhang, 1994). Cytoplasmic male-sterility trait has been found in Japanese radish by Ogura (1968). Ogura-type male-sterile cytoplasm was originated in wild radishes (Yamagishi and Terachi, 1994). Sterility was induced by the interaction between homozygous recessive nuclear genes *ms* and the *S*-cytoplasm. *ORF138* is the mitochondrion-encoded protein responsible for Ogura cytoplasmic male- sterility (*cms*) in radish and *Brassica* species (Duroc *et al*, 2006).

Breeding objectives

Main breeding objectives of radish are mentioned below.

- High marketable yield with white, long/stumpy, non-pithy roots (longer standing period in field without becoming pithy) with thin tap root and non-branching habit
- Uniformity, dark green leaf without hair, cylindrical shape, smooth skin and firm root texture with slow bolting and late-flowering habit
- Early-rooting and short duration suitable for fitting in different vegetable-based cropping sequence
- Superior edible quality such as desired pungency, low nitrate content, less pore formation, etc. and for varieties eaten fresh as salad, increased sweetness and root colour are important quality parameters
- Characteristic cylindrical root of appropriate length for mechanical harvesting
- Wider adaptability to different agro-climatic conditions; particularly tolerance to heat, low temperature, drought, excessive rain and saline, alkaline and acid soils
- Resistance to important diseases *viz*., alternaria leaf blight (*Alternaria dauci*), white rust (*Albugo candida*), yellow disease (*Fusarium oxysporum* f.sp. *conglutinans*) and club root (*Plasmodiophora brassicae*)
- Resistance to insect pest particularly mustard saw fly (*Athalia lugens proxima*) and aphids (*Myzus persiae*)

Breeding methods and development of variety

The temperate or European radish has been originated in Europe and is basically temperate in nature. Most of the cultivars of European radish group grown in India are exotic introductions. However, one variety "Pusa Himani" was bred through hybridization of Temperate x Asiatic /Oriental type followed by selection. In this breeding scheme, a semi-domesticate type, "Black Radish" possessing temperate habit, which is found in the North Indian hills has been utilized.

Cultivars under the "Asiatic" group have been developed in India. Large range of tropicalized cultivars suitable for growing in subtropical and tropical condition with heat tolerance and capable of forming roots even under day temperature of 35°C is the significant feature of adaptation of Asiatic races in India. These adaptable Asiatic types, called the "Desi" type in India have been utilized in breeding many Asiatic type varieties.

Radish is highly cross-pollinating with strong sporophytic self-incompatibility. It is an insect-pollinated crop. It shows considerable inbreeding depression on selfing by bud pollination. The commonly adopted breeding method include different population improvement schemes particularly mass selection, recurrent selection and family breeding utilizing wide variability of tropical radish group to develop many tropical varieties. Hybridization followed by population improvement has also been followed to develop varieties. Adapted cultivars have been utilized in hybridization programme between Tropical type x Temperate type or Oriental type x Temperate type followed by different population improvement schemes to develop many tropical type varieties in India.

The cultivars grown in India can conveniently be described under two broad groups *viz.*, European or temperate type and Asiatic or tropical and subtropical type according to their temperature sensitivity, root characters and seed production behaviour.

- **European or Temperate Type:** White Icicle, Rapid Red White Tipped, Scarlet Globe, Scarlet Long, Chinese Pink
- **Asiatic or Tropical and Sub-tropical type:** Japanese White, Pusa Desi, Pusa Chetki, Pusa Reshmi, Pusa Mridula, Pusa Sweta, Pusa Jamuni, Pusa Gulabi, Punjab Safaid, Punjab Ageti, Kalyanpur No-9, CO-1, Kashi Sweta, Kashi Hans, Arka Nishant, Punjab Pasand, Kashi Mooli-40 , Kashi Lohit , Hisar Sel.-1, Palam Hriday

Germplasm resources and Plant introduction

The radish collection at the Plant Genetic Resources unit, USDA, New York, USA consists of more than 710 accessions chiefly belong to *Raphanus sativus* species. China is rich in radish germplasm resources, and has collected and preserved 2073 domestic and foreign radish germplasm resources, with 95% of the preserved resources being domestic local varieties. Therefore, China occupies an important position in the collection and preservation of radish germplasm resources and research.

In India, tropical radish germplasm, in particular is maintained at NBPGR, New Delhi, ICAR-IARI, New Delhi, ICAR-IIVR, Varanasi, IIHR, Bengaluru, PAU, Ludhiana, Punjab, HPKV, Palampur, Himachal Pradesh and other places.

Most of the cultivars belonging to temperate European type grown in India are exotic introductions mainly from Europe.

Table 3: Exotic variety directly released for cultivation

Name of the variety	Country from where introduced
White Icicle	Europe
Rapid Red White Tipped	Europe
Scarlet Globe	Europe
Scarlet Long	Europe
Chinese Pink	China
Japanese White	Japan (introduction from Japan in the name Shiroaguri-Kyo)

Variability and population improvement

Large range of tropicalized Asiatic cultivars have been developed in India which are suitable for growing in subtropical and tropical condition with heat tolerance and capable of forming roots even under day temperature of even 35°C. Wide variation in colour, shape and size of roots of these types are available in India. Some of the distinct and popular cultivars of this group are Jaunpuri Giant (Very big radish, about 1 m long, 50-60 cm in girth and 10-15 kg in weight, cultivated in and around Jaunpur district of U.P.); Bombay Long White (Big, white cultivar, pungent and juicy); Bombay Long Red (Big, red skinned cultivar, pungent and juicy, grown during winter in Northern and Eastern India); Kontai Long (Big, red skinned local cultivar of West Bengal; juicy and pungent, weighing about 2-4 kg, grown in winter); Aus Radish (Small-rooted and mostly white or pinkish cultivar with dense top is grown in high temperature condition of rainy-autumn season in West Bengal).

Radish is highly cross-pollinating with strong sporophytic self-incompatibility. It is an insect-pollinated crop. It shows considerable inbreeding depression on selfing by bud pollination. The commonly adopted breeding include different population improvement schemes particularly mass selection, recurrent selection and family breeding utilizing wide variability of tropical radish group to develop many tropical varieties *viz.*, Pusa Desi, Pusa Mridula, Pusa Sweta, Pusa Gulabi, Punjab Ageti, etc. Exotic introductions have also been utilized in the population improvement programme to develop very few tropical type variety *viz.*, Pusa Chetki (material from Denmark), Kashi Sweta (material from Japan), Arka Nishant (material from Singapore), etc which favour mild temperature condition for optimum growth.

Table 4: Varieties developed through population improvement schemes

Name of variety	Indigenous / Exotic material
Pusa Desi	Indigenous (old sabji mandi, Delhi), Mass selection
Pusa Chetki	Exotic (Denmark), mass selection
Pusa Mridula	Indigenous , Mass selection
Pusa Sweta	Indigenous , Mass selection
Pusa Jamuni	Indigenous , Mass selection
Pusa Gulabi	Indigenous , Mass selection
Punjab Ageti	Indigenous , Mass selection
Kalyanpur No-9	Indigenous , Mass selection

Name of variety	Indigenous / Exotic material
CO-1	Indigenous , Mass selection)from a germplasm (RS-44-2-1)
Kashi Sweta	Exotic (Japan), Mass selection
Kashi Hans	Indigenous , Mass selection
Arka Nishant	Exotic (Singapore), 10 cycles of Mass selection
Punjab Pasand	Repeated crossing of selected plants from a base population of Pusa Chetki and recurrent selection
Kashi Mooli-40	Indigenous , Mass selection
Kashi Lohit	Indigenous , Mass selection
Hisar Sel.-1	Indigenous , Mass selection
Palam Hriday	Indigenous , Mass selection

Hybridization and population improvement

Hybridization followed by population improvement is also adopted to develop few varieties. Adapted cultivars of tropical type have been utilized in hybridization programme between Tropical x Temperate or Oriental x Temperate type followed by different population improvement schemes to develop some varieties in India. The varieties developed by combining temperate and tropical types behave like tropical types as far as seed production is concerned however, they prefer mild temperature for optimum growth and yield *viz.*, Pusa Reshmi, Punjab Safaid, etc. The variety, "Pusa Himani" developed through Temperate x Oriental type cross followed by population improvement scheme behave like temperate type as far as seed production is concerned. This variety also prefers mild temperature for optimum growth and yield.

Transgression of useful traits is possible by wide interspecific and intergeneric hybridization between *Raphanus* and *Brassica* species with the utilization of embryo rescue technique or ovule culture however, to obtain a hybrid between *Brassica* and *Raphanus, Brassica* should be the maternal parent (Been and Park, 1983). Production of both diploid and allopolyploid hybrids have proved useful in breeding different *Brassica* vegetables and radish.

Table 5: Varieties bred through hybridization and population improvement

Name of variety	Parentage
Pusa Himani (Temperate)	Black Radish (temperate semi-domesticated type) × Japanese White (Asiatic/ Oriental)
Pusa Reshmi (Tropical)	Green Top × Desi type
Punjab Safaid (Tropical)	White 5 × Japanese White

Heterosis breeding

Heterosis breeding has been an important method of radish breeding mainly because of uniformity of roots in the hybrids. Commercial hybrid seed production of radish is done utilizing sporophytic self-incompatibility and cytoplasmic male sterility system.

Self-incompatibility

Radishes are normally self-incompatible and insect pollinated. The incompatibility system is of sporophytic type with multiple S-alleles located at a single locus in which the papillae on the stigma surface form the barrier to pollen tube penetration. Generally, low seed set ratio is employed to indentify self-incompatible lines. Pollen tube growth behaviour revealed that self-incompatible lines had a relatively lower number of germinated pollen grains per stigma and pollen tubes per style than self-compatible lines. These two self-incompatibility test methods gave comparable results. However, Horisaki *et al.* (2003) opined that high level of *SI* in radish is controlled by a dominant gene and insect-pollination of self-incompatibility (*SI*) test can be considered to be a more reliable and stricter method than the artificial self-pollination test with respect to an evaluation of SI levels.

Development of homozygous self-incompatible lines

Suitable lines from indigenous or well adapted cultivars are grown for 2-3 generations to attain stability in important traits. By disrupting the stigma surface or by applying pollen immediately to the conducting tissue of the style or bud pollination, 2-3 days before anthesis when the stigma has already become receptive due to protogyny, it is possible to get normal seed set by selfing. Homozygous self-incompatible lines are developed from the desirable plants of these well-adapted line/cultivars by selfing. Treatment of the stigma with small amount of ether or KOH (10%) solution also reduce self-incompatibility. The self-incompatible homozygous lines are maintained by normal selfing through bud pollination or 4% CO_2 treatment under closed poly house. Self-incompatible lines can also be maintained vegetatively by tissue culture method.

Male sterility

Male sterility has been reported in Japanese and Chinese radish. Cytoplasmic male sterility (CMS) is a maternally-inherited trait that is characterized by the inability of a plant to produce functional pollen. Male-sterile cytoplasms have long been of interest because of their use in hybrid seed production and also as an invaluable material to study mitochondrial function and interaction between mitochondrial and nuclear genes. The CMS trait is mitochondrially encoded and always found associated with mitochondrial DNA (mtDNA) rearrangements. Ogura (1968) described a type of male sterility in radish, controlled by the interaction of a recessive gene, *ms* and *S*-cytoplasm. The restorer *R* allele is frequent in European and Chinese cultivars and the maintainer *r* allele is found in Japanese cultivars. In *Ogura* radish, the abortion of male gametogenesis occurs at the early microspore stage due to degeneration of the tapetal tissues (Ogura, 1968). Ogura CMS has been successfully transferred to the cultivars developed and grown in India *viz*., Japanese White, Pusa Chetki and Pusa Desi. The *Ogura* CMS lines are utilized in hybrid seed production.

Production of hybrid seeds

Roots of either self incompatible but cross compatible parental lines or male sterile (female line) and the male fertile lines are planted in isolation for cross pollination in open condition to produce the hybrid seeds. In European radish type, roots of both the self-incompatible lines and seed and pollen parents of the male sterility system are vernalized for 6-8 weeks at 4° to 8°C temperature before planting. Stecklings of 35-40 g weight which gives higher seed yield are prepared from the vernalized roots of the parental lines. Before commercial hybrid seed production, combining ability of the inbred lines are tested to select the parental lines. In the CMS based hybrid seed production, the fertile plants should not differ from the CMS plants for height, flowering period and flower odour. In both self-incompatibility and cytoplasmic male sterility based system, yield of hybrid seeds is heavily dependent on the bee pollinator activity for pollination. So, keeping strong bee hives in the hybrid seed production field is mandatory for getting adequate bee foraging and at the same time, cross- pollination. The F_1 hybrids are produced by single crosses, three-way crosses or double crosses using self-incompatible but cross-compatible lines.

Breeding for disease resistance

Alternaria leaf blight caused by *Alternaria brassicae* is one of the most important foliar diseases, considerably reducing the yield. Several tropical radish cultivars or accessions developed in India are good source of resistance genes for this disease. The breeding line IIHR-1-1 from ICAR-IIHR, Bengaluru shows resistance against white rust caused by *Albugo candida.*

Table 6: Source of disease resistance in Indian radish

Biotic stress	Source of resistance
Alternaria blight (*Alternaria brassicae*)	Accessions 7401, 6802, 8801, 8803, 8805, Pusa Desi and Jaunpuri (Resistant) Accessions 7212, 7110, 7108, 7208, 7218, 8804, S-271 and IHR-1 (Moderately resistant)
White rust (*Albugo candida*)	IIHR-1-1 (Resistant)

Resistance to club root disease caused by *Plasmodiophora brassicae* is controlled by a dominant gene and resistance to white rust caused by *Albugo candida* is also controlled a dominant gene. The genetics of resistance to yellow disease caused by *Fusarium oxysporum* f.sp. *conglutinans* race 2 is conditioned by several genes with varying levels of disease severity.

Table 7: Genetic control of resistance to diseases in radish

Disease	Inheritance pattern/ Gene symbol
Yellow disease caused by *Fusarium oxysporum* f.sp. *conglutinans* race 2	Polygenic control
Club root caused by *Plasmodiophora brassicae*	Dominant gene (s)
White rust caused by *Albugo candida*	Single dominant gene

Source: Akhtar (2015)

Breeding for quality

In radish, quality is related to pungency, sugar content, water content, pore extent, degree of bolting and also external appearance like root length, shape, colour, gloss, smoothness of skin, etc. Good salad type radishes are non-pungent, white or bright yellow-skinned, crispy with a low water content.

The red, pink, violet and purple colour of radish is due to the presence of anthocyanin pigment. They have nutraceutical, colourant and antioxidant properties. Among the various anthocyanins, pelargonidine (acylated pelargonidin-3-sophoroside-5-glucoside) and cyanidine (acylated cyaniding-3-sophoroside-5-glucoside) are responsible for red/pink and purple/violet colour of the roots in radish, respectively (Tatsuzawa *et al.*, 2010). Absence of anthocyanins resulted in white colour. In recent years, the use of coloured radish in the salads or as garnish is gaining popularity in India.

Generally, pigmented radish contains higher amounts of anthocyanins, phenolic compounds and antioxidant activity than non-pigmented radish. Singh *et al.* (2017) studied 24 lines of coloured radish and reported that they differed by 4.98 fold for total phenolics, 36.16 fold for anthocyanins content, 4.96 fold for FRAP activity (Ferric reducing ability of plasma also called Ferric ion reducing antioxidant power) and 4.03 fold for CUPRAC activity (Cupric Reducing Antioxidant Capacity). They estimated 20-250% higher antioxidants in coloured-rooted radishes as compared to the white-rooted cultivars. Improved lines of coloured radishes possessed 80-250% higher ascorbic acid (18.5–26.5.0 mg/ 100g fresh weight), total phenolics (32.5–65.0 mg/100 g fresh weight), anthocyanin content (90–175 µg/g fresh weight), antioxidant-FRAP value (3.15–5.90 µmol/g fresh weight) and antioxidant-CUPRAC value (5.25–11.50 µmol/g fresh weight as compared to white-coloured commercial cultivars (Singh *et al.*, 2017).

Pusa Gulabi is the first entire pink-fleshed, nutritional and anthocynin rich tropical variety which has been developed at IARI, New Delhi by mass selection from indigenous germplasm. Kashi Lohit is also a tropical variety characterized by attractive red coloured and iciclical roots. This ascorbic acid, anthocynin and phytochemical-rich variety has been developed at ICAR-IIVR, Varanasi by mass selection from the indigenous germplasm.

References

Akhtar, S. 2015. Radish. In: *Hand book of Vegetables* Vol.III (Eds. Peter, K. V. and Hazra, P.), Studium Press LLC, Houston, Texas, USA, pp. 311-341.

Been, C.G. and Park, H.G. 1983. Application of ovule culture to production of intergeneric hybrids between *Brassica* and *Raphanus*. *J. Kor. Soc. Hort. Sci.*, **25**:100-108

Duroc, Y., Gaillard, C., Hiard, S., Tinchant, C., Berthome, R., Pelletier, G. and Budar, F. 2006. Nuclear expression of a cytoplasmic male sterility gene modifies mitochondrial morphology in yeast and plant cells. *Plant Sci.*, **170**: 755-767.

Hazra, P., Banerjee, M.K. and Chattopadhyay, A. 2022. *Varieties of Vegetable crops in India*, Kalyani Publishers, New Delhi, Ludhiana, 204p.

Horisaki, A., Tanaka, N. and Niikura, S. 2003. The effectiveness of insectpollination test to evaluate the level of self-incompatibility and their genetic analysis in radish (*Raphanus sativus* L.). *Theor. Appl. Genet.*, **107**: 1009-1013.

Niikura, S. and Matsuura, S. 2001. Genetic variation of the self-incompatibility alleles (S-alleles) in the cultivated radish (*Raphanus sativus* L.) by the PCR-RFLP method. *Acta Hort.*, **546**: 359-365.

Ogura, H. 1968. Studies of the male sterility in Japanese radish with special reference to the utilization of the sterility towards the practical raising of hybrid seeds. *Meun. Fac. Agric. Kagoshima Univ.*, **6**: 39-78.

Singh, B.K., Koley, T.K., Karmakar, P., Tripathi, A., Singh, B. and Singh, M. 2017. Pigmented radish (*Raphanus sativus*): Genetic variability, heritability and interrelationships of total phenolics, anthocyanins and antioxidant activity. *Indian J. Agric. Sci.*, **87**: 1600-1606.

Tatsuzawa, F., Saito, N., Toki, K., Shinoda, K., Shigihara, A. and Honda, T. 2010. Acylated Cyanidin 3-sophoroside-5-glucosides from the Purple Roots of Red Radish (*Raphanus sativus* L.) 'Benikanmi'. *J. Jap. Soc. Hortic. Sci.*, **79**: 103–107.

Zhou, C. J. and Zhang, Y. L. 1994. Studies on several properties of radish male sterility. *Acta Hort. Sinica*, 21: 65

28

Garden pea

Garden pea, *Pisum sativum* var. *hortense* under the family Fabaceae (Leguminosae) having chromosome No. 2n = 2x = 14 is an all-time favourite vegetable in the world. As a cool season crop, it is extensively grown in early spring season in temperate zone and winter season in the tropics and subtropics. The immature/green peas are used as fresh vegetable as well as frozen or canned. Green peas are quite nutritious and contain high amount of protein and a fair amount of vitamin C, zinc, fibre and antioxidants. Diet containing green pea may improve the functionality of the gut.

Origin, domestication and adaptation in India

Discovery of the carbonized pea seed remains indicated that the smooth-seeded field pea was cultivated in the Near East and Europe as early as 7000 B. C. The earliest archaeological finds of peas come from Neolithic Syria, Turkey and Jordan. In Egypt, evidence of peas dates from ca. 4800–4400 BC. The wild pea was restricted to the Mediterranean basin and the Near East. Central Asia (Kazakhstan, Uzbekistan, Tazikistan, Turkmenistan, Kyrgyzstan, Afghanistan), Abyssinia (Ethiopia) and the Mediterranean basin are the primary centres of origin of pea with the near-east (Syria, Lebanon, Israel, Jordon) as the secondary centre of origin. In another view, *Pisum* is considered to have been originated in Ethiopia and from where it spread during pre-historic times to Mediterranean region, Central Europe, the Near East and subsequently to rest of the world. The wild progenitor of *Pisum sativum* is unknown.

Several genetic studies were carried out to understand the genetic changes in pea during the process of domestication and many of the traits were modified typically as parts of the domestication syndrome (Weeden, 2007).

Table 1: Genetic studies on traits forming the domestication syndrome of pea

Wild type phenotype	Cultivated phenotype	Number of loci identified	Gene symbol
Dehiscent pods	Indehiscent pods	1	*Dpo*
Dormancy present	Dormancy lacking	1	*A*
Tall	Dwarf	1	Le
Many basal branches	Few basal branches	1 (minimum)	*Rms* series
Small seeds	Large seeds	Several	*QTL*
Poor seed quality	Good seed quality	4 (minimum)	*R, A, Pl, Gty*
Long day flowering	Day neutral flowering	4	*Sn, Hr, Lf, E*

Source: Weeden (2007), Ram and Choudhary (2012)

In India it was used in pre-historic period during the Indus Valley Civilization (c. 2500 – 2000 B. C.) as evidenced by the carbonized pea seeds discovered in Mohan-ja-Daro and Harappan excavations. Archaeological evidences suggest the existence of pea in India before the advent of the Aryans. This pea was round and non-sweet pulse type (*Pisum sativum var. arvense*). Pea was mentioned in the Vedas (c.1500–1000 B. C.) after the arrival of the Aryans in India. The Aryans introduced it into Europe. It was introduced into the USA in early 1600's. Smooth-seeded field pea was first brought into cultivation and later came the garden pea varieties having white flowers with mostly wrinkled seeds however, some garden pea varieties also have smooth seeds.

The sweet garden pea (*Pisum sativum* var. *hortense*) varieties developed in temperate regions of Europe might have been introduced in India during the British regime. Some early varieties like, Bonia, Kaip, Asauji, etc. have been developed from the early introductions of hardy and round seeded garden pea varieties by the Britishers. These are less sensitive to wilt even under warmer conditions and have later been used in different breeding programmes.

Crop botany and floral biology

- Garden pea is a self pollinated annual herb, bushy or climbing, glabrous, usually glaucous (bluish-green); stem is weak, round and slender, 30-150 cm long
- Leaves are alternate, pinnate with 1-3 pairs of leaflets and a terminal branched tendril, leaflets are ovate or elliptic and 1.5-6 cm long; leaf size in most cases increases up to the first node bearing the first flower.
- The leaf type may be conventional, semi-leafless and leafless.
- Stipules are large, leaf-like and up to 10 cm long; the node at which the first flower emerges is characteristic of a given variety; in temperate regions, the number of nodes at which the first flower emerges varies from 4 in the earliest to about 25 in late maturing types under field conditions.
- Flowers borne on the same peduncle that produce pods; pods mature at different times, the youngest being at the tip; on a whole plant basis, flowering is sequential and upwards from node to node.
- The inflorescence is racemose, solitary, axillary or up to 3 flowers per raceme; flowers are papilionaceous, pedicellate, hermaphrodite and zygomorphic, each with a short peduncle.
- Calyx is oblique; five sepals in the calyx which are united at the base.
- The corolla is papilionaceous with an erect petal spreading at anthesis; petals are five, one standard, two wings and two keels; corolla may be white, pink or purple; the wings are adherent to the boat shaped keel, enclosing the androecium and gynoecium; aestivation is axillary.
- The androecium consists of 10 diadelphous (9 + 1) stamens; filaments are broad, anthers are uniform.
- Gynoecium is monocarpellary and has a superior ovary; style is long, curved, and flattened, stigma is minute, terminal and slightly hairy; placentation is marginal.

- The pods are botanically fruit; pods are swollen or compressed, short-stalked, straight or curved; 4-15 cm long, 1.5-2.5 cm wide, monocarpellary, containing 2 – 10 seeds per pod, pendulous, varying in size and colour, slightly angular, 2-valved, pod dehiscence from both the sutures; colour of the pod is most commonly green, occasionally golden yellow or infrequently purple.
- Seeds are angular or globose, smooth or wrinkled; seeds consist of two large cotyledons surrounding the embryo and enclosed by the seed testa, which may be yellow, green, brownish or other coloured; average 100 seed weight is 12 – 18 gram; wrinkle-seeded garden peas are sweeter than smooth seeded types.

Reproductive behaviour

It is highly self-pollinated crop due to hermaphrodite sex form, homogamy i.e. maturation of stigma and anthers at the same time. Stamens and pistil in opened flower remain enveloped together inside the tube like structure of jointed petals called "keel" leading to self-pollination. Flowers open only once between 5.00 a.m. and 11.00 a.m. Though the flowers open in the morning, the anthers dehisce much earlier even 24 hours before anthesis however, dehiscence is maximum at anthesis. Stigma becomes receptive and pollens are fertile maximum on the day of anthesis. Stigma remains receptive for 3 days after emasculation at 16°C and for one day at 20° to 24°C. Pollens are viable from the time of anther dehiscence until few days thereafter. Since the dehiscence of anthers is much in advance of the blooming, emasculation needs to be carried out in mature flower buds in the preceding afternoon hours. Emasculation of matured buds in the afternoon hours followed by hand pollination in the next day morning between 8.00 to 10.00 a.m. is the best method of hybridization. The flower buds, which are to open in one or two days, are selected for emasculation. After giving a slit in the keel with a pair of forceps, the stamens are removed. Hand pollination is done in the next morning by rubbing the matured anthers of the male parent against the stigma of the female parent. Pollens are collected in the morning from the freshly opened flower. The standard and wings of the flower from the intended male parent are removed and by slight depression of the keel, stigma laden with the pollen grains protrudes out which can be used as a brush for pollination.

Genetics of important characters

Pea belongs to the family Fabaceae, the third largest family of the flowering plants having more than 450 genera and over 12000 species. Pea is a diploid species (2n = 2x= 14) with the basic chromosome number, x = 7. The translocations and other chromosome arrangements are common. Five species comes under the genus *Pisum viz*., *P. fulvum* (red-yellow pea), *P. abyssinicum* (Abyssinian pea), *P. sativum* (garden pea), *P. humile* (dwarf pea) and *P. elatius* (Mediterranean pea). Later, close relationship was found among *P. sativum*, *P. humile* and *P. elatius* and the genus is divided into three major groups *viz*., *P. fulvum*, *P. abyssinicum* and the complex of *P. sativum*, *P. humile* and *P. elatius* (Vershinin *et al*., 2003).

Long cultivation of peas has led to gradual separation of types for different uses *viz*., vegetable use, dry seed use and for use of edible pods.

Table 2: Basic characters of different types of pea, *Pisum sativum*

Type of pea	Basic characters
Garden pea (*Pisum sativum* var. *hortense*)	These are basically sweet-seeded pea universally used for vegetable purpose. Matured seeds of this type are normally wrinkled. It has tough inedible pods that protects the sweet, tender green peas inside. These are delicious and nutritious. The sugar in peas begins to turn starch as soon as they are picked.
Edible podded pea (*Pisum sativum* var. *saccharatum*)	The Edible podded peas lack pod parchment or fibre, unlike field or garden peas that have fibrous pods. This group includes both "Snow pea" and "Snap pea". Snow peas have flat, tender pods with small, undeveloped peas inside. Snap peas have crunchy, delicious plump pods with sweet and full sized peas inside. The most important trait that separates snap peas from snow pea is the thickened pod wall of snap peas.
Field pea (*Pisum sativum* var. *hortense*)	Field peas are usually round seeded, less sweet and are consumed as dry seeds in various preparations. Dry pea seeds have high levels of essential amino acids, lysine and tryptophan which are normally low in cereal grains.

Source: Ram and Choudhary (2012)

The garden pea was the model system employed by Gregor Mendel, father of Genetics to formulate the famous laws of inheritance. The seven characters studied by Mendel have later been mapped.

Table 3: The characters Gregor Mendel used to formulate the famous laws of inheritance

Phenotype of plant character	Genetic control and gene symbol	Chromosome
Shape of matured seeds	Smooth dominant over wrinkled (*R/r*)	7
Seed colour	Yellow dominant over green (*I/i*)	1
Flower colour	Purple/ violet dominant over white (*A/a*)	1
Mature pods	Smooth and expanded dominant over wrinkled and indented (*V/v*)	4
Colour of unripe pods	Green dominant over yellow (*Gp/ gp*)	5
Inflorescence	Axillary dominant over terminal (*Fa/ fa*)	4
Plant height	Tall dominant over dwarf (*Le /le*)	4

Source: Ram and Choudhary (2012)

Note: Mendel dealt with the genes, *A/a* and *I/I* on chromosome 1, *V/v*, *Fa/ fa*, *Le /le* on chromosome 4 , *Gp/ gp* on chromosome 5 and *R/r* on chromosome 7. However, Mendel did not encounter linkage while formulating the law of independent assortment. This has happened because the genes *A/a* and *I/I* on chromosome 1 and the genes *V/v*, *Fa/ fa*, *Le /le* on chromosome 4 was situated far apart hence, crossing over occurred between these genes freely. For this reason, linkage was not realized.

Several qualitative characters of pea are under the control of single gene.

Table 4. Some important monogenic characters of pea

Gene symbol	**Phenotype of plant character**
cry	Influences internode length and plant height along with *la* and *le*
la	Internode length and plant height along with *cry* and *le*
ls	Internode length and plant height along with *cry* and *la*
ram	Increases number of branches
af	Leaflets converted into tendrils
cri	Leaves, stipules, flowers and pod folded and crisp
lat	Double leaflet and stipule area
st	Stipules reduced to strap like structure
tac	Tendrils present on acacia leaves
tl	Leaves with extra leaflets and no tendrils
wa	Without wax on pods, upper and lower stipule surfaces and underside of leaflets
was	Reduced wax as with *wa*
wb	Pods without wax, little wax on rest of plant
wel	Wax absent from all parts of the plant
wex	Extra wax on all aerial plant parts
a	Absence of anthocyanin; dominant allele for anthocyanin production in plant, flower and seed
alb	Plant without chlorophyll, albino, lethal
fn	With *fna* determines number of flowers on the inflorescence; greatly influenced by environment
fna	With *fn* determines number of flowers on the inflorescence greatly influenced by environment
fa	Stem fasciation with *fas*
fas	Stem fasciation with *fa*
yp	Yellow pollen
com	Sides of seeds flattened
r, rb	Seed cotyledons wrinkled
gty	Gritty seed surface
i	Green cotyledons; dominant allele *I* produces yellow cotyledons
it	Increases pod width by 25%
Bt	Apex of pods blunt
Con	Affects curvature of pods
Dpo	Pods tough and leathery; readily dehisce at maturity
gp	Young pod yellow
Pur	With *A, Pu* pod colour purple
ce	Flower rose : dependent on *A* for manifestation of colour
PI	Hilum black with *A Ar B*
di	Small dimpled depression in seed; observable only with a *R* seeds
P	Reduces or eliminates sclerenchymatous membrane on inner pod walls
n	Pod wall thick

Source: Gritton (1986), Kalloo (1993), Swarup (2006), Ram and Choudhary (2012)

Three single recessive genes, *cry, la* and *le* influence internode length and plant height. Each gene governs these characters along with other two genes. Similarly, branching is controlled by two single recessive genes, *fr* and *fru* in presence of each other. A single recessive gene, *ram* is responsible for increasing the number of branches. The leaf, leaflet, stipule and tendril characters are governed by single recessive genes. The leaflets are converted into tendrils due to the presence of the gene, *af*, double leaflet and stipule area due to the gene *lat*, tendrils present on acacia leaves is due to the gene, *tac* and leaves with extra leaflets and no tendrils is governed by the gene "*tl*". Stem fasciation is controlled by two single recessive genes, *fa* and *fas* along with each other. The wax or bloom trait is inherited by single recessive genes, such as, *wa* for absence of wax on pods, upper and lower stipule surfaces and underside of leaflets, *wb* for pods without wax, little wax on rest of the plant and *wel* for absence of wax in all parts of the plant.

Colour of plant and its parts, like foliage, flower, and seed are also governed by different single recessive genes *viz.*, the gene "*a*" controls absence of anthocyanin in plant, flower and seed, *ch-1* is responsible for light yellowish green plant, "*d*" for green leaf axil, *pa* for dark green immature seed and foliage and *vm* with effect similar to *pa*. Number of flowers on the inflorescence is controlled by two different recessive single genes, *fm* and *fna* in the presence of each other.

The gene *b* controls pink flower and *ce* for rose coloured flower and both are dependent on the dominant gene, *A* for manifestation of colour. Single recessive genes determine various seed characteristics, like flattened seed sides (*com*), dimpled seed (*di*), wrinkled seed cotyledon (*r, rb*), gritty seed surface (*gty*) and green cotyledons (*i*). Black hilum character is controlled by dominant gene *PI* along with *A, Ar* and *B*. A single recessive gene "*it*" increases pod width by 25 per cent, the dominant gene *Con* controls curvature of pods, dominant gene *Bt* for blunt apex of pods and recessive gene "*n*" is responsible for thick pod wall. Tough and leathery pods that dehisce readily at maturity are due to the presence of a dominant single gene *Dpo*. Two different single recessive genes, *p* and *v* are responsible for reducing or eliminating sclerenchymatous membrane on inner pod walls. The purple pod colour is governed by two different dominant genes *Pu* and *Pur*, along with the dominant gene *A* and yellow colour of young pods is controlled by a recessive gene *gp*.

The highly heritable polygenic characters are plant height, earliness, number of pods per plant, pod length, seeds per pod and 100 seed weight. Pod yield generally shows low heritability. Number of pods per plant, number of seeds per plant, 100 seed weight, number of days to maturity and plant height had much direct effect on yield. Number of pods per inflorescence, number of seeds per pod, and seed weight, besides plant height, leaf area, number of branches, early flowering and number of pods have direct association with yield. Nature of gene action, degree of dominance, and interallelic gene effects were studied for different plant characters. Seed yield per plant had significant additive genetic variance and positive epistatic gene action for seed yield per plant. Days to flowering showed non-additive gene action with partial dominance and over-dominance. Partial dominance or over-dominance were important for plant height.

Breeding objectives

The followings are the main objectives of pea breeding

- Early maturity, preferably 50 – 60 days after seed sowing particularly in short-lived winter condition.
- Mid-season (80 – 90 days after sowing) and late (100–120 days after sowing) maturing variety for prolonged winter condition.
- High number of straight or slightly curved, well filled and dark green pods
- Bold, dimpled or wrinkled, greenish seed with high sugar percentage.
- Ratio between weight of green pods and weight of seeds in mature green pods (shelling percentage) should be about 40 per cent or more.
- High yield of green and matured pods
- Suitability for processing, mainly canning, dehydration, and freezing.
- Resistance to disease, like powdery mildew, *(Erysiphe polygoni)*, wilt (*Fusarium oxysporum* f. *pisi*), rust (*Uromyces fabae*), and virus diseases (pea seed-borne mosaic, pea enation mosaic, bean yellow mosaic).
- Resistance to insect pests, like leaf miner (*Phytomyza atricornis*), stemfly (*Ophiomyia phaseoli*), pod borer (*Etiella zinckenella*), etc.
- Resistance to high temperature stresses

Breeding methods and development of variety

Pea is highly self-pollinating because stamens and pistil in opened flower remain enveloped together inside the tube like structure of jointed petals called "keel" leading to self-pollination. The breeding procedures adopted for pea improvement include plant introduction, individual plant selection, single seed descent method of selection, hybridization followed by pedigree or bulk method of selection, backcross breeding and induced mutation.

However, introduction of exotic varieties forms the mainstay of breeding garden pea in India. They were either used as a variety as such or utilized in hybridization programmes to develop new varieties. Some varieties have been developed through single plant or pure line selection from the exotic varieties or few well-adapted indigenous genotypes. However, most of the varieties have been developed by inter-varietal hybridization followed by pedigree, bulk-pedigree or some modified methods of handling the segregating generations. The varieties grown in India have been described under two groups according to their maturity period, *viz*., early and mid-season and late varieties. The early varieties can again be sub-divided to smooth-seeded and wrinkle-seeded varieties. Edible podded types form a separate varietal group.

- **Early variety (smooth seeded):** Asauji, Early Superb, Meteor, Lucknow Boniya, Harbhajan, Kashi Nandini, Kashi Udai, Kashi Mukti, Pant Sabji Matar-3, Palam Triloki

- **Early variety (wrinkled seeded):** Arkel, Early Badger, Little Marvel, Alaska, Yates Early Crop, Kelvedon Wonder, VL-Ageti Matar-7 (VL-7), Jawahar Matar 3, Jawahar Matar 4, Pant Matar 2, Jawahar Peas 54, Mattar Ageta 6, Hisar Harit, Jawahar Peas-4, Kashi Nandini, Narendra Sabji Matar-6, Narendra Sabji Matar-4, VRP-2, Pusa Shree, Pusa Pragati, Arka Mayur, Pant Sabji Matar -5
- **Midseason and late varieties (mostly smooth seeded):** Bonneville, Thomas Laxton, Alderman, Perfection New Line, T-19, Lincon, Delwiche Commando, Khapar Kheda, NP 29, Jawahar Matar 1, GC-195, Jawahar Matar 2, VL-Matar-3, Pant Uphar, Punjab 88, Azad P-2, Vivek-6, Ooty-1, Jawahar Pea 83, Jawahar Peas 15, NDVP-8, VL-8, Vivek Matar-9, Narendra Sabji Matar-5, Kashi Shakti, Azad P-5, Swarna Mukti, Swarna Amar, Arka Ajit, NDVP-10, VRP-3, Priya, DPP-9411, Palam Priya, Arka Karthik, Arka Priya, Arka Chaitra, Arka Tapas, Arka Uttam, Vivek Matar 8, Kashi Samridhi, Punjab 89, Phule Priya , Solan Nirog, Palam Sumool
- **Edible podded variety (snow pea or sugar pea or snap pea):** Sylvia, Oregon Sweet Podded, Arka Sampoorna, Arka Apoorva, Mithi Phali

Germplasm resources and Plant introduction

A large number of *ex situ* germplasm collections are maintained in different countries *viz.*, Australia, Bulgaria, China, Germany, Italy, The Netherlands, Poland, Russia, Syria, Sweden, UK, USA, and other countries (Ambrose, 2008). Perhaps one of the best world collection of pea germplasm is being maintained in the Nordic Gene Bank at Lund in Sweden. Exhaustive germplasm collections are also being maintained in USA, Germany, Italy and U. K.

In India, garden pea germplasm is maintained at NBPGR, New Delhi, ICAR-IARI, New Delhi, ICAR-IIVR, Varanasi, CSAUAT, Kanpur and other places. Wide range of field pea germplasm is also maintained at ICRISAT, Hyderabad, and ICAR- Indian Institute of Pulses Research, Kanpur. Good number of exotic varieties of garden pea have been released directly for cultivation in India (Hazra et al (2022). However only few varieties, like Arkel, Bonneville and Lincoln are still being grown by the farmers while most of the other exotic cultivars have been replaced by several improved and popular varieties which have been developed later.

Table 5: Exotic variety directly released for cultivation

Name of the variety	Country from where introduced
Early Superb	England
Meteor	England
Arkel	France
Early Badger	USA
Little Marvel	England
Alaska	England
Yates Early Crop	Australia
Kelvedon Wonder	England
Bonneville	USA

Thomas Laxton	USA
Alderman	USA
Perfection New Line	USA
Lincon	USA
Sylvia	Sweden
Oregon Sweet Podded	USA

Variability and selection

Several local cultivars of pea were grown in the country, such as, Boniya, Lucknow-Boniya, Local Yellow Batri, Hara Boniya, Asauji (Amritsar), Hoshiarpuri, Kap, Kanawari, Khaparkheda (Madhya Pradesh), Kalanagini, Kinnauri and a few others. These local varieties are basically smooth-seeded field pea types. Single plant selection or pure line selection method has been followed to develop some old varieties mostly from these local varieties *viz.*, Asauji, T-19, Mithi Phali, etc. Few varieties *viz.*, Pusa Shree, Pusa Pragati, Priya, etc. have also been developed through selection from different exotic genetic materials. Selection should effectively be exercised based on pod yield and most important pod yield components *viz.*, plant height, earliness, pods per plant, seeds per pod, pod length and 100-seed weight.

Table 6: Varieties developed through selection

Name of variety	Indigenous / Exotic material
Asauji	Indigenous well adapted genotype from early introduction (Amritsar)
Lucknow Boniya	Indigenous well adapted genotype from early introduction (around Lucknow and Meerut in Uttar Pradesh)
Harbhajan	Exotic (EC 33866)
Pusa Shree	Exotic
Pusa Pragati	Exotic
T-19	Indigenous well adapted genotype from early introduction (Amritsar)
Khapar Kheda	Indigenous well adapted genotype from early introduction (Madhya Pradesh)
NP 29	Indigenous well adapted genotype from early introduction (U.P)
Pant Uphar	Exotic
Ooty-1	Exotic (Accession PS 33)
Priya	Exotic
Vivek Matar 8	Exotic
Mithi Phali	Indigenous well adapted genotype (Punjab)

Hybridization and selection

The main emphasis in pea improvement has been on early maturity, yield, quality, and resistance to diseases and insect pests. Hybridization and selection and backcross breeding methods have been adopted in breeding for disease resistance. The most often practiced method of breeding in peas is pedigree method capitalizing on better recombinants and transgressive segregates. The crosses are made normally between better and contrasting type of parents. Selection for highly heritable traits like, resistance to disease and other monogenic characters in early generations followed by line sowing in small plots in F_4/F_5 generation where superior lines are selected and then superior plants are selected from the superior lines in F_6 generation. The superior lines consisting of superior plants based on the desirable characters are evaluated through multilocation trial to select the best pure line to be released as a new variety. Single seed descent method of breeding (modified bulk method) in the early generations to maintain wide range of variability followed by pedigree method breeding is also employed to develop new varieties. Most of the varieties in India have been developed by hybridization followed by selection.

Table 7: Varieties bred through hybridization followed by selection

Name of variety	Parentage
Kashi Nandini	P-1542 x VT-2-1
Kashi Udai	Arkel x FC-1
Kashi Mukti	No.7 x PM-5
Pant Sabji Matar-3	Arkel x GC 141
VL-Ageti Matar-7	Pant Uphar x Arkel
Jawahar Matar 3	T19 × Early Badger
Jawahar Matar 4	T19 × Little Marvel
Pant Matar 2	Early Badger × Pant Uphar
Jawahar Peas 54	(Arkel × JM5) × ('4bc' × JP 501)
Mattar Ageta 6	Massey Gem × Harabona
Hisar Harit	Bonneville × P23
Jawahar Peas-4	Local Yellow Batri × (6588 × 46C)
Narendra Sabji Matar-6	KS-123 x Arkel
Arka Mayur	IIHR 105 × Arka Pramodh
Jawahar Matar 1	T19 × Greater Progress
GC-195	T19 x Little Marvel
Jawahar Matar 2	Greater Progress × Russian-2
VL-Matar-3	Old Sugar × Early Wrinkled Dwarf 2-2-1
Punjab 88	Pusa-2 × Morrasis-55
Azad P-2	Bonneville x 6587
Vivek-6	Pant Uphar × VL Matar-3
Jawahar Pea 83	(JMI × JP 829) × (46 c × JP 501)
Jawahar Peas 15	(JMI × R 98 B) × JP 501 N2
Kashi Shakti	Hara Bona x NDVP-8
Arka Ajit	(Bonneville x IIHR-209) x Freezer-656
Arka Karthik	Arka Ajit × IIHR 554
Arka Priya	Arka Ajit × IIHR 562
Arka Chaitra	Arka Ajit × Arka Sampoorna
Arka Tapas	(Arka Pramodh × Oregon Sugar) × Arka Priya)
Arka Uttam	(Arka Ajit × Arka Sampoorna) × (Arka Pramodh × Oregon Sugar)
Arka Sampoorna	(Bonneville × IIHR 209) × Freezer 656 × Oregon Sugar)
Arka Apoorva	Arka Pramodh × Oregon Sugar

Utilization of mutants and applied mutagenesis

Historically, mutation breeding has occupied a centre stage in pea breeding where chemical mutagens (EMS, NEU, EI, NMU) or radioactive mutagenic agents (mainly gamma radiation) are used to create random mutations in commercially accepted cultivars in many countries. Garden pea is a model for understanding compound leaf development. Noteworthy mutation breeding researches in garden pea have been reported in Sweden, Italy, Germany, Poland and Bulgaria. Mutants have played a major role in ideotype / plant type breeding in peas. Two useful recessive mutant genes, *st* (reduced stipule size) and *af* (tendrils substituted for leaflets) were utilized to produce the 'leafless' pea genotype, *afafstst*. The "*le*" and "*afila*" mutations were introduced to produce different field pea varieties adapted to mechanical harvesting.

At JNKVV, Jabalpur, some induced mutations are early flowering *viz.*, 46 C and JP 829 (flowering from 4th to 6th node) and fasciated mutants, *viz.*, R 701, R 710, JP 625, JP 67, 251 A, 997, 999 and others. *Acacia mutants* (tendrils converted into leaflets) are *Acacia Long*, *Acacia Batri*, *Acacia Purple* and a few other types of mutants. Research works have been done to combine mutant characters with other desirable traits including multiple disease and pest resistance and quality characters.

Breeding for resistance to diseases

Single dominant genes confer resistance to several diseases, like enation mosaic virus (*En*), Near Wilt, *Fusarium oxysporum* f. *pisi* race 2 (*Fnw*), Fusarium wilt, Fusarium *oxysporum* f. *pisi* race 1 (*Fw*), brown root of peas, *Fusarium solani* f. sp. *pisi*, rust, *Uromyces fabae*, downy mildew *Peronospora pisci* and bacterial blight. *Pseudomonas syringe* pv. *pisi* race 1.

Single recessive genes confer resistance to pea seed-borne mosaic virus (*sbm*), powdery mildew (*er; er-2*), bean yellow mosaic virus (*mo*), top yellow virus, pea streak virus, pea mosaic virus (*pmv*) and bean virus 2. Resistance to Ascochyta blight (*Ascochyta pisi*) is governed by duplicate factors or single dominant gene. Linked polygenes with a recessive gene (*le* or *pl*) or single dominant gene govern resistance to pea root rot, *Aphanomyces euteiches*. Resistance to bacterial brown spot of pea, *Pseudomonas syringe* pv. *pisi*, is due to a complex system of inheritance.

Table 8: Genes for resistance in peas

Gene symbol	Resistance to disease
Single dominant gene, *En*	Resistance to enation mosaic virus
Single dominant gene, *Fnw*	Resistance to near wilt , *Fusarium oxysporium* f. *pisi* race 2,
Single dominant gene, *Fw*	Resistance to Fusarium wilt, *Fusarium oxysporium* f. *pisi* race 1
Single recessive gene, *sbm*	Resistance to pea seed borne mosaic virus
Single recessive gene, *er-1*	Resistance to powdery mildew, *Erysiphe polygoni*
Monogenic, dominant	Resistance to rust, *Uromyces fabae*
Monogenic, dominant	Resistance to brown root of peas, *Fusarium solani* f. sp. *pisi*
Monogenic, dominant	Resistance to downy mildew, *Peronospora pisi*
Monogenic, dominant	Resistance to Asochyta blight, *Asochyta pisi*

Gene symbol	Resistance to disease
Duplicate factor; or Single dominant gene	Resistance to pea root rot, *Aphanomyces euteiches*
Single dominant gene, *mo*	Resistance to Bean yellow mosaic virus
Single recessive gene	Resistance to Pea leaf-roll virus
Single recessive gene	Resistance to Pea mosaic virus
Complex inheritance	Resistance to brown spot of pea, *Pseudomonas syringe* pv. *pisi*

Source: Gritton (1986), Kalloo (1993), Swarup (2006), Ram and Choudhary (2012)

In India, breeding for disease and pest resistance in pea was started at the Jawaharlal Nehru Krishi Vishawa Vidyalaya, Jabalpur about three decades ago and several varieties were developed having resistance to a few important diseases, like powdery mildew, Fusarium wilt and rust along with other desired attributes such as, maturity, yield of green pods, length of pods, and seed quality *viz*., 100 seed-weight, wrinkled and seed colour (Narasinghani and Tiwari, 1993). A few multiple disease resistant varieties with insect resistance have also been developed. There were many sources of powdery mildew resistance, such as, P185, P388, P6585, P6587, P6588, Sel. 1, Morasis-55, T-10, T-56, JP501 and a few others. There are only two pea lines, JB Batri 3 and JP Batri Brown 4 show resistance to rust however, both these lines are highly susceptible to powdery mildew.

Important powdery mildew resistant varieties developed in India are Jawahar Pea 83 from a double cross (JMI × JP 829) × (46 c × JP 501), Jawahar Pea 4 from a triple cross Local Yellow Batri × (6588 × 46C), Jawahar Peas 54 from a double cross (Arkel × JM5) × ('4bc' × JP 501), Kashi Mukti from the cross (No.7 x PM-5), VL-Matar-3 from the cross Old Sugar × Early Wrinkled Dwarf 2-2-1, VL-Matar-3 from the cross Old Sugar × Early Wrinkled Dwarf 2-2-1 and Azad P-2 from the cross Bonneville x 6587.

A duel resistant (powdery mildew and Fusarium wilt) variety Jawahar Peas 15 was developed through advanced generation selections from the triple cross (JMI × R 98 B) × JP 501 N2. The variety JP 179 was developed which is resistant to Fusarium wilt and powdery mildew and shows tolerance to rust. The variety Arka Karthik, Arka Priya, Arka Ajit developed at IIHR, Bengaluru has combined resistance to rust and powdery mildew. The snap pea variety, Arka Sampoorna is also resistant to rust and powdery mildew.

Several exotic pea varieties, like Alaska, Early Giant, Sylvia, Kelvedon Monarch and a few others as well as some Indian varieties, like T17, Selection 1, GC66, GC468, Lokar, Pusa Vipasha, and the local cultivar, Kalanagini were found to be resistant to *Fusarium* wilt when tested under epiphytotic conditions. However, wilt resistance varied at different locations due to prevalence of different races. A *Fusarium* resistant source, JP 501 A/2, was identified at JNKVV, Jabalpur. It was also resistant to powdery mildew however, the powdery mildew resistant varieties are generally susceptible to rust.

Breeding for resistance to insect pests

Breeding for resistance to leaf miner was undertaken at JNKVV, Jabalpur and HAU, Hisar. Sources of resistance to leaf miner were JP 179, JP 169-1, JP 747 and a few others at Jabalpur and LMR 4, LMR 10 and LMR 20 at Hisar. However, improved varieties having resistance to leaf miner have not yet been developed.

The two pea lines, JP9 and JP179 show resistance to pea weevil, *Bruchus pisorum*. The crosses of *Pisum fulvum* with *Pisum sativum* hold promise for resistance to the pea weevil. A few lines having multiple resistance to diseases and pests were developed at JNKVV, Jabalpur. These multiple resistant lines are, JP 179 (highly resistant to powdery mildew, resistant to Fusarium wilt, tolerant to rust and resistant to leaf miner and pea weevil), JP9 (resistant to powdery mildew and pea weevil), JP 501 A/2 (resistant to powdery mildew, Fusarium wilt and pea weevil) and JP Batri Brown 3 and JP Batri Brown 4 (resistant to rust and pea weevil). Attempts have been made to combine the multiple disease and pest resistance of JP 179 with some important varieties for developing improved cultivars for commercial cultivation. Some improved varieties have been found tolerant against important insect pests *viz*., Pant Uphar shows tolerance against pea stem fly, Kashi Nandini is tolerant to leaf miner and pod borer.

Breeding for abiotic stress

The varieties, Arka Uttam, Arka Tapas and Arka Chaitra, all developed from the segregating generation of a double cross, (Arka Ajit × Arka Sampoorna) × (Arka Pramodh × Oregon Sugar) show tolerance to high temperature (upto 35^0C). The variety, Vivek-6 is moderately tolerant to moisture stress. Some research reports suggest that cold resistance is controlled by intermediate dominance while other reports indicated being under the control of many recessive genes.

Breeding for quality

The wrinkled seeded peas contain 26 – 33 % protein content while, it is 23–31 % in smooth seeds peas. Inheritance of protein content is polygenically controlled, and high protein content is mainly governed by recessive genes. The varieties GC 195 and the local cultivar, Kinnauri have high soluble protein content which can be used as effective donor parents.

Dehydration, canning and freezing are the most common methods of processing peas. Large sized wrinkled and dark green peas as found in the variety Arkel are suitable for dehydration. For canning, both round and wrinkled seeded varieties like, T19 and Bonneville can be used. However, for freezing, wrinkled and green seeded varieties like, Arkel , Kashi Nandini, Jawahar Matar 4, etc. are suitable.

References

Ambrose, M. 2008. Garden Pea. In: *Handbook of Plant Breeding* (Eds. Prohens, J., Nuez, F. and Carena, M.J.) VolI, Springer, Heidelberg. Pp. 3-26.

Gritton, E.T. 1986. Pea breeding. In: *Breeding Vegetable crops* (Ed. Bassett, M.J.), AVI Publishing Co. Inc. West Port, Connecticut, USA pp. 283-319.

Hazra, P., Banerjee, M.K. and Chattopadhyay, A. 2022. *Varieties of Vegetable crops in India*, Kalyani Publishers, New Delhi, Ludhiana, 204p.

Kalloo, G. 1993. Pea- *Pisum sativum*. In: *Genetic Improvement of Vegetable crops* (Eds. Kalloo, G and Bergh, B.O.), Pergamon Press, Oxford, New York, pp. 409-425.

Narasinghani, V.G. and Tiwari, A. 1993. Improvement of garden pea. In: *Advances in Horticulture*, Vol V, Part 1(Eds. Chadha, K.L. and Kalloo, G.), Malhotra Publishing House, New Delhi, pp. 217-233.

Ram, H.H. and Choudhary, H. 2012. Pea. In: *Hand book of Vegetables* (Eds. Peter, K. V. and Hazra, P.), Studium Press LLC, Houston, Texas, USA, pp. 391-420.

Swarup, V. 2006. *Vegetable Science and Technology in India*. Kalyani Publishers, New Delhi, Ludhiana, 656p.

Vershinin, A.V., Allnutt, T.R., Knox, M.R., Ambrose, M.J. and Ellis, T.H.N. 2003. Transposable elements reveal the impact of introgression, rather transposition in *Pisum* diversity, evolution and domestication. *Mol. Biol. Evol.*, **20**: 2067-2075.

Weeden, N.F. 2007. Genetic changes accompanying the domestication of *Pisum sativum*: Is there a common genetic basis to the "Domestication Syndrome" for legumes? *Ann. Bot.*, **100**: 1017-1025.

29

French bean

French bean, *Phaseolus vulgaris* under the family Fabaceae (Leguminosae) and having chromosome No.2n = 2x = 22 is an important leguminous vegetable crop grown worldwide for its tender fruit (pod) as vegetable and dry seeds as pulse. French beans are unique among plant foods due to their high protein content. It is a highly variable species that has a long history of cultivation. Most of the bean varieties for processing (canned, frozen, and freeze-dried) are round podded, while fresh market cultivars are often flat or oval shaped. French bean is a valuable source of the amino acids lysine and tryptophan; minerals likes, iron, copper and zinc; and beneficial phytochemicals, antioxidants and flavonoids.

Origin, domestication and adaptation in India

French bean is thought to have been originated in South and Central America. Southern Mexico and Central America are considered to be the primary centre of origin, while secondary centres lies in Peru-Bolivia-Ecuador region of American continent. It was actually domesticated in Mesoamerica first and travelled south, probably along with squash (*Cucurbita* species) and maize. Multi-locus sequence data have indicated that the domestication of common bean was initiated 8000 years ago. Cultivation of French beans started by the American Indian tribes settled in Tehuacan Valley of Mexico and in Callejon de Huaylas, Peru. When Christopher Columbus returned from his second voyage to the New World in the year 1493 he brought French beans with him in the Mediterranean region. French beans were considered to be rare to find and expensive but soon became one of the commonly used beans in the 19th century. In France, French beans were introduced in the year 1597 by the Conquistador (Spanish explorers). In Europe, the French bean spread rapidly in the 16th and 17th centuries and reached England by 1594. *Phaseolus aborigineus* is known as the progenitor of French bean and the domestication of this species might have occurred in Brazil and northern America. During the process of domestication of common beans, several morphological changes like, growth habit, testa colours, pod structure, etc. have occurred. Marked phenotypic differences for these morphological and physiological traits that distinguish wild progenitors and cultivated descendants are the "domestication syndrome". However, lack of information about their genetic control have limited the utilization of wild germplasm for crop involvement.

Determinate or bushy cultivars which are utilized both for the tender green pods as vegetable and dry seeds as pulse (Rajmah) has been introduced in India comparatively in recent times from Europe. Most of the local types indigenous to North-eastern region are indeterminate, pole types.

Crop botany and floral biology

- All wild members of French bean have climbing habit, but many cultivars are determinate in growth habit which are classified as "bush type" while cultivars with indeterminate growth habit are grouped "pole type".
- Leaves are alternate, green or purple which are divided into three oval, smooth-edged leaflets, each 6–15 cm long and 3–11 cm wide.
- Inflorescence is unbranched, axillary short raceme bearing several flowers.
- Bracts on the rachis of the inflorescences are persistent, and the size and shape of the bracteoles are distinguishing characteristics of bean cultivars.
- Flowers are hermaphrodite and of about 1.0 cm long; stigma and anthers mature at the same time, medium sized, few in numbers and borne on the short raceme.
- The floral structure contributes to the high rate of self-pollination: anther dehiscence and stigma receptivity occur at the same time, before the flower is fully open, and the anthers and stigma are positioned near one another at the time of anther dehiscence and stigma receptivity
- Calyx is gamosepalous, sepals are 5 in number; corolla is papilionaceous and exerted with a recurved standard petal; petals are five, one standard, two wings and two keels; wings are large; keel is spirally coiled in a characteristic manner around stamens and pistil; flower colour ranges from white, yellow, bluish-purple, lilac, deep violet, pink, rose or lavender depending on the type and variety.
- Stamens are diadelphous, 9 united +1 free.
- Ovary is monocarpellary, unilocular, placentation is marginal with many ovules borne in two alternating rows along the ventral suture; style is enclosed inside the keel and hairy on the inner side; stigma is oblique.
- The flowers give way to fruits (pods, beans) 8–20 cm long and 1–1.5 cm wide, bivalved pod that bear seeds alternately on both sides of the dorsal margin of the placentae; pods may be green, light green, yellow, black or purple in colour, each containing 4–6 beans.
- The pods which may be round or flat are of two types: fibrous "string" running along the length of the pod and stringless which lack the tough, fibrous "string".
- As the name implies, string or snap beans break easily when the pod is bent, giving off a distinct audible snap sound.
- The seeds are smooth, plump, kidney-shaped, up to 1.5 cm long, range widely in colour, and are often mottled in two or more colours; flowering is reduced once the seeds begin to form inside the pods.
- Average 100 seed weight is about 25 gram

Reproductive behaviour

It is a pre-dominantly self-pollinated crop due to hermaphrodite sex form and homogamy i.e., maturation of stigma and anthers at the same time although, dehiscence of anthers takes place much earlier than anthesis in the morning. However, natural crossing some time occurs because of high population pressure of the pollinators, principally bumblebee and to a lesser extent by the honeybee in the field. Stamens and pistil in opened flower remain enveloped together inside the tube like structure of jointed petals called "keel" leading to self-pollination. Flowers open only once between 7.00 and 9.00 a.m. depending on the weather condition and never closes upon fertilization. However, the corolla is shed within few days of blooming. The stigma becomes receptive two days before anthesis however, receptivity is maximum on the day of anthesis. Even though the pollens are shed from the anthers much before the anthesis, the pollens are not transferred to the stigma at this time unless the flower is opened. Agitation or tripping of the flower by bees helps to transfer the pollens to the stigma. A short time after anthesis, the stigma retracts into the keel and meets the pollens and self-pollination is accomplished. After pollination, the pollen tube grows toward the ovary and fertilization takes place about 8 to 9 hours later. Pollen tubes from the foreign pollen may grow more quickly than self-pollen, leading to a higher incidence of cross-fertilization if cross-pollination occurs. Pollen viability is at maximum between 5.00 to 9.00 a.m. on the day of anthesis. Since the dehiscence of anthers is much in advance of the blooming, emasculation needs to be carried out in mature flower buds in the preceding afternoon hours. The flowers are highly sensitive and drop off easily with slight mechanical disturbance or injury. To prevent drying out of the emasculated bud, leaflet may be folded and pinned around the emasculated bud. A tissue paper can also be used to cover and protect the bud. Being a large flowered leguminous vegetable crop, it is sufficient to open only one side of the flower bud by careful removal of the keel, which remains spirally around the stamen and style. Emasculation of the anthers is then done carefully. Emasculation of matured buds in the afternoon hours followed by hand pollination in the next day morning between 7.00 to 9.00 a.m is the best method of hybridization. Pollens are collected in the morning from the freshly opened flowers. The standard and wings of the flower from the intended male parent are removed. By careful removal of the spiral keel, stigma and hairs beneath covered with the pollen grains itself can be used as a brush for pollination.

Genetics of important characters

French bean (*Phaseolus vulgaris*) is a diploid species with chromosome number 2n =2x= 22. Polyploidy did not play any role in evolution of *Phaseolus*, but structural chromosome changes in the somatic chromosomes like, inversions, heterozygosity, presence of univalent and pollen sterility were involved in the evolution of this genus.

In French bean, the qualitative characters include growth habit (pole type and bush type), green hypocotyl, dwarf plant, unifoliate leaf, leaf variegation, short and narrow pod, stringy pod, cell wall fibre, pod membrane, pod shape and size, straight pod, interlocular cavity, hard seeds, seed eye pattern, seed colour, etc.

Table 1: Some important simply inherited characters of French bean

Character/ Gene symbol	Phenotype / inheritance pattern
ace	Shiny pod
aph	Plants have only two leaves, both unifoliate; sterile
Beg	With "*P*" gives begonia red flower colour
Blu	Blue flower colour mutant
C^r	Indistinct, inconsistent mottling on seed coat
cry	Crypto-dwarf; a dwarfing gene
Da	Straight pod; dominant
do	Dwarf out-crossing mutant
$dt\text{-}1^a$ $dt\text{-}2^a$	Early, day length neutral flowering
dw1 dw2	Dwarf plant due to duplicate genes
Fa	Basic gene for pod membrane
Fb Fc	Supplementary genes for pod membrane
fd	Delayed flowering response under long days
Fin	Indeterminate dominant over determinate
gas	Gamete sterile causing both male and female sterility
glb	Glossy bronzing leaf mutant
la lb	Parchment pod dominant over tender pod
Li	Long internode dominant over short internode
lo	Short inflorescence
Mf	Brownish-violet blotch on the base of standard petal
ms-1	Induced genic male sterile mutant
nie	An induced mutation for effective nodulation by *Rhizobium* sp.
nnd (sun-1)	An induced mutation for non- nodulation by *Rhizobium* sp.
nnd-2	An induced mutation for non- nodulation by *Rhizobium* sp.
nts	An induced mutation for abundant nodulation by *Rhizobium* sp. In presence of high nitrogen
ol	Overlapping leaflet mutant
P	Basic colour gene; "*P*" without colour genes is colourless as is "*p*"
pa	Pale green leaves
pc	Persistent green pod colour
pg	Pale green foliage mutant
prc	Progressive chlorosis mutant
ram	Branched inflorescence
rnd	Round leaf mutant
St	Stringy pod; dominant character
T	Self-coloured seed coat and coloured flowers
Th-1 Th-2	Genes for equal value for seed thickness
To	Cell wall fibre
te	Short and narrow pod
uni	Unifoliate leaves; complete sterility; recessive character

Uni-2	Dominant mutation for unifoliate true leaves
Growth habit	Three genes control pole and bushy growth habit Viny type (*T*) dominant over bush type (*t*) Long stem (*L*) dominant over short stem (*l*) Indeterminate growth habit (*A*) dominant over determinate growth habit (*a*) Pole type: *L _ A –T* Bush type: *ll aa tt*
Stringy pod	Single dominant gene; Stringy dominant over non-stringy
Seed colour	At least 8 genes control seed colour and "*P*" is basic to coloured seeds White seeds are homozygous recessive *pp* Black seed is epistatic to all other colour
Interlocular cavitation	Single dominant gene
Seed eye pattern	Single gene

Source: Yarnell (1965), Bassett (1993), Ram (1998), Swarup (2006), Singh and Dutta (2015)

Generally, pod yield and yield components are governed by polygenes and gene action controlling these characters is complex in nature. Important quantitative characters governed by polygenes are plant height, internode length, number of primary and secondary branches, days to flowering, days to first harvest, pod weight, pod length, number of pods per plant, number of seeds per pod and 100-seed weight.

Breeding objectives

In French bean, breeding objectives for improvement of vegetable types (for fresh market and processed products) and dry bean type are different. The followings are the main objectives of French bean breeding for green pod used as vegetable for fresh market and processing purpose.

- Bush type of determinate growth habit with upright leaves and good number of branches per plant (plant height 50 to 65 cm), early flowering
- Short vine for pole type of indeterminate growth habit, early flowering
- Pods should be attractive, bright green, round or oval, straight or slightly curved, string less, crisp, fleshy and smooth, free from interlocular space, pod wall fibre just enough to retain shape, suitable for steam cooking.
- Pods should be 12-15 cm long, bright green, round or oval, smooth with small and few seeds, free from interlocular space with concentrate fruiting at mid height of the plant.
- Pods may be flat, green, and fleshy and having good cooking quality (for fresh market)
- Good processing quality for canning (light or medium green, round, fleshy, stringless pod with small and few seeds) and freezing (small sized, bright green, round, fleshy, stringless pod for freezing whole pod)
- Higher pod number and pod yield per plant

- Resistance to diseases *viz*., powdery mildew (*Erysiphe polygoni*), rust (*Uromyces appendiculatus*), root rot (*Rhizoctonia solani*), anthracnose (*Colletotrichum lindemuthianum*), ashy stem blight (*Macrophomina phaseolina*), bacterial blight (*Xanthomonas campestris pv. phaseoli),* bean common mosaic virus, etc.
- Resistance to insect pests viz., aphids *(Aphis craccivora),* stem fly *(Ophiomyia phaseoli),* pod borer *(Helicoverpa armigera),* etc.
- Resistance to abiotic stresses like, high temperature, soil salinity, cold and frost

Important characters for breeding French bean for dry seed as pulse (Rajmah) include, pod number, pod yield, high pod wall fibre content for easy dehiscence of dry pods, high dry seed weight and light seed colour.

Breeding methods and development of variety

Introduction plays very important role in breeding French bean in India. In fact, most of the popular varieties grown in India are exotic varieties. Selections from exotic cultivars or lines also developed some varieties. French bean is highly self-pollinating species because stamens and pistil in opened flower remain enveloped together inside the tube like structure of jointed petals called "keel" leading to self-pollination. Apart from plant introduction, the breeding methods adopted for French bean improvement include individual plant selection, pure line selection, hybridization followed by single seed descent, pedigree or bulk method of selection, backcross breeding and induced mutation. Recombination breeding approaches through planned hybridization and selection have been followed to incorporate desirable characters including disease resistance. Varieties of French bean developed in India can conveniently be classified according to growth habit *viz*., bush type and pole type.

- **Bush type variety:** Contender, Giant Stringless, Bountiful, Premier, Jampa, Pusa Parbati, VL-Bauni-1, Arka Komal, Arka Suvidha, Arka Anoop, Arka Sharath, Arka Arjun, Pant Anupama, YCD-1, Phule Surekha, Kashi Sampann , Kashi Rajhans , Kashi Param, Azad Rajmah-1, Phule Suyash, Azad Rajmah-1, Swarna Priya, Swarna Lata
- **Pole type variety:** Kentacky Wonder, Pusa Himlata, TKD-1, KKL-1, Arka Sukomal

Germplasm resources and Plant introduction

Large number of germplasm collections of French bean or common bean are maintained in different countries *viz*., Brazil, CIAT (Columbia), China, ICARDA (Syria), USA, Denmark, Ukraine, Turkey, Norway, Iran, etc.

In India, French bean germplasm is maintained at NBPGR, New Delhi and its regional stations at Phagli, Shimla, Shillong, Akola and Trichur, ICAR-IIVR, Varanasi, ICAR-IARI, New Delhi, ICAR-IIHR, Bengaluru and other places. In the early period of French bean breeding in India, several exotic varieties / germplasm were introduced as breeding material. Several exotic varieties were released directly as a variety for cultivation in India (Hazra et al (2022). Few varieties, like Contender, Premier, etc. are still being grown by the farmers.

Table 2: Exotic variety directly released for cultivation

Name of the variety	Country from where introduced
Contender	USA
Giant Stringless	USA
Bountiful	USA
Premier	USA
Jampa	Mexico
Kentacky Wonder	USA

Variability and selection

In India, main stay of developing varieties of French bean is selection from the exotic varieties / germplasm. Different studies suggested that selection index based on the characters *viz*., green pod yield per plant, number of branches per plant, number of nodes bearing pods, number of pods per plant, pod length, and pod weight recorded highest genetic advance and relative efficiency. For yield of dry seeds, important yield components are number of pods per plant, pod length and 100-seed weight.

French bean is highly self-pollinating species due to hermaphrodite sex form and homogamy however, anthers dehisce much earlier than anthesis. Stamens and pistil in opened flower remain enveloped together inside the tube like structure of jointed petals called "keel" which favours self-pollination. Both pure line selection and single plant selection from the heterogeneous and heterozygous genetic resources are employed in French bean breeding. Although, single plant selection is mostly followed due to convenience, pure line selection is more effective in selecting desirable individuals from the exotic genetic resources. Maximum number of varieties have been developed through selection.

Table 3: Varieties developed through selection

Name of variety	Indigenous / Exotic material
VL-Bauni-1	Exotic
Arka Komal	Exotic (Australia)
Arka Anoop	Exotic
Arka Sharath	Exotic
Arka Arjun	Exotic
Pant Anupama	Exotic
YCD-1	Indigenous (local type from Shevory hills)
Phule Surekha	Exotic (Jampa Improved from Mexico)
Kashi Sampann	Exotic
Kashi Rajhans	Exotic
Kashi Param	Exotic
Azad Rajmah-1	Exotic
Phule Suyash	Exotic
Azad Rajmah-1	Exotic
Swarna Priya	Exotic (pure line selection)
Swarna Lata	Exotic (pure line selection)

Name of variety	Indigenous / Exotic material
Pusa Himlata	Exotic
KKL-1	Indigenous (local type from Kodaikanal)
Arka Sukomal	Exotic

Hybridization and selection

Hybridization and selection method of breeding has also been followed employing mostly exotic genetic resources to increase the genetic variability and to combine the desired characters from two or more genetic materials into the newly developed variety. Being highly self-pollinated crops, pedigree, bulk, bulk-pedigree or back cross-pedigree method of breeding up to at least F_6 and later segregating generation is followed to isolate the desirable segregates. Backcrossing followed by pedigree method of selection is efficient in resistance breeding when resistance is governed by more than one gene. Hybridization and selection method of breeding has been followed to develop only two varieties *viz*., Arka Suvidha and TKD-1

Table 4: Varieties bred through hybridization followed by selection

Name of variety	Parentage
Arka Suvidha	Blue Crop x Contender
TKD-1	Selection-1 × PVl18

Utilization of mutants and applied mutagenesis

Seed-coat colour is an important horticultural character to determine the marketability of a dry bean variety and there is regional preference for certain colors of beans. It is known that many black bean varieties are superior in disease resistance and yielding capacity but this colour is not preferred in many bean growing areas. Moh (1971) long before advocated applied mutagenesis of black bean varieties with EMS or gamma rays to induce a range of seed-coat colours from white, yellow to grey brown. Coupled with the screening technique, it is practicable to isolate the seed-coat colour mutants which is correlated with green hypocotyl colour at a very early stage of seedling development. Genetic tests showed that the induced mutants are due to single genetic locus changes from dominant to recessive. Thus, by using mutation breeding technique, homozygous lines of seed-coat colour mutants can be achieved in the second selfed generation. In India, only one variety "Pusa Parbati" has so far been developed through X-ray irradiation of the American variety "Wax Pod".

Table 5: Varieties/ mutants developed through applied mutagenesis

Variety/ Mutant	Applied mutagenesis / Remark
Pusa Parbati	X-ray irradiation of the American variety "Wax Pod"

Resistance breeding

On screening the germplasm including improved varieties, number of genotypes/ varieties could be identified which showed resistance against different disease-causing fungi, bacteria and virus. These genetic materials can be used as resistant variety as such or used as donor parent in further breeding programme.

Table 6: Varieties showing resistance to different diseases

Resistance to disease	Variety showing resistance
Powdery mildew (*Erisiphe polygoni*)	Contender, Pusa Parbati, Jampa, Swarna Priya and Swarna Lata
Rust (*Uromyces appendiculatus*)	Pant Anupama, VL Bauni-1, Arka Bold, Pant Bean 2, Kentucky Wonder, Arka Sukomal, Arka Anoop, YCD-1
Angular leaf spot (*Pseudocercospora griseola*)	Lakshmi, Pant Anupama
Bean common mosaic virus	Pant Anupama, Pant Bean 2, Pusa Parbati
French bean golden mosaic virus	Arka Arjoon, Kashi Rajhans and Kashi Sampann
Bacterial blight (*Xanthomonas campestris pv. phaseoli*)	Arka Anoop
Root rot (*Rhizoctonia solani*),	YCD-1

Source: Swarup (2006), Hazra *et al.* (2022)

Table 7: Genes for resistance to disease, insect pest and nematode in French bean

Disease/ insect pest	Inheritance pattern
Powdery mildew	Resistance controlled by single dominant gene
Bean rust	Resistance controlled by single dominant or incompletely dominant gene
Anthracnose(*Colletotrichum lindemuthianum*)	Resistance controlled by single dominant gene "*A*", "*Are*", "*F*" for different strains
Angular leaf spot	One recessive gene from *Phaseolus coccinues*
Bacterial halo blight	Single recessive gene *pt pt* for race 1 and 2; Single dominant gene *Pt Pt* for race 1
Fusarium yellow	Incomplete dominant gene
Bean common mosaic virus	Single dominant gene "*I*"
Curly top virus	Two peristaltic dominant genes
Root rot (*Fusarium solani*)	Several quantitative genes
Alfalfa mosaic virus	Resistance controlled by single dominant gene, *Amv-1*
Bean common mosaic virus; strain unspecific	Resistance controlled by single recessive gene, *bc-u*
Bean pod mottle virus	Resistance controlled by single dominant gene, *Bpm*
Peanut mottle virus	Resistance controlled by single incompletely dominant gene, *Pmv*
Tobacco mosaic virus	Resistance controlled by single dominant gene, *Tm*
Root knot nematode, *Meloidogyne incognita*	Resistance controlled by single dominant gene, *Mel*

Source: Silbernagel (1986), Ram (1998), Swarup (2006), Singh and Dutta (2015)

Resistance to most of the diseases are controlled by single gene. Back-cross breeding is commonly adopted method of breeding for transferring disease resistance to a cultivar susceptible to that particular disease but having other desired characters. Backcrossing followed by pedigree method of selection is efficient in resistance breeding when resistance is governed by more than one gene. Additive genetic variance was evident for percent disease index of anthracnose caused by *Colletotrichum lindemuthianum*. Hence rapid genetic gain can be achieved in the early generation for tolerance to this disease through simple breeding methods (Chatterjee *et al*., 2025)

Breeding for quality

French beans are unique among plant foods due to their high protein content. For this reason, they are considered an important protein source for the vegetarians. It also contains a rich variety of phytochemicals with antioxidant activity, such as phenolic acids, flavonoids, flavonols, isoflavones, anthocyanins and pro-anthocyanins. Enhancement of different quality traits particularly, protein content and quality along with different phytochemical contents need to be taken into consideration in French bean breeding programme for increasing yield, quality and disease resistance.

References

Bassett, M.J. 1993. List of genes- *Phaseolus vulgaris* L. *Beans Improv. Coop*., **36**: 6-23.

Chatterjee, S., Mukherjee, D., Choudhuri, P., Maurya, P.K. Maji, A., Mandal, A.K. and Chattopadhyay, A. 2025. Breeding strategies for simultaneous improvement in anthracnose disease resistance and economically important traits in French bean. *Legume Res*., **48**:183-189.

Hazra, P., Banerjee, M.K. and Chattopadhyay, A. 2022. *Varieties of Vegetable crops in India*, Kalyani Publishers, New Delhi, Ludhiana, 204p.

Moh, C.C. 1971. Mutation breeding in seed coat colors of beans (*Phaseolus vulgaris*). Euphytica, **20**: 119-125.

Ram, H.H. 1998. *Vegetable Breeding: Principles and Practices*, Kalyani Publishers, New Delhi, Ludhiana.

Silbernagel, M.J. 1986. Snap bean breeding. In: *Breeding Vegetable crops* (Ed. Bassett, M.J.), AVI Publishing Co., Westport, Connecticut, USA, pp. 243-282.

Singh, D.K. and Dutta, A.K. 2015. French bean. In: *Hand book of Vegetables* Vol III (Eds. Peter, K. V. and Hazra, P.), Studium Press LLC, Houston, Texas, USA, pp. 150-191.

Swarup, V. 2006. *Vegetable Science and Technology in India*. Kalyani Publishers, New Delhi, Ludhiana, 656p.

Yarnell, S.H. 1965. Cytogenetics of the vegetable crops, IV: Legumes. *Bot. Rev*., **31**: 247-330.

30

Vegetable cowpea

Cowpea, *Vigna unguiculata* under the family Fabaceae (Leguminosae) and having chromosome no. 2n = 2x =22 is in cultivation from ancient times in the tropics of old world. Cowpea is grown for green edible pods, immature seeds, dry grains and edible leaves as leafy greens. The immature pods and seeds as well as dry seeds besides being widely used as fresh vegetable and pulse are also frozen and canned. Immature pods are good source of protein, fibre, minerals, calcium and vitamins particularly vitamin A, B and C. It covers large areas in South Asia including India, Bangladesh, Pakistan, Myanmar and Sri Lanka besides Southeast Asian countries, like Indonesia, Malaysia, Thailand and Philippines. It is also grown extensively in Brazil and other countries of Latin America. In Africa, it is commonly eaten as a pulse along with sorghum and millets while in India it is used both as pulse and vegetable almost throughout the country. Cowpea is also grown for hay or forage and green manure crop.

Origin, domestication and adaptation in India

Cowpea is one of the most ancient human food sources and has probably been used as a crop plant since Neolithic times. All evidence points to its originating in Africa because of the presence of conspecific wild forms, although where the crop was first domesticated is uncertain. Among the African countries, Ethiopia, Central and South Africa and West Africa are considered to be the probable centre of origin. Although, there is no archaeological evidence for early cowpea cultivation in the centre of diversity, there was early consensus regarding West Africa being the centre of origin and place of early domestication. However, later research using molecular markers has suggested that domestication may have instead occurred in East Africa and currently both the theories carry equal weight.

Remains of charred cowpeas from rock shelters in Central Ghana have been dated to the second millennium BC. In 2300 BC, cowpea is believed to have made its way into South East Asia where secondary domestication events may have occurred. In India, cowpea is known since Vedic times. However, it is generally agreed that the cowpea is of African origin as conspecific wild forms are found in Africa but are absent in Asia. However, India is considered the modern centre of diversity of cowpea cultivars. From South East Asia, cowpeas travelled north to the Mediterranean, where they were used by the Greeks and Romans. The first written references to the cowpea were in 300 BC and they probably reached Central and North America during the slave trade through the 17th to early 19th centuries. The first written reference of the word 'cowpea' appeared in 1798 in the United States. The name was most likely acquired due to their use as a fodder crop for cows.

The earlier view was that the progenitor of cowpea might be the wild *Vigna unguiculata* sub-species *dekindtiana* but later it was suggested that *Vigna unguiculata* sub-species *mensensis* was the most likely progenitor of cowpea. Both these sub-species are prostrate in growth habit and little differences are there between these two subspecies except for the habitat that sub-species *mensensis* occurs in forests and sub-species *dekindtiana* in the savannah (Steele and Mehra, 1980).

In the African context, the role of cowpea is predominantly that of pulse and African use of pulse continues, but after the introduction of *Vigna unguiculata* sub-species *unguiculata* (cultigroup Unguiculata) to southeast Asia and India, two other subspecies (cultigroup) were evolved: *Vigna unguiculata* sub-species *sesquipedalis* (Cultigroup Sesquipedalis , long-podded vegetable type) mainly in southeast Asia and India and *Vigna unguiculata* sub-species *cylindrica* (Cultigroup Biflora, short-podded fodder type) in India and southeast Asia under the predominant influence of human selection. The other sub-species *Vigna unguiculata* sub-species *textilis* (Cultigroup Textilis, fibre type) was probably selected from *unguiculata* forms in West Africa for fibre from its long peduncle.

Crop botany and floral biology

- Cowpea is an annual herbaceous legume and of the four subspecies of cowpea, three are cultivated.
- Black-eyed pea, a common name of the *unguiculata* /cultivar group Unguiculata, describes the presence of a distinctive black spot at the hilum of the seed; *sesquipedalis* in Latin means "foot and a half long", and this subspecies/cultivar group Sesquipedalis is characterized by unusually long pods, leading to the common names of yard long bean, asparagus bean and Chinese long-bean; sub-species *cylindrica* /cultivar group Biflora is characterized by very small podded fodder type.
- There is a high level of morphological diversity found within the species with large variations in the size, shape and structure of the plant.
- The plants are either be short and bushy (as short as 20 cm), prostrate or climbing annuals; climbing vine may grow up to a height of 2 metres.
- It has a well-developed root system and the taproot can penetrate to a depth of 2.4 m after eight weeks.
- All cowpeas are virtually glabrous in nature; size and shape of the leaves varies greatly, making this an important feature for classifying and distinguishing the varieties.
- Inflorescence is unbranched, axillary racemes bearing several flowers at the terminal end of peduncles, which vary greatly from 5-60 cm in length; one peduncle can support 1-4 or more pods.

- Flowers are bisexual, zygomorphic, solitary in the raceme, large and showy, and typically papilonaceous,
- Sepals are 5 in number, connate in an irregular calyx tube; corolla is papilionaceous with an erect and exerted standard petal spreading at anthesis; petals are five, one standard, two wings and two keels.
- Pigmentation pattern of corolla varies from white to solid mauve with yellow spot near the base of the standard petal; the wings are adherent to the boat shaped keel, enclosing the androecium and gynoecium.
- Stamens are diadelphous, 9 united +1 free; anthers uniform and bright yellow.
- Ovary is monocarpellary, unilocular, placentation marginal with many ovules are borne in two alternating rows along the ventral suture; style incurved, persistent and hairy on the inner side, stigma terminal, glabrous and capitate.
- Fruits or pods are either pendent or vertically attached to the raceme axis; fruits are mostly linear, although curved and coiled shaped pods are also found; fruits (pods) are long, narrow and more or less circular in cross section; length of the pods may vary from less than 11 to more than 100 cm depending on the sub-species (cultigroups). Pods of the vegetable cowpea belonging to true cultivar group Sesquipedalis and integrades or intermediate types of Unguiculata and Sesquipedalis are pendent, 30-90 cm long, fleshy and inflated, tending to shrink when dry.
- Seeds are small, exalbuminus with large embryo and fleshy cotyledon, varying in size from square, globular or long (8-12 mm) and kidney shaped; texture and colour of the seeds is very diverse; seeds can have a smooth or rough coat and be speckled, mottled or blotchy; different colours of seeds are black, grey, blue, white, brown, reddish, crimson, yellow, creamy, holstein, etc.

Reproductive behaviour

It is highly self-pollinated crop due to hermaphrodite sex form and homogamy i.e. maturation of stigma and anthers at the same time. However, anthers dehisce much earlier than anthesis. Stamens and pistil in opened flower remain enveloped together inside the tube like structure of jointed petals called "keel" which favours self-pollination. Infestation of some beetles, which bore through the keel tube, leads to some cross-pollination in hot and humid environment. Flowers open only once between 7.00 and 9.00 a.m. On cloudy days, the flowers may open even in the afternoon. Though the flowers open late in the morning, anthers dehisce much earlier even at night. Anther dehiscence is influenced by temperature and presence of moonlight. Dry and warm atmosphere and dark night tend to delay anther dehiscence. Pollens remain viable for 42 hours at room temperature of 28-29°C and 91% relative humidity. Since the dehiscence of anthers is much in advance of the blooming, emasculation needs to be carried out in mature flower buds in the preceding afternoon. Flowers are highly sensitive and drop off easily with slight mechanical disturbance or injury. To prevent drying out emasculated bud, leaflet may be folded and pinned around the bud. A tissue paper can also be used to cover and protect the bud. Emasculation of matured buds in the afternoon hours followed by hand pollination in the next day morning between 7.00 to 9.00 a.m is the best method of hybridization. Pollens are collected in the morning from a freshly opened flower. Standard and wings of the flower from the intended male parent are removed. By slight depression of the keel, stigma covered with the pollen grains protrudes out which can be used as a brush for pollination.

Genetics of important characters

Cowpea is a diploid species (2n = 2x= 22) with the basic chromosome number, x = 11. The pachytene chromosome study showed that the 11 bivalent complement had one short (18 µm), seven medium (26–36 µm) and three long (41–45 µm) chromosomes. The autotetraploids had meiotic abnormalities, like quadravalents, trivalents and univalents and laggards and unequal distributions in anaphase, resulting in sterility.

The genus *Vigna* is a pantropical genus with 22 species in India and Southeast Asia, 120 species in Africa and a few in America and Australia. In earlier classification, four cultivated subspecies under *Vigna unguiculata* was recognized *viz*., sub-species *unguiculata*, *cylindrica* , *sesquipedalis* and *textilis*. Later, the taxonomy of cowpea has been refined with the down-gradation of the sub-species level to cultigroups *viz*., Cv.gr Unguiculata (sub-species *unguiculata* , pulse or vegetable type), Cv-gr Biflora (sub-species *cylindrica*, fodder type), Cv-gr Sesquipedalis (sub-species *sesquipedalis,* vegetable type) and Cv-gr. Textilis (sub-species *textilis*, fibre type) and all these four cultigroups were lumped under one sub-species, *Vigna unguiculata* ssp. *unguiculata* (Baudoin and Marechal, 1985). However, the sub-species level classification of cowpea is still mentioned in different literatures.

Table 1: Basic characters of three cultivated sub-species / cultigroups of cowpea

Sub-species/ Cultigroup	Growth habit	Pod character	Seed character
V.unguiculata ssp. *unguiculata* / Cv.gr. Unguiculata	Prostrate, erect, semi-erect, climbing habit	Small to medium podded pulse or vegetable type; linear, crescent, round, coiled, pendant or semi-pendant pod orientation; more or less circular in cross section; 20–30 cm in length.	Small, 5-7 mm, cylindrical, square and slightly elongated; presence of a distinctive black spot at the hilum of the seed; colours of seeds are black, grey, blue, white, brown, reddish, crimson, yellow, creamy, holstein, etc.
V.unguiculata ssp. *cylindrica* / Cv.gr. Biflora	Mainly prostrate, sometimes climbing habit	Very small podded fodder type, non-fleshy pod and more or less circular in cross section; erect pod orientation; 7–13 cm in length.	Small, 5–6 mm square or globular shaped; different seed colours

Sub-species/ Cultigroup	Growth habit	Pod character	Seed character
V.unguiculata ssp. *sesquipedalis* / Cv.gr. Sesquipedalis	True Sesquipedalis types are climbing; integrades are less viny	True Sesquipedalis types have pendent, 80-90 cm unusually long, fleshy and inflated pod, tending to shrink when dry; integrades of Unguiculata and Sesquipedalis have pendent, 25-30 cm long, fleshy and inflated pods	Big, 8–12 mm, flat, elongated and kidney shaped; colours of seeds are generally black or brown

Source: Pandit and Hazra (2012)

In cowpea, several simple inherited qualitative characters include axillary buds, branch, peduncle, growth habit, leaf shape, petiole character, inflorescence, flower colour, flowering, pod shape, colour, pod attachment, pod surface character, seed colour, etc.

Table 2: Some important simply inherited characters of cowpea

Character/ Gene symbol	Phenotype / inheritance pattern
A	Alfalfa – like pod shape; dominant character
B	Blue seed coat; dominant character
Bl	Black seed coat, also for anthocyanin production in pod tip, calyx and peduncle; heterozygote produces mottled seeds
Bk	Black pod; dominant to white
Bp	Brown pod; dominant to straw colour
ax	Axillary buds; recessive character
Bcy	Brown calyx colour; dominant to green
Ci	Compound inflorescence; dominant character
Cy	Cylindrical pod; dominant character
crpt	Crumpled petal; recessive character
df	Dwarf (slow growth, dark green leaves, short internodes); recessive
D	Dark flower colour; dominant to light colour
Er	Erect pod attachment; dominant to drooping or pendant pod attachment
Ef-1	Early flowering-1; dominant character
Ef-2	Early flowering-2; dominant character
Gr	Green bud; dominant to white bud
Gp	Green pod; dominant to cream pod
La	Lanceolate leaf; dominant character
Llf	Long leaf; dominant to small leaf
ls	Small leaf size; recessive to large leaf

Character/ Gene symbol	Phenotype / inheritance pattern
lg	Light green pod; recessive to dark green pod
ms-1, ms-2, ms-3, ms-4	Male sterility-1, 2, 3, 4; recessive character
Nlf	Narrow leaf; dominant to broad leaf
O	Hilum ring seed coat pattern; dominant character
Pbr	Purple branch base; dominant to green
Pg	Pale green plant; dominant character
Pn	Peduncle length; long dominant to short
Pf	Purple flower; dominant to white
Pb	Purple petiole base; dominant to green
P	Purple pod; dominant to green, also causes anthocyanin production in calyx and peduncle
pa-1	Pod appearance-1; wrinkled dry pod recessive to smooth
pa-2	Pod appearance-2; wrinkled dry pod recessive to smooth
Pu	Purple pod, stem and petiole; dominant character
R	Red seed coat; dominant character
Sh	Spindly growth habit, main stem elongated with few side branches
stx	Sesquipedalis-like soft texture of pod; recessive character
S	Spotting pattern, black patches on seed coat; dominant character
un	Unifoliate leaf, two lateral leaflets and their petioles missing; recessive
V	Seed coat mottling; dominant character
Growth habit	Viny growth habit dominant over bushy habit
Flower colour	Violet flower colour dominant over white Violet colour is controlled by two dominant genes, D and L, and it shows dominance over light violet colour
Pod length	Short pod partially dominant over long pod
Pod weight	Light pod partially dominant over heavy pod
Seed shape	Elongated and reniform shape dominant over shorter and cylindrical shape

Source: Fery (1985), Ram (1998), Swarup (2006), Pandit and Hazra (2012)

Several studies suggested that viny growth habit was dominant over bushy growth habit. One study suggested two duplicate genes, V_1 and V_2 controlled growth habit while other study indicated the character to be monogenic. Anthocyanin pigmentation was dominant to non-pigmentation and purple pigmentation was controlled by two duplicate dominant genes. Pod pigmentation was controlled by two complementary genes with pleiotropy. Violet and light violet flower colours were due to the presence of two dominant genes, *D* and *L*, and violet colour was dominant over light violet colour. One study indicated calyx colour being monogenic dominant while other study suggested four genes for the control of this character. Different studies suggested the inheritance of seed colour to be monogenic, digenic and trigenic. Such variation in the genetic control of a character in different studies generally happens due to utilization of different genetic materials for the study.

The characters conditioned by polygenes include flowering and maturity, pod number per plant, pod length, pod weight, seed weight and shape, pod and seed protein content and pod yield per plant. Inheritance of pod length and weight was studied employing different *V. unguiculata* cv-gr. Unguiculata x *V. unguiculata* cv-gr. Sesquipedalis crosses where small and light weight pod of *V.unguiculata* cv-gr. Unguiculata showed partial dominance over long and heavy pod of *V. unguiculata* cv-gr. Sesquipedalis (Hazra *et al.*, 1993; Hazra *et al.*, 1994). Genetic studies of pod yield and seed protein content suggested the encouraging possibility of enhancing pod yield without concomitant decrease in protein content (Hazra *et al.,* 1996).

Breeding objectives

The followings are the main objectives of vegetable cowpea breeding

- Climbing type for the yard-long bean type under *Vigna unguiculata* Cv-gr. Sesquipedalis
- Plant should be dwarf and bushy or semi-viny and photoinsensitive which can be grown in both summer-rainy and autumn-winter season
- For the dwarf and bushy or semi-viny and photoinsensitive type, early maturity of 70 to 80 days is preferred with the completion of harvesting of green pods by 3-4 pickings.
- For the yard-long bean type under *Vigna unguiculata* Cv-gr. Sesquipedalis early flowering and harvesting of pods are also preferred character however pods are harvested by several picking due to indeterminate growth habit.
- Pods of the yard-long bean type under *Vigna unguiculata* Cv-gr. Sesquipedalis should be long to very long (60 to 80 cm), green, light green, stringless, round, tender and of good cooking quality
- Pods of the dwarf and bushy or semi-viny and photoinsensitive type should be medium long (25 to 30 cm), green, light green, stringless, round, tender, meaty with crisp texture and of good cooking quality
- High green pod yield with high number of pods per plant
- The dwarf and bushy or semi-viny and photoinsensitive type should be suitable for both summer-rainy and autumn-winter season
- Resistance to different diseases *viz.*, leaf spot (*Pseudocercospora cruenta*), powdery mildew (*Erysiphe pisi*), rust (*Uromyces phaseoli var. vignae*), Cercospora leaf spot (*Cercospora* sp.), bacterial blight (*Xanthomonas axonopodis* pv. *vignicola),* cowpea mosaic and golden mosaic virus disease, etc.
- Resistance to different insect pests *viz.*, pod borer (*Helicoverpa armigera*), aphids (*Aphis* sp.), bean weevil (*Bruchus rufimanus*), etc.
- Resistance to abiotic stresses *viz.*, high temperature, drought and water logging

Breeding methods and development of variety

In India, most of the cowpeas grown for tender pod as vegetable are intermediate types between the cv-gr. Unguiculata and cv-gr. Sesquipedalis along with few true cv-gr. Sesquipedalis (long podded vegetable cowpea) types. After the introduction of Unguiculata forms from Africa to southeast Asia and India, Cultigroup Sesquipedalis were evolved in southeast Asia and India under the predominant influence of human selection. For this reason, wide array of such intermediate types are available in India. India is also considered as modern centre of diversity of cowpea cultivars.

In India, initially a few improved varieties were developed through selection from the exotic varieties / germplasm. Vegetable cowpea breeding in India mainly aims at developing bushy plant stature, photoinsensitive nature, determinate habit, tender pod and disease resistance. Several introduced lines which are determinate and photo-insensitive in nature have been utilized in breeding. Combination of characters of two cultigroups *viz*., Unguiculata (mainly grain type) or cv-gr. Biflora (mainly fodder type) with the cv-gr. Sesquipedalis (long podded vegetable type) has also been done to develop suitable non-viny or semi-viny and photoinsensitive type vegetable cowpea varieties.

Varieties of vegetable cowpea can conveniently be classified according to the growth habit namely, determinate and bush type and indeterminate and semi-viny type.

- **Determinate and bush type:** Philippine Early, Pusa Phalguni, Pusa Dofasli, Pusa Komal, Cowpea 263, Pusa Rituraj, Arka Samrudhi, Bidhan Barbati-1 , Bidhan Barbati-2, Bidhan Sadabahar, Kashi Shyamal, Kashi Gauri, Arka Suman, Swarna Suphala, Swarna Sweta, Swarna Mukut, Narendra Lobia-2 , Kashi Kanchan, Kashi Nidhi, Kashi Unnati, Kashi Sudha, Pant Lobia-1 , Pant Lobia- 2, Pant lobia-3, Pant Lobia-4
- **Indeterminate, viny /semi-viny type:** Pusa Barsati, CO-2, Vamban-2, Arka Garima, Sel 2-1, Arka Mangala, Anaswara, Swarna Harita, Konkan Wali

Germplasm resources and Plant introduction

The International Institute of Tropical Agriculture (IITA), Ibadan, Nigeria maintains the world's largest collection of cowpea germplasm of over 15,000 accessions. Genetic resources of long podded vegetable type under cultigroup Sesquipedalis is maintained at AVRDC, Taiwan and other countries like, Malaysia, Indonesia, Philippines, etc.

In India, cowpea germplasm is maintained at NBPGR, New Delhi, ICAR-IARI, New Delhi, ICAR-IIHR, Bengaluru, ICAR-IIVR, Varanasi, ICAR- Indian Institute of Pulses Research, Kanpur and other places. In India, germplasm collections were made from Rajasthan, Gujarat, Maharashtra, Madhya Pradesh, Tamil Nadu, West Bengal and Uttar Pradesh.

In the early period of cowpea breeding in India, several exotic varieties / germplasm were introduced as breeding material. However, only one variety "Philippine Early" introduced from Philippines was released directly as a variety for cultivation.

Table 3: Exotic variety directly released for cultivation

Name of the variety	Country from where introduced
Philippine Early	Philippines

Variability and selection

In India, initially a few improved varieties were developed through selection from the exotic varieties / germplasm. For example, the variety "Pusa Phalguni" was selected from the variety "Delaque du Tonkin" introduced from Canada and the variety "Pusa Barsati" was selected from a genotype introduced from Phillippines. Later, large number of varieties of both determinate / bushy and viny / semi-viny type *viz.*, Arka Mangala, Kashi Shyamal, Swarna Sweta, Kashi Kanchan, Kashi Sudha etc. have been developed by single plant selection from the wide range of indigenous germplasm. Different studies suggested that selection index based on five characters *viz.*, green pod yield per plant, pod length, number of pods per cluster, number of pods per plant and ten pod weight recorded highest genetic advance and relative efficiency. Cowpea is highly self-pollinating species due to hermaphrodite sex form and homogamy however, anthers dehisce much earlier than anthesis. Stamens and pistil in opened flower remain enveloped together inside the tube like structure of jointed petals called "keel" which favours self-pollination. Both pure line selection and single plant selection are employed in cowpea breeding. Although, single plant selection is mostly followed due to convenience, pure line selection is more effective in selecting desirable individuals from the heterozygous and heterogeneous genetic resources. Maximum number of varieties have been developed in India through selection from predominantly indigenous genetic materials.

Table 4: Varieties developed through selection

Name of variety	Indigenous / Exotic material
Pusa Phalguni	Exotic (Delaque du Tonkin from Canada)
Pusa Barsati	Exotic (Philippines)
Vamban-2	Exotic (pure line selection from IT81-D-1228-10)
Cowpea 263	Indigenous (Bangalore local)
Sel 2-1	Indigenous
Arka Mangala	Indigenous
Kashi Shyamal	Indigenous
Kashi Gauri	Indigenous
Anaswara	Indigenous
Swarna Suphala	Exotic
Swarna Harita	Indigenous
Swarna Sweta	Indigenous
Kashi Kanchan	Indigenous
Kashi Nidhi	Indigenous
Kashi Unnati	Indigenous
Kashi Sudha	Indigenous
Konkan Wali	Indigenous

Name of variety	Indigenous / Exotic material
Pant Lobia-1	Indigenous
Pant Lobia- 2	Indigenous
Pant lobia-3	Indigenous
Pant Lobia-4	Indigenous

Hybridization and selection

Hybridization and selection method of breeding has also been followed employing both indigenous / exotic genetic resources to increase the genetic variability and to combine the desired characters from two or more genetic materials into the newly developed variety. Being highly self-pollinated crops, pedigree, bulk method, bulk-pedigree or back cross-pedigree method of breeding up to at least F_6 and later segregating generation is followed to isolate the desirable segregates. Backcrossing followed by pedigree method of selection is efficient in resistance breeding when resistance is governed by more than one gene. This method has been followed to develop the variety "Arka Garima" at IIHR, Bengaluru. Backcross-pedigree method of breeding is also efficient to isolate desirable segregates from the segregating population of inter sub-specific / inter-cultigroup cross. This method has been followed to develop two varieties *viz*., "Bidhan Barbati-1" and Bidhan Barbati-2 at Bidhan Chandra Krishi Viswavidyalaya, West Bengal (Hazra *et al*., 2001).

Breeding strategy has been framed to develop determinate / less-viny and photo insensitive types through combination of erect habit, short pod length, early maturity and photo-insensitivity of the genotypes belonging to the cultigroup Unguiculata or Biflora with the viny genotypes having long pod length, succulence and fleshiness of the cultigroup Sesquipedalis (Som and Hazra ,1993; Hazra *et al*., 2007). Although all the sub-species / cultigroup can be crossed with each other, conspicuous differences in crossability exists depending on the genetic diversity of parental lines (Hazra *et al*., 2000). Inheritance studies showed that viny character was dominant to dwarf and bushy type and small and light weight pod of cultigroup Unguiculata / Biflora showed partial dominance over long and heavy pod of cultigroup Sesquipedalis (Hazra *et al*., 1993; Hazra *et al*., 1994). Hence, advancement of generations from highly heterotic inter sub-species/ inter-cultigroup cross combinations (Unguiculata × Sesquipedalis or Biflora × Sesquipedalis) by pedigree method was not effective to isolate desirable segregates. Hence, modified backcross-pedigree method in that the F_1 was backcrossed once with the Sesquipedalis parent followed by pedigree method of selection up to F_7 generations has been found effective to develop relatively short statured plant types with photo insensitivity, synchrony in flowering and fruit set and medium-long succulent pods. Several varieties have been bred following this inter sub-species/ inter-cultigroup crossing and selection scheme *viz*., Bidhan Barbati-1, Bidhan Barbati-2, CO-2, Arka Garima, etc.

Inter-varietal hybridization and pedigree method of selection in the advanced generation has also been followed to develop several photo-insensitive varieties having bushy or semi-viny growth habit, medium long, meaty and fleshy pods *viz*.,

Bidhan Sadabahar (Bidhan Barbati-1 x Kashi Kanchan), Arka Suman (Pusa Komal x Arka Garima), Narendra Lobia-2 (Sel-2-1 x Red Seeded). Good number of varieties have been developed following this method of breeding (Chattopadhyay *et al.*, 2020; Hazra *et al.*, 2022).

Table 5: Varieties bred through hybridization and selection

Name of variety	Parentage
CO-2	C 521 (grain type) × C 419 (vegetable type)
Pusa Dofasli	Pusa Phalguni × Philippines Bush (EC 21662)
Pusa Komal	(Pusa Dofasli × EC 26410) x P-426, bacterial blight resistant line
Pusa Rituraj	Pusa Dofasli x EC 26410
Arka Garima	TUV-762 (*Vigna unguiculata* subsp. *unguiculata* × *Vigna unguiculata* subsp. *sesquipedalis*
Arka Samrudhi	Arka Garima × Pusa Komal
Bidhan Barbati-1	EC 243954 (subsp. *unguiculata*) × EC 305827 (subsp. *sesquipedalis*)
Bidhan Barbati-2	V-70 (subsp. *cylindrica*) × Sel TM-3 (subsp. *sesquidalis*)
Bidhan Sadabahar	Bidhan Barbati-1 x Kashi Kanchan
Arka Suman	Pusa Komal x Arka Garima
Narendra Lobia-2	Sel-2-1 x Red Seeded

Breeding for resistance to diseases

In India, screening of germplasm against some diseases like, bacterial blight, leaf spot, cowpea mosaic virus, golden mosaic and yellow mosaic virus has identified several promising genotypes which can be used in resistance breeding programme.

Table 6: Varieties showing resistance to different diseases

Resistance to disease	Variety
Cowpea golden mosaic virus	Bidhan Barbati-1, Kashi Kanchan, Kashi Nidhi, Kashi Unnati , Kashi Sudha
Cowpea mosaic virus	Bidhan Barbati-1, Swarna Sweta, Swarna Harita
Yellow mosaic virus	Pant Lobia 1, Pant Lobia 2, Pant Lobia 3, Pant Lobia 4
Rust	Swarna Sweta, Swarna Harita
Leaf spot	Kashi Kanchan, Kashi Nidhi, Kashi Unnati , Kashi Sudha
Bacterial blight	Pant Lobia 3

Source: Hazra *et al.* (2022)

Table 7: Genes for resistance to disease, insect pest and nematode in cowpeas

Gene symbol	Resistance to disease
Bc-1	Bacterial canker, *Xanthomonas vignicola* resistance-1
Bc-2	Bacterial canker, *Xanthomonas vignicola* resistance-2
bcm	Cowpea mosaic virus resistance, may be synonym to *blc*
blc	Cowpea mosaic virus resistance, may be synonym to *bcm*
by	Resistance to bean yellow mosaic virus
CC	Resistance to cowpea chlorotic mottle virus; the recessive allele at the *Mvi* locus which may be the synonym

Gene symbol	Resistance to disease
Cm	Resistance to cucumber mosaic virus
rh	Resistance to beetle, *Callosobruchus maculatus*
Rk	Resistance to root knot nematode; allelic to rk^i
rk^i	Intermediate resistance to root knot nematode; allelic to *Rk*
Sr	Resistance to stem rot, *Phytophthora vignae*
Sbm	Resistance to southern bean mosaic virus
Vw	Resistance to verticillium wilt, *Verticillium albo-atrum*
Ymr	Resistance to cowpea yellow mosaic virus

Source: Fery (1985), Ram (1998), Swarup (2006), Pandit and Hazra (2012)

Resistance to most of the diseases are controlled by single gene. Back-crossing is commonly adopted method of breeding for transferring disease resistance to a cultivar susceptible to that particular disease but having other desired attributes. Backcrossing followed by pedigree method of selection is efficient in resistance breeding when resistance is governed by more than one gene.

The widely adaptable, photo-insensitive and bacterial blight resistant variety "Pusa Komal" was developed through the series of planned hybridizations and subsequent selections at IARI, New Delhi. Three parents were involved in the breeding programme. In the advanced generation of the three parent crossing scheme (Pusa Dofasli × EC 26410) x bacterial blight resistant line P-426, a bushy, determinate and bacterial blight resistant segregate "P-85-2" having medium long (22-23 cm), light green, cylindrical and tender pod could be selected which was later released as the variety "Pusa Komal".

Breeding for quality

Cowpea is a good source of protein, iron and dietary fibre. Addition of even small amount of cowpea ensures nutritional balance of the diet and enhances protein quality by the synergistic effect of high lysine from the protein of cowpea and high methionine from the protein of cereals. Genetic studies of pod yield and seed protein content suggested the encouraging possibility of enhancing pod yield without concomitant decrease in protein content (Hazra *et al.,* 1996). Widely divergent genotypes should be screened for protein content and quality along with iron content of both pod and seed to develop a gene pool to be utilized in cowpea breeding programme for quality. However, cowpea contains some anti-nutritional factors like oxalate, phytate, tannin, lectin and stachyose (Ileke, 2014). Enhancement of nutrient particularly, protein and iron and at the same time, lowering of anti-nutrient compounds, particularly, oxalic acid and phytate contents should be taken into consideration during breeding vegetable cowpea for green pod yield and quality

References

Baudin, J.P. and Marechal, R. 1985. Genetic diversity in *Vigna.* In: *Cowpea Research, Production and Utilization* (Eds. Singh, S.R. and Rachie, K.O.), John Wiley and Sons Ltd., New York, pp.3-9.

Chattopadhyay, A., Dutta, S., Mandal, A.K., Maurya, P.K., Banerjee, S., Bhattacharjee, T. and Hazra, P. 2020. Bidhan Sadabahar: A bushy cowpea variety resistant to cowpea aphid borne mosaic virus. *Indian Hort.*, **65**: 3-4.

Fery, R.L. 1985. The genetics of cowpea – a review of the world literature. In: *Cowpea Research, Production and Utilization* (Eds. Singh, S.R. and Rachie, K.O.), John Wiley and Sons Ltd., New York.

Hazra, P., Banerjee, M.K. and Chattopadhyay, A. 2022. *Varieties of Vegetable crops in India*, Kalyani Publishers, New Delhi, Ludhiana, 204p.

Hazra, P., Chattopadhyay, A., Dasgupta, T. and Som, M. G. 2001. Bidhan Barbati-1 and Bidhan Barbati-2: new high yielding and non-viny vegetable cowpeas. *Indian Hort.,* **46**:13-15.

Hazra, P., Chattopadhyay, A., Som, M. G. and Das, P.K. 2000. Effects of season and cultivar group on the crossability of cowpea (*Vigna unguiculata).* SABRAO *J. Breed. Genet.*, **32**: 23-29.

Hazra, P., Chattopadhyaya, A., Dasgupta, T., Kar, N., Das, P. K. and Som, M. G. 2007. Breeding strategy for improving plant type, pod characters and protein content in pods and seeds in vegetable cowpea (*Vigna unguiculata* L. Walp.)., *Acta Hort.*, **752**: 275-280.

Hazra, P., Som, M. G. and Das, P. K.1994. Inheritance of pod weight in cowpea. *Indian J. Genet. Plant Breed.*, **54**: 175-178.

Hazra, P., Som, M. G.and Das, P. K.1993. Inheritance of pod length in cowpea. *Indian J. Hort.*, **50**: 358-363.

Hazra, P., Som, M. G.and Das, P. K.1996. Combining ability for pod yield and seed protein in cowpea. *Indian J. Genet. Plant Breed.*, **56**: 553-556.

Ileke, K. 2014. Anti-nutritional factors in cowpea cultivars and their effects on susceptibility to *Callosobruchus maculatus* (Fab.) [Coleoptera: Bruchidae] infestation. *Bioscience Methods*, **5**: 1-8, DOI:10.5376/bm.2014.05.0002

Pandit, M.K. and Hazra, P. 2012. Cowpea. In: *Hand book of Vegetables* (Eds. Peter, K. V. and Hazra, P.), Studium Press LLC, Houston, Texas, USA, pp. 189-229.

Ram, H.H. 1998. *Vegetable Breeding: Principles and Practices*, Kalyani Publishers, New Delhi, Ludhiana.

Som, M. G. and Hazra, P. 1993. Cowpea. In: *Genetic Improvement of Vegetable Crops* (Eds. Kalloo, G. and Bergh, B.O.). Pergamon Press, Oxford, New York, Seoul, Tokyo, pp.339-354.

Steele, W.M. and Mehra, K.L. 1980. Structure, evolution and adaptation to farming system and environments in *Vigna*. In: *Advances in Legume Science* (Eds. Summerfield, R.J. and Bunting, A.H.), Her Majesty's Stationary office, London, U.K., pp. 393-404.

Swarup, V. 2006. *Vegetable Science and Technology in India.* Kalyani Publishers, New Delhi, Ludhiana, 656p.

31

Dolichos bean or Hyacinth bean

Dolichos bean or Hyacinth bean, *Lablab purpureus* (L).Sweet. (Syn. *Dolichos lablab* Roxb.) under the family Fabaceae (Leguminosae) and having Chromosome No.: 2n = 2x = 22 is one of the most ancient among the cultivated plants. Young, immature pods are cooked and eaten like green beans. It is grown throughout the tropical regions of Asia, Africa and America. In India, it is grown extensively in the cool season in Tamil Nadu, Andhra Pradesh, Karnataka, Madhya Pradesh, West Bengal and Maharashtra. The pods are one of the rich sources of protein and it also provides fair amount of carbohydrates, calcium, phosphorus sodium, magnesium and iron, β-carotene, riboflavin, thiamine and nicotinic acid.

Origin, domestication and adaptation in India

Dolichos bean or Hyacinth bean has been originated in India as wild forms of this bean are found in this country. From India, it was introduced to China, Western Asia, Egypt and other tropical countries of South and South-east Asia and Africa. Vegetable and pulse type dolichos bean belong to two distinct botanical varieties under *Lablab purpureus viz*., *Lablab purpureus* var. *typicus* and *Lablab purpureus* var. *lignosus*, respectively. Various forms and cultivars of both the cultivated types are available in India.

Crop botany and floral biology

- It is a short-lived perennial herb, frequently grown as an annual; thick stem is usually twining and may reach to 1.5 to 18.0 m in length; however, bushy, semi-erect, and prostrate forms are also available.
- It shows huge variation in form and habit; two types of distinct hyacinth bean forms are recognized under two botanical varieties *viz*., *Lablab purpureus* var. *typicus* and *Lablab purpureus* var. *lignosus*
- The taproot is well developed with many laterals and well-developed adventitious roots.
- The trifoliate leaves are tinged purple, each up to 15 cm long and may be hairy on the undersides.
- Inflorescence is axillary, erect, long racemes; flowers are borne on a straight upright stalk, often a foot high on which they open in succession.

- Flowers are bisexual, zygomorphic, complete, cyclic, pentamerous, dichlamydeous, heterochlamydeous, bracteate, bracteolate and pedicellate; flowers are borne solitary in the raceme on short pedicels, generally white, violet or purple and typically papilonaceous.
- Calyx is campanulate; sepals are 5 in number, imbricate and gamosepalous; corolla is papilionaceous, white, violet or purple in colour with five petals; polypatalous, descendingly imbricate; of the five petals, two keels forming a tube and enclosing the androecium and gynoecium, two wings and one big standard petal.
- Androecium consists of 10 stamens, diadelphous, 9 united +1 free; filaments are alternately short and long; anthers are dithecous, introse and uniform.
- Ovary is sub-sessile, superior, mono-carpellary, unilocular with few ovules on marginal placentation; ovules are borne in two alternating rows along the ventral suture; style is incurved, persistent, terminal and hairy, stigma is terminal, glabrous and capitate.
- Fruits are botanically pods; pods are variable in size (5-15 x 1-5cm) and generally flat and glossy; 5-6 seeded; often curved with a strongly curved beak; pod shape may be extra-long and flat, long-flat, short-flat, long and slightly curved, tubular, slightly curved and bloated and narrow with different colour (green, light green, purple, white, green with purple sutures, etc.); fruits are usually tipped with the persistent style.
- Seeds are non-endospermous, large, thick, ovoided, slightly flattened with large embryo and fleshy cotyledon; seeds vary in size and colour and have a prominent white hilum along one third of the edge; different colours of seeds are black, grey, white, brown, reddish, crimson, yellow, creamy, etc.; average 100 seed weight is generally 3.20 gram.
- Mature, dried seeds are reportedly toxic due to high levels of cyanogenic glucosides and should be boiled with at least two changes of water before eating to remove the toxins

Reproductive behaviour

It is highly self-pollinated crop due to hermaphrodite sex form and homogamy *i.e.* maturation of stigma and anthers at the same time. Stamens and pistil in opened flower remain enveloped together inside the tube like structure of jointed petals called "keel" which favours self-pollination. The extrovert stigma through the keel tube in few varieties leads to cross-pollination by insects up to 6-10%. Anthesis starts in the morning by 7.00 a.m. and continues up to 5.00 p.m. Anther dehiscence generally occurs before anthesis and continues from 5.00 a.m. to 2.00 p.m. Stigma becomes receptive and pollens are fertile on the day of anthesis. Emasculation of matured buds in the afternoon hours followed by hand pollination in the next day morning between 8.00 to 10.00 a.m is the best method of hybridization. It needs good care for emasculation because the flowers, though large are prone to injury.

Genetics of important characters

Dolichos bean is a diploid species (2n = 2x= 22) with the basic chromosome number, x = 11. Vegetable and pulse type dolichus bean belong to two distinct botanical varieties under *Lablab purpureus viz., Lablab purpureus* var. *typicus* (vegetable type) and *Lablab purpureus* var. *lignosus* (pulse type).

Table 1: Basic characters of two botanical varieties of dolichos bean

Botanical variety	Growth habit	Pod character	Seed character	Photo-sensitivity
Lablab purpureus var. *typicus* (Vegetable type) It is predominantly grown for soft and fleshy whole pods for use as a vegetable.	Perennial twining herb, cultivated mostly as an annual; due to its twinning habit it is trained on a trellis.	Pods are soft, flat, glossy, longer and more tapering with long axis of seeds parallel to the suture of the pod; thick and pendant in orientation; pod shape may be extra-long and flat, long-flat, short-flat, long and slightly curved, tubular and bloated and narrow with different colour (green, light green, purple, white, green with purple sutures, etc.)	Seeds are bold, white, yellow, brownish, purple or black	Photo-sensitive, cool season and short day legume crop
Lablab purpureus var. *lignosus* (Pulse type) It is grown predominantly for fresh beans for use as a vegetable and to a limited extent for split pulse for use in various food preparations.	Semi- erect or bushy, perennial herb, cultivated as an annual; it shows little or no tendency to climb; leaflets are innately trifoliate and smaller than those of var. *typicus*	Pods are tough, firm-walled and parchmented which are relatively shorter and more abruptly truncated and long axis of the seeds is perpendicular to the suture of the pod; fibrous and erect in orientation; pod colour is mostly white or cream; pods exude oily substances that emit characteristic fragrance	Seeds are comparatively smaller in size; white or brownish, in colour	Photo-insensitive legume crop which can be grown much earlier in the season

Source: Ramesh and Byregowda (2016), Hazra *et al.* (2022)

Most of the plant characters like, growth habit, inflorescence type, colour of leaf margin, leaf vein, stem, petiole, nodes and internodes, pod characters and colour are conditioned by single or two genes with complementary action or by three genes.

Table 2: Some important simply inherited characters of dolichos bean

Character	Phenotype / inheritance pattern/ Gene symbol
Growth habit; Spreading / compact	Monogenic dominant, *Sg*; Spreading > compact
Growth habit; Erect /Prostrate	Three genes, one basic and two complementary; Erect > Prostrate
Growth habit; Determinate / Indeterminate	Two / Three genes; Indeterminate > determinate
Growth habit and inflorescence type	Three genes
Photoperiod sensitivity	Monogenic dominant, *Se*
Leaf margin colour; Purple / green	Monogenic dominant ; Purple > green
Leaf vein colour; Purple / green	Digenic; complementary gene action; Purple > green
Petiole colour; Purple / green	Two complementary genes, P_{p1} and P_{p2}; Purple > green
Stem colour; Purple / green	Three complementary genes, P_{s1}, P_{s2} and P_{s3}; Purple > green
Leaf colour; Dark green / green	Three complementary genes; Dark green > green
Leaf texture; Rough / smooth	Two complementary genes; Rough > smooth
Photoperiod sensitivity to flowering; Sensitive / insensitive	Monogenic dominant; Sensitivity > insensitivity
Flower colour; Purple / white	Monogenic dominant; Purple > white
Orientation of dry pods to branches ; Erect / drooping	Monogenic dominant; Erect > dropping
Pod width; Broad / narrow	Monogenic dominant; Broad > narrow
Nature of pod surface; Smooth / shrivelled	Two genes, Duplicate dominant; Smooth > shrivelled
Pod colour; Green / light green	Monogenic dominant; Green > light green
Pod colour	Three duplicate genes
Pod curvature; Straight / curved	Two genes, complementary, epistasis; Straight > curved
Pod fragrance; High / low	Two genes, Inhibitory epistasis; High > low
Pod form	Monogenic dominant and two complementary genes
Purple colour of nodes, flowers and pod sutures and black seed coat	Monogenic dominant (closely associated character)
Purple internodes	Monogenic dominant; *Pint*
Purple pod	Monogenic dominant; *Ppd*
Non-tapering pod tips	Monogenic dominant; *Ntpt*
Seed coat colour Khaki / Chocolate / black / brown	Two genes, Supplementary epistasis; Khaki > chocolate > black > buff
Seed coat colour; Red / white	Monogenic dominant ; Red > 1white
Seed coat colour; Chocolate / brown	Monogenic dominant; Chocolate > brown
Seed shape; Round / flat	Monogenic dominant; Flat > Round

Source: Kalloo and Pandey (1993), Swarup (2006), Ramesh and Byregowda (2016)

The polygenic characters include plant height, vine length, plant spread, days to flowering, pod length and width, pod number per plant, green pod yield, seed length, width, number of seeds per pod, etc.

Breeding objectives

The followings are the main objectives of dolichos bean breeding

- Short-viny and early flowering type with early harvesting of pods under *Lablab purpureus* var. *typicus*
- Pods of viny, pole type under *Lablab purpureus* var. *typicus* should be soft, thick, long, dark green or green, shiny, slightly curved, tender and of good cooking quality
- Bushy and compact or semi-viny growth habit, early maturity, photo-insensitive and intermediate types between *Lablab purpureus* var. *typicus* and *Lablab purpureus* var. *typicus* with high number of soft, medium long, dark green or green, tender pods of good cooking quality
- The bushy, compact, semi-viny and photo-insensitive types should be suitable for both rainy-autumn and autumn-winter season
- High green pod yield with high number of pods per plant
- Resistance to different diseases *viz.*, anthracnose (*Colletotrichum lindemuthianum*), powdery mildew (*Erysiphe polygoni*), rust (*Uromyces fabae*), ashy stem blight (*Macrophomina phaseolina*), bacterial leaf spot (*Xanthomonas phaseoli),* dolichos yellow mosaic virus disease, etc.
- Resistance to different insect pests *viz.*, stem fly (*Ophiomyia phaseoli),* pod borer (*Helicoverpa armigera*), aphids (*Aphis* sp.) , bean weevil (*Bruchus rufimanus*), etc.
- Resistance to abiotic stresses *viz.*, high temperature, drought and water logging

Breeding methods and development of variety

Wide variability in various forms and cultivars of both the cultivated types under *Lablab purpureus* var. *typicus* and *Lablab purpureus* var. *lignosus* are available in India. Hence, selection from indigenous variability is the most important breeding method for improving vegetable type dolichos bean. Vegetable podded, medium viny or bush types have mostly been developed through hybridization and selection approach involving photo-insensitive and bushy pulse type and viny, vegetable type as the parents in the hybridization programme. The varieties developed in India may be described under three broad class *viz.*, Long-viny and photosensitive type, Semi-viny and photo-insensitive type and Bushy and photo-insensitive type

Long-viny and photo-sensitive type variety

These varieties basically belong to viny, vegetable type (*Lablab purpureus* var. *typicus*) which are photo-sensitive (short day) in nature and developed through selection from the indigenous germplasm.

Pusa Early Prolific, JDL-79, CO-1, CO-2, CO-3, CO-4, CO-5, Rajni, Dasarawal, JDL 53, Deepaliwal, KDB 403, KDB 405, Pusa Sem-2, Pusa Sem-3, Phule Gouri, JDL-79, Kashi Sheetal , Kashi Harittima, Swarna Utkrisht, Swarna Rituvar

Semi-viny and photo-insensitive type

These soft-podded, semi-viny and photo-insensitive types have been developed through hybridization and selection method involving photo-insensitive and bushy pulse type, *Lablab purpureus* var. *lignosus* and viny, vegetable type, *Lablab purpureus* var. *typicus* as two parents in the hybridization programme.

Arka Adarsh, Arka Krishna, Arka Pradhan, Arka Visthar, Arka Bhavani, Arka Prasidhi, Arka Swagath, Arka Amogh, Arka Sambhram, Arka Soumya

Bushy and photo-insensitive type

Most of the soft-podded, bushy and photo-insensitive types have been developed through hybridization and selection method involving photo-insensitive and bushy pulse type, *Lablab purpureus* var. *lignosus* and viny, vegetable type, *Lablab purpureus* var. *typicus* as two parents in the hybridization programme. Pods of these varieties are not as long and tender as that of pole type, but their photo-insensitive nature enable them to grow much earlier in the season. Few semi-viny/bush type varieties have also been developed by selection from the germplasm of *Lablab purpureus* var. *lignosus*. One variety was a spontaneous mutant (CO-9) and the other (CO-10) was isolated in the M_2 generation generation of induced mutation.

Hebbel Avare-1, Hebbel Avare-3, Hebbal Avare-4, CO-6, CO-7, CO-8, CO-9, CO-10, CO-11, CO-13, Wal Konkan-1, Arka Jay, Arka Vijay, Konkan Bhushan, CO-12

Germplasm resources and Plant introduction

Dolichos bean germplasm is maintained in different countries or institutes *viz.*, International Institute of Tropical agriculture (IITA), Nigeria, Wealth Scientific and Industrial Research Organization (CSIRO), Australia, AVRDC, Taiwan, China, Philippines, South America, Ethiopia, Kenya, etc.

In India, largest collection of dolichos bean genetic resources is held at the University of Agricultural Sciences (UAS), Bengaluru. Wide range of germplasm is also maintained at NBPGR, New Delhi, ICAR-IIVR, Varanasi, ICAR-IARI, New Delhi, JNKVV, Jabalpur and other places. In India, germplasm collections were made from Madhya Pradesh, Haryana, Odisha, West Bengal and Andhra Pradesh.

No exotic genetic materials have so far been utilized in breeding dolichos bean in India.

Variability and selection

Wide variability of both viny, vegetable type (*Lablab purpureus* var. *typicus*) and bushy pulse type, *Lablab purpureus* var. *lignosus* is available in India which favours effective selection of desirable genotypes. There are reports of positive correlation between number of pods per plant and pod and seed yield; number inflorescence per plant with pods per plant and pod weight; Different studies suggested that selection index based on plant height, plant spread, number of pods per plant, length and breadth of pod, weight of pod, seed length and width,100 dry seed weight, protein content of pod and pod yield/plant should be considered while formulating selection indices in the improvement of pole type dolichos bean (Das *et al.*, 1987).

Dolichos bean is highly self-pollinating species due to hermaphrodite sex form and homogamy however, anthers dehisce much earlier than anthesis. Stamens and pistil in opened flower remain enveloped together inside the tube like structure of jointed petals called "keel" which favours self-pollination. Both pure line selection and single plant selection are employed in the breeding programme. Although, single plant selection is mostly followed due to convenience, pure line selection is more effective in selecting desirable individuals from the indigenous / exotic genetic resources. Large number of varieties of the long-viny and soft-podded, photo-sensitive (short day) vegetable types belonging to *Lablab purpureus* var. *typicus* have been developed by single plant selection from the wide range of indigenous germplasm *viz.*, Pusa Early Prolific, JDL-79, CO-1, CO-2, CO-3, CO-4, Rajni, JDL-53, Kashi Sheetal, etc. Few semi-viny / bush type varieties *viz.*, CO2, CO3, CO4 have also been developed by selection from the germplasm of *Lablab purpureus* var. *lignosus*. Maximum number of varieties have been developed in India through selection from indigenous genetic materials.

Table 3: Varieties developed through selection

Name of variety	Indigenous / Exotic material
Pusa Early Prolific	Indigenous
JDL-79	Indigenous (Madhya Pradesh)
CO-1	Indigenous (selection from HD-18)
CO-2	Indigenous (Tamil Nadu named "Chinna Avarai")
CO-3	Indigenous (Tamil Nadu named "Yanaikathu Avarai")
CO-4	Indigenous (Tamil Nadu named "Sivappa Avarai")
CO-5	Indigenous (Tamil Nadu named "Kozhikkal Avarai")
Rajni	Indigenous (U.P)
Dasarawal	Indigenous (Maharashtra)
JDL 53	Indigenous (Madhya Pradesh)
Deepaliwal	Indigenous (Maharashtra)
KDB 403	Indigenous (U.P)
KDB 405	Indigenous (U.P)
Pusa Sem-2	Indigenous
Pusa Sem-3	Indigenous
Phule Gouri	Indigenous (Maharashtra)
Arka Swagath	Indigenous (IC 556736)
JDL-79	Indigenous (Madhya Pradesh)
Kashi Sheetal	Indigenous (U.P)
Kashi Harittima	Indigenous (U.P)
Swarna Utkrisht	Indigenous
Swarna Rituvar	Indigenous

Hybridization and selection

Considerable heterosis was manifested in the intervarietal hybrids belonging to *Lablab purpureus* var. *typicus* for pod setting, yield components and pod yield (Hazra and Som, 1991). Heterosis has not yet been exploited commercially, However, good number of short-viny or bushy and photo-insensitive, vegetable type varieties having soft but comparatively short-pod have been developed through hybridization and selection method involving photo-insensitive and bushy pulse type, *Lablab purpureus* var. *lignosus* and photo-sensitive, viny, soft-podded, vegetable type, *Lablab purpureus* var. *typicus* as two parents in the hybridization programme. In the advanced generations, selection of the segregates has been exercised on growth habit, photo-sensitivity and soft pod type. Being highly self-pollinated crops, pedigree, bulk, bulk-pedigree or back cross-pedigree method of breeding up to at least F_{7-8} generation is followed to isolate the desirable segregates. This method has been followed to develop good number of varieties *viz*., Arka Adarsh, Arka Krishna, Arka Pradhan, Arka Visthar, Arka Jay, Arka Vijay, etc. Pods of these varieties are not as long and tender as that of pole type but are sufficiently soft to be used as vegetable type. However, photo-insensitive nature of these varieties enables them to grow much earlier in the season.

Table 4: Varieties bred through hybridization and selection

Name of variety	Parentage
Arka Adarsh	IIHR 178 × Arka Swagath
Arka Krishna	IIHR 178 × Arka Swagath
Arka Pradhan	IC 556824 IPS-2 × Arka Swagath
Arka Visthar	IIHR 178 × Arka Swagath
Arka Bhavani	IIHR 178 × Arka Swagath
Arka Prasidhi	IC 556824 IPS-2 × Arka Swagath
Arka Amogh	(Arka Jay × Arka Vijay) × Konkan Bhushan)
Arka Sambhram	(Arka Jay × Arka Vijay) × Konkan Bhushan)
Arka Soumya	(Arka Jay × Arka Vijay) × Konkan Bhushan)
Hebbel Avare-1	Local Avare × Red Typicus
Hebbel Avare-3	Hebbal Avare 1 × US 67-31 (USA)
Hebbal Avare-4	Hebbal Avare × CO 80
CO-6	DL 3169 × CO-5 (pole type)
CO-7	DL 3169 × CO-5 (pole type)
CO-8	CO-5 × DL 3169
CO-11	CO-9 × White Yanai Kathu (pole type)
CO-13	CO-9 × Florikifield (pole type)
Wal Konkan-1	Wal-2-K2 × WaL 125-136
Arka Jay	Hebbel Avare-3 × IHR-99 (pole type)
Arka Vijay	Hebbel Avare-3 × Pusa Early Prolific (pole type)
Konkan Bhushan	Hebbel Avare 3 × Wall
CO-12	CO-9 × CO-4 (pole type)

Utilization of mutants and applied mutagenesis

Very little works have so far been done on applied mutagenesis in dolichos bean. In a preliminary study on the effect of gamma radiations on seed germination and growth of a viny variety, Deepali and a dwarf variety, Konkan Bhushan, Harish Kumar *et al.* (2016) reported that seed germination and plant growth progressively decreased with increased doses of gamma radiations from 250 gray to 550gray. One bushy type variety was developed from a spontaneous mutant (CO-9) and the other of such type (CO-10) was isolated in the M_2 generation of the variety CO-6 irradiated with 240 Gray gamma radiation.

Table 5: Varieties/ mutants developed through applied mutagenesis

Variety/ Mutant	Applied mutagenesis / Remark
CO-9	Spontaneous mutant with bushy habit
CO-10	240 Gray gamma radiation to CO-6

Breeding for resistance to diseases

Resistance to bacterial leaf spot (*Xanthomonas campestris* pv. *phaseoli*) is controlled by three genes. Dolichos yellow mosaic virus (DYMV) is caused by Gemini virus and transmitted by whitefly. This serious virus disease affects the production of dolichos bean in South Asia. Diseased plants produce characteristic bright yellow mosaic patches on the leaves and early infections cause reductions in yield. Singh *et al.* (2012) identified VRSEM 894, VRSEM 887 and VRSEM 860 as resistance to DYMV among 300 germplasm accessions based on initial field screening under natural infection followed by screening of 34 symptomless accessions using sap inoculation in field condition (Rai *et al.*, 2015). Backcrossing followed by pedigree method of selection is efficient in resistance breeding when resistance is governed by more than one gene.

Breeding for quality

Fresh pods of dolichos bean are rich in proteins, minerals and other nutrients. However, phytic acid content – a major nutrient inhibitor should be low (1.8–2.4 mg per 100g of tender pods) to enhance bioavailability of minerals, making the immature pods a potential functional food and a valuable asset in overcoming malnutrition, especially in the developing countries (Pandey *et al.*, 2023). Several reports suggest that pod yield is negatively correlated with crude fibre and seed protein content is negatively correlated with pod weight, seeds per pod and seed yield per plant. Enhancement of different quality traits *viz.*, protein content and quality, different minerals, phytochemical contents and crude fibre and at the same time, lowering of the anti-nutrient compounds particularly, phytic acid content should be taken into consideration in dolichos breeding programme for improving plant stature, pod yield and quality.

References

Das, A.K., Hazra, P. and Som, M.G. 1987. Genetic variability and Heritability studies in dolichos bean (*Dolichos lablab* (Roxb.L.) *Veg. Sci.*, **14**:169-173.

Harish Kumar, Ghawade, S.M., Meghwal, M. and Shivaputra. 2016. Mutagenic effect of gamma radiation on seed germination, plant growth and mortality of dolichos bean (*Lablab purpureus* L.), *Prog. Res.*, **11**: 4844-4847

Hazra, P., Banerjee, M.K. and Chattopadhyay, A. 2022. *Varieties of Vegetable crops in India*, Kalyani Publishers, New Delhi, Ludhiana, 204p.

Hazra, P. and Som, M.G. 1991. Study on flower shedding and pod setting in dolichos bean *(Lablab purpureus (L.)* Sweet. Bangladesh J.Bot., **20**: 85-86.

Kalloo, G. and Pandey, S.C. 1993. Hyacinth bean Lablab purpureus L. Sweet. In: Genetic Improvement of Vegetable crops (Kalloo, G. and Bergh, B.O.), Pe*rgamon Press, O*xford, New York, *Seoul, Tokyo, pp.* 387-389.

Pandey, D.K., Singh , S., Dubey , S.K., Mehra , T.S., Dixit , S. and Sawargaonkar , G. 2023. Nutrient profiling of lablab bean (*Lablab purpureus*) from north-eastern India: A potential legume for plant-based meat alternatives. *J. Food Comp. Anal.*, https://doi.org/10.1016/j.jfca.2023.105252.

Rai, N., Rai, K.K., Venkataramanappa, V. and Saha, S. 2015, Molecular approach coupled with biochemical attributes to elucidate the presence of DYMV in leaf samples of *Lablab purpureus* L. genotypes. *Appl. Biochem. Biotechnol.*, 176: 1 - 15. DOI:10.1007/s12010-015-1915-5

Ramesh, S. and Byregowda, M. 2016. Dolichos bean (*Lablab purpureus* L. Sweet., var. *lignosus*) Genetics and Breeding- Present status and future prospects. *Mysore J. Agric. Sci.*, **50**: 481- 500.

Singh, P.K., Ashish Kumar, Rai, N. and Singh, D.V. 2012, Identification of host plant resistance to Dolichos Mosaic Virus (DYMV) in Dolichos bean (*Lablab purpureus*). *J. Plant Path. Microbiol.*, **3**: 1 - 3.

Swarup, V. 2006. *Vegetable Science and Technology in India*. Kalyani Publishers, New Delhi, Ludhiana, 656p.

32

Vegetable amaranth

The genus *Amaranthus* is rather unique in having species which are used for vegetable, grain and ornamental purposes. Vegetable amaranth, *Amaranthus* spp. under the family Amaranthaceae (Chromosome No.: $2n = 2x = 32, 34$) is the most common tropical leafy vegetable in Southeast Asian countries especially India, Malaysia, Indonesia, southern China and hot and humid regions of Africa. It is an excellent source of β carotene, vitamin C, iron, calcium, potassium, magnesium, phosphorus, folic acid and dietary fibre. Its protein content (20-38% based on dry matter) includes methionine and other sulphur-containing amino acids. It is cultivated throughout India however, it is a common leafy vegetable in southern states *viz*., Tamil Nadu, Kerala, Karnataka, Telangana and Andhra Pradesh.

Origin, domestication and adaptation in India

The phylogenetic relationship and taxonomic delimitation in the genus *Amaranthus* under the family Amaranthaceae are still not resolved with extreme clarity. Phylogenetic analysis revealed clear separation of vegetable and grain amaranths. The derivation of grain and vegetable amaranths represent two lines of descent from the weed progenitor. *Amaranthus hybridus* L. This species has spread to different tropical countries like Mexico, Thailand, Malaysia, Indonesia and the Philippines. *Amaranthus* is a cosmopolitan genus of annual or short-lived perennial plants collectively known as amaranths. Vegetable amaranth is the most common leafy vegetable crop grown in India. Several kinds of vegetable amaranths under the genus *Amaranthus* and belonging to different species are grown in India.

- *Amaranthus tricolor*: It is the most important vegetable amaranth which is native to South and South-East Asia. Most probably it has been originated in India and from where it spread to the neighboring countries like, Indonesia, Malaysia, Hong Kong, Taiwan and Southern China. It is one of several species of amaranth now cultivated in warm regions across the world. It occupies the predominant position in India with different morphological forms, colour and shape of leaves.
- *Amaranthus blitum*: It is native to the Mediterranean region however, it is naturalized in other parts of the world, including much of eastern North America, much of tropical Africa, Western Europe and Japan. It is commonly grown in Greece, Taiwan and India.

- *Amaranthus dubius*: It is native to Central America however, it is widely introduced throughout the world particularly, Central America, Indonesia and India. The species occurs locally in France and Germany and is naturalized or invasive in tropical and subtropical regions of the United States (Florida and Hawaii), Africa, Asia, Australia and the Pacific.
- *Amaranthus tristris*: It is native of tropical America but now occurs almost throughout the tropics of the world.
- *Amaranthus hypochondriacus*: It has been originated in Argentina and Peru of South America.
- *Amaranthus lividus*: It occurs as a weed in Southern and Central Europe where it is also a popular leafy vegetable.
- *Amaranthus viridis*: It is possibly of Asian origin but now it is a cosmopolitan weed in the tropical and subtropical regions of the world; it is also penetrating far into temperate regions (e.g. in Europe, North America, Asia and Australia).
- *Amaranthus caudatus*: It has been originated in cold regions of high Andes in South America
- *Amaranthus polygonoid* : It is endemic in North America and the Caribbean islands.
- *Amaranthus viridis*: It has been originated in East Asia and widespread in tropical and subtropical regions of the world and in the upland rice area.

Crop botany and floral biology

- *Amaranthus* is a cosmopolitan genus of annual or short-lived perennial plants, with inflorescences and foliage ranging from purple, through red and green to gold; this genus shows a wide variety of morphological diversity among and even within certain species.
- Members of this genus share many characteristics and uses with members of the closely related genus *Celosia*.
- Members of this family have simple leaves that are opposite or alternate, with margins entire or coarsely toothed and without stipules.
- Depending on the cultivar, photoperiod and cultural practices, flowering may start 4-8 weeks after sowing, making the plant less suitable for consumption as leafy greens.
- In *A. tricolor* and *A. cruentus,* the seeds mature after 3-4 months and then the plant dies; *A. dubius* continues its generative stage for a much longer period and when cut regularly, the plant may become shrubby and perennial, but even at its mature stage, the leaves are succulent enough for consumption.
- Inflorescence is terminal and axillary with catkin-like cymes of densely packed flowers; flowers are solitary or aggregated in cymes, spikes or panicles and unisexual; inflorescence is differently coloured from green to deep purple and orange and with different shades of these colours; the basic unit of the inflorescence is dichasial cymes called glomerule.

- At least four times as many female flowers as male flowers are borne in the inflorescence; flowers are small, regular and mostly unisexual, monoecious, pentamerous, bracteate and bracteolate which are scaly, perianth parts are 4-5, imbricate, membraneous, often persistent, stamens are 2-5, placed opposite to perianth parts, often some reduced to staminodes; filaments are either free or united below, anthers one or two-celled; carpels are 2-3.
- Ovary is superior with commonly one campylotropus ovule; style is single to 2 or 3 branched.
- Fruit is botanically the utricle, indehiscent, enclosed by the persistent perianth parts.
- Seeds varying in colour, black, brown or white, compressed, smooth, shiny with floury endosperm.

Reproductive behaviour

Vegetable amaranth is a monoecious crop but predominantly self-pollinated with percent cross-pollination ranging from 0– 40 per cent. In spite of monoecious sex form, predominant self-pollination in vegetable amaranth occurs due to presence of high percentage of male flowers (10-25%) per glomerule and more number axillary glomerules. Grain amaranth is also predominantly self-pollinated with somewhat higher 16 – 40 % out-crossing. It is characteristically wind pollinated but abundant pollen production, especially in the higher flowers, causes a high degree of self-pollination. Bees occasionally visit the grain amaranth species having colourful inflorescence leading to more cross-pollination in this species than the vegetable type. From the breeding point of view, the small, closely grouped flowers in glomerules make emasculation extremely difficult. The most satisfactory method of hybridization is to pollinate heavily as soon as the stigma is receptive and to remove the staminate flowers by hand.

Genetics of important characters

Amaranthus genus belongs to Amaranthaceae family and 298 species of this genus have been identified. Species in the *Amaranthus* genus generally consist of annual and herbaceous plants. There are also *Amaranthus* species such as *Amaranthus deflexus*, *Amaranthus tricolor* and *Amaranthus spinosus*, which are short-lived perennial herbaceous, shrub and woody.

The genus *Amaranthus* has two basic chromosome number, x = 16 and x = 17. All the *Amaranthus* species are diploids with chromosome number 2n = 32 or 2n = 34 and only *Amaranthus dubious* is a tetraploid species (Mallika *et al*, 1988). However, Srivastava and Roy (2014) reported new chromosome number in *Amaranthus blitum* (2n=28) which indicated that it must have been derived from the new basic number (x =14) and show similarity with the earlier findings of Pal *et al.* (2000) for *Amaranthus tenuifolius* (2n=28).

Table 1: Chromosome number of different *Amaranthus* species

Chromosome number	***Amaranthus* species**
2n = 28	*Amaranthus blitum, Amaranthus tenuifolius*
2n = 32	*Amaranthus hypochondriacus, Amaranthus caudatus, Amaranthus hybridus*
2n = 32 or 34	*Amaranthus tricolor, Amaranthus cruentus, Amaranthus viridis, Amaranthus lividus, Amaranthus graecizans , Amaranthus tristris, Amaranthus polygonoid, Amaranthus lividus*
2n = 64	*Amaranthus dubius*

Among the vegetable amaranth, *Amaranthus tricolor* is most popular in South Asia, Southeast Asia and India; *Amaranthus dubius* is popular in Central America, India, Indonesia, Africa and Caribbean islands; *Amaranthus lividus* in Greece, Taiwan and India; *Amaranthus dubius* is widely available in Central America, Indonesia and India; *Amaranthus viridis* is particularly popular in India. Other *Amaranthus* species used as leafy vegetable, although to a lesser extent, are *Amaranthus tristris, Amaranthus hypochondriacus, Amaranthus polygonoid, Amaranthus caudatus*, etc.

Amaranthus hypochondriacus, Amaranthus caudatus, Amaranthus cruentus are also used for grain purpose. Important characters of different *Amaranthus* species are mentioned below.

Table 2: Important characters of different *Amaranthus* species

***Amaranthus* species**	**Origin**	**Characters**
Amaranthus tricolor	South and South-East Asia	Erect, annual or short-lived perennial herb, up to 1.5 m tall; some cultivars are shrub and woody; red or light to dark green leaves are elliptical to lanceolate (oval to lance-shaped) or broadly ovate (egg-shaped). Flowers occur in rounded clusters in the leaf axils (the space between the upper surface of the leaf and the stem); It has thick erect stem and larger leaves; It responds to cutting and is distinct by its much longer growing period. Some varieties flower at least 4 months after sowing. It can be sown early in summer and it covers summer, rainy season and part of autumn before it flowers. Leaves, tender twigs and even matured main stem of some particular varieties of this species are used as vegetable.
Amaranthus blitum	Mediterra-nean region	It is an erect or semi-prostrate annual plant; cotyledons with rounded to truncate apex; single or branched stem can grow to one-meter-tall; green or purplish leaves are up to 10 cm long on stalks of a similar length and are arranged spirally; leaves are simple, roughly triangular in shape and have entire margins; it has early flowering habit within 45 days after sowing; it responds to cutting; Inflorescence is mostly axillary but may also arise as short spike terminally with tiny male and female flowers clustered together; fruits are small globular capsules containing disc-shaped seeds; seed with minutely punctiform surface and diameter is 1.1-1.2 mm

***Amaranthus* species**	**Origin**	**Characters**
Amaranthus dubius	Central America	It usually grows to a size of 80–120 cm; it is more branching and having cone shaped plant frame with straight and upright main stem; branching starts from the base at near 90° angle; inflorescence emerges at terminal of main branches as well as in the axils; it has both green and red varieties, as well as some with mixed colours; the green variety is practically indistinguishable from *Amaranthus viridis*; plants continue its generative stage for a much longer period and when cut regularly, the plant may become shrubby and perennial, but even at its mature stage, the leaves are succulent enough for consumption; it flowers from summer to fall in the tropics, but can flower throughout the year in subtropical conditions.
Amaranthus cruentus	Guatemala in Central America	It is an annual herbaceous plant and has a short growing period of 4–6 weeks; it produces one dominant, large, central root (tap root); thick stems are often straight and branched, 0.1 to 2.0 m in height, ribbed, and red dyed; topped with clusters of dark pink flowers; the seeds mature after 3-4 months and then the plant dies; it shares many morphological features of *Amaranthus hybridus.*
Amaranthus tristris	Tropical America	It is erect annual herb which mostly grows up to 90 cm tall; stem rather slender to stout, usually branched, angular, glabrous or increasingly furnished upwards (especially in the inflorescence) with short to rather long, multicellular hairs; leaves are glabrous, or thinly and shortly pilose on the inferior surface of the primary venation, long-petiolate (petioles up to 8.5 cm. long, sometimes longer than the lamina), lamina ovate or rhomboid-ovate; flowers are green, in the lower part of the plant in axillary clusters of 4–10 cm in diameter; the plants respond favourably to cuttings.
Amaranthus hypochondriacus	Argentina and Peru	It is an herbaceous annual or short-lived perennial plant; vigorous, upright that typically reaches 40–200 cm; leaves are simple and alternately arranged, with entire margins, rhombic-ovate to broadly lanceolate in shape, about 4–12 cm long and 2–7 cm wide; leaves and seeds are very nutritious and have a mild flavour; seeds also contain phenolic compounds
Amaranthus viridis	East Asia	It is an annual herb with an upright, light green stem that grows to about 60–80 cm in height; numerous branches emerge from the base, and the leaves are ovate, 3–6 cm long, 2–4 cm wide, with long petioles of about 5 cm; plant has terminal panicles with few branches, and small

Amaranthus species	Origin	Characters
		green flowers with 3 stamens; it is eaten as a boiled green or as a leafy vegetable in many parts of the world; in India, it is a common vegetable in Bengali cuisine; plants are small and cone shaped; main stem is straight and branches arise in less than 90° angle; leaves are small and broad; inflorescence is both terminal and axillary.
Amaranthus polygonoid	North America and the Caribbean	It is a terrestrial and annual herb with prostrate growth habit; branched mostly at the base with soft spikes on young shoots; leaves are ovate, leaf margin entire, obovate-rhombic to narrowly ovate, sometimes lanceolate, long ptioled, apex deeply notched; flowers are simple in axillary spikes; it is a less grown leaf amaranth species.
Amaranthus lividus	Southern and Central Europe	Plants are annual herb, glabrous, prostrate, decumbent or ascending; stem erect, simple or branching from the base, more or less angular, mostly glabrous; leaves are alternate on long petioles, ovate, mostly glabrous, distinctly emarginated at the apex. Inflorescences are slender terminal spikes or panicles and also axillary clusters; in some forms only axillary clusters are present; spikes are erect or sometimes reflexed and green; flowers are borne in terminal and axillary spikes, green, male and female flowers are intermixed; it is also a less grown species.
Amaranthus caudatus	High Andes in South America	It is most recognizable for its striking flowering panicles that can reach up to 90 cm long; colour of these highly dense flowering panicles ranges from black to red and more commonly white; the red varieties of *A. caudatus* are due to a high content of betacyanins. It can grow anywhere and grows best in full sunlight; it is grown in a wide variety of conditions, both humid and arid; it is also less grown vegetable amaranth species.

Source: Swarup (2006), Hazra et al (2022)

There are only few reports on the inheritance of qualitative and quantitative characters in vegetable amaranth. Most of the genetical studies were carried out in the *Amaranthus* species for grain purpose “grain amaranth”. Inheritance of seed colour has been found to be under polygenic control and complex in nature. Black seed colour is dominant to white or gold. In grain amaranth, green perianth colour shows dominance which could be employed as a marker character to determine the extent of cross-pollination.

Breeding objectives

The followings are the main objectives of vegetable amaranth breeding

- Higher leaf yield, broad leaves, longer harvest duration and late bolting

- Plants with uniform growth habit, medium height with broader leaves, medium-thick stem and leaf / stem ratio should be above one, high rejuvenating ability which responds to multiple cutting.
- Stable yield in different environments and late bolting type
- Good nutritional profile that includes high protein, β carotene, vitamin C, iron, calcium, potassium, magnesium, phosphorus, folic acid contents and low in oxalate and nitrate contents
- Resistance to lodging and tolerance to low and high temperature
- Resistance to important diseases *viz*., damping off (*Pythium aphanidermatum*), white rust (*Albugo bliti*), leaf spot (*Cercospora* sp.), etc.
- Resistance to insect pests and nematodes *viz*., stem borer (*Lixus truncatulus*), leaf caterpillar (*Hymeria recurvalis*) , root knot nematode (*Meloidogyne incognita*), etc

Breeding methods and development of variety

Several kinds of vegetable amaranth under the genus *Amaranthus* and belonging to different species are grown in India. The cultivars vary considerably in plant height, size and colour of leaves, branching habit, flowering time and response to cutting. Natural interspecific and intervarietal hybridization might have caused such wide variation. Mostly 'green' and 'red' types are cultivated and both these types are nutritionally almost similar.

Since vegetable amaranths are predominantly self-pollinating species, improvement programme in India has been done by single plant selection, pure line selection or mass selection of superior types from indigenous germplasm. Interspecific hybridization has not yet been successfully utilized in the breeding programme. The varieties can be grouped according to the species under which they belong.

- ***Amaranthus tricolor***

  CO-2, CO-5, Pusa Badi Chaulai, Pusa Kiran, Pusa Kirti, Pusa Lal Chauli, Arka Suguna, Arka Arunima, Arka Varna, Arka Samraksha, Annapurna, Arun, Mohini, Krishnasree, Renusree, Kashi Suhavaani

- ***Amaranthus blitum***

  Pusa Chhoti Chaulai, Pusa Kirti

- ***Amaranthus dubius***

  CO-1

- ***Amaranthus tristris***

  CO-3

- ***Amaranthus hypochondriacus***

  CO-4

- ***Amaranthus polygonoid***

  Sirukeerai

- ***Amaranthus viridis***

  No improved variety under this species has so far been developed
- ***Amaranthus lividus***

  No improved variety under this species has so far been developed
- ***Amaranthus caudatus***

  No improved variety under this species has so far been developed

Germplasm resources and Plant introduction

Wide range of germplasm are maintained at the National Laboratory for Genetic Resource Preservation, Fort Collins, Colorado State University, USA (global collection), National Horticultural Research Institute, Ibadan, Nigeria (African collection), El Instituto Nacional de Investigaciones Forestales, Agrícolas y Pecuarias (INIFAP), Mexico (new World collection) and National Bureau of Plant Genetic Resources (NBPGR), New Delhi (Asian collection).

In India, wide range of germplasm is also maintained at ICAR-IIVR, Varanasi, ICAR-IARI, New Delhi, ICAR-IIHR, Bengaluru, Tamil Nadu Agricultural University, Coimbatore, Kerala Agricultural University, Vellanikkara, Kerala and other places.

Very rarely exotic genetic materials have been utilized in breeding vegetable amaranth in India.

Variability and selection

Collection, evaluation and characterization of germplasm or lines is the first step in genetic improvement of any crop. Remarkable germplasm diversity, adaptability to diverse growing conditions and unique reproductive behaviour ranging from outcrossing to greater self-pollination has led this crop to exhibit wide genetic variability for many important quantitative and quality characters. It offers substantial scope to identify suitable types for different regions. Performance of the *Amaranthus* species depends on its origin and location for evaluation. *A. caudatus* genotypes originating from Asian countries produced more leaves (25.8±8.9), while *A. caudatus* from the US only produced 12.2 leaves. Similarly, *A. cruentus*, genotypes from South America tended to have more leaves (71.2±33.7) than others. The *A. hybridus* genotypes from the US had fewest leaves (30.0±8.) and those from Africa had the most (74.8±11.7) (Wu *et al.*,2000). Leaves of *A. tricolor* have considerable quantities of protein (average 1.24±0.03 mg/100 mg fresh), carotenoid (average 0.83±0.02 mg/g fresh), ascorbic acid (112.33±5.00 mg/100g fresh) and fibre (8.39±0.10%) contents (Shukla *et al.*, 2006). The anti-nutrient, oxalate content widely ranged between 0.6 % and 2.1% and nitrate content varied from 0.04 % to 1.6 % which offer ample scope for selection of genotypes having low in the anti-nutrient compounds (Celine *et al.*, 2007).

Indirect selection for the traits *viz.*, higher leaf number per plant, longer leaves with more leaf area and weight, higher diameter of stem base, high stem weight, fibre content and more harvest index could result in high leaf yield per plant. Since vegetable amaranths are predominantly self-pollinating species, single plant selection (e.g., Pusa

Kirti, Arka Samraksha, etc.), pure line selection (e.g., Arka Suguna , Arka Arunima, etc.) or mass selection (Pusa Badi Chaulai, Pusa Lal Chauli, etc.) of superior types from mostly indigenous germplasm is the most prominent method of improvement.

Table 4: Varieties developed through different methods of selection

Name of variety	Indigenous / Exotic material
Single plant selection	
CO-5	Indigenous
Pusa Kirti	Indigenous
Arka Samraksha	Indigenous
Krishnasree	Indigenous
Renusree	Indigenous
Kashi Suhavaani	Indigenous
Pusa Chhoti Chaulai	Indigenous
Pure line selection	
Arka Suguna	Exotic from Taiwan (IIHR 13560)
Arka Arunima	Indigenous (IIHR 49)
Arka Varna	Indigenous
Annapurna	Indigenous; from Pauri (Uttaranchal)
Mass selection	
CO-2	Indigenous
Pusa Badi Chaulai	Indigenous
Pusa Lal Chauli	Indigenous
Arun	Indigenous
Mohini	Indigenous
Pusa Kirti	Indigenous
CO-1	Indigenous
CO-3	Indigenous (A-83)
CO-4	Indigenous
Sirukeerai	Indigenous

Hybridization and selection

Yield potential of most of the crops has been improved by the adoption of hybridization in the breeding programme. In vegetable amaranths, interspecific hybridization though has not yet been successfully utilized yet, it has great potential for transgression of useful genes from the allied species. Homozygous parental lines which are needed for the development of hybrids can be produced by selfing or through double haploid technology. Many tissue culture-based direct shoot regeneration protocols are also available in *Amaranthus* species (Pal *et al.,* 2013).

However, development of efficient crossing methods is important for genetic research and improvement programme of amaranth. Stetter *et al.* (2016) evaluated three different crossing methods (open pollination, hot water emasculation and hand emasculation) and reported that all three methods successfully produced F_1 hybrid, but with different success rates. Open pollination had the lowest (10%), hand emasculation had the highest success rate (74%) and hot water emasculation showed an intermediate success rate (26%). Substantial advances in the understanding of the genome and modern genetic marker systems has been made in amaranth and these resources can be used to improve breeding methodology for amaranth (Stetter *et al.,* 2016). Male sterility identified in grain amaranth under *A. hypochondriacus* can simplify the hybridization process (Das, 2016).

In vegetable amaranth breeding, negligible efforts have been made for its genetic improvement through hybridization programme due to variability and versatility of inflorescence, ratio and distribution of male and pistillate flowers in the inflorescence. However, promising crosses of *Amaranthus tricolor* having high foliage yield with increased plant height, branches per plant and leaf size may be advanced to isolate transgressive segregates to obtain highest returns from vegetable amaranth breeding. The cross combinations generally do not show inbreeding depression hence, pedigree method of breeding can be adopted in advanced generations to develop nutritionally rich and high yielding varieties (Rastogi *et al.*, 2015). However, only one natural *A. tricolor* x *A. tristis* cross, showing predominantly the characters of *A. tricolor* has been advanced for the development of a variety, Pusa Kiran.

Table : Varieties bred through hybridization and selection

Name of variety	Parentage
Pusa Kiran	Natural cross, *A. tricolor* x *A. tristis*; the selected line possessed predominantly the characters of *A. tricolor*

Utilization of mutants and applied mutagenesis

In grain amaranth breeding, induced mutagenesis was used through physical mutagenesis based on gamma radiation and chemical mutagenesis by alkylating substances, such as dimethyl sulfate, ethyl methane sulfonate, ethyleneimine, and others. Several promising mutants could be selected which showed dwarf plant stature, broad leaves, late bolting, basal branching, high rejuvenating ability, bold seed size and high green yield (reviewed by Pylypets, 2023). Various methods of applied mutagenesis can be applied to isolate desirable mutants of vegetable amaranth for onward development of new variety. The varieties *viz*., 'Centenario' (improved grain yield) in Peru, 'New Asutake' (early maturity) in Japan, 'Sterk' (tolerance to moisture and heat stress) in Russia, 'Pribina' in Slovakia have been developed by utilizing applied mutagenesis followed by isolation of promising mutants (Joshi *et al.,* 2018).

Alteration of ploidy level

Induction of polyploidy has been proved to be a handy tool in tampering with the genotype of the organisms and in widening the amplitude of variation in the gene pool. Though various methods of induction of polyploidy have been developed, induction of polyploidy through the application of colchicine has been met with remarkable success especially with many herbaceous plants including two species of *Amaranthus viz. Amaranthus hypochondriacus* L. and *Amaranthus dubius* Mart. ex Thell. As the vegetative parts such as the leaf and stem of *Amaranthus* species are edible and autopolyploidy very often magnifies the vegetative characters, this method of increasing the ploidy level can be utilized for development of a new variety

(Behera and Patnaik, 1975). Autotetraploidy was also induced in *Amaranthus blitum* by colchicine. The induced tetraploids had higher number and bigger leaves with prolonged vegetative phase. However, no variety has yet been developed through this method of breeding.

Breeding for quality

Vegetable amaranths are unique leafy vegetables due to good source of protein, sulphur-containing amino acids like, methionine and other vitamins and minerals like, β carotene, ascorbic acid, iron, calcium, potassium, magnesium, phosphorus, folic acid and dietary fibre. However, some nutrient inhibitor particularly, oxalic acid and nitrate contents should be lowered to enhance the bioavailability of minerals, making this leafy vegetable a potential functional food.

References

Behera , B. and Patnaik , S.N. 1975. Induced Polyploidy in *Amaranthus hypochondriacus* L. and *Amaranthus dubius* Mart. ex Thell. *Cytologia*, **40**: 157-168.

Celine, V.A., Shankaran, S.S., Seema, S., Deepa, S.N., Sreelathakumary, I. and Abdul Vahab, M. 2007. Characterization and evaluation of vegetable amaranthus (*Amaranthus tricolor* L.) for high yield, quality and resistance to *Rhizoctonia solani. Acta Hort.*,**752:** 447-452.

Das, S. 2016. *Amaranthus: A Promising Crop of Future*, Springer, Singapore, 208p.

Hazra, P., Banerjee, M.K. and Chattopadhyay, A. 2022. *Varieties of Vegetable crops in India*, Kalyani Publishers, New Delhi, Ludhiana, 204p.

Mallika, V.K., Ramachandran, C. and Peter, K.V. 1988. Genome analysis in *Amaranthus* I. Meiotic studies in species and hybrids. Genome, 30: 285.

Pal, M., Ohri, D. and Subrahmanyam. 2000. A new basic chromosome number for *Amaranthus* (Amaranthaceae). *Cytologia*, **65**: 13-16

Pal. A., Swain, S.S., Das, A.B., Mukherjee, A.K. and Chand, P.K. 2013.Stable germ line transformation of a leafy vegetable crop amaranth (*Amaranthus tricolor* L.) mediated by *Agrobacterium tumefaciens. In Vitro Cell Dev. Biol. Plant,* **49:** 114–128.

Pylypets, S.O. 2023. The use of induced mutagenesis in grain amaranth breeding (Review). *Bull. Sumy Nat. Agrarian Univ*., Vol 52. https://doi.org/10.32782/agrobio.2023.2.11.

Rastogi, A., Mishra, B.K., Srivastava, M., Siddiqui, A., Pandey, R., Verma, N. and Shukla, S. 2015. Identification of heterotic crosses based on combining ability in vegetable amaranthus (*Amaranthus tricolor* L.). *Asian J. Agril. Res*., **9**: 84-94.

Shukla, S., Bhargava, A., Chatterjee, A., Srivastava, A. and Singh, S.P. 2006.Genotypic variability in vegetable amaranth (*Amaranthus tricolor* L.) for foliage yield and its contributing traits over successive cuttings and years. *Euphytica,* **151:** 103–110.

Srivastava , R. and Roy , B.K. 2014. A new chromosome number for *Amaranthus blitum* . J. *New Biol. Rep*., **3**: 111 – 114.

Stetter, M. G., Zeitler, L., Steinhaus, A., Kroener, K., Biljecki, M. and Schmid, K. J. 2016. Crossing methods and cultivation conditions for rapid production of segregating populations in three grain amaranth species. *Front. Plant Sci.,* **7:** https://doi.org/10.3389/fpls.2016.00816

Swarup, V. 2006. *Vegetable Science and Technology in India*. Kalyani Publishers, New Delhi, Ludhiana, 656p.

Wu, H., Sun, M., Yue, S., Sun, H., Cai, Y., Huang, R., Brenner, D. and Corke, H. 2000.Field evaluation of an *Amaranthus* genetic resource collection in China. *Genet. Resour. Crop Evol.,***47:** 43–53.

33

Palak or Spinach beet

Palak or spinach beet or beet leaf, *Beta vulgaris var bengalensis* under the family Amaranthaceae (previously under Chenopodiaceae) (Chromosome No. 2n = 2x = 18) is one of the most common leafy vegetables of the tropical and subtropical regions of the world. Its tender, soft and succulent leaves are used as a leafy vegetable. It is a rich and cheap source of β carotene, iron, essential amino acids, ascorbic acid, thiamine, riboflavin, pyridoxine, minerals like, calcium, iron, phosphorus, magnesium, potassium and zinc and fibre. It is very common vegetable in India, particularly in eastern and northern plains of Uttar Pradesh, Bihar, West Bengal, Rajasthan, Punjab and Haryana. It is also grown in Maharashtra, Gujarat and Madhya Pradesh however, it is not common in southern states.

Origin, domestication and adaptation in India

Palak originated from Indo-Chinese region. The Romans used this crop as food for humans and animals. The crop was introduced to USA in 1800 AD (Campbell, 1976). However, it is not known when it was introduced into India. Mentioning of palak in the medical treatise, Charak Samhita, c. 600 BC is the testimony of its long history of cultivation in India. Perhaps it was first cultivated in Bengal. In India, earlier improvement through natural selection by the farmers might have been in the direction of larger succulent leaves, late bolters and ability for regeneration after repeated cuttings. However, despite such a long history of cultivation genetic variability recorded in this species is not wide enough, commensurate with outcrossing breeding behaviour.

Crop botany and floral biology

- The plant is an annual herb; has large, leafy, thick-stemmed green which looks like garden beet without swollen taproot.
- Basal leaves are few not forming a dense rosette, with long strong petiole, ovate to oblong-ovate, up to 20 cm long, apex obtuse, passing above into smaller narrower leaves.
- Infloresence is a terminal open panicle bearing small sessile flowers either singly or in clusters of 2 to 7 in the axil of linear bracts of infloresence axis and its secondary branches.
- Flowers are bisexual, two or three in a cluster on simple or branched terminal spike.

- Flowers are small, inconspicuous without corolla but with green calyx (perianth), nearly epigynous, actinomorphous and pentamerous.
- Perianth segments 5, united at base, narrowly oblong, incurved in fruit, perianth becomes thicker towards the base as the fruit ripens, utricle adnate to perianth at base.
- Stamens are five, opposite to perianth lobes and slightly united to them; anthers are dithecous.
- Ovary is embeded in the fleshy unilocular receptacles and three carpelled, syncarpous with single ovule arising just above the base; style and stigma are short.
- Fruits are mostly aggregate which are formed by the cohesion of two or more fruits and held together by sowllen perianth base and thus forming an irregular dry cork like body known as seed ball or the so-called seed; it is called "multiple seed" or multi germ; if the fruit is formed from a single ovary, it is termed as 'single germ seed' or monogerm.
- Seed is kidney shaped and brown to reddish-brown in colour; average 100 seed weight is about 1.7 gram.

Reproductive behaviour

All cultivars of palak are annual type and their seeds can be produced freely in the plains of India. Palak bears hermaphrodite flowers but varying levels of cross-fertilization takes place in this crop due to long anthesis and anther dehiscence time, open inflorescence like a large spike, which sways in wind, prolonged stigma receptivity and conspicuous production of very fine pollens, which drift in air for long time. It is anemophilous crop and cross-pollination is accomplished by blowing air although considerable self-pollination also occurs. Anthesis starts in the morning at about 7.00 a.m. and continues up to 5.00 p.m. depending on the prevailing weather conditions. Anther deshiscence occurs at about 8.00 a.m. and continues upto 6.30 p.m. The stigma becomes receptive 6 hours before and 8 hours after anthesis. Pollen remains viable on the day of anthesis. For experimental hybridization purpose, matured buds of suitable size on a spike are selected one day prior to anthesis. Emasculation is done with forceps and the whole spike is bagged. Pollination is carried out next day morning by dusting the pollens from the spike of the intended male parent on the emasculated buds. Spike from the male parent may also be bagged along with the emasculated spike for effecting pollination.

Genetics of important characters

Palak or spinach beet belongs to same species in which garden beet, sugar beet and Swiss chard are included. It is a diploid species, with chromosome number, $2n = 2x = 18$, the basic chromosome number being $x= 9$. *Beta vulgaris* has 8 isobrachial chromosome pairs and one, the longest (2-3 µm) heterobrachial chromosome pair. Ploidy levels could be altered to tetraploids, triploids and aneuploids due to application of colchicine.

Palak or Spinach beet is sometimes erroneously called spinach which not only belongs to different genus and species of the same family but also different in morphpology and sex form.

Table 1: Basic difference between palak and spinach

Name of the crop	Scientific name	Chromosome number	Sex form	Leaf morphology	Temperature sensitivity
Palak	*Beta vulgaris var. bengalensis*	Diploid (2n = 2x = 18)	Hermaphrodite sex form; perianth not hardened in fruit	Succulent leaves with entire margin; leaf texture only "Smooth" type	It tolerates high temperature and grow well in hot weather condition
Spinach	*Spinacia oleracea L.*	Diploid (2n = 2x = 12)	Basically dioecious crop however, few monoecious plants with varying proportion of male and female flowers are also found; fruiting perianth enlarge, become hardened and often spiny	Lobed leaf margin; leaf texture may be "Savoy", "Semi-savoy" or "Smooth" type	It is purely a cool season crop and cannot tolerate high temperature; in warm weather and long day condition, it tend to flower quickly

Source: Sadhankumar *et al* (2015), Hazra *et al* (2022)

There are very few reports on the genetics and breeding of palak. The red or purple colour of stems and leaves is dominant over green and governed by two dominant genes with a segregation ratio of 9 purple and 7 green in F_2 population. The monogerm seed character is controlled by a single recessive gene with multigene was partially dominant perhaps with some modifier genes. The genic male sterility manifested in palak is controlled by two recessive genes with a segregation ratio of 15 male fertile and 1 male sterile in F_2 population. In the cross between garden beet and palak, both early bolting character and higher dry matter content is controlled by dominant genes (Swarup, 2006).

Breeding objectives

The followings are the main objectives of palak breeding

- Vigorous plant growth, quick growing, strong, succulent stem, larger, succulent leaves and late bolters
- Plants with uniform growth habit, medium height, preferably green coloured stems and leaves, high rejuvenating ability which responds to multiple cutting.
- Stable yield and wide adaptability to varying environments

- Good taste, flavour and nutritional profile that includes high β carotene, vitamin C, iron, calcium, potassium, magnesium, phosphorus, and low in oxalate and nitrate contents
- Resistance to important diseases *viz.*, damping off (*Pythium sp.*), leaf spot (*Cercospora beticola*), etc.
- Resistance to insect pests *viz.*, aphids (*Aphis* sp.), leaf eating caterpillar (*Laphygma exigua*), etc

Breeding methods and development of variety

The basic breeding methods employed in India for the development of palak include mass selection from the indigenous variability, hybridization followed by bulk method of selection, alteration of ploidy level and applied mutagenesis.

- **Improved varieties:** All Green, Pusa Palak, Pusa Jyoti, Pusa Harit, Kashi Baramasi, Jobner Green, HS-23, Palak No. 51-56, Banerjee's Giant, Ooty-1, Pant Composite, Arka Anupama, Pusa Bharati, Punjab Selection, Punjab Green

Germplasm resources and Plant introduction

In India, wide range of germplasm of palak is maintained at National Bureau of Plant Genetic Resources (NBPGR), New Delhi, ICAR-IARI, New Delhi, ICAR-IIVR, Varanasi, G. B. Pant University of Agriculture and Technology, Pantnagar, Punjab Agricultural University, Ludhiana, Chaudhary Charan Singh Haryana Agricultural University, Hisar, Chandra Shekhar Azad University of Agriculture & Technology, Kanpur and other places.

No exotic genetic materials have been utilized in breeding palak in India.

Variability and selection

Despite long history of cultivation of palak in India, genetic variability observed in this species is not wide enough, commensurate with appreciable percentage of outcrossing. In India, earlier improvement through natural selection by the farmers might have been in the direction of larger succulent leaves, late bolters and ability for regeneration after repeated cuttings. Palak bears hermaphrodite flowers and is basically a cross-pollinating species due to wind pollination. However, considerable self-pollination also occurs due to long anthesis and anther dehiscence time, open inflorescence like a large spike, which sways in wind, prolonged stigma receptivity and conspicuous production of very fine pollens, which drift in air for long time. Since palak shows inbreeding depression hence, much selfing should be avoided and instead sib-mating will be more effective. One variety, "All Green" has so far been developed by inbred selection from indigenous material. Generally, mass selection is adopted for the improvement of the heterozygous and heterogeneous population and most of the varieties have been developed following this method. *viz.*, Kashi Baramasi, HS-23, Pusa Bharati, Punjab Green, etc. One variety, "Pant Composite" was developed through mass selection and subsequent compositing of the finally selected lines.

Table 2: Varieties developed through different methods of selection

Name of variety	Indigenous / Exotic material
Mass selection	
Kashi Baramasi	Indigenous
HS-23	Indigenous
Palak No. 51-56	Indigenous
Ooty-1	Indigenous (local type of Himachal Pradesh)
Pant Composite	Mass selection and subsequent composite of the finally selected lines
Pusa Bharati	Indigenous; Improvement over Pusa Jyoti through selection
Punjab Selection	Indigenous
Punjab Green	Indigenous
Inbred selection	
All Green	Indigenous

Hybridization and selection

Palak generally shows inbreeding depression hence, much selfing should be avoided and instead sib-mating will be more effective. Hence, pedigree method of breeding may be less effective for selection of desirable segregates in the advanced generations of the hybrids. However, one variety, Arka Anupama was developed by pedigree method of selection from the cross between two lines of palak, IIHR 10 × IIHR8. Generally, hybridization followed by bulk method of selection would be more effective. Hybridization between palak x Swiss chard, palak x garden beet, palak x sugar beet followed by selection in the advanced generations have been successfully adopted to develop improved varieties having late bolting habit, wider adaptability to varying climate and better rejuvenation ability *viz*., Pusa Palak (Swiss chard × palak), Pusa Harit (Sugar beet x palak) and Banerjee's Giant (palak x garden beet).

Table 3: Varieties bred through hybridization and selection

Name of variety	Parentage
Pusa Palak	Swiss chard (*Beta vulgaris* var. *cicla*) × local palak cultivar and subsequent advanced generation selections by pedigree method
Pusa Harit	Sugar beet x local palak and subsequent advanced generation selections by pedigree method
Banerjee's Giant	Hybridization of Local palak x garden beet followed by subsequent selections in advanced generations
Arka Anupama	IIHR 10 × IIHR8 followed by pedigree method of selection

Utilization of mutants and Mutation breeding

In palak breeding, induced mutagenesis by gamma radiation and chemical mutagens has been found effective in isolating promising high yielding mutants having late bolting habit, broad leaves and enhanced nutrient contents. However, no variety has so far been developed through applied mutagenesis. Only one variety "Jobner Green" was developed through isolation of spontaneous mutant from the population of a local collection (SI. No.5).

Table 4: Varieties developed through applied mutagenesis

Variety	Applied mutagenesis
Jobner Green	Isolation of spontaneous mutant from the population of a local collection (SI. No.5)

Alteration of ploidy level

Induction of polyploidy has been proved to be a handy tool in tampering with the genotype of the organisms and in widening the amplitude of variation in the gene pool. In the leafy vegetables, polyploidy increases size of the leaves and dry matter as an important economic value. Delay in the flowering time in the polyploid plants increase the fresh weight, dry weight, and β carotene and ascorbic acid contents. Induction of polyploidy is achieved through the traditional method of increasing the ploidy levels. Generally, un-germinated seeds are immersed in the aqueous solutions of different mitotic inhibitor *viz.*, colchicine, trifluralin, oryzalin, etc. However, induction of polyploidy through the application of colchicine is convenient and successful in many herbaceous plants including leaf amaranth, spinach and palak.

The induced tetraploids had higher number and bigger leaves with prolonged vegetative phase. However, only one tetraploid variety, Pusa Jyoti has so far been developed through polyplodization of the variety "All Green" by the application of 2% colchicine solution for 24 hours

Table 5: Varieties developed through alteration of ploidy level

Variety	Alteration of ploidy
Pusa Jyoti	Tetraploid of the variety, All Green by applying 2% colchicine solution for 24 hours

Breeding for quality

Palak or spinach beet is unique leafy vegetable due to good and cheap source of β carotene, ascorbic acid, thiamine, riboflavin, pyridoxine and minerals like, calcium, iron, phosphorus, magnesium, potassium and zinc and dietary fibre. Enhancement of nutrients particularly, β carotene, ascorbic acid, methionine and minerals and at the same time lowering of anti-nutrient compounds, particularly, oxalic acid and nitrate contents should be taken into consideration during breeding palak for leaf yield and quality.

References

Campbell, G.K.G. 1976. Sugar beet. In: *Evolution of Crop plants* (Ed. Simmonds, N.W.), Longman, London

Hazra, P., Banerjee, M.K. and Chattopadhyay, A. 2022. *Varieties of Vegetable crops in India*, Kalyani Publishers, New Delhi, Ludhiana, 204p.

Sadhankumar, P.G., Nirmaladevi, S. and Peter, K.V. 201. Tropical leafy vegetables. In: *Hand book of Vegetables* (Eds. Peter, K. V. and Hazra, P.), Studium Press LLC, Houston, Texas, USA, pp. 641-655.

Swarup, V. 2006. *Vegetable Science and Technology in India.* Kalyani Publishers, New Delhi, Ludhiana, 656p.

34

Okra

Okra, *Abelmoschus esculentus* under the family Malvaceae (Chromosome No.: 2n = 126, 130, 132, 144) is an important vegetable in the tropical and sub-tropical parts of the world. Okra is mainly used for its tender green fruits as vegetable in many countries. In India, tender fruits are used as fried and boiled or in curry and sambar. It is also frozen, dehydrated and canned. Tender fruit contains appreciably high potassium, vitamin B, vitamin C, folic acid and calcium. It is a popular vegetable crop of low in calories and high in dietary fibre and protein contents, rich in essential amino acids, lysine and tryptophan.

Origin, domestication and adaptation in India

Okra has been regarded as of African origin. Perhaps it has been originated in Ethiopia and upper Nile region of Sudan. However, according to Zeven and Zhukovsky (1975), the genus *Abelmoschus* is considered to be of Asiatic origin. In fact, its cytotaxonomy confused that *Abelmoschus* might be of Asiatic origin and at least part of the cultigen. Joshi and Hardas (1976) believed that either the cultigen might have originated in Asia or it was present in Africa and India as a polyphyletic species. Significant native diversity of okra and its allied species in India points towards its secondary diversity here. Indian geneticists and cytologists reported that the wild relative *A. tuberculatus* which grows wild in India is one of the ancestors of cultivated okra, *A. esculentus* which is considered to be a natural amphidiploid species. Okra has been mentioned in ancient Sanskrit literature in the name "tindisha" and "gandhamula" which indicated very old association of this genus and its different species with India.

Crop botany and floral biology

- Okra is a tall, annual plant with thick, stiff upright stem; plant shows indeterminate growth habit; stem is robust, variable in height and branching depending on the variety; height varies widely between 50 and 400 cm depending on the genotype however, most of the varieties generally reach 180- 200 cm height.
- Leaves are alternate and can appear very different at different stages of the plant growth; very young plants have rounded leaves with serrated edges, they look more like small saucers and as they mature, they become larger and many lobed sometimes as few as three lobes other usually palmately five-lobed.
- In the early stages, leaves have a large surface area and slightly lighter veins, often slightly yellow and serrated edges and as the plant become taller, the

leaves become more lobed with much less leaf area however, in all the stages, the leaves are covered with fine bristles or spines that make them coarse to touch.

- Flowering is continuous and produced individually at a leaf node near the top of the stem; flower bud appears in the axil of each leaf above 6th to 8th leaf depending upon the cultivar; the crown of the stem at this time bears 3-4 underdeveloped flowers but later on, during the period of profuse flowering of the plant there may be as many as 10 undeveloped flowers on a single crown.
- As the stem elongates, the lower most flower buds open into flowers; there may be a period of 2, 3 or more days between the time of development of each flower but never does more than one flower appear on a single stem.
- Flower is axillary and solitary, 3.75-8.75 cm diameter, with five yellow to pale yellow petals arranged in a cone.
- Flowers are solitary with about 2.0 cm long peduncle, actinomorphic (regular), hypogynous, bisexual, self-pollinating and bracteates.
- Flower bud takes about 22-26 days from initiation to full bloom; flowers open only once in the morning and close after pollination on the same day and the following morning the corolla withers.
- Sepals subtended by an involucre of connate bracts; epicalyx 10, narrow, hairy bracteoles, deciduous and fall before fruit reaches maturity; the cone shaped calyx sheath cover the bud at the early stage before opening; calyx split longitudinally as flower opens.
- Flowers are showy, as much as 2.0 inches across, resembles the cotton flower, with its wide corolla usually made up of five petals; 5-7 cm long, free from one another but adnate to staminal column at the base, twisted, yellow with crimson spot on claw.
- The erect style consists of a 5-9 parts, each part with a capitate and burgundy coloured stigma, surrounded by the staminal column which is united to the base of the petals with numerous stamens.
- Stamens are many, epipetalous, monoadelphous i.e., united by the filaments into a single bundle, anther one celled i.e., one lobed, reniform, transversely attached to the filament; filaments are free and branched; pollens garins are large with spiny exine.
- Ovary superior multi-chambered (5-7), carpel many (usually 5-7), united, stigma sticky; ovule many in each chamber, placentation axile.
- The fruit is a capsule, oblong and endocarpic - schizocarpic; fruit grows quickly after flowering; the greatest increase in fruit length and diameter occurs during 4th to 6th day after pollination and at this stage, fruit is most often plucked; fruit colour may be green, light green and even purple containing numerous white seeds; as the fruits ripe, seed colour changes to brownish-black.

Reproductive behaviour

Okra is mainly self-pollinated crop but better termed as often-cross pollinated crop considering comparatively high of 8-12% natural cross pollination. Cross pollination is mainly accomplished by the honey bees which frequently visit the flowers due to the presence mucilage, sticky pollens and showy petals of the flower. Anthesis starts from 7.00 am, continues up to 10.30 am. Dehiscence of anthers is transverse and occurs 15 to 20 minutes after anthesis and it completes within 5-10 minutes liberating somewhat heavy and sticky pollens. Pollen fertility is the maximum between an hour before and after opening of the flower. Pollens remain viable for 30 days at 25°C temperature and 50% relative humidity condition. Stigma is receptive at the opening of the flower hence, cross-pollination at bud stage is not successful. For manual cross pollination, buds which will open next day morning should be collected in the afternoon hours and kept over-night wrapped with wet cloth. Emasculated flowers are cross-pollinated in the next day morning by gently rubbing the staminal colomn of the flowers of intended male parent on the stigma head.

Genetics of important characters

The cultivated okra, *Abelmoschus esculentus* has chromosome number, 2n ranging from 66 to 144. In India, the cytological studies indicated a series of polyploids with x = 12 and chromosome numbers, 2n = 72, 108, 120, 132 and 144. However, the most commonly reported chromosome number of cultivated okra is 2n = 120-130. Most of the cytological studies suggest that okra is a natural amphidiploid. The cytological studies undertaken in India and Japan have indicated that cultivated okra (*Abelmoschus esculentus*), 2n =130 has been evolved through hybridization between *A. tuberculatus* (n=29) and another species (n=36) which is still unknown, followed by doubling of the chromosome number. It is quite certain that *A. tuberculatus* (n=29, 2n = 58) is one of the progenitor species which took part in the natural hybridization process during evolution of cultivated okra.

Wild relatives of okra give a wide diversity in okra germplasm. The number of species recognized varies from 6 to 15. The genus has generally high chromosome numbers, and includes several allopolyploids. There is considerable variation in chromosome counts from single species, and it is not clear whether this is the result of the difficulty in counting of chromosomes, polymorphism in chromosome count, or the presence of cryptic species (Narkhede *et al.*, 2015). Chromosome number of other *Abelmoschus* species are given below.

Table 1: Chromosome number of different *Abelmoschus* species

***Abelmoschus* species**	**Chromosome number**
Abelmoschus esculentus	66, 72, 108, 118, 120, 122, 124, 130, 131, 132, 134, 144
Abelmoschus manihot ssp. *manihot*	60, 66, 68
Abelmoschus manihot ssp. *tetraphyllus* var. *tetraphullus*	130
Abelmoschus manihot ssp. *tetraphyllus* var. *pungens*	138

Abelmoschus moschatus	72
Abelmoschus ficulneus	72, 78
Abelmoschus angulosus	56
Abelmoschus sagittifolius	36
Abelmoschus tuberculatus	58
Abelmoschus crinitus	138
Abelmoschus cailli	185-199

Source: Sureshbabu (2012)

Table. 2: Brief information regarding the 12 wild relatives of okra

Wild *Abelmoschus* sp.	Brief information regarding the species
Abelmoschus angulosus	It is a wild species recorded from Sri Lanka, India (Gujrat, Kerala, Maharshtra and Tamil Nadu), Pakistan, Cambodia, Laos, Vietnam, Indonesia. Hooker's Flora of British India describes two varieties – var. *grandifloras* (with scabrous (not hispid) petioles, yellow petals, and sub-pyriform seeds) and var. *purpureus* (with rigid spreading hairs on the petiole, purple petals and globose seeds. This species shows resistance to YVMV and powdery mildew consistently.
Abelmoschus tuberculatus	It is a wild species from India, morphologically similar to, and long confused with *Abelmoschus esculentus*. However, it has a different karyotype (2n = 58 versus 2n =130), and hybrid between it and *Abelmoschus esculentus* are sterile. Morphologically, it can be distinguished by the small and tuberculate fruits (warty fruit).
Abelmoschus crinitus	It is a wild species from southern China (Guangxi, Guizhou, Hainan, Yunnan), India, Nepal, Myanmar, Thailand and Vietnam. It is a perennial herb, growing between 0.5 to 2.0 m in height. The root is fusiform. The plant has an indumentum composed of 5-6 mm long yellow hairs and a minute grey puberulence (fine hair). The corolla is around 10 to 13 cm diameter, and is yellow with a purple centre. The staminal column is around 2.0 cm long, with scattered anthers. The style is branched, with 5 arms, each with a flat stigma. The chromosome number of is 2n=138.
Abelmoschus tetraphyllus	This species is sometimes classified as a subspecies of *Abelmoschus manihot*. However, it has a different cytotype (2n=138 versus 2n=68). There are two subspecies of *Abelmoschus manihot* var. *tetraphyllus*: *Abelmoscus manihot* var. *tetraphyllus*: Chromosome number: 2n=130; originated from Southern Asia (Including India, Myanmar and Thialand) and southwest pacific (including Philippines, Moluccas and Papua New Guinea); *Abelmoschus manihot* ssp *tetraphyllus* var. *pungens*: Chromosome number: 2n=138; originated from Indonesia, Philippines, Christmas Islands, Japan, China, India, Nepal, the Indo-Chinese peninsula and Myanmar.
Abelmoschus manihot	It is native to India, Nepal and southern China. It is an annual or perennial herb growing 1 to 2 m. The corolla is yellow with a purple centre, and about 12 cm in diameter. It is sometimes grown as an ornamental; fruit is 8-ridged; Chromosome number: 2n = 68

Wild *Abelmoschus* sp.	Brief information regarding the species
Abelmoschus moschatus	It is native to India, south east Asia and southern China. It is an annual or perennial herb growing 1 to 2 m. The corolla is yellow with purple centre, and about 7 to 12 cm in diameter. It is grown commercially for its musk scented seeds. It is also sometimes grown as an ornamental plant.
Abelmoschus ficulneus	It is found in the Indian subcontinent, Australia, Madagascar, and in the Sahel zone of Africa, including Niger and northen Nigeria. It is prickly annual herb, with palmately 3-5-lobed glabrous leaves. Flowers white in colour turning pink. Chromosome number: 2n =72.
Abelmoschus caillei	It is cultivated species from West Africa, where it is originated as an allopolyploid hybrid between *Abelmoschus manihot* and *Abelmoschus esculentus*. The same allopolyploid has been produced experimentally in Japan.
Abelmoschus muliensis	It is endemic to southwest Sichuan (a province of South-western China), where it occurs on grassy slopes between 1200 m to 2100 m altitude. It is a velutinous (velvety) herb reaching 1 m in height, with a stem about 0.5 cm in diameter, and most parts with a dense covering of yellow hairs. The foliage is alternate, stipulate and shortly petiolate. The stipules are narrow, and 1 to 1.5 cm long. The petiole is 1 to 4 cm long. The blades of the lower leaves are 6-9 cm in diameter, and have a rounded cordate base, and obtuse or acute apex and a broadly dentate margin. Those of the upper leaves are ovate-sagittate, 7-10 cm long and 5-7 cm wide, with a broadly dentate margin and an obtuse apex. Flowers are borne singly or in clusters in the leaf axils, and that the epicalyx is composed 12 filiform lobes which are 2-3 mm wide and 1.5-3 cm long
Abelmoschus sagittifolius	It is often classified as subspecies (*tubrosus*) of *Abelmoschus moschatus*. However, it has different karyotype (2n=36 versus 2n=72), and for this reason, it is considered as a separate species. It is found in southern China (Guangdong, Guangxi, Guizhou, Hainan, Yunnan), mainland south east Asia, extending to eastern India, and also in the Kimberely and Arnhem Land areas of northen Australia. In China it is found in pine forests, on grassy slopes, and in other habitats. It is also grown as an ornamental. It is perennial herb, growing to 40 to 100 cm, or rarely 200 cm, tall. Most vegetative parts are densely pubescent, and are sometimes also setose or stellate-hairy. The flowers are white, pale yellow or pink, and 4 or 5 cm in diameter. They are borne solitary in the laef axils, on slender but densely scabrous, 4 to 7 cm long petioles. The epicalyx composed of 6 to 12 filiform, sparsely hirsute, spreading or reflexed, bracteoles. These are around 15 mm long, and 1 to 1.7 mm wide. The petals are 3 to 4 cm long with an oblong shape.
Abelmoschus enbeepeegearensis	It is morphologically allied to *Abelmoschus crinitus* and *Abelmoschus moschatus*, but it is not freely crossable. It is found in some pocket of Western Ghat (Kerala, Karnataka and Tamil nadu). It is a rare species with distinct features.

Source: Siemonsma (1982); Narkhede *et al* (2015)

Simply inherited characters of okra include earliness, plant pigmentation, petiole and leaf characters, corolla and calyx colour. There are several polygenic characters *viz*., plant height, days to flower, days to marketable maturity of fruits, fruits per plant, fruit length, fruit weight, fruit diameter, ridges on fruit, intermodal length, seed weight, seeds per fruit, fruit yield, etc.

Table. 2: Inheritance of important qualitative characters in okra

Character	**Inheritance pattern**
Plant colour	Monogenic: Dark green dominant over light green and green recessive to greenish-red
Stem colour	Monogenic: Purple dominant over green
Leaf margin	Monogenic: Cut leaves dominant over lobed leaves
Yellow corolla	Monogenic: Yellow dominant over cream
Calyx pigmentation	Monogenic: Pigmentation dominant over green
Fruit shape	Digenic with epistatic effect: Angular dominant over roundish shape
Fruit colour	Both Monogenic and Digenic
Fruit hairness	Monogenic, incompletely dominant
Resistance to Yellow vein mosaic virus	Two recessive genes, *yv1* and *yv2* in *Abelmoschus esculentus* x *Abelmoschus esculentus* cross Two complementary dominant genes in *Abelmoschus esculentus* x *Abelmoschus esculentus* cross
	Two complementary dominant genes in *Abelmoschus esculentus* x *Abelmoschus manihot* ssp. *manihot* cross Single dominant gene in *Abelmoschus esculentus* x *Abelmoschus manihot* ssp. *manihot* cross
	Polygenic with additive effect

Source: Swarup (2006); Ali *et al*. (2000) ; Seth *et al* (2017)

As resistance to YVMV is a complex trait, it should be studied comprehensively utilizing an integrated approach like transcriptomics, proteomics and metabolomics (Mishra *et al*., 2017). With respect to the genetics of OELCV resistance, pattern of segregation in the F_2 generation of two cross combinations showed that the resistance to OELCV disease was controlled by two duplicate recessive genes, while generation mean analysis has shown the involvement of both additive and non-additive effects for the inheritance of this disease in okra.

Breeding objectives

Some of the important traits that have been the main focus for different purpose of okra are outlined below

- Short plant height with shorter length of internodes and more number of short nodes
- High and early pod yield and prolonged harvest
- Dark green or shining green, tender, thin, medium long smooth with 5-ridged fruits, fruits free from conspicuous hairs

- Fruits suitable for processing industry and export
- For dehydration purpose, fruits should have less fibre, less mucilage, high protein, vitamins, dry matter, and minerals, and high dehydration/rehydration ratio
- For canning and freezing purpose, the fruits should be small, tender, dark green with low dry matter and rich in nutrients
- For export market, the fruits should be 6-7 cm long, dark green, 5-ridged, smooth pod with pointed tip
- Resistant to important diseases *viz.*, Bhendi Yellow Vein Mosaic Virus (BYVMV), Okra Enation Leaf Curl Virus (OELCV), Cercospora leaf blight (*Cercospora abelmoschii*), Alternaria leaf spot (*Alternaria hibiscinum*), powdery mildew (*Erysiphe cichoracearum*), etc.
- Resistant to insect pests *viz.*, fruit and shoot borer (*Earias vittella*), jassids (*Amrasca biguttula biguttula*), etc. and root knot nematodes (*Meloidogyne* spp.)
- Tolerant to abiotic stresses (drought, salinity, high temperature).

Breeding methods and development of varieties

Okra breeding started in India through selection from indigenous material and some introductions. Since okra is an autogamous but often cross-pollinated crop (8-12% outcrossing), the breeding methods suitable for improving the self-pollinated crops can usefully be employed in this crop also. The most important methods are pure line selection, single plant selection and intervarietal or inter-specific hybridization followed by pedigree method or back-cross pedigree method of selection and mutation breeding.

The wild *Abelmoschus* genepool has many desirable traits including biotic stress resistances such as Bhendi Yellow Vein Mosaic Virus (BYVMV) and Okra Enation Leaf Curl Virus (OELCV) diseases and abiotic stress tolerances such as drought and high temperature (Suma *et al.*, 2023). During the last three decades concerted efforts were put forth to develop BYVMV resistant varieties through transfer of resistance gene from the wild species (*A. manihot* ssp. *manihot*, *A. tetraphyllus*) or some already developed resistant varieties. Interspecific hybridization in okra is of immense value to exploit species diversity in the genus *Abelmoschus* for developing male sterile lines and advanced breeding lines that possess resistance to BYVMV and OELCV diseases. Important varieties developed in India are described below (Hazra *et al*, 2022).

High-yielding varieties with good fruit quality

Pusa Makhmali, Pusa Sawani, CO-1, Gujarat Bhendi-1, Gujarat Bhendi-2, Gujarat Okra-3, Perkins Long Green (Harbhajan Bhindi), MDU-1, Azad Bhindi-1, Utkal Gaurav, Phule Kirti, Punjab No.13, Salkeerthi, Kiran

BYVMV resistant varieties

Hisar Naveen, Selection-2, Punjab-7, Punjab Padmini, Punjab 8 (EMS 8), Punjab Suhawani, Parbhani Kranti, Arka Anamika, ArkaAbhay, Pusa A-4, Varsha Uphar, Hisar Unnat, Azad Kranti, Kashi Saatdhari, Kashi Mohini (VRO-3), Kashi Pragati (VRO-6), Kashi Vibhuti (VRO-5), Kashi Lila (IIVR-11), Kashi Mangali (VRO-4), Kashi Kranti (VRO-22), Kashi Vardaan (VRO-25), Kashi Lalima (VROR-157), Kashi Chaman (VRO-109), Phule Utkarsha, Susthira, Aruna, NDO-10, Bidhan Sahebmukto

Germplasm resources and Plant introduction

Okra germplasm collections are maintained in Southern Regional Plant Introduction Station, USDA-ARS-SAA, Griffin, USA, AVRDC, Taiwan, Institut International de Research Scientifique pour le development in Afrique, Wageningen, The Netherlands, Genetiques et Amelior. des Plantes Tropicales, ORSTOM F-34032 Montpellier Cedex, France, National Plant Genetic Resources Laboratory, IPB/UPLB College 4031 Laguna, Philippines, Horticultural Research Section Agricultural Research Corporation Wad Medani, Sudan. Genetico e Prod. Sementi DIVRAPA, Universita di Torino Via Pietro Giuria 15,10126 Torino, Italy, Agril. Rehabilitation Programme, Dept. of Agric. & Livestock, Konedobu, Papua New Guinea, Mount Makulu Agric. Research Station, Chilanga, Zambia and other organizations/ countries.

In India, okra germplasm including cultivars and wild species is maintained at National Bureau of Plant genetic resources (NBPGR), New Delhi, ICAR-IIVR, Varanasi, ICAR-IARI, New Delhi, ICAR-IIHR, Bengaluru and other places.

Only one variety, Perkins Long Green introduced from USA which was later named as "Harbhajan Bhindi" has been released for cultivation in India.

Table 2: Exotic variety directly released for cultivation

Name of the variety	Country from where introduced
Perkins Long Green (Harbhajan Bhindi)	USA

Variability and selection

In cultivated okra, genetic diversity exists for number of ridges, extent of hairiness and pigmentation on plant parts including fruits, ratio between length and width (diameter) of fruits, attachment of fruits on plant, lobing and lamina surface of leaves and also branching and height of the plant. In India, maximum variability in plant type and fruit type is available in Eastern part (Odisha, West Bengal, Bihar) and North-Eastern part (Assam, Meghalaya, Sikkim) where land races and primitive types are still found which are generally poor yielder but possessing varying degree of tolerance to different stresses. In hilly areas (warmer parts), smooth fruited variety may develop hard hairs and sometimes non-pigmented ones develop partial pigmentation.

A. esculentus forms are adapted to different agro-climates. Plant and fruit types from different countries may look different but these are all *A. esculentus*, barring few. In West Africa, Soudanian (*A. esculentus*) and Guinnean (*A.caillei*) forms have got established. In The Philippines, Sounde (extra early) and Gbogboligbo types (both *A. esculentus*) have got established. In Papua New Guinnea, Fiji and Solomon Islands forms of *A. manihot*, both flowering types (Aidiba) as well as non-flowering types (Aibika) have got established as leafy vegetables suitable for rejuvenation after multiple cuttings and perpetuated through easily rooted stem cuttings. Presence of some form of *A.esculentus* having round non-lobed leaves, short fruits and probably with 2n =72 in Saurashtra and Dwarka region of India and some part of Japan indicate genetic variability within *A.esculentus* itself.

Magnitude of genetic variability has been found to be high for number of fruits per plant, plant height, onset of flowering, number of nodes, stem diameter, leaf colour and leaf shape. Complex characteristics such as yield must be related to many individually distinguishable characteristics. Yield of okra have a positive and significant correlation with plant height, number of branches/plant, stem diameter, leaves/plant, number of fruits/plant, fruit weight and fruit length. Total seed yield is correlated with number of pods per plant, seeds per pod and seed size. High heritability of these yield components makes it possible to select these characters from the very early generation onwards for breeding high yielding genotypes. High correlation is found in case of colour characteristics of stem, leaves (vein base, petiole) and fruits which suggested that anthocyanin (red colour) tends to be produced in various plant parts at the same time.

Okra is mainly self-pollinated crop but better termed as often-cross pollinated crop considering comparatively high of 8-12% natural cross pollination. However, much variability in okra cultivars was not recorded in India. Some of the old local cultivars enlisted in the catalogue of seed companies are Lucknow Dwarf, Long Green, Velvet Green, Smooth Velvet, Red Wonder, Samalkota, Satdhari, Panchdhari, etc.

Selection methods for self-pollinated crops *viz*., single plant selection and pure line selection are applicable for okra. Maximum number of varieties have been developed in India through selection from the variability available in indigenous germplasm (Hazra *et al*, 2022).

Table 3: Varieties developed through selection

Name of variety	Indigenous / Exotic material
Pusa Makhmali	Indigenous; pure line selection
CO-1	Indigenous, Red Wonder collected from Hyderabad; single plant selection
Gujarat Bhendi-1	Indigenous; pure line selection
Gujarat Bhendi-2	Indigenous; pure line selection
Azad Bhindi-1	Indigenous; single plant selection
Phule Kirti	Indigenous; single plant selection
Punjab No.13	Indigenous; single plant selection
Salkeerthi	Indigenous; single plant selection
Kiran	Indigenous; single plant selection
Hisar Naveen	Indigenous; single plant selection (moderately tolerant to BYVMV)
Punjab Suhawani	Indigenous; single plant selection (moderately tolerant to BYVMV)
Kashi Mangali	Indigenous; pure line selection (moderately tolerant to BYVMV)
Kashi Vardaan	Indigenous; single plant selection (moderately tolerant to BYVMV)
Kashi Lalima	Indigenous; single plant selection (red fruit)

Name of variety	Indigenous / Exotic material
Kashi Chaman	Indigenous; single plant selection (moderately tolerant to BYVMV)
Susthira	Indigenous; single plant selection (moderately tolerant to BYVMV)
Aruna	Indigenous; single plant selection (moderately tolerant to BYVMV)
NDO-10	Indigenous; single plant selection (moderately tolerant to BYVMV)

Hybridization and selection

Inter-varietal hybridization followed by pedigree or back cross-pedigree method of selection in segregating generations are followed for the development of good number of high yielding varieties having tolerance to BYVMV.

Table 4: Varieties bred through hybridization followed by selection

Name of variety	Parentage
Pusa Sawani	Pusa Makhmali × IC1541 (showed resistance to BYVMV earlier)
Gujarat Okra-3	HRB 55 × Arka Anamika (tolerance to BYVMV)
Utkal Gaurav	Pusa Sawani × Goudiapada Local (moderate tolerance to BYVMV)
Selection-2	(Pusa Sawani × Best one) x (Pusa Swani × IC 7194) (moderate tolerance to BYVMV)
Pusa A-4	Inter-varietal cross (moderate tolerance to BYVMV)
Varsha Uphar	Lam Selection-l × Parvani Kranti (moderate tolerance to BYVMV)
Hisar Unnat	Selection 2-2 × Parbhani Kranti (moderate tolerance to BYVMV)
Azad Kranti	Inter-varietal cross (moderate tolerance to BYVMV)
Kashi Saatdhari	(PK × IC111542) × IIVR-20 (resistant to YVMV and OLCV)
Kashi Mohini (VRO-3)	Inter-varietal cross (moderate tolerance to BYVMV and OLCV)
Kashi Kranti	VRO-6 × 67-10-02 (moderate tolerance to BYVMV)
Phule Utkarsha	Parbhani Kranti × Varsha Vphar (moderate tolerance to BYVMV)
Bidhan Sahebmukto	VRO-6 × Mukta (moderate tolerance to BYVMV)

Crop wild relatives and inter-specific hybridization for resistance breeding

Cultivation of okra is facing serious threats due to two major viral diseases, namely, Bhendi yellow vein mosaic virus (BYVMV) and Okra enation leaf curl virus (OELCV). Host-plant tolerance to the viruses is one of the most practical, economical and environment friendly strategies for reducing yield loss in okra. Undomesticated crop wild relatives are rich sources of genes providing resistance to various diseases, pests, and unfavourable environmental conditions. Wild species like *Abelmoschus manihot, A. manihot* ssp. *manihot*, *A. manihot* ssp. *tetraphyllus, A. tetraphyllus*, *A. tuberculatus, A. caillei, A. crinitus, A. angulosus var. grandiflorus A. pungens var. mizoramensis, A. vitifolius* and *A. panduraeformis* show resistance to BYVMV and OELCV.

Table 5: Inter-specific crossability of *Abelmoschus* species

Inter-specific hybrid	**Crossability status**
Abelmoschus caillei x Abelmoschus esculentus	Successful hybrid
Abelmoschus caillei x Abelmoschus manihot	Successful hybrid; The cross is only successful when *Abelmoschus caillei* is the seed parent
Abelmoschus esculentus x Abelmoschus angulosus	Successful hybrid
Abelmoschus esculentus x Abelmoschus tuberculatus	Crossable without the need for any intervention other than emasculation and hand pollination. The hybrids show hybrid vigour, but are sterile. A fertile allopolyploid, which retains hybrid vigour, has been created by colchicine treatment.
Abelmoschus esculentus x Abelmoschus manihot	Crossable. The hybrid shows hybrid vigour, but are partial sterile. A fertile allopolyploid has been produced.
Abelmoschus esculentus x Abelmoschus tetraphyllus	Crossable. An allopolyploid can be produced
Abelmoshus esculentus x Abelmoschus moschatus	These two species can be crossed, with difficulty
Abelmoschus esculentus x Abelmoschus tetraphyllus	The 72 chromosome race of *Abelmoscus esculentus* can be crossed with *Abelmoschus tetraphyllus.*
Abelmoschus ficulneus x Abelmoschus tuberculatus	Crossable but the hybrid is sterile. The experimental allopolyploid is also sterile.
Abelmoschus manihot x Abelmoschus tuberculatus	Crossable but the hybrid is sterile

Source: Siemonsaa (1981); Narkhede *et al.* (2015)

The wild relatives of okra particularly *Abelmoschus manihot* ssp. *manihot*, *Abelmoschus caillei* and *A. tetraphyllus* have been utilized in the breeding programmes for the transfer of BYVMV and OELCV resistant genes into the cultivated okra background in India, despite the occurrence of unilateral incompatibility (set fruits and seed only when cultivated okra used as female parent) and sterility problem in interspecific hybrids (Singh *et al.,* 2007).

With respect to OELCV, absence of characterized resistance sources in the cultivated background has forced the plant breeders to look beyond the cultivated gene pool into the wild relatives like *A. crinitus*, *A. ficulneus*, *A. angulosus* and *A. manihot* as stable and reliable sources of resistance to OELCV (Singh *et al.*, 2007)

Introgression of genetic materials from the wild species to the cultivated okra through wide hybridization and repeated back-crossing plays an important role in genetic modification and enriching the cultivated gene-pool with novel genetic variations. The wild *Abelmoschus* gene pool has many other desirable traits like short internodal length, more number of productive branches, extended bearing, perennation tendency, reduced fruit length (more consumer preferred trait), high mucilage content (medicinal value), abiotic stress tolerances such as drought, high temperature (Suma *et al.*, 2023).

Inter-specific hybridization and back cross method of breeding or inter-specific hybridization followed by back cross- pedigree method of selection in segregating generations have been followed for the development of good number of BYVMV resistant varieties in India.

Table 4: Resistant varieties developed through inter-specific hybridization

Variety	Inter-specific hybridization
Punjab-7	*A. esculentus* cv. Pusa Sawani × *A. manihot* ssp. *manihot* (resistant to BYVMV)
Punjab Padmini	*A. esculentus* cv. Reshmi × *A. manihot* ssp. *manihot* x a F_2 segregate of the cross *A. esculentus* cv. Pusa Sawani x *A. manihot* ssp. *manihot* followed by selection (resistant to BYVMV)
Parbhani Kranti	*A. esculentus* cv. Pusa Sawani × *A. manihot* ssp. *manihot*(resistant to BYVMV)
Arka Anamika	*Abelmoschus esculentus* × *A. manihot* ssp. *tetraphyllus* followed by back cross method of breeding (resistant to BYVMV)
Arka Abhay	*Abelmoschus esculentus* × *A. manihot* ssp. *tetraphyllus* followed by back cross method of breeding (resistant to BYVMV)
Kashi Pragati (VRO-6)	(NIC 9303 × PK 20) (resistant to BYVMV and OLCV)
Kashi Vibhuti (VRO-5)	(NIC 9303 × IC111547) × IIVR 20 (resistant to BYVMV and OLCV)
Kashi Lila	(NIC 9305 × HRB 9-2) × IIVR-50 (resistant to BYVMV and OLCV)

Utilization of mutants and Mutation breeding

Mutation breeding is also a viable alternative for breeding resistance to BYVMV disease. Phadvibulya *et al.* (2009) irradiated seeds of okra varieties, Annie and Okura, with Gamma rays at 400 and 600Gy. One M_4 plant of Okura (B-21) irradiated at 400Gy was found to be highly resistant to BYVMV disease. At Bidhan Chandra Krishi Viswavidyalaya, West Bengal, optimum gamma irradiation dose determined through the calculation of LD_{50} and GR_{50} was 350 Gy and 450 Gy (Hazra *et al.*, 2021). Two selected putative mutant families, namely, 350//10///3-9////28 and 450//66///2-4////39, isolated in the M_5 generation of Pusa Sawani showed consistent resistance against BYVMV disease. DNA fingerprinting further validated the alterations occurred in these two isolated mutants compared to the parental genotype (Hazra *et al.*, 2023). In earlier study, induced mutation using chemical mutagen, Ethyl methane sulfonate (EMS) has been successfully utilized at Punjab Agricultural University, Ludhiana to develop a BYVMV resistant improved variety, EMS-8. At Madurai, Tamil Nadu Agricultural University, a gamma radiation induced mutant variety resistant to BYVMV in the name, MDU-1 was developed. Another gamma radiation induced mutant variety suitable for processing, Parvani Tillu was developed at Maharashtra Agricultural University, Parvani.

Table 5: Varieties developed through applied mutagenesis

Variety/ Mutant	Applied mutagenesis / remarks
MDU-1	Gamma irradiated mutant of Pusa Sawani (resistant to BYVMV)
Punjab 8 (EMS 8)	EMS mutant of Pusa Sawani (resistant to BYVMV)
Line 350//10///3-9////28 and 450//66///2-4////39	Gamma irradiated mutant of Pusa Sawani (resistant to BYVMV)

Heterosis breeding

Hybrid vigour to the extent of 60 -70 % increased yield coupled with resistance to BYVMV has been commercially exploited in okra. Hand emasculation and pollination has become the commercial method of hybrid seed production because of high rate of crossing success although number of hybrid seeds per fruit by cross pollination are not many.

Hybridization technique

Emasculation

A plump bud that will open next day should be emasculated in the afternoon hours. The cone shaped calyx sheath covering the bud along with the closely packed overlapping corolla is cut with a sharp blade by a ring cut at the base of the bud. Utmost care should be taken so that the staminal column and style inside do not get injured. The cone shaped calyx sheath is pulled off upwardly. The petals are removed one by one with the forceps. Anthers are then removed from the staminal column by downward scraping from stigma head end with blunt scalpel or forceps. Stigma head is very sticky so care should be taken during scrapping so that no anthers remain adhered to the stigma head. Paper bagging protects the emasculated buds. All open and un emasculated flowers should be removed.

Pollen collection

Anthesis generally takes place between 7.30 to 10.30 a.m. and anthers dehisce after 15 to 20 minutes. Dehiscence is completed within 5-10 minutes liberating somewhat heavy and sticky pollens. Pollens are collected in the morning from the dehisced anthers of the previously bagged flowers by gently knocking with a blunt scalpel. Pollen production is higher in the anthers of apical region compared to those of the basal region. In another methods of manual cross pollination, bud which will open next day morning are collected in the afternoon hours and kept overnight wrapped with wet cloth. Emasculated flowers are cross pollinated by gently rubbing the staminal column of these flower collected in the previous day.

Pollen storage

Use of fresh pollens gives higher set of fruits and seeds. Pollens remain viable at 5°C temperature and 50% relative humidity condition for 30 days.

Pollination

Pollination is done within 8.00 to 11.00 a.m. in the morning by touching the stigma with a camel hair brush dipped in the pollen mass. For commercial hybridization purpose, plumpy and matured buds from the plants of the male parent which will open in the next morning are collected in the afternoon hours and kept overnight wrapped with moist cloth. In the morning, the cone shaped calyx sheath and petals are removed exposing the staminal column with mature and dehiscing anthers. The staminal columns of the flower buds are rubbed gently over the stigma of the emasculated flower of the female line to effect cross- pollination. Maximum pollens must be deposited on the stigma to get maximum hybrid seeds per fruit. A thread is tied around the peduncle to identify the crossed fruits. Pollination may continue for 40 to 50 days.

Male sterility in hybrid seed production

Use of male sterility would be very effective for commercial hybrid seed production. Induction of male sterility by use of chemical (Naveen *et al.*, 2017) and by irradiation (Phadvibulya *et al.*, 2009) has indicated its possibility in commercial hybrid seed production.

In okra, the male sterility is controlled by a single recessive gene (*ms1*) and is expressed in stable manner when the gene is present in homozygous condition (*ms1ms1*). Anthesis is normal in genic male sterile (GMS) plants but anther dehiscence is partial and microsporogenesis is normal up to the tetrad stage (Pitchaimuthu *et al.*, 2012). Under open field, alternate planting of 2 rows of male sterile plants and 1 row of fertile plants (2:1 planting ratio) can accommodate about 4166 male sterile plants/ha and produce 52.56% higher seed yield (5.66 q/ha of hybrid seed) as compared to fertile control. It has been found that for production of 1 kg of hybrid seed using male sterile line, it requires 3 hours to hand pollinate 274 flowers, whereas, by hand emasculation and hand pollination it took over 9 hours, around 70% saving in time and manual labour can be achieved.

Breeding for resistance to disease

Bhendi Yellow vein mosaic virus

Intensive efforts have been made in India to develop virus-resistant/tolerant varieties with the utilization of undomesticated wild relatives of *A. esculentus*. Inter-varietal hybridization utilizing one resistant variety has also been employed to develop resistant variety with good fruit quality. However, resistance to the begomoviruses is found unstable with frequent breakdowns. Appearance of new viral strains, high recombination rate of begomoviruses, emergence of the cryptic species (morphologically similar but genetically different) of the insect vector, *Bemisia tabaci* with its increased host range are the main hindrance in BYVMV resistant breeding programme (Chowda-Reddy *et al.*, 2012).

Hence, an intensive and systematic wide hybridization or pre-breeding programme is required to identify and transfer the genes for tolerance to BYVMV disease from the wild *Abelmoschus* species to the cultivated species. Several earlier attempts of wide hybridization in okra have been reported, which include *A. esculentus* x *A. manihot* ssp. *manihot*; *Abelmoschus esculentus* x *A. manihot* ssp. *tetraphyllus*, *A. esculentus* x *A. moschatus*; *A. esculentus* x *A. angulosus* and *A. esculentus* x *A. caillei*.

Presence of pre- and post-zygotic barriers to interspecific geneflow, differences in ploidy levels and genotype specific variations in chromosome numbers led to varying degrees of sterility in F_1 plants of interspecific crosses. Joseph *et al.* (2013) advocated colchicine treatment at the seedling stage for fertility restoration in the interspecific crosses of *A. esculentus* with *A. tuberculatus* and *A. tetraphyllus*. They also reported that *A. caillei* can be used as a potential bridging parent to transfer desirable genes in okra breeding programmes.

Rattan and Kumar (2020) developed the protocol for embryo rescue of interspecific hybrids involving *A. esculentus* × *A. tetraphyllus*. Badiger *et al.* (2023) elaborated the pollen germination, pollen–pistil interaction, and crossability studies in interspecific and induced colchiploid populations involving crosses between *A. esculentus* and wild species, *A. manihot* var. *tetraphyllus* and *A. moschatus*. Majority of the wide cross derivatives had an intermediate fruit morphology and dominance of wild characters *viz*., hispid fruits, stem, leaves, tough fruit fibre, vigorous perennial growth habit and prolonged flowering and fruiting.

Screening and identification of resistant wild species

↓

Development of interspecific hybrids
(Emasculation and pollination)

↓

Hybrid fruits harvested, seed extracted and stored
(Fertility restoration)
Seeds of F_1 sown in polybags

↓

Colchicine @ 0.1% applied to the apical meristem
of seedlings at 2 leaf stage

↓

Surviving seedlings develop tumour like growth followed
by new sprout developmentat colchicines treated site

↓

Amphidiploid generation advancement

↓

Generation advancement and backcrossing of amphidiploids
with the cultivated okra variety

Fig. 1: Flowchart of wide hybridization, *Abelmoschus* amphidiploids development, and generation advancement (Suma *et al.*, 2023)

About 27 begomoviruses were reported to infect okra of which, bhendi yellow vein mosaic virus disease (BYVMV) and okra enation leaf curl virus disease (OELCV) most severely affect its production in terms of yield and fruit quality (Sanwal *et al.*, 2014). Bhendi yellow vein mosaic virus (BYVMV) belongs to the genus Begomovirus and family Geminiviridae. This disease is caused by a complex consisting of the monopartite begomo virus BYVMV encapsulated within a typical geminated particle (Jose and Usha, 2003) and a small satellite DNA with additional molecules known as beta satellites and/or alpa satellites (Briddon and Stanley, 2006). The begomoviruses native to the New World have only bipartite genomes (having DNA-A and DNA-B components) whereas, of Old World have both bipartite and monopartite (have DNA-A homolog and lacks DNA-B) genomes (Brown *et al.*, 2012). Alpha satellites (DNA-A) are self-replicating (autonomous) circular ssDNA molecules, encodes factors required for the viral encapsidation, replication, suppression of host defence and play no role in symptom induction (Briddon *et al.*, 2004). Whereas, beta satellites (DNA-B) associated with the monopartite viruses, encodes factors essential for viral systemic movement, host range determination and induce typical disease symptoms in their original hosts (Venkataravanappa *et al.*, 2012).

The biochemical parameters also impart resistance to BYVMV disease in okra. Resistant cultivar contains lesser amount of soluble sugars, higher amount phenols, orthodihydroxy phenol, flavonols, ascorbic acid and total chlorophyll content and enhanced activity of enzymes like, peroxidase, polyphenol oxidase, phenylalanine ammonialyase (PAL) than the susceptible cultivars (Seth *et al.*, 2017). The wild parents resistant to BYVMV had maximum phenolics, peroxidase, polyphenol oxidase activity and lower reducing, non-reducing, total sugars and total nitrogen content compared to cultivated okra (Prabu *et al.*, 2009).

Enation leaf curl virus

Singh *et al.* (2007) reported that *A. esculentus* lines VRO-5, VRO-6, KS-410, VRO-4, 111-8-9 (Red), 131-8-1 (Red) and IIVR-10 were free from infection. Among wild taxa, infection was absent in *A. pungens*, *A. manihot*, *A. crinitus*, *A. ficulneus*, *A. moschatus* and *A. angulosus.* Nine accessions of *A. tetraphyllus* and eight accessions of *A. tuberculatus* expressed resistance (less than 10% infection) against okra enation leaf curl virus disease infection. Under West Bengal condition, among 35 okra genotypes, only 3 genotypes (285-1-3A-1-17-1, VRO-178 and 285-1-3-1-15-2-1) expressed appreciable OELCV disease tolerance during rainy season (Yadav *et al.*, 2018).

Devi *et al.* (2018) recorded two genotypes 10/OKYVRES-5 and 10/OKYVRES-10 which showed high tolerance against enation leaf curl virus disease over two consecutive years. Jamir *et al.* (2020) found that two genotypes IC111551 and IC433616 were moderately susceptible against ELCV disease up to 60 days after sowing with disease severity (15.87 PDI).

Inheritance pattern of resistance to OELCV disease is complex. The segregation pattern for OELCV disease in the F_2 generation of Tolerant × Susceptible crosses indicated existence of 2 genes with duplicate recessive epistasis (Yadav *et al.*, 2020; Devi *et al.*,

2020). Peroxidase activity and total phenol content in okra leaf exhibited consistent and significant negative correlation with the severity of OELCV disease over the growth stages suggesting their implication as biochemical 'markers' for identification of okra genotype resistant to this disease (Yadav *et al.*, 2020).

Fungal diseases

Singh *et al.* (2007) recorded complete resistance to powdery mildew (*Erysiphe cichoracearum*) in *A. tetraphyllus*, *A. moschatus*, *A. crinitus* and moderate resistance to *A. angulosus*, *A. pungens* and *A. manihot*. Under Rahuri, Maharashtra conditions, the wild species *A. caillei* and *A. moschatus* were found to be immune to powdery mildew (Prabu *et al.*, 2009). Vidhate *et al.* (2014) evaluated 20 lines of cultivated okra during rainy season in Parbhani, Maharashtra and 8 lines (PBN OK 4, PBN OK 5, PBN OK 6, PBN OK 8, PBN OK 9, AOH 23, Surya and NOH 100) showed moderate resistance to powdery mildew disease. Resistance to *Cercospora* leaf spot was also identified in wild species *viz.*, *A. crinitus*, *A. moschatus*, *A. pungens*, *A. manihot* and *A. angulosus* (Singh *et al.*, 2007).

Breeding resistance to insect-pests

Jassid

The wild relatives *A. moschatus*, *A. ficulneus* and *A. crinitus* showed resistance to jassid infestation (Singh *et al.*, 2007). Different lines of the wild species *A. moschatus* recorded least jassid (nymph) population per leaf (Prabu *et al.*, 2009). Halder *et al.* (2016) studied the physical and biochemical basis of resistance against *Amrasca bigutulla bigutulla.* They identified one tolerant genotype VROB-181 which had relatively higher number of trichomes on leaf lamina (11.85/cm^2), mid-rib (9.17cm) and vein (9.95 cm), mid-rib thickness (1.58 mm), leaf length (22.43 cm) and total phenol content in green leaves (75.04 mg/100 g).

White fly

In some lines of the wild taxa *viz*, *A. angulosus* and *A. moschatus* minimum mean whitefly (adult) population per leaf was recorded which could be utilized in the breeding programme to develop YVMV resistant okra variety.

Shoot and fruit borer

Singh *et al.* (2007) screened 119 lines of wild okra and observed that 16 accessions of *A. tuberculatus* recorded less than 10% fruit damage. It may be due to the presence of tubercles on the surface of the fruit wall and toughness of the fruit wall which does not permit the neonate larvae to feed and survive. Prabu *et al.* (2009) reported that 3 lines of *A. tuberculatus* were found to be immune to fruit borer infestation while 4 lines of *A. tetraphyllus* and *A. manihot* ssp. *tetraphyllus* were completely free from shoot borer infestation. Halder *et al.* (2015) screened 10 genotypes to study the relationship between physico-morphic and biochemical characters against *Earias vittella.* They identified one tolerant genotype SB 6 which had dense trichomes (51.8/cm^2), wider fruit angle (26.4°), thick stem (1.28 cm), fewer leaves (20/plant), fewer fruits (13.67/plant) and branches (1.4/plant) and higher total phenol content (34.78 mg/100 g).

Red Spider mite

Singh *et al.* (2007) reported that three accessions of wild species *A. caillei* (EC-305656, EC-305664, EC-305696) did not harbour any mite population exhibiting complete resistance.

Root-knot nematodes

Meloidogyne javanica and *M. incognita* are two important species of nematodes which infest okra and reduce its yield considerably. Unfortunately, no source of resistance against both the species is so far available. Five lines *viz.*, IC-111472, IC-128372, IC-169482, IC-411698 and IC-433686 exhibited moderately resistant reaction with minimum gall index under Odisha condition (Mohanta and Mohanty, 2012). Mukhtar *et al.* (2014) identified five cultivars *viz.*, Sanam, Dikshah, Arka Anamika, Ikra-1 and Ikra-2 and Kedarnath *et al.* (2017) identified two cultivars, Arka Abhay and Anoop as moderately resistant to *M. incognita.*

Source of resistance to *M. hapla* and *M. arenaria* was reported in the cultivar Clemson Spineless of *A. esculentus* (Sasser and Kirby, 1979) however, *M. arenaria* has not yet been reported in India. The species *M. hapla* has however, been found in the hills of North India.

Breeding for quality

Fruits should be dark, velvety-green, less fibrous and low in mucilage content. At the same time, fruits should have high dry matter content, being suitable for dehydration. Okra fruits are also highlighted for their contribution of phenolic compounds as potential anti-oxidants.

References

Ali, M., Hossain, M.Z. and Sarker, N.C. 2000. Inheritance of Yellow Vein Mosaic Virus (YVMV) tolerance in a cultivar of okra (*Abelmoschus esculentus* (L.) Moench). *Euphytica*, **111** : 205-209.

Badiger, M., Yadav, R.K., Sharma, B.B., Bhat, K.V., Tomar, B.S., Lata S.,Vinay N.D. and Das A. 2023. Pollen germination, pollen–pistil interaction and crossability studies in interspecific and induced colchiploid population of *Abelmoschus* species. *Springer Sci. Rev.*,**71**: 107–127.

Briddon, R.W. and Stanley, J. 2006. Subviral agents associated with plant single-stranded DNA viruses. *Virology*, **344** : 198-210.

Briddon, R.W., Bull, S.E., Amin, I., Mansoor, S., Bedford, I.D., Rishi, N., Siwatch, S.S., Zafar, M.Y., Abdel-Salam, A.M. and Markham, P.G. 2004. Diversity of DNA 1: a satellite-like molecule associated with monopartite begomovirus–DNA β complexes. *Virology*, **324** : 462-474.

Brown, J.K., Fauquet, C.M., Briddon, R.W., Zerbini, M., Moriones, E., and Navas-Castillo, J. 2012. Family Geminiviridae, In: *Virus Taxonomy* (Eds. King, A.M.Q. , Adams, M.J. , Carstens, E.B. and Lefkowitz, E.J.), Elsevier Academic Press, London, pp. 351-373.

Chowda-Reddy, R.V., Kirankumar, M., Seal, S.E., Muniyappa, V., Valand, G.B., Govindappa, M.R. and Colvin, J. 2012. *Bemisia tabaci* phylogenetic groups in India and the relative transmission efficacy of Tomato leaf curl Bangalore virus by an indigenous and an exotic population. *J. Integr. Agric.*, **11** : 235-248.

Devi, R., Kanitkar, K., Narendhar, R., Sehmi, K. and Subramaniam, K. 2020. A narrative review of the patient journey through the lens of non-communicable diseases in low- and middle-income countries. *Adv. Ther.*, **37**: 4808–4830.

Halder, J., Sanwal, S.K., Deb, D., Rai, A.B. and Singh, B. 2016. Mechanisms of physical and biochemical basis of resistance against leaf-hopper (*Amrasca biguttula biguttula*) in different okra (*Abelmoschus esculentus*) genotypes. *Indian J. Agric. Sci.*, **86**:481-484.

Halder, J., Sanwal, S.K., Rai, A.K., Rai, A.B. Singh, B. and Singh, B.K. 2015. Role of physico-morphic and biochemical characters of different okra genotypes in relation to population of okra shoot and fruit borer, *Earias vittella* (Noctuidae: Lepidoptera). *Indian J. Agric. Sci*, **85** : 278-282.

Hazra, S., Gorai, S., Kumar, V. U., Bhattacharya, S., Maji, A., Jambhulkar, S., Ali, N., and Chattopadhyay, A. 2021. Optimization of gamma radiation dose for induction of mutations in okra. *Intern. J. Veg. Sci.*, **27**: 574–584.

Hazra, S., Gorai, S., Roy, S., Bose, S., Hazra, P., Chattopadhyay, A., Nasimali, Md., Jambhulkar, S. and Maji, A. 2023. Isolation of YVMV resistant mutants of okra (*Abelmoschus esculentus* L.) through applied mutagenesis. *Plant Breeding*, 2023;1–14, DOI: 10.1111/pbr.13151

Jamir, I., Mandal, A.K., Devi, A.P., Bhattacharjee, T., Maurya, P.K., Dutta, S., Chattopadhyay, A., Pramanik, K. and Banik, S. 2020. Screening of genotypes against viral diseases and assessment of yield loss due to yellow vein mosaic virus in okra grown in the eastern part of India. *Indian Phytopat.*, **73**: 125–133.

Jose, J. and Usha, R. 2003. Bhendi Yellow Vein Mosaic Disease in India is caused by association of a DNA β Satellite with a Begomovirus. *Virology*, **205**;310-317.

Joseph, J.K., Nissar, V.A.M., Latha, M., Shrikant, S., Patil, P., Malik, S.K., Negi, K.S., Keisham, M., Rao, S.R., Yadav, S.R. and Bhat, K.V. 2013. Genetic resources and crossability relationship among various species of *Abelmoschus*. *Curr. Hortic.*, **1**; 35-46.

Joshi, A.B. and Hardas, M.W. 1976. Okra. In: *Evolution of crop plants* (Ed. Simmonds, N.W.), Longman, London, pp. 184-195.

Kedarnath, Ravichandra, N.G., Preethi, D.M., Reddy, B.M.R., Pavithra, B.S. and Pavithra, R.S. 2017. Screening of okra (*Abelmoschus esculentus*) cultivars for resistance against root knot nematode (*Meloidogyne incognita*) under field condition in Karnataka, India. *Int. J. Curr. Microbiol. App. Sci.*, **6** : 3420-3426.

Mishra, G. P., Singh, B., Seth, T., Singh, A. K., Halder, J., Krishnan, N., Tiwari, S. K. and Singh, P. M. 2017. Biotechnological advancements and begomovirus management in okra (*Abelmoschus esculentus* L.): status and perspectives. *Front. Plant Sci.*, **8**(360). doi: 10.3389/fpls.2017.00360

Mohanta, S. and Mohanty, K.C. 2012. Screening of okra germplasms/varieties for resistance against *Meloidogyne incognita*. *J. Plant Prot. Environ.*, **9** : 66-68.

Mukhtar, T., Hussain, M.A., Kayani, M.Z. and Aslam, M.N. 2014. Evaluation of resistance to root-knot nematodes (*Meloidogyne incognita*) in okra cultivars. *Crop Prot.*, **56** : 25-30.

Narkhede, G.W., Deshmukh, S.B. and Shinde, S.M. 2015. Wild relatives of Okra (*Abelmoschus* spp.) – An Overview. *Eco. Env. & Cons.* **21**: 301-306.

Naveen, N., Pallavi, H.M., Satish, D., Cholin, S.S., Ganiger, V.M. and Babu, A.G. 2017. Microsporogenesis manipulation by cha's to eliminate emasculation in hybrid seed production of okra (*Albemoschus esculemtus* L.). *J. Exp. Biol. Agric. Sci.*, **5** : 623-630.

Phadvibulya, V., Boonsirichai, K., Adthalungrong, A. and Srithongchai, W. 2009. *Induced Plant Mutations in the Genomics Era* (Ed. Shu, Q.Y.), Food and Agriculture Organization of the United Nations, Rome, pp. 349-351.

Pitchaimuthu, M., Dutta, O.P. and Swamy, K.R.M. 2012. Studies on inheritance of geneic male sterility (GMS) and hybrid seed production in okra [*Abelmoschus esculentus* (L.) Moench.]. *J. Hort. Sci*., **7**: 199-202.

Prabu, T., Warade, S.D., Saidi, M. and Baheti, H.S. 2009. Screening wild and cultivated okra species for resistance to important pests. *Indian J. Plant Protec*., **37** : 87-91.

Rattan, P. and Kumar, S. 2020. Development of interspecific hybrids (*Abelmoschus esculentus* × *A. tetraphyllus*) in okra using embryo rescue approach. *Biosci. Biotechnol. Res. Asia*, **17**:517-523.

Sanwal, S.K., Singh, M., Singh, B. and Naik, P.S. 2014 Resistance to yellow vein mosaic virus and okra enation leaf curl virus: Challenges and future strategies. *Curr. Sci*., **106** : 1470-1471.

Sasser, J.M. and Kirby, M.F. 1979. *Crop Cultivars Resistant to Root-knot Nematodes Meloidogyne species*, Raligh, North Carolina, USA.

Seth, T., Chattopadhyay, A., Dutta, S., Hazra, P. and Singh, B. 2017. Genetic control of yellow vein mosaic virus disease in okra and its relationship with biochemical parameters. *Euphytica*, **213** : 1- 13.

Siemonsma, J. 1982. West African Okra – Morphological and cytogenetical Indications for the Existence of a Natural Amphiploid of *Abelmoschus esculentus* (L.) Monech and *Abelmoschus manihot* (L). Medikus. *Euphytica*, **31** : 242-252

Singh, B., Rai, M., Kalloo, G., Satpathy, S., and Pandey, K. K. 2007. Wild taxa of okra (*Abelmoschus* species): reservoir of genes for resistance to biotic stresses. *Acta Hort*., **752** : 323-328.

Suma, A., Joseph, J.K., Bhat, V.K., Latha, M., Jayasundaran, L.C., Pitchaimuthu, M., Nissar, V., Thirumalaisamy, P.P., Pandey, C.D., Pandey, S., Kumar, A., Gautam, R.K. and Singh, G.P. 2023.Genetic enhancement of okra [Abelmoschus esculentus (L.) Moench] germplasm through wide hybridization. *Front. Plant Sci*., **3**:1284070

Sureshbabu, K.V. 2012. Okra. In: *Hand book of Vegetables* (Eds. Peter, K. V. and Hazra, P.), Studium Press LLC, Houston, Texas, USA, pp. 333-353

Swarup, V. 2006. *Vegetable Science and Technology in India*. Kalyani Publishers, Ludhiana, 656p.

Venkataravanappa, V., Reddy, C.N.L., Jalali, S. and Reddy, M.K. 2012. Molecular characterization of distinct bipartite begomovirus infecting bhendi (L.) in India. *Virus Genes*, **44** : 522-535.

Vidhate, A.B., Badgujar, S.L. and Wagh, S.S. 2014. Screening of elite material against powdery mildew disease of okra under field conditions. *Trends Biosci*., **7** : 3010-3012.

Yadav, Y., Maurya, P. K., and Bhattacharjee, T., Banerjee, S., Dutta, S., Mandal, A.K., Chattopadhyay, A. and Hazra, P. 2020. Inheritance pattern of okra enation leaf curl disease among cultivated species and its relationship with biochemical parameters. *J. Genet*. **99**: 1–12.

Yadav, Y., Maurya, P.K., Banerjee, S., Bhattacharjee, T., Jamir, I., Chatterjee, S., Mandal, A.K., Dutta, S. and Chattopadhyay, A. 2018. Enation leaf curl virus (ELCV): A real threat in major okra production belts of India. *Int. J. Veg. Sci*., DOI: 10.1080/19315260.2018.1540025

Zeven, A.C. and Zhukovsky, P.M. 1975. *Dictionary of Cultivated plants and their Centres of diversity: excluding ornamentals, forest trees and lower plants*. Centre for Agricultural Publication and Documentation, Wageningen, The Netherlands.

35

Muskmelon

Muskmelon, *Cucumis melo* under the family Cucurbitaceae (Chromosome number: $2n = 2\times = 24$) is a highly relished crop because of its attractive flavour, sweet taste and refreshing effect. It is an important crop of Africa, Middle East, Asia, Southeast Asia, Japan, USA, South and Central America, France, Spain, Israel and many eastern European countries. Mature fruits are eaten fresh as a dessert fruit, canned or used for syrup or jam. Muskmelon is quite rich in ß carotene, vitamin B9, vitamin C and minerals like potassium and low in calorie content.

Origin, domestication and adaptation in India

The genus *Cucumis*, which includes cucumber (*Cucumis sativus*) and melon (*Cucumis melo*), has numerous wild African species, and it has therefore been assumed that melon originated in tropical Africa in eastern region, south of Sahara Desert. This proposition is supported by the presence of true wild melons which can still be found in some African countries. Some botanists consider melons native to the Levant and Egypt, while others place their origin in Iran, India, Afghanistan or Central Asia. Sebastian *et al* (2010) reported that the wild progenitor of *C. melo* occurs in India, and they confirmed that the Southeast Asian *Cucumis hystrix* is the closest relative of cucumber and most surprisingly, the closest relative of melon, *Cucumis picrocarpus* was from Australia. The wild forms of muskmelon reported from India are feral escapes derived from local cultivars.

Melon was cultivated in Egypt during 2400 BC and in Greece during 300 BC. It was introduced into the USA by Columbus in 1494. Now, it is grown in both Old and New World. Different cultivars of muskmelon were introduced in India probably during Mughal invasion around 14th century from central Asian region. Since then, it has spread to different parts of the country as far down to Andhra Pradesh and Karnataka.

Crop botany and floral biology

- Plants are prostrate, highly branched with unbranched tendrils; stem is softly hairy, striate or angled, leaves are orbicular to ovate to reniform, usually five angled, sometimes shallowly three to seven – lobed, hairy or somewhat scrabrous, 8-12 cm across.
- It is a polymorphic species where most cultivars grown particularly in India show andromonoecious sex form (staminate and bisexual flowers on the same plant). The hermaphrodite flowers are borne at the first one or two nodes of

the branches and one or more staminate (male) flowers at all other nodes; Less common sex form is monoecious, with pistillate flowers being produced instead of hermaphroditic flowers.

- Gynoecious and hermaphrodite sex form are rarely found in some recessive mutants, producing perfect flowers or pistillate flowers at every node.
- The staminate flowers are clustered; pistillate or bisexual flowers are solitary on short stout pedicels.
- In monoecious sex form, staminate flowers are clustered and borne on long pedicels, while the pistillate flowers are usually solitary with comparatively stout, short pedicels.
- Sepals are 5-lobed, 6-8 mm in length, forming a perianth tube; calyx alternate with corolla lobes.
- Petals are 5, rotate, deeply parted and bright, lemon-yellow to deep orange in colour, small, approximately 3 cm in diameter.
- Stamens are usually 3, two stamens having two lobes (thecae) and one unilocular.
- Carpels are 1-5, usually 3, syncarpous, ovary inferior, thick short style terminated by three bilobed or divided papillate stigma.
- Fruits are pepo and widely divergent in sizes, shapes, colours and rind thickness; fruit shape is usually spherical or oval-oblong; fruit surface is smooth, glabrous, some are deeply ridged and others are covered with corcky (reticulate) netting; vein tracts (sutures) are areas of longitudinal indentations or stripes on the fruit surface.
- Immature fruits are not sweet, instead bland, similar to cucumbers; sweetness is developed with the maturity of the fruit; proper stage of maturity is very important to ensure maximum sweetness and quality.
- Seeds are numerous, flat, slender, spindle shaped, white or buff coloured and smooth; seed ranges in length from 5-15 mm; average 100 seed weight is generally 2.5 gram.

Reproductive behaviour

Andromonoecious sex form is predominant in muskmelon cultivars grown in India. The cultivars of monoecious and gynoecious sex form is highly cross-pollinated in nature however, andromonoecious sex form favour a higher degree of natural self-pollination. It is entomophillous and major pollinators are bees (*Apis florea, Apis dorsata, Nomioides* sp., *Halictine* sp., etc.).Beetles like *Conopophilus* sp. and moths like *Planidia* sp, *Pygargonia* sp also help in natural cross pollination. Muskmelon show negligible inbreeding depression.

Anthesis of both staminate and hermaphrodite / pistillate flowers generally starts in the morning hours from 5.30 to 7.00 a.m. Anther dehiscence in the staminate or hermaphrodite flower generally starts after anthesis. Pollens remain viable from

5.00 a.m. to 2.00 p.m. in ambient condition. Stigma receptivity is at its maximum at the time of anthesis although the stigma becomes receptive 2 hours earlier and 2-4 hours thereafter. Best time of hybridization is between 6.00 to 8.00 a.m. on the day of anthesis. In androemonocious sex form, the flowers are emasculated a day prior to anthesis in the afternoon hours and bagged followed by hand pollination in the next morning and subsequent bagging for two more days is the commercial method of hybrid seed production. In the parental lines of monoecious sex form, bagging the staminate and pistillate flowers prior to opening followed by hand pollination and subsequent bagging for two more days is the method of hybrid seed production. Harvesting the fully matured hermaphrodite / staminate flower buds in the afternoon hours that will open in next morning and keeping them covered with thin moist cloth over night for their use in cross-pollination by gentle rubbing on the stigma head in the next day is the conventional practice. Male sterility has been successfully utilized in developing F_1 hybrids. Muskmelons do not cross with watermelon, cucumber, pumpkin, or squash, but varieties within the species intercross frequently.

Genetics of important characters

Melon is a diploid species (2n = 2x= 24) and a polymorphic taxon comprising large number of botanical and horticultural groups, which includes both desserts as well as cooking and salad types. Cytological characteristics indicate that melon (*Cucumis melo*) having basic chromosome number (x = 12) and cucumber (*Cucumis sativus*) having basic chromosome number (x = 7) are separate and unrelated species. Muskmelon is crossable with *Cucumis melo* var *momordica, Cucumis melo* var *flexuosus* and *Cucumis callosus.*

Several qualitative characters of muskmelon are under the control of single gene.

Table 1: Some important monogenic characters of muskmelon

Gene symbol	Phenotype of plant character
a	Androecious; interact with *g*
ab	Abrachiate; lacking lateral branches. Interacts with *a* and *g* (e.g. *abab aa g+* plants produce only staminate flowers)
b	Bush, short internodes, compact plant habit
Bi	*Bitter.* Bitter seedling; dominant to non-bitter
g	Gynomonoecious (*gynomonoecious.* Mostly pistillate, fewer perfect flowers. Epistatic to *a* : a+_*g*+_monoecious*; a+_ gg* gynoecious; *aa g+* _ andromonoecious; *aa gg* hermaphrodite.
gf	Green flesh colour; recessive to salmon flesh
gf-1	Green flesh colour; recessive to white flesh
gl	Glabrous, lacking in trichome
gp	Green petal; Corolla leaf like in colour and venation
y	*Halo* cotyledons; yellow cotyledons, later turning green
jf	Juicy flesh
Me	Mealy flesh texture
Mt	Mottled leaf pattern
Mu	Musky flavour

Gene symbol	Phenotype of plant character
ms-1	*Male sterile-1;* Indehiscent anthers with empty pollen walls in tetrad stage
ms-2	*Male sterile-2;* Anthers indehiscent, containing mostly empty pollen walls
ms-3	*Male sterile-3.* Waxy and translucent indehiscent anthers, containing two types of empty pollen sacs
n	*Nectarless;* nectaries lacking in flowers
O	*Oval fruit;* dominant over round, associated with a
p	Pentamerous. Five carpels and stamens; recessive to trimerous
Pa	*Pale* green foliage. *PaPa* plants are white (lethal); *PaPa+ are yellow*
r	*red* stem. Red pigment under epidermis of stems, especially at nodes; tan seed colour
ri	Ridged fruit surface; recessive to ridge less
s	Presence of sutures; recessive to ribless
si-1, si-2	Short internode-1, Short intrnode-1
So	Sour taste; dominant to sweet taste
sp	Spherical fruit shape; recessive to obtuse, incomplete dominance
st	Striped fruit; recessive to non-striped fruit
v	Virescent; chlorotic foliage
w	White colour of matured fruit; recessive to dark green fruit skin
wf	White flesh; recessive to orange flesh
Wi	White colour of immature fruit; dominant to green
Wt	White testa; dominant to yellow or tan coloured seed coat
Y	Yellow epicarp; dominant to white fruit skin
yg	Yellow-green leaves; reduced chlorophyll content

Source: Pitrat (2006)

Many important quantitative characters under polygenic control are vine length, earliness, fruit size, fruit shape, fruit weight, fruits per plant, rind thicknes, flesh thickness, yield per plant, etc.

Table 2: Gene action for different quantitative characters

Quantitative character	Gene action
Earliness	Additive, dominance, over- dominance Partially dominant over lateness Both additive and dominance, dominance being more important Additive, dominance, epistasis
Fruit size	Both additive and non-additive gene action
Fruit shape	Additive
Rind thickness	Additive, dominance, epistasis
Fruit weight	Both additive and non-additive gene action
Fruits per plant	Both additive and non-additive gene action Dominance Additive, dominance, epistasis

Total soluble solids	Additive, partial dominance Over-dominance Additive, dominance, over-dominance
Vine length	Both additive and dominance, dominance being more important
Blossom scar	Additive
Seed size	Additive
Total fruit yield	Both additive and dominance, dominance being more important, over-dominance Additive Dominance

Source: Swarup (2006), Munshi and Choudhary (2015)

Breeding objectives

Some of the important characters that have been the main focus of muskmelon breeding are outlined below

- Preferably short vines with high hermaphrodite / female to male flower ratio in the vine
- High total and early yield potential in consistent manner and better adaptability
- Lower number of node at which first hermaphrodite / pistillate flower appears which gives a fair indication of earliness
- Early maturing, spherical, flattish round or oblong fruit with small seed cavity and high flesh recovery per volume
- Netted cantaloupe, honeydew, sutured melon and Galia types
- Epicarp / skin colour and surface should be smooth with sutures and may be netted
- Flesh colour may be white, greenish or salmon-orange.
- Fruit should be sweet (TSS content should be minimum 10° Brix) with high musky aroma
- Good keeping quality and transportability
- Resistance/tolerance to different diseases like, fusarium wilt (*Fusarium oxysporum* f.sp. *nivenum*), downy mildew (*Pseudoperonospora cubensis*), powdery mildew (*Podosphaera xanthii*), cucumber masaic virus, etc.
- Resistance/tolerance to insect pests like, fruit fly (*Bactrocera cucurbitae*), red pumpkin beetle (*Aulacophora foveicollis*), etc.
- Resistance to abiotic stresses like cold temperature, seed germination at low temperature, soil salinity, etc.

Breeding methods and development of variety

Melon fruit is botanically a pepo. The flesh is either sweet or bland, with or without a musky aroma, and the rind may be smooth, ribbed, wrinkled, or netted. In North America, the sweet-flesh varieties are often collectively called muskmelon, including the musky netted-rind varieties and the inodorous smooth-rind varieties, and cantaloupe usually means the former type. However, muskmelon in a narrow sense only refers to the musky netted-rind type, while the true cantaloupe is the European type with ribbed and often warty rind that is seldom grown in North America.

Existence of extensive variability of this crop amply supports the proposition that India is either the place of origin or secondary centre of diversity of muskmelon. Different local cultivars are quite popular in various parts of the country and they are being cultivated in different names like, Sharbat-e-Anar and Hingan in Kurnool, Lucknow Safeda, Kanpuria, Jogia, Mathuria, Batti, Kajra, Jaunpuri, Mau (of Azamgarh), Bhagpat (of Meerut), Netted (Jametha), Mahaban, Faizabadi, Nawabganji in different areas of Uttar Pradesh; Tonk melon of Rajasthan, Sankheda and Galteshwar (Panam) in Gujarat; Kharri melon of Hoshangabad; Jalgaon, Kabri, Gurbeli, Peeli and Kavit in different other areas of Madhya Pradesh; Sanganer, Haragola, Safeda Gola, Motia and Sharda in different areas of Rajasthan; Bathesa, Papay or Papaya, Chiranji and Ladoo in Andhra Pradesh; Kalapa in Karnataka; Batli, Goose and Jam (Neel) in Maharashtra; Kutana and Bhagpat in Haryana; Hari Dhari, Musa (Sharda melon), Chamirai and Sweet Chitkabra in Punjab, etc.

Table 3: Different botanical varieties of *Cucumis melo*

Botanical variety	**Characteristics**
Cucumis melo var. *agrestis*	Wild melon, very small, inedible fruits with thin mesocarp
Cucumis melo var. *cantaloupensis*	Medium-large fruits, netted rind, aromatic, andromonoecious sex form, Galia, Ananas type
Cucumis melo var. *inodorus*	Large sized fruit, winter melon, non-aromatic, andromonoecious sex form, Cassaba type
Cucumis melo var. *flexuosus*	Highly elongated fruit, non-sweet, eaten immature as cucumber, monoecious sex form
Cucumis melo var. *conomon*	White fleshed, thin rind fruit, generally eated as pickle, andromonoecious sex form
Cucumis melo var. *chito and dudain*	Small plum-size fruits, aromatic, both monoecious and andromonoecious sex form
Cucumis melo var. *momordica*	Less sweet fruit which split at maturity, monoecious sex form

Source: Munshi and Choudhary (2015)

The dessert type muskmelon cultivars grown in India comprise different local cultivars and improved varieties developed through individual plant selection and inbreeding or inbred selection from different locally adapted types and exotic introductions. Good number of the varieties has also been developed by inter-varietal hybridization followed by pedigree, backcross or some modified methods of handling the segregating populations. Muskmelon cultivars grown in India may be classified according to colour of flesh (mesocarp) and netting on fruit surface.

- **White/ creamy white/ greenish-white flesh:** Lucknow Safeda, Pusa Sarda , Arka Rajhans, Arka Jeet

- **Green /Light-green flesh:** Hara Madhu, Durgapura Madhu, Pusa Madhurima, Punjab Rasila, RM-43, RM-50, MHY-3, MHY-5, Gujarat Muskmelon-1, Gujarat Muskmelon-3
- **Salmon-orange/ orange flesh:** Punjab Sunheri, Hisar Madhur, Pusa Sharbati, Pusa Madhuras, Arka Siri, MH-10, Gujarat Muskmelon-2, Kashi Madhu, Narendra Muskmelon-1, Narendra Muskmelon-2
- **Netting on fruit surface:** Arka Siri, Punjab Rasila, Gujarat Muskmelon-2, Pusa Madhurima, Pusa Sharbati

Germplasm resources and Plant introduction

Melon exhibits great dissimilarity in fruit characters such as size, shape, colour, taste, texture and harvesting period. Most melons cultivated in India are in the Cantalupensis, Reticulatus, Momordica, Acidulus, Flexuosus, Ameri, and Dudain Groups. Genetic variability study of several collections from different parts of India exhibited that the Indian germplasm contains unique genetic variability (Malik *et al.*, 2014). Indian melon germplasm may, therefore, be considered as an important reservoir of genetic variability for melon breeding. Global gene banks house Indian melon accessions that originated from North Rajasthan, Uttar Pradesh, and Madhya Pradesh (McCreight *et al.*, 2013). The melon collection in the U.S. National Plant Germplasm System (NPGS) is a valuable resource to conserve natural genetic diversity and provide novel traits for melon breeding. International Board for Plant Genetic Resources had initiated melon germplasm from southwestern Asia (Turkey to Afganisthan), India (particularly, western ghats and Indo-Gangetic plains), Mayanmar, southern China, tropical Africa (Sudan) and southern Africa.

In India, muskmelon germplasm is maintained at NBPGR, New Delhi, ICAR-IIVR, Varanasi, ICAR-IARI, New Delhi, ICAR-IIHR, Bengaluru, SKNAU, Jobner, Durgapura, Rajasthan, PAU, Ludhiana, Punjab and other places.

No exotic variety has so far been released directly for cultivation in India.

Individual plant selection and inbreeding

Huge genetic diversity exists in India for the dessert type muskmelon cultivars. In this method, an individual plant from the source population consisting of heterozygous and non-uniform materials is selected based on the selection indices and unlike mass selection, the seeds of the selected plants are not mixed even if they are equally good and look similar. This method of selection is based on the principle that the actual worth of the selected plant may be masked by environmental variation and, therefore, the only way to recognize it is by growing the progeny of each selected individual plant for re-evaluation. Progenies of individual plant for evaluation can be developed by selfing as muskmelon do not show loss of vigour due to inbreeding. Homogzygosity for the concerned characters can also be attained in the individuals of the progeny by such repeated selfings. Maximum number of varieties have been developed by this method of breeding from different indigenous genetic resources which amply suggested the worth of Indian melon germplasm. Only one variety, Arka Siri has so far been developed from exotic germplasm (Tuscan cantaloupe).

Table 4: Varieties developed through selection and inbreeding

Name of variety	Indigenous / Exotic material
Lucknow Safeda	Indigenous (U.P)
Hara Madhu	Indigenous (U.P., Kutana type)
Durgapura Madhu	Indigenous (Rajaasthan)
Pusa Madhuras	Indigenous (Rajaasthan)
Pusa Madhurima	Indigenous
Pusa Sarda	Indigenous
Arka Rajhans	Indigenous (Rajasthan)
Arka Jeet	Indigenous (Bati strain, U.P)
Arka Siri	Exotic (Tuscan cantaloupe)
RM-43	Indigenous (Rajasthan)
Annamalai	Indigenous
MH-10	Indigenous (Punjab)
Gujarat Muskmelon-1	Indigenous (Sabarkantha district, Gujarat)
Gujarat Muskmelon-3	Indigenous (Gujarat)
Gujarat Muskmelon-2	Indigenous (Nagpur, Maharashtra)
Kashi Madhu	Indigenous (U.P.)
Narendra Muskmelon-1	Indigenous (U.P.)
Narendra Muskmelon-2	Indigenous (U.P.)

Hybridization and selection

Planned hybridization involving both indigenous and exotic genetic materials followed by pedigree method of selection in segregating generations or modified backcross pedigree method of breeding has been followed for the development of good number of varieties in India (Hazra *et al*, 2022). Parents to be employed in the hybridization programme should possesses specific characters which are expected to be combined in the segregating generations. The parental lines should also be genetically diverse for obtaining high heterotic F_1 hybrids and broad spectrum of variability in the segregating generations. Selection in segregating generations should be based on the specific characters to be incorporated in the recombined lines and important selection indices for fruit yield and quality.

In a mass pedigree system of selection, Andrus and Bohn (1967) obtained several advantages in cantaloupe breeding. They advocated that inbreeding be withheld until after some period of mass selection with or perhaps without conscious selection. The principle here is to allow natural simulated selection with a resultant opportunity to allow for more genetic recombination. Emphasis in melon breeding, being dessert fruits, should be on getting a cultivar that has dependable quality of a standard grade rather than outstandingly sweet or extraordinary flavour. By inbreeding and pure line selection a cultivar can be selected with uniformity in one character, but only by sacrificing some other indispensable quality characteristic. Invariably, more homozygous inbreds lack earliness, productivity or one or the other quality trait. Thus far, in no case, a wholly satisfactory combination of horticultural characters in an apparently homozygous line has been obtained. The behavioural characteristics which result in good overall performance are based on polygenic systems and it is not uncommon to harvest two fruits of muskmelon from the same vine on different days with varying quality of sweetness.

Table 5: Varieties bred through hybridization followed by selection

Name of variety	Parentage
Punjab Sunheri	Hara Madhu x Edisto (American introduction)
Hisar Madhur	Pusa Sharbati × 75-34
Pusa Sharbati	Kutana × PMR6 (U.S.A.)
Punjab Rasila	WMR-29 x Hara Madhu
RM-50	Durgapura Madhu x Sel-1
MHY-3	Durgapura Madhu x Pusa Madhuras
MHY-5	Durgapura Madhu x Hara Madhu

Heterosis breeding

In muskmelon, heterosis breeding has different requirements. Being dessert fruits, quality characteristics of the hybrids should be superior in respect of uniformity and stability over pure bred cultivar. Earlier maturity is another major advantage in the acceptance of melon hybrids. Muskmelon hybrids has become very popular among the farmer mainly due to high hermaphrodite/ female to male ratio, good quality fruit like small cavity size, thick flesh, high aroma and crispy flesh, high number of fruits per plant, uniformity, extended fruiting span, wider adaptability and early and high yield. Muskmelon is extensively cultivated in India with several high quality indigenous cultivars as well as line-bred varieties. For this reason, area under muskmelon hybrids in India is comparatively low. It has been estimated that about 20-25 % muskmelon area in India is at present under different hybrids. The andromonoecious condition of muskmelon necessitates emasculation. Emasculation of the hermaphrodite flower and bagging of staminate and emasculated flowers in the afternoon hours and hand pollination of the emasculated hermaphrodite flower in the morning hours of the next day with the pollens from the staminate flower of the intended male parent is the commercial method of hybrid seed production.

Use of male-sterility

Male sterility governed by recessive gene (*ms-1*) was first reported by Bohn and Whitakar (1949). Since then, at least four additional male sterile recessive alleles, *viz.* *ms-2*, *ms-3*, *ms-4* and *ms-5* have been identified. A spontaneous recessive mutation for male sterility has been found in which the male flowers abort at an early stage (Bytschvarov *et al*, 1978). The hybrid seed is produced in an isolated field in which generally three rows of female (male-sterile line) or seed-parent line is alternated with one row of male line (monoecious sex form). However, Foster (1968) in a series of experiments highlighted the importance of monoecious sex form in producing hybrids being cheaper method of hybrid seed production and opined that male sterility has not been of much of an advantage in melon hybrids.

Use of gynoecious lines

Gynoecious lines have opened the opportunity for hybrid seed production of muskmelon in open and insect-pollinated condition. Gynoecious sex form is based on digenic inheritance. Hermaphrodite is designated as double recessive (*aagg*) with monoecious having the two dominant genes (*AAGG*). The andromonoecious (+a) and gynomonoecious (g+) are heterozygotes. Gynoecious sex expression is controlled by multiple modifying genes in addition to the major genotype *AAgg* or *Aagg*. Kubicki (1966) described a method for producing 100% genetically pure gynoecious lines. Such lines are derived from a cross of hermaphrodite (*aagg*) with a monoecious (*AAGG*) line and isolating gynoecious (*AAgg*) and hermaphrodite (*aagg*) lines. Crossing of gynoecious (*AAgg*) and hermaphrodite (*aagg*) lines produce all gynoecious (*Aagg*) plants. Gynoecious lines are generally maintained through induction of hermaphrodite or male flowers by applying 200-250 ppm silver thiosulphate at 2-4 leaf stage.

Breeding for resistance to diseases

Breeding melons for disease resistance has resulted in several accomplishments. Multiple resistant lines in cantaloupe have been developed carrying resistance to powdery and downy mildew, *Alternaria* and watermelon mosaic virus (Thomas and Web, 1981). Ivanova (1986) investigated sources of resistance to powdery mildew among melons in Uzbekistan and reported highest resistance in Super Market, Perlita, Kurume 1 and Joknean. Kenigsbuch and Cohen (1987) found that inheritance of resistance to powdery mildew in muskmelon in PI124111 to race 1 was conferred by a single dominant gene and to race 2 by an incompletely dominant gene. The downy mildew resistance in muskmelon line of MR-1 was conferred by 2 incompletely dominant genes designated *Pc1* and *Pc2* (Thomas *et al.*, 1988).

The genetics and mode of gene action for resistance to tomato leaf curl New Delhi virus (ToLCNDV) in muskmelon was studied under conditions of natural, epiphytotic and whitefly-mediated challenge inoculation. In order to accomplish this investigation, susceptible parent (DOM 115) and resistant parent (DSM 132) were utilized to produce the F_1 , F_2 , BC_1 P_1 and BC_1 P_2 populations. Under natural epiphytotic and whitefly mediated challenge inoculation screening procedures, complementary types of epistasis have been found because both components, dominance [h] and dominant × dominance [l], were oriented in the same direction. The results indicated that the recombination breeding, and hybridization followed by selection at later generations could be used to improve resistance to ToLCNDV (Padmanabha *et al*, 2022). Several cucumber green mottle mosaic virus (*CGMMV*) and downy mildew resistance lines DVRM-1, DVRM-2, DMDR-1, DMDR-2 and DMDR-3 were developed by Munshi and Verma (2000) by crossing Phoot x muskmelon *viz*., (*C. melo* var. *momordica* (resistant) × M4 (susceptible) and FM1 (resistant) × Pusa Madhuras (susceptible).

Table 6: Genes for resistance to disease and insect pests in muskmelon

Gene symbol	Resistance to the disease/ insect pest
Af	Resistance to red pumpkin beetle (*Aulacophora foveicollis*); dominant control
Ag	Tolerance to *Aphis gossypii* ; dominant control
Ac	Resistance to *Alternaria cucumerina* ; dominant control
dc-1	Resistance to *Bactrocera cucurbitae-1 ;* recessive control
dc-2	Resistance to *Bactrocera cucurbitae-2 ;* recessive control
Fom-1	Resistance to *Fusarium oxysporum* f. *melonis*, race 1; dominant control
Fom-2	Resistance to *Fusarium oxysporum* f. *melonis*, race 2; dominant control
Fom-3	Resistance to *Fusarium oxysporum* f. *melonis*, race 3; dominant control
Me	Resistance to gummy stem blight, *Mycosphaerella citronella*; dominant control
Pm-1	Resistance to powdery mildew, *Podosphaera xanthii* race-1; dominant control
Pm-2	Resistance to powdery mildew, *Podosphaera xanthii* race-2; dominant control
Pm-3	Resistance to powdery mildew, *Podosphaera xanthii* race-3; dominant control
Pm-4	Resistance to powdery mildew, *Podosphaera xanthii* race-4; dominant control
Pm-5	Resistance to powdery mildew, *Podosphaera xanthii* race-5; dominant control
Pm-6	Resistance to powdery mildew, *Podosphaera xanthii* race-6; dominant control
Pc-1, Pc-2	Resistance to downy mildew, *Pseudoperonospora cubensis*; dominant control
Pc-1	Resistance to downy mildew, *Pseudoperonospora cubensis*; incompletely dominant partial resistance

Source: Pitrat (2006)

Distant hybridization was carried out to introgress disease resistance from the wild relatives of the cultivated species. Some successful inter-specific hybridization in melon are *Cucumis melo* × *C. figarei, C. melo* ×*C. zeyheri, C. melo* × *C. meeusii, C. melo* × *C. myriocarpus* for transferring resistance to *Fusarium oxysporum* ssp. *melonis* and cucumber green mottle mosaic virus and *C. melo* × *C. africanus* for transferring cucumber green mottle mosaic virus (*CGMMV*) in melons. However, it is very difficult to obtain F_1 hybrids from most of such interspecific crossings. Crossability depends on the particular accessions of the two species used as parents however, such hybrids are normally highly sterile because of impaired ability of the staminate flowers to produce functional pollens. Inter-specific crosses among different species is generally done by repeated pollination during the bud and flowering stage although percentage fruit set and embryo size are greatly influenced by the pollen parent. However, the use of *in vitro* culture of the immature embryo by embryo rescue technique is generally employed for obtaining the inter-specific F_1 hybrid.

Breeding for quality

Inheritance of flavonoid, carotenoids and chlorophyll pigments has been studied extensively in melon. The pigments accumulating in rinds of different melon genotypes during fruit development were analyzed. It was found that melon rind colour is based on different combinations of chlorophyll, carotenoids and flavonoids according to the cultivar tested and their ratios changed during fruit maturation. Moreover, in "canary yellow" type melons, naringenin chalcone, a yellow flavonoid pigment previously unknown to occur in melons, has been identified as the major fruit colorant in mature rinds. Naringenin chalcone is also prominent in other melon types, occurring together with carotenoids (mainly β-carotene) and chlorophyll. Both chlorophyll and carotenoid pigments segregate jointly in an F_2 population originating from a cross between a yellow canary line and a line with green rind. In contrast, the content of naringenin, chalcone segregates as a monogenic trait independently to carotenoids and chlorophyll (Tadmor *et al.*, 2010).

In melon, α-carotene and β-carotene are found in the fruit. There is an opportunity through traditional plant breeding to increase the amount of these and other carotenes to improve the nutrition of melon. Some carotenes can be observed in melon fruit as an orange colour (the higher the orange colour the higher the concentration of β-carotene). Genetic knowledge regarding the inheritance (how genes are transmitted from generation to generation) about these compounds is critical for effective and efficient plant improvement through breeding. Several genes were identified which control β-carotene in fruit and are related to the colour of the fruit. These genes can be used to breed higher level of β-carotene in melon varieties. Cuevas *et al.* (2008) suggested that accumulation of β-carotene in melon is under complex genetic control, where epistasis plays an important role in trait expression. Although the inheritance of β-carotene content and fruit maturity is complex, introgression of early fruit maturity genes from Chinese germplasm might solve the problem. Such introgression may lead to increased yield potential in US market types while retaining relatively high β-carotene content (i.e., orange mesocarp), if stringent, multiple location and early generation family selection (F_{3-4}) is practiced for fruit maturity with concomitant selection for quantity of β-carotene (Cuevas *et al.*, 2010).

Identification of molecular markers and orange flesh colour associated with major QTL for high β-carotene content in muskmelon was reported by Napier (2006). Presence or absence β-carotene in muskmelon is controlled by two genes, green flesh *gf* and white flesh *wf*. In its dominant form the *wf* gene is responsible for orange flesh colour; however, the epistatic interactions of *gf* and *wf* can create three flesh colours: orange, white and green. Bulk segregent analysis was used with RAPD markers to identify molecular markers associated with high β-carotene content. Flesh colour and colour intensity both had significant relationships with β-carotene content. A single QTL was also found to be linked with the *WF* locus. The identified QTL can be used to screen potential breeding lines for high β-carotene. It was also confirmed that the visual ratings of flesh colour intensity can be reliably used to select high β-carotene content melons.

References

Andrus, C.F. and Bohn, G.W. 1967. Cantaloupe breeding: shifts in population means and variability under mass selection. *Proc. Amer. Soc. Hort. Sci.*, **90**: 209-222.

Bohn, G.W. and Whitaker, T.A. 1949. A gene for male sterility in the muskmelon. *Proc. Amer. Soc. Hort. Sci.*, **53**: 309-14.

Bytschvarov, S., Petkov, T. and Losanov, P. 1978. New varieties of onions, cucumbers and other cucurbits. *Internationale Zeitschrift der Landwirtschaft*, **2**: 155-158.

Cuevas, H.E., Staub, J.E. and Simon, P.W. 2010. Inheritance of beta-carotene-associated mesocarp color and fruit maturity of melon (*Cucumis melo* L.). *Euphytica.,***173**: 129–140.

Cuevas, H.E., Staub, J.E., Simon, P.W., Zalapa, J.E. and McCreight, J.D. 2008. Mapping of genetic loci that regulate quantity of beta-carotene in fruit of US Western Shipping melon (*Cucumis melo* L.). *Theor. Appl. Genet.*, **117**: 1345-1359.

Foster, R.E. 1968. F_1 hybrid muskmelons. V. Monoecism and male sterility in commercial seed production. *J. Hered.*, **59**: 205-207.

Hazra, P., Banerjee, M.K. and Chattopadhyay, A. 2022. *Varieties of Vegetable crops in India*, Kalyani Publishers, New Delhi, Ludhiana, 204p.

Ivanova, A.E. 1986.Sources of resistance to powdery mildew among melons in Uzbekistan. *Nauchno – TeknicheskiiByulletenVsesoyuznogo Ordena Leninai Ordena Druzby Narodov Nauc/1110 0 Issledovatel'skogoInstitutaRastonievodstvaimeniN.I. Vavilova.*, No. 165, pp. 78-80.

Kenigsbuch, D. and Cohen, Y. 1987. Inheritance of gynoecious sex type in muskmelon. *Cucurbit Genet. Coop. Rep.*, **10**: 47-48.

Kubicki, B. 1966. Genetic basis for obtaining gynoecious muskmelon lines and the possibility of their use for hybrid seed production. Genet. *Polonica*, **7**: 27- 29.

Malik, A., Vashisth, V.K., Singh, K. and Sharma, A. 2014. Diversity among melon (*Cucumis melo* L.) landraces from the Indo-Gangetic plains of India and their genetic relationship with USA melon cultivars. *Genet. Resour. Crop Evol.*, **61**: 1189-1208.

McCreight, J.D., Staub, J.E., Wehner, T.C. and Dhillon, N.P.S. 2013. Gone global: familiar and exotic cucurbits have Asian origins. *Hort Sci.*, **48**:1078–1089

Munshi, A. D. and Verma, V. K. 2000. Field screening of muskmelon germplasm against cucumber green mottled mosaic virus. *Indian J. Pl. Genet. Resour.*, **13**: 72-74.

Munshi, A.D. and Choudhary, H. 2015. Muskmelon. In: *Hand book of Vegetables* Vol.III (Eds. Peter , K. V.and Hazra, P.), Studium Press LLC, Houston, Texas, USA, pp. 271-310.

Napier, A.B. 2006. Identification and confirmation of molecular markers and orange flesh color associated with major QTL for high beta-carotene content in muskmelon. *M.Sc. Thesis*, Texas A and M University, USA.

Padmanabha, K., Choudhary, H., Mishra, G.P., , Mandal, B., , Solanke, A.U., , Mishra, D.C. and Yadav, R.K. 2022. Genetic analysis of resistance to tomato leaf curl New Delhi virus in muskmelon (*Cucumis melo* L.) *Indian J. Genet. Plant Breed.*, **82**: 499-506

Pitrat, M. 2006. Gene list for melon. *Cucurbit Genet. Coop. Rep.*, **28-29**: 142-163

Sebastian , P., Schaefer, H. and Renner, S.S. 2010. Cucumber (*Cucumis sativus*) and melon (*C. melo*) have numerous wild relatives in Asia and Australia, and the sister species of melon is from Australia. *Proc. Nat. acad. Sci.* USA, **107**:14269-14273

Swarup, V. 2006. *Vegetable Science and Technology in India.* Kalyani Publishers, New Delhi, Ludhiana, 656p.

Tadmor, Y., Burger, J., Yaakov, I., Feder, A., Libhaber, S.E., Portnoy, V., Meir, A., Tzuri, G., Sa'ar, U., Rogachev, I., Aharoni, A., Abeliovich, H., Schaffer, A.A., Lewinsohn, E. and Katzir, N. 2010. Genetics of flavonoid, carotenoid, and chlorophyll pigments in melon fruit rinds. *J. Agric. Food Chem.*, **58**: 10722–10728.

Thomas, C. E. and Webb, R. E. 1981. W1, W3, W4, W5, and W6 multidisease-resistant muskmelon breeding lines. *HortSci.*,**16**: 96.

Thomas, C.E., Cohe, Y. and McCreight, J.D., Jourdain, E.L. and Cohen, S. 1988. Inheritance of resistance to downy mildew in *Cucumis melo*. *Plant Dis.*, **72**: 33-35.

36

Cucumber

Cucumber, *Cucumis sativus* under the family Cucurbitaceae (Chromosome No: 2n = 2× =14) is grown throughout the world for consumption of fresh and tender fruits as slicing cucumber, and for pickles as pickling cucumber. Slicing cucumbers are usually larger and have thicker skins, while pickling cucumbers are usually smaller and have thinner skins. The flesh of cucumbers is rich vitamin C, and folic acid, fibre and a range of minerals including magnesium, molybdenum and potassium.

Origin, domestication and adaptation in India

The genus *Cucumis*, which includes cucumber (*Cucumis sativus*) and melon (*Cucumis melo*), has numerous wild African species, and it has therefore been assumed that melon originated in Africa. For cucumber, this seemed less likely because wild cucumbers exist in India and a closely related species, *Cucumis hystrix* is available in the Eastern Himalayas. Using DNA sequences from plastid and nuclear markers for some 100 *Cucumis* accessions from Africa, Australia, and Asia, Sebastian *et al* (2010) concluded that melon and cucumber are of Asian origin and have numerous previously overlooked species-level relatives in Australia and around the Indian Ocean. Cultivated for at least 3,000 years, the cultivated cucumbers *Cucumis sativus* were domesticated in India from wild *C. sativus* var. *hardwickii*. A great many varieties are available, along with its closest relative, *Cucumis hystrix*. Its Sanskit equivalent name "Urvaruka" and Ervaruka" as mentioned in the ancient treatises of India "Rigveda" and Charaka Samhita" amply justifies its cultivation in India dates back 3000 years.

Table 1: Distribution of wild relatives of cucumber in India

Wild species	Distribution in India
C. sativus var. *hardwickii* (Syn. *C. hardwickii*	Western Himalayas
Cucumis hystrix	Plains of Eastern region, north-eastern hills and Assam valley
Cucumis prophetarum	Dry north-western plains, Rajasthan and adjoining areas
Cucumis setosus	Eastern India, upper Gangetic plains, West Bengal and peninsular region
Cucumis trigonus	Eastern and Western Ghats, north-eastern region, extending to the Himalayas

Source: Swarup (2006)

It was probably introduced to Europe by the Greeks or Romans. Records of cucumber cultivation appear in France in the 9th century, England in the 14th century, and in North America by the mid-16th century. Cucumber now grows in most continents and many different types of cucumber are traded on the global market.

Crop botany and floral biology

- The plant is a morphologically variable annual herbaceous climber; stems are prostrate, angular, and covered in white pubescence; stipules are absent, and the plant bears unbranched axillary tendrils up to 30cm long; vine uses their simple tendrils to climb over structures or other vegetation.
- The pubescent leaves are alternately arranged on 10-16 cm long petioles, simple, basally cordate, and apically acute with 3-7 palmate lobes; the palmately-veined leaves are nearly orbicular, 7-20 cm long and broad.
- Plants are generally monoecious in sex form with staminate and pistillate flowers are borne separately in the same plant; its axillary flowers are actinomorphic; gynoecious, andromonoecious and hermaphrodite sex forms are also found in some genotypes.
- Inflorescence is axillary, solitary or clustered or racemose.
- Both staminate and pistillate flowers have a pubescent 5-parted calyx composed of 0.5-1cm long white pubescent sepals.
- The sepals are long, narrow, and acute; on pistillate flowers, the calyx is fused to the ovary and forms a hypanthium.
- The corolla is approximately 2cm long, yellow, fused, less than half of its length, campanulate, 5-parted with oblong to lanceolate lobes.
- Staminate flowers are borne in clusters of 3-7 with short, slender pedicels (0.5-2cm long) whereas the pistillate flowers are usually solitary with stout, short pedicels.
- Staminate flowers bear 3 stamens, of which two have 2-celled anthers (0.3-0.4cm long) and one a 1-celled anther.
- Pistillate flowers are solitary or paired on pedicels shorter than the staminate flower (<0.5cm long) before fruit development, whereas during fruit development the pedicels can elongate up to 5 cm long; pistilate flowers have 3 staminodes, a pubescent, ellipsoid, unilocular ovary 2-5cm long, with a short style and 3 stigmas.
- Sepals are 5, united, often deeply lobed, campanulate and inserting into corolla tube.
- Petals are 5, united, often deeply 5 lobed, campanulate, inserting into the calyx tube; corolla colour is generally yellow but cream, orange and white flowers are also found.
- Stamens are usually 3, synandrous or syngenesious; anthers are single or double lobed.

- Carpels are 3, syncarpous, ovary inferior, unilocular, parietal placentation and falsely trilocular; ovules are many.
- The pistillate flowers are pollinated mainly by bees that are attracted by the nectar produced by the flowers; honey bees are more effective than wild bees.; pollen is sticky, which makes wind pollination improbable.
- Botanically, the fruits are many seeded fleshy pepo with no internal divisions; mesocarp is fleshy, usually cylindrical or oval in shape but variable in size, skin colour and thickness and spine colour with many seeds.
- The fruits are glabrous, and can be smooth or warty, yellow or green, ranging from 5-100 cm long and weigh 50 g to 4 kg.
- Seeds are flat, slender, spindle shaped, creamy-white in colour, non-endospermic with straight embryo and large oily cotyledons; 100 seed weight is generally 3.2 to 3.3 gram.

Reproductive behaviour

Cucumber is predominantly monoecious in sex form. however, gynoecious sex form is found in some genotypes. Two other sex forms *viz*., andromonoecious, and hermaphrodite sex form are rarely found. Cucumber is highly cross pollinated in nature and major pollinators are bees (*Apis florea, Apis dorsata, Nomioides* sp., *Halictine* sp., etc.). Although, 100% natural cross pollination occur in monoecious sex form, yet reasonable extent of self fertilization happens because of foraging of insect pollinators in the staminate flowers of the same vine. Cucumber show negligible inbreeding depression. Anthesis generally starts in the morning hours from 5.30 to 7.00 a.m. Pollen dehiscence generally starts after anthesis and pollens remain fertile in ambient condition up to 2.00 p.m. Stigma receptivity is the maximum at the time of anthesis. Best time of hybridization is between 7.00 to 10.00 a.m on the day of anthesis. Bagging the staminate and pistillate flowers in the afternoon hours prior to opening followed by hand pollination in the next day morning and subsequent bagging for two more days is the commercial method of hybrid seed production of cucumber. The temperate gynoecious lines are unstable under high temperature and long photoperiodic conditions. Tropical gynoecious lines are suitable for hybrid seed production in the tropical and sub-tropical regions under open pollination.

Genetics of important characters

Cucumber is a diploid species ($2n = 2x = 14$) with the basic chromosome number, $x = 7$. The progenitor form of cultivated cucumber, *C. sativus* var. *hardwickii* (Sy. *C. hardwickii*) is also diploid ($2n = 14$). The genus *Cucumis* comprises 30 species which are distributed in Africa and Asia. Cytological characteristics indicate that the cucumber (*Cucumis sativus*) having basic chromosome number ($x = 7$) and basically of Asiatic origin and melon (*Cucumis melo*) having basic chromosome number ($x = 12$) and basically of African origin are separate and unrelated species.

Several qualitative characters of cucumber are under the control of single gene. However, sex expression is manifested through interaction between two major genes.

Table 2: Some important monogenic characters of cucumber

Gene symbol	Phenotype of plant character
a	androecious; recessive at *Acr/acr* locus
Acr	Female
al	albino cotyledon
ap	apetalous. Male-sterile. Anthers become sepal-like
B	Black or brown spines; dominant to white spine
B-2	Black spine-2; Interacts with B to produce F_2 of 15 black: 1 white spine
B-3	Black spine-3; Interacts with B-4 to produce an F_2 of nine black: 7 white spine
B-4	Black spine-4; Interacts conversely with B-3
bi	Bitter-free; All plant parts lacking cucurbitacins
Bt	Bitter fruit; Fruit with extreme bitter flavour
by	bushy. Short internodes
c	cream mature fruit colour; Interaction with R is evident in the F2 ratio of 9 red (*RC*) : 3 orange (*Rc*) : 3 yellow (*rC*) : 1 cream (*rc*).
cd	chlorophyll deficient
cl	closed flower
co	green corolla
cp	compact; reduced internode length
D	dull fruit skin
de	determinate habit. Short vine with stem terminating in flowers
df	delayed flowering
dvl	divided leaf
dw	dwarf; Short internodes
F	Female; High degree of pistillate sex expression; interacts with a and M; strongly modified by environment
Fba	Flower bud abortion
Fl	Fruit length; Expressed in an additive fashion
g	golden leaves
gig	gigantism; First leaf larger than normal
gl	glabrous; Foliage lacking trichomes
gn	green mature fruit
gy	gynoecious; Recessive gene for high degree of pistillate sex expression
H	Heavy netting of fruit; dominant to no netting
I	locule number; Many fruit locules and pentamerous androecium
m	andromonoecious; Plants are andromonoecious if (*mf*); monoecious if (*Mf*); gynoecious if (*MF*) and hermaphroditic if (*mF*)
m-2	andromonoecious-2. Bisexual flowers with normal ovaries
ms-1	male sterile-1; Staminate flowers abort before anthesis
ms-2	male sterile-2; pollen abortion occurs after first mitotic division

Gene symbol	Phenotype of plant character
n	negative geotropic peduncle response; Pistillate flowers grow upright
ns	numerous spines
O	Orange-yellow corolla
P	Prominent tubercles
Pc	Parthenocarpy; Sets fruit without pollination
pf+	single pistillate flower; incompletely dominant over multiple pistillate flower
pfd	double pistillate flower
pfm	multiple pistillate flower
R	Red mature fruit; Interacts with c; linked or pleiotropic with B and H
s	spine size and frequency; Many small fruit spines
ss	small spines; Large, coarse fruit spines is dominant to small, fine fruit spines
T	Tall plant; Tall incompletely dominant to short
td	Tendrilless
te	tender skin of fruit; of some European cultivars
Tr	Trimonoecious; Producing staminate, perfect, and pistillate flowers in this sequence
Tu	Tuberculate fruit; Warty fruit characteristic of American cultivars is dominant to smooth
u	uniform immature fruit colour
w	white immature fruit colour
wf	white flesh; Intense white flesh color is recessive to dingy white
yc-1, yc-2	yellow cotyledons-1, 2
yf	yellow flesh; Interacts with *wf* to produce F_2 of 12 white (*Wf Yf* and *wf Yf*) : 3 yellow (*Wf yf*) : 1 orange (*wf yf*).
yg	yellow-green immature fruit colour; Recessive to dark green and epistatic to light green
yp	yellow plant; Light yellow-green foliage; slow growth

Source: Wehner (2005), Swarup (2006), Pandey and Choudhary (2015)

Many important quantitative characters under polygenic control are vine length, earliness, fruit size, fruit weight, fruit shape, fruits per plant, fruit firmness and crispiness, fruit yield per plant, etc.

Breeding objectives

Some of the important characters that have been the main focus of cucumber breeding are outlined below

- Vines medium-long, with high female to male flower ratio in the vine
- High total and early yield potential and wide adaptability
- Lower number of node at which first pistillate flower appears which gives a fair indication of earliness

- Early maturing, uniform, long, cylindrical shaped fruit without crook-neck
- Attractive dark green skin colour with minimum of prickles in slicing cucumber
- Early maturing, nice green or dark green "Poinsett type" fruit, uniform long cylindrical shape without crook-neck
- Fruits tender, without carpel separation showing hollow spots at edible maturity
- Texture should be firm and crisp and the flavour should be without bitterness
- Fruits should have high TSS, β carotene and vitamin C contents
- Parthenocarpic - gynoecious and indeterminate types for protected cultivation
- Bunching or concentrated fruit set producing multiple pistillate flowers on individual nodes to suit machine harvesting of finger size fruits to suit whole fruit canning for export especially in case of pickling cucumber.
- Resistance/tolerance to different diseases like, powdery mildew (*Podosphaera xanthii*), downy mildew (*Pseudoperonospora cubensis*), cucumber mosaic virus, etc.
- Resistance/tolerance to insect pests like, fruit fly (*Bactrocera cucurbitae*), red pumpkin beetle (*Aulacophora foveicollis*), etc.
- Resistance to abiotic stresses like cold temperature, soil salinity, etc.

Breeding methods and development of variety

Cucumber is mostly grown in India utilizing wide array of indigenous cultivars and some exotic varieties. Some improved varieties have also been developed through individual plant selection and inbreeding or inbred selection from different locally adapted types. Genetically controlled gynoecious sex form and parthenocarpy character has also been introgressed in some varieties. Cucumber cultivars are basically classified into two broad groups; slicing cucumber (used for fresh market for different uses) and pickling cucumber (used for processing). The slicing type cultivars are much larger than pickling type and generally cylindrical in shape. Parthenocarpic and parthenocarpic-gynoecious types form a separate group because fruits are set without pollination.

Slicing type

Japanese Long Green, Straight Eight, Poinsett, China, Sheetal, Himangi, Phule Shubhangi, Phule Prachi, Solan Green, Kalyanpur Green, Gujarat Cucumber-1, Swarna Poorna, Swarna Ageti, Swarna Sheetal, Pant Khira-1, Pusa Uday , CO 1, Punjab Naveen

Pickling type

No variety has so far been developed in India

Parthenocarpic / Parthenocarpic-gynoecious type

Pusa Parthenocarpic Cucumber-6, Pant Parthenocarpic Khira-2, Pant Parthenocarpic Khira-3, Punjab Kheera-1

Germplasm resources and Plant introduction

Germplasm of cucumber includes different cultivars as well as wild species including *C. hystrix*, and *C. sativus* L. var. *hardwickii*. High levels of phenotypic and genetic diversity of cucumber along with different wild relatives are found in Asia, and India. Primary and secondary centers of diversity can serve as useful sources of variation, and have been widely used to incorporate traits such as disease resistance into cultivated materials. Germplasm collections of cucumber is maintained in many countries like, USDA, USA, Bulgaria, the Netherlands, Germany, Czechoslovakia, Hungary, Russia, Turkey, Phillippines, China and Japan.

In India, cucumber germplasm is maintained at NBPGR, New Delhi, ICAR-IARI, New Delhi, ICAR-IIVR, Varanasi, GBPUAT, Pantnagar, MPKV, Rahuri and other places. Good number of long-fruited exotic varieties of slicing cucumber have been imported mainly from USA, Japan and China. Gherkin (*Cucumis anguria*) and small-sized pickling cucumbers have also been imported because they are now in good demand mainly for their processed product for export.

Table 3: Exotic variety directly released for cultivation

Name of the variety	Country from where introduced
Japanese Long Green	Japan
Straight Eight	USA
Poinsett	USA
China	China

Individual plant selection and inbreeding

Huge genetic diversity exists in India for cucumber cultivars. In this method, individual plant from this indigenous source population consisting of heterozygous and non-uniform materials is selected and unlike mass selection, the seeds of the selected plants are not mixed even if they are equally good and look similar. This method of selection is based on the principle that the actual worth of the selected plant may be masked by the environmental variation and, therefore, the only way to recognize it is by growing the progeny of each selected individual plant for re-evaluation. Progenies of individual plant for evaluation can be developed by selfing as cucumber does not show loss of vigour due to inbreeding. Homogzygosity for the concerned characters can also be attained in the individuals of the progeny by such repeated selfings. Fruit quality traits, including external appearance (shape and colour) and texture (described by characteristics such as firmness and crispness) are the most important traits in cucumber because they are directly related to the product's commercial value. Significant phenotypic and genotypic correlation was observed between fruit yield and fruit weight. number of days to 50% flowering and fruit width which could be employed as selection indices for improvement of fruit yield in cucumber.

Maximum number of varieties have been developed by this method of breeding from different indigenous genetic resources which amply suggested the worth of Indian cucumber germplasm.

Table 4: Varieties developed through selection and inbreeding

Name of variety	Indigenous / Exotic material
Sheetal	Indigenous (Maharashtra)
Himangi	Indigenous (Maharashtra)
Phule Shubhangi	Indigenous (Maharashtra)
Phule Prachi	Indigenous (Maharashtra)
Solan Green	Indigenous
Kalyanpur Green	Indigenous (U.P)
Gujarat Cucumber-1	Indigenous
Swarna Poorna	Indigenous
Swarna Sheetal	Indigenous
Pant Khira-1	Indigenous
Pusa Uday	Indigenous

Hybridization and selection

Planned hybridization involving both indigenous and exotic genetic materials followed by pedigree method of selection in segregating generations or modified backcross pedigree method of breeding can be followed for the development of varieties having recombined characters. However, no variety has so far been developed in India following hybridization and selection method of breeding (Hazra *et al*, 2022). Parents to be employed in the hybridization programme should possesses specific characters which are expected to be combined in the segregating generations. The parental lines should also be genetically diverse for obtaining high heterotic F_1 hybrids and broad spectrum of variability in the segregating generations. Selection in segregating generations should be based on the specific characters to be incorporated in the recombined lines and important selection indices for fruit yield and quality.

In cucumber, the first fruits forming seeds inhibit further fruit set on the plant. The wild relative, *Cucumis hardwickii* has the unique ability to have simultaneous setting of several fruits in the vine (Munshi *et al*, 2007) which can be utilized to develop varieties which can set fruits simultaneously.

Development of tropical gynoecious lines

In temperate locations, Munger (1979) found that the gynoecious lines Gy14, SR551F, Gy3, Gy57, and Table Green 68 are the best for producing F_1 hybrids in slicing and pickled cucumbers. However, temperature and photoperiod influence the expression of gonoecious sex form (More and Munger, 1987). Unfortunately, in the high temperatures and long photoperiodic conditions in the tropical zones, those lines have been proven to be unstable for gynoecy. As a result, there is a need to develop gynoecious lines, which are suitable for tropical condition. Temperate gynoecious lines were crossed with tropical monoecious lines. Silver-nitrate was used to maintain gynoecious aggregates (250 ppm, twice). As a result, several tropical gynoecious lines could be isolated at ICAR-IARI, New Delhi (More and Seshadri,1988)

Development of parthenocarpic and gynoecious variety

Gynoecious-parthenocarpic hybrids in cucumber have great importance for its successful cultivation under protected conditions to enhance productivity and quality. Two fundamental steps in conventional breeding to develop parthenocarpic and gynoecious lines are i) creation of a breeding population that segregates for the parthenocarpic trait of one parental genotype and ii) selecting progeny from a segregated population with parthenocarpy and favourable non-parthenocarpic parent characteristics.

In India, research works was undertaken to develop and maintain the gynoecious inbred lines from the segregating generation of several private sector gynoecious-parthenocarpic cucumber hybrids. At ICAR-IARI, New Delhi, segregation of F_2 population and test crosses for parthenocarpic fruit development suggested that parthenocarpy in gynoecious and parthenocarpic cucumber line Pusa Parthenocarpic Cucumber -2 is under the control of incomplete dominant gene (Jat *et al*, 2017). Further study through development of inbred backcross lines (IBLs) in Indian slicing cucumber with the gynoecious trait confirmed the dominance of *F* alleles manifesting gynoecious trait (Behra *et al.*, 2022). The F_2 mapping population was developed using the Indian monoecious cultivar of light-green-fruited Pusa Uday as the recurrent parent and the American gynoecious pickling line G421 as a donor. Two markers, SSR13251 and SSR15516, were found to be closely linked to the *F* locus at 1.5 and 4.5 cM, respectively, by marker analysis of F_2 individuals. The amplified product of 430 bp generated using a SCAR marker (*CsACS1G*) in gynoecious (*MMFF*) and sub-gynoecious (*MMFf*) plants of the F_2 population confirmed the dominance of *F* alleles. The IBLs were developed by backcrossing Pusa Uday with BC_1 and BC_2 progenies followed by foreground and background selection of plants displaying the gynoecious habit in the background of Indian slicing cucumber. These IBLs could be utilized for the development of hybrids with higher yield, earliness, and quality.

At ICAR-IIHR, Bengaluru, gynoecious-parthenocarpic lines could be isolated from the segregating generations of the gynoecious parthenocarpic cucumber hybrids *viz.*, Silyon hybrid, Pickling Cucumber-1, Pickling Cucumber-2 and Pune Cucumber Hybrid. Four slicing cucumber lines, namely IIHR-434, IIHR-435, IIHR-436 and IIHR-437, were raised to develop segregating populations. The individual plant selections were made in F_2 population and through the pedigree method, these lines were forwarded to F_4 generation based on the gynoecy sex expression. Mean performance of all the four gynoecious lines was recorded. The F_4 populations were validated at phenotypic as well as through SSR markers linked to gynoecious trait. The SSR-02021 and SSR-18718 genotypic data showed that most of the plants were gynoecious in all four advance gynoecious lines and these gynoecious inbred lines were further utilized to develop gynoecious-parthenocarpic cucumber F_1 hybrids (Bommesh *et al*, 2020).

In India, several gynoecious-parthenocarpic varieties could be developed from these gynoecious-parthenocarpic lines which were isolated from the segregating generations of the gynoecious parthenocarpic cucumber hybrids (Hazra *et al*, 2022).

Table 5: Parthenocarpic and gynoecious variety

Name of variety	Parentage
Pusa Parthenocarpic Cucumber-2	Variety with parthenocarpic and gynoecious character
Pusa Parthenocarpic Cucumber-6	Variety with parthenocarpic and gynoecious character
Pant Parthenocarpic Khira-2	Variety with parthenocarpic and gynoecious character
Pant Parthenocarpic Khira-3	Variety with parthenocarpic and gynoecious character
Punjab Kheera-1	Variety with parthenocarpic character

Mutation breeding

Wang (2014) studied applied mutagenesis of cucumber cv. Shannong No. 5 with the exposure of Ethyl methanesulfonate (EMS). The seeds were treated by 1% EMS for 12 h, 24 h and 48 h. Median lethal dose was 1% EMS for 24 h. After treated by 1% EMS for 24 h, 541 M_1 plants were grown in greenhouse for phenotype investigation. The fertility of M_1 plants was very low, and only 79 lines produced seeds after selfing. Sixty independent M_2 families comprising 600 M_2 plants were investigated for phenotypic alteration, and 11 individual mutant lines were isolated into six mutant groups:

- *Short-fruit* mutants,
- *Long-fruit* mutants,
- *Small-flower* mutants,
- *Big-flower* mutants,
- *Opposite-tendril* mutants
- *Clustered-leaf* mutants.

The mutation frequency was 18.3%. Two selected representatives, *short-fruit* mutants and *clustered-leaf* mutants, showed 1:3 segregation ratio in M_2 populations. This ratio is consistent with classic Mendelian model, indicating that the two kinds of mutants may be controlled by a single recessive gene, respectively. *Long-fruit* phenotype was stably inherited and no segregation was observed in M_3 generation, indicating that this mutant line may be homozygous. In India, only one variety has so far been developed through gamma radiation induced mutant which showed earliness.

Table 7: Varieties/ mutants developed through applied mutagenesis

Variety/ Mutant	Applied mutagenesis / Remark
Swarna Ageti	Gamma radiation mutation

Heterosis breeding

Hayes and Jones (1916) reported heterosis in cucumber for the first time. Since then a large number of hybrids have been developed in different countries including India. Cucumber is extensively cultivated in India with several high quality indigenous cultivars as well as line-bred varieties. For this reason, area under cucumber hybrids in India is comparatively low. It has been estimated that about 30-35 % cucumber area in India is at present under different hybrids as against almost 90% of the total area under cucumber is covered by hybrids in Western countries. Hybrid cucumber is highly valued for earliness, high yield, extended harvest span and external attributes like straight fruits with rounded ends, smooth skin of homogeneous green colour, crispy-textured flesh with refreshing flavour especially in slicing cucumber, adaptability to diverse climatic conditions and resistance to diseases, particularly, downy mildew, powdery mildew and cucumber mosaic virus. Most of the hybrids carry multiple resistances to diseases due to the fact that dominant genes control resistance to some diseases. In cucumber, per cent pistillate nodes registered moderate to highly significant positive correlations with yield. In protected growing condition, not only high yield but also sustaining yield over atleast four months is very important for making the poly house / glass house culture competitive to produce superior quality fruits. Earliness is another character which is manifested in the hybrids, besides lesser percentage of unmarketable culls. The latter character is important in pickling hybrids because misshapen fruits are rejected at the time of grading for processing. In poly house / glass house culture, parthenocarpy and multiple disease resistance have been incorporated in the gynoecious hybrids to obviate the insect pollinators for fruit set. In pickling hybrids, machine harvesting has come into vogue requiring development of hybrids with a concentrated fruit set. Multiple pistillate lines producing more than one fruit at each flowering node has become an important character in pickling types to suit once-over harvest.

Being monoecious in sex expression and cross pollinated, it provides ample scope for the utilization of hybrid vigour. Bagging of staminate and pistillate flowers in the afternoon hours and hand pollination of the pistillate flower in the next day morning hours after opening of the pistillate flower with the pollens from the staminate flower of the intended male parent is the commercial method of hybrid seed production in India and many other countries. However, high propensity of hybrid seed production particularly in the Western countries could be achieved mainly due to the exploitation of gynoecious sex form in the development of hybrids.

Use of gynoecious lines in hybrid seed production

According to Yamasaki *et al.* (2002), sex determination in cucumber plants is genetically controlled by the F and M loci. The F and M loci interact to produce three different sex phenotypes viz., gynoecious (*M-F-*), monoecious (*M-ff*) and andromonoecious (*mmff*). Spraying of GA3 (2,000 ppm) or silver nitrate or silver thiosulphate (250 ppm) at first true leaf stage and again 14 days later, induced the production of staminate flower in the gynoecious line. The female flowers of the gynoecious plants can be selfed using the pollen from the induced male flowers and thus maintained in a homozygous condition. The hybrid seeds are produced in an isolated field in which generally three rows of female (gynoecious) or seed-parent line is alternated with one row of male line (monoecious sex form). Hybrid seeds are collected from the female lines. The female line is propagated in another isolated field, in which one row out of three or four is sprayed with GA3, silver nitrate or silver thiosulphate to induce male flowers. The natural sib and self-pollinated seeds are obtained for growing the female lines in hybrid seed production programme. In hybrid-seed production field, it is necessary to keep at least four beehives/ ha to assure cross-pollination. The quality and yield of cucumber hybrids using female (gynoecious) parents and monoecious male parents were indistinguishable from those obtained using near isogenic gynoecious male parents with the induction of male flower by application silver nitrate at 2-4 leaf stage (Staub *et al.*, 1986).

Characterization of gynoecious-parthenocarpic and monoecious cucumber lines along with their hybrids indicated that the monoecious × gynoecious hybrids expressed predominantly gynoecious sex form and late parthenocarpy. The monoecious parent recorded higher yield as compared to the gynoecious parent. Acute fruit shape at peduncle end was the sole qualitative character which positively influenced the ascorbic acid content of fruit. Regression modelling revealed fruit length, crispy fruit texture and total chlorophyll content to be most influential towards yield in a positive direction (Bose *et al*, 2023).

Breeding for resistance to biotic stresses

Distant hybridization has been carried out to introgress very specific characters like disease resistance from the wild relatives of the cultivated species. Some successful inter-specific hybridization in cucumber are *Cucumis sativus* × *C. trigonus* (source of fruit fly resistance) and *C. sativus* × *C. hardwicki* and *Cucumis sativus* × *Cucumis hystrix* F_1 hybrid is both male and female sterile and fertility could be restored by using *C. hystrix* as the seed parent (Chen and Adelberg, 2000). In all the distant hybridization, it is very difficult to obtain F_1 hybrids from most of such interspecific crossings. Crossability depends on the particular accessions of the two species used as parents however, such hybrids are normally highly sterile because of impaired ability of the staminate flowers to produce functional pollens. Inter-specific crosses among different species is generally done by repeated pollination during the bud and flowering stage although percentage fruit set and embryo size are greatly influenced by the pollen parent. However, the use of *in vitro* culture of the immature embryo by embryo rescue technique is generally employed for obtaining the inter-specific F_1 hybrid.

Table 8: Genes for resistance to biotic stresses in cucumber

Gene symbol	Resistance to disease
Ar	Anthracnose, *Colletotrichum lagenarium* resistance
Bw	Bacterial wilt resistance, *Erwinia tracheiphila* resistance
Cca	*Corynespora cassiicola* resistance
Cm	*Corynespora melonis* resistance
Ccu	Scab, *Cladosporium cucumerinum* resistance
cla	*Colletotrichum lagenarium* resistance
Cmv	Cucumber mosaic virus resistance
dm-1, dm-2, dm-3	downy mildew resistance-1, 2, 3
Foc	*Fusarium oxysporum* f. sp. *cucumerinum* resistance
mj	Root-knot nematode (*Meloidogyne javanica*) resistance from *Cucumis sativus* var. *hardwickii*
pm-1, pm-2, pm-3	powdery mildew resistance-1, 2, 3. Resistance to *Sphaerotheca fuliginia*
psl	*Pseudomonas lachrymans* resistance
Wmv-1, Wmv-2, Wmv-3, Wmv-4	Watermelon mosaic virus resistance
zym	zucchini yellow mosaic virus resistance

Source: Wehner (2005), Swarup (2006), Pandey and Choudhary (2015)

Study on the genetics of powdery mildew resistance revealed that it was controlled by one major dominant and 2 semi-dominant independent genes. In order to analyze the inheritance of powdery mildew resistance Morishita *et al.* (2000) crossed PI 197088-5 (resistant) parent and Sharp One (susceptible) in Japan. They, however, found that resistance was controlled by two recessive genes, and plants with *aabb* showed resistance at both 26 and 20°C, and plants with *aaBB* or *aaBb* showed resistance at 26°C and concluded that PI 197088-5 is a novel gene source for powdery mildew resistance in winter cucumbers in Japan. According to Provvidenti (1987), zucchini yellow mosaic virus (ZYMV) in cucumber was controlled by a single recessive gene, designated *zym*. Pershin *et al.* (1988) suggested that resistance to downy mildew in cucumber was controlled by at least 3 major recessive genes. A full length cDNA clone of the RNA genome of the ZYMV was constructed downstream from a bacteriophage T 7 RNA polymerase promoter, a single extra guanosine residue not present in ZYMV RNA was added to the 5' and 3'ends. Capped (m7GppG) ZYMV RNA transcripts were infectious in 10 of 91 *C. pepo* test plants and uncapped RNA transcripts were not infectious (Gal-On *et al.*, 1991). The 3' terminal sequences of the ZYMV genome from Beijing, China was characterized by RT-PCR from the total RNA of infected leaves and cloned into the pMD18-T vector, and the sequences were 1269bp long and contained a coat protein (CP) gene, which consisted of 837 nucleotides encoding 279 amino acids (Zhang *et al.*, 2006).

Back cross method of breeding is generally employed to transfer the highly heritable character like disease resistance into the superior variety / inbred, which is referred to as the recurrent parent. Often six generations of selection and backcrossing to the recurrent parent are required to recover the desired genotype i.e., the recurrent parent with resistance to specific disease.

Breeding for quality

A distinct group of cultivars of cucumber, known as *Xishuangbanna* gourd, is classified as *C. sativus* var. *xishuangbannesis* Qi & Yuan is grown by the Hani people of southwestern China at elevations of 1000 m or higher. These land races are largely unknown outside of eastern Asia. Rind of this variety is white, yellow or brownish-orange, sometimes with distinct netting, smooth, with yellow to orange flesh colour rich in β carotenes which were used by the breeder to introgress β-carotene into the commercial cultivars. Three varieties namely Early Orange Mass 400 (EOM 400), Early Orange Mass 402 (EOM 402) and Late Orange Mass 404 (EOM 404), were derived from the cross between U.S pickling cucumber and orange fruited Xishuangbanna cucumber PI 509549, from Peoples Republic of China (Simon and Navazio, 1997). Further the inheritance of β-carotene-associated flesh colour in cucumber was studied and it was found that mesocarp and endocarp F_2 segregation adequately fit a 15:1 [low β-carotene (0.01–0.34 lg g^{-1}): high-β-carotene (1.90–2.72 lg g^{-1})] and 3:1 (low-β-carotene: high β-carotene) ratio, respectively (Cuevas *et al.*, 2010). Likewise, segregation of β-carotene concentration in mesocarp and endocarp tissues in BC_1P_2 progeny adequately fit a 3:1 (low β -carotene: high-β-carotene) and 1:1 (low-β-carotene: high-β-carotene) ratio, respectively. Progeny segregations indicate that two recessive genes control the β -carotene content in the mesocarp, while one recessive gene controls β-carotene content in the endocarp.

A large number of orange fleshed cucumber accessions were collected from Mizoram of North-east region of India which had significantly high carotenoid content than Chinese type and thus, appeared to be promising for utilization in breeding programme for enhancing β -carotene content. A unique accession namely IC420405, showing orange flesh (mesocarp and endocarp) colour and lemon yellow skin colour was identified at ICAR-NBPGR, New Delhi (Ranjan *et al.*, 2019a) and subsequently registered with ICAR-NBPGR, New Delhi (INGR18029). The germplasm having common character *viz*., orange flesh colour in the adjoining regions of Myanmar and southern China are indicative of close linkages of continuous variation in this region (Ranjan *et al.*, 2019b).

References

Behera, T.K., Boopalakrishnan, G.,Jat, G. S., Munshi, A.D., Choudhary, H., Ravindran, A., Shilpi Kumari and Rajni Kumari. 2022. Deriving stable tropical gynoecious inbred lines of slicing cucumber from American pickling cucumber using MABB. *Hortic. Environ. Biotechnol.* , **63**: 263–274

Bommesh, J.C., Pitchaimuthu, M., Ravishankar, K.V. and Varalakshmi, B. 2020. Development of tropical gynoecious inbred lines in cucumber (*Cucumis sativus*) and validation by DNA markers. *Agril. Res.*,**9**: 161-168.

Bose S., Hazra, S., Hazra· P., Chattopadhyay, A., Maji, A., Basfore· S.and Karak, C. 2023. Characterization of gynoecious-parthenocarpic and monoecious cucumber lines (*Cucumis sativus* L.) and regression modelling to obtain high yielding and functionally rich genotypes. *Hort. Environ.Biotechnol.*, https://doi.org/10.1007/s13580-023-00579-y

Chen, J.F. and Adelberg, J. 2000. Interspecific hybridization in Cucumis-progress, problems, and perspectives. *Hort Sci.*, **35**: 11-15.

Cuevas, H.E., Song, H., Staub, J. E. and Simon, P.W. 2010. Inheritance of beta-carotene-associated flesh color in cucumber (*Cucumis sativus* L.) fruit. *Euphytica.*,**171**: 301–311.

Gal-On, A., Antignus, Y., Rosner, A. and Raccah, B. 1991. Infectious *in vitro* RNA transcripts derived from cloned cDNA of the cucurbit potyvirus, zucchini yellow mosaic virus. *J. Gen. Virol.*, **72**: 2639-2643.

Hayes H K and Jones D F. 1916. First generation crosses in cucumber. *Republic Conferance of Agricultural Experiment Statistics*, **5**: 319–22

Hazra, P., Banerjee, M.K. and Chattopadhyay, A. 2022. *Varieties of Vegetable crops in India*, Kalyani Publishers, New Delhi, Ludhiana, 204p.

Jat, G.S., Munshi, A.D., Behera, T,K. and Bhardwaj, C. 2017. Inheritance of parthenocarpy in gynoecious cucumber(*Cucumis sativus* L.) cultivar PPC-2. *J. Hort. Sci.*, **12**: 193-197.

More, T.A. and Munger, H.M. 1987. Effect of temperature and photoperiod on gynoecious sex expression in cucumber. *Veg. Sci.*, **14**:42–50

More, T.A. and Seshadri, V.S. 1988. Development of tropical gynoecious lines in cucumber. *Rep. Cucurbit Genet. Coop.*, **11**:17–18

Morishita, M., Sugiyama, K. and Saito, T. 2000.Gene sources of powdery mildew resistance in cucumber. *Acta Hort.*, **521**: 83-90.

Munger, H.M. 1979. A summary of cucumbers released from Cornell breeding program. *Vegetable Improvement Newsl. Cornell Univ*. USA, **24**: 3–4

Munshi, A.D., Panda, B., Kumar, R., Bisht, I.S. and Behera, T.K. 2007. Genetic variability in *Cucumis sativus* var. *hardwickii* R. germplasm. *Cucurbit Genet Coop Rep.*, **30**: 5–10

Pandey, S. and Choudhary, B.R. 2015. Cucumber. In: *Hand book of Vegetables* Vol.III (Eds. Peter, K. V. and Hazra, P.), Studium Press LLC, Houston, Texas, USA, pp. 121-140.

Pershin, A. F., Medvedeva, N. I. and Medvedev, A. V. 1988. Quantitative approach to studying the genetics of disease resistance. IV. Interaction of the genetic systems for resistance to powdery and downy mildews in cucumber. *Genetika*, **24**: 484-493.

Provvidenti, R. 1997. New american summer squash cultivars possessing a high level of resistance to a strata of zucchini yellow mosaic virus from china. *Rep. Cucurbit Genet. Coop.*, **20**: 57-57.

Ranjan, P.**,** Gangopadhyay, K.K., Bhardwaj, R., Pandey, C.D., Srivastava, R., Bag, M.K., Prasad, T.V., Meena, B.L., Dutta, M., Bansal, K.C., Pal, A.K., Koley, T.K., Pandey, S., Singh, B., Munshi, A.D. and Harish, G.D. 2019a. IC420405 (IC0420405; INGR18029), a Cucumber (*Cucumis sativus*) Germplasm with High Carotenoid Content. Orange Flesh Colour. *Indian J. Plant Genet. Resour.*, **32**: 437.

Ranjan, P., Pandey, A., Munshi, A.D., Bhardwaj, R., Gangopadhyay, K.K., Malav, P.K, Pandey, C.D., Pradheep, K., Tomar, B.S. and Kumar, A. 2019b. Orange-fleshed cucumber (*Cucumis sativus* var. *sativus* L.) germplasm from North-East India: agro-morphological, biochemical and evolutionary studies. *Genet. Resour. Crop. Evol.*, **66**: 1217–1230.

Sebastian , P., Schaefer, H. and Renner, S.S. 2010. Cucumber (*Cucumis sativus*) and melon (*C. melo*) have numerous wild relatives in Asia and Australia, and the sister species of melon is from Australia. *Proc. Nat. acad. Sci.* USA, 107:14269-14273

Simon, P.W. and Navazio, J.P. 1997. Early orange mass 400, early orange mass 402, and late orange mass 404: high-carotene cucumber germplasm. *Hort Sci.*.,**32**: 144-145.

Staub, J.E., Balgooyen, B. and Tolla, G.E. 1986. Quality and yield of cucumber hybrids using gynoecious and bisexual parents. *Hort Sci.*, **21**: 510-512.

Swarup, V. 2006. *Vegetable Science and Technology in India.* Kalyani Publishers, New Delhi, Ludhiana, 656p.

Wang, L. , Zhang, B. , Li, J. , Yang, X. and Ren, Z. 2014. Ethyl Methanesulfonate (EMS)-Mediated Mutagenesis of Cucumber (*Cucumis sativus* L.). *Agril. Sci.*, **5**: 716-721.

Wehner, T.C. 2005. Gene List 2005 for Cucumber. *Rep. Cucurbit Genet. Coop.*, **28-29**:105-141.

Yamasaki, S., Fujii, N., Takahashi, H., Mizusawa, H. and Matsuura, S. 2002. Molecular approach to the study of sex determination in cucumber plants. *Acta Hort.*, **588**: 309-312.

Zhang, Y.J., Li, M.F., Xiang, N. and Li, G.F. 2006. Identification and the analysis on characterization of the 3′ terminal sequences of the genome of Zucchini yellow mosaic virus from Beijing. *Acta Phytophylacica Sin.*, **33**: 32-36

37

Watermelon

Watermelon, *Citrullus lanatus* under the family Cucurbitaceae (Chromosome number: $2n = 2\times = 22$) is an important and popular summer season crop of the USA, Middle east, Japan, Africa and India. Watermelon is primarily consumed fresh for their sweet, juicy and slightly grainy flesh and are often eaten as desserts. Watermelon is actually a nutrient dense food. It provides high levels of vitamins, minerals, andantioxidants particularly lycopene and relatively low in calories.

Origin, domestication and adaptation in India

The origin of watermelon was in tropical Africa perhaps in Kalahari Desert, now in Botswana. The progenitor species, *Citrullus colocynthis* is endemic to Africa. Wide variation of this progenitor species is also found in Rajasthan and Gujarat in India. Wild watermelon seeds have been found in Uan Muhuggiag, a prehistoric site in Libya that dates to approximately 3500 BC. Watermelon was domesticated in Africa and later it was introduced in India. From India it spread to Near East and Asia. The watermelon which was domesticated in Egypt by 2000 BC were not the sweet modern variety. The fruit flesh of wild watermelons is watery, but typically hard-textured, pale-coloured and bland or bitter. The familiar sweet dessert watermelons, *C. lanatus*, featuring non-bitter, tender, well-coloured flesh have a narrow genetic base, suggesting that they originated from a series of selection events in a single ancestral population. North-eastern Africa is the centre of origin of the dessert watermelon which were domesticated for water and food there over 4000 years ago, and the sweet dessert watermelons emerged in Mediterranean lands by approximately 2000 years ago (Paris, 2015). Sweet dessert watermelons spread across the Mediterranean world during Roman times. It spread to China in 10th or 11th century AD and was grown in the Americas before or after the migration of Europeans there.

In India it is known from ancient times. It was mentioned in the name "Kalingam" in ancient Sanskrit literatures. Wide cultivation from very early period resulted evolution of extensive variability and this designated India as the secondary centre of diversity of this crop.

Crop botany and floral biology

- It is a trailing annual with stems, as long as 400 cm; roots are shallow (40-50 cm) and extensive (60-90 cm), with taproot and many lateral roots.

- The stems are hairy, rounded to angular in cross-section, and have branched tendrils at each node which are pinnately divided into three or four pairs of lobes.
- Leaves are ovate (oval but broader towards the base) to obovate (oval but broader at the apex), 8-20 cm long, scabrid (rough to touch) and deeply pinnatified and the lobes are pinnately divided into three or four pairs of lobes.
- Plants are generally monoecious in sex form which bears separate staminate and pistillate flowers on the same plant. The flowers are unisexual, solitary, axillary and pedicillate.
- Hermaphrodite sex form is rarely found in some genotypes; flowers are small, yellow, not very showy and borne in the leaf axils.
- Staminate flowers are also borne singly and have 12-30 cm long pedicels, whereas the pistillate flowers are usually solitary with comparatively short and stout peduncles.
- Calyx lobes alternate with corolla lobes.
- Flowers are small, 4-5 cm across, sulphur-yellow in colour, and less showy than other cucurbits.
- Petals are 5, pasted, rotate or broadly campanulate, greenish yellow and united in a tube.
- Stamens are usually 3, inserted at the base of corolla; filaments are free, two stamens having two locules (thecae); anthers are flexuous.
- Ovary is inferior; carpels are 1-5, usually 3, thick, short styled, terminated by 3 bilobed or divided papillate stigma. The stigma is wet consisting of multicellular papillae invested by copious extra cellular secretions.
- The fruit, botanically called "pepo" consists of a firm outer rind, a layer of white inner rind, flesh of 0.5-1.4 cm thick, and an interior coloured edible pulp embedded with seeds; edible part of the fruit is the endocarp (placenta), which contrasts with muskmelon (*Cucumis melo*), where the edible part of the fruit is the mesocarp.
- Fruit is round to oblong or cylindrical in shape; skin is smooth with colour ranging from yellow, orange, light green to almost black, either solid or striped with a paler green or marbled; flesh (endocarp) may be white, cream, yellow, pale red, red or dark red; fruit rind varies from thin to thick and from brittle to tough.
- The seeds are embedded in the edible pulp, small in size (4-6 cm), flat, smooth and of many colours such as white, tan, greenish-red and black; average 100 seed weight is generally11.3 gram.

Reproductive behaviour

Monoecious sex form is predominant in watermelon and rarely hermaphrodite sex form is found. It is entomophillous and major pollinators are bees (*Apis florea, Apis dorsata, Nomioides sp., Halictine* sp., etc.). Beetles like *Conopophilus* sp., moths like, *Planidia* sp., *Pygargonia* sp. also helps in natural cross-pollination. Although 100% natural cross-pollination occur in monoecious sex form, yet reasonable extent of self-fertilization happens because of geitenogamy due to foraging of insect pollinators in the staminate flowers of the same vine. Watermelon like other members of the Cucurbitaceae family show negligible inbreeding depression. Anthesis of both staminate and pistillate flowers generally starts in the morning hours from 5.30 to 8.00 a.m. Anther dehiscence in the staminate flower generally starts one hour before anthesis and continues till 8.00 a.m. Pollen viability is maximum at anthesis and it decreases drastically if kept at room temperature for five hours. Pollen from the staminate/hermaphrodite flowers at the optimum stage can be stored for a few days in a desiccator. Stigma receptivity is maximum at the time of anthesis. At high temperature condition, stigmatic fluid starts drying and the stigma becomes non-receptive. Best time of hybridization is between 6.00 to 9.00 a.m on the day of anthesis. Bagging the staminate and pistillate flowers in the afternoon hours prior to opening followed by hand pollination in the next morning and subsequent bagging for two more days is the commercial method of hybrid seed production of watermelon. Fully matured staminate flower buds that will open in the next morning should be plucked in the afternoon hours and kept them covered with thin moist cloth over night for their use in cross-pollination in the morning hours of the next day.

Genetics of important characters

The basic and haploid chromosome numbers of watermelon (*C. lanatus*) and its wild relatives viz., *C. colocynthis*, *C. ecirrhosus*, *C. naudianus* and *C. mucosospermus* are same, x = n =11 (2n = 22). Many important qualitative characters are controlled by single gene *viz*., andromonoecius sex form, fruit skin colour, fruit shape, flesh colour, bitterness in fruit, seed size, seed colour, etc.

Table 1: Some important monogenic characters of watermelon

Gene symbol	Phenotype of plant character
a	Androecious; recessive to monoecious
C	Canary yellow flesh; dominant to pink flesh
d	Dotted seed coat; flat black seed coat colour is monogenically dominant to the dotted black seed coat colour
dw-1, dw-2	Dwarf; short internodes
e	Explosive rind; thin, tender rind which burst when cut
f	Furrowed fruit surface
g	Light green skin; recessive to dark green skin
g^s	Striped green skin; recessive to dark skin, dominant to light-green skin
go	Golden yellow colour of matured and older leaves
gms	Glabrous male sterility
I	The *s* gene is epistatic for *l*, and *LL SS* is assigned to medium-size seeds; *ll SS*, large-size seeds; and *LL ss* or *ll ss* is assigned to small-size seeds
m	Greenish-white mottling on fruit skin
nl	Non-lobed leaves; lobed leaf show incomplete dominance over non-lobed

Gene symbol	Phenotype of plant character
O	Oval fruit; incompletely dominant to spherical shape
p	Pencilled lines on fruit skin; recessive to netted fruit skin
R	Dotted black seed coat colour (*R_*) is dominant to green seed coat colour (*rr*)
r	Red seed coat (*rrttWW)*; interacts with *w* and *t*
s	Short seeds; epistatic to *l*
Scr	Scarlet flesh colour
Sp	Spotted cotyledon, leaves and fruits
su	Non-bitter fruit character
Wf	White flesh; dominant to yellow and red
y	Yellow flesh; recessive to red
Yb	Yellow belly (ground spot)

Source: Ram (1998), Gusmini and Wehner (2006), Swarup (2006), Paudel *et al* (2019)

Many quantitative characters *viz*., maturity, earliness, fruit shape index, fruit weight, rind thickness, total soluble solids, etc. are governed by polygenes.

Table 2: Gene action for different quantitative characters

Quantitative character	Gene action
Earliness	Additive, dominance, epistasis; Intermediate in maturity
Fruit shape index	Partial dominance
Fruit weight	Dominance, epistasis (dominance x dominance)
Rind thickness	Partial dominance
Total soluble solids	Additive; both additive and dominance variance

Source: Swarup (2006)

Breeding objectives

Some of the important characters that have been the main focus of watermelon breeding are mentioned below

- Medium sized, strong plant with more number of branches and high female to male flower ratio in the vine
- High number of fruits per plant
- High total and early yield potential and wider adaptability
- Early maturing; lower number of node at which first pistillate flower appears which gives a fair indication of earliness
- Jubilee pattern type, large, elongated melons with finely-textured, bright pink-red flesh and sweet
- Sugar Baby and Icebox type with 3 to 3.5 Kg fruit weight, blackish-green colour skin, bright pink-red flesh
- Early maturing, spherical or oblong (ice box type) shaped fruit
- Rind colour solid dark to light green or striped and tough, resistance to cracking which are useful for transportable quality

- Flesh should be firm and colour should preferably be deep pink or pink or yellow with fewer and smaller seeds.
- Fruit should be sweet and 10°Brix TSS may be taken as a minimum standard for selection of fruits to meet the consumer acceptance
- High fructose and lycopene contents in the flesh
- Resistance /tolerance to different diseases like, watermelon bud necrosis virus, anthracnose (*Colletotrichum orbiculare*), fusarium wilt (*Fusarium oxysporum* f.sp. *nivenum*), powdery mildew (*Podosphaera xanthii*), etc.
- Resistance/tolerance to insect pests like, fruit fly (*Bactrocera cucurbitae*), red pumpkin beetle (*Aulacophora foveicollis*), etc.
- Resistance to abiotic stresses like cold temperature, soil salinity, etc.

Breeding methods and development of variety

Watermelon was introduced in India in very early times, which resulted evolution of extensive variability and this designated India as the secondary centre of diversity of this crop. Even today, different local cultivars are available in India and are grown, though in minimum scale, in different names like, Farrukhabadi, Moradabadi, Faizabadi in Uttar Pradesh; Mateera in Bikaner, Rajasthan; Gangasagar in West Bengal; Katagolan which is cultivated in the riverbed of Jamuna keeps well at ambient temperature. However, these indigenous cultivars have mostly been replaced by improved varieties, particularly hybrids. Watermelon cultivars grown in India comprise of exotic varieties, selection from different locally adapted types and exotic introductions. Good number of the varieties have been developed through inter-varietal hybridization followed by pedigree, backcross or some modified methods of handling the segregating generations. Triploid seedless watermelon hybrids have also been developed through alteration of ploidy method of breeding. Watermelon cultivars grown in India may be classified into different groups.

- **Exotic varieties:** Asahi Yamato, Sugar Baby, New Hampshire Midget, Improved Shipper, Dixielce, Fuken, Charleston Grey
- **Early maturing variety:** Special No.1, PKM-1, Thar Manak, AHW-19, AHW-65
- **Late maturing variety:** Durgapura Meetha, Durgapura Kesar
- **Ice-box type variety:** Arka Muthu
- **Disease resistant variety:** Arka Manik
- **Seedless triploid hybrid:** Pusa Bedana, Arka Madhura, Swarna, Shonima

Germplasm resources and Plant introduction

Watermelon germplasm collections are maintained in China, South Africa and Zimbabwe. The United States Department of Agriculture (USDA), Agricultural Research Service (ARS), National Plant Germplasm System (NPGS), maintains a large collection of watermelon and related *Citrullus* spp. germplasm. The USDA/

ARS/NPGS, Germplasm Resources Information Network (GRIN) http:// www. ars-grin. gov/ npgs contains general information on accessions held within the USDA/ NPGS collection. Other countries *viz*., Bulgaria, Czechoslovakia, Germany, Hungary, India, Turkey, Italy, Nigeria, Pakistan, Peru, Phillipines and Russia also maintains wide range of germplasm collections of watermelon.

In India, watermelon germplasm is maintained at NBPGR, New Delhi, ICAR-IIHR, Bengaluru, ICAR-IIVR, Varanasi, SKNAU, Jobner, Durgapura, Rajasthan, KAU, Vellanikkara, Kerala and other places. Maximum varieties grown in India are either exotic variety mainly from USA and Japan or different private sector hybrids.

Table 3: Exotic variety directly released for cultivation

Name of the variety	Country from where introduced
Asahi Yamato	Japan
Sugar Baby	USA
New Hampshire Midget	USA
Improved Shipper	USA
Dixielce	USA
Fuken	Japan
Charleston Grey	USA

Individual plant selection and inbreeding

In this method, an individual plant from the source population consisting of heterozygous and non-uniform materials is selected based on the selection indices and unlike mass selection, the seeds of the selected plants are not mixed even if they are equally good and look similar. This method of selection is based on the principle that the actual worth of the selected plant may be masked by the environmental variation and, therefore, the only way to recognize it is by growing the progeny of each selected individual plant for re-evaluation. Progenies of the individual plant for evaluation can be developed by selfing as watermelon do not show loss of vigour due to inbreeding. Homogzygosity for the concerned characters can also be attained in the individuals of the progeny by such repeated selfings. Good number of varieties have been developed by this method of breeding from different indigenous genetic resources.

Table 4: Varieties developed through selection and inbreeding

Name of variety	Indigenous / Exotic material
Durgapura Meetha	Indigenous
Durgapura Kesar	Indigenous
Special No.1	Indigenous
Arka Muthu	Indigenous (Attur, Tamil Nadu)
PKM-1	Indigenous
Thar Manak	Indigenous
AHW-19	Indigenous (Bikner, Rajasthan)
AHW-65	Indigenous (Bikner, Rajasthan)

Hybridization and selection

Planned hybridization involving both indigenous and exotic genetic materials followed by pedigree method of selection in segregating generations or modified backcross pedigree method of breeding has been followed for the development of very few varieties in India (Hazra *et al*, 2022). Parents to be employed in the hybridization programme should possesses specific characters which are expected to be combined in the segregating generations. The parental lines should also be genetically diverse for obtaining high heterotic F_1 hybrids and broad spectrum of variability in segregating generations. Selection in segregating generations should be based on the specific characters to be incorporated in the recombined lines and important selection indices for fruit yield and quality. Multiple disease resistant variety, Arka Manik has been developed following this method breeding.

Table 5: Varieties bred through hybridization followed by selection

Name of variety	Parentage
Arka Manik	HR21 (a collection from Rajasthan) × Crimson Sweet (USA)

Heterosis breeding

In watermelon, heterosis breeding has widely been adopted in crop improvement in different countries including India. Being dessert fruits, quality characteristics of the hybrids should be superior in respect of uniformity and stability over pure bred cultivar. Earlier maturity is another major advantage in the acceptance of melon hybrids. Watermelon hybrids has become very popular among the farmer mainly due to high female to male flower ratio resulting high number of fruits per plant, fruits are of attractive shape, very sweet, crispy, deep pink fleshed and dark green skin, good transportability, wide adaptability, extended fruiting span and high yield. It has been estimated that about 55-60 % watermelon area in India is at present under different hybrids. The monoecious sex form of watermelon only necessitates bagging of both staminate and pistillate flowers on the intended male and female parents before opening followed by hand pollination for the production of hybrid seeds. Bagging of staminate and pistillate flowers in the afternoon hours and hand pollination of the pistillate flower in the morning hours of the next day with the pollens from the staminate flower of the intended male parent is the commercial method of hybrid seed production. The glabrous male-sterile (gms) mutant was male sterile due to the absence of synapsis in meiosis (Ray and Sherman, 1988). Use of male sterile line may reduce the cost of hybrid seed.

Development of triploids and seedless hybrid

Triploids contain 3 complete sets of chromosome and can be developed by hybridizing normal diploid (2x=22) with induced tetraploid (4x=44). Triploidy arise by the fusion of gametes containing haploid (n) and dihaploid (2n) chromosomes. In triploids, only trivalents are produced in meiosis instead of normal bivalents in the diploids. These trivalents do not disjoin in equal number of chromosomes to the two poles. Therefore, the number of chromosomes in the gametes would vary from n to 2n. Most of these gametes are unbalanced leading to high degree of sterility. In triploid watermelon, frequency of viable gametes is only 0.1 percent. So seeds are not produced in the fruits of the triploids. This phenomenon has been commercially utilized in developing four seedless triploids. It is essential to artificially pollinate the flower of 3x plant with diploid viable pollens to induce fruit set.

Seedless watermelon is one classic example of the practical use of colchicine and application of autopolyploidy in crop improvement. There are several steps in producing seedless watermelon fruits (Kihara, 1951; Kondo, 1955).

Production of tetraploid lines

Application of colchicine (0.2 % to 0.4 %) to the growing points of young seedlings (at 1 to 2 true-leaf stage), for 2 to 3 days successively, gives satisfactory results. The treatment is done under controlled conditions avoiding direct sunlight. Growth of tetraploids is slow and terminal bud should be carefully promoted to grow. The 4x plants are characterized by "gigantism" characters like thick and broad leaves, bigger stomata and pollen grain. When the plants start flowering, the pollen grains may be tested for fertility and also the size of pollen grains. Fruiting will be less in first generation with very few seeds, around 30 to 40 due to high pollen sterility. The seeds will be broad and bigger. The 4x plants should be checked for chromosome number, both in somatic cells and in meiosis. Perfect 11 quadrivalents is not generally found rather, it ranges from 4 to 10 with varying number of univalents, bivalents and trivalents.

Maintenance of tetraploid lines at stable level is important and continuous selection for improvement of quality and vigour in tetraploid lines has to be done for at least 4 or 5 generations. Chance reversion to diploidy can be detected by numerous small seeds in the fruits. Tetraploid fruits have characteristic large blossom scar. Higher fertility during continuous selection is possibly related to lesser multivalent frequency, which probably leads to the true breeding behaviour. Attainment of superior fruit quality (firm flesh, deep colour, stable TSS content of 12 to 13 %) in the tetraploids is an important requirement for the production of high quality seedless fruits.

Adequate caution is needed during seed extraction from the fruits of tetraploids. Fermentation of the pulp should be avoided which will affect the seed germination. Similarly, at the time of sowing of seeds of tetraploids, overnight pre-soaking should be avoided. Seeds need to be germinated under controlled conditions ensuring slow and continuous absorption of moisture.

Production of 3x seeds

Triploid (3x) seeds can be produced by the following methods

(i) Hand pollination of the pistillate flower of the tetraploid line by the pollens of the diploid male parent

(ii) Growing adjacent rows of tetraploid female and diploid male lines in isolated field, pinching of male buds of tetraploid plants and leaving the pollination to the insects and fruit set to natural condition of insect pollination

In 1:1 planting of tetraploid female and diploid male, the percentage of 4x seeds (by sib/selfs) under open pollinated conditions comes to about 30 and the triploid and tetraploid seeds cannot be distinguished. The production of 3x seeds is, therefore, expensive partly because of low seed content of the tetraploid lines and mixture of 4x seeds. Under open pollinated conditions hardly 4.5 to 6.8 kg per acre of triploid seeds can be produced.

Another difficulty arises at the time of fruiting, when 2x, 3x and 4x fruits have to be distinguished. Hence, at the time of choice of parents for crossing 4x plants with 2x male, contrasting fruit characters are chosen. If 4x plants produce fruits with light green rind and stripes and 2x male chosen, would produce fruits of dark green rind, the 3x fruits will have intermediate green rind with stripes. The 3x fruits will have triangular configuration, in contrast to round fruits of 4x and 2x plants.

Gene transfer system may be useful in the improvement of tetraploid lines and in the production of seedless triploid watermelon. Resistance to fusarium wilt disease has been incorporated in tetraploid lines in the USA and Japan so that triploids carry resistance.

Germination of 3x seeds

Good seed germination is important because of high cost of production of 3x seeds. Poor germination is caused by hard seed-coat. Seeds will have to be sown in flat or peat cups with moisture carefully maintained at a temperature of 24-29°C in the first 72 hours. Naturally, germination of seedlings under controlled conditions and their subsequent transplantation increase the cost of production of seedless fruits. Transplantation of seedlings should be done in 2-true-leaf stage.

Production of 3x fruits

There will be no fruit set in triploid plants, if there are only 3x plants in the field. Even a few tetraploid plants occurring as a mixture would not be enough for good fruit set. Stimulation of fruit development has to be brought about by 2x pollen through bee pollination. Hence a pollinizer, a diploid cultivar has to be planted in the field. Ne Smith and Duval (2001) recorded that greatest number of triploid fruit per unit land area was in the harvest row 3.0 m from the pollinizer row. When the distance from the pollinizer row was 6.0 m or greater, triploid fruit numbers diminished substantially. Yield estimates made each year using the fruit density data suggested that a 1 pollinizer: 4 triploid ratio gave the maximum total triploid fruit yield per hectare for 1.5 m row spacing. These results should prove useful in designing field planting strategies to optimize triploid watermelon production.

The 3x plants are of long duration, fruits maturing in 120 to 130 days in contrast to 85-90 days in diploid cultivars. The firm texture of the rind is an attractive feature, but there may be few empty ovular cavities in the fruit. Sweetness should be uniformly distributed from the rind to the centre. Triploids are sterile both on male and female sides and unequal chromosomal distribution in meiosis causes pollen and ovular abortion. However, few ovules may develop into a hardened seed-coat and these are called 'ex-embryonate' or "false' seeds. In India, four triploid hybrids have so far been developed at ICAR-IARI, New Delhi, ICAR-IIHR, Bengaluru and Kerala Agricultural University, Vellanikkara, Kerala.

Table 6: Development of triploid hybrids

Name of variety	Parentage
Pusa Bedana	Tetra 2 (tetraploid) × Pusa Rasal (diploid)
Arka Madhura	Tetraploid × Diploid
Swarna	Tetraploid × Diploid
Shonima	Tetraploid × Diploid

Breeding for resistance to diseases and insect-pests

Resistance to different diseases *viz*., anthracnose, gummy stem blight, fusrium wilt, powdery mildew and to insects *viz*., fruit fly and red pumpkin beetle are governed by single gene. Some cultivars show some level of tolerance to anthracnose, powdery mildew and bacterial fruit blotch. Few wild relatives of watermelon *viz*., citron (*Citrullus lanatus* var. *citroides*), *C. colocynthis* and *C. naudianus* are resistant to several diseases and insect pests. The citron is resistant to *Fusarium* wilt, *C. colocynthis* shows resistance to watermelon mosaic virus (WMV-1, WMV-2) and zucchini yellow mosaic virus and *C. naudianus* shows immunity to anthracnose. These wild species are cross-compatible among themselves as well as with watermelon. These wild species have been utilized in resistance breeding of watermelon.

Table 7: Genes for resistance to disease and insect pests in watermelon

Gene symbol	Resistance to the disease/ insect pest
Af	Resistance to red pumpkin beetle (*Aulacophora foveicollis*); dominant control
Ar-1	Resistance to anthracnose (*Colletotrichum orbiculare), race 1;* dominant control
Ar-2	Resistance to anthracnose (*Colletotrichum orbiculare), race 2;* dominant control
dh	Resistance to gummy stem blight (*Didymella bryoniae*); recessive control
Fo-1	Resistance to *Fusarium oxysporum* race 1; dominant control
Fwr	Resistance to fruit fly (*Bactrocera cucurbitae*); dominant control
pm	Resistance to powdery mildew (*Podosphaera xanthii*); recessive control

Source: Swarup (2006)

Breeding for quality

Fruit quality traits of sweet watermelon are crucial for new product development and commercialization. Sweet watermelon fruit quality traits are affected by the compositions of phytochemical compounds, phytohormones, and fruit flesh firmness which are affected by genes, the growing environment and their interaction. These compositions determine fruit ripening, eating quality, and postharvest shelf-life. Knowledge of the genetic profile and analyses of quality traits in watermelon is vital to develop improved cultivars with enhanced nutritional compositions, consumer-preferred traits, and extended storage life. Genetic mechanisms involved in the biosynthesis of phytochemical compounds (i.e., sugars, carotenoids, amino acids, organic acids, and volatile compounds), phytohormones (i.e., ethylene and abscisic acid) and fruit flesh structural components (i.e., cellulose, hemicellulose, and pectin) elicited during fruit development and ripening need to be critically considered for selection of genetic materials and framing of breeding method. Gene-editing technology and innovations associated with fruit quality traits for selection and accelerated cultivar development is useful for watermelon variety design, product profiling and development to serve the diverse value chains of the crop (Mashilo *et al*, 2022).

References

Gusmini , G. and Wehner , T.C. 2006. Qualitative Inheritance of Rind Pattern and Flesh Color in Watermelon. *J. Hered.*, **97**: 177-185.

Hazra, P., Banerjee, M.K. and Chattopadhyay, A. 2022. *Varieties of Vegetable crops in India*, Kalyani Publishers, New Delhi, Ludhiana, 204p.

Kihara, H. 1951. Triploid watermelons. *Proc. Amer. Soc. Hort. Sci.*, **58**: 217-230.

Kondo, N. 1955. *Studies on the triploid watermelon*. The Institute for Breeding Research, Tokyo Agricultural University, 52p.

Mashilo, J., Shimelis, H., Mantlo, R., Nqwepe, R.M. and Thungo, Z. 2022. Genetic analysis of fruit quality traits in sweet watermelon (*Citrullus lanatus* var. *lanatus*): A Review. *Front. Plant Sci.*, Volume 13 - 2022 , https: // doi.org /10. 3389 / fpls. 2022. 834696

NeSmith, D. S. and Duval, J. R. 2001. Fruit set of triploid watermelons as a function of distance from a diploid pollinizer. *Hort Sci.*, **36**: 60-61.

Paris, H.S. 2015. Origin and emergence of the sweet dessert watermelon, *Citrullus lanatus. Ann Bot.*, **116**: 133-148.

Paudel, L., Clevenger, J. and McGregor , C. 2019. Chromosomal locations and interactions of four loci associated with seed coat color in watermelon. *Front. Plant Sci.*, doi: 10.3389/fpls.2019.00788

Ram, H.H. 1998. *Vegetable Breeding. Principles and Practices*. Kalyani Publishers, New Delhi, Ludhiana.

Ray, D. T., and Sherman, J. D. 1988. Desynaptic chromosome behavior of the GMS mutant in watermelon. *J. Hered.* **79**: 397–399.

Swarup, V. 2006. *Vegetable Science and Technology in India*. Kalyani Publishers, New Delhi, Ludhiana, 656p.

38

Pumpkin and Summer squash

Pumpkin (*Cucurbita moschata*) Duch. ex Poir. and summer squash (*Cucurbita pepo*) under the family Cucurbitaceae (Chromosome No. $2n = 2\times = 40$) are important tropical vegetable crops grown all around the world. The term "squash" is generally employed to designate the forms of *C. pepo* for immature fruits, all baking cultivars of *C. maxima* and the cushaw-type cultivars of *C. mixta* whose matured fruits are consumed. The term "pumpkin" is normally applied to the edible fruit of *Cucurbita moschata*, *C. maxima* and *C. argyrosperma* utilized when ripe as boiled or cooked vegetable or in pie. However, pumpkin is mostly associated with *Cucurbita moschata* Duch. ex Poir. Pumpkin cultivars are generally more tolerant to hot, humid weather than cultivars of *C. maxima* or *C. pepo*. It is extensively cultivated in India, Africa, Latin America, Southern Asia and United States. Among the five domesticated *cucurbita* species, *C. moschata*, *C. pepo* and *C. maxima* are economically most important and among them summer squash, *C. pepo* is the most important in the world. However, of the different species of *Cucurbita*, pumpkin (*C. moschata)* is the most popular and grown throughout India.

Pumpkin fruits are rich in ß carotene, flavonoid, polyphenolic antioxidants such as lutein, xanthin and different minerals *viz*., calcium, potassium and magnesium. Pumpkins help in keeping the heartbeat regular and blood pressure low. The fibre in pumpkin can also play a part in lowering blood pressure as well as cholesterol. Pumpkin occupies a prominent place among the vegetables owing to its high productivity, nutritive value, good storability, long period of availability and better transport potentialities.

Origin, domestication and adaptation in India

Cucurbita species under the family, Cucurbitaceae are native to the Andes and Mesoamerica (southern North America and most of Central America which extends from approximately central Mexico through Belize, Guatemala, El Salvador, Honduras, Nicaragua, and northern Costa Rica). Not surprisingly, the wild species, *C. lundelliana* and *C. martinezii*, which are evidently closely related to the cultivated species, occur in this area. The five domesticated *species of Cucurbita viz*., *Cucurbita moschata*, *C. pepo*, *C. maxima*, *C. argyrosperma* and *C. ficifolia* are grown worldwide for their edible fruits, variously known as squash, pumpkin, or gourd, depending on species, variety, and local parlance. Native areas of the cultivated species of *Cucurbita* appear to be North Mexico of Mexico City for *C. pepo*, South America for *C. maxima*, South America, Central America and Mexico for *C. moschata*, Mexico and South Mexico

City for *C. mixata* and Central America and Mexico for *C. ficifolia* (Whitaker and Robinson 1986).

All the three crops of *Cucurbita* sp. are introduced crop in India. Huge genetic variability in pumpkin, in particular, suggests its early introduction and is favoured due to its wide acceptability throughout India.

Table 1: Botanical description of three *Cucurbita* species

Character	**Pumpkin (*C. moschata*)**	**Winter squash (*C. maxima*)**	**Summer squash (*C. pepo*)**
Growth habit, sex expression, foliage texture	Annual vine, Monoecious, Foliage soft hairy	Annual vine, Monoecious, Foliage is stiffly hairy	Annual vine or bush, Monoecious, Foliage is stiff, upright, harsh, and prickly to touch
Foliage shape	Foliage 5-6 lobed, lobes acute or rarely obtuse, often with whitish blotches on the upper surface	Foliage more or less reniform with 5 rounded shallow lobes, margin minutely dented	Foliage broad, obtuse or acute, usually with 5 deep lobes, margin dentate, upper surface glabrous and lower with glandular club shaped hairs
Tendril	Tendril many fid	Tendril 2-6 fid	Tendril branched
Flower	Solitary	Solitary	Solitary
Calyx	Calyx tube of staminate flower short, lobes 3, linear, very often leafy	Calyx tube club shaped, lobes 5, short, linear and hairy	Calyx tube obscurely 5-angled
Flower	Corolla campanulate, pale yellow to yellow, mostly reflexed, lobes 5, stamen 3, filamentous, swollen at the base, glanduliform	Corolla campanulate, light to deep yellow, lobes 5, reflexed, stamen 3, filaments thick, glanduliform	Corolla campanulate, bright yellow to organge with erect or spreading lobes, stamens 3, filaments swollen below
Peduncle	Smooth, 5-angled, expanded or flared at fruit attachment	Spongy, cylindrical, soft, corky	Five angled with little or no expansion at fruit attachment
Fruits	Usually large of variable shape and size, globuler, cylindrical, flattened	Usually large, oblong-cylindrical or flattened-cylindrical	Usually of variable size, shape and colour
Seeds	Yellow with a thin or rugged margin, scalloped or shredded in appearance, the margin more deeply coloured than the body of the seed	Seed white or yellowish with margins of different colour	Seeds whitish-yellow, broadly or narrowly ovate, usually with raised, smooth, well-differentiated margin

Source: Whitaker and Davis (1962), Chakravarty (1982)

Crop botany and floral biology

- Pumpkin is an annual crop with creeping or climbing stems (growing up to 5 m) bearing tendrils. The stems are strong, cylindrical or pentangular, with petioles measuring 12–30 cm; stems and leaves are mildly hairy.
- The leaves are circular, kidney-shaped, heart-shaped or triangular, often deeply indented at the base, weakly lobed, wavy and toothed, more or less white spotted, up to 20 cm long and 30 cm wide; broad leaves are attached to the thick stems and, in most cultivars, are covered in trichomes that may induce dermatitis on contact.
- Plants are monoecious in sex form with unisexual flower; after about 6-8 weeks of growth, the plant produces flowers; flowering and anthesis is influenced by temperature.
- Flowers are solitary in leaf axil; corolla of the staminate flowers are generally larger than pistillate flowers and with long pedicels.
- Corolla campanulate (bell-shaped), pale yellow, bright yellow to orange in colour, 5 lobed half way down and up to 12 cm long, mostly reflexed and gamopetalous in nature; petals follow inner circle, each petal alternating with a paired sepal.
- The staminate flowers produce pollen with large granules and the pistillate flowers are pollinated by honey bee and bumble bee.
- Calyx with 5 sepals adnate to the ovary, long stalked; the paired sepals arise first around the margin of receptacle; calyx tube of staminate flower is shorter than that of the pistillate flower.
- Stamens are short and usually 3, filaments are free and swollen at the base, glanduliform; anthers are united into a column like mass. Pollen grains are heavy and sticky.
- Ovary is typically inferior, tricarpellary and occasionally 4 or 5, syncarpous, unilocular and ovules on parietal placentation; the reflexed margins of the carpels form the placentae which appear as longitudinal parietal ridges to which the ovules are attached; ovary has a thick fleshy style topped by three large, forked stigmas.
- Fruits are many seeded pepo with fleshy mesocarp, indehiscent and vary in shape, colour, flavour and taste; fruits are round, oblate, oval, oblong, or pear-shaped, variously ribbed, 15– 60 cm in diameter, and weigh up to 45 kg; pumpkin fruits are generally harvested as mature and ripe stage, which has a hard rind and firm mesocarp while fruits of summer squash is generally harvested at immature stage.
- In pumpkin, the peduncle is strong, smooth, normally 5- angled, flared at the fruit attachment with a rounded pentangular base and large apex while in summer squash, peduncle is five angled with little or no expansion at fruit attachment.

- Flesh of pumpkin is deep yellow, orange, pale green, or white, and the hollow centre contains pulpy loose fibres and numerous seeds.
- Seeds are numerous, encased in a lignified seed coat attached to the fibrous endocarp adjacent to a hollow central cavity; oval, flattened, whitish-yellow to yellow in colour with a thin or ragged margin, shredded in appearance and the margin is more deeply coloured than the body of the seed, with a meaty kernel.

Reproductive behaviour

Both pumpkin and summer squash are highly cross-pollinated in nature due to monoecious sex form and entomophilies pollination. The staminate flowers produce pollens with large granules and the pistillate flowers are pollinated by honeybee and bumble bee. Although it is 100% cross pollinated in nature yet high extent of self-fertilization happens because of bee foraging in the staminate flowers of the same vine. Pumpkin and summer squash generally show negligible inbreeding depression. Anthesis starts in the early morning at 3:30 a.m. and maximum number of flowers opens between 4:00 to 5.00 a.m. Flowers remain open for about 8 to 10 hours after which they start losing their yellow colour and wither. Pollen dehiscent generally starts from 5:30 am. in the summer months and it is somewhat delayed with the advent of low temperature. Stigma receptivity is maximum at the time of anthesis although stigma becomes receptive one day prior to and remains one day after anthesis. Best time of hybridization is within 6.00 am on the day of anthesis. Bagging the staminate and pistillate flowers in the afternoon hours prior to opening followed by hand pollination in the next morning and subsequent bagging for two more days is the commercial method of hybrid seed production. Harvesting the fully matured staminate flowers in the afternoon hours that will open in the next day morning and keeping them covered with thin moist cloth overnight for their use in cross pollination in the next early morning hours is also practiced. Pumpkin can be crossed with *C. pepo*, *C. maxima* and *C. mixta* with difficulty however, F_1 sterility is usually encountered in these interspecific crosses. Embryo culture and/or amphidiploidy is generally employed to overcome this problem.

Genetics of important characters

The basic and haploid chromosome numbers of the five domesticated *species of Cucurbita viz*., *Cucurbita moschata*, *C. pepo*, *C. maxima*, *C. argyrosperma* and *C. ficifolia* are are same, x = n = 20 (2n = 40). However, it is not easy to cross the *Cucurbita* species and the interspecific hybrids are mostly sterile. However, in some cases a few viable seeds can be obtained.

The three economically important species, *C. pepo*, *C. moschata*, and *C. maxima* are highly polymorphic in fruit characteristics, inspiring much research into their inheritance although most of such studies have been done in *C. pepo* and *C. maxima*. The new gene list for pumpkin and squash includes descriptions of gene interactions and the genetic background of the parents that had been used for crossing to allow easy confirmation of previous work and provide a sound foundation for further investigation. Information on simply inherited characters documented in the gene list is almost lacking for *C. moschata* and among the few that have been reported so far are Trifluralin (a herbicide) tolerance controlled by a dominant gene "*T*", naked kernel (absence of tough seed coat) trait controlled by a pair of recessive genes and mottled light and dark green fruit colour, monogenic dominant (*Mldg*) over dark green fruit.

Table 2: Some important monogenic characters of *Cucurbita* species

Gene symbol	Phenotype of plant character
a	Super male plant, androecious sex form
B	Yellow fruit pigmentation before anthesis
Bi	Bitter fruit containing high cucubitacin content
bl	Blue fruit colour, incompletely dominant over green
Bu	Bushy plant , dominant over viny in early stage but recessive at later stage
C	Coloured fruit, green fruit epistatic to white, *r*
D	Dark green stem, dominant over light green stem
Di	Disc fruit shape, dominant to spherical
G	Gynoecious sex form
Gr	Green rind
Hr	Hard fruit rind, dominant to soft rind
I	Light fruit rind
lo	Lobed leaves
lt	Leafy tendril
ly	Light yellow corolla, recessive to orange-yellow
M	Mottled leaves, silvery grey areas in the axils of leaf veins
Mldg	Mottled light and dark green fruit colour, monogenic dominant over dark green fruit, *C. moschata*
ms-1	Male sterility-1
ms-2	Male sterility-2
n	Naked seed, with a lignified seed coat, single recessive gene Pair of recessive genes in *C. moschata*
r	White fruit colour, hypostatic to *C*
Rd	Red skin of fruit, dominant to green, white, yellow and grey
ro	Rosette leaf
s	Both male and female sterility
St	Longitudinal striped fruit
T	Trifluralin (a herbicide) tolerance, dominant gene in *C. moschata*
W	White matured fruit, dominant to green, partially epistatic to yellow, *Y*
Y	Yellow fruit colour, dominant to green
Ygp	Yellow-green placenta
ys	Yellow seedling lacking in chlorophyll, lethal

Source: Whitaker and Robinson (1986), Paris and Brown (2005), Thapa and Hazra (2012)

Polygenic inheritance of quantitative characters has mostly been studied in summer squash (*C. pepo*) however such information on pumpkin (*C. moschata*) is meagre.

Table 3. Gene action for different quantitative characters of *Cucurbita* species

Quantitative character	**Gene action**
Cucurbita moschata	
Vine length, number of branches, node of first staminate and pistillate flower, days to first staminate and pistillate flower, seeds per fruit, 100-seed weight, seed size index	Non-additive gene action Dominance gene action Over-dominance gene action
Fruit weight	Dominance gene action
Fruits/ plant, flesh thickness, fruit shape index, fruit yield/ plant	Over-dominance gene action
Total soluble solids, carotenoids, ascorbic acid content, calcium, iron contents	Over-dominance gene action
Cucurbita pepo	
Fruit shape, vine length	Additive gene action
Number of branches/plant	Both additive and non-additive gene action
Fruits/plant, fruit length, early yield, total yield	Non-additive gene action
Early maturity and fruit yield/plant	Both additive and non-additive gene action

Source: Swarup (2006), Thapa and Hazra (2012)

Breeding objectives

The main objectives of pumpkin and summer squash breeding are mentioned below.

- Medium or short vines with high degree of female sex expression for pumpkin and bushy growth habit for summer squash
- High early and total fruit yield
- Earliness as indicated by first pistillate flower at early node number
- Round, oblong, flattish-round shaped fruit with preferably non-ridged fruit surface for pumpkin
- Thick, yellow, yellowish-orange or deep yellow flesh of good quality with high β-carotene, sugar contents and high flesh recovery for pumpkin
- Cylindrical fruit with non-ridged surface, thick fruit flesh and small seed cavity containing high β-carotene for summer squash
- Resistance/tolerance to different diseases like, powdery mildew (*Podosphaera xanthii*), downy mildew (*Pseudoperonospora cubensis*), zucchini yellow mosaic virus, watermelon mosaic virus, cucumber mosaic virus, etc.
- Resistance/tolerance to insect pests like, fruit fly (*Bactrocera cucurbitae*), red pumpkin beetle (*Aulacophora foveicollis*), etc.
- Resistance to abiotic stresses like cold temperature, soil salinity, etc.

Breeding methods and development of variety

Pumpkin was introduced in India in very early times, which resulted evolution of extensive variability. In India, pumpkin area is dominated by different local cultivars under different regional name. On the other hand, exotic varieties dominate summer squash growing areas in India. As in other cucurbitaceous vegetable crops, inbreeding depression is not manifested in pumpkin and summer squash. So selection from heterogeneous and heterozygous open pollinated indigenous cultivars (as in pumpkin) or exotic varieties (as in summer squash) followed by selfings to develop uniform lines are the chief breeding method. So, cultivars of these three *Cucurbita* crops mainly constitute different local cultivars and improved selections. In summer squash, some exotic varieties have been released directly as varieties in India. Very few varieties of pumpkin have been developed through inter-varietal hybridization followed by selection in the segregating generations by pedigree method or modified back cross-pedigree method of breeding.

The varieties grown in India can be grouped under three *Cucurbita* species.

- **Pumpkin (*Cucurbita moschata*):** Arka Chandan, CO-1, CO-2, Ambili, Sooraj, Saras, Pusa Viswas, Pusa Vikas, Solan Badami, Kashi Harit, Narendra Agrim, Narendra Amrit, Narendra Abhooshan, Azad Pumpkin-1, Thar Kavi, PAU Magaz Kadoo -1, Punjab Samrat , Anand Pumpkin-1,
- **Summer squash (*Cucurbita pepo*):** Early Yellow Prolific, Australian Green, Patty Pan, Punjab Chappan Kaddu, Pusa Pasand, Kashi Subhangi
- **Winter squash (*Cucurbita maxima*):** Arka Suryamukhi

Germplasm resources and Plant introduction

Domesticated *Cucurbita* species has been remarked as one of the plant genera with the highest diversity in colour, shape and fruit dimensions. Their economic and cultural values are related to the consumption of the mature or immature fruits, seeds, flowers, and to the use as decoration. The genus, *Cucurbita* containing a wealth of genetic variation in fruit size, shape, colour, flavour and nutritional value have originated in the Americas. *Cucurbita* genetic resources are maintained in large collections in a dozen countries including Mexico, USDA, USA, Bulgaria, Czechoslovakia, Hungary, The Netherlands, Portugal, Russia, Spain, Japan and Turkey. Importance of these collections as a critical resource for future world food security, along with the need to share genetic information and germplasm, has been formally recognized in the United Nations International Treaty on Plant Genetic Resources for Food and Agriculture. A central clearing house for this information, GeneSys (https://www.genesys-pgr.org,) was established in 2011 as an outcome of this recognition. Among the sources of information included in the GeneSys database are the European Cooperative Programme for Plant Genetic Resources (EURISCO), Consultative Group on International Agricultural Research (CGIAR), and the United States Department of Agriculture National Plant Germplasm System (USDA NPGS).

In India, pumpkin germplasm is maintained at NBPGR, New Delhi, ICAR-IARI, New Delhi, ICAR-IIHR, Bengaluru, ICAR-IIVR, Varanasi, KAU, Vellanikkara, Kerala and other places. Summer squash collections are maintained at NBPGR, New Delhi, ICAR-IIHR, Bengaluru, PAU, Ludhiana and IARI Regional station, Katrain.

In pumpkin, no exotic varieties have so far been released directly for cultivation in India. However, good number of varieties of summer squash grown in India are exotic variety mainly from USA and Australia.

Table 4: Exotic variety directly released for cultivation

Name of the variety	Country from where introduced
Early Yellow Prolific (Summer squash)	USA
Australian Green (Summer squash)	Australia
Patty Pan (Summer squash)	USA

Individual plant selection and inbreeding

In this method, an individual plant from the source population consisting of heterozygous and non-uniform materials is selected based on the selection indices and unlike mass selection, the seeds of the selected plants are not mixed even if they are equally good and look similar. This method of selection is based on the principle that the actual worth of the selected plant may be masked by the environmental variation and, therefore, the only way to recognize it is by growing the progeny of each selected individual plant for re-evaluation. Progenies of individual plant for evaluation can be developed by selfing as both pumpkin and summer squash do not show loss of vigour due to inbreeding. Homogzygosity for the concerned characters can also be attained in the individuals of the progeny by such repeated selfings.

In pumpkin and summer squash, inbreeding and individual plant selection through pedigree breeding can be practiced as effectively as in self-pollinated crops. Different studies suggested that 3-4 cycles of pedigree breeding and selfing is adequate to develop improved genetic stock of different *Cucurbita* species, particularly pumpkin (Koutsika- Sotiriou *et al*., 2004; Hazra *et al*, 2007a). Different studies clearly suggested that fruit weight, particularly weight of first matured fruit, fruits/plant, equatorial and polar diameter of fruit, pericarp thickness of fruit and days to first pistillate flower are the key fruit yield components of pumpkin. Different studies also indicated the possibility of simultaneous selection for sugar, ascorbic acid and β carotene content of pulp without a concomitant depression in fruit yield (Hazra *et al*, 2007a). High correlations exist between sensory sweetness and total ethanol- soluble sugars, and between sensory texture and solids content of fruit pulp of *C. moschata.* Early generation selections of the inbred as well as breeding lines should be made with preference given to reduced vine length, high ratio of pistillate to staminate blossoms, earliness, high fruit yield, bright-orange colour of flesh with high total solids content. Almost all the varieties of pumpkin and few of summer and winter squash have been developed by this method of breeding.

Table 5: Varieties developed through individual selection and inbreeding

Name of variety	Indigenous / Exotic material
Arka Chandan	Indigenous (Rajasthan)
CO-1	Indigenous (Tamil Nadu)
CO-2	Indigenous (Tamil Nadu)
Ambili	Indigenous (Thrissur, Kerala)
Sooraj	Indigenous (Kerala)
Saras	Indigenous (Kerala)
Pusa Viswas	Indigenous (Sabzi Mandi, Delhi)
Pusa Vikas	Indigenous
Solan Badami	Indigenous (Himachal Pradesh)
Narendra Agrim	Indigenous (UP)
Narendra Amrit	Indigenous (UP)
Narendra Abhooshan	Indigenous (UP)
Azad Pumpkin-1	Indigenous (UP)
Thar Kavi	Indigenous (UP)
PAU Magaz Kadoo -1	Indigenous (seeds are without testa)
Punjab Samrat	Exotic (Russia)
Punjab Nawab	Indigenous (Punjab)
Anand Pumpkin-1	Indigenous (Gujarat)
Punjab Chappan Kaddu (Summer squash)	Indigenous (Punjab)
Pusa Pasand (Summer squash)	Indigenous
Kashi Subhangi (Summer squash)	Indigenous
Arka Suryamukhi (Winter squash)	Exotic

Hybridization and selection

Controlled hybridization between selected parental lines followed by pedigree breeding is also effective in developing improved varieties through selection of desirable segregates. Pedigree method of selection in segregating generations or modified backcross-pedigree method of breeding has been followed for the development of new variety. After crossing between the selected parents, individual plants possessing specific characters are selected in several generations through inbreeding. When one of the parents is of much better horticultural type than the other, it may be desirable to backcross to the better parent for one or more generations, followed by several generations of selfing till uniformity is attained in the breeding lines. Ultimately selfed or sibbed seeds from a number of selected plants are bulked. However, very few varieties of pumpkin have been developed in India through this method (Hazra *et al*, 2022). Huge requirement of space in the field for the evaluation of a reasonably large segregating population to identify the desirable one becomes the main practical difficulty of this breeding method.

Table 6 . Varieties bred through hybridization followed by selection

Name of variety	Parentage
Kashi Harit	NDPK-24 x PKM

Interspecific hybridization

Crossing among different *Cucurbita* species has been well documented by a number of researchers. Whitaker and Robinson (1986) from different studies on interspecific hybridization of *C. moschata*, *C. pepo*, *C. maxima* and *C. mixta* arrived at three obvious conclusions:

i) F_1 hybrids can be obtained from most interspecific crossings, usually with difficulty and crossability depends on the particular accessions of the two species used as parents however, such hybrids are normally highly sterile because of impaired ability of the staminate flowers to produce functional pollen

ii) The breeding data suggest that these four species are arranged like a wheel with *C. moschata* as the hub, the others species as spokes

iii) There is no evidence for spontaneous hybridization among these four species.

Table 7: Interspecific crossability among different *Cucurbita* species

Species	(1)	(2)	(3)	(4)	(5)	(6)
C. moschata (1)	SC	***	*	*	***	CC (*)
C. pepo (2)	***	SC	***	*	*	CC +
C. mixta (3)	*	***	SC	*	***	CC +
C. maxima (4)	*	*	+	SC	**	CC (x)
C. ficifolia (5)	*	**	***	**	SC	CC (x)
C. lundelliana (6)	CC (x)	CC +	CC +	CC (x)	CC (x)	SC

Source : Rai *et al* (2006), Thapa and Hazra (2012)

Note: SC: Self-compatible, CC: Cross-compatible, *: F_1 viable but self-sterile, **: Few viable F_1 seeds and weak F_1 plants, ***: F_1 sparingly viable and self-sterile, +: Small F_1 seeds and very weak seedlings, (x): Self-fertile F_1 plant

In all the interspecific hybrid combinations, *C. maxima* × *C. pepo* (and *vice versa*), *C. pepo* × *C. moschata*, *C. ficifolia* × *C. maxima*, (*C. maxima* × *C. moschata*) × *C. pepo*, *C. argyrosperma* × *C. moschata* (and *vice versa*), the relative nuclear DNA contents was intermediate and differed significantly from those determined in the parental species suggesting that differences in genome size within the *Cucurbita* genus are big enough to be used efficiently in determination of interspecific hybrids (Sisko *et al,* 2003).

Basic objectives of interspecific crosses are to transfer one or few simply inherited desirable characters like disease resistance, good quality flesh, etc. between species. For example, good quality flesh from *C. maxima* to *C. moschata*, bushy growth habit from *C. pepo* to *C. moschata*, disease resistance from *C. lundelliana* or *C. martinezii* to *C. moschata*. Transfer of resistance to powdery mildew from *C. lundelliana* to *C. pepo* was successful through bridge cross between these two species.

Heterosis breeding

Different studies reported dominance and over-dominance gene action for most of the fruit yield components and fruit quality characters *viz.*, days to first pistillate flower, fruit weight, fruits/plant, pericarp thickness, seeds/plant, total soluble solids, β carotene, ascorbic acid, calcium and iron contents of pumpkin (*C. moschata*) and summer squash (*C. pepo*) which clearly suggested the importance of exploitation of heterosis commercially. Several studies suggested different advantages of the F_1 hybrids of pumpkin over open-pollinated cultivars include increased vigour, enhanced fruit yield, earlier and more uniform maturity. Several research reports suggested exploitable positive heterosis for earliness, fruit number, fruit weight, fruit flesh thickness, fruit yield, total soluble solids, total sugar and βcarotene content of the pulp. In summer squash, earliness and high number of fruits per plant are the prime advantages of the hybrids.

Sex form of both pumpkin and summer squash are ubiquitously monoecious hence, bagging of staminate and pistillate flowers in the afternoon hours and hand pollination of the pistillate flower in next day morning after opening of the pistillate flower with the pollens from the staminate flower of the intended male parent is the commercial method of hybrid seed production. Application of ethephon or ethrel to the young plants at 2-3-leaf stage would suppress male flower formation for 2-3 weeks and suppression for considerable period of time can be achieved at higher concentrations of 600 ppm applied twice at 2- and 4-leaf stage has made hybrid summer squash seed production comparatively easier. It has been estimated that nearly 56% of total summer squash seed produced in the USA is of' F_1 hybrids.

Development of gynoecious line or cytoplasmic male sterility in pumpkin will improve commercial hybrid seed production in pumpkin. Identification of gynoecious sex form in *C. maxima*, *C. pepo* and *C. foetidissima* has opened a possibility of transferring this monogenic dominant character to pumpkin.

Breeding for resistance to diseases

Disease resistance has become the major objective in breeding programmes of pumpkin, squashes and other *Cucurbita* species. Pumpkin and squashes are particularly susceptible to different fungal diseases particularly powdery mildew (*Sphaerotheca fulliginea*), downy mildew (*Pseudoperonospora cubensis*) and root rot (*Phytopthora capsici*) and several virus diseases *viz.*, watermelon mosaic virus, cucumber mosaic virus, squash mosaic virus, bottle gourd mosaic virus, cucurbit latent virus, squash leaf curl virus, zucchini yellow mosaic, papaya ring spot virus, pumpkin mosaic virus, etc. Pumpkin yellow vein mosaic (PYVMV) disease symptoms occur frequently in pumpkins in India where diseased plants show vein yellowing, which sometimes coalesces to form chlorotic patches and infected plants are stunted with premature dropping of flowers and PYVMV is most similar to the tomato leaf curl New Delhi virus from India.

The summer squash variety, Punjab Chappan Kaddu-1 has been found to have field tolerance to downy mildew, powdery mildew, cucumber mosaic virus and red pumpkin beetle. Different accessions of *C. moschata* viz., Minina, originating from Portugal (Gilbert *et al.*, 1993), Nigerian Local (Brown *et al.*, 2003), BGH 1934, BGH 1937 and BGH 1943 (Moura *et al.*, 2005) and wild species, *C. ecuadorensis* (Provvidenti *et al.*, 1997) are known to carry resistance against ZYMV. Transfer of ZYMV resistance from the resistant *C. moschata* cultivars and *C. ecuadorensis* to the varieties of *C. moschata, C. pepo* and *C. maxima* was successful in some cases resulting in the development of new cultivars showing high level of resistance (Desbiez *et al.*, 2003). However, no RAPD markers linked to ZYMV resistance were found (Brown *et al.*, 2003).

Different feral species of *Cucurbita* are found resistant to many diseases *viz.*, *C. lundelliana* and *C. martinezii* are resistant to powdery mildew and different viruses such as watermelon mosaic, cucumber mosaic virus and zucchini yellow mosaic virus, *C. equadorensis* is resistant to papaya ringspot virus and *C. andreana*, *C. lundelliana*, *C. equadorensis* and *C. foetidissima* are resistant to bacterial wilt caused by *Erwinia trachiiphila* (Robinson *et al.* 1988). Generally, backcross breeding method is employed for the transfer of resistance from the donor parent to the commercial cultivar.

Breeding for resistance to insect pests

Breeding for insect resistance is difficult because of the intricate difficulties of three phases of work (Whitaker and Robinson 1986):

i) Rearing the insects or obtaining natural infestation
ii) Exposing test plants at the appropriate stage to a uniform number of insects
iii) Accurately assaying the response

Very limited work has been done so far on breeding *C. moschata* cultivars resistant to insect pests. Few reports on the identification of source of resistance in pumpkin include combined resistance to prickle worm, striped cucumber beetle, and serpentine leaf miner, resistance to red pumpkin beetle and resistance to leaf silvering disorder in response to feeding by *Bemisia argentifolii* at immature stage (Wessel-Beaver and Katzir, 2000). In summer squash, inheritance of resistance to red pumpkin beetle was controlled by both additive and non-additive gene action, non-additive gene action being predominant. The resistance to fruit fly is controlled by a single dominant gene *Fr*.

Table 8: Inheritance of resistance to different diseases in *Cucurbita* species

Resistance to disease	**Inheritance pattern**
Powdery mildew in *C. lundellina*	Single dominant gene, *Pm*
Powdery mildew in *C. martinezii*	Single dominant gene with modifier genes
Papaya ring spot virus in *C. ecuadorensis*	Polygene, additive gene action
Zucchini yellow mosaic virus in *C. ecuadorensis*	Polygene, additive gene action
Watermelon mosaic virus in *C. maxima*	Polygene, additive gene action
Watermelon mosaic virus and Zucchini yellow mosaic virus in *C. moschata*	Single Dominant gene "*Zym*" or two closely linked gene
Cucumber mosaic virus in *C. pepo*	Two unlinked recessive genes

Fruit fly in *C. pepo*	Single dominant gene *Fr*
Watermelon mosaic virus in *C. moschata*	Single dominant gene "*Wmv*"
Cucumber mosaic virus in *C. moschata*	Single dominant gene "*Cmv*"
Papaya ring spot virus in *C. moschata*	Single recessive gene "*prv*"
Pumpkin mosaic virus in *C. moschata*	Single recessive gene
Yellow vein mosaic virus in *C. moschata*	Single dominant gene
Whitefly (*Bemisia argentifolii*) induced silver leaf disorder *C. moschata*	Single dominant gene

Source: Thapa and Hazra (2012)

Breeding for fruit quality

Pumpkins are one of the most overlooked and underutilized food and medicinal plants. It is an extraordinary vegetable with the potential to be used as a medicinal as well as a nutritious multifunctional food. Pumpkin flesh contain essential minerals as well as phytochemicals (β-carotene, total flavonoids, and total phenolics) that can help with antiaging and the immune system. Particularly, pumpkin seeds are rich in zinc. Since, low-cost powders made from pumpkin components could be employed in the food and pharmaceutical industries as a source of functional ingredients and nutraceuticals, pumpkin paste can be manufactured ahead of time and frozen for later use. It has the potential to improve people's nutritional status, particularly among vulnerable groups, in terms of vitamin A-requirements. Several studies emphasized the importance of hybrid breeding to improve different quality traits particularly β carotene content in pumpkin (Hazra *et al.*, 2007a,b).

The '*B*' gene, in *C. pepo* which conditions precocious yellow fruit pigmentation before anthesis produce deep yellow coloured fruit with high vitamin A content. It is a chlorophyll deficiency character in different plant parts and controlled by two unlinked genes, *B1* and *B2*. This gene from *C. pepo* was successfully transferred to *C. moschata* (Paris *et al.*,1985).

References

Brown, R.N., Bolanos, H.A., Myers, J.R. and Jahn, M.M. 2003. Inheritance of resistance to four cucurbit viruses in *Cucurbita moschata. Euphytica.*,**129**: 253-258.

Chakravarty, H.L. 1982. *Fascicles of Flora of India Fascicle II, Cucurbitaceae*, Botanical Survey of India, Howrah, India, 136p.

Desbiez, C., Gal-On, A., Girard, M., Wipf-Scheibel, C. and Lecoq, H. 2003. Increase in Zucchini yellow mosaic virus symptom severity in tolerant zucchini cultivars is related to a point mutation in P3 protein and is associated with a loss of relative fitness on susceptible plants. *Phytopathology*, **93**: 1478-1484.

Gilbert-Albertini, F., Lecoq, H. and Pitrat, M. 1993. Resistance of *Cucurbita moschata* to watermelon mosaic virus type 2 and its genetic relation to resistance to zucchini yellow mosaic virus. *Euphytica*, **69**: 231–237.

Hazra, P., Mandal, A. K., Dutta, A. K., Sikder, D. and Pandit, M. K. 2007a. Breeding pumpkin (*Cucurbita moschata* Duch ex Poir.) for high yield and carotene content., *Acta Hort.*, **752**: 431- 435.

Hazra, P., Mandal, A. K., Dutta, A. K. and Ram, H.H. 2007b. Breeding pumpkin (*Cucurbita moschata* Duch ex Poir.) for fruit yield and other characters. *Intern. J. Plant Breed.*, **1**: 51-64.

Koutsika-Sotiriou, M., Traka-Mavrona, E., Tsivelikas, A.L., Mpardas, G., Mpeis, A. and Klonari, E. 2004. Use of genetic resources in a dual approach toward selecting improved scion/rootstock grafting combinations of melon (*Cucumis melo* L.) on *Cucurbita spp.* In: *Progress in Cucurbit Genetics and Breeding Research,* (Eds. Lebeda, A. and Paris, H.S.), *Proc. 8th EUCARPIA meeting on Cucurbit Genetics and Breeding,* Olomouc, Czech Republic., July 12-17, 2004, pp. 163-167.

Moura, M.D.C., Zerbini, F.M., Silva, D.J.H.D. and Queiroz, M.A.D. 2005. Identification of pumpkin populations as resistance sources against the zucchini yellow mosaic virus (ZYMV). *Hortic. Bras*., **23**: 206-210.

Nayar, N.M. and More, T.A. 1998. *Cucurbits*. Oxford &IBH Publishing Co. (Pvt.) Ltd. New Delhi, Calcutta.

Paris, H.S. and Brown, R.N. 2005. The genes of pumpkin and squash. *Hort Sci.,* **40**: 1620-1630

Paris, H.S., Norson, H. and Burger, Y. 1985. Precocious PI 165561 and precocious PI 165561 R pumpkin breeding lines. *Hort Sci.,* **20**: 778-779

Provvidenti, R. 1997. New american summer squash cultivars possessing a high level of resistance to a strata of zucchini yellow mosaic virus from china. *Rep. Cucurbit Genet. Coop*., **20**, 57-57.

Rai, M., Pandey, S., Ram, D., Kumar, S. and Singh, M. 2006. Cucurbit Research in India: An overview In: *Cucurbits Breeding and Production Technology* (Eds. Gautam, P. L. , Ram, H. H. and Singh, H. P.) *Proc. Natl. Seminar onCucurbits*. G.B. Pant Univ. Agril and Tech., Pantnagar, 22-23 September, 2005, pp.1-20.

Robinson, R.W., Weeden, N.F. and Provvidenti, R. 1988. Inheritance of resistance to zucchini yellow mosaic virus in the interspecific cross *C. maxima* × *C. ecuadorensis*. *Rep. Cucurbit Genet. Coop.,* **11:** 74-75

Sisko, M., Ivancic, A. and Bohanec, B. 2003. Genome size analysis in the genus *Cucurbita* and its use for determination of interspecific hybrids obtained using the embryo-rescue technique. *Plant Sci.,* **165**: 663-669.

Thapa, U. and Hazra, P. 2012. Pumpkin and Squash. In: *Hand book of Vegetables* (Eds. Peter, K. V. and Hazra, P.), Studium Press LLC, Houston, Texas, USA, pp. 491-531.

Wessel-Beaver, L. and Katzir, N. 2000. Inheritance of silver leaf resistance in *Cucurbita moschata. Acta Hort.,* **510:** 289-295

Whitakar, T.W. and Davis, G.N. 1962. *Cucurbits, Botany, Cultivation, Utilization.* World Crops Books, Interscience Publishers, New York, 250p.

Whitakar, T.W. and Robinson, R.W. 1986. Squash Breeding. In: *Breeding Vegetable crops* (Ed. Bassett, M.J.), AVI Publishing Company, Westport, Connecticut, USA, pp. 209-242.

39

Ridge gourd and Sponge gourd

Ridge gourd, *Luffa acutangula* and sponge gourd, *Luffa cylindrica* under the family Cucurbitaceae (Chromosome number: $2n = 2\times = 26$) is an important warm season fruit vegetable crop. Ridge gourd is widely cultivated in India, Southeast Asia, China, Japan, Egypt and other parts of Africa. Sponge gourd is now cultivated in all tropical regions of the world. It was an important crop used for industrial filters until World War II and Japan cultivated the best crop for this purpose. The main commercial production areas of sponge gourd today are China, Korea, India, Japan, Central America and Brazil.

Immature fruits of ridge gourd and sponge gourd are eaten as a vegetable and at maturity both become very fibrous. The word luffa or 'loofah' is of Arabic origin and the sponge characteristic of the crop has been described in old Egyptian writings.

Fruits are good sources of vitamin C, riboflavin, zinc and antioxidants. It is effective in balancing blood sugar levels, help eliminate toxic waste and prevent fat accumulation. Ridge gourd also helps reduce excessive body heat during summers which can prevent many ailments.

Origin, domestication and adaptation in India

Both ridge gourd and sponge gourd has been originated in subtropical regions of Asia, most probably from India as the primary centre. They were domesticated and known in India from ancient times. *Luffa* crop was mentioned in Kautilya's Arthashastra, dating back 350-300 B.C. It was also described in early Egyptian and Chinese literature.

The wild species, *Luffa graveolens* is the progenitor of the cultivated ridge gourd and sponge gourd. The wild dioecious species, *L. echinata* and monoecious *L. operculata* were evolved from the wild species *L. graveolens*. Centre of diversity of these species are in South and Southeast Asia.

Luffa, an old world genus comprise nine species including *Luffa acutangula, Luffa cylindrica, Luffa hermaphrodita, Luffa graveolens , Luffa echinata, Luffa tuberosa* and *Luffa umbellate* are native to India. Great diversity of *Luffa* gourds, particularly ridge gourd occurs in India. The species, *L. echinata* var. *longistylis* is widely distributed in north-western Himalaya and upper Gangetic plains. *Luffa graveolens* occurs in north-eastern plains, extending south to Tamil Nadu and sporadic in Eastern Himalayas. The species *L. umbellata* is confined to the eastern coast and *L. acutangula* var. *amara* is mainly distributed in peninsular tract. *Luffa hermaphrodita* having trimonoecious (male, female and hermaphrodite flowers in the same plant) or hermaphrodite sex form is cultivated in the name "Satputia" in some places of Bihar and West Bengal.

A large number of local cultivars of ridge gourd and sponge gourd, particularly ridge gourd are grown in India. In ridge gourd very wide variability in shape and size (one metre to a few centimeters long) are available in India and some of them are even slightly bitter in taste.

Crop botany and floral biology

- Vine is annual, vigorous with slender, five angled stem; leaves are sub-circular, membranous, deltoid to nearly orbicular in outline, but acutely pointed at the apex, usually five to seven lobed, scrabrous, nearly glabrous and dentate margins; tendrils are robust, often 3- fid, puberulous.
- Plants are generally monoecious in sex form with staminate and pistillate flowers are borne separately in the same plant; hermaphrodite sex form is also found in some genotypes of ridge gourd.
- Inflorescence is axillary, solitary or clustered or racemose; many staminate flowers, even upto 20 are borne in racemes with long stalk which open in succession (in sponge gourd, staminate flower is solitary); peduncles 10-15 cm long, pedicels 1-4 cm long, white-puberulous, probracts 3-7 x 2-4 mm, fleshy, green, ovate with 3-10 glistening glands on the upper surface
- The pistillate flowers are single which may or may not develop on the same leaf axil; in sponge gourd both the inflorescences appear on the same leaf axil; peduncles 5-10 cm long, ovary elongate, 10-angular, apex constricted, style short, stigmas 3, expanded, 2-lipped
- In hermaphrodite sex form of ridge gourd, the flowers appear in the leaf axil in cluster
- Calyx tube campanulate; sepals are 5- partite, glandular; lobes lanceolate, 4-6 x 2-3 mm, apex acuminate, slightly reflexed, densely white-pubescent, 1- nerved.
- Corolla obovate, petals are 5, 2.0 to 2.5 cm in length and lemon-yellow to yellow in colour in ridge gourd; corolla orange-yellow with green veins in sponge gourd, lobes obcordate, 15- 25 x 10-20 mm, both surfaces are sub-glabrous.
- Stamens are usually 3, synandrous or syngenesious, free, 1 unilocular, 2 bilocular, filaments 3-4 mm long, bearded at the base, anthers puberulous; stamens are 5 and free in sponge gourd.
- Carpels are 3, syncarpous in ridge gourd; ovary inferior, filiform with 10 longitudinal ridges on which swollen glands are borne; ridges on the ovary is absent in sponge gourd; ovules are many with one short style and trifid stigma.
- Flowers of ridge gourd open in the afternoon hours and remain open throughout the night while the flowers of sponge gourd open in early morning hours and remain open for 3-4 hours after anthesis.
- Fruits are many seeded pepo, club shaped, and variable in size, crowned by large sepals, stipule persistent; fruits are ribbed with 10 ridges; fruits are smooth in sponge gourd; fibre obtained from the mature dry fruit is xylem fibers and arranged in a cell like structure.

- Seeds are numerous, black and pitted in ridge gourd; seeds are black on white and non-pitted in sponge gourd; average 100 seed weight is generally 9.5 gram in ridge gourd and 10.6 g in sponge gourd.

Reproductive behaviour

Monoecious sex form is predominant in ridge gourd and sponge gourd however, other sex forms namely, hermaphrodite, andromonoecious, androecious, gynomonoecious are rarely found. Hermaphrodite sex form is found in an old indigenous cultivar of ridge gourd named "Satputia". Both these *Luffa* gourds are highly cross-pollinated in nature due to the predominantly monoecious sex form. It is entomophillous and major pollinators are bees (*Apis florea,Apis dorsata, Nomioides* sp., *Halictine* sp., etc.). Ridge gourd and sponge gourd like other members of the family Cucurbitaceae shows negligible inbreeding depression. Anthesis of both staminate and pistillate flowers of ridge gourd generally starts in the afternoon hours from 5.00 to 8.00 p.m. and remain open throughout night. On the other hand, anthesis of both staminate and pistillate flowers of sponge gourd generally starts in the morning hours from 6.00 to 10.00 a.m. Anther dehiscence in the staminate flower of both the *Luffa* gourds generally coincides with the anthesis. Pollen fertility is the maximum on the day of anthesis which may remain viable up to 1 to 3 days after anthesis in ambient condition depending on the season. Stigma receptivity is maximum at the time of anthesis although the stigma becomes receptive 6 hours earlier to 48 hours after anthesis. In ridge gourd, best time of hybridization is between 4.00 to 6.00 p.m. on the day of anthesis. However, flowers are ready for selfing and cross-pollination in the early morning of next day also. In sponge gourd, best time of hybridization is between 6.00 to 8.00 a.m. in the morning hours of the day of anthesis. Bagging the staminate and pistillate flowers prior to opening in the morning hours followed by hand pollination in the afternoon to evening hours and subsequent bagging for two more days is the commercial method of hybrid seed production of ridge gourd. In sponge gourd, bagging the staminate and pistillate flowers prior to opening in the afternoon hours followed by hand pollination in the next day morning hours and subsequent bagging for two more days is the commercial method of hybrid seed production.

Genetics of important characters

Ridge gourd, *Luffa acutangula* , sponge gourd, *Luffa cylindrica* and "Satputia" (*Luffa hermaphrodita*) are diploid species (2n = 2x= 26) and are cross-compatible. However, all these species remain isolated in nature. The isolation of species may be genic and not chromosomal as indicated by cytological investigations. The amphidiploids produced between *Luffa acutangula* x *Luffa graveolens* (2n = 52) had slight pollen sterility, partly due to univalent.

Few reports are there regarding qualitative characters of ridge gourd and sponge gourd which are under the control of single gene.

Table 1: Some important monogenic characters of ridge gourd and sponge gourd

Character	**Genetic control**
Stamen character	Five free stamens of *Luffa cylindrica* is monogenically dominant over (2+2+1) type androecium of *Luffa acutangula*
Corolla colour	Orange-yellow with green veins of *Luffa cylindrica* is governed by single dominant gene
Fruit surface	Smooth fruit surface of *Luffa cylindrica* is governed by single dominant gene
Seed surface	Non-pitted seed surface of *Luffa cylindrica* is governed by single dominant gene
Bitterness in fruit	Bitterness in fruit monogenically dominant (*Bi*) over non-bitter fruit
Sex form in *Luffa acutangula*	Sex form is controlled by two suppressor genes, A and G, the former (A) supress femaleness (G). Digenic segregation ratio in F_2, with 9 Monoecious (*AG*), 3 andromoncecious (*aG*), 3 gynomonoecious (*Ag*) and 1 hermaphrodite (*ag*); The gynomoncecious type is susceptible to environmental fluctuation, giving rise occasionally to gynoecious and trimoncecious forms. Interspecific hybrid *Luffa cylindrica* x *Luffa hermaphrodita*, Monoecious x hermaphrodite is monoecious
Male sterility	Cytoplasmic male sterility in ridge gourd with two dominant fertility restorer gene (*Rf1* and *Rf2*)

Source: Singh *et al.* (1948), Ram (2006), Pradeep Kumar and Bhardwaj (2015)

Many important quantitative characters of both ridge gourd and sponge gourd under polygenic control are vine length, days to flowering, node number to first female flower, fruit weight, fruits per plant, yield per plant, etc.

Table 2: Gene action for different quantitative characters of ridge and sponge gourd

Quantitative character	**Gene action**
Ridge gourd	
Fruit number per plant	Non-additive gene action
Fruit weight	Additive gene action Non-additive gene action Both additive and non-additive gene action
Fruit length	Additive gene action Both additive and non-additive gene action
Fruit girth	Non-additive gene action
Fruit yield	Non-additive gene action
Sponge gourd	
Days to opening of first male flower	Both additive and non-additive gene action, most of the epistatic components
Days to opening of first female flower	Both additive and non-additive gene action, most of the epistatic components
Number of nodes for opening of first male flower	Both additive and non-additive gene action, most of the epistatic components

Source: Pradeep Kumar and Bhardwaj (2015)

Breeding objectives

Some of the important characters that have been the main focus of ridge gourd and sponge gourd breeding are outlined below

- Vines medium-long, vigorous, well branched, 2.0 to 2.5 m with high female to male ratio
- Earliness; lower number of node at which first pistillate (or hermaphrodite in case of "Satputia" cultivar) flower appears which gives a fair indication of earliness
- Early maturing, medium-sized, non-fibrous at marketable stage, more or less cylindrical, straight, green to dark green fruits; fruits with bulged blossom end and thin neck is not preferred; extra-long fruits often break and get damaged during transportation.
- High early and total yield potential
- Fruit of ridge gourd should be green or dark green; fruit of smooth gourd should be whitish-green and light green.
- Endocarp (pulp) soft, light greenish, non-bitter and contain less fibre
- Resistance/tolerance to different diseases like, powdery mildew (*Podosphaera xanthii*), downy mildew ((*Pseudoperonospora cubensis*), cucumber mosaic virus, etc.
- Resistance/tolerance to insect pests like, fruit fly (*Bactrocera cucurbitae*), red pumpkin beetle (*Aulacophora foveicollis*), leaf miner (*Liriomyza trifoli*), etc.
- Resistance to downy mildew, powdery mildew, leaf miner and fruit fly

Breeding methods and development of variety

Improved varieties of both ridge gourd and sponge gourd have been developed in India mainly by single plant selection from the existing genetic variability followed by inbreeding to stabilize the genotype. Hybridization between the indigenous genotype followed by selection method of breeding and applied mutagenesis have also been followed to develop few varieties of ridge gourd.

Table 3: Morphological and flowering characters of ridge gourd and smooth gourd

Characters	**Ridge gourd (*Luffa acutangula*)**	**Sponge gourd (*Luffa cylindrica*)**
Sex form	Six sex form are found (monoecious, trimonoecious, andromonoecious, gynomonoecious, gynoecious, hermaphrodite)	Generally monoecious sex form is found
Leaf character	Leaves are shallowly lobbed and larger in size; no white patches are found on the leaf	Leaves are deeply lobed and comparatively smaller in size; white patches are found on the leaves
Staminate flower	Smaller in size	Larger in size
Stamen character	Stamens are usually 3 (2+2+1)	Stamens are five and free

Corolla colour	Lemon-yellow to yellow in colour	Orange-yellow with green veins
Shape of ovary	Ovary is disc shaped and not tomentose	Ovary is cylindrical and tomentose
Anthesis	Afternoon hours and remain open throughout the night	Early morning hours and remain open for 3-4 hours
Fruit shape	Fruits are angled and 10-ribbed	Fruits are smooth surfaced
Seed colour and shape	Seeds are black and pitted	Seeds are black, white, flat and not pitted

Source: Pradeep Kumar and Bhardwaj (2015), Hazra (2020)

The varieties of ridge gourd can conveniently be grouped according to fruit size while those of the sponge gourd can be grouped according to fruit colour.

Ridge gourd

- **Fruit size medium to medium-long:** Pusa Nasdar, Punjab Sadabahar, Arka Sujat, Swarna Manjari, Swarna Uphar, Deepthi, Pant Torai-1, Kashi Jyoti, Thar Karni, Gujarat Anand Ridge Gourd-1, Phule Sucheta
- **Fruit size long to very long:** Pusa Nutan, CO-1, CO-2, PKM-1, Konkan Harita, Arka Prasan, Arka Sumeet, Kashi Shivaani, GJRGH-1, Hisar Kalitori, Swarna Manjhari, Swarna Uphar

Sponge gourd

- **Fruit green / dark green:** Pusa Chikni, Pusa Sneha, Kashi Rakshita, Kashi Shreya, Kashi Divya, Kalyanpur Hari Chikni, Azad Torai-1, Azad Torai-2, Thar Tapish
- **Fruit medium green, pale green/ yellowish-green:** Pusa Supriya, Phule Prajakta, Gujarat Sponge Gourd-1, Swarna Prabha

Germplasm resources and Plant introduction

Rich diversity of ridge gourd and sponge gourd occurs in Indo-Gangetic plains, terai regions and northeastern plains in India. Diversity of *Luffa* genotypes is maintained in India, Thailand, Vietnam, the Philippines, Indonesia, China, etc.

In India, ridge gourd and sponge gourd germplasm is maintained at NBPGR, New Delhi, ICAR-IIVR, Varanasi, ICAR-IIHR, Bengaluru ICAR-IARI, New Delhi and other places.

No exotic variety has so far been released directly for cultivation in India.

Individual plant selection and inbreeding

Huge genetic diversity exists in India for both ridge gourd and sponge gourd cultivars. In this method of breeding, individual plant from the indigenous source population consisting of heterozygous and non-uniform materials is selected and unlike mass selection, the seeds of the selected plants are not mixed even if they are equally good and look similar. This method of selection is based on the principle that the actual worth of the selected plant may be masked by the environmental variation and, therefore, the only way to recognize it is by growing the progeny of each selected individual plant for re-evaluation. In sponge gourd, reevaluation of the bulk population of the variety, Pusa Supriya has led to the development of another variety, Pusa Sneha. Progenies of individual plant for evaluation can be developed by selfing as both ridge gourd and sponge gourd do not show loss of vigour due to inbreeding. Homogzygosity for the concerned characters can also be attained in the individuals of the progeny by such repeated selfings. Most of the varieties of ridge gourd and all the varieties of sponge gourd have been developed by this method of breeding.

Table 4: Varieties of ridge and sponge gourd developed through selection and inbreeding

Name of variety	Indigenous / Exotic material
Ridge gourd	
Puss Nasdar	Indigenous (Madhya Pradesh)
Pusa Nutan	Indigenous
CO-1	Indigenous (Tamil Nadu)
CO-2	Indigenous (Tamil Nadu)
Konkan Harita	Indigenous (Maharashtra)
Punjab Sadabahar	Indigenous (Punjab)
Arka Prasan	Indigenous (Karnataka)
Swarna Manjari	Indigenous (Jharkhand)
Swarna Uphar	Indigenous (Jharkhand)
Deepthi	Indigenous (Kerala)
Pant Torai-1	Indigenous
Kashi Jyoti	Indigenous (U.P)
Kashi Shivaani	Indigenous (U.P)
Thar Karni	Indigenous (Rajasthan)
GJRGH-1	Indigenous (Gujarat)
Gujarat Anand Ridge Gourd-1	Indigenous (Gujarat)
Hisar Kalitori	Indigenous (Haryana)
Phule Sucheta	Indigenous (Maharashtra)
Sponge gourd	
Pusa Chikni	Indigenous (Bihar)
Pusa Supriya	Indigenous (Delhi)
Pusa Sneha	Bulk population of Pusa Supriya
Phule Prajakta	Indigenous (Maharashtra)
Kashi Rakshita	Indigenous (U.P)
Kashi Shreya	Indigenous (U.P)
Kashi Divya	Indigenous (U.P)
Kalyanpur Hari Chikni	Indigenous (U.P)
Gujarat Sponge Gourd-1	Indigenous (Gujarat)
Swarna Prabha	Indigenous (Jharkhand)
Azad Torai-1	Indigenous (U.P)
Azad Torai-2	Indigenous (U.P)
Thar Tapish	Indigenous (Rajasthan)

Hybridization and selection

Planned hybridization between indigenous parental materials followed by pedigree method of selection in segregating generations has been followed for the development of few varieties of ridge gourd in India (Hazra *et al*, 2022). Parents to be employed in the hybridization programme should possesses specific characters which are expected to be combined in the segregating generations. The parental lines should also be genetically diverse for obtaining high heterotic F_1 hybrids and broad spectrum of variability in segregating generations.

Interspecific hybridization between *Luffa hermaphrodita* 'Satputia' (bisexual sex form and cluster bearing habit) and sponge gourd, *Luffa cylindrica* (monoecious and solitary bearing) can be highly useful for the introgression of cluster bearing, high yield and gynoecism in sponge gourd (Sidhu and Kaur,2021). However, the occurrence of self-pollination in closed flowers of 'Satputia' and anthesis of two species at different time intervals creates hindrance in interspecific hybridization. Pre-anthesis emasculation (28 h before anthesis) of buds of Satputia in the evening and pollination with sponge gourd in the morning on the day of anthesis resulted in high fruit set. Interspecific hybrids were monoecious and morphologically intermediate for most of the vegetative, flower and fruit traits. Although pollen size was less than both the parents, but the pollen density improved in F_1 vines. Fertile hybrids could be easily used to generate F_2 generations. In F_2 generation, gynoecious, adroecious, andromonoecoius, monoecious and cluster bearing vines of variable length and fruit size were observed. In back cross and triple cross generations, most of the vines were monoecious except a few adroecious and gynoecious with improved fruit size, vine growth and bearing capacity. Backcross and triple cross with sponge gourd displayed a shift towards this species.

Such interspecific hybridization between *Luffa hermaphrodita* 'Satputia' and ridge gourd, *Luffa acutangula* (monoecious and solitary bearing) has also been highly successful. Anthesis of both these species takes place in the afternoon hours which is convenient for interspecific hybridization. Emasculation of the bud of the hermaphrodite flowers of *Luffa hermaphrodita* in the afternoon about 24 hours before anthesis and pollination with the pollens from the staminate flowers of the ridge gourd (monoecious sex form) in the afternoon hours on the day of anthesis (next day) resulted in high fruit set (Das, 2020). Interspecific hybrids were monoecious and morphologically intermediate for most of the vegetative, flower and fruit characters. In the F_2 generation, cluster bearing vines having different sex form *viz*., monoecious, andromonoecoius, hermaphrodite and vines of variable length and fruit size segregated which has given ample scope for the selection of suitable, cluster bearing, medium-long fruited, soft-fleshed and high yielding varieties of ridge gourd (Chatterjee, 2023). However, no variety has so far been developed through this interspecific hybridization programme. Only two varieties of ridge gourd have so far been developed following hybridization and selection method of breeding.

Table 5: Varieties of ridge gourd bred through hybridization and selection

Name of variety	Parentage
Arka Sujat	IIHR 54 (an early maturing and medium fruited line) × IIHR 18 (a long fruited line)
Arka Sumeet	IIHR 54 × IIHR 24 (a long fruited line)

Mutation breeding

Very little work has so far been done on applied mutagenesis in ridge gourd and sponge gourd however, one variety of ridge gourd, PKM-1 has so far been developed through induced mutation by gamma radiation of an indigenous material (H 160).

Table: Varieties/ mutants developed through applied mutagenesis

Variety/ Mutant	Applied mutagenesis / Remark
PKM-1	Gamma radiation mutation of H 160

Heterosis breeding

In ridge gourd and sponge gourd, fruit yield per plant and most of the important fruit yield components *viz.,* earliness (days to opening of first female flower), days to first and last picking, number of fruit pickings, fruit weight, fruit girth, fruit length are governed predominantly by non-additive genetic variance which amply suggest exploitation of heterosis for improvement of yield and yield components in both ridge gourd and sponge gourd.

Ridge gourd hybrids, in particular, has become very popular among the farmer mainly due to high female to male ratio in the vine, good quality fruit like dark green, ridged, pretty shape, white pulp with seeds embedded in soft flesh, extended fruiting span, wider adaptability, early and high yield and resistance to diseases. However, ridge gourd is extensively cultivated in India with several high quality indigenous cultivars as well as few line-bred varieties. For this reason, area under ridge gourd hybrids in India is comparatively low. It has been estimated that about 25-30 % ridge gourd area in India is at present under different hybrids.

Being monoecious in sex expression and cross pollinated, it provides ample scope for the utilization of hybrid vigour. Bagging of the staminate and pistillate flowers in the afternoon about 24 hours before anthesis and cross-pollination with the pollens from the staminate flowers of the intended male parent in the afternoon hours on the day of anthesis (next day) is the commercial method of hybrid seed production.

The three-line breeding approach using cytoplasmic-genic male sterility (CGMS) is an easy system to adopt among different male sterility systems in hybrid seed production. In cucurbits, this system was reported for the first time from India in ridge gourd (Pradeepkumar *et al.* 2012). The male-sterile lines produce rudimentary male flowers either in clusters and bear normal fertile female flowers. Pollinators can be employed to produce hybrid seeds in open or protected structures once the male sterile female lines are identified.

Breeding for resistance to diseases

Inheritance a resistance to downy mildew in ridge gourd could be explained by the additive dominance model, involving at least 4 additive genes (Hua and Sen, 1999). Based on screening against natural infection and challenge inoculation study and PCR detection of Tomato Leaf Curl New Delhi Virus two lines of sponge gourd (*Luffa cylindrica*), DSG-6 and DSG-7 were found highly resistant to the virus (Islam *et al.*, 2010). The frequency distribution of F_2 based on chi-square test revealed monogenic dominant nature of resistance of the disease. This finding was further confirmed by identification of SRAP markers linked to the single dominant resistance gene (Islam *et al.*, 2011). Further DSG-6(IC0588956; INGR12013), was registered with NBPGR, New Delhi (Munshi *et al.*, 2012). The variety, Kashi Shreya is resistant to downy mildew, powdery mildew and sponge gourd mosaic virus under field condition and it is tolerant to leaf minor, fruit fly and red pumpkin beetle. The other variety, Swarna Prabha is tolerant to powdery and downy mildew and resistant to leaf miner under field conditions.

Breeding for quality

All the nutritional parameters such as ascorbic acid, total carotenoids, total phenolics content and *in vitro* antioxidant activity exhibited higher values of dominance variance over the additive genetic variance, more than unity value of average degree of dominance and low narrow-sense heritability (<50 %). The antioxidants and mineral content in the fresh fruits are predominantly attributed by the non-additive genetic component. Hence, the various tools of hybrid breeding would be useful to breed antioxidants- and minerals-rich genotypes of ridge gourd (Karmakar *et al*, 2013).

References

Chatterjee, G. 2023. *Expression of flowering and fruiting behaviour in F_1 and F_2 generation of ridge gourd.* M.Sc. thesis (unpublished), Department of Vegetable Science, Bidhan Chandra Krishi Viswavidyalaya, Mohanpur, West Bengal, 61p.

Das, S. 2020. *Studies on flowering and fruiting behaviour in ridge gourd.* M.Sc. thesis (unpublished), Department of Vegetable Science, Bidhan Chandra Krishi Viswavidyalaya, Mohanpur, West Bengal, 74p.

Hazra, P. 2020. *Seed production Technology for Vegetable, Tuber and Spice crops* , Brillion publishing, New Delhi, 651p

Hua, X. W. and Sen, X. D. 1999. Analysis of resistance heredity in ridge gourd to downy mildew. *J. South China Agri. Univ.*, **20**: 20-23.

Islam, S., Munshi, A.D, Verma, M., Arya, L., Mandal, B., Behera, T.K., Kumar, R. and Lal, S.K. 2011. Screening of *Luffa cylindrica* Roem. for resistance against Tomato Leaf Curl New Delhi Virus, inheritance of resistance, and identification of SRAP markers linked to the single dominant resistance gene. *J. Hortic. Sci. Biotech.*, **86**: 661-667.

Islam, S., Munshi, A.D., Mandal, B., Kumar, R. and Behera, T.K. 2010. Genetics of resistance in *Luffa cylindrica* Roem. against Tomato leaf curl New Delhi virus. *Euphytica.*, **174**: 83-89.

Karmakar,P., Munshi, A. D. Behera, T. K., Kumar, R., Sureja, A.K., Kaur, C. and Singh, B.K. 2013. Quantification and inheritance of antioxidant properties and mineral content in ridge gourd (*Luffa acutangula*). *Agril. Res.*, **2**: 222-228.

Munshi, A.D., Islam, S., Kumar, R., Mandal, B., Behera, T.K. and Sureja, A.K. 2012. DSG-6 (IC0588956; INGR12013), a Sponge gourd (*Luffa cylindrica*) Germplasm with highly resistant to Tomato Leaf Curl New Delhi Virus. *Indian J. Plant Genet. Resour.*, **25**: 321-322.

Pradeepkumar, T., Hegade, V.C., Kannan, D., Sujatha, R., George, T.E. and Nirmaladevi, S. 2012. Inheritance of male sterility and presence of dominant fertility restorer gene in ridge gourd (*Luffa acutangula* (Roxb.) L.). *Sci. Hort.*, **144**: 60-64.

Pradeepkumar, T. and Bhardwaj, A. 2015. Ridge and Smooth gourd. In: *Hand book of Vegetables* Vol.III (Eds. Peter, K. V. and Hazra, P.), Studium Press LLC, Houston, Texas, USA, pp. 343-363.

Ram, H.H. 2006. *Vegetable Breeding: Principles and Practices*, Kalyani Publishers, Ludhiana, pp. 423-425.

Sidhu, M. K. and Kaur, J. 2021. Interspecific hybridization between sponge gourd (*Luffa cylindrica* L.) and 'Satputia' (*Luffa hermaphrodita* Singh & Bhandari) for pre-introgression of cluster bearing, high yield and gynoecism. *J. Genet.*, 100:73 https://doi.org/10.1007/s12041-021-01323-0

Singh, H.B., Ramanujam, S. and Pal, B.P. 1948. Inheritance of Sex Forms in *Luffa acutangula* Roxb. *Nature*, **161**: 775-776.

40

Bitter Gourd

Bitter gourd or Bitter melon, *Momordica charantia* L under the family Cucurbitaceae (Chromosome no. 2n = 2× = 22) is widely grown in India, Indonesia, Malaysia, Singapore, Thailand, China, Japan, East Africa, the Caribbean and South America for its edible fruit. Tender fruits are widely used as cooked vegetables in a number of ways. Bitter gourd is an excellent source of vitamin C, iron, magnesium, potassium, calcium and ß carotene. The fruits contain non-toxic glycoside momordicin, which gives it bitterness and also health benefits. It also contains an insulin-like compound called Polypeptide-p or p-insulin and steroidal compound "Charantin" having hypoglycemic property.

Origin, domestication and adaptation in India

Wild or semi-domesticated variants spread across Asia in prehistorical period, and it was probably domesticated in Southeast Asia. It has probably originated in eastern India or Indo-Burma area or southern China. Uncarbonized seed coat fragments have been tentatively identified from the spirit cave in northern Thailand however, there have been no archeological reports of bitter gourd remains in China (Marr *et al*, 2004).Population resequencing revealed the divergence between wild and South Asian cultivars about 6,000 years ago, followed by the separation of the Southeast Asian cultivars about 800 years ago, with the latter exhibiting more extreme trait divergence from wild progenitors and stronger signs of selection on fruit traits. Unlike some crops where the largest phenotypic changes and traces of selection happened between wild and cultivar groups, in bitter gourd large differences exist between two regional cultivar groups, likely reflecting the distinct consumer preferences in different countries (Matsumura, 2020). Hazra *et al*., (2022) also recorded huge divergence between the predominant bitter gourd types of Indian origin (dark green, spindle shaped fruits with sharp tubercles) and the predominant types in the region of China, Japan, Indonesia, Korea (whitish yellow, yellowish white, obtuse or roundish-oblong fruit with light ridges or roundish blunt tubercles).

Crop botany and floral biology

- Plant is a tropical, fast growing and climbing vine that may grows up to 2-3 m; stem may be either hairless or slightly hairy.
- From the apex of the central taproot, the stems spread to climb over any available support; plant with thin stems and tendrils require a trellis to support their climbing vines.

- The well-branched, slender, green stems are usually slightly 5-angled or ridged, and carry unbranched tendrils in the leaf axils.
- The leaves are carried singly along the stems on 3-5 cm long stalks, and each leaf is 4-10 cm long, rounded in outline, and deeply 5-9 lobed; foliage has an unpleasant smell when crushed.
- Bitter gourd is generally monoecious in sex form where staminate and pistillate flowers are borne on separate nodes in the same vine; gynoecious sex form is rarely found in some genotypes.
- Flowers are solitary in leaf axil, actinomorphic, pedicillate and are yellow in colour with prominent sessile bract at or near the base of slender peduncle; staminate flowers are generally having comparatively long peduncle than the pistillate flowers with conspicuous ovary.
- Staminate flowers have a slender basal swelling, which is continuous with the base of the sepal tube, which ends in five blunt sepals.
- Petals are five, oval, yellow, 10-20 cm long, and five central stamens; filaments are free but anthers are united; flower opens at sunrise and remains open for only one day.
- Anthesis typically occurs in the early morning hours between 3:30 and 7:30 a.m. when flowers are completely open; pollen viability is maximum at anthesis but lose relatively rapidly.
- Staminate flowers appear first and usually exceed the number of pistillate flowers by about 20:1.
- Corolla rotate, parted nearly to the base, light to deep yellow in colour with five petals; plants with white flowers are rarely seen.
- The pistillate flowers are single; styles short, 3 in number and terminated by 3 bilobed or divided stigma; ovary is typically inferior, tricarpellary, syncarpous and unilocular with many ovules; calyx with deeply 5- fid; sepals adnate to the ovary.
- The stigma is usually receptive for one day before or after flower opening, after which it dries and turns brown.
- The pistillate flower consists of an inferior ovary and a three-lobed, wet stigma that is attached to a columnar, hollow style.
- The ovary contains three carpels typical of many cucurbits, each with 14 to 18 ovules, surrounded by an ovary wall; number of ovules in an ovary can be up to 60, the average being 40; anatropous ovules are attached to parietal placenta in two irregularly aligned rows in each carpel; unlike other cucurbits, no more than four ovules can be seen in ovary cross-section.
- Typically, pollen tubes penetrate papillae tissue within 1 hour of pollination arriving at the ovary cavities about 6 hours after pollination, and thus fertilization is accomplished within 18 to 24 hours post pollination.

- Fruits are many-seeded pepo with fleshy mesocarp, fusiform, pendulous, 5-25 cm long, ribbed, with numerous tubercles; fruits turn to orange-yellow colour when ripe.
- Fruits come in a variety of size (medium to very big), shape (spindle, obtuse, ovate, elongated obtuse, oblong, etc.), tubercle character (deep and sharp tubercles, medium tubercles, smooth ridges, smooth surface, etc.) and fruit colour (dark green, green, light green, yellowish-white and white). The cultivar common in China, Indonesia, Korea is 20–30 cm long, oblong with bluntly tapering ends and yellowish-white and white in colour, with a gently undulating, warty surface while the cultivars more typical of India has a narrower shape with pointed ends, and a surface covered with jagged, triangular "teeth" and ridges; dark green, green to white in colour. Between these two extremes, there are any numbers of intermediate forms.
- Seeds are many, flattened, 1.5 to 2.0 cm long, brownish with scarlet aril and average 100 seeds weight is about 20 to 21 g.

Reproductive behaviour

Bitter gourd is highly cross pollinated in nature due to monoecious sex form. It is entomophilies and major pollinators are honeybees (*Apis* sp.). It shows negligible inbreeding depression. Hence, inbred selection is an important breeding method for this crop. Anthesis generally starts by 4.00 a.m and gets completed by 9.00 a.m. Anther dehiscence starts by 5.00 am and gets completed by 8.00 am. Stigma receptivity is maximum at the time of anthesis although stigma becomes receptive one day prior to and remains one day after anthesis also. Best time of hybridization is between 5.00 to 7.00 a.m. on the day of anthesis. Bagging both the staminate and pistillate flowers prior to opening in the afternoon hours followed by hand pollination next morning and subsequent bagging for two more days is the commercial method of hybrid seed production. Hand pollination is carried out by rubbing the anthers of the freshly opened or prior harvested matured staminate flowers against the stigma of the protected pistillate flowers. Harvesting the fully matured staminate flowers in the afternoon hours that will open in the next morning and keeping them covered with thin moist cloth overnight for their use in cross pollination is also practiced for hybrid seed production.

Genetics of important characters

Bitter gourd is a diploid species (2n = 2x= 22). The genus *Momordica* has about 80 species and chromosome number of *Momordica* species ranges between 2n = 22 (*M. charantia*, *M. charantia* var. *muricata*, *M. balsamina*, *M. cymbalaria*, syn. *M. tuberosa*) and 2n = 44 (*M. foetida*) and some species are having chromosome number in between *viz*., *M. dioica* and *M. cochinchinensis* (2n =28). Bitter gourd does not cross with *M. dioica*, *M. balsamina* and *M. cymbalaria*. Though *Momordica* species are cross-pollinated, they have remained isolated in nature, perhaps because of aneuploidy.

Fruit skin colour, fruit surface, tubercles, bitterness, fruit curviness, seed size, seed colour, etc. are monogenically inherited characters. Simple inheritance of these fruit traits will enable the breeder to incorporate these traits in the line-bred varieties and hybrids. The black seed coat colour may be utilized as a morphological marker for hybrid seed purity test.

Table 1: Some simply inherited characters of bitter gourd

Character	**Inheritance pattern**
Fruit surface	Monogenic, tubercles dominant over smooth surface Monogenic, continuous ridge dominant over smooth surface
Fruit curviness	Monogenic, curviness dominant over straight fruit
Seed colour	Monogenic, Black seed dominant over light brown seed Monogenic, Dark brown seed dominant over light brown seed
Fruit skin colour	Green dominant over white; F_2 shows continuous variation Controlled by two genes Controlled by three genes
Bitterness	Monogenic; bitterness dominant over mild bitterness
Seed size	Monogenic; small seed size dominant over large seed size

Source: Swarup (2006), Behera and Rajashree (2015), Huang and Hsieh (2017), Saha *et al*, 2020, Rao *et al* (2022)

Many important quantitative and fruit quality characters under polygenic control are vine length, earliness, fruit length, diameter, weight, fruits per plant, total chlorophyll content, β-carotene, total phenol, flavonoid and ascorbic acid contents, fruit yield per plant, etc.

Generation mean analysis was employed to understand the nature of gene action involved in conditioning of fruit length, diameter, weight, total chlorophyll content, β-carotene, and ascorbic acid contents, utilizing two contrasting crosses involving dark green (Meghna-2) and whitish-yellow fruited (BG 1346501) varieties and very small fruited feral species, *M. charantia* var. *muricata.* In both crosses, dominance, additive, and most of the epistatic components were significant indicating a complex nature of inheritance for all the characters. The dominance × dominance interaction effect was larger than the additive × additive effect. Duplicate epistasis in both the crosses indicated hindrance in advancement through selection. Most characters are under control of non-additive gene action, and the magnitude of additive component was lower than dominance × dominance component hence, selection for most characters might be difficult (Saha *et al*, 2020).

Huang and Hsieh (2017) suggested that green fruit colour was dominant over the white colour. But the fruit colour of the F_2 population showed a continuous distribution from white to dark green indicating that this trait was controlled by quantitative genes. Genetic analyses fitted the three parameters model ($\chi^2 = 1.59$, $P = 0.6607$) and showed significant to highly significant additive and dominance effects, respectively. The generation mean analysis results of the six parameters model revealed a significant additive effect on chlorophyll b concentration. Chlorophyll a and total chlorophyll concentrations were non-significant in all parameters. The magnitudes of additive and additive × additive gene effects were larger than those of the dominance and dominance × dominance gene effects, which revealed that additive genes were more important in the control of chlorophyll concentrations.

Breeding objectives

Some of the important characters that have been the main focus of bitter gourd breeding are mentioned below

- Preferably long, well branched and vigorous vines having large sized leaves with high female to male flower ratio in the vine
- Lower number of node at which first pistillate flower appears which gives a fair indication of earliness
- Fruit size vary widely from very small (4-5 cm) to very big (40-50 cm); generally, 10-15 cm long, medium-sized fruits are preferred
- Fruit should be spindle shaped, egg-shaped, elongated or club-shaped with thicker blossom end
- Spindle shaped green fruit with medium prickle or continuous ridges and blunt tubercles on the fruit surface is preferred for export.
- Fruits should be uniformly medium-thick to thick; thick fruit is preferred for stuffing and pickles
- Fruit surface may be warty with tubercles or smooth with smooth continuous or broken ridges; fruits should not have too many tubercles which may break during transportation.
- Fruit colour may be dark green, green, light green, yellowish-white and white depending on regional preference.
- Fruits with medium bitterness is preferred in many parts of the country; small sized very bitter fruits are preferred in West Bengal and Bihar.
- Fruit should contain different compounds having medicinal properties such as natural cucurbitane type triterpenoid (charantin), polypeptides (polypeptide-K), glycoalkaloids, phenolics and antioxidants
- Fruits should contain small seeds which remain soft in the edible stage of the fruit
- Extended harvest span, high total and early yield
- Better adaptability to different climatic condition
- Resistance/tolerance to different diseases like, powdery mildew (*Podosphaera xanthii*), downy mildew ((*Pseudoperonospora cubensis*), cucumber mosaic virus, etc.
- Resistance/tolerance to insect pests like, fruit fly (*Bactrocera cucurbitae*), red pumpkin beetle (*Aulacophora foveicollis*), etc.
- Resistance to abiotic stresses like cold temperature, soil salinity, etc.

Breeding methods and development of variety

In bitter gourd wide genotypic variation exists in India for fruit shape and size ranging from 5 to 400 g because of its probable centre of origin and extensive cultivation. Large number of indigenous cultivars are grown in India under different names like, Small Green, Large White, Long Green, Karala Guchhedar, Karala Long, Kerala Jhalarder, Baramasi, Khandesh Mali, Panipat Local, Ambala Local, Washim Local, Coimbatore Local, Gorakhpur Local, Jaunpuri, Faizabadi, Kareli, Kalyanpur Baramasi , Pride of Surat, Pride of Gujarat (cultivated widely in Gujarat), Gangajali, Meghna (cultivated widely in West Bengal), etc. Obviously, improved varieties have been developed in India chiefly through selection and subsequent purification by inbreeding because bitter gourd like other cucurbits does not show inbreeding depression. Indigenous genotypes have also been utilized in hybridization programmes to develop new variety and hybrid.

Bitter gourd cultivars grown in India may be classified according to fruit size and colour.

Long / extra-long fruit

Coimbatore Long, Coimbatore Green, Coimbatore Long White, VK-1 Priya, MDU-1, Preethi, Phule Green Gold, Punjab Kareli-1, Kalyanpur Baramasi, Kashi Urvasi

Medium/ medium-small sized fruit

Pusa Do Mousumi, Arka Harit, Co-1, MC-23, Pusa Vishesh, Priyanka, Hirkani, Phule Ujwala, Konkan Tara, Punjab 14, Punjab Karela-15, Punjab Jhaar Karela-1, Kalyanpur Sona, Pant Karela-1, Pant Karela-2

Dark green / Green / light green fruit

Coimbatore Green, Pusa Do Mousumi, Arka Harit, VK-1 Priya, Co-1, MC-23, Pusa Vishesh, Hirkani, Phule Green Gold, Phule Ujwala, Konkan Tara, Punjab 14, Punjab Kareli-1, Punjab Karela-15, Punjab Jhaar Karela-1, Kalyanpur Baramasi, Kalyanpur Sona, Kashi Urvasi, Pant Karela-1, Pant Karela-2

While/ whitish-green / yellowish-white fruit

Coimbatore Long, Coimbatore Long White, MDU-1, Preethi, Priyanka

Germplasm resources and Plant introduction

In India, apart from bitter gourd, the genus *Momordica* comprises a number of small and often poorly known species but bear edible fruits esteemed for their medicinal, nutritional properties and taste. Taxonomic confusion exists in the genus because of lack of familiarity with taxa and widespread use of common names and is region specific and poorly represented in herbaria. Analysis of morphological data for determining the genetic variation within seven *Momordica* species lead to the recognition of three groups.

Table 2: Variation in *Momordica* species

Group	Species	Characters
Group- 1	*M. charantia* and *M. balsamina*	n = 11, annual, monoecious, non-tuberous roots and muricate - tubercled fruit surface
Group-2	*M. dioica*, *M. sahyadrica*, *M. subangulata* subsp. *renigera* and *M. cochinchinensis*	n = 14, perennial, dioecious, tuberous tap roots and echinate-soft papillate fruit surface
Group-3	*M. cymbalaria*	n = 9, perennial, monoecious, tuberous tap roots, and ribbed fruit surface

Source: Bharathi *et al* (2013)

Germplasm collections of bitter gourd are being maintained in Germany, USA, Japan, Phillippines, South Africa, Malayasia, Thailand and many other countries.

In India, bitter gourd germplasm is maintained at NBPGR, New Delhi, ICAR-IARI, New Delhi, ICAR-IIVR, Varanasi, IIHR, Bengaluru, KAU, Vellanikkara, Kerala, PAU, Ludhiana, Punjab and other places.

No exotic variety has so far been released directly for cultivation in India.

Individual plant selection and inbreeding

Huge genetic diversity exists in India for bitter gourd cultivars. In this method of breeding, individual plant from this indigenous source population consisting of heterozygous and non-uniform materials is selected and the seeds of the selected plants are not mixed even if they are equally good and look similar. Selfed seeds of the selected lines are grown and evaluated separately. This method of selection is based on the principle that the actual worth of the selected plant may be masked by the environmental variation and, therefore, the only way to recognize it is by growing the progeny of each selected individual plant for re-evaluation. Progenies of individual plant for evaluation can be developed by selfing as bitter gourd does not show loss of vigour due to inbreeding. Homogzygosity for the concerned characters can also be attained in the individuals of the progeny by such repeated selfings.

Fruit quality traits, including external appearance are the most important traits in bitter gourd because they are directly related to the product's commercial value. Spindle shaped green fruit with medium prickle or continuous ridges and blunt tubercles on fruit surface is preferred for export. In addition to appropriate internal colour, fruit must be firm, without excessive seed development and free from internal defects. Fruit colour also governs its marketability although, regional preference is there for the fruit colour. For example, green-fruited types are in demand in southern China while white-fruited types are preferred in central China. Similarly, dark green to glossy green fruits are favoured in northern and eastern India whereas white fruits are preferred in southern India (Behera and Rajashree, 2015).

It emerged from the study of trait linking that the predominant bitter gourd types of Indian origin (dark green, spindle shaped fruits with sharp tubercles) were associated with high nutrient and most of the nutraceutical contents (β carotene, total chlorophyll and ascorbic acid contents) but comparatively low in charantin content and the predominant types in the region of China, Japan, Indonesia, Korea (whitish-yellow, yellowish-white, obtuse or roundish-oblong fruit with light ridges or roundish-blunt tubercles) were associated with comparatively low nutrient and most of the nutraceutical contents but high in charantin content. This study confirmed the effectiveness of combining morphological characters with nutritional and nutraceutical contents of fruit in future breeding programmes (Hazra *et al*., 2022).

Most of the varieties have been developed in India by this method of breeding.

Table 3: Varieties developed through selection and inbreeding

Name of variety	Indigenous / Exotic material
Coimbatore Long	Indigenous (Tamil Nadu)
Coimbatore Green	Indigenous (Tamil Nadu)
Coimbatore Long White	Indigenous (Tamil Nadu)
Pusa Do Mousumi	Indigenous
Arka Harit	Indigenous (Rajasthan)
VK-1 Priya	Indigenous (Kerala)
Co-1	Indigenous (Tamil Nadu)
MC-23	Indigenous (Tamil Nadu)
Pusa Vishesh	Indigenous (Hapur, U.P)
Preethi	Indigenous (Thrissur, Kerala)
Priyanka	Indigenous (Thiruvalla, Kerala)
Hirkani	Indigenous (Maharashtra)
Phule Ujwala	Indigenous (Maharashtra)
Konkan Tara	Indigenous (Maharashtra)
Punjab 14	Indigenous (Punjab)
Punjab Karela-1	Indigenous (Khanna of Ludhiana , Punjab)
Punjab Karela-15	Indigenous (Punjab)
Punjab Jhaar Karela-1	Indigenous (Punjab)
Kalyanpur Baramasi	Indigenous (U.P)
Kalyanpur Sona	Indigenous (U.P)
Pant Karela-1	Indigenous
Pant Karela-2	Indigenous

Hybridization and selection

It emerged from the generation mean analysis involving two contrasting varieties of bitter gourd and a very small fruited feral species, *M. charantia* var. *muricata* that all the characters *viz*., fruit length, diameter, weight, total chlorophyll content, β-carotene, and ascorbic acid contents exhibited complex genetic behaviour. Simple selection in early segregating generations may not be effective for improvement of these traits. Complex genetic behaviour, particularly additive and dominance components, could be exploited in later generations. Selection for improvement of all the traits should be delayed to later segregating generations. The modified bulk method of selection is recommended for improving fruit physico-chemical characters of bitter gourd (Saha *et al*, 2020). Inter-varietal hybridization involving indigenous parental materials and selection in the segregating generations has been followed for the development of very few varieties in India (Hazra *et al*, 2022).

Table 4: Varieties bred through hybridization followed by selection

Name of variety	Parentage
Phule Green Gold	Green long × Delhi Local
Kashi Urvasi	IC-85650B × IC-44435A

Heterosis breeding

Bitter gourd being cross-pollinated due to monoecious sex form holds ample scope for exploitation of heterosis. It amply responds to heterosis breeding due to prevalence of high magnitude of non-additive gene effects and over-dominance reactions for the inheritance of most of fruit and fruit quality characters which indicated that heterosis breeding might be advantageous for improvement of different quantitative and quality traits in bitter gourd (Tiwari *et al.*, 2018; Saha *et al*, 2020). However, more pronounced hybrid vigour could be observed with the inclusion of diverse parents based on number of fruits, fruit weight, fruit length, fruit shape and fruit yield. Bitter gourd hybrids have become very popular among the farmers mainly due to high female to male ratio in the vine, earliness, attractive, medium thick, spindle shaped or elongated fruit, high number of fruits per plant, uniformity, extended fruiting span, wider adaptability and early and high yield. Bitter gourd is extensively cultivated in India with several high quality indigenous farmers' variety as well as some line-bred varieties. For this reason, area under bitter gourd hybrids in India is comparatively low. It has been estimated that about 25-30 % bitter gourd area in India is at present under different hybrids. The monoecious sex form of bitter gourd only necessitates bagging of both staminate and pistillate flowers on the intended male and female parents before opening followed by hand pollination for the production of hybrid seeds. Bagging of staminate and pistillate flowers in the afternoon hours and hand pollination of the pistillate flower in the morning hours of the next day with the pollens from the staminate flower of the intended male parent is the commercial method of hybrid seed production.

Gynoeous sex form in hybrid seed production

High propensity of hybrid seed production could be achieved with the exploitation of gynoecious sex form in the development of hybrids. In the Indian germplasm of bitter gourd, gynoecy (*gy1*) is recessive to monoecy (Ram *et al*, 2006; Behera *et al*, 2009; Vinay *et al*, 2023). Protein CHROMATIN REMODELING 4 were identified with possible role in gynoecious sex expression (Vinay *et al*, 2023). Such gynoecious or predominantly female line hold ample promise for facilitating hybrid breeding programme.

Mutation breeding

In bitter gourd, gamma irradiation and EMS (ethylmethane sulfonate) treatment led to an increased genetic variability in many quantitative characters (Mallaiah and Zafar, 1986). One variety, MDU 1 could be developed by gamma radiation of seeds of the landrace MC-103 (Rajasasekharan and Shanmugavelu, 1984). Research works on applied mutagenesis in bitter gourd under DAE, Bhabha Atomic Research Centre, Govt. of India sponsored Regional Nuclear Agriculture Research Centre at Bidhan Chandra Krishi Viswavidyalaya, Mohanpur revealed very low mutation frequency ranging between 0.40 to 0.67 %. However, exposure of dry seeds to 200-250 Gy dose gamma radiation was most effective in inducing desirable macro-mutation which could be isolated based on variation in sex form, fruit colour, fruit shape, prickliness, fruit length, girth, weight and charantin content (Dutta *et al.*, 2021). Radicle scavenging activity in two mutant isolated from the M_2 population of the white fruited cultivar, BG-1346501 irradiated by 100 Gy gamma radiation (BG-1346501- 100 Gy) and 200 Gy gamma radiation (BG-1346501-200 Gy) improved significantly by 9.12 and 16.12 percent, respectively and total charantin content was maximum (1.26 mg /100 g dry weight) in the mutant line BG-1346501-100 Gy (Dutta *et al.*, 2021; Hazra *et al.*, 2022).

Table 5: Varieties/ mutants developed through applied mutagenesis

Variety/ Mutant	Applied mutagenesis / Remark
MDU-1	500 gray gamma radiation mutation of MC-103

Breeding for resistance to diseases

Identification and incorporation of resistance to economically important pests such as fruit fly and various disease causing pathogens, particularly downy mildew and viruses is important in resistance breeding programme. However, genes for resistance to several important pests and pathogens are generally not found in cultivated species, *Momordica charantia* var. *charantia.* The variety Pant Karela-1, developed by single plant selection followed by inbreeding at GBPUAT, Pantnagar shows tolerance to red pumpkin beetle. This finding amply suggest planned screening of wide diversity of bitter gourd available in the country in resistance breeding programme.

Very small fruited, wild feral species *Momordica charantia* var. *muricata* shows high level of tolerance to fruit fly. It crosses with the cultivated bitter gourd. However, small fruit size and roundish shape of *Momordica charantia* var. *muricata* show dominance hence, it is unlikely to isolate segregates with desirable fruit size from the *Momordica charantia* var. *charantia* x *Momordica charantia* var. *muricata* cross. Wild African types *viz.*, *M. foetida* and *M. rostrata* can be exploited as potential source of disease resistance (Njoroge and van Luijk, 2004). *M. balsamina* may contribute tolerance to red pumpkin beetle but it generally does not cross with the cultivated bitter gourd.

Rathod *et al.* (2021) investigated the interspecific hybridization programme of *Momordica charantia* x *Momordica balsamina* in detail. Hybridization success rate was very low with a higher rate of seedling mortality. A total of 116 accessions of *Momordica charantia* (female parent) were used for artificial cross-pollination with *Momordica balsamina* (IC-467683). Among these interspecific crosses, only one cross combination (Pusa Aushadhi × *Momordica balsamina*) produced the higher percentage of normal seeds and only one seed was germinated followed by development into a full plant. The pollen-pistil interaction study confirmed the existence of post dynamic barriers in all these unsuccessful crosses. A single plant progeny of *M. charantia* × *M. balsamina* was derived and its advanced generations were studied at morphological, cytological, and molecular levels. The plant phenotype predominately resembled the female parent with variation in a few traits. The microsatellite (SSRs) markers analysis showed that amplified fragments of interspecific hybrids exhibited highly conserved similarity with the fragments of female parents with a very low frequency of male fragments. The F_2 and backcross generations were produced from the single partial interspecific hybrid (Pusa Aushadhi × *M. balsamina*). Twenty advanced generation F_2 plants showed variation with majority traits resembles the female parent. The microsatellite analysis revealed that the progenies of backcross generation amplified the fragments with high similarity to both female and the male parent which confirmed the partial introgression of chromosome segments of *M. balsamina*.

Breeding for quality

Fruits of bitter gourd are a rich source of carbohydrates, proteins, vitamins and minerals and possess the highest nutritive and medicinal value among the gourds (Upadhyay *et al*., 2015). Considerable variation in nutrients particularly, β carotene and ascorbic acid and phytochemicals has been observed in bitter gourd due to genotypic variation (Yuwai *et al*., 1991; Hazra *et al*, 2022). In fact, bitter gourd fruits have higher crude protein content (11.4-20.9 g kg-1) than that of tomato and cucumber (Xiang *et al*., 2000). In phytochemical, bitter gourd mainly consists of proteins, steroidal compound "Charantin", poly phenols, flavonoids, glycosides, fatty acids and volatile constituents (Lee *et al*., 2009; Haque *et al*., 2011; Hazra *et al*, 2022). Breeding for quality traits are typically focusing on accessions with relatively high vitamin C content, and moderate bitterness (Dey *et al*., 2006). Bitter gourd lines like DBTG-3, DBTG-8, DBTG-6, and DBTG-9 contain>1000 mg.kg-1 vitamin C in fruits as compared to 500 mg.kg-1 in standard cultivars (Dey *et al*., 2006). Many Indian bitter gourd varieties already reported rich in traits like high total soluble solid content (>3.1°Brix; MC-84, Preethi, RHRBG-5, and PBIG-1), vitamin C (>950mg.kg-1; Konkan Tara and Hirkani) and protein (>1.5%; DVBTG-1, Preethi, Hirkani and Konkan Tara) content (Kore *et al*., 2003). These nutritionally rich cultivars /lines are used in breeding programs for development of cultivars with high quality traits along with yield.

References

Behera, T.K., Dey, S.S., Munshi, A.D., Gaikwad, A.B., Pal, A. and Singh, I. 2009. Sex inheritance and development of gynoecious hybrids in bitter gourd (*Momordica charantia* L.). *Sci. Hort*., **120**: 130-133.

Bharathi, L. K., Munshi, A. D., Behera, T. K., Joseph John, K., Bhat, K. V. and Sidhu, A. S.2013. Morphological relationship among the *Momordica* species of Indian occurrence. *Indian J. Genet Plant Breed*., **73**: 278-286

Dey, S., Singh, A. K., Chandel, D. and Behera, T. K. 2006. Genetic diversity of bitter gourd (*Momordica charantia* L.) genotypes revealed by RAPD markers and agronomic traits. *Sci. Hort*., **109**:21-28

Dutta, S., Hazra, P., Saha, S., Acharya, B., Bhattacharjee, T., Maurya, P.K., Banerjee, S., Chakraborty,I. and Chattopadhyay, A. 2021. Applied mutagenesis could improve economically important traits in bitter gourd (*Momordica charantia* L.). *J.Genet*., 100:43 https://doi.org/10.1007/s12041-021-01283-5

Haque, M. E., Alam, M. B. and Hossain, M. S. 2011. The efficacy of cucurbitane type triterpenoids, glycosides and phenolic compounds isolated from *Momordica charantia*: A review. *Intern. J. Pharmaceu. Sci. Res.*, **2**: 1135-1146

Huang , H.Y. and Hsieh , C.H. 2017. and Genetic Research on fruit color traits of the bitter gourd (*Momordica charantia* L.). *The Hort. J.*, **86**: 238-243.

Kore, V. N., Dhanawate, H. P., Thorat, S. T., Mahajan, T. S., Patil, R. S. and Mane, A. V. 2003. Comparative studies on chemical composition of fruits and fruit yield of improved bitter gourd (*Momordica charantia* L.) genotypes. *J. Soils Crops*, **13**:91-94

Lee, S. Y., Eom, S. H., Kim, Y. K., Park, N. I. and Park, S. U. 2009. Cucurbitane-type triterpenoids in *Momordica charantia* Linn. *J. Med. Plants Res.*, **3**: 1264-1269.

Mallaiah, B. and Nizam, Z. 1986. Induced polygenic variability in *Momordica charantia* L. *J. Indian Bot. Soc.*, **65**: 80.

Marr, K.L., Xia, Y.M., Bhattarai, N.K. 2004. Allozyme, morphological and nutritional analysis bearing on the domestication of *Momordica charantia* L. (Cucurbitaceae). *Econ. Bot.*, **58**: 435-455.

Matsumura, H., Hsiao, M.C. Linb, Y.P., Toyoda, A., Taniai, N., Tarorad, K., Urasakid, N., Anand, S.A., Dhillone, N.P.S., Schafleitner, R. and Ruei Lee, C. 2020. Long-read bitter gourd (*Momordica charantia*) genome and the genomic architecture of non-classic domestication. *Proc. Nat. Acad. Sci.*, 117: 14543-14551

Rajasekharan, K.R. and Shanmugavelu, K.G. 1984. MDU. 1 bitter gourd. *South Indian Hort.*, **32**: 47.

Ram, D., Kumar, S., Singh, M., Rai, M. and Kalloo, G. 2006. Inheritance of gynoecism in bitter gourd (*Momordica charantia* L.). *J. Hered.*, **97**: 294-295..

Rao, G., Behera, T.K., Munshi, A.D., Jat, G.S. and Boopalakrishnan, G. 2022. Genetic inheritance of fruit traits and seed coat colour in bitter gourd (*Momordica charantia*). *Indian J. Agril. Sci.*, **92**: 357–61

Rathod , V., Behera, T.K., Munshi, A.D., Gaikwad, A.B., Singh, S., Vinay, N.D., Boopalakrishnan, G. and Jat, G.S. 2021. Developing partial interspecific hybrids of *Momordica charantia* × *Momordica balsamina* and their advance generations. *Sci. Hort.*, **281**: https://doi.org/10.1016/j.scienta.2021.109985

Saha, S., Hazra, P., Dutta, S., Acharya, B., Maurya, P. K., Banerjee, S, Banerjee, T., Mukherjee, D. and Chattopadhyay, A. 2020. Genetic expression of fruit physico-chemical characters in bitter gourd inter-specific crosses. *Intern. J. Veg. Sci.*, **27**: 131–143

Tiwari, N.K., Pandey, A.K., Singh, U.N. and Singh, V.B. 2018. Gene action involved in expression of various traits in bitter gourd (*Momordica charantia* L.). *J. Pharmacogn. Phytochem.*, **7**: 3703–3706.

Upadhyay, A., Agrahari, P. and Singh, D. K. 2015. A review on salient pharmacological features of *Momordica charantia*. *Intern. J. Pharmacol.*, **11**: 405-413

Vinay, N.D., Matsumura, H., Munshi, A.D., Ellur, R.K., Chinnusamy, V., Singh , A., Iquebal, M.A., Jaiswal, S., Jat, G.S., Panigrahi, I., Gaikwad, A.B., Rao, A.R., Dey , S.S. and Behera, T.K. 2023. Molecular mapping of genomic regions and identification of possible candidate genes associated with gynoecious sex expression in bitter gourd. *Front. Plant Sci.*, 14 - 2023 | https://doi.org/10.3389/fpls.2023.1071648

Yuwai, K. E., Rao, K. S., Kaluwin, C., Jones, G. P. and Rivett, D. E. 1991. Chemical composition of *Momordica charantia* L. fruits. *J. Agril. Food Chem.*, **39**: 1762-1763.

41

Bottle gourd

Bottle gourd, *Lagenaria siceraria* under the family Cucurbitaceae (Chromosome No: $2n = 2\times = 22$) is commonly grown vegetable crop of India, Africa, Central America and other warmer regions of the world. Tender fruits are widely used as cooked vegetables in a number of ways. Bottle gourd is a good source of vitamin C and zinc that may offer several skin benefits. It also contains minerals like calcium, magnesium, iron, and zinc. Its dietary fibre enhances gastro-intestinal health and promotes bowel movements and prevents constipation.

Origin, domestication and adaptation in India

It is the only crop known to have been cultivated in pre-Columbian times in both the Old and New World. Bottle gourd is indigenous to the tropical Africa. It has bihemispheric distribution in Africa, the old world and America, the new World. Its utilization by man is about 16,000 years in the Old World and 12,000 years in the New World. It was independently domesticated in the two hemispheres. It was one of the world's first cultivated plants grown not primarily for food, but for the use of hard-shelled fruit as containers. The apparent wild plant produces thinner-walled fruit that, when dried, would not endure the rigors of use on long journeys as a water container. Today's bottle gourd may owe it's tough, waterproof wall to selection pressures over its long history of domestication. The bottle gourd was theorized to have drifted across the Atlantic Ocean from Africa to South America. Plastid genomes of a broad sample of bottle gourds was thoroughly studied, and it was concluded that North and South American specimens were most closely related to wild African variants and could have drifted over the ocean several or many times, as long as 10,000 years ago. Archaeological evidences indicated its presence in Egyptian Tombs during 3000 to 5000 B.C., Spirit caves of Thailand during 6000-10000 B.C., Mexico during 5000 to 7000 B.C. and Peru during 3000 to 4000 B.C.

Bottlegourd had pre-historic existence in India. Several wild and semi-wild forms of bottle gourd are still found in Africa and India. Mentioning of bottle gourd in different names like, "Alaabu", Katutumbi, Alobu, etc. in ancient Sanskrit literatures amply justified its antiquity in India. Its wide acceptability in India favours significant diversification of this crop in the country.

Crop botany and floral biology

- Plant is a vigorous, quick-growing, annual and climbing vine with large leaves and a lush appearance; vine is branched and climbs by means of long forked tendrils along the stem.
- The thick stem is furrowed longitudinally; foliage is covered with soft hairs and has a foul musky odour when crushed; leaves have a velvety texture because of the fine hairs, especially on the under surface; size of the leaves may be up to 35 cm wide, circular in overall shape, with smooth margins, a few broad lobes or with undulate margins.
- Plant grows fast and may begin to flower within 40-50 days after seeding and plants are monoecious in sex form with unisexual flower; an andromonoecious form has been reported from India.
- The flowers are generally borne singly on the axils of the leaves; the staminate flowers are borne on about 10 cm long peduncles and the pistillate flowers on short peduncles and conspicuous ovary; very few cultivars bear more than one flower (both staminate and pistillate); flowers are white and attractive, up to 10 cm in diameter, with spreading petals; the staminate and pistillate flowers are similar in appearance.
- Calyx with five sepals adnate to the ovary; calyx tube of staminate flower is shorter than that of the pistillate flower; corolla campanulate, 5 lobed and free
- The anthers are borne on short filaments grouped at the centre of the flower; stamens are short, woolly and usually three, two as compound and one as single; filaments are free and swollen at the base, glanduliform; anthers are united into a column like mass; Pollen grains are heavy and sticky.
- The stigmas are short, thickened, and branched.
- Ovary is typically inferior; ovary has a thick style topped by three thick, woolly, and bitubed stigmas.
- Fruits are many-seeded pepo with fleshy mesocarp, variable in size and shape: cylindrical, oval, egg-shaped, club shaped, round, etc. The fruits at maturity are hard-shelled, smooth surfaced, green to whitish-green or tan in colour, and variously striped or mottled.
- Seeds are numerous, flattened, whitish or brown in colour embedded in whitish green pulp; each seed is rectangular with grooved notches near the attached end; average 100 seeds weight is about 15g.

Reproductive behaviour

Bottle gourd is highly cross-pollinated in nature due to monoecious sex form and entomophilies. Major pollinators are honeybees (*Apis* sp.). Although it is 100% cross-pollinated in nature yet reasonable extent of self-fertilization happens because of bee foraging in the staminate flowers of the same vine. Bottle gourd also shows negligible inbreeding depression. Anthesis generally starts in the afternoon hours from 2.00 p.m.

onwards in north and eastern Indian condition. Anther dehiscence generally starts in the afternoon hours. Stigma receptivity is maximum at the time of anthesis although stigma becomes receptive one day prior to and remains one day after anthesis. Best time of hybridization is between 3.00 to 5.00 p.m. on the day of anthesis. Bagging the staminate and pistillate flowers prior to opening in the morning hours followed by hand pollination in the afternoon hours of the same day and subsequent bagging for two more days is the commercial method of hybrid seed production.

Genetics of important characters

Bottle gourd is a diploid species (2n = 2x= 22) with normal meiosis of 11 bivalents. Modern forms of bottle gourd have also been considered secondary polyploidy and derived from the ancestor with a basic chromosome number x = 5. Although polyploidy could not be detected in bottle gourd, differences in DNA content within accessions was observed (Enonch *et al.*, 2008). Bottle gourd crosses with the wild species *Lagenaria sphaerica* but the F_1 hybrid is sterile. It has been found that cultivated bottle gourd is not crossable with other five wild relatives *viz.*, *L. breviflora, L. abyssinica, L. rufa, L. sphaerica* and *L. guineensis*.

Information on genetics of different qualitative characters of bottle gourd is very limited mainly because of viny and spreading nature which requires large space to accommodate plant population for genetic studies.

Table 1: Some important simply inherited characters of bottle gourd

Character	**Genetic control and gene symbol**
Leaf shape	Segmented leaf governed by single dominant gene
Fruit colour	Patchy dominant over white; Green fruit colour monogenically dominant over white
Bitterness	*Bi*; governed by single dominant gene
Fruit shape	Governed by two major gene Long fruit shape (*A-B-*) Oval fruit shape (*A-bb*) Pear fruit shape (*aa B-*)
Andromonoecious sex form	Governed by single recessive gene
Fruit shape	Long fruit partially dominant (*AA*) over round fruit (*aa*); disc shape is dominant to round fruit
Seed colour	Dark brown monogenically dominant over white

Source: Swarup (2006), Dhatt and Khosa (2015)

Tiwari and Ram (2009) suggested that long fruit shape of PBOG 54 was found to be conditioned by two dominant genes (*A-B-*). The intermediate fruit shape i.e. oval (*A-bb*) and pear (*aa B-*) were determined by a combination of one dominant and one recessive gene and round fruit shape of PBOG 13 (*aabb*) by two recessive genes.

Many important quantitative characters under polygenic control are length of vine, number of branches per plant, days to first pistillate flower, fruit length, fruit girth, fruits per plant, days to maturity, seeds per fruit, seed size, seed weight, yield per plant, etc. In most of the characters both additive and non-additive gene action are important for the control of these characters. However, fruit yield per plant is pre-dominantly controlled by dominant alleles which indicated the importance of heterosis breeding for enhancement of yield in bottle gourd.

Breeding objectives

Some of the important characters that have been the main focus of bottle gourd breeding are mentioned below

- Plant vigorous, vines medium-long, with high female to male flower ratio in the vine
- Lower number of node at which first pistillate flower appears which gives a fair indication of earliness
- Early maturing, medium-sized, round, flattish-round, long, uniformly cylindrical or club shaped green to light green fruits without neck
- Attractive glossy green fruit of round/cylindrical shape, pubescent with sparse hairs persisting on the skin at immature stage and non-fibrous flesh at edible stage
- Flesh soft and should not become fibrous quickly in optimum sized fruits at edible maturity.
- Wider adaptability to different climatic condition
- High total and early yield potential
- Resistance/tolerance to different diseases like, powdery mildew (*Podosphaera xanthii*), downy mildew (*Pseudoperonospora cubensis*), cucumber mosaic virus, watermelon mosaic virus, etc.
- Resistance/tolerance to insect pests like, fruit fly (*Bactrocera cucurbitae*), red pumpkin beetle (*Aulacophora foveicollis*), etc.
- Resistance to abiotic stresses like cold temperature, soil salinity, etc.

Breeding methods and development of variety

Huge genetic diversity of bottle gourd in India has been utilized to develop several improved varieties of different fruit shape through single plant selection followed by inbreeding to stabilize the genotype. Few varieties have also been developed by inter-varietal hybridization followed by pedigree, backcross or some modified methods of handling the segregating populations. Several hybrids have also been developed utilizing the indigenous germplasm. The varieties developed in India can conveniently be grouped according to the fruit shape.

- **Long /cylindrical fruited variety:** Pusa Summer Prolific Long, Punjab Long, Arka Bahar, Arka Nutan, Pusa Naveen, Azad Nutan, Kalyanpur Long Green, Samrat, Rajendra Chamatkar, Kashi Bahar, Narendra Jyoti, Pant Lauki-3, NDBG-132, Pusa Samridhi, Kashi Ganga
- **Round/oblong/ pear fruited variety:** Pusa Summer Prolific Round, Punjab Round, Punjab Komal, Pusa Sandesh, Pusa Santushti, Arka Ganga, Kashi Kiran, Kashi Kundal, Thar Samridhi

- **Bottle / club shaped variety:** Narendra Dharidar, Narendra Rashmi, Arka Shreyas
- **Pot shaped/ small fruited variety:** CO 1, Kashi Kirti

Germplasm resources and Plant introduction

Bottle gourd is amongst the widely grown yet under-researched cucurbits genetic resources in sub-Saharan Africa. Large germplasm collections of bottle gourd mainly sourced from sub-Saharan African countries (i.e., South Africa, Zambia, Zimbabwe, Swaziland, Namibia, Botswana, Egypt, Ethiopia, Nigeria, and Zaire) are held by the United States Department of Agriculture (USDA), Plant Genetic Resource Conservation Unit (USDA-PGRCU). Germplasm collections have also been made in Spain.

In India, bottle gourd germplasm is maintained at NBPGR, New Delhi, ICAR-IARI, New Delhi, ICAR-IIVR, Varanasi, ICAR-IIHR, Bengaluru, ANDUAT, Ayodhya, PAU, Ludhiana, Punjab and other places.

No exotic variety has so far been released directly for cultivation in India.

Individual plant selection and inbreeding

Huge genetic diversity exists in India for bottle gourd cultivars of different shape and size. In this method of breeding, individual plant from the indigenous source population consisting of heterozygous and non-uniform materials is selected and unlike mass selection, the seeds of the selected plants are not mixed even if they are equally good and look similar. This method of selection is based on the principle that the actual worth of the selected plant may be masked by the environmental variation and, therefore, the only way to recognize it is by the growing the progeny of each selected individual plant for re-evaluation. Progenies of individual plant for evaluation can be developed by selfing as bottle gourd does not show loss of vigour due to inbreeding. Homogzygosity for the concerned characters can also be attained in the individuals of the progeny by such repeated selfings. Fruit yield per plant recorded positive and significant correlation coefficients with days to first female flower, vine length, number of primary branches/plant, number of fruits/plant and fruit weight which suggested their importance as selection indices during exercising selection. Most of the varieties in India have been developed by this method of breeding from wide range of germplasm available in India.

Table 2: Varieties developed through individual selection and inbreeding

Name of variety	Indigenous / Exotic material
Pusa Summer Prolific Long	Indigenous (Delhi)
Punjab Long	Indigenous (Punjab)
Arka Bahar	Indigenous (Karnataka); inbred selection
Arka Nutan	Indigenous (Karnataka)
Pusa Naveen	Indigenous
Azad Nutan	Indigenous (U.P)
Kalyanpur Long Green	Indigenous (U.P)

Name of variety	Indigenous / Exotic material
Samrat	Indigenous (Maharashtra)
Rajendra Chamatkar	Indigenous (Bihar)
Kashi Bahar	Indigenous (U.P)
Narendra Jyoti	Indigenous (U.P)
Pant Lauki-3	Indigenous (U.P); inbred selection
NDBG-132	Indigenous (U.P)
Pusa Samridhi	Indigenous (Delhi)
Kashi Ganga	Indigenous (U.P)
Pusa Summer Prolific Round	Indigenous (Delhi)
Punjab Round	Indigenous (Punjab)
Pusa Sandesh	Indigenous (Delhi)
Pusa Santushti	Indigenous (Delhi)
Arka Ganga	Indigenous (Karnataka)
Kashi Kiran	Indigenous (U.P)
Kashi Kundal	Indigenous (U.P)
Thar Samridhi	Indigenous (Rajasthan)
Narendra Rashmi	Indigenous (U.P)
Arka Shreyas	Indigenous (Karnataka)
CO 1	Indigenous (Tamil Nadu)
Kashi Kirti	Indigenous (U.P)

Hybridization and selection

Planned hybridization involving indigenous genetic materials followed by pedigree method of selection in segregating generations or modified backcross-pedigree method of breeding has been followed for the development of few varieties in India (Hazra *et al*, 2022). Parents to be employed in the hybridization programme should possesses specific characters which are expected to be combined in the segregating generations. The parental lines should also be genetically diverse for obtaining high heterotic F_1 hybrids and broad spectrum of variability in the segregating generations. Selection in segregating generations should be based on the specific characters to be incorporated in the recombined lines and important selection indices for fruit yield and quality.

Table 3: Varieties bred through hybridization and selection

Name of variety	Parentage
Punjab Komal	LC 11 (round) × LC 5 (oblong)
Narendra Dharidar	NDBG-108 x NDBG-1

Heterosis breeding

Heterosis breeding has high potentialities for most of the cucurbits including bottle gourd. Bottle gourd hybrids has become very popular among the farmer mainly due to high female to male flower ratio, attractive, green, long and straight, cylindrical or club-shaped fruit shape without crooked neck suitable for both domestic and export market, high number of fruits per plant, few seeds, uniformity, extended fruiting span, wider adaptability, high yield and resistance to biotic stresses. It has been estimated that about 40-45% bottle gourd area in India is at present under different hybrids. Heterosis breeding offers good opportunity in bottle gourd due to monoecious sex form, high number of seeds per cross pollination and requirement of less number of plants per unit area.

Bagging of staminate and pistillate flowers in the morning hours and hand pollination of the pistillate flower in the afternoon hours of the same day after opening of the pistillate flower with the pollens from the staminate flower of the intended male parent is the commercial method of hybrid seed production. Genetically controlled male sterility due to single recessive *ms* gene (Dutta, 1983) has not yet been utilized for the development of hybrids.

Breeding for resistance to biotic stresses

In bottle gourd several Indian and exotic accession showed resistance to different diseases and insect-pests.

Table 4: Germplasm showing resistance to different diseases and insect pests

Germplasm	Resistance to the disease/ insect-pest
PI 271353 (India)	Cucumber mosaic virus, tobacco ringspot virus, tomato ringspot virus, watermelon mosaic virus
Doodhi Long Green, Three Feet Long Green, PI 271353, PI 271357, PI 381804, PI 273663(India)	Powdery mildew
Renshi (Taiwan)	Fusarium wilt
Punjab Komal (India)	Cucumber mosaic virus
PI 381825, PI 321831, PI381834	Zucchini yellow mosaic virus
PI 391602 (China), PI 181913 (Syria), PI 188809 (Phillippines), Pi 273663 (Ethiopia), PI 270456 (Mexico)	Squash mosaic virus
U-10-315, NIC-1225, NB-28 (India)	Red pumpkin beetle
NB-29 (India)	Fruit fly

Source: Swarup (2006), Dhatt and Khosa (2015)

Availability of resistant source and determination of genetic control of resistance character follows the transfer of resistance character in the commercial variety possessing superior horticultural traits. Inheritance of resistance to zucchini yellow mosaic virus was found to be under dominant genetic control (Ling and Levi, 2007). Backcross breeding method is generally followed for the transfer of resistance controlled by one or two genes while, recurrent selection is followed in case the resistance is governed by poly genes. Development of mapping population can help finding of marker associated with the resistance character.

Breeding for fruit quality

Bottle gourd is an important multi-purpose cucurbit crop grown for its leaf, fruit, and seed. It is widely cultivated in the tropical region providing vital human nutrition and serving as food security crop. There is wide genetic variation among bottle gourd genetic resources for diverse qualitative and quantitative attributes for effective variety design, product development, and marketing. However, adequate research works has not yet been carried out in this crop. Focus should be given on genetic improvement and genetic analysis targeting agronomic and horticultural attributes, nutritional composition combined with biotic and abiotic stress tolerance. Several studies emphasized the importance of hybrid breeding to improve different quality traits *viz.*, dry matter content, ascorbic acid, reducing sugar, non-reducing sugar, total sugar and total soluble solids contents in bottle gourd.

References

Dhatt, A.S. and Khosa, J.S. 2015. Bottle gourd. In: *Hand book of Vegetables* Vol.III (Eds. Peter, K. V. and Hazra, P.), Studium Press LLC, Houston, Texas, USA, pp. 49-78.

Dutta, O.P. 1983. Male sterility in okra (*Abelmoschus esculentus*) and bottle gourd (*Lagenaria siceraria*) and its utilization in hybrid seed production. *Thesis Abst.*, **9**: 341-342.

Enonch, G., Dako, A., Fuchus, J. Ahanchede, A. and Blattner, F.R. 2008. Flow citometric analysis in *Lagenaria siceraria* (Cucurbitaceae) indicates correlation of genome size with usage types and growing elevation. *Plant Syst. Evol.*, **276**: 9-19.

Hazra, P., Banerjee, M.K. and Chattopadhyay, A. 2022. *Varieties of Vegetable crops in India*, Kalyani Publishers, New Delhi, Ludhiana, 204p.

Ling, S.K. and Levi, A. 2007. Sources of resistance to zucchini yellow mosaic virus in *Lagenaria siceraria* germplasm. *Hort Sci.*, **42**: 1124-1126.

Swarup, V. 2006. *Vegetable Science and Technology in India.* Kalyani Publishers, New Delhi, Ludhiana, 656p.

Tiwari, A. and Ram, H.H. 2009. Inheritance of fruit shape in bottle gourd (*Lagenaria siceraria* (Mol.) Standl). *Veg. Sci.*, **36**: 147-149.

Index